LINEAR INTEGRATED CIRCUITS

Er. Rishabh Anand

KHANNA BOOK PUBLISHING CO. (P) LTD.

::: Publisher of Engineering and Computer Books :::

Head Office	:	4C/4344, Ansari Road, Darya Ganj, New Delhi-110002 Phone: 011-23244447/48 Fax: 011-23244447
Regd. Office	:	1694-95, Nai Sarak, Delhi-110006 Phone: 011-23283211 Fax: 011-23256658
E-Mail	:	sales@khannapublishers.com, contact@khannabooks.com
Website	:	www.khannapublishers.com

Dear Readers,

To prevent the piracy, this book is secured with HIGH SECURITY HOLOGRAM on the front title cover. In case you don't find the hologram on the front cover title, please write us to at contact@khannabooks.com or whatsapp us at +91-99109 09320 and avail special gift voucher for yourself.

Specimen of Hologram on front Cover title:

Moreover, there is a SPECIAL DISCOUNT COUPON for you with EVERY HOLOGRAM.

How to avail this SPECIAL DISCOUNT:

Step 1: Scratch the hologram
Step 2: Under the scratch area, your "coupon code" is available
Step 3: Logon to www.khannabooks.com
Step 4: Use your "coupon code" in the shopping cart and get your copy at a special discount
Step 5: Enjoy your reading!

Thanks,
Khanna Publishing House

Price: ₹ 450.00

LINEAR INTEGRATED CIRCUITS

Rishabh Anand

ISBN: 978-93-81068-93-9

Edition: 2014
Reprint: 2021

Published by:

Khanna Book Publishing Co. (P) Ltd.
4C/4344, Ansari Road, Darya Ganj, New Delhi-110 002
Phone: 011-23244447-48 **Mobile:** +91-9910909320
E-mail: contact@khannabooks.com

Printed by: SPS Printers

Preface

Make things as simple as possible, but not simpler

-Albert Einstein

The term Integrated Circuit (IC) reflects the capabilities of semiconductor industry to fabricate complex electric circuit consisting of a large number of components on a single substrate. The operational amplifier or op-amp is the most versatile active element among linear ICs. The present text covers the design and application of op-amp and other ICs. The emphasis is on the fundamental design concepts.

In the last two years, digital electronics has seen revolutionary changes. This, in turn, has both demanded and enabled an evolution of analog integrated circuits. More and more functions are being integrated into a single IC. This increased level of integration, when applied to the analog world, has produced integrated circuits that provide widely ranging functions. These ICs are amazingly easy to use. Because of this simplicity, and because of a high performance-to-cost ratio, analog integrated circuits have replaced most discrete transistor amplifier circuits. The *operational amplifier* (Op Amp) is the most widely used analog IC and has become the true building block of analog electronics. The field of analog integrated circuits has developed and matured tremendously. The initial groundwork was laid in bipolar technology, followed by a rapid evolution of MOS analog integrated circuits.

Plenty of good books on *Linear Integrated Circuits* are available in the market. But they are either too concise or too elaborate, either too simple or too advanced, either devoid of mathematics or purely analytical. A few of these books also do not cover the entire syllabus of various universities. This book is a comprehensive treatment of the analysis, design, and fabrication of analog integrated circuits.

Since this is a teaching text dedicated to those new to the field, the reader is provided with a "soft start". The book follows a 'tall and thin' approach. The book is intended to be useful both as a text for students and as a reference book for practicing engineers. Considerable amount of material is also included on applications. In practice, analysis, design and applications are closely linked and knowledge of each of these three aspects is essential for designers and users of ICs. The book provides logical method of explaining various complicated concepts and stepwise methods to explain the important topics. Each chapter is well supported with necessary illustrations and practical examples. All care has been taken to make students comfortable in understanding the basic concepts of the subject.

The book not only covers the entire scope of the subject but explains the philosophy of the subject. This makes the understanding of this subject clearer and makes it more interesting. The book will be very useful not only to the students but also to the subject teachers.

The author hopes that the book will fulfill the pressing need of interested readers and welcome any suggestions towards the improvement of the book.

Er. Rishabh Anand

Acknowledgement

We all marvel at the beautiful rose. The rose, with all its beauty and grandeur. But seldom we pause and thank the kind of gardener who patiently manured and watered it. The book I have written is the fruit of such unseen hands.

I am especially grateful to my teachers and my colleagues specially Mrs. Aparna Mahajan, Mr. Sanjeev Kumar, Mrs. Rashmi Mahajan, Wg.cdr. P. Patnayak, GpCapt. H.S. Dua, Wg.cdr. Onkar Singh, Mrs. Dimple Saproo, Mrs. Naresh Kumari, Mrs. Chitra Kaul, Director M.M Jha, Mrs.Savita Khazanchi, Mrs. Meenakshi and other faculty members who influenced my academic grown during my study.

There was, in and always will be the love of my students whom I taught, for what is written in this book has been the fruit of experience that I taught them. Finally, thanks to my mother and to rest of my family for their patience and support during the long hours of writing this book and above all there is the one almighty whose humble children we are. It is his blessings we cherish and pray for. It is the blessing I wish for you.

I express my deep thanks to *Mr. Puneet Khanna and Mr. Mukul Seth* from Khanna Book Publishing Co., and entire team for publishing the book in such a beautiful get-up and well in time.

Er. Rishabh Anand

Message to My Students

As you have signed up for this course and probably your teacher has adopted this book! You had no say about the choice, although you can form and express your opinion on the book after reading it. What I really meant is that although this is an easy book to study from (in my opinion), it is no-nonsense book. It does not have cartoons or nice-looking photographs to amuse you. From here on it is all business and hard work.

This book has many examples. Some of these are simple, for the purpose of illustrations of new ideas and subject matter. Some examples are made quite elaborate, in order to develop desired technical skill and confidence. Furthermore, the objective of this book is to present the complex subject in a clear and thorough manner.

I hope that you will enjoy this book, and it will be remain as major text book in your college career. My advice to you is not sell it back at the end of semester. If you do, not only will you be violating the "buy low and sell high" axiom of finance but definitely be missing it during your higher semester or project assignments.

Er. Rishabh Anand

Brief Table of Contents

Detailed Table of Contents

Chapter

1

Integrated Circuit Fabrication

1.1 INTRODUCTION

An integrated circuit is one in which the active components (e.g., transistors, diodes etc.) and passive elements (e.g., resistors, capacitors, etc.) are automatically part of a semiconductor chip.

Integrated electronics, generally referred as **microelectronics**, is the outcome of continuous research and improvements in the characteristics and miniaturizations employed in the fabrication of the solid state devices and components. Thus the integrated electronics can be considered as logical extension of the silicon device technology. Today the integrated electronics is used to realize complete circuits over a single semiconductor chip of silicon. Such circuits are commonly known as **Integrated Circuits (ICs)**. The main semiconductor material utilized is silicon, but for special applications, gallium arsenide and other compound semiconductor materials may also be utilized. In the IC Fabrication technology, the circuit designer faces some limitations and opportunities both.

Since the development the first IC, the circuit design has become more and more sophisticated and the integrated circuit more complex. A single silicon chip may be of the order of one square centimeter, which encloses over a million transistors and other component. These components are not

Inside the Chapter

similar to the conventional component but the application of each component is similar to that discrete component of similar type. Intense research on silicon processing and increased automation in design and manufacturing have led the integrated circuits to the lower costs and higher fabrication yields.

There are lot of advantages of integrated circuits (ICs) over the circuits with discrete components which have made ICs very popular worldwide. The various advantages of the integrated circuits are as follows :

(*i*) Practically **size** of an IC is thousands of times smaller than the discrete circuits, which increases the equipment density.

(*ii*) Thousands of silicon wafers consisting individually millions of components, can be produced or manufactured simultaneously, known as **mass production**. Due to this, the **cost** of IC is very very low.

(*iii*) As the number of components are fabricated on a single silicon wafer, which are not similar to the conventional one, the **weight** of IC reduces as compared to the discrete circuits with same number of components.

(*iv*) ICs operate at low voltages. The **power consumption** of ICs is very low.

(*v*) In ICs, soldered joints are absent and thus the high **reliability** of the system is the main feature of ICs.

(*vi*) As power consumption is less, the temperature rise is also low which increases **reliability, accuracy** and **life** of the ICs.

(*vii*) As number of components are fabricated on a same silicon wafer with same fabrication processes simultaneously, thus the **matching of devices** is excellent.

(*viii*) Effects of the parasitic capacitances are minimised in ICs, hence the **operating speeds** are higher.

The ICs are available with wide variety of characteristics and different types of packages with very complex circuits integrated.

In this chapter we will discuss first classification of ICs and then different IC fabrication processes. At the end we will discuss the realization of various active and passive devices.

1.2 CLASSIFICATION OF INTEGRATED CIRCUITS

Depending upon the functional utility, the integrated circuits are classified as **linear ICs** and **digital ICs**. From the point of view of structural considerations, ICs can be divided as **monolithic ICs, thick-thin film ICs** and **hybrid ICs**. The monolithic ICs are most common type of IC. In the monolithic ICs, all the active and passive components alongwith their interconnections are manufactured on a single silicon chip. In general, the monolithic ICs has power limitations; hence they are preferred only in low power applications.

In general, **thick-thin film ICs** are used to produce only the passive elements. The film technology is used to develop thick-thin film ICs. By combining the passive elements with pre-fabricated active components. Generally the film circuits are made by depositing various film capacitors and resistors on a separate substrate such as glass or ceramic used as passive substrate. And then the pre-fabricated active components are added into the substrate. The **thick film technology** is easy and inexpensive. The thin film technology produces components with greater precision, but it is expensive as compared with the other technology.

For high power applications, hybrid ICs are used. Hybrid ICs combine two or more monolithic ICs in one package. They may combine monolithic ICs with thick-thin film circuits.

On the basis of the active devices used, the ICs are classified as **bipolar ICs** and **unipolar ICs**. The bipolar ICs use bipolar junction transistors (BJTs), while the unipolar ICs use field effect transistors (FETs).

Depending upon the isolation technique used, the bipolar ICs are further classified as **p-n junction isolation ICs** and **dielectric isolation ICs**.

Depending upon the type of FET used, the unipolar ICs are further classified as **MOSFET unipolar ICs** and **JFET unipolar ICs**.

Based upon the above requirements, two distinctly different IC technology namely, Monolithic technology and Hybrid technology have been developed.

Table 1.1 gives the total information about the IC families on the basis of substrate used, structure of IC, active and passive devices used and finally the application of the IC.

Table 1.1

Properties \ IC family	Monolithic IC	Thick-thin film	Hybrid IC
Substrate	Silicon	Glass, Ceramic	(*i*) Glass, Ceramic (*ii*) Silicon
Structure	Active and passive devices along with interconnections on a single chip.	Passive components are fabricated on a substrate and then prefabricated active components are added along with interconnections.	(*i*) Passive devices and interconnections on one insulating substrate with active devices wire wound. (*ii*) Active devices on a single chip while passive devices along with interconnections on thick-thin film.
Active devices	(*i*) BJT (*ii*) MOSFET	MOSFET	(*i*) BJT, MOSFET (*ii*) BJT, MOSFET
Passive devices	(*i*) Diffused resistors, oxide capacitor (*ii*) MOS resistor. oxide capacitor	Metal film resistor, oxide capacitor	(*i*) Metal film resistor, cement resistor, oxide capacitor (*ii*) Metal film resistor, oxide capacitor
Application	Linear and digital IC	Digital IC	(*i*) Linear and digital IC (*ii*) Linear and digital IC

1.3 INTEGRATED CIRCUIT CHIP SIZE AND LEVELS OF INTEGRATION

In 1947, **Prof. Bardeen** and **Prof. Brattain**, the members of a research group at **Bell Laboratories**, invented first germanium transistor. Actually the team was attempting to make a device resembling a field effect transistor. Even though the basic principle behind the operation of MOSFET was proposed in 1935, it took long period to materialize this device. By the year of 1951, the bipolar junction transistors (BJTs) and the junction field effect transistors (JFETs) came in the market. Immediately in 1954, silicon took the place of germanium in BJTs. In late 1960, first time MOSFET was produced.

Meanwhile the interconnection of electric circuits with the help of the printed Circuit assembly was under progress. In 1952, the scientist found the possibility of semiconductor integrated circuits. Finally in 1958 **Jack Kilby of Texas Instruments** invented first integrated circuit. Shortly after this invention **Prof. Robert Noyce** and **Gorden Moore** of **Fairchild semiconductor**, USA, announced the planar process for silicon bipolar transistors which stepped ahead to the development of monolithic IC After realizing the strength of the IC, it become the most demanded device. Then progressively inventions were made in increasing the number of active devices per chip by reducing device dimensions.

Depending upon the number of active devices per chip, there are different levels of integration as explained below.

When the active devices per chip are less than 100, then it is referred as **small scale integration (SSI)**. Most of the SSI chips use integrated resistors, diodes and bipolar transistors.

When the count of active devices per chip is between 100 to 1000, then it is referred as **medium scale integration (MSI)**. In most of the MSI chips, BJTs and enhancement mode MOSFETs are integrated.

In large scale integration (LSI) ICs, the number of active devices per chip ranges between 1000 to 100,000. In general, LSI chips use MOS transistors; as it requires less number of steps for integration. Thus more number of components can be produced on the chip with MOS transistors than with the bipolar transistors.

When the active devices per chip are over hundreds of thousands, then it is referred as **very large scale integration (VLSI)**. Almost all modem chips employ VLSI technique.

In general, from **small scale integration (SSI) level to the very large scale integration (VLSI) level,** the device density has been increased tremendously. Also the chip area is increased from 1 mm^2 (for SSI level) to 1 cm^2 (for VLSI level). The variation in the chip area for SSI chip to VLSI chip is a shown in the Fig. 1.1.

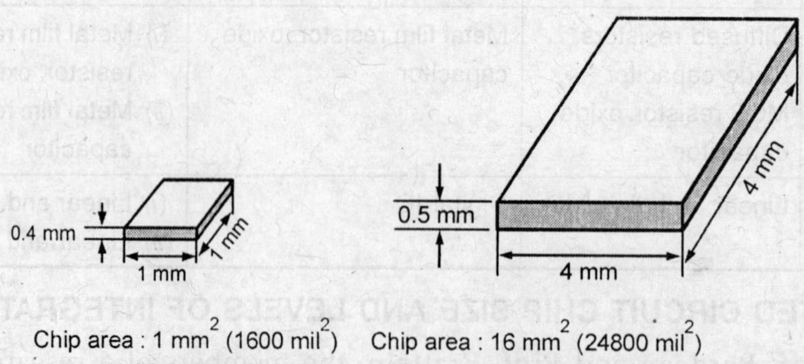

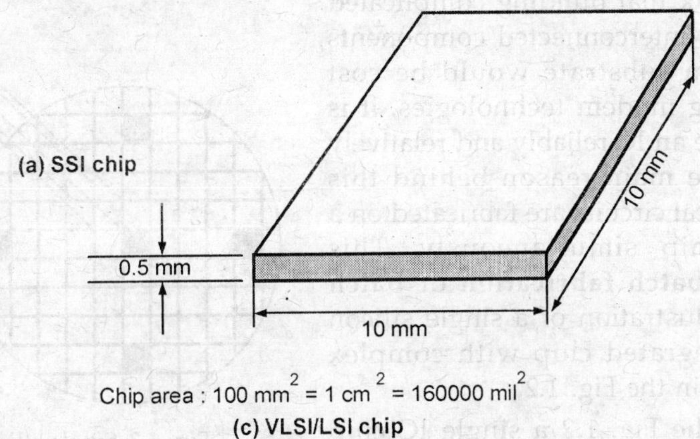

Fig. 1.1 Chip are for various integrated circuits levels
(1 mil = 0.0254 mm = 2.54 m cm = 0.001 in)

Recently a new level of integration has been introduced which is known as **ultra large scale integration (ULSI)**. In ULSI technique, more than one million active devices are integrated on a single chip. Pentium microprocessors use ULSI technology.

Table 1.2 gives the categories of ICs on the basis of number of active devices. Note that the number of active devices per chip is nothing but the number of transistors per chip. Sometimes an alternative parameter for transistor count per chip may be used. It is referred as gate count.

Table 1.2

Level of integration	Number of active devices per chip
1. Small scale integration (SSI)	less than 100
2. Medium scale integration (MSI)	100 - 10000
3. Large scale integration (LSI)	1000 - 100,000
4. Very large scale integration (VLSI)	over 100,000
5. Ultra large scale integration (ULSI)	over 1 million

1.4 FUNDAMENTALS OF MONOLITHIC IC TECHNOLOGY

We have already studied that in monolithic ICs all the active as well as passive components along with interconnections are integrated on a single crystal. The word 'monolithic' is originated from Greek word and means 'single stone' or 'one stone'. Thus this word is appropriate as the components are integrated on a single piece of silicon crystal. The fabrication of different discrete devices such as diodes, transistors and integrated circuits is carried out by the same technology. The various processes involved in the fabrication of different devices are carried out in a single plane. Hence this is also referred as a **planar technology**.

One might think that building complicated circuits with many interconnected components on a single silicon substrate would be cost effective. But using modem technologies, it is possible to fabricate an IC reliably and relatively inexpensively. The main reason behind this is that many identical circuits are fabricated on a single silicon chip simultaneously. This process is called **batch fabrication** or **batch production**. The illustration of a single silicon wafer and an integrated chip with complex circuit is as shown in the Fig. 1.2.

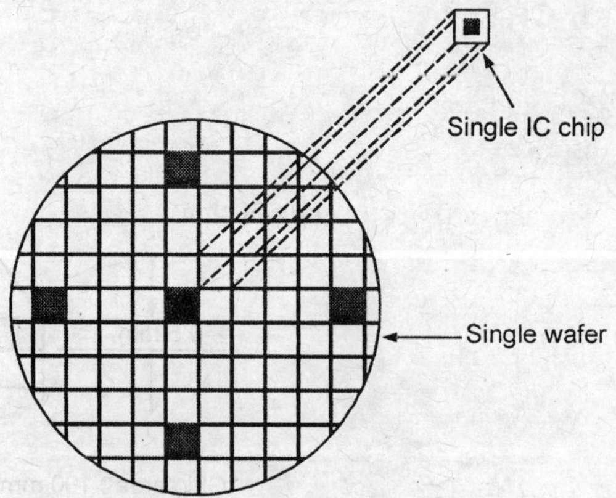

Fig. 1.2 Single silicon wafer and individual IC chip

As shown in the Fig. 1.2 a single IC chip which may vary in area 1 mm^2 to 100 mm^2, contains upto several million devices. Eventhough the fabrication processes are complex and expensive, the resulting large number of the integrated circuits make ultimate cost of the individual IC fairly low.

Another important consideration is the **percentage of usable devices**. i.e. **fault free devices**. Generally the faulty devices occur due to some defects in Si or defects in the fabrication steps. The percentage of the usable devices per chip area is called **yield**. Now let us consider a silicon wafer of 150 mm diameter consisting approximately 12,000 chips of rectangular shape of sides 1 mm. Each IC may contain the components varying in number from few tens to millions. And in one batch, 15 such wafers are fabricated. Hence, in one batch, simultaneously we can get 1,80,000 ICs. How assume that the percentage of the usable devices (i.e. fault free devices) per chip area is 25% only. Then also we can have as many of 45,000 perfect fault free IC chips in one batch.

1.4.1 Basic Planar Processes in IC Fabrication

The basic planar processes involved during the fabrication of ICs are as listed below.

1. Crystal growth and wafer preparation.
2. Epitaxial growth.
3. Oxidation.
4. Lithography.
5. Reactive plasma etching.
6. Diffusion.
7. Ion Implantation.
8. Metallization.
9. Assembly techniques and packaging.

Let us study each process in detail one by one.

1.5 CRYSTAL GROWTH AND WAFER PREPARATION

The first important process in IC fabrication is the formation of a silicon wafer through the crystal growth. This process can be subdivided into number of sub processes as given below.

1. Silicon crystal ingots growing.
2. Ingot trimming and grinding.
3. Ingot slicing.
4. Wafer etching.
5. Wafer polishing.
6. Wafer cleaning.

Let us consider subprocess one by one.

1.5.1 Silicon Crystal Ingots Growing

The important semiconductor material for the fabrication of semiconductor devices and integrated circuits is silicon. Other semiconductor materials are **germanium** and **gallium arsenide**. At present 95% of the semiconductor devices and integrated circuits are manufactured using silicon only and for very special applications gallium arsenide is preferred. We have already discussed that the bipolar junction transistor (BJT) was developed first in 1948, with germanium as a basic semiconductor material. But it was observed that silicon is better option than germanium. The comparison between silicon and germanium is as given in Table 1.3.

Table 1.3

Germanium	Silicon
1. It is unsuitable for certain application due to high junction leakage currents as it has relatively narrow energy band gap (0.66 eV).	It is comparatively suitable for all applications as junction leakage currents are negligible as the energy band gap is comparatively broader (1.1 eV).
2. Germanium devices can be operated upto 100°C temperatures.	Silicon devices can be operated upto 200°C temperature.
3. Germanium oxides are unsuitable for certain device applications.	Silicon dioxides are the most wanted for the planar processes.
4. The intrinsic resistivity without any dopant is 47 Ω-cm, hence not suitable for high voltage rectifying devices.	The intrinsic resistivity without any dopant is 230,000 Ω-cm hence most suitable for high voltage rectifying devices as well as infrared sensing devices.
5. Germanium is costlier as compared to silicon.	Silicon is cheaper as compared to germanium.

Basically silicon is the element, found in nature in the form of **silica** and **silicates**. It is found abundantly in the nature in the form of **silicon dioxide**. So this silicon dioxide constitutes almost 20% of the earth's crust. So obviously one can not start fabricating integrating circuits using silicon dioxide with the earth's crust. The sand can be converted into pure silica on through number of processes. For the fabrication of ICs, the **silicon** must be in **crystalline form**. The crystalline form of silicon is the pure silicon with no deflects and no contaminations.

To obtain this purest form of silicon, first **metallurgical grade silicon (MGS)** is produced in a submerged electron arc furnace. Using this step, MGS is solidified with a purity of about 98%. Then the silicon is pulverized mechanically. Then it is added with unhydrous hydrogen chloride to form trichlorosilane ($SiHCl_3$). This process is carried out with a catalyst in a fluidized bed at 350°C. After this reaction, the trichlorosilane obtained is liquid at room temperature. Then by using fractional distillation, purification of trichlorosilane is carried out. After this purified trichlorosilane is applied with chemical vapour deposition process. With the help of chemical reaction, hydrogen is reduced from trichlorosilane. It results in the rods of silicon. This multistep process continues for many hours and finally results in polycrystalline structured EGS rods with 0.2 m diameter and several meter length. This process is advantageous because the costing of the process is low and the by-products of the reactions arc less harmful. The **electronic grade silicon (EGS)** is also known as **semiconductor grade silicon (SGS)** which the highly purified form of silicon i.e., polycrystalline silicon. This form of silicon consists many small crystals. However for fabrication of ICs a crystalline silicon is needed. The single crystal silicon can be obtained by using method known as **crystal growth**.

The primary method of the crystal growth is **Czochralski (CZ) method**. In practice all the silicon required for integrated circuits is prepared by using this method only. In general, a phase change from solid, liquid or gas phases to crystalline solid phases to crystalline solid phase is nothing but growing crystal. Czochralski method is used for silicon crystal growth from which ultimately silicon wafers are produced. The apparatus used for the crystal growth is called **Czochralski crystal growth apparatus** or puller. The puller has four important subsystems namely furnace, crystal pulling mechanism, ambient control and control systems. The simplified version of Czochralski crystal puller is as shown in the Fig. 1.3.

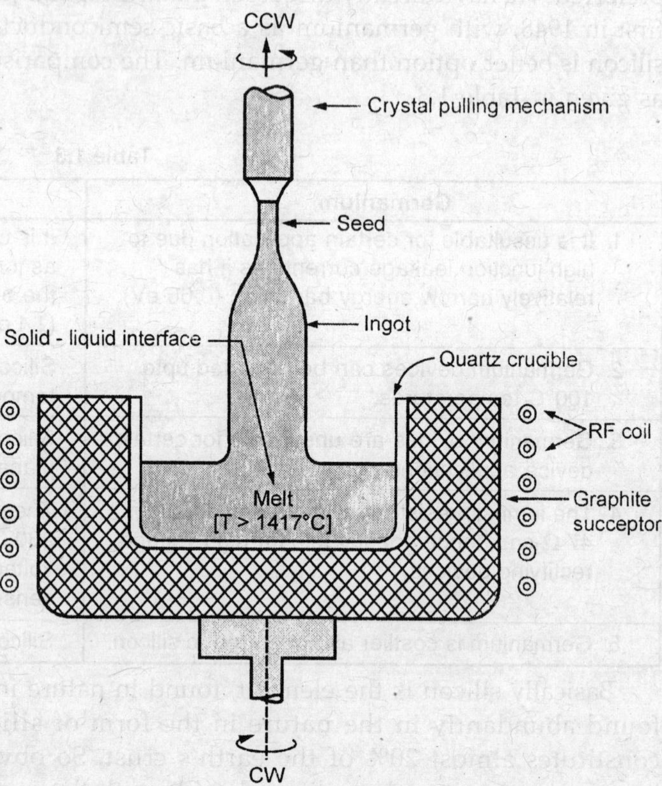

Fig. 1.3 Czochralski crystal grower

The furnace consists a crucible, crucible support, rotation mechanism and heating element housed in a chamber. The crucible is made up of fused silica (SiO_2) as this material is chemically unreactive with molten silicon. An EGS block is heated in a fused Silica crucible with the appropriate dopant using a heating element. The material in the crucible is heated to a temperature which is greater than the melting point of silicon, i.e. 1417°C. A resistance heating is preferred for large pullers; while induction heating is used for small melt sizes.

Then a small single crystal rod of silicon is immersed into the molten material. This rod is called **seed crystal**. This seed crystal is located at the crystal pulling assembly. In this assembly, using rotating mechanism seed shaft and seed chuck are rotated. Lowering the seed crystal in molten silicon allows the crystal ingot to form on the seed by solidification. The main function of crystal pulling assembly, is to control the pull rate of seed crystal and crystal rotation both, with minimum vibrations and precision. During the crystal growth, the crystal is rotated slowly, by stirring the molten and averaging out temperature gradients leading to inhomogeneous solidification. To get the ingots of circular cross-section, the crucible and the seed crystal are rotated in opposite direction. When the seed crystal is pulled out of the molten material, due to solidification, silicon ingot gets formed exactly same as seed crystal.

In general, the diameter is controlled by the pull rate. The standard diameter of the ingot is about 150 mm and the length is about 2 m. The ingot with such dimensions weighs about 60 kg.

The silicon growth is generally conducted in a vacuum or in an inert gas like helium or argon.

To control the process parameters such as temperature, crystal diameter, pull rate and rotation speed etc., control system is used which works under either open loop control or closed loop control.

1.5.2 Ingot Trimming and Grinding

First of all, the seed which initiated the crystal growth is separated from the circular ingot. The top and bottom ends are also cut off. As silicon is hard and brittle material, industrial grade diamond is used for shaping and cutting it. This process is called **ingot trimming**. After completion of the crystal growth, it is generally tested for resistivity and perfection evaluation. So the portions of the ingot failed in the above tests are also cut.

Note that these cuttings can be recycled for new crystal growth after cleaning.

After trimming of the ingot, the **surface grinding of the ingot** is carried out. Actually the ingots are slightly oversized. Hence with the help of lathelike diamond tool, the ingot is ground to a precise diameter.

After grinding the ingot to a precise diameter, generally two flats are ground along the length of the ingot. The larger flat is called **major** or **primary** flat and it is positioned relative to the crystal direction. The X-ray technique is used to locate the primary flat. The primary flat is very important as

(*i*) it serves for mechanical alignment of the wafer in automatic processing, and

(*ii*) it serves for orienting ICs on the wafer relative to the crystal.

The smaller flat is called secondary flat which is used to identify the orientation (< 100 > or < 111 >) and conductivity (p or n) of the wafer.

1.5.3 Ingot Slicing

After completing ingot trimming and grinding process, the ingot is ready for next process i.e. **ingot slicing**. The slices of the ingot are called **wafers** and typically the thickness of after may very from 0.4 mm to 1 mm. This process is very important as it is necessary to maintain the flat plane and desired surface orientations.

The slicing also determines the orientation of the surface. In general, there are two orientations < 100 > and < 111 >. Out of these orientations, the wafers with < 100 > orientations are cut **'on orientation'**; while wafers with < 111 > orientations are cut 'off orientations'. The position of the flats can be identified according to the standards laid by the Semiconductor Equipment and Materials Institute (SEMI) as shown in the Fig. 1.4.

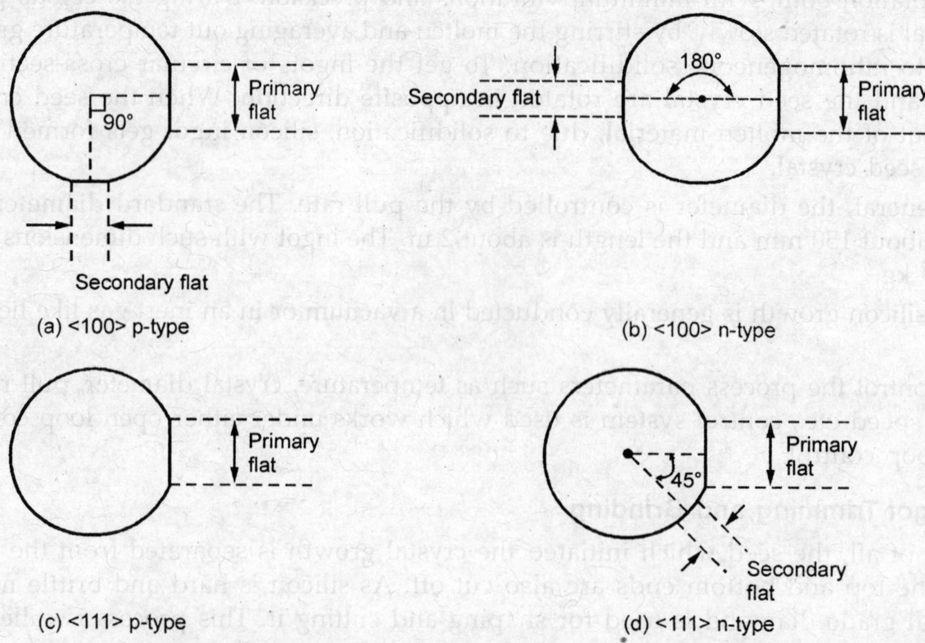

Fig. 1.4 Identifying flats on < 100 > and < 111 > silicon wafer

The ingot is sliced using a circular cutting blade kept in tension on the outer edge while having the cutting edge on the inner diameter.

The thickness of the wafer is determined by the slicing. It is another important wafer parameter because thicker wafers can easily withstand the stresses of subsequent thermal processes. The higher quality of slicing is achieved by using capacitive sensing device near the blade which helps the blade to be positioned correctly to achieve exactly flat plane cut.

1.5.4 Wafer Etching

If the sliced wafers are to be used for VLSI application, then before etching process two sided mechanical **lapping** process is carried out. Using this process, wafers with uniform flatness are achieved which are mostly required for photolithography.

Due to the machining operations during trimming, grinding and slicing, the surface and edges of the wafers get contaminated and even damaged. The depth of damage depends on the mechanical operations carried previously. It is observed that the damaged and contaminated regions are not more than 10 μm deep. Even by using chemical etching process, all the damaged and contaminated edges can be removed. Practically mixture of hydrofluoric, acetic and nitric acids is used in chemical etching. This is called **acidic etching**. The other alternative is to use **alkaline etching** using potassium hydroxide or sodium hydroxide. By the etching process, typically 10 μm to 30 μm of wafer is removed from both the sides.

1.5.5 Wafer Polishing

After etching, the wafer is polished to eliminate the microcracks and debris. The main intension of polishing a wafer is to provide a smooth and perfect flat surface such that the device features can be engraved. The polishing is done with the help of a polishing machine. Note that the polishing process removes further 10 μm to 30 μm of the wafer surface typically.

The three steps of lapping, etching and polishing reduce the wafer thickness by 40 to 150 μm. Thus by considering the typical figures, to have the wafer of 200 μm thickness, the thickness of the substrate required is 560 μm. In practice, the processed 6 inch wafers are typically 250 μm to 500 μm thick.

1.5.6 Wafer Cleaning

The silicon wafers are cleaned using chemicals. Generally organic films, heavy metals are deposited on the surface of the wafers. Hence by using HCl - H_2O_2 aqueous solution, metallic impurities can be removed. First the wafer is cleaned by using HCl - H_2O_2. Then wafer is rinsed in water to deionize. Again the wafer is dipped in hydrofluoric acid. Then again the wafer is rinsed in water.

After cleaning process, the wafer is ready for the formation of the dies.

1.6 EPITAXIAL GROWTH

The epitaxy means **'arranged upon'**. In epitaxy a monocrystalline film is formed on the top of a monocrystalline surface. Thus epitaxy is crystalline growth process in which the foundation layer i.e. substrate works as seed crystal. The epitaxial layer formed on the substrate may be either n-doped, p-doped or intrinsic. The polarity and the concentration of the dopant to be used is not determined by the foundation layer. In general for p-type doping bi-borane (B_2H_6) and for n-type doping phosphine (PH_3) are used with the steam of silicon tetrachloride hydrogen gas.

The epitaxial growth of pure silicon can be represented with the help of following chemical reaction.

$$SiCl_4 + 2H_2 \xrightarrow{1250^\circ C} Si + 4HCl$$

There are two types of epitaxy as given below.

1. **Homoepitaxy:** When the epitaxial layer and the substrate on which the epitaxial layer is to be formed, are of same materials, then the process is called homo epitaxy. The silicon process in which silicon is grown or formed over silicon substrate is an example of homoepitaxy.

2. **Heteroepitaxy:** When the epitaxial layer and the substrate on which the epitaxial layer is to be formed are not of identical material, then the process is called heteroepitaxy. But for crystalline growth with heteroepitaxy, the materials must have identical crystal structure.

While studying epitaxy one must note that this process is not restricted to a single layer formation on the substrate. But multiple epitaxial layers are also possible which are widely found in various applications such as microwave p-i-n diodes.

In general bipolar integrated circuits use epitaxial layer process in which high resistivity epitaxial layer is formed over a low resistivity substrate. To provide isolation between the epitaxial growth and the substrate, the doping used in both layers is of opposite type. Due to this a heavily doped buried layer is formed. The **buried layer** is also called **diffusion layer** as shown in the Fig. 1.5.

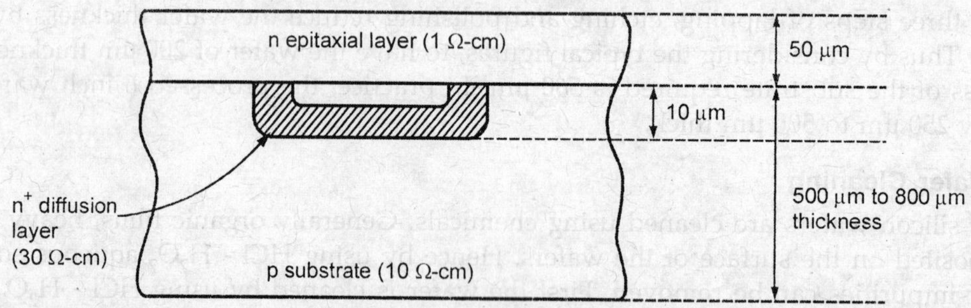

Fig. 1.5 Cross sectional view of bipolar integrated circuit

The advantages of epitaxy are as follows -

1. Because of one or more buried layers, the designer can control the doping in the structure.
2. The properties of the epitaxial layer and the bulk material are different.
3. Using epitaxial structures the performances of RAMs (Random access memories) and CMOS ICs can be improved.

The two main epitaxial processes are

1. Chemical vapour deposition (CVD) and
2. Molecular beam epitaxy (MBE)

1.6.1 Vapour Phase Epitaxy

In chemical vapour deposition (CVD), the film is formed on the surface of the substrate by thermal decomposition and or the reaction of various gaseous compounds. As in CVD, the epitaxial layer is formed from the gaseous vapour phase, hence it is called **vapour phase epitaxy**.

The process of CVD growth takes place in the reaction chamber in which a long quartz tabe with RF induction coil wound on it. The silicon wafers are placed in the graphite rod which acts as base. It is also called graphite **suscepter** which supports the wafer. Then the suscepter is heated by RF induction to temperature between 1100°C to 1300°C. The four important compounds for the reaction are silicon tetrachloride ($SiCl_4$), trichlorosilane ($SiHCl_3$), dichlorosilane (SiH_2Cl_2) and silane (SiH_4). Out of these compounds, silicon tetrachloride ($SiCl_4$) is the compound used widely in the industries. The temperature for $SiCl_4$ reaction is controlled between 1150 to 1250°C and respectively lower temperatures for remaining vapours. The different gases required for epitaxial layer formation are allowed in the chamber through different channels. This doping is done by using hydrides of arsenic, boron, phosphorus. These are Arsine (A_5H_3) diborane (B_2H_6) and phosphine (PH_3). In general, the film grows at 1 μm/min typically. Now a days, instead of RF induction, heating with quartz halogen lamps is preferred as it provides more uniform heating. The doping capability of an epitaxial layer lies in the range 10^{12} to 10^{22} atoms/cm^3.

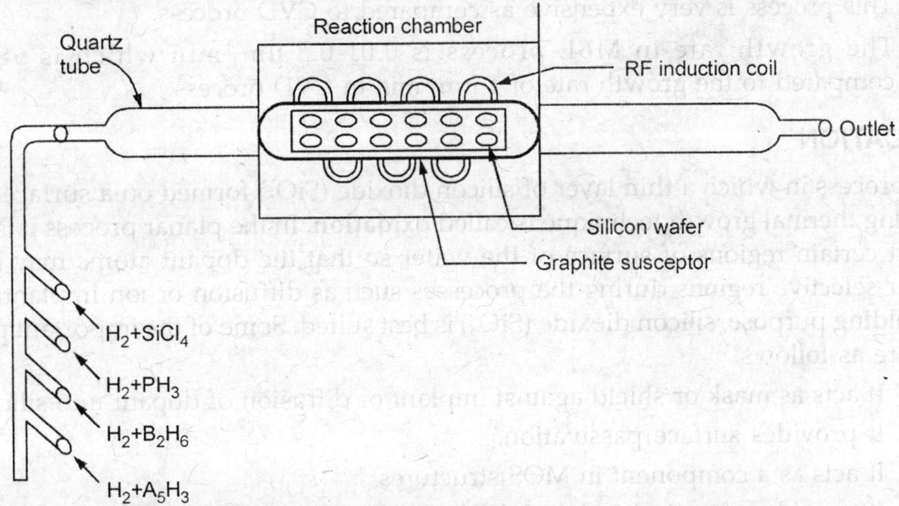

Fig. 1.6 Schematic representation of vapour phase epitaxy process

1.6.2 Molecular Beam Epitaxy

The **molecular beam epitaxy (MBE)** is based on evaporation. In MBE, the film is evaporated and deposited one layer at a time. In the process, no chemical reactions are considered. Instead of that the evaporation of silicon and other dopant is carried out under ultra high vacuum (UHV) pressures of the orders of 10^{-8} to 10^{-10} Torr. The through put is slow. The growth rate is also very slow typically, 0.01 to 0.3 μm/min. The molecular beam epitaxy is carried out under temperature ranging between 600°C to 900°C which is comparatively low temperature. As this process is very expensive, it is extensively used in only special applications such as **Ga As technology, silicon on insulator (SIC)** and **silicon on sapphire (SOS).**

1.6.3 Advantages of MBE Over CVD Process

1. MBE is low temperature process which is advantageous for VLSI technology.
2. While preparing thin layers using MBE process, autodoping and autodiffusion both are minimized.

3. The MBE process can be used for generating complicated doping profiles as it regulates the amount of dopant.
4. As MBE process is based on the evaporation of silicon and the dopants, hence no chemical reactions are involved in it.
5. For MBE process safety precautions are not required extensively as compared to those required in CVD process.

1.6.4 Disadvantages of MBE Process

1. For overall perfect and pure film, it is necessary to maintain a very low pressure of the order of 10^{-10} Torr which is slightly difficult.
2. This process is very expensive as compared to CVD process.
3. The growth rate in MBE process is 0.01-0.3 μm/min which is very small compared to the growth rate of 1 μm/min in CVD process.

1.7 OXIDATION

The process in which a thin layer of silicon dioxide (SiO_2) formed on a surface of silicon wafer using thermal growth technique is called **oxidation**. In the planar process it is essential to protect certain regions of surface of the wafer so that the dopant atoms may be driven into other selective regions during the processes such as diffusion or ion implantation. For such shielding purpose, silicon dioxide (SiO_2) is best suited. Some of the important properties of SiO_2 are as follows.

(*i*) It acts as mask or shield against implant or diffusion of dopant into silicon wafer.

(*ii*) It provides surface passivation.

(*iii*) It acts as a component in MOS structures.

(*iv*) It provides electrical isolation between multilevel interconnected layers.

(*v*) The most commonly used silicon dopants such as phosphorus, boron, arsenic and antimony diffuse with difficulty in SiO_2. In other words, all these dopants have low diffusion coefficient in SiO_2. Hence SiO_2 acts as shield against all the above mentioned impurities.

Following are the different techniques developed for forming oxide layers.

(*i*) Thermal oxidation.

(*ii*) Wet anodization or oxidation.

(*iii*) Vapour-phase technique (CVD).

(*iv*) Plasma anodization or oxidation.

Thermal oxidation is the basic process in the fabrication of semiconductor devices. When the interface between oxide and silicon requires a low charge density level, this type of oxidation is proffered.

When the oxide layer is required on the top of a metal layer such as in multilevel metallization structure, the **vapour phase technique** i.e. **chemical vapour deposition (CVD)** is used.

Plasma oxidation is a low temperature vacuum process carried out in a pure oxygen discharge. This process allows to grow high quality oxides on the surface of silicon wafer as compared with other techniques. The plasma oxidation can grow thick oxide layers at comparatively lower temperatures with higher growth rate. The growth rate depends on temperature of substrate, plasma density and dopant concentration in substrate.

1.7.1 Thermal Oxidation

The oxidation processes is called thermal oxidation because to grow the oxide layer, the temperature maintained is high.

The selection of the oxidation technique depends upon the thickness and the oxide properties required. For thin oxides with low charge density at interface, the oxides are grown in dry oxygen. This is also called dry oxidation explained by the reaction given below.

$$\underset{\text{(solid)}}{Si} + O_2 \longrightarrow \underset{\text{(solid)}}{SiO_2} \text{ (dry oxidation)}$$

For thick layers of oxides, steam or water vapour is used at high pressure for oxidation. This is called **wet oxidation**. The chemical reaction for wet oxidation is as follows.

$$\underset{\text{(solid)}}{Si} + 2H_2O \longrightarrow \underset{\text{(solid)}}{SiO_2} + \underset{\text{(gas)}}{2H_2} \text{ (wet oxidation)}$$

In the most commonly used oxidation technique, the wafers are stacked vertically in a slotted paddle or boat which is open ended. This is made up of quartz and it is placed in quartz tube. This quartz tube is slowly passed through resistance heated horizontal furnace. Typically the temperature maintained is 700°C to 1200°C. The oxidizing agent used may be dry oxygen or a mixture of oxygen and water vapour. The schematic for the thermal oxidation equipment is as shown in the Fig. 1.7.

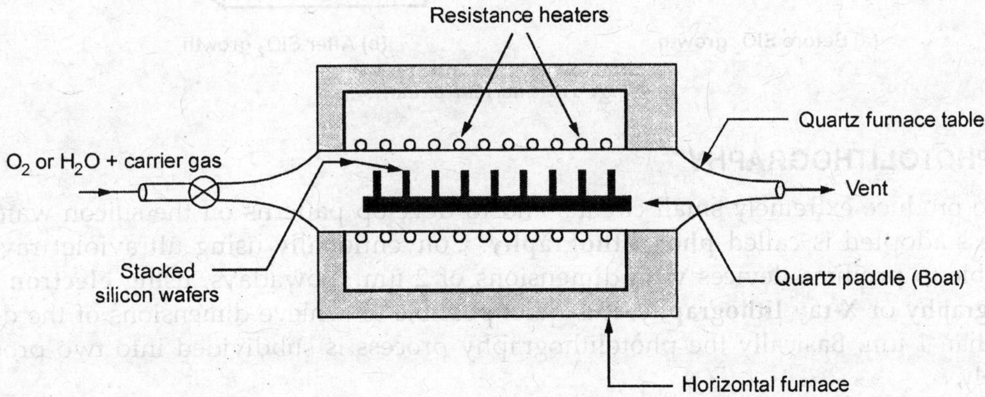

Fig. 1.7 Thermal oxidation equipment

Basically an oxidation is high temperature process. Due to the oxidation process, the layer departs from its original location. To overcome this undesired result, the oxidation process is carried out at low temperatures. But this increases the growth time. To overcome this, pressure is increased because an increase by 1 atm pressure, decreases temperature by 20°C for same growth rate. Thus high pressure oxidations with pressures upto 25 atm are used in the industrial applications at the temperatures in the range of 700 to 900°C.

Initially the oxide growth formation rate is very fast and then gradually it slows down, as oxygen atoms have to diffuse through the oxide to reach the interface between silicon substrate and SiO_2. As compared to dry oxidation process, wet oxidation process is faster at a given temperature. Typically to grow 1 μm thick oxide layer, dry oxidation process takes 2 hour 30 minutes, while wet oxidation takes only 1 hour and 30 minutes. Eventhough the wet oxidation process is time saving, it has a drawback of higher impurity contents of the oxides. Generally MOS ICs require a very pure oxide for reliable performance. For this purpose specially dry oxidation process is preferred.

The process is known as **thermal oxidation** because the oxide layer is formed on silicon wafer using high temperatures. The silicon surface has very high affinity for oxygen, an oxide layer rapidly forms when silicon is exposed to a oxidising agent even at room temperature. During the oxidation process, the Si-SiO_2 interface pierces into the silicon substrate. It is observed that after completion of oxidation process, the oxide of thickness d consumes 0.44 d thickness of silicon. This is illustrated in the Fig. 1.8.

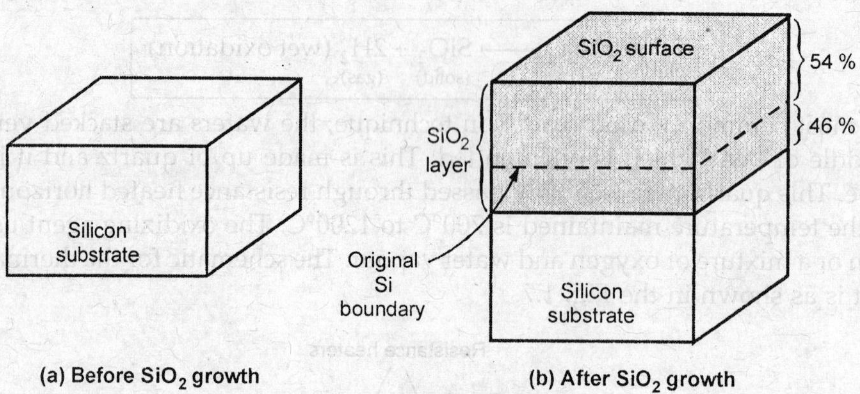

Fig. 1.8 Thermal oxidation

1.8 PHOTOLITHOGRAPHY

To produce extremely small circuits and to develop patterns on the silicon wafer, the process adopted is called **photolithography**. Conventionally using ultraviolet rays it is possible to produce devices with dimensions of 2 μm. Nowadays, using **electron beam lithography** or **X-ray lithography**, it is quiet possible to achieve dimensions of the devices less than 1 μm. Basically the photolithography process is subdivided into two processes namely,

(*i*) Photomask generation.

(*ii*) Lithography i.e. pattern generation.

(*iii*) Etching.

1.8.1 Photomask Generation

During the fabrication process of semiconductor devices, it is necessary to identify the regions of each circuit on the wafer so as to carry out selective doping process. Basically selective doping means in certain regions of the wafer, identical dopant is deposited or interconnections are made and at the same time remaining portion of the wafer is protected against doping. In general, a wafer consists of hundreds of identical circuits and each circuit may consist millions of devices. Thus identical steps are carried out simultaneously which require selected regions are exposed only; while the others are protected. This requirement can be fulfilled by using **mask** (protective layer pattern). During fabrication process, for one of the processes such as oxidation, diffusion, ion implantation and epitaxial growth, separate mask is used. First entire wafer is covered by a protective layer with certain pattern generated on it and then protective layer is removed from the selected regions.

A generation of photomask means designing a desired pattern for each device. A **mask** is simply a glass plate consisting a pattern for the complete wafer. The mask is generated using computer controlled system. The pattern to be generated is stored in the digital form. With the proper commands executed by the computer, a pattern generator is activated which uses an electron beam to draw pattern. This is commonly known as photoengraving. This pattern is drawn on high quality glass which is coated with material such as chromium.

After obtaining mask for each layer, individual mask is photographed. Then using step and repeat camera, photorepeating is done. Photorepeating means obtaining successive images in array, on the photographic plate which is moved in equal steps during exposure. After developing the exposed plate, final mask is prepared. This mask is usually called master mask. From this master mask, working copies are made which are called reticles. Such reticles are used for actual pattern transfer on a semiconductor wafer.

In general, reticle is about 5 or 10 times the actual size of the chip. The reticle consists two types of the regions namely transparent region (with no metal) and opaque region (with metal) when the light is passed through reticle placed on a photoresist coated wafer, only the shadow of the reticle is projected onto the surface of the wafer.

The mask for each level of fabrication process consists a multiple pattern with exactly identical boundaries but different patterns as shown in the Fig. 1.9.

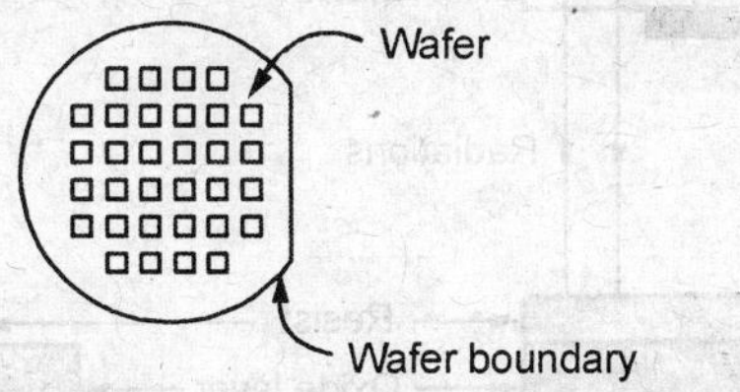

Fig. 1.9 Multiple images formed by lithography

For larger wafers accommodating number of individual integrated circuits or sites, a **wafer stepper** is used. A **wafer stepper** is an apparatus which holds the wafer and controls accurate movement to align optics to each site. This sequence produces large number of identical sites n the wafer shown in the Fig. 1.10. The test site locations contain various test structures and circuits. These test sites are very much essential to test wafer electrically during various fabrication processes.

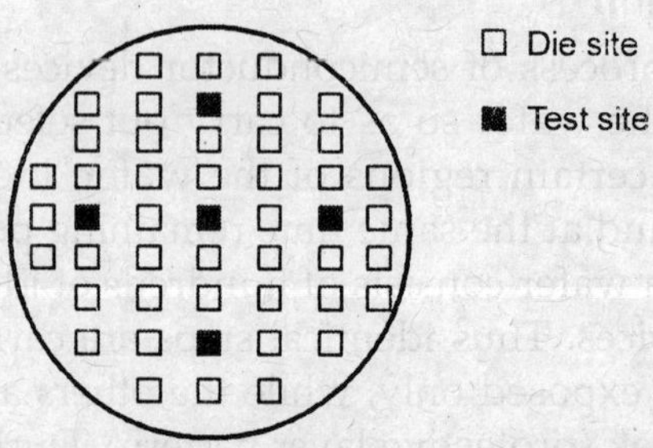

Fig. 1.10 Wafer sites for wafer stepper

1.8.2 Lithography

Lithography is a process in which the pattern generated on the photomask is transferred or imaged on the wafer covered with photoresist. A lithography process can be realized with subprocesses given as,

(*i*) photomask generation and deposition of photoresist on the wafer,
(*ii*) pattern transferring or imaging on the surface of wafer, and
(*iii*) etching of silicon oxide.

The lithographic process actually states with the generation of a photomask. To transfer this pattern present on the photomask, the wafer is coated with light sensitive liquid called **photoresist** (or simply **resist**). To have uniform coating of photoresist on the surface of the wafer, the photoresist liquid is sprayed on a spinning wafer held in a place by a chuck. Then the radiations from the radiating source are directed to the wafer through the photomask. The radiations travel through the clear or transparent portions of the pattern only. While the opaque portions block the radiations. To remove the material from the surface of the wafer which is not masked, the etching process is used. The schematic representation of the lithographic process is as shown in the Fig. 1.11.

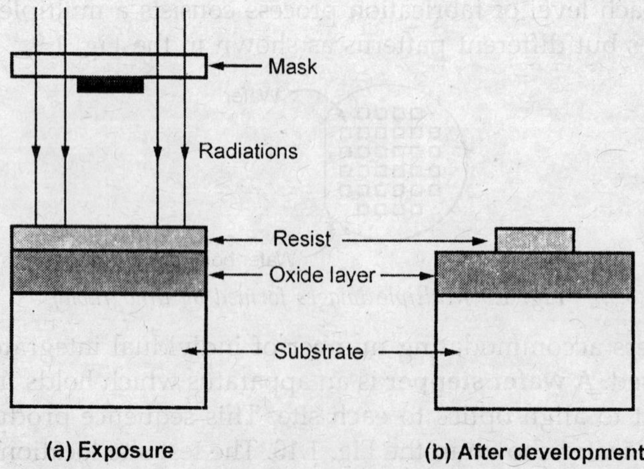

Fig. 1.11 Schematic representation of lithographic process

1.8.2.1 Optical Lithography

The lithographic process carried out using ultraviolet light as radiating source is called **optical lithography**. As explained earlier, a light sensitive liquid called a photoresist. The photoresist in a liquid form is sprayed over the surface of the wafer. The wafer is spined at high speed to achieve uniform thin coating of the photoresist over the surface of the wafer. To remove excess solvent, the wafer is baked in an oven. This is called **prebaking**. Then the wafer is cooled at room temperature. Then the wafer is exposed to U.V. light through the mask and then it is developed using developer solution (e.g. trichloroethylene) which dissolves the unexposed regions on the photoresist. After developing, the wafer is baked again, commonly called **postbaking**.

In general, a photoresist can be classified as positive photoresist and negative photoresist. With the positive photoresist, the window is opened wherever the ultraviolet light is passed through the transparent portion of the photomask. The common examples of positive photoresist are **MP-2400, HPR-206** etc. With the **negative photoresist**, the window is opened only under the opaque parts of the photomask. The common example of negative photoresist is **kodac microneg 747**. The negative photoresist on exposure to the light becomes less soluble in a developer solution, while positive photoresist becomes more soluble.

In the optical lithographic process, we have studied that the photomask is placed in contact with the wafer through a photoresist. At the surface of crystal, there is a possibility of irregular particles along with the dust particles. During process these particles stick to the mask. When such a mask is used for further operations, it causes defects in the surface. To overcome this, a proximity printing process is used. In this process, the distance of 10 to 20 μm is maintained between mask and the wafer.

1.8.2.2 Electron Beam Lithography (EBL)

When the feature size required is less than 1 μm, electron beam lithography is preferred as against optical lithography. The EBL offers higher resolution as compared with optical lithography. In this system, the photomask is generated using either **electron beam pattern generator or electron beam projection system**. The exposure time is longer in EBL as compared with optical lithography. Also the equipment is very costly. Hence this method is used only when very small device dimensions are demanded.

1.8.2.3 X-Ray lithography

All optical materials become opaque, at a reduced wavelength, because of fundamental absorption but the transmission increases in the X-ray region. The X-ray wavelength is small, hence the diffraction effects are reduced which gives high quality image. Using **proximity printing** technique, a pattern is generated on the wafer. The feature size of 0.1 μm is also possible with X-ray lithography. But similar to the EBL, the proces is too slow and the cost of the equipment is very high.

1.8.3 Etching

Etching is a process in which the material which is not masked by the lithographic process is removed. The etching technique is useful in removing unmasked material uniformly or selectively. The most widely used etching a techniques as wet etching, reactive plasma etching and ion etching.

In **wet etching**, the unmasked regions on the wafer are etched using wet etchants (i.e. chemicals) such as nitric acid, hydrofluoric acid. The main advantage of wet etching is high throughput. That means large number of wafers can be immersed in the etchants simultaneously to etch off unmasked regions. The etching rate in wet etching process is dependent on temperature. The selectivity is higher in wet etching process. This process minimises lateral removal of resist and overetching of the substrate material.

The **reactive plasma** etching process is dry etching process. The wafer under consideration is kept in a reaction chamber. The reaction chamber is filled with reactive gas. This chamber is excited by high value R.F. field producing a plasma discharge in the reaction chamber. The plasma is a collection of electrons, positive ions, negative ions and molecules generated due to the interaction of gases introduced in the chamber. These ions react with the wafer and remove the unmasked material. The reactive plasma etching is preferred over wet etching because it is less sensitive to temperature changes and the process can be controlled easily. The pattern obtained after reactive plasma etching is superior than obtained using wet etching. The commonly used gas for etching silicides, silicon, silicon dioxide is CF_4 and its compounds. For SiO_2 (silicon dioxide) $SiCl_4$ is most commonly used gas while for silicides and silicon etching NF_3 is also proffered.

In the **reactive ion etching** process, the wafers are kept in a reaction chamber between two electrodes, anode and cathode. Using high electric field between two electrodes, the plasma is produced between these electrodes. The plasma consists different ions, which partly react with wafer and partly remove the material by ion bombardment. This process is advantageous as it provides highest selectivity and uniformity.

The plasma reactor used in the reactive plasma etching process is as shown in the Fig. 1.12.

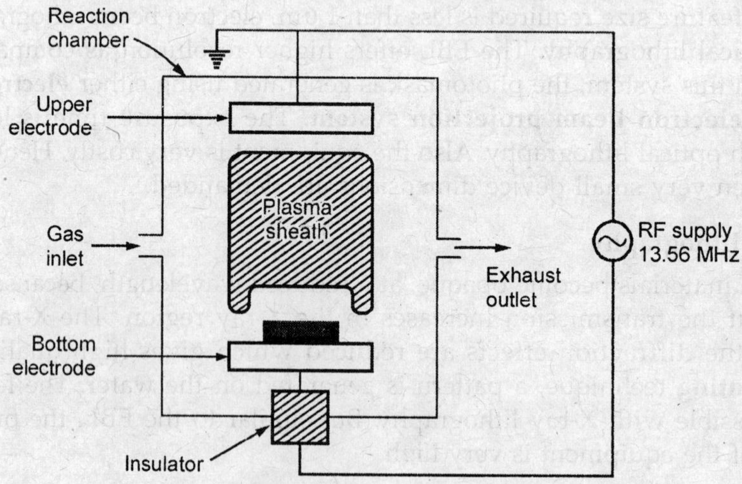

Fig. 1.12 Plasma reactor

The Fig. 1.13 illustrates the window opening using the lithographic process for positive and negative photoresist.

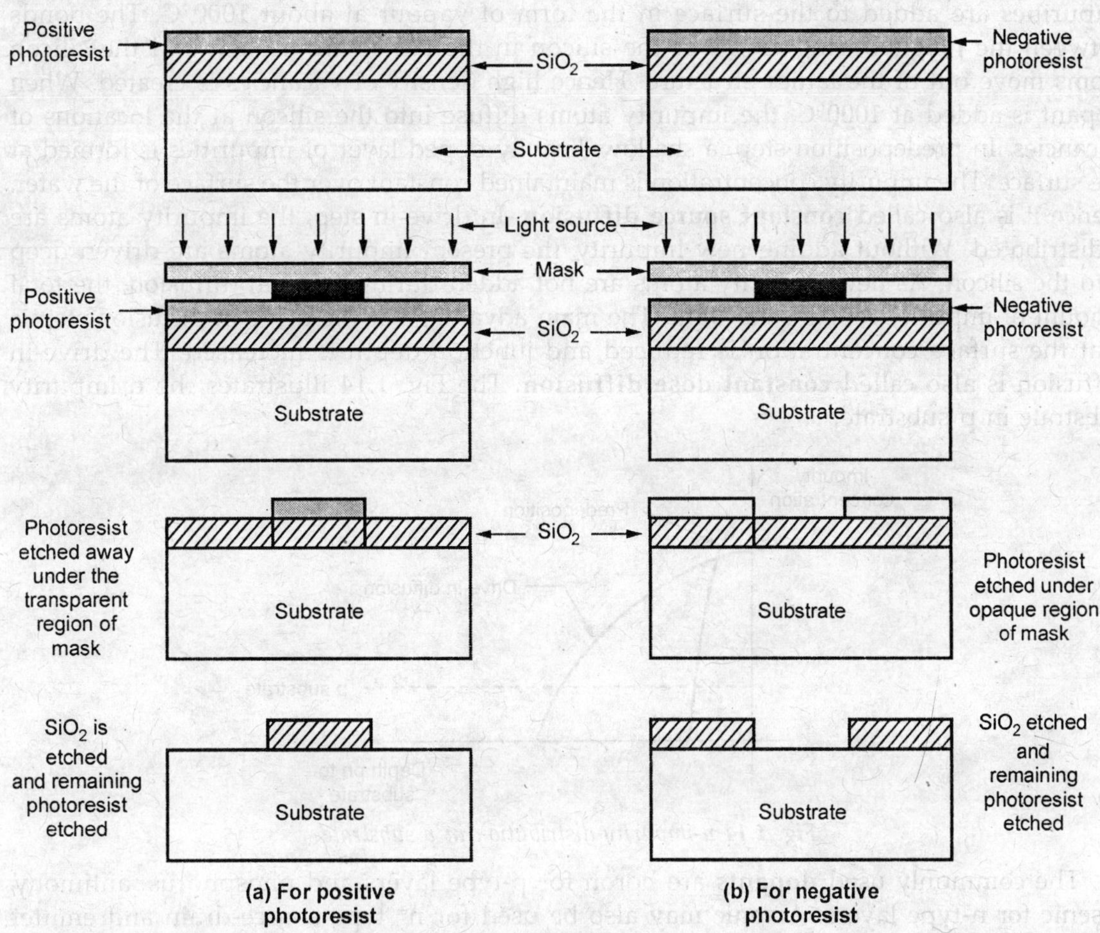

Fig. 1.13 Window opening using lithographic process using positive and negative photoresist

1.9 DIFFUSION

The process of doping i.e. adding impurity to the silicon wafer is called **diffusion.** Diffusion is used to form bases, emitters and resistors in bipolar device technology and to form source and drain regions in MOS device technology. The dopant can be added into the silicon wafer by using one of the most commonly used methods.

(*i*) Diffusion from a **chemical source** in a vapour form at very high temperature.

(*ii*) Diffusion from **doped-oxide** source, and

(*iii*) Diffusion from **ion-implanted** layer.

In general, diffusion process takes place in two steps namely **pre-deposition** diffusion and drive-in **diffusion**. In the first step, i.e. **pre-deposition** diffusion step, very high concentration of dopant atoms are implanted on the surface of silicon. These dopants or impurities are added to the surface in the form of vapour at about 1000°C. The bonds between the neighbouring atoms of the silicon in the lattice are broken and the silicon atoms move out of the lattice structure. Hence high density of vacancies is created. When dopant is added at 1000°C, the impurity atoms diffuse into the silicon at the locations of vacancies. In predeposition step, a shallow, heavily doped layer of impurities is formed at the surface. The impurity concentration is maintained constant over the surface of the wafer. Hence it is also called **constant source diffusion**. In drive-in step, the impurity atoms are redistributed. Without adding new impurity, the present impurity atoms are driven deep into the silicon. As new impurity atoms are not added during drive-in diffusion, the total amount of impurity remains constant. The main advantage of the drive-in diffusion step is that the surface concentration is reduced and junction depth is increased. The drive-in diffusion is also called **constant dose diffusion**. The Fig. 1.14 illustrates the n impurity substrate in p substrate.

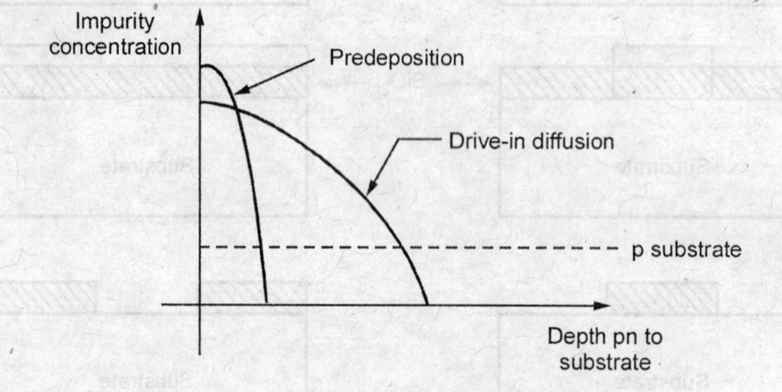

Fig. 1.14 n-impurity distribution in p substrate

The commonly used **dopants** are boron for p-type layers and phosphorus, antimony, arsenic for n-type layers. Arsenic may also be used for n^+ type source-drain and emitter diffusions. It is observed that practically, instead of above mentioned element, the compounds of these elements are more suitable. Such compounds may be derived from solid, liquid or gaseous sources.

A schematic representation of the diffusion process is as shown in the Fig. 1.15. The silicon wafers are stacked vertically on a movable quartz boat inside quartz furnace tube. The temperature of the quartz furnace tube is increased with the help of resistance heaters. The dopant to be introduced is kept in a container. The dopant in the liquid form e.g. $POCl_3$ may be placed inside the quartz tube in low temperature region or preferably outside the tube with precisely controlled temperature to maintain dopant in the liquid form. Typically nitrogen and oxygen are used as carrier gases.

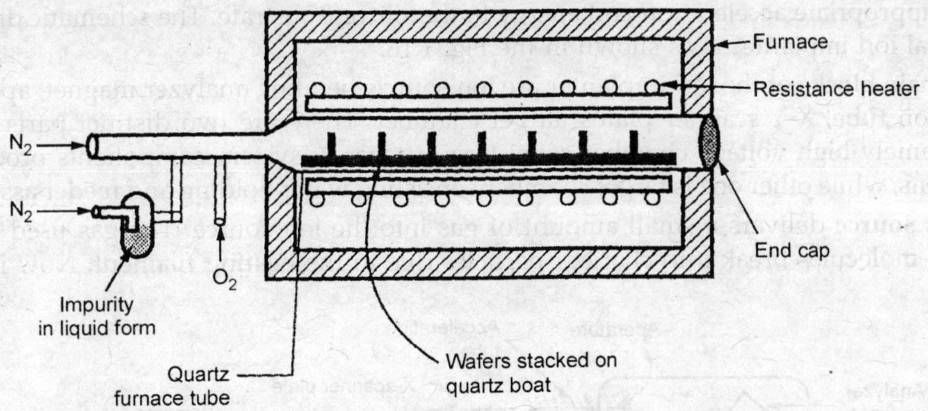

Fig. 1.15 Schematic representation of diffusion process

When the carrier gases pass over the container, they carry dopant vapour into furnace. In furnace, the gases are deposited on the silicon surface. The layer containing silicon, oxygen, phosphorus is formed on the surface. At high temperature such as 900°C to 1000°C, the dopant gets diffused into silicon. For drive-in diffusion, the temperature is further increased to 1100°C. So that the dopant gets diffused with more depth in silicon. For this step, with a proper control circuitry, the impurity supply is stopped. The proper depth can be achieved by controlling parameters such as time and temperature of the process.

1.10 ION IMPLANTATION

As we know, the conductivity of the semiconductor increases when small impurity is added to it. The process of adding impurity is called **doping** while the impurity to be added is called **dopant**. So **ion implantation** is a process of adding dopant to the silicon substrate. The ion implantation process is controllable, reproducible and also there are no unwanted side effects. The ion implantation process is proffered over diffusion because of following reasons.

(*i*) The impurity concentration is highly uniform typically within 1%, over the wafer,

(*ii*) The degree of uniformity is maintained same from wafer to wafer,

(*iii*) The layer can be formed anywhere within substrate,

(*iv*) The lateral spread is very small.

Basically the ion implantation process is low temperature process. In this process, the dopant atoms are vapourized. They are accelerated by an accelerator and then bombarded on silicon substrate. The entire wafer or selected part of it, is exposed to the beam of vapourized, accelerated dopant atoms. The beam injects the dopant atoms into unmasked sections of the substrate. The depant atoms directly enter the crystal lattice of the silicon. In the lattice, due to the collisions with silicon atoms the dopant atoms starts loosing energy. When the energy is totally lost, the dopant atoms are found at some depth within the lattice itself. The depth of penetrations is controlled by the acceleration energy of the incident beam and the doping concentration. In general, the ion implantation is made through thin oxide, as compared to masking which is made through thicker layer of oxide.

The main objective of the basic ion implantation is to direct a beam of dopant atoms with the appropriate acceleration and energy to the silicon substrate. The schematic diagram of a typical ion-implanter is as shown in the Fig. 1.16.

The main blocks of the ion-implanter are ion source, bending analyzer magnet, aperture, acceleration tube, X-Y scanner plates, target chamber. There are two distinct parts of the system namely high voltage chamber consisting number of system components producing desired ions, while other one is target chamber consisting wafer holding and feeder assembly.

A **gas source** delivers a small amount of gas into the ion source. The gas used is BF_3. There are molecules break into charged particles due to the heating filament. Now in the

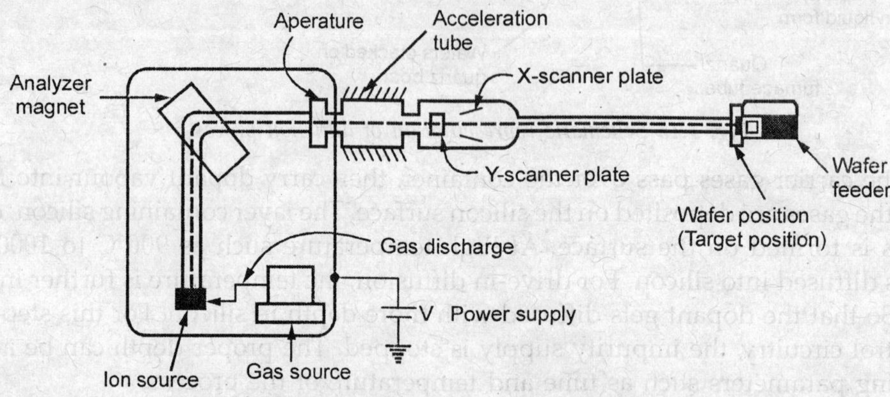

Fig. 1.16 Schematic diagram of typical ion-implanter

ion source, there are desired ions along with other charged particles. Due to the high voltage (about 20 kV), the charged ions are pulled out of the ion source into the bending magnet analyzer. Note that the pressure in the system is maintained very low (of the order of 10^{-6} Torr) so as to avoid scattering of ions due to gas molecules. The **bending analyzer magnet** selects the ions with desired charge to mass ratio with the help of properly applied magnetic field. Thus the desired ions only can travel through the analyzer, while the others impinge on the analyzer walls. In the acceleration tube, the ions are accelerated to the sufficiently high implantation energy.

The aperture focuses the beam of ions. The X-Y scanner plates adjust the sweep of the beam over the wafer placed in target chamber. The wafer is slightly offset to the axis of the acceleration tube so as to avoid deflection of ions on to the wafer. In typical ion-implanter, accelerator voltages range from few kV to 250 kV for medium energy implanters, while upto 2 MV for high energy implanters. Typically a medium energy implanter is 6 m long, 3 m wide and 2 m high. It process 200 wafers per hour. The total number of ions entering the target is called **dose**. The medium energy implantation dosage extends from about 10^{10} to 10^{17} atoms/cm^2.

1.11 ISOLATION TECHNIQUES

We have studied that on a single silicon wafer, large number of components are fabricated. So it is important to provide isolation between these components electrically. The most commonly used isolation techniques are, (*i*) p-n junction isolation

(*ii*) Dielectric isolation.

The selection of the isolation technique in fabrication of any IC depends on the application of the IC i.e. for general purpose applications or specialized military applications.

1.11.1 p-n Junction Isolation Technique

Consider p-substrate with n-type epitaxial layer grown over it. To provide isolation, a p-type impurity with high concentration (i.e. p^+ type impurity) is diffused selectively into an epitaxial layer such that it reaches to the p-substrate as shown in the Fig. 1.17.

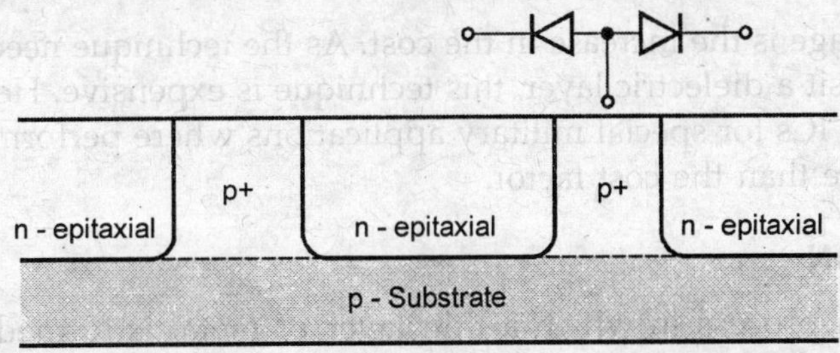

Fig. 1.17 p-n junction isolation technique

From the Fig. 1.17 it is clear that n-epitaxial region forms a region which is surrounded by p-type regions. This region is called **island**. Two regions are separated by two back to back diodes as shown in the Fig. 1.17. These two back to back diodes serve as **isolation regions** if both are reverse biased. So when very high negative voltage is applied to p-type substrate, both the diodes are reverse biased and thus the electrical isolation is observed between two isolation islands.

The important consideration for the isolation is that the concentration of the acceptor atoms must be higher than the concentration of the p-type substrate. If this condition is not maintained, then the depletion layer of the reverse biased diode penetrates into the p + region and may connect isolation islands together.

The main advantage of p-n junction isolation is that different components can be fabricated within the isolation islands. But the disadvantage of this technique is the presence of undesirable and unavoidable parasitic capacitances at the isolation p-n junctions. These capacitances limit high frequency performance of the circuit. Hence this technique is generally proffered for the general purpose ICs only.

1.11.2 Dielectric Isolation Technique

In the dielectric isolation technique, the electrical isolation to each component is provided by surrounding them with a solid dielectric. The most commonly used dielectric are silicon dioxide, ruby etc. It also fulfils the purpose of physical isolation. The dielectric layer thickness is large enough so that the parasitic capacitances are negligible. The dielectric Isolation technique is illustrated in the Fig. 1.18.

The main advantage of the technique is that within same silicon substrate p-n-p as well as n-p-n transistors can be fabricated.

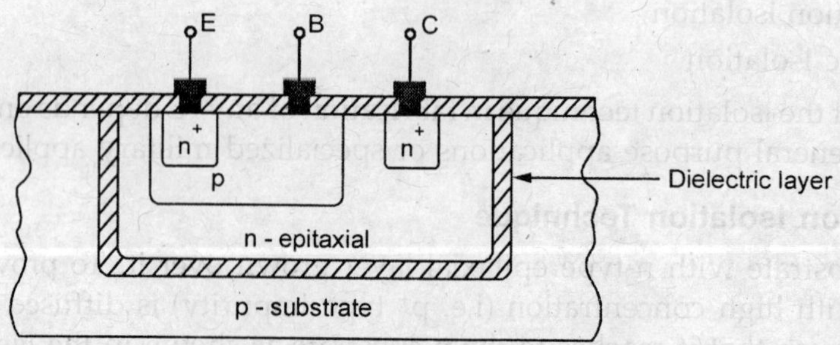

Fig. 1.18 Dielectric isolation technique

But the disadvantage is the increase in the cost. As the technique needs additional steps in fabrication to deposit a dielectric layer, this technique is expensive. Hence this technique is used for fabricating ICs for special military applications where performance of the circuit is of prime importance than the cost factor.

1.12 METALLIZATION

Metallization is a process in which a thin layer of metal is formed which is used to make interconnections between the components on the chip as well as interconnections between the components and the outside world.

In general metallization applications are divided into three groups, (*i*) gate, (*ii*) contact and (*iii*) interconnects. In VLSI, different new metallization schemes for gate, interconnections, ohmic contacts are introduced. But the application of any new metallization scheme is tested in accordance with certain requirements as given below.

(*i*) The metal layer should be of low resistivity.

(*ii*) The formation of layer should be easy.

(*iii*) The layer should be easy for pattern generation to etched off.

(*iv*) The layer should be stable in oxidizing ambients.

(*v*) The layer should have surface smoothness, mechanically stable with low stress and good adherence.

(*vi*) The layer should not react with final metal.

(*vii*) The metal should not contaminate devices, wafers.

(*viii*) The device characteristics should be good enough.

Based on the types of applications there are three types of metallizations namely gate metallization, contact metallization and interconnection metallization. The metallization which connects a base (in bipolar transistors) or gate (in MOSFETs) to the neighbouring two regions is called **gate metallization.** The metallization which is directly in contact with semiconductor is called **contact metallization.** The metallization which connects number of bipolar devices or MOSFETs is called **interconnection metallization.**

1.12.1 Metals and Alloys for Metallization

In most of the IC's, aluminium is the widely used metal for metallization because,

(*i*) It is a good conductor.

(*ii*) It can form mechanical bonds with silicon.

(*iii*) It can form low resistance, ohmic contacts with heavily doped n-type and p-type silicon.

But now a days alongwith aluminium, platinum silicide (Pt Si) has been used as a schottky barrier contact. It may be used as an ohmic contact simply for deep junctions. For high reliable connections to the outside, platinum/titanium/gold beam lead technology is preferred.

Actually no metal satisfies all the desired properties. Eventhough aluminium is most widely used metal, it has certain drawbacks such as :

(*i*) Low melting point

(*ii*) Undesirable electromigration behaviour.

In MOS devices, polysilicon is generally used for gate metallization. With this metal in contact with gate oxides, devices characteristics and processes are unaltered. To have low resistance at gate and interconnection level, polysilicon is replaced by polysilicon/refractory silicides.

In the production of microprocessors and RAMs, refractor silicides with highest compatibility are used. The common examples are molybdenum ($MOSi_2$), tantalum ($TaSi_2$) and tungston (WSi_2) which are all disilicides.

Because of junction spiking, high contact resistance, electromigration resistance, contact failure takes place. The contact problems can be overcome by using (*i*) Si-Alloy, (*ii*) polysilicon layers gate and aluminium, (*iii*) selectively deposited tungston. Some times self aligned silicide such as platinum silicide (PtSi) is used in-between silicon and silicide to ensure high metallurgical contact.

When aluminium is used as a top metal there is a possibility of interaction of silicide with aluminium in the 200 to 500°C temperature range. To overcome this, transition metal nitrides, carbides and borides are used as barrier between silicide and aluminium.

Lastly the most important characteristics of a metallization is that it should be good adhesive. From that point of view, the metals, forming oxides, such as Al, Ta, Tl etc. are most commonly used. The titanium is the most commonly used metal providing good adherence. The important property of titanium is that it forms a good bond with SiO_2 at two temperatures and acts as a glue layer.

The following Table 1.4 gives the list of metals and alloys used for different metallization applications.

Table 1.4

Metallization	Metals / Alloys
1. Gates, contacts and interconnections.	Polysilicon, silicides, nitrides, carbides, borides, refractory metals, aluminium and alloys of two or three of above.
2. Top level.	Aluminium.
3. Diffusion barrier.	Nitrides, carbides, silicides, borides.
4. Selectively formed.	Tungston, aluminium and silicides.

The important properties of the metallization are,

(*i*) Resistivity (μ Ω-cm),

(*ii*) Melting point (°C), and

(*iii*) Linear thermal expansion coefficient (ppm/°C).

The resistivity ρ for the pure, thick and single crystalline film is lower; while for impure and grain boundaries, it is higher. Due to chemical or metallurgical interactions, the resistivity of metallization either increases or decreases.

The second important property is the melting point. The solid state diffusion controls grain growth, annealing of defects as well as interactions in solid state. It is observed that the solid state is effective only at a temperature larger than one third of the melting point of the solid in which diffusion takes place.

The last property is the thermal expansion coefficient α. If there is a difference between the thermal expansion coefficients of the film and substrate, the stress conditions are observed on thin films. The stress will be greater, if the difference between the thermal expansions coefficients is more. The silicides show large stress conditions. This internal stress is balanced by a substrate producing opposite stress. Thus for proper formation of metal layer on substrate, the film stresses and factors affecting must be studied in detail.

1.12.2 Metallization Process

The process takes place in a chamber which is called **vacuum evaporation chamber**. The chamber pressure is adjusted to the range of 10^{-6} to 10^{-7} Torr. The material to be evaporated is placed in a basket. Then using electron gun, high power density electron beam is focussed at the Substrate surface of the material. Due to this, material starts heating up and vaporizing. These vapours hit substrate and condense there to form a thing film coating. After the metallization process is over, the thin film is patterned to form required interconnections. By using proper etching process, aluminium is removed form unwanted places.

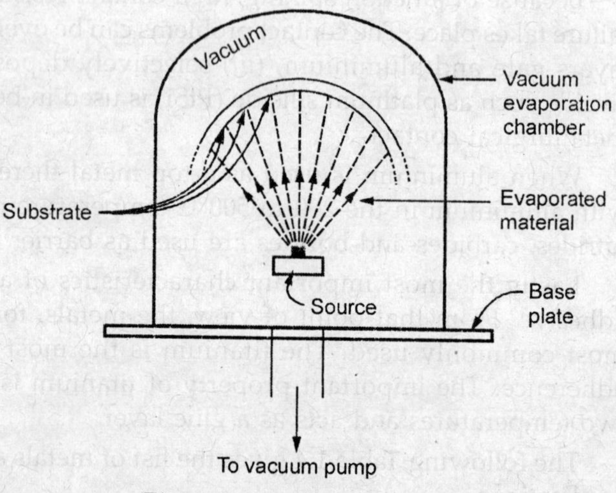

Fig. 1.19 Metallization process

In general, there are two important deposition processes.

1. Chemical Vapour Deposition: This process has number of advantages over other process such as, (*i*) low temperature process, (*ii*) high throughput, and (*iii*) excellent step coverage.

Using chemical vapour deposition technique only Molybdenum and tungston is deposited. This process is carried at low pressure, hence also known as LPCVD. Using laser CVD, selective area deposition or direct writing throughout, both is possible.

2. Physical Vapour Deposition: There are two types of physical vapour deposition processes namely **evaporation** and **sputtering**. In both the processes, first the condensed phase of the solid is converted into gaseous or vapour phase. Then the gaseous phase of the solid is transported to the substrate. And then lastly the gaseous source is condensed on the substrate followed by film growth.

The **evaporation method** is the simplest method of film deposition by the condensation of a vapour on a substrate. In this method, the temperature of the substrate is maintained lower than that of the vapour. When any metal is heated to sufficiently high temperature, it vapourizes. To heat a metal to high temperature different methods of heating are used such as resistance heating, inductive heating, electron bombardment and laser heating. For aluminium, which is the most common metal used for metallization, any of the methods can be used.

In **sputter deposition**, unlike evaporation method, energetic ions are bombarded on the target material. Due to this process, some atoms of the target materials are released. These released atoms are then condensed on the substrate. The sputtering deposition process is applicable to any type of materials such as insulators, semiconductors, metals or alloys. As compared to evaporation process, the sputtering process is well controlled. This is carried at relatively high pressures like 1 Pa.

1.12.3 Metallization Applications

1. Gate and interconnection metallization controls the speed of the circuit by controlling the resistance of the interconnection lines. For high speed operation, such resistance should be as small as possible.
2. The gate and interconnection metallization also controls flat band voltage V_{FB}, which is essential to maintain a flat band condition in the semiconductor.
3. The contact metallization gives electrically and mechanically stable ohmic contact having contact resistance negligibly small compared to the device resistance.
4. The top level metal is thick as it carries current which provides connection to the outside world.
5. The metallization is used to produce rectifying contacts and diffusion barriers between reacting metallic films.

1.13 PACKAGING OF ICs

After completing all the fabrication processes, several chips are ready on a wafer. Each of the chip is nothing but a complete circuit. Now the next step is to separate out these chips and package them individually. Using a diamond tipped tool, lines are scribed along the rectangular grides on the surface of the wafer. The wafer is cut off along the lines drawn using the sharp diamond tipped tool. Thus individual chips are separated from each other. Then each chip is assembled on a suitable package.

For comparing and assessing different packages, features are considered (*i*) maximum pin count (*ii*) dimensions (*iii*) pitch (spacing between the centres of adjacent pins) (*iv*) encapsulating material (ceramic or plastic) (*v*) mode of mounting (plated through hole-TH or surface mount - SM) (*vi*) maximum power dissipation.

For TH mounting DIP (dual in line package) and PGA (pin grid array) are the only standard packages. For small scale integration (SSI) and medium scale integration (MSI), the different packages available are SIP (single in-line package), ZIP (zig-zag-in-line package) and QIPC quad-in-line package with TH mounting type. For low pin counts the packages available are SO (small out line package), SSOP (Shrunk small out line package). Both these are with SM mounting type.

With SM mounting types there are two more packages namely chip carrier and TQFP (thin quad flat pack). The chip carrier uses either plastic or ceramic as encapsulating material. The TQFP has very low profile and small weight.

Table 1.5 gives the different package types with their abbreviations.

Table 1.5

Package type	Abbreviation
Dual-in-line	DIP
Small outline	SO_1 SOIC
Shrunk small outline package	SSOP
Single-in-line	SIP
Zig zag-in-line	ZIP
Quad-in-line	QIP
Plastic leaded chip carrier	PLCC
Leadless ceramic chip carrier	LCCC
Leadless chip carrier	LLCC or LCC
Leaded chip carrier	LDCC
Flat pack	FP
Quad flat pack	QFP
Ceramic quad flat pack	CQFP
Plastic quad flat pack	PQFP
Thin quad flat pack	TQFP
Pin grid array	PGA
Plastic pin grid array	PPGA
Ceramic pin grid array	CPGA

Table 1.6 summarizes typical package types with typical pin counts and mounting type.

Table 1.6

Package	Pin count range	Mounting type
Dual-in-line (DIP)	8 - 64	TH
Single-in-line (SIP)	5 - 40	TH
Zig-zag-in-line (ZIP)	14 - 28	TH
Quad-in-line (QIP)	14 - 64	TH
Small outline (SO)	8 - 32	SM
Chip carrier (CC)	16 - 200	SM
Flat Pack (FP)	10 - 300	SM
Pin grid array (PGA)	68 - 500	SM

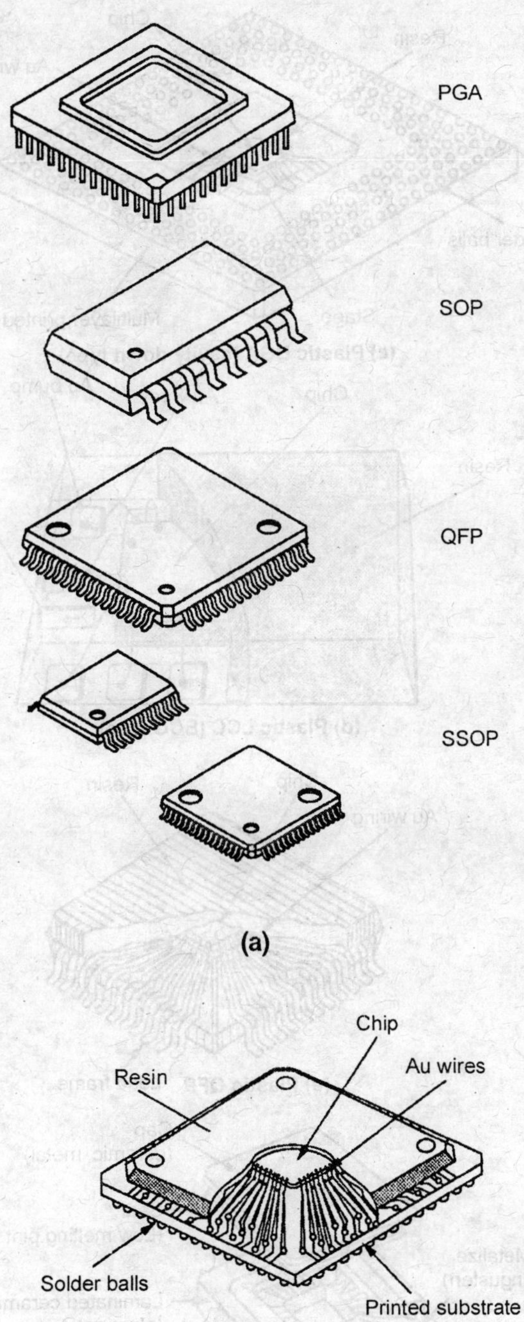

(b) Plastic BGA (mold type)

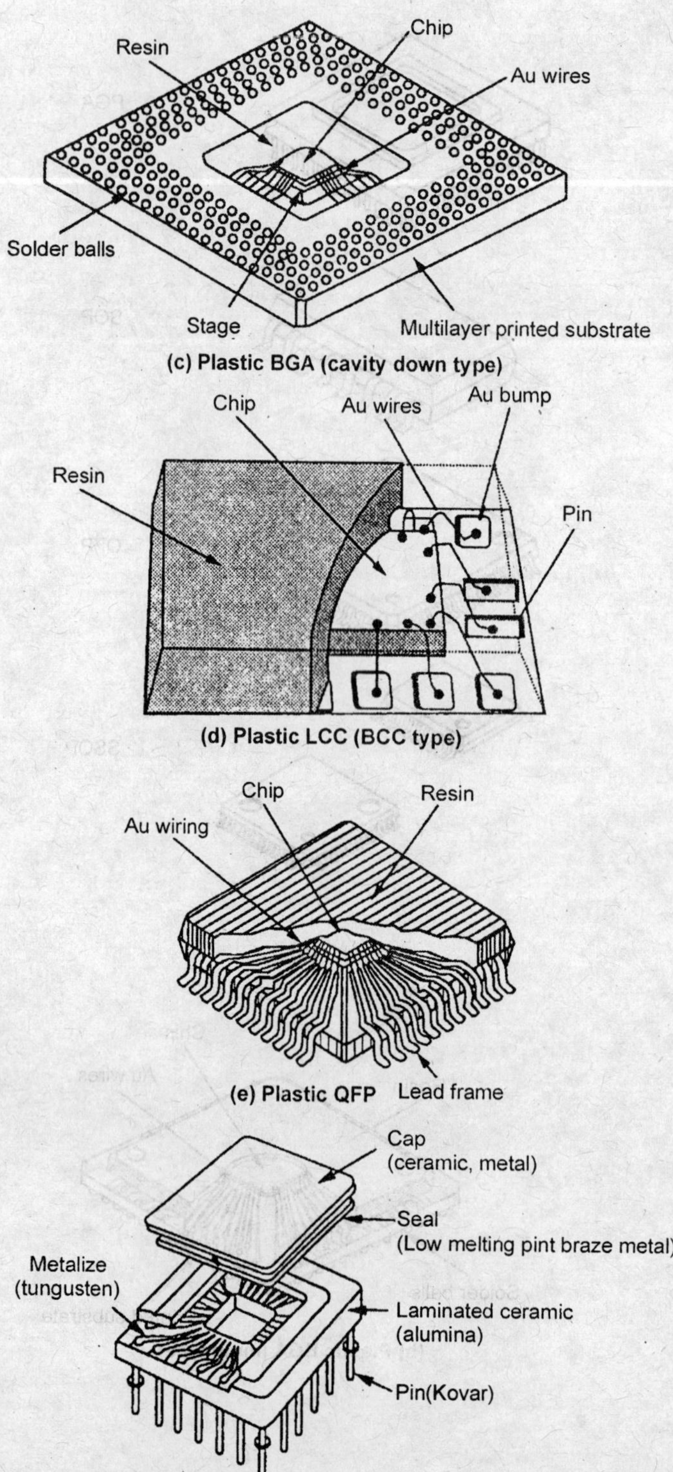

(c) Plastic BGA (cavity down type)

(d) Plastic LCC (BCC type)

(e) Plastic QFP

(f) Ceramic PGA (laminated)

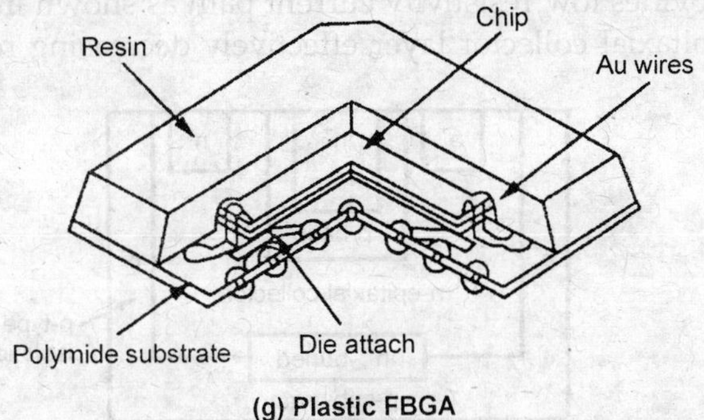

(g) Plastic FBGA

Fig. 1.20 Different IC packages

1.14 REALIZATION OF ACTIVE AND PASSIVE DEVICES IN INTEGRATED CIRCUITS

In this section, we shall study how to realize active and passive devices such as transistors, diodes, resistors, capacitors and inductors etc. in integrated circuits.

1.14.1 Monolithic Transistors

The cross-sectional view of an integrated monolithic transistor and discrete planar transistor are as shown in the Fig. 1.21 (*a*) and (*b*) respectively.

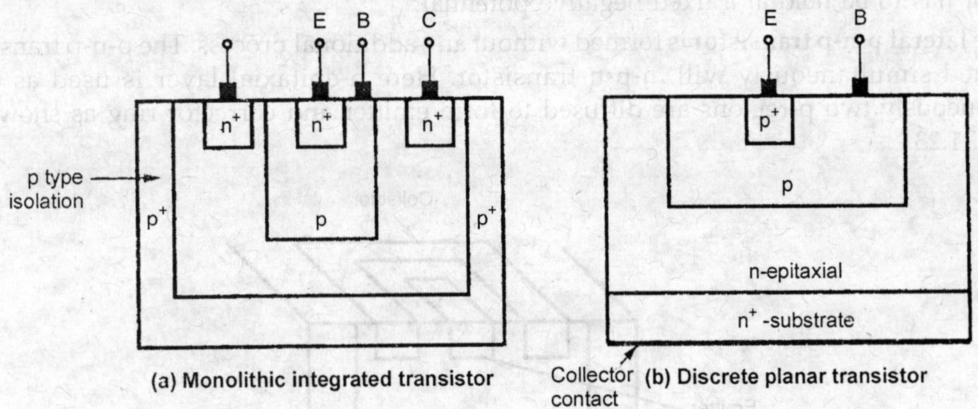

(a) Monolithic integrated transistor **(b) Discrete planar transistor**

Fig. 1.21

The main difference between monolithic integrated transistor and discrete planar transistor is that collector contact in monolithic integrated transistor is at top, while in the discrete planar transistor it is at bottom. Because of this the collector series resistance of the collector current path increases. This effectively increases the collector to emitter voltage $V_{CE(sat)}$ of the device. Also in monolithic integrated transistor as substrate is held at negative potential, additional parasitic capacitance appears between collector and substrate.

To overcome the increase in collector series resistance, buried n^+ layer is incorporated by using additional process step. The buried layer can be processed with heavily doped n^+ region in between p-type substrate and n-type epitaxial collector. The advantage of buried

n^+ layer is that it provides low resistivity current path as shown in the Fig. 1.22. The buried n^+ layer shunts n-epitaxial collector layer effectively decreasing resistance.

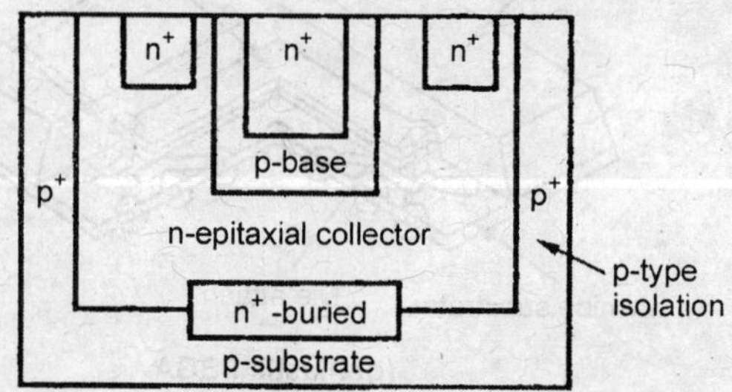

Fig. 1.22 Monolithic integrated transistor with n^+ buried layer

1.14.1.1 p-n-p Transistor

There are different ways of fabricating p-n-p transistor in integrated circuits. The important means of integrating p-n-p transistor are (*i*) vertical p-n-p, (*ii*) lateral p-n-p and (*iii*) triple diffused p-n-p.

In **vertical p-n-p transistor**, the p type substrate is used as p-type collector while the n-epitaxial layer is used as base. Obviously the next diffusion layer of p-type is used as an emitter. This type of p-n-p is called as **substrate p-n-p transistor**. The main drawback is that collector has to be held at a fixed negative potential.

The **lateral p-n-p transistor** is formed without any additional process. The p-n-p transistor is formed simultaneously will, n-p-n transistor. Here n-epitaxial layer is used as base simultaneously two p-regions are diffused to form emitter and corrector ring as shown in the Fig. 1.23.

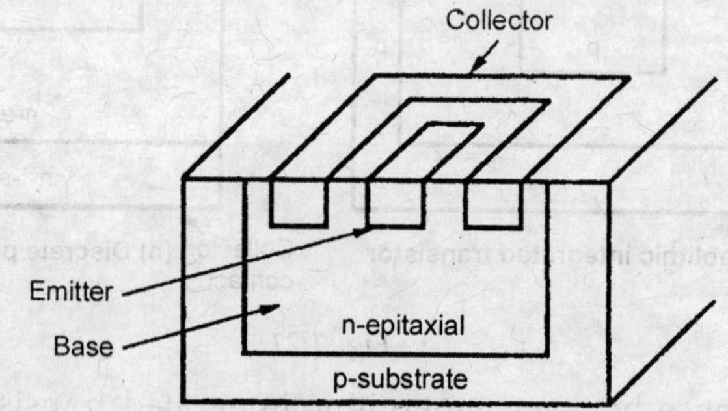

Fig. 1.23 Lateral p-n-p transistor

If an extra p-type diffusion is added after n-diffusion then we can get p-n-p transistor starting for standard n-p-n transistor. The p-n-p transistor thus obtained is called **triple diffused p-n-p** transistor. This type of transistor is fabricated only if there is a special need because of complicated process calculations and increase in cost due to addition in fabrication process.

1.14.2 Planar p-n Junction Diode Fabrication

Diodes are used extensively in the integrated circuits, for various applications such as digital applications. Note that in the integrated circuits, a p-n junction diode is formed from the bipolar transistor. Generally any two terminals of the transistor are connected together to get one terminal of diode, while the remaining terminal of the transistor serves as the second terminal of diode. Different transistor connections to utilize it as a diode are as shown in the Fig. 1.24.

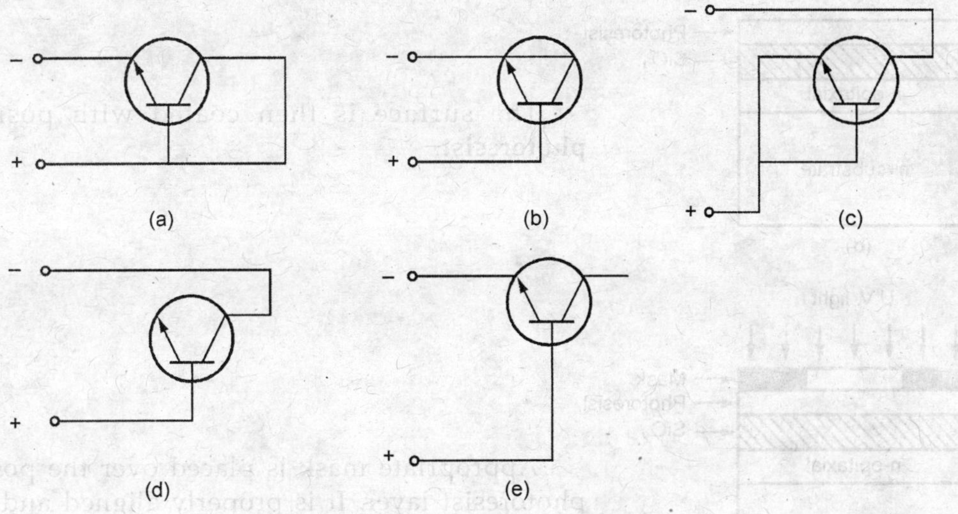

Fig. 1.24 Transistor utilized as diode in IC

Depending upon desired circuit performance and application, diode connection is selected. The diode represented in the Fig. 1.24 (*a*) is used in digital circuits for high speed applications because of its lowest storage time and lowest forward voltage drop. The diodes represented in the Fig. 1.24 (*b*) and (*e*) are used as stored charged devices. The diodes represented in the Fig. 1.24 (*c*) and (*d*) have highest breakdown voltage.

Fabrication of a planar p-n junction diode is illustrated in the Fig. 1.25.

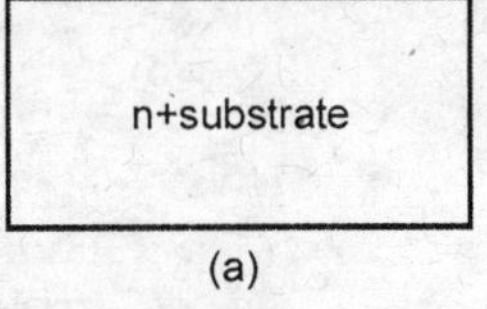

(a)

The starting material for the planar p-n junction diode is n^+ substrate which is grown by using **Czochralski growing process**. The substrate is about 150 μm thick.

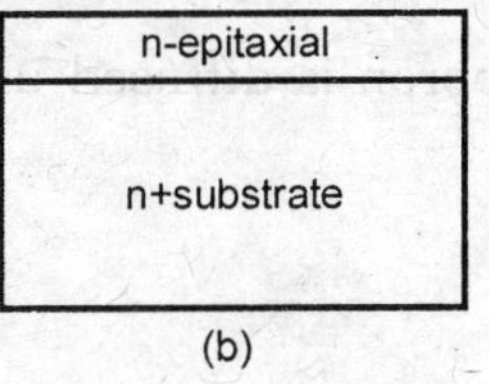

(b)

Using **epitaxial growth process,** a layer of n-type silicon is deposited on the substrate. The layer is about 1 to 5 μm thick.

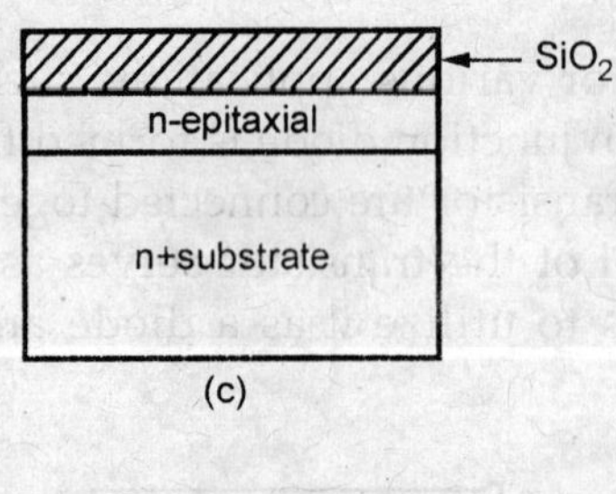

(c)

Using **oxidation process**, silicon dioxide (SiO_2) is deposited.

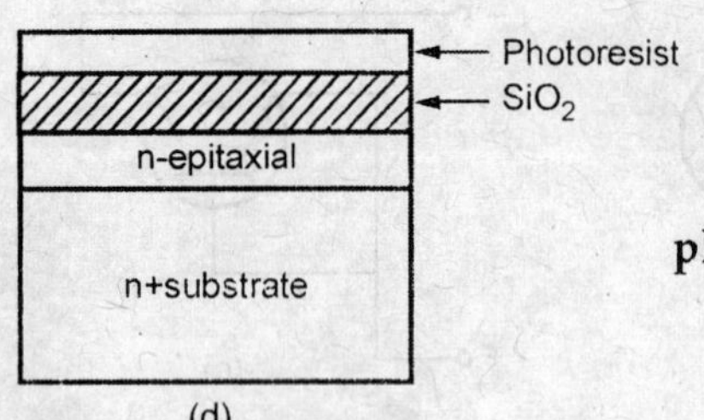

(d)

The surface is then coated with positive **photoresist**.

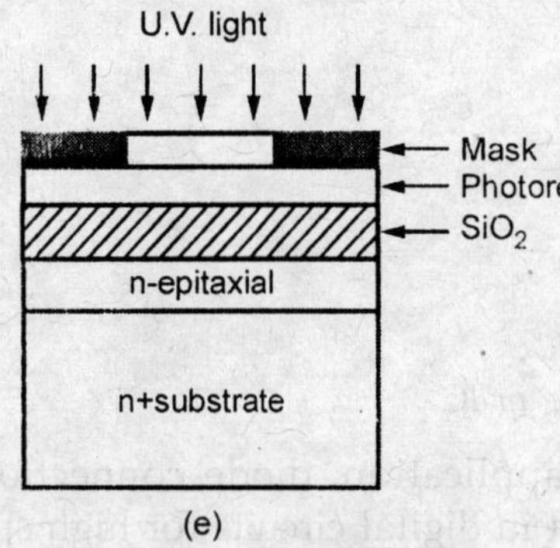

(e)

Appropriate mask is placed over the positive photoresist layer. It is properly aligned and then exposed to the ultraviolet light.

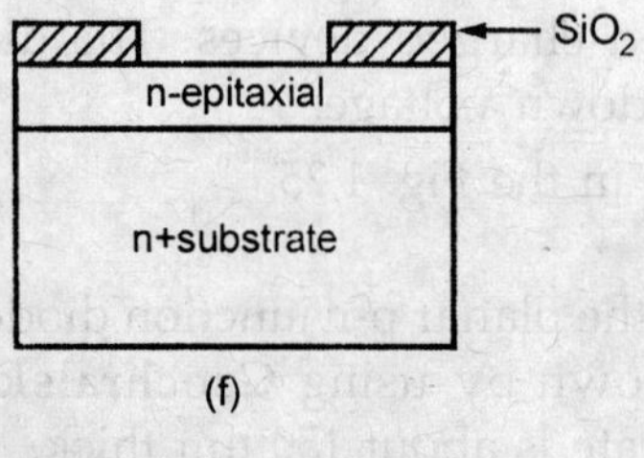

(f)

Then the mask is removed and photoresist is removed. Using **etching process**, only silicon dioxide layer under the exposed resist is etched.

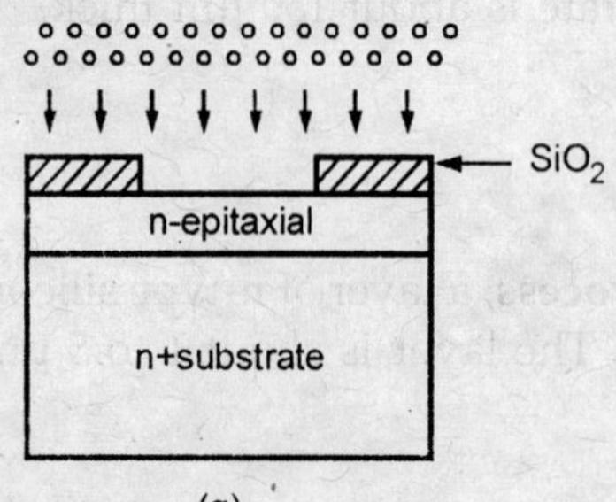

(g)

Then to form p-region, boron is diffused using **ion-implantation** process.

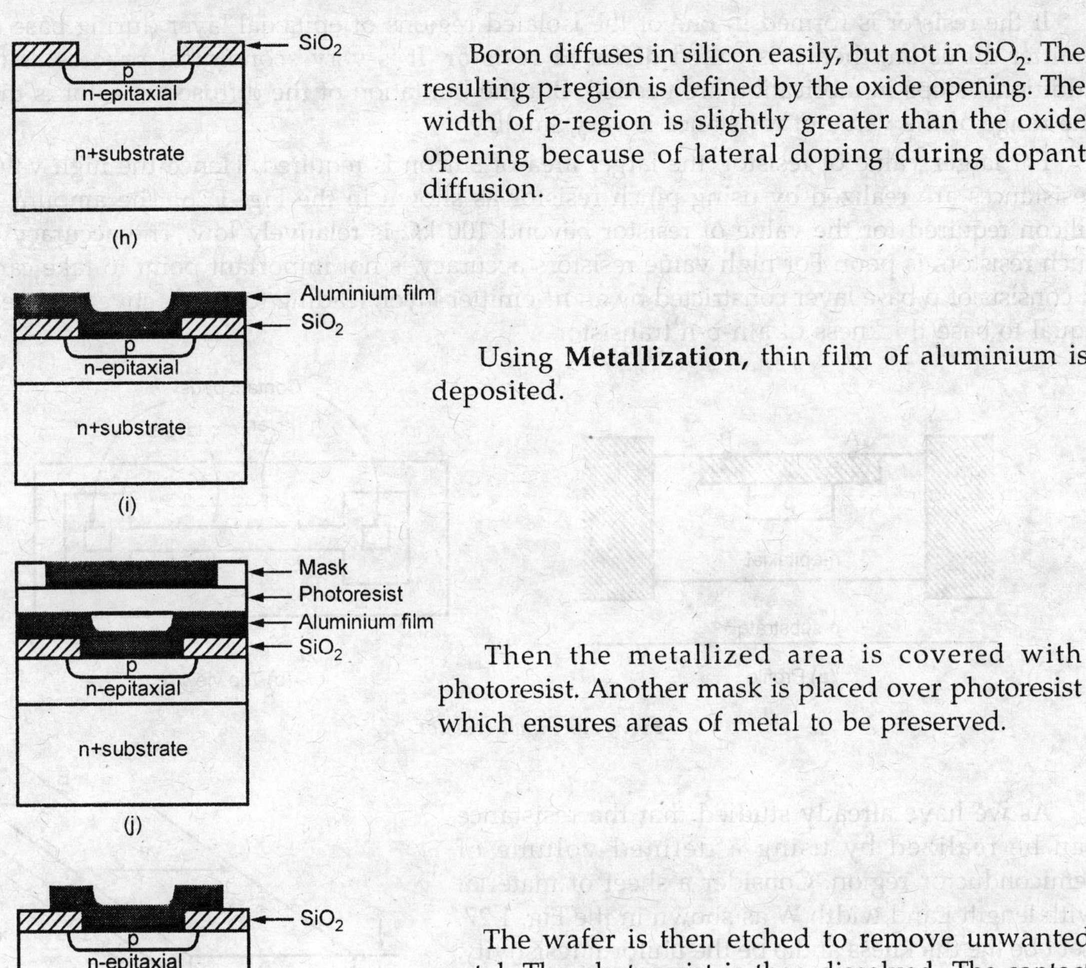

Boron diffuses in silicon easily, but not in SiO_2. The resulting p-region is defined by the oxide opening. The width of p-region is slightly greater than the oxide opening because of lateral doping during dopant diffusion.

Using **Metallization,** thin film of aluminium is deposited.

Then the metallized area is covered with photoresist. Another mask is placed over photoresist which ensures areas of metal to be preserved.

The wafer is then etched to remove unwanted metal. The photoresist is then dissolved. The contact metal is deposited on the back surface and using heat treatment ohmic contacts are made.

Fig. 1.25 Fabrication of planar p-n junction diode

1.14.3 Integrated Resistors

In integrate circuit design, the importance is given to the maximum usage of transistors. For example digital CMOS, nMOS and GaAs circuits are fabricated entirely with transistors and diodes. The resistors are grouped into two groups; one formed within monolithic IC and other composed of film resistors.

The monolithic IC resistors consist suitably dimensioned layers which would form part of the transistor normally. Obviously the resistivity of such layers is determined from transistor characteristics.

If the resistor is formed in one of the isolated regions of epitaxial layer during base or emitter diffusion, then it is called **diffused resistor**. It is very economical process as no additional steps in fabrication are needed. But the limitation of the diffused resistor is that the range of the value of resistance is very small.

For larger value of resistor, the larger area of silicon is required. Hence the high value resistances are realized by using pinch resistor as shown in the Fig. 1.26. The amount of silicon required for the value of resistor beyond 100 kΩ is relatively low. The accuracy of such resistors is poor. For high value resistors accuracy is not important point to take care. It consists of p base layer constricted by an n^+ emitter layer, leading to an effective thickness equal to base thickness of a n-p-n transistor.

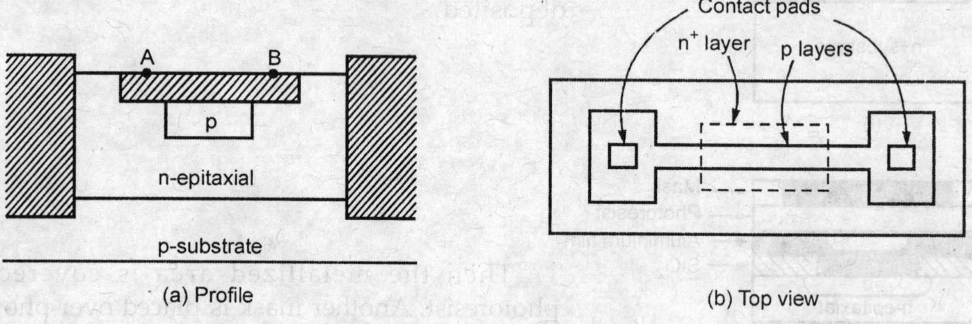

Fig. 1.26 Pinch resistor

As we have already studied that the resistance can be realized by using a defined volume of semiconductor region. Consider a sheet of material with length Land width W as shown in the Fig. 1.27. Let t be the thickness and ρ be the uniform resistivity.

The resistance R between layers A and B is given by

Fig. 1.27 Sheet resistor

$$A = \frac{\rho L}{A} = \frac{\rho L}{tW} \quad \text{...(1)}$$

For square surface area, W = L, then the resistance of the material is given by

$$R_S = \frac{\rho L}{(tL)} = \frac{\rho}{t} \quad \text{...(2)}$$

Thus rearranging equation (1) using equation (2), we can write

$$R = R_S\left(\frac{L}{W}\right) \quad \text{... (3)}$$

Here ratio $\frac{L}{W}$ is called **aspect ratio**. Thus using this technique base resistor in the range 20 Ω to 300 Ω can be fabricated. Similarly the resistance of the emitter diffusion can be fabricated as sheet resistor but the range of resistance is only 10 to 1 kΩ.

By using n-epitaxial collector layer, the large value resistances than base and emitter diffusion can be achieved. Such resistors are called epitaxial resistors as shown in the Fig. 1.28.

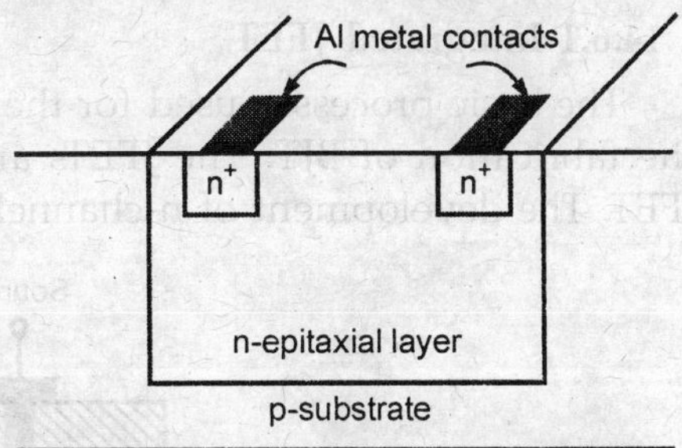

Fig. 1.28 Epitaxial resistor

1.14.4 Integrated Capacitors

The common parallel plate capacitor structures are as shown in the Fig. 1.29.

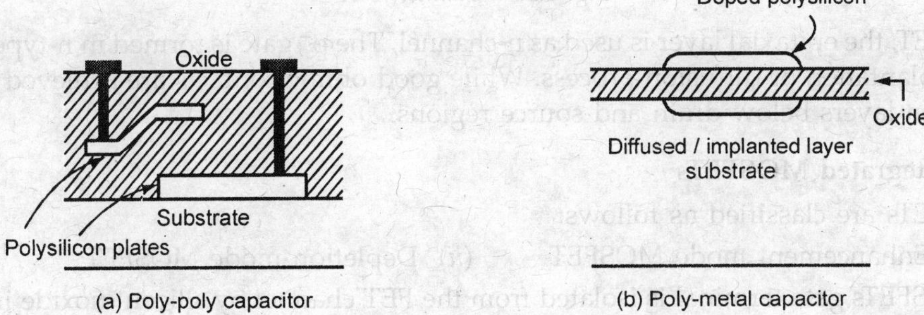

Fig. 1.29 Capacitor structures

In most widely used type shown in Fig. 1.29 (*a*), the two polysilicon plates are separated by silicon dioxide (SiO_2). Here the lower plate rests on the top of the substrate.

The capacitor shown in Fig. 1.29 (*b*) is MOS capacitor. It consists a implanted or diffused heavily doped layer within substrate while a polysilicon or metal plate on the top of a thin oxide layer. For MOS capacitors, generally gate oxide is used with no extra processing step.

1.14.5 Integrated Inductors

In ICs, inductors, transformers and chokes are not integrated because of bulkiness.

The inductors can be fabricated on a chip in the form of a thin film spirals by successively depositing the conducting patterns. But using fabricated inductor, a very small value of inductor is possible which is of the order of few nano-henries. But the practical approach is not to integrate inductors. The inductors are simulated on a chip with the help of R-C networks or some other type of a network. In case of RF and IF circuits, the use of inductor is unavoidable. Under such conditions, inductors are used externally with the integrated circuit.

1.14.6 Integrated FETs

In general the field effect transistors are of two types namely

(*i*) Junction field effect transistor (JFET)

(*ii*) Metal oxide semiconductor field effect transistor (MOSFET)

1.14.6.1 Integrated JFET

The basic processes used for the fabrication of JFET are exactly similar to those used in the fabrication of BJT. The JFETs are further classified as n-channel JFET and p-channel JFET. The development of n-channel JFET is as shown in the Fig. 1.30.

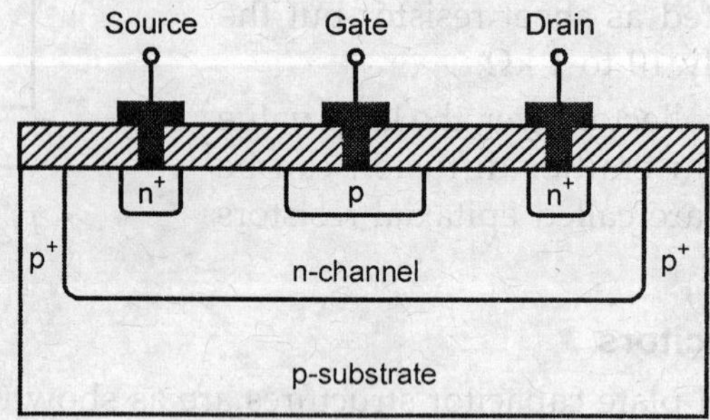

Fig. 1.30 n-channel JFET

The JFET, the epitaxial layer is used as n-channel, The p^+ gate is formed in n-type channel by ion-implantation or diffusion process. While good ohmic contacts are achieved by using n^+ diffusion layers below drain and source regions.

1.14.6.2 Integrated MOSFETs

MOSFETs are classified as follows:

(*i*) Enhancement mode MOSFET (*ii*) Depletion mode MOSFET

In MOSFETs gate terminal is isolated from the FET channel by silicon dioxide insulating layer. As the layer is insulating type, it provides very high input resistance. In providing superior barrier for impurities penetrating SiO_2 layer, silicon nitride (Si_3N_4) is sandwiched between two silicon dioxide (SiO_2) layers. This helps in increasing overall dielectric constant.

The n-channel MOSFET of enhancement and depletion mode are as shown in the Fig. 1.31 (*a*) and (*b*) respectively.

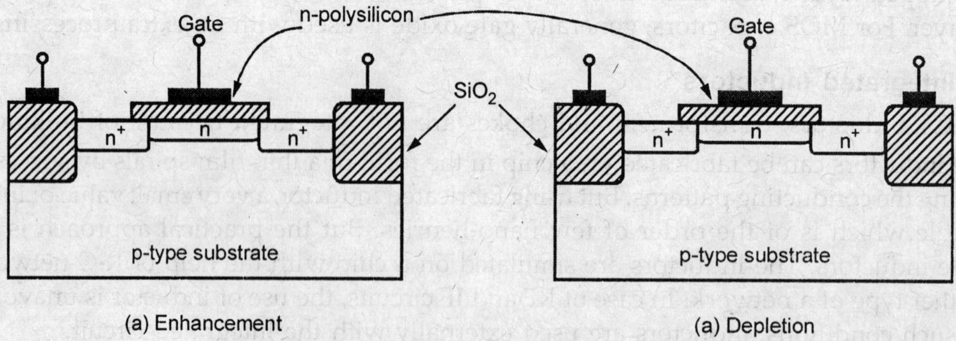

Fig. 1.31 n-channel MOSFET

In the enhancement mode, MOSFET is in OFF state when gate-source bias is zero, while MOSFET turns ON by positive gate source voltage. In depletion mode, because of n-implanted channel conduction is possible in ON state for zero gate-source voltage. While negative gate source voltage is required to turn it OFF.

1.14.6.3 Integrated CMOSs

When n-channel MOSFET and p-channel MOSFET both are integrated on same chip, the device is termed as complementary CMOS. In CMOS fabrication, n-type well is diffused in p-type substrate. Also p-channel MOSFET is fabricated within this n-well. Basically this n-well forms substrate for p-channel MOSFET. In the fabrication of p-channel MOSFET two additional steps are required as compared to n-channel MOSFET fabrication. The additional steps are formation of n-well and ion-implantation of p-type source and drain regions. The cross-section of CMOS IC is as shown in the Fig. 1.32.

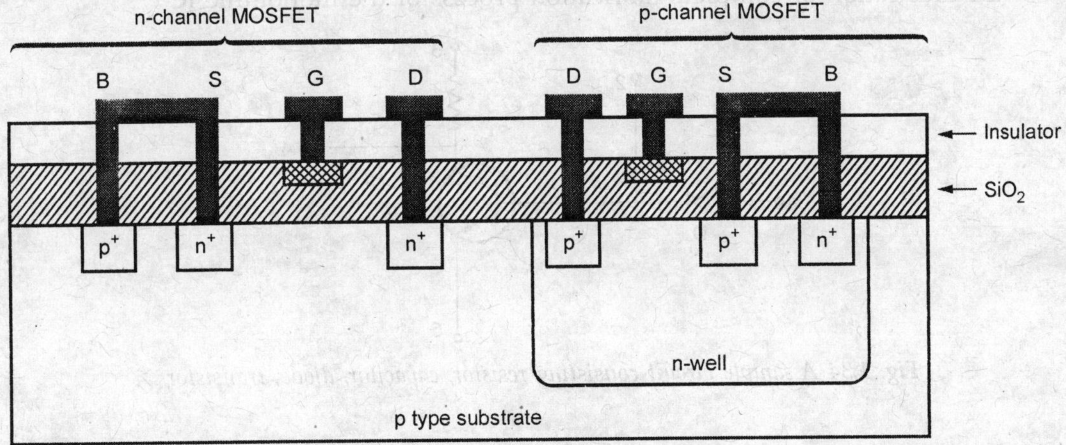

Fig. 1.32 Cross-section of CMOS

1.15 REALIZATION OF MONOLITHIC INTEGRATED CIRCUIT

After the discussion of the basic fabrication processes let us now consider how circuit is converted into monolithic integrated circuit. The key element in the IC is the **transistor.** Along with transistors, diodes, resistors and capacitors are also integrated over a single silicon wafer which is called **monolithic IC.** The interrelationship between various processes steps in the fabrication of IC is illustrated in the Fig. 1.33.

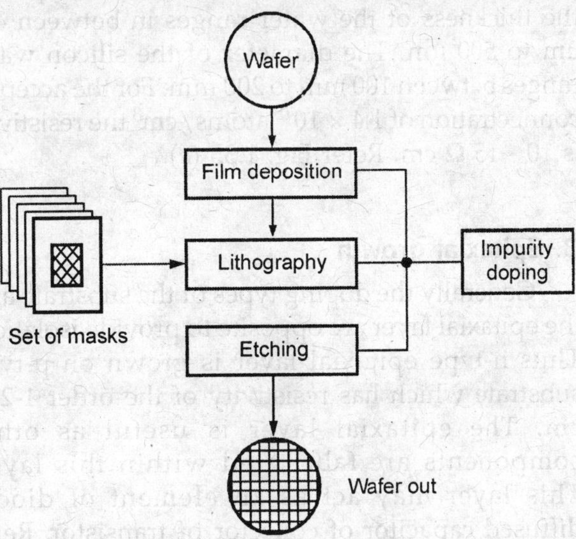

Fig. 1.33 IC fabrication schematic flow diagram

The starting material for the IC fabrication is the **polished silicon wafer** with a specific orientation and resistivity. The formation of oxide films using **thermal oxidation** and deposition of polysilicon, dielectric for isolation and metal films for interconnections are included in film deposition. Then it is followed by the lithography or ion implantation process. Then it is subjected to the etching process. The final IC is made by sequentially transferring the patterns from set of masks onto the surface of a wafer.

After processing, the wafer consists thousands of rectangular chips which all are identical. Then chips are separated by using laser cutting. The chips are tested electrically. The faulty chips are marked and removed from the batch, while the good ICs are packaged to provide proper termal, electrical, interconnection environment for different electric applications.

Consider a simple circuit as shown in Fig. 1.34.

For the illustration of complete fabrication process of the monolithic IC

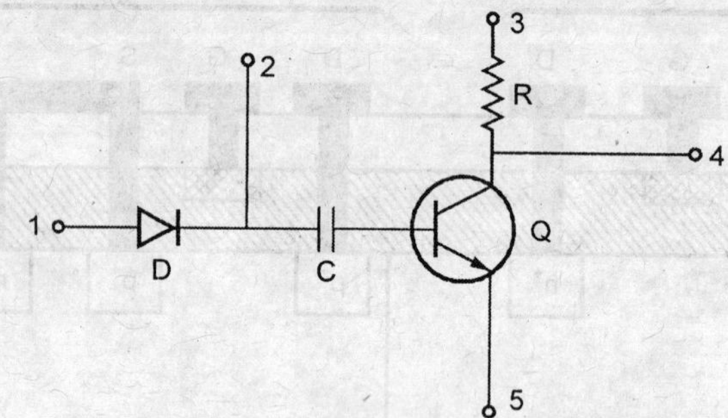

Fig. 1.34 A sample circuit consisting resistor, capacitor, diode, transistor

A. Preparation of wafer

The starting material for the integrated circuit is p-type silicon which is called substrate. Typically the thickness of the wafer ranges in between 400 μm to 500 μm. The diameter of the silicon wafer ranges between 100 mm to 200 mm. For the acceptor connectration of 1.4×10^{15} atoms/cm^3 the resistivity is 10 - 15 Ω cm. Refer Fig. 1.35 (*a*).

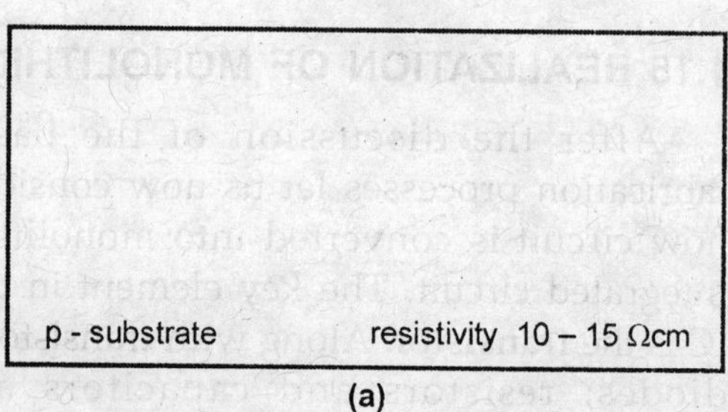

(a)

B. Epitaxial growth

Generally the doping types of the substrate and the epitaxial layer are opposite to provide isolation. Thus n-type epitaxial layer is grown on p-type substrate which has resistivity of the order 1-2 Ω cm. The epitaxial layer is useful as other components are fabricated within this layer. This layer may act as an element of diode, diffused capacitor of collector of transistor. Refer Fig. 1.35 (*b*).

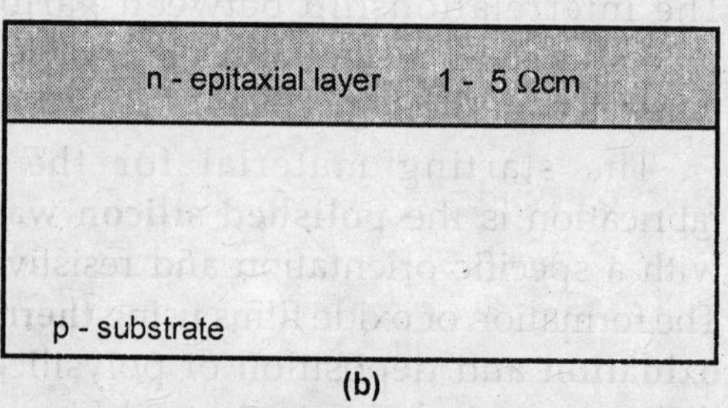

(b)

C. Oxidation

After the growth of an epitaxial layer, SiO_2 layer is grown on the n-epitaxial layer. This oxide layer is grown by using **thermal oxidation method.** The thickness of the SiO_2 layer is smaller as compared to previous layers. Typically it ranges between 0.05 to 2 µm. Refer Fig. 1.35 (*c*).

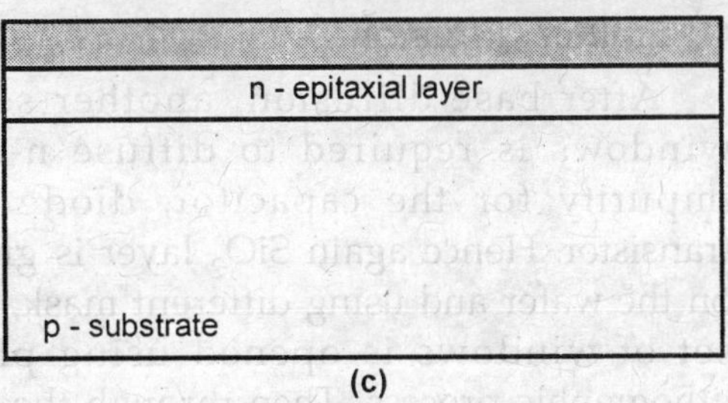

(c)

D. Lithographic process

Then the wafer is coated with negative photoresist. To isolated the four components of the circuit, p^+ type layer is diffused. For this a opening has to be made using the proper mask. So proper mask is kept on a wafer and then u.v. light is passed. After that it is photoetched to get a wafer with openings for isolation diffusion. Refer Fig. 1.35 (*d*). Note that we will have to use this process frequently with different pattern of the mask.

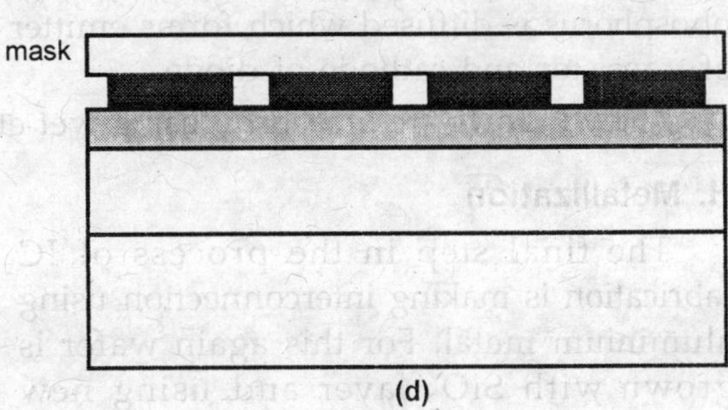

(d)

E. Isolation diffusion using p-n junction isolation technique

After the lithographic process, from the openings from where SiO_2 is etched out, heavy doping of p-type is diffused for very long interval such that the impurities reach p-substrate penetrating n-type epitaxial layer. Thus we get for isolation island for four components. Generally the concentration of acceptor atoms between isolation island is kept higher than p-substrate which ensures perfect electrical isolation. Refer Fig. 1.35 (*e*).

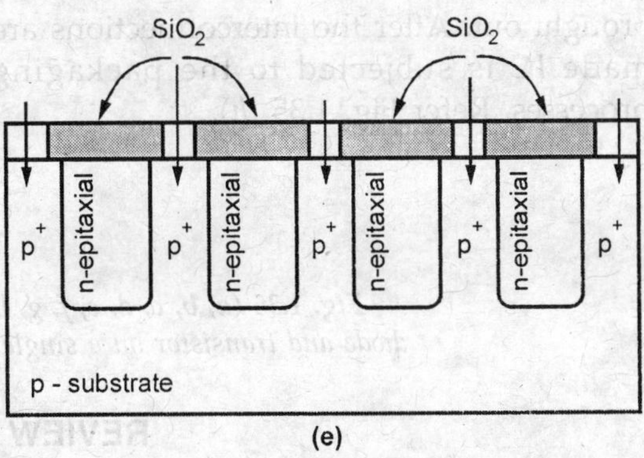

(e)

F. Base diffusion

After isolation diffusion once again a layer of SiO_2 is grown over a wafer and then using again photo lithographic technique different pattern is marked on wafer to have opening for the diffusion of p-type impurity such as boron. The impurity diffusion depth is controlled so that it can not penetrate epitaxial layer to reach p-substrate. Refer Fig. 1.35 (*f*). This serves as base of transistor, anode of diode etc.

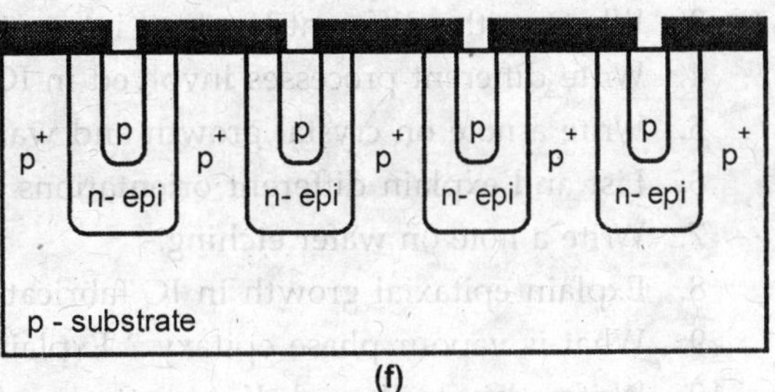

(f)

G. Emitter diffusion

After base diffusion, another set of windows is required to diffuse n-type impurity for the capacitor, diode and transistor. Hence again SiO_2 layer is grown on the wafer and using different mask, new set of windows is opened using photo lithographic process. Then through the new set of windows, n-type impurity e.g. phosphorus is diffused which forms emitter of transistor and cathode of diode.

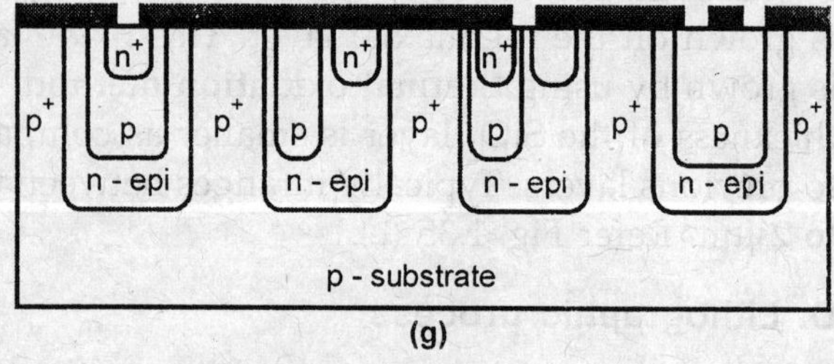

The windows are etched by using wet etching technique. Refer Fig. 1.35 (*g*).

H. Metallization

The final step in the process of IC fabrication is making interconnection using aluminium metal. For this again wafer is grown with SiO_2 layer and using new photomask, new set of windows is opened at points from where terminals are to be brought out. After the interconnections are made IC is subjected to the packaging processes. Refer Fig. 1.35 (*h*).

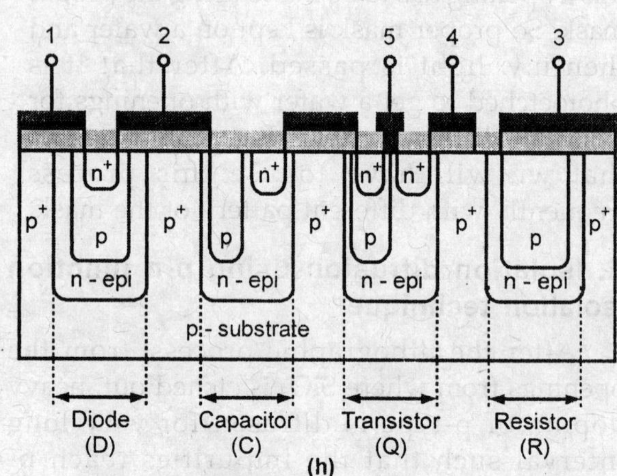

Fig. 1.35 (a, b, c, d, e, f, g, h) Fabrication of resistor, capacitor, diode and transistor on a single chip in monolithic integrated circuit.

REVIEW QUESTIONS

1. Give classification of integrated circuits.
2. Compare IC families on the basis of substrate used, structure, active devices, passive devices and applications.
3. What are the different levels of integration ?
4. Write different processes involved in IC fabrication.
5. Write a note on crystal growth and wafer preparation.
6. List and explain different orientations in silicon.
7. Write a note on wafer etching.
8. Explain epitaxial growth in IC fabrication.
9. What is vapour phase epitaxy ? Explain the same with suitable diagram.
10. Write advantages and disadvantages of molecular beam epitaxy.
11. Write the important uses of silicon dioxide.

12. Write a note on different techniques of oxidation.
13. What is lithography ? Explain in brief lithographic process with schematic representation.
14. Write a note on optical lithography.
15. Explain in brief photo etching process with suitable diagrams.
16. What is reactive plasma etching ?
17. Write a note on reactive plasma etching process.
18. Explain in short - Different pattern transfer techniques.
19. What is diffusion ? Give different techniques which are commonly used.
20. Why Ion Implantation is preferred over diffusion ?
21. Explain with schematic diagram - Ion implantation.
22. Write a note on Isolation techniques.
23. Compare p-n junction Isolation technique with dielectric isolation technique.
24. What is metallization ?
25. What are the different types of metallization ?
26. Write a note on - Alloys and metals used for different types of metallization.
27. Write a note on metallization process.
28. What are different types of physical vapour deposition process ?
29. Write applications of metallization.
30. What are different packages of ICs ? List common packages with their abbreviations, pin count and mounting type.
31. Explain the various methods of fabrication of transistor in monolithic integrated circuits.
32. Write a note on fabrication of resistor and capacitor on monolithic IC.
33. Explain in brief - Integration of inductor and capacitor.
34. With the help of neat sketches, explain the fabrication process of a monolithic IC.
35. Explain the word "Epitaxy".
36. Explain why "Buried Layer" is used.
37. To what voltage is the substrate connected and why?

❖❖❖

Chapter

2

Differential Amplifiers

2.1 INTRODUCTION

A device which accepts an input signal and produces an output signal proportional to the input, is called an **amplifier**. An amplifier which amplifies the difference between the two input signals is called **differential amplifier**. The differential amplifier configuration is very much popular and is used in variety of analog circuits. The differential amplifier is an essential and basic building block in modern IC amplifiers.

The term IC means integrated circuit where all the components are fabricated on the same chip. Most of the ICs are produced by the monolithic process. In this process, all the active elements like transistors and all the passive elements like resistances, capacitances are fabricated on a single piece of semiconductor material, generally silicon. This is called silicon wafer or a substrate. Such a monolithic process makes low cost mass production of ICs possible.

With the help of ICs the circuit design becomes very simple. The variety of useful circuits can be built without the necessity of knowing about the complex internal circuitry. The various advantages of linear integrated circuits are,

1. Small size

Practically size of an IC is thousands of times smaller than the discrete circuit.

Inside the Chapter

2. Low cost

Thousands of silicon wafers consisting of number of components individually, are produced simultaneously, called mass production. Due to this, cost of IC is very very low and hence millions of ICs are now in use, annually.

3. Less weight

Number of components are fabricated on a single silicon wafer hence weight of IC is very much less compared to a discrete circuit consisting of same number of components.

4. Low supply voltages

ICs work at lower voltages which avoids the need of the large supply voltages.

5. Low power consumption

The power consumption of ICs is very very low.

6. Highly reliable

Due to absence of soldered connections and less interconnections, the ICs are highly reliable from the performance point of view. Due to low power consumption, the temperature rise of ICs is also low which further increases the reliability, accuracy and the life of ICs.

In addition to these important advantages, the ICs are used popularly because of versatility, flexibility and dependability.

The IC version of operational amplifiers is also popular which is inexpensive, takes up less space and consumes less power. The basic building block of modem linear IC op-amp is **differential amplifier**.

The differential amplifier has certain advantages like excellent stability, high versatility, immune to interference signals, due to which it is used as a basic circuit for most of the analog circuits, ranging from d.c. to high frequency range applications. The discussion and analysis of differential amplifier prepares the background for analysis and design procedures for the operational amplifiers. It helps in understanding the characteristics of the operational amplifiers.

Thus, this chapter includes the detail analysis of differential amplifier circuit. The chapter also provides the discussion of various supporting circuits which are used to improve the performance of a basic differential amplifier. Before beginning the actual study of the differential amplifier, let us review the fundamental characteristics of a voltage amplifier.

2.2 AMPLIFIER FUNDAMENTALS

A two port device which accepts an input signal and produces an output signal proportional to the input, is called an **amplifier**. The proportionality constant between input and output is called a **gain of the amplifier**. There are variety of amplifiers available depending upon the nature of the input and output signals. These types are the **voltage amplifier** (voltage input, voltage output), the **current amplifier** (current input, current output), the **transconductance amplifier** (voltage input, current output) and the **transresistance amplifier** (current input, voltage output).

***Key Point:** The differential amplifier is basically a voltage amplifier is basically a voltage amplifier which amplifies the difference between the two input voltage signals.*

Consider a voltage amplifier having a constant gain A as shown in the Fig. 2.1, supplying the power to the load resistance R_L.

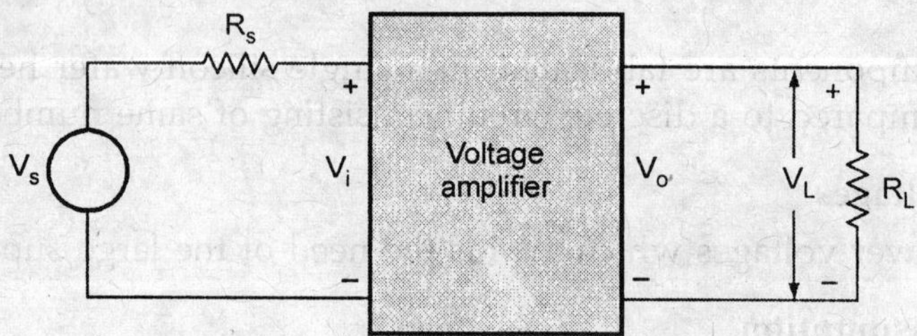

Fig. 2.1 Voltage amplifier supplying load R_L.

Let R_s = Source resistance

R_i = Input resistance of an amplifier

R_o = Output resistance of an amplifier

The equivalent circuit of the input side of an amplifier can be obtained as shown in the Fig. 2.2.

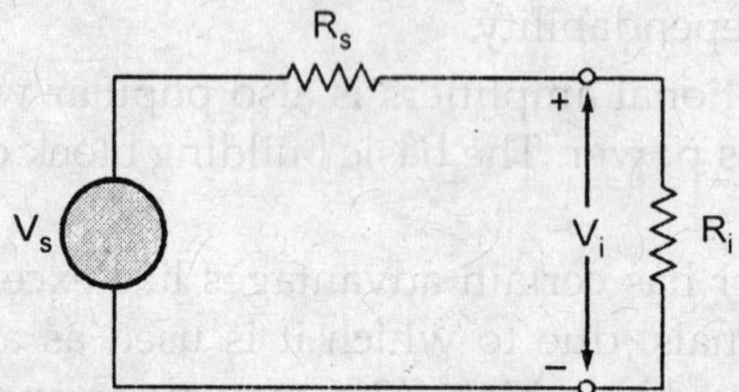

Fig. 2.2 Equivalent circuit of input.

From the Fig. 2.2, we can write the expression for the input voltage V_i, received by the amplifier for the amplification, from the source voltage V_s.

$$V_i = \left[\frac{R_i}{R_s + R_i}\right] V_s \quad \text{... by voltage divider rule}$$

Let us analyse the value of V_i for the values of R_i and R_s.

Let $R_i = 500\ \Omega$ and $R_s = 50\ \Omega$

$$\therefore \quad V_i = \left[\frac{500}{50 + 500}\right] V_s = 0.909 V_s$$

Thus only 90.9% of source voltage is available for amplification, as the input voltage V_i of the amplifier.

Now if the input resistance is increased by some means, then with $R_i = 50\ \Omega$ and $R_s = 50\ \Omega$ as before we get,

$$\therefore \qquad V_i = \left[\frac{50 \times 10^3}{50 + 50 \times 10^3}\right] = 0.999 V_s$$

Thus now 99.9% of source voltage is available as the input V_i to the amplifier.

In practice the input voltage to the amplifier must be same as that of source voltage.

***Key Point:** To achieve this, the **input resistance** (generally referred as input impedance) must be as large as possible i.e. ideally infinite.*

This is one of the basic requirement of a differential amplifier that its input impedance must be as large as possible.

The amplifier, amplifies the input voltage V_i and produces the output voltage V_o equal to AV_i where A is called gain of the amplifier. Is this voltage, entirely available to the load R_L ? Let us find the answer for this question.

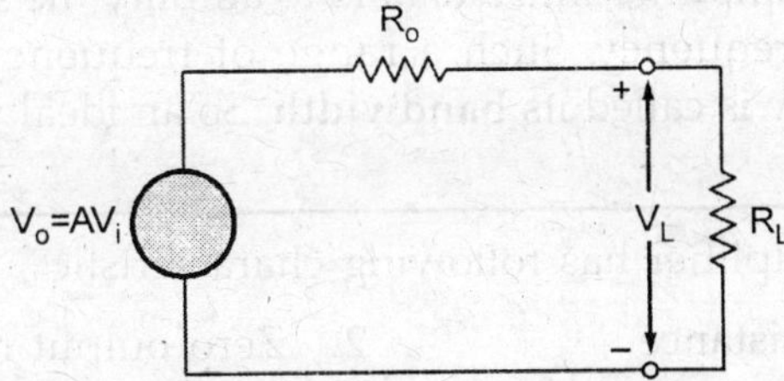

Fig. 2.3 Equivalent circuit of output.

Consider the equivalent circuit of the output side of the amplifier as shown in the Fig. 2.3.

The resistance R_o represents the **output resistance** (generally referred as output impedance) of the amplifier.

By applying voltage divider rule we can write,

$$V_L = \left[\frac{R_L}{R_o + R_L}\right] V_o$$

Now if $\qquad R_L = 50\ \Omega$ and $R_o = 100\ \Omega$ then,

$$V_L = \left[\frac{50}{100 + 50}\right] V_o = 0.333 V_o$$

Thus only 98.03% of output voltage is available to the load as the load voltage V_L. This is mainly because of drop across the resistance R_o.

***Key Point:** Thus as current demanded by the load increases, drop across output resistance increases and the voltage available to the load decreases.*

Such an effect is called loading effect. In other words it can be said that the amplifier output is loaded. Loading complicates the life because every time as we change the source or the load there is change in the loading effect which alters the calculation of the overall gain of the amplifier. Hence it is necessary to reduce the loading effect and if possible to eliminate it.

Consider $R_L = 50\ \Omega$ and $R_o = 1\ \Omega$ then,

$$V_L = \left[\frac{50}{1+50}\right] V_o = 0.9803 V_o$$

Thus now 98.03% of output is available to the load. This shows that for a lower value of R_o, the loading effect reduces considerably and we can get load voltage almost same as the output voltage.

Key Point: *Hence the output impedance of the amplifier must be as small as possible, ideally zero to eliminate the loading effect.*

This is another basic requirement of a differential amplifier.

The main purpose of a differential amplifier is to amplify the difference between the two voltages which might be very small. So it should provide a maximum voltage gain, ideally infinite. Also, the amplifier must be able to amplify the signal without any distortion for the entire range of frequency. Such a range of frequency over which the amplifier performance is satisfactory is called its **bandwidth**. So an ideal voltage amplifier has infinite bandwidth.

Thus ideal voltage amplifier has following characteristics,

1. Infinite input resistance.
2. Zero output resistance.
3. Infinite gain
4. Infinite bandwidth

Let us study now the differential amplifier and its characteristics based on the basic requirements of a voltage amplifier.

2.3 BASICS OF DIFFERENTIAL AMPLIFIER

Key Point: *The differential amplifier amplifies the difference between two input voltage signals. Hence it is also called difference amplifier.*

Consider an ideal differential amplifier shown in the Fig. 2.4.

V_1 and V_2 are the two input signals while V_o is the single ended output. Each signal is measured with respect to the ground.

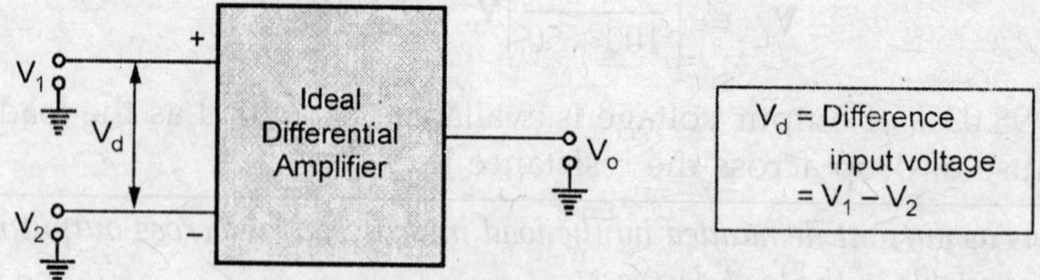

Fig. 2.4 Ideal differential amplifier.

In an ideal differential amplifier, the output voltage V_o is proportional to the difference between the two input signals. Hence we can write,

$$V_o \propto (V_1 - V_2) \quad \text{...(1)}$$

2.3.1 Differential Gain A_d

From the equation (1) we can write,

$$\therefore \quad \boxed{V_o = A_d(V_1 - V_2)} \quad ...(2)$$

where A_d is the constant of proportionality. The A_d is the gain with which differential amplifier amplifies the difference between two input signals. Hence it is called differential gain of the differential amplifier.

Thus, A_d = **differential gain**

The difference between the two inputs $(V_1 - V_2)$ is generally called **difference voltage** and denoted as V_d.

$$\therefore \quad \boxed{V_o = A_d V_d} \quad ...(3)$$

Hence the differential gain can be expressed as,

$$\therefore \quad \boxed{A_d = \frac{V_o}{V_d}} \quad ...(4)$$

Generally the differential gain is expressed in its decibel (dB) value as,

$$\therefore \quad \boxed{A_d = 20 \log_{10} (A_d) \text{ in dB}} \quad ...(5)$$

2.3.2 Common Mode Gain A_c

If we apply two input voltages which are equal in all the respects to the differential amplifier i.e. $V_1 = V_2$ then ideally the output voltage $V_o = (V_1 - V_2)A_d$, must be zero.

But the output voltage of the practical differential amplifier not only depends on the difference voltage but also depends on the average common level of the two inputs. Such an average level of the two input signals is called **common mode signal** denoted as V_c.

$$\therefore \quad \boxed{V_c = \frac{V_1 + V_2}{2}} \quad ...(6)$$

Practically, the differential amplifier produces the output voltage proportional to such common mode signal, also.

Key Point: *The gain with which it amplifies the common mode signal to produce the output is called* ***common mode gain*** *of the differential amplifier denoted as A_c.*

$$\therefore \quad \boxed{V_o = A_c V_c} \quad ...(7)$$

Thus there exists some finite output for $V_1 = V_2$ due to such common mode gain A_c, in case of practical differential amplifiers.

So the total output of any differential amplifier can be expressed as,

$$\therefore \quad \boxed{V_o = A_d V_d + A_c V_c} \quad ...(8)$$

This shows that if one input is + 25 μV and other is – 25 μV then the output of the amplifier will not be same, with the inputs as 600 μV and 650 μV, though the difference between the two sets of the inputs is 50 μV.

***Key Point:** For an ideal differential amplifier, the differential gain A_d must be infinite while the common mode gain must be zero. This ensures zero output for $V_1 = V_2$.*

But due to mismatch in the internal circuitry, there is some output available for $V_1 = V_2$ and gain A_c is not practically zero. The value of such common mode gain A_c is very very small while the value of the differential gain A_d is always very large.

At this stage, we can define one important parameter of the differential amplifier known as common mode rejection ratio (CMRR).

2.3.3 Common Mode Rejection Ratio (CMRR)

When the same voltage is applied to both the inputs, the differential amplifier is said to be operated in a common mode configuration. Many disturbance signals, noise signals appear as a common input signal to both the input terminals of the differential amplifier. Such a common signal should be rejected by the differential amplifier.

The ability of a differential amplifier to reject a common mode signal is expressed by a ratio called **common mode rejection** ratio denoted as CMRR.

It is defined as the ratio of the differential voltage gain A_d to common mode voltage gain A_c.

$$\therefore \quad \text{CMRR} = \rho = \left|\frac{A_d}{A_c}\right| \qquad \text{... (9)}$$

***Key Point:** Ideally the common mode voltage gain is zero, hence the ideal value of CMRR is infinite.*

For a practical differential amplifier A_d is large and A_c is small hence the value of CMRR is also very large.

Many a times, CMRR is also expressed in dB, as

$$\therefore \quad \text{CMRR in dB} = 20\log\left|\frac{A_d}{A_c}\right| \text{dB} \qquad \text{...(10)}$$

The output voltage can be expressed in terms of CMRR as below:

$$\therefore \quad V_o = A_d V_d + A_c V_c = A_d V_d\left[1 + \frac{A_c V_c}{A_d V_d}\right]$$

$$\therefore \quad V_o = A_d V_d\left[1 + \frac{1}{\left(\frac{A_d}{A_c}\right)}\frac{V_c}{V_d}\right]$$

$$\therefore \quad V_o = A_d V_d\left[1 + \frac{1}{\text{CMRR}}\cdot\frac{V_c}{V_d}\right] \qquad \text{...(11)}$$

This equation explains that as CMRR is practically very large, though both V_c and V_d components are present, the output is mostly proportional to the difference signal only. The common mode component is greatly rejected.

2.3.4 Features of Differential Amplifier

The various features of a differential amplifier are:

1. High differential voltage gain.
2. Low common mode gain.
3. High CMRR.
4. Two input terminals.
5. High input impedance.
6. Large bandwidth.
7. Low offset voltages and currents.
8. Low output impedance.

Example 2.1: Determine the output voltage of a differential amplifier for the input voltages of 300 μV and 240 μV. The differential gain of the amplifier is 5000 and the value of the CMRR is (*i*) 100 and (*ii*) 10^5.

Solution: The differential amplifier is represented as shown in the Fig. 2.5.

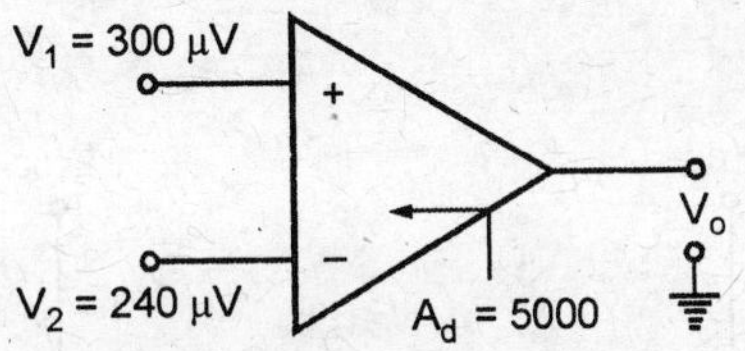

Fig. 2.5

(*i*)

$$\text{CMRR} = 100$$

$$V_D = V_1 - V_2 = 300 - 240$$

$$= 60\ \mu V$$

$$V_c = \frac{V_1 + V_2}{2} = \frac{300 + 240}{2}$$

$$= 270\ \mu V$$

$$\text{CMRR} = \frac{A_d}{A_c}$$

$$\therefore \quad 100 = \frac{5000}{A_c}$$

$$\therefore \quad A_c = A_d V_d + A_c V_c = 5000 \times 60 + 50 \times 270 = 313500\ \mu V$$

$$= \mathbf{313.5\ \mu V}$$

(*ii*) CMRR $= 10^5$

$$\therefore \quad A_c = \frac{A_d}{CMRR} = \frac{5000}{10^5} = 0.05$$

$$\therefore \quad V_o = A_d V_d + A_c V_c = 5000 \times 60 + 50 \times 270 = 313500\ \mu V$$

$$= \mathbf{313.5\ \mu V}$$

(*iii*) CMRR $= 10^5$

$$\therefore \quad A_c = \frac{A_d}{CMRR} = \frac{5000}{10^5} = 0.05$$

$$\therefore \quad V_o = A_d V_d + A_c V_c = 5000 \times 60 + 0.05 \times 270$$

$$= 300013.5\ \mu V = \mathbf{300.0135\ \mu V}$$

Ideally A_c must be zero and output should be only $A_d V_d$ which is $5000 \times 60 \times 10^{-6}$ i.e. 300 μV.

Key Point: *It can be seen that higher the value of CMRR, the output is almost proportional to the difference voltage V_d, rejecting the common mode signal.*

2.4 TRANSISTORISED DIFFERENTIAL AMPLIFIER

The transistorised differential amplifier basically uses the emitter biased circuits which are identical in characteristics. Such two identical emitter biased circuits are shown in the Fig. 2.6.

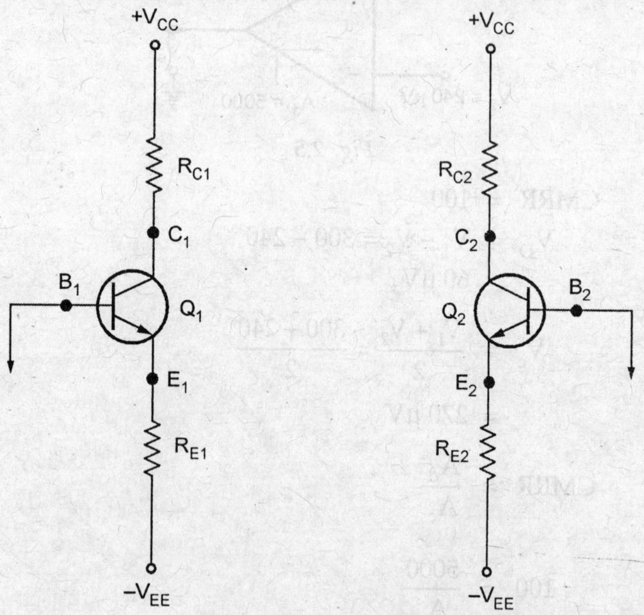

Fig. 2.6 Emitter biased circuits.

The two transistors Q_1 and Q_2 have exactly matched characteristics. The two collector resistances R_{C1} and R_{C2} are equal while the two emitter resistances R_{E1} and R_{E2} are also equal.

Thus $R_{C1} = R_{C2}$ and $R_{E1} = R_{E2}$

The magnitudes of + V_{CC} and − V_{EE} are also same. The differential amplifier can be obtained by using such two emitter biased circuits. This is achieved by connecting emitter E_1 of Q_1 to the emitter E_2 of Q_2. Due to this, R_{E1} appears in parallel with R_{E2} and the combination can be replaced by a single resistance denoted as R_E. The base B_1 of Q_1 is connected to the input 1 which is V_{S1} while the base B_2 of Q_2 is connected to the input 2 which is V_{S2}. The supply voltages are measured with respect to ground. The balanced output is taken between the collector C_1 of Q_1 and the collector C_2 of Q_2. Such an amplifier is called **emitter coupled differential amplifier**. The two collector resistances are same hence can be denoted as R_C.

The output can be taken between two collectors or in between one of the two collectors and the ground. When the output is taken between the two collectors, none of them is grounded then it is called **balanced output, double ended output** or **floating output**.

When the output is taken between any of the collectors and the ground, it is called unbalanced output or single ended output.

The complete circuit diagram of such a basic dual input, balanced output differential amplifier is shown in the Fig. 2.7.

As the output is taken between two output terminals, none of them is grounded, it is called balanced output differential amplifier.

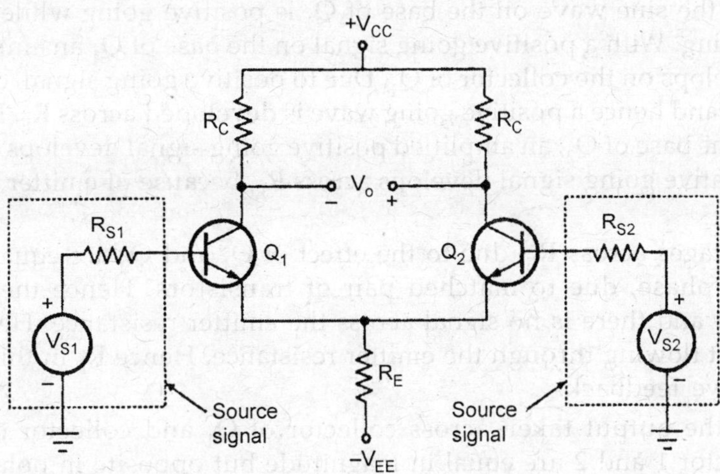

Fig. 2.7 Dual input, balanced output differential amplifier.

Let us study the circuit operation in the two modes namely

(*i*) Differential mode operation

(*ii*) Common mode operation.

2.4.1 Differential Mode Operation

In the differential mode, the two input signals are different from each other. Consider the two input signals which are same in magnitude but 180° out of phase. These signals, with opposite phase can be obtained from the center tap transformer. The circuit used in differential mode operation is shown in the Fig. 2.8.

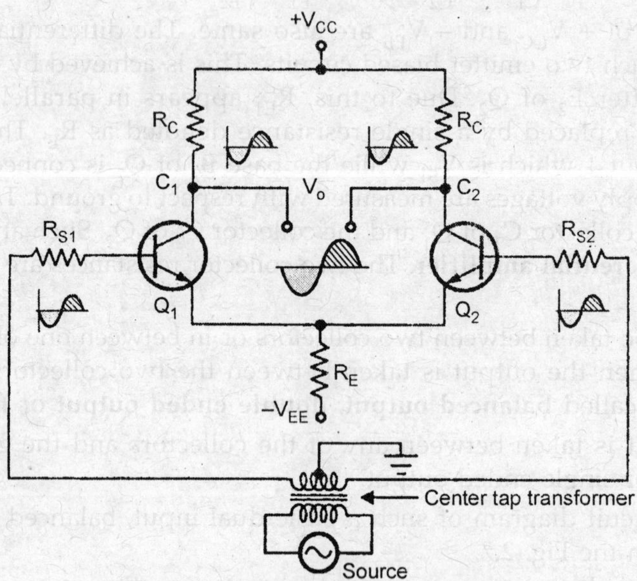

Fig. 2.8 Differential mode operation.

Assume that the sine wave on the base of Q_1 is positive going while on the base of Q_2 is negative going. With a positive going signal on the base of Q_1 an amplified negative going signal develops on the collector of Q_1. Due to positive going signal, current through R_E also increases and hence a positive going wave is developed across R_E. Due to negative going signal on the base of Q_2, an amplified positive going signal develops on the collector of Q_2. And a negative going signal develops across R_E, because of emitter follower action of Q_2.

So signal voltages across R_E, due to the effect of Q_1 and Q_2 are equal in magnitude and 180° out of phase, due to matched pair of transistors. Hence these two signals cancel each other and there is no signal across the emitter resistance. Hence there is no a.c. signal current flowing through the emitter resistance. Hence R_E in this case does not introduce negative feedback.

While V_o is the output taken across collector of Q_1 and collector of Q_2. The two outputs on collector 1 and 2 are equal in magnitude but opposite in polarity. And V_o is the difference between these two signals, e.g. + 10 – (– 10) = + 20.

***Key Point:** Hence the difference output V_o is twice as large as the signal voltage from either collector to ground.*

2.4.2 Common Mode Operation

In this mode, the signals applied to the base of Q_1 and Q_2 are derived from the same source. So the two signals are equal in magnitude as well as in phase. The circuit diagram is shown in the Fig. 2.9.

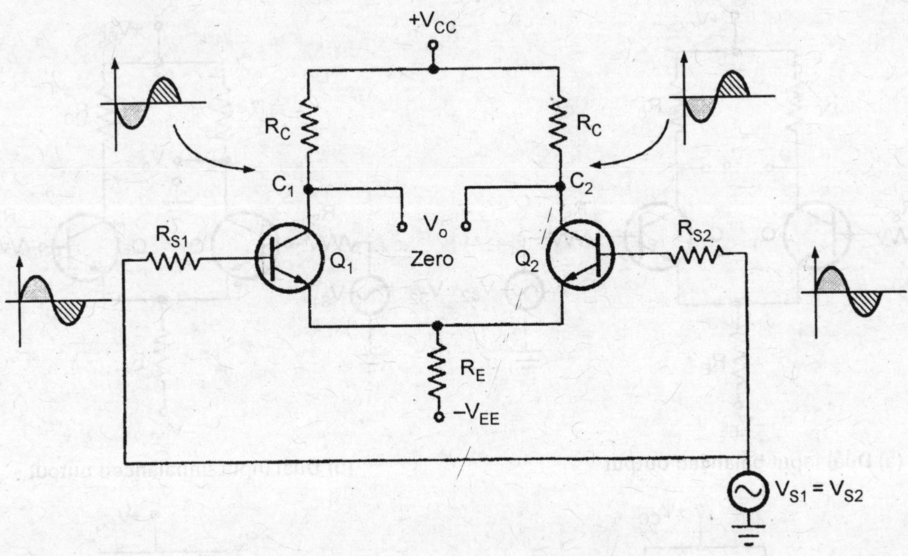

Fig. 2.9 Common mode operation.

In phase signal voltages at the bases of Q_1 and Q_2 causes in phase signal voltages to appear across R_E, which add together. Hence R_E carries a signal current and provides a negative feedback. This feedback reduces the common mode gain of differential amplifier.

While the two signals causes in phase signal voltages of equal magnitude to appear across the two collectors of Q_1 and Q_2. Now the output voltage is the difference between the two collector voltages, which are equal and also same in phase, e.g. (10) – (10) = 0. Thus the difference output V_o is almost zero, negligibly small. **Ideally it should be zero.**

2.5 DIFFERENTIAL AMPLIFIER CIRCUIT CONFIGURATIONS

The differential amplifier, in the difference amplifier stage in the op-amp, can be used in four configurations:

(*i*) Dual input, balanced output differential amplifier.
(*ii*) Dual input, unbalanced output differential amplifier.
(*iii*) Single input, balanced output differential amplifier.
(*iv*) Single input, unbalanced output differential amplifier.

The differential amplifier uses two transistors in common emitter configuration. If output is taken between the two collectors it is called **balanced output or double ended output**. While if the output is taken between one collector with respect to ground it is called **unbalanced output or single ended output.** If the signal is given to both the input terminals it is called **dual input,** while if the signal is given to only one input terminal and other terminal is grounded it is called **single input or single ended input.**

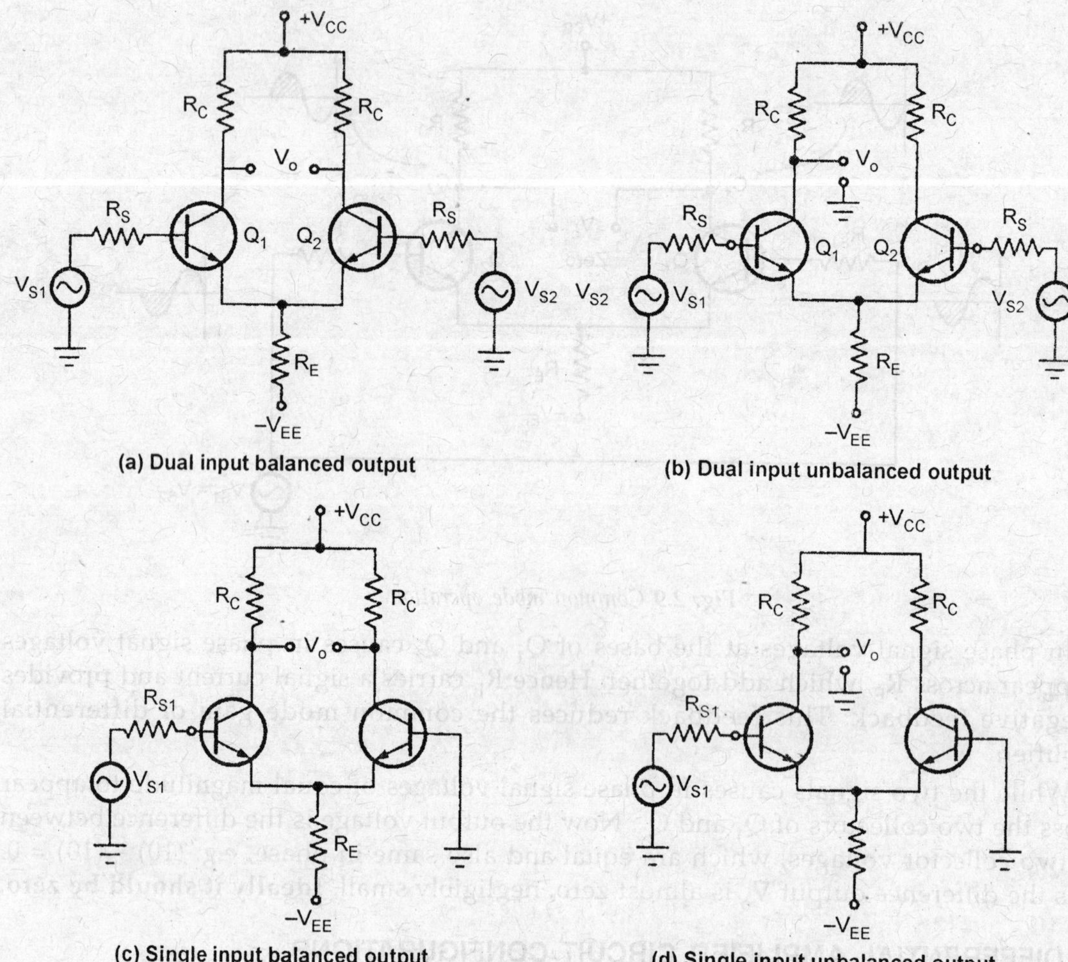

Fig. 2.10

Out of these four configurations the dual input, balanced output is the basic differential amplifier configuration. This is shown in the Fig. 2.10 (*a*). The dual input, unbalanced output differential amplifier is shown in the Fig. 2.10 (*b*). The single input, balanced output differential amplifier is shown in the Fig. 2.10 (*c*) and the single input, unbalanced output differential amplifier is shown in the Fig. 2.10 (*d*).

The operation of dual input balanced output differential amplifier in differential mode and common mode is already discussed in the last section. This configuration is also called symmetrical differential amplifier.

2.6 D.C. ANALYSIS OF DIFFERENTIAL AMPLIFIER

The d.c. analysis means to obtain the operating point values i.e. I_{CQ} and V_{CEQ} for the transistors used. The supply voltages are d.c. while the input signals are a.c., so d.c.

equivalent circuit can be obtained simply by reducing the input a.c. signals to zero. The d.c. equivalent circuit thus obtained is shown in the Fig. 2.11. Assuming $R_{S1} = R_{S2}$, the source resistance is simply denoted by R_S.

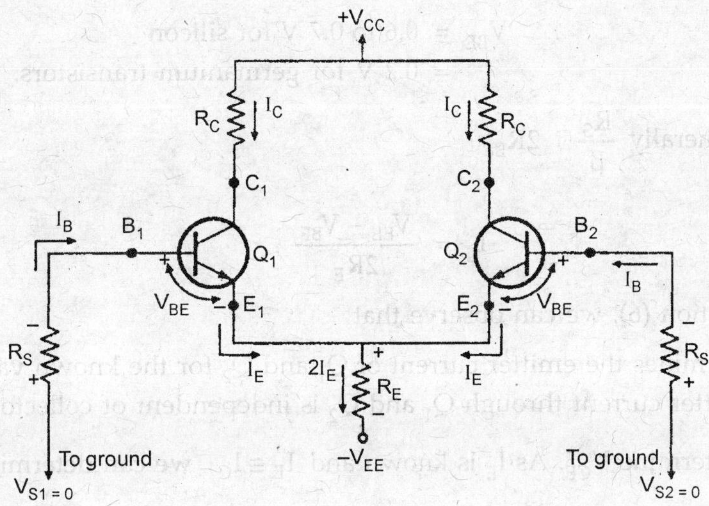

Fig. 2.11 D.C. equivalent circuit.

The transistors Q_1 and Q_2 are matched transistors and hence for such a matched pair we can assume:

(*i*) Both the transistors have the same characteristics.

(*ii*) $R_{E1} = R_{E2}$ hence $R_E = R_{E1} \| R_{E2}$.

(*iii*) $R_{C1} = R_{C2}$ hence denoted as R_C.

(*iv*) $|V_{CC}| = |V_{EE}|$ and both are measured with respect to ground.

As the two transistors are matched and circuit is symmetrical, it is enough to find out operating point I_{CQ} and V_{CEQ}, for anyone of the two transistors. The same is applicable for the other transistor.

Applying KVL to base-emitter loop of the transistor Q_1,

$$-I_B R_S - V_{BE} - 2I_E R_E + V_{EE} = 0 \qquad \text{...(1)}$$

But $\quad I_C = b\, I_B$ and $I_C = I_E$

$$I_B = \frac{I_E}{\beta} \qquad \text{...(2)}$$

Substituting in equation (1), we get

$$\frac{-I_E R_S}{\beta} - V_{BE} - 2I_E R_E + V_{EE} = 0 \qquad \text{...(3)}$$

$$\therefore \quad I_E\left[\frac{-R_S}{\beta} - 2R_E\right] + V_{EE} - V_{BE} = 0 \qquad \text{...(4)}$$

$$\therefore \quad I_E = \frac{V_{EE} - V_{BE}}{\frac{R_S}{\beta} + 2R_E} \quad \text{...(5)}$$

where V_{BE} = 0.6 to 0.7 V for silicon
= 0.2 V for germanium transistors.

In practice, generally $\frac{R_S}{\beta} \ll 2R_E$

$$\therefore \quad I_E = \frac{V_{EE} - V_{BE}}{2R_E} \quad \text{...(6)}$$

From the equation (6), we can observe that

(*i*) RE determines the emitter current of Q_1 and Q_2 for the known value of V_{E1}.
(*ii*) The emitter current through Q_1 and Q_2 is independent of collector resistance R_C.

Now let us determine V_{CE}. As I_E is known and $I_E \cong I_C$, we can determine the collector voltage of Q_1 as

$$V_C = V_{CC} - I_C R_C \quad \text{...(7)}$$

Neglecting the drop across R_S, we can say that the voltage at the emitter of Q_1 is approximately equal to $- V_{BE}$. Hence the collector to emitter voltage is

$$\therefore \quad V_{CE} = V_C - V_E = (V_{CC} - I_C R_C) - (-V_{BE})$$

$$\therefore \quad V_{CE} = V_{CC} + V_{BE} - I_C R_C \quad \text{... (8)}$$

Hence $I_E = I_C = I_{CQ}$ while $V_{CE} = V_{CEQ}$ for given values of V_{CC} and V_{EE}.

Key Point: *In the equation (6), the sign of V_{EE} is already considered to be negative, while deriving it. Hence while using this equation to solve the problem, only the magnitude of V_{EE} should be used and negative sign of V_{EE} should not be used again.*

Thus for both the transistors, we can determine operating point values, using equations (6) and (8). With the same biasing arrangement, the d.c. analysis remains same for all the four possible configurations of differential amplifier.

$$I_E = \frac{V_{EE} - V_{BE}}{\frac{R_S}{\beta} + 2R_E} = \frac{V_{EE} - V_{BE}}{2R_E} = I_{CQ}$$

$$V_{CEQ} = V_{CC} + V_{BE} - I_{CQ} R_C$$

Key Point: *The d.c. analysis and expressions for I_{CQ} and V_{CEQ} remain same for all the differential amplifier circuit configurations.*

Example 2.2: Calculate the operating point values for the circuit shown in the Fig. 2.12.

Solution: From the Fig. 2.12 we can write

RC = 4.7 kΩ, RE = 3.3 kΩ, Vcc = 12 V, V_{EE} = – 12 V

From equation (6),

$$I_E = \frac{V_{EE} - V_{BE}}{2R_E}$$

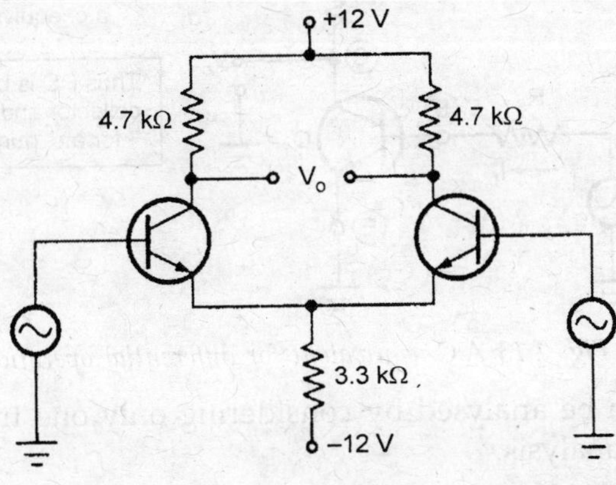

Fig. 2.12

The sign of V_{EE} is already considered while deriving the equation hence V_{EE} must be replaced by 12 V. And assume V_{BE} = 0.7 V.

$$\therefore \quad I_E = \frac{12 - 0.7}{2 \times 3.3 \times 10^3} = 1.712 \text{ mA}$$

Now $\quad I_C = I_E = 1.712$ mA

$\therefore \quad I_{CQ} = 1.712$ mA

$\therefore \quad V_{CEQ} = V_{CC} + V_{BE} - I_{CQ}R_C$

$= 12 + 0.7 - 1.712 \times 10^{-3} \times 4.7 \times 10^3 = 4.653$ V.

$\therefore$ Q point is **(1.712 mA, 4.653 V)**.

2.7 A.C. ANALYSIS OF DIFFERENTIAL AMPLIFIER USING h-PARAMETERS

In the a.c. analysis, we will calculate the differential gain A_d, common mode gain A_c, input resistance R_i and the output resistance R_o of the differential amplifier circuit, using the h-parameters.

2.7.1 Differential Gain (A_d)

For the differential gain calculation, the two input signals must be different from each other. Let the two a.c. input signals be equal in magnitude but having 180° phase difference in between them. The magnitude of each a.c. input voltage V_{S1} and V_{S2} be $V_S/2$.

The two a.c. emitter currents I_{e1} and I_{e2} are equal in magnitude and 180° out of phase. Hence they cancel each other to get resultant a.c. current through the emitter as zero. This is already discussed in the section 2.4.1. Hence for the a.c. purposes emitter terminal can be grounded. The a.c. small signal differential amplifier circuit with grounded emitter terminal is shown in the Fig. 2.13. As the two transistors are matched, the a.c. equivalent circuit for the other transistor is identical to the one shown in the Fig. 2.13.

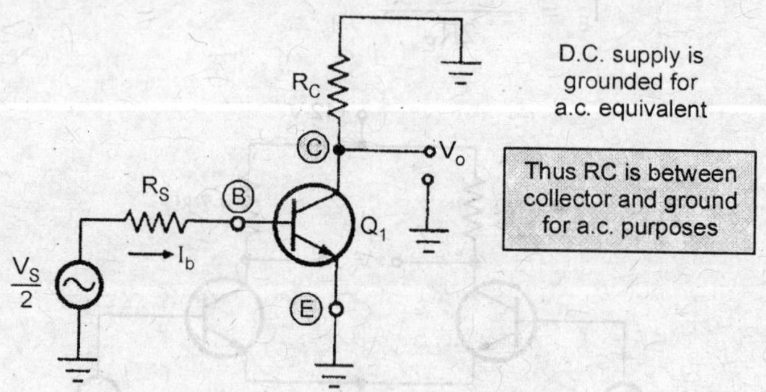

Fig. 2.13 A.C. equivalent for differential operation.

Thus the circuit can be analysed by considering only one transistor. This is called as **half circuit concept** of analysis.

The approximate hybrid model for the above circuit can be shown as in the Fig. 2.14, neglecting h_{oe},

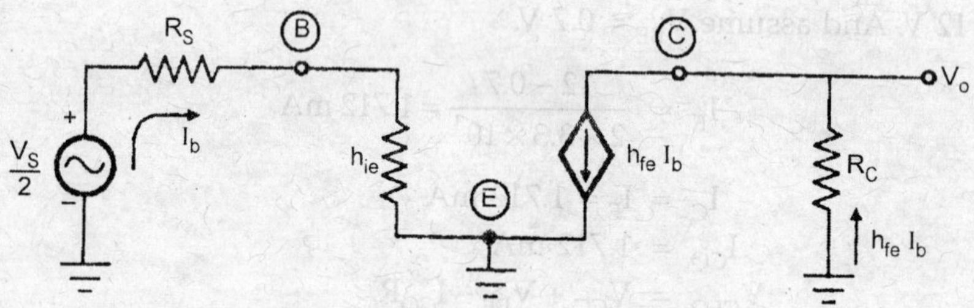

Fig. 2.14 Approximate hybrid model.

Applying KVL to the input loop,

$$-I_bR_S - I_bh_{ie} + \frac{V_S}{2} = 0 \quad \text{... (1)}$$

$$\therefore \quad -I_b(R_S + h_{ie}) = -\frac{V_S}{2}$$

$$\therefore \quad I_b = \frac{V_S}{2(R_S + h_{ie})} \quad \text{... (2)}$$

Applying KVL to the output loop,

$$V_o = -h_{fe}I_bR_C \quad \text{...(3)}$$

Substituting equation (2) in equation (3),

$$V_o = -h_{fe} R_C \frac{V_S}{2(R_S + h_{ie})}$$

$$\boxed{\frac{V_o}{V_S} = \frac{-h_{fe} R_C}{2(R_S + h_{ie})}} \quad ...(4)$$

The negative sign indicates the phase difference between input and output.

Now two input signal magnitudes are $\frac{V_S}{2}$ but they are opposite in polarity, as 180° out of phase.

$$\therefore \quad V_d = V_1 - V_2 = \frac{V_S}{2} - \left(-\frac{V_S}{2}\right) = V_S$$

The **magnitude** of the differential gain A_d is

$$\boxed{A_d = \frac{V_o}{V_S} = \frac{h_{fe} R_C}{2(R_S + h_{ie})}} \quad \text{(For unbalanced output)} \quad ...(5)$$

Where V_S = Differential input

Key Point: *The differential gain obtained in the equation (5) is for the circuit where output is measured with respect to ground. So it is unbalanced output.*

We are interested to obtain A_d for the differential amplifier with the balanced output. Balanced output is across the two collectors of the transistors Q_1 and Q_2, which are perfectly matched. Such balanced output is double than that obtained above, with unbalanced output. Hence the expression for A_d with balanced output changes as

$$A_d = 2 \times \frac{h_{fe} R_C}{2(R_S + h_{ie})}$$

$$\therefore \quad \boxed{A_d = \frac{h_{fe} R_C}{(R_S + h_{ie})} \text{ (magnitude)}} \quad ...(6)$$

This is the differential gain for balanced output dual input differential amplifier circuit.

2.7.2 Common Mode Gain (A_c)

Let the magnitude of both the a.c. input signals be V_S and are in phase with each other. Hence the differential input $V_d = 0$ while the common mode input V_c is the average value of the two.

$$\therefore \quad V_c = \frac{V_1 + V_2}{2} = \frac{V_S + V_S}{2}$$

$$= V_S \quad ...(7)$$

While the output can be expressed as

$$V_o = A_c V_S \quad ...(8)$$

$$A_c = \frac{V_o}{V_S} \qquad \text{...(8a)}$$

But now both the emitter currents $I_{e1} = I_{e2} = I_e$, flows through R_E in the same direction.

Hence the total current flowing through R_E is $2I_e$.

As the two transistors are matched, a.c. equivalent of common mode operation can be shown, considering only one transistor, as in the Fig. 2.15.

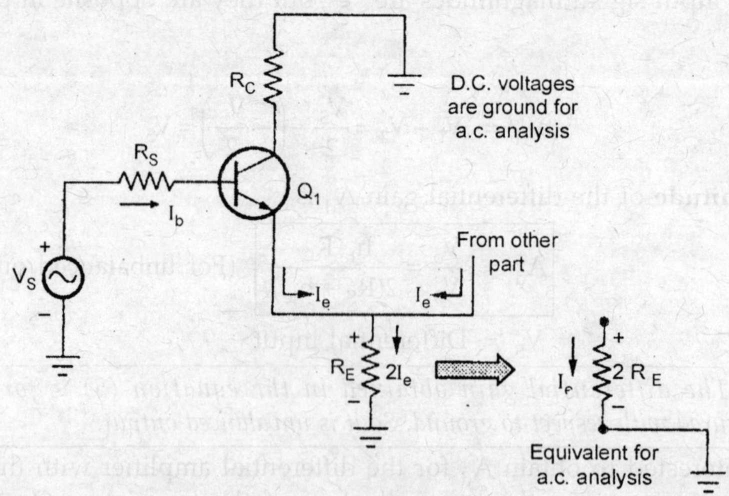

Fig. 2.15 A.C. equivalent for common mode operation.

The approximate hybrid model for the above circuit can be shown as in the Fig. 2.16.

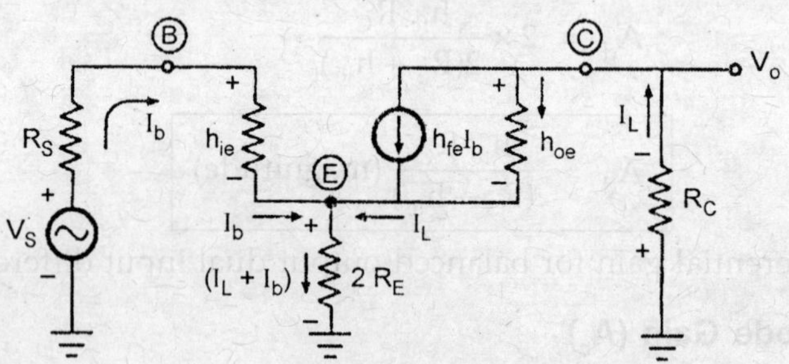

Fig. 2.16 Approximate hybrid model.

As the current through R_E is $2I_e$, for simplicity of assumed to be I_e and effective emitter resistance as $2R_E$. Hence the emitter resistance is shown $2R_E$ in the Fig. 2.16.

So current through R_C = load current I_L

effective emitter resistance = $2R_E$

current through emitter resistance = $I_L + I_b$

current through h_{oe} = $(I_L - h_{fe} I_b)$

Applying KVL to the input side,

$$-I_bR_S - I_bh_{ie} - 2R_E(I_L + I_b) + V_S = 0$$

$$\therefore \quad V_S = I_b(R_S + h_{ie} + 2R_E) + I_L(2R_E) \quad \text{...(8b)}$$

While $$V_o = -I_LR_C \quad \text{...(9)}$$

Negative sign due to the assumed direction of current. Applying KVL to the output loop,

$$\frac{-(I_L - h_{fe}I_b)}{h_{oe}} - 2R_E(I_L + I_b) - I_LR_C = 0$$

$$\therefore \quad -\frac{I_L}{h_{oe}} + \frac{h_{fe}}{h_{oe}}I_b - 2R_EI_L - 2R_EI_b - I_LR_C = 0$$

$$\therefore \quad I_b\left[\frac{h_{fe}}{h_{oe}} - 2R_E\right] = I_L\left[\frac{1}{h_{oe}} + 2R_E + R_C\right]$$

$$I_b[h_{fe} - 2R_E\,h_{oe}] = I_L[1 + h_{oe}(2R_E + R_C)]$$

$$\therefore \quad \frac{I_L}{I_b} = \frac{[h_{fe} - 2R_E\,h_{oe}]}{[1 + h_{oe}(2R_E + R_C)]} \quad \text{...(10)}$$

Substituting the value I_b, into the equation (8*b*), we get

$$V_S = \frac{I_L[1 + h_{oe}(2R_E + R_C)](R_S + h_{ie} + 2R_E)}{[h_{fe} - 2R_E\,h_{oe}]} + I_L(2R_E)$$

$$\therefore \quad \frac{V_S}{I_L} = \frac{[1 + h_{oe}(2R_E + R_C)](R_S \;\; h_{ie} + 2R_E)}{[h_{fe} - 2R_E\,h_{oe}]} + (2R_E)$$

Finding L.C.M. and adjusting the terms, we get

$$\therefore \quad \frac{V_S}{I_L} = \frac{2R_E(1 + h_{fe}) + R_S(1 + 2R_Eh_{oe}) + h_{ie}(1 + 2R_Eh_{oe}) + h_{oe}R_C[2R_E + R_S + h_{ie}]}{[h_{fe} - 2R_E\,h_{oe}]}$$

$$\therefore \quad \frac{V_S}{I_L} = \frac{2R_E(1 + h_{fe}) + (R_S + h_{ie})(1 + 2R_Eh_{oe}) + h_{oe}R_C[2R_E + R_S + h_{ie}]}{[h_{fe} - 2R_E\,h_{oe}]} \quad \text{...(11)}$$

Neglecting the terms of $h_{oe}R_C$ as practically $h_{oe}R_C \ll 1$.

$$\therefore \quad \frac{V_S}{I_L} = \frac{2R_E(1 + h_{fe}) + (R_S + h_{ie})(1 + 2R_Eh_{oe})}{[h_{fe} - 2R_Eh_{oe}]} \quad \text{...(12)}$$

Substituting the value of I_L, in the equation (9)

$$V_o = \frac{-V_S(h_{fe} - 2R_Eh_{oe})R_C}{2R_E(1 + h_{fe}) + (R_S + h_{ie})(1 + 2R_Eh_{oe})}$$

Hence the common mode gain can be written as (absorbing negative sign),

$$A_c = \frac{V_o}{V_S} = \frac{(2R_E h_{oe} - h_{fe})R_C}{2R_E(1+h_{fe}) + (R_S + h_{ie})(1+2R_E h_{oe})} \quad \text{... (13)}$$

In practice h_{oe} is generally neglected hence the expression for A_c can be further modified as

$$\therefore \quad A_c = \frac{-h_{fe} R_C}{R_S + h_{ie} + 2R_E(1+h_{fe})} \quad \text{... (14)}$$

***Key Point:** The expression for A_c remains same whether the output is balanced or unbalanced.*

2.7.3 Common Mode Rejection Ratio (CMRR)

Once the differential and common mode gains are obtained, the expression for the CMRR can be obtained as,

$$\text{CMRR} = \left| \frac{A_d}{A_c} \right|$$

$$\therefore \quad \text{CMRR} = \frac{R_S + h_{ie} + 2R_E(1+h_{fe})}{(R_S + h_{ie})} \quad \text{... (15)}$$

This is CMRR for **dual input balanced output** differential amplifier circuit.

2.7.4 Differential Input Impedance (R_i)

It is the equivalent resistance between one of the input and the ground when the other input terminal is grounded. Hence

$$R_i = \frac{V_S}{I_b} \quad \text{... (16)}$$

But referring to Fig. 2.14 and equation (1) we can write,

$$\therefore \quad R_i = 2(R_S + h_{ie}) \quad \text{... (17)}$$

For one transistor and input pair, the input resistance is $R_S + h_{ie}$ hence for dual input circuit the total input resistance is $2(R_S + h_{ie})$, as the two circuits are perfectly matched.

***Key Point:** This input resistance is not dependent on whether output is balanced or unbalanced.*

2.7.5. Output Impedance (R_o)

If is defined as the equivalent resistance between one of the output terminals with respect to ground. As seen from the Fig. 2.14, the resistance between output terminal with respect to ground is R_C.

$$\therefore \quad R_o = R_C$$

In the a.c. analysis of dual input, balanced output differential amplifier, it can be observed that the results are derived by considering only one transistor and one output at the time of analysis. And then due to matched pair of transistors, the results are derived for dual input, balanced output configuration. Hence the results for the remaining configurations can be directly obtained from the results derived earlier. And hence the summary of the differential amplifier circuit configurations is directly provided in the Table 2.1.

Table 2.1

Configuration gain	Circuit Resistance	A_d voltage Resistance	R_{in} Input	R_o Output
Dual Input Balanced Output	V_{CC}, R_C, R_C, V_o, R_{S1}, R_{S2}, V_{S1}, V_{S2}, R_E, $-V_{EE}$	$\dfrac{h_{fe}R_C}{R_S + h_{ie}}$	$2(R_S + h_{ie})$	R_C
Dual Input Unbalanced Output	V_{CC}, R_C, R_C, V_o, R_{S1}, R_{S2}, V_{S1}, V_{S2}, R_E, $-V_{EE}$	$\dfrac{h_{fe}R_C}{2(R_S + h_{ie})}$	$2(R_S + h_{ie})$	R_C
Single Input Balanced Output	V_{CC}, R_C, R_C, V_o, R_{S1}, R_{S2}, V_{S1}, R_E, $-V_{EE}$	$\dfrac{h_{fe}R_C}{R_S + h_{ie}}$	$2(R_S + h_{ie})$	R_C

Single Input Unbalanced Output	V_{CC}, R_C, R_C, V_o, R_{S1}, V_{S1}, R_E, $-V_{EE}$	$\frac{h_{fe}R_C}{2(R_S+h_{ie})}$	$2(R_S+h_{ie})$	R_C

Example 2.3: The Fig. 2.17 shows dual input, balanced output differential amplifier configuration. Assuming silicon transistors with h_{ie} = 2.8 kΩ. Calculate

(*i*) Operating point values
(*ii*) Differential gain
(*iii*) Common mode gain
(*iv*) CMRR
(*v*) Output if V_{S1} = 70 mV peak to peak at 1 kHz
and V_{S2} = 40 mV peak to peak at 1 kHz

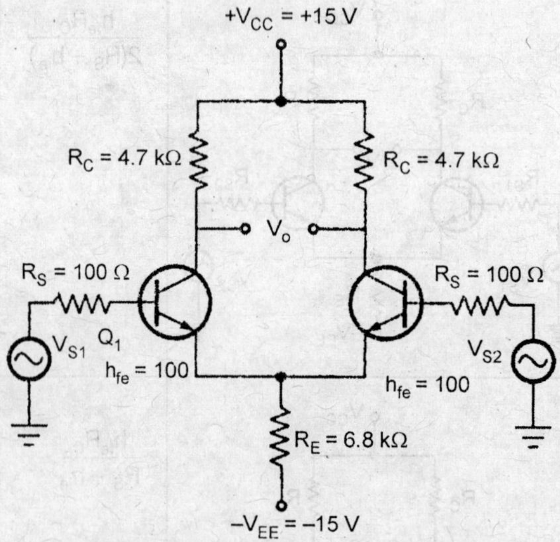

Fig. 2.17

Solution: As the transistors are silicon,

$$V_{BE} = 0.7 \text{ V}$$

(*i*) Operating point values are I_{CQ} and V_{CEQ}

$$I_E = \frac{V_{EE} - V_{BE}}{2R_E + \frac{R_S}{\beta}}$$

$$\beta = h_{fe} = 100$$

$$\therefore \quad I_E = \frac{15 - 0.7}{2 \times 6.8 \times 10^3 + \frac{100}{100}} = 1.051 \text{ mA}$$

Now $\quad I_C = I_E = 1.051$ mA

$\therefore \quad I_{CQ} =$ **1.051 mA**

Now $\quad V_{CE} = V_{CC} + V_{BE} - I_C R_C = 15 + 0.7 - 1.051 \times 10^{-3} \times 4.7 \times 10^3$

$= 10.758$ V

$\therefore \quad V_{CEQ} =$ **10.758 mA**

(*ii*) Differential gain

$$A_d = \frac{h_{fe} R_C}{R_S + h_{ie}} = \frac{100 \times 4.7 \times 10^3}{100 + 2.8 \times 10^3} = \mathbf{162.068}$$

(*iii*) Common mode gain

$$A_c = \frac{h_{fe} R_C}{2R_E(1 + h_{fe}) + R_S + h_{ie}}$$

$$= \frac{100 \times 4.7 \times 10^3}{2 \times 6.8 \times 10^3 (1 + 100) + 100 + 2.8 \times 10^3}$$

$= \mathbf{0.3414}$

(*iv*) $\quad CMRR = \frac{A_d}{A_c} = \frac{162.068}{0.3414} = \mathbf{474.652}$

i.e. $\quad$ CMRR = 20 log (474.652) = **53.527 dB**

(*v*) Output voltage

$$V_o = A_d V_d + A_c V_c$$

Now $\quad V_d = V_{S1} + V_{S2} = 70 - 40 = 30$ mV peak to peak

$$V_c = \frac{V_{S1} + V_{S2}}{2} = \frac{70 + 40}{2} = 55 \text{ mV peak to peak}$$

$\therefore \quad V_o = 162.068 \times 30 \times 10^{-3} + 55 \times 10^{-3} \times 0.3414 = 4.86204 + 0.0187$

$=$ **4.88 V peak to peak**

Key Point: *In practice $A_c V_c$ can be neglected as compared to $A_d V_d$ as common mode gain is always very small compared to differential gain.*

2.8 A.C. ANALYSIS OF DIFFERENTIAL AMPLIFIER USING R-PARAMETERS

The a.c. analysis can be obtained using r-parameters of transistors instead of h-parameters. An important r-parameter of a transistor is called a.c. emitter resistance denoted as r_e. The mathematical expression to obtain r_e is,

$$r_e = \frac{26mV}{I_E} = \frac{26 \times 10^{-3}}{I_E} \Omega$$

where I_E = D.C. emitter current of differential amplifier

Using this r_e, the expressions for A_d, R, and R_o can be obtained. The detail analysis is not expected in this course hence directly the expressions for A_d, R_i and R_o in terms of r_e are given in the Table 2.2.

Table 2.2

No.	Configuration	A_d	R_i	R_o
1.	Dual input balanced output	$\frac{R_C}{r_e}$	$2r_e\beta_{ac}$	R_C
2.	Dual input unbalanced output	$\frac{R_C}{2r_e}$	$2r_e\beta_{ac}$	R_C
3.	Single input balanced output	$\frac{R_C}{r_e}$	$2r_e\beta_{ac}$	R_C
4.	Single input unbalanced output	$\frac{R_C}{2r_e}$	$2r_e\beta_{ac}$	R_C

Alternatively one can find out h_{ie} from r_e and use the expressions in terms of h-parameters. The approximate relation between r_e and h_{ie} is given by,

$$h_{ie} = h_{fe}\, r_e \quad \text{where } h_{fe} = \beta \text{ for transistor}$$

Note that the above relation is approximate hence there is little difference in the value of A_d obtained using r-parameters and h-parameters. But this difference is small and practically acceptable.

While solving problems

If h_{ie} of the transistor is not given in the problem, find r_e and use the expressions in terms of r_e to obtain A_d, r_o and r_i.

2.8.1 Small Signal T-Model

The important r-parameter of the transistor is the a.c. emitter resistance r_e. Based on this, the small signal T-model of the transistor can be obtained as shown in the Fig. 2.18.

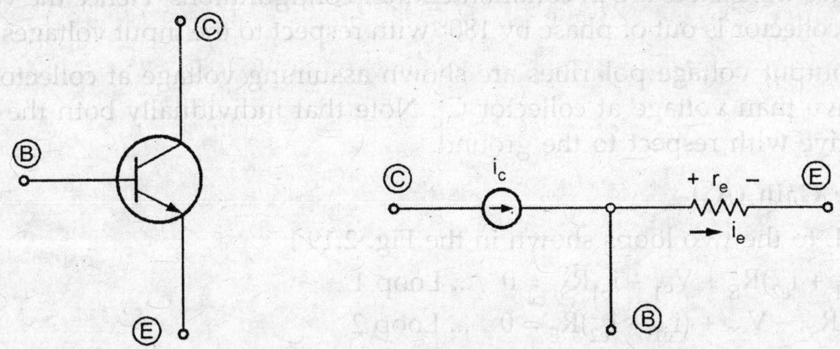

Fig. 2.18

The analysis of the differential amplifier using r-parameter is based on the small signal T-model of the transistor.

$$\therefore \qquad r_e = \frac{26\text{ mV}}{I_E}\,\Omega$$

The steps to perform the a.c. analysis using r-parameters are,

1. Set the d.c. voltages + V_{CC} and – V_{EE} to zero, by grounding the supply terminals.
2. Use the small signal T-models for the two transistors to obtain the equivalent circuit.

The d.c. analysis of all the differential amplifier configuration remains same irrespective of use of any parameters.

2.8.2 A.C. Analysis of Dual Input Balanced Output Differential Amplifier

The a.c. equivalent circuit of dual input, balanced output differential amplifier is shown in the Fig. 2.19.

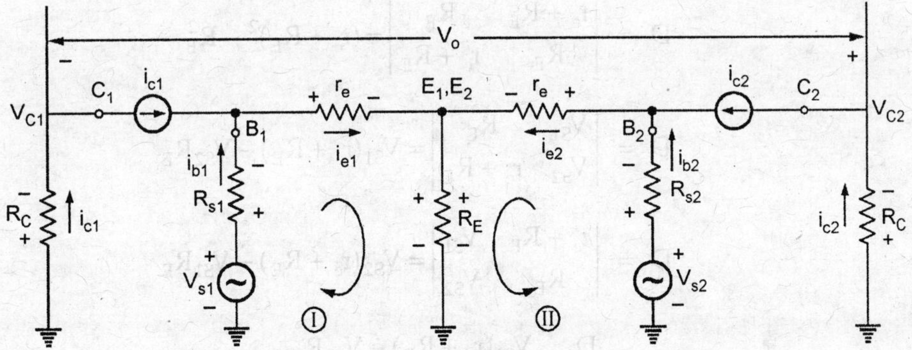

Fig. 2.19 A.C equivalent circuit.

Points to Remember for a.c. Analysis

1. As the two transistors are matched, $I_{E1} = I_{E2}$ and hence $r_{e1} = r_{e2}$. Hence the a.c. emitter resistance of both the transistors Q_1 and Q_2 is indicated as $r_{e1} = r_{e2} = r_e$.

2. The two transistors are in common-emitter configurations. Hence the voltage across each collector is out of phase by 180° with respect to the input voltages V_{S1} and V_{S2}.
3. The output voltage polarities are shown assuming voltage at collector C_2 is more positive than voltage at collector C_1. Note that individually both the voltages are negative with respect to the ground.

2.8.2.1 Voltage Gain (A_d)

Apply KVL to the two loops shown in the Fig. 2.19

$$-i_{e1}r_e - (i_{e1} + i_{e2})R_E + V_{S1} - i_{b1}R_{S1} = 0 \quad \text{... Loop 1} \qquad \text{...(1)}$$

$$-i_{e2}r_e + i_{b2}R_{S2} - V_{S2} + (i_{e1} + i_{e2})R_E = 0 \quad \text{... Loop 2} \qquad \text{...(2)}$$

For the transistors we can write,

$$i_{b1} = \frac{i_{e1}}{\beta_{ac}} \quad \text{and} \quad i_{b2} = \frac{i_{e2}}{\beta_{ac}}$$

Substituting in above,

$$V_{S1} - \frac{R_{S1}}{\beta_{ac}} i_{e1} - r_e i_{e1} - R_E(i_{e1} + i_{e2}) = 0 \qquad \text{...(3)}$$

$$V_{S2} - \frac{R_{S2}}{\beta_{ac}} i_{e2} - r_e i_{e2} - R_E(i_{e1} + i_{e2}) = 0 \qquad \text{...(4)}$$

As β_{ac} value is large, $\frac{R_{S1}}{\beta_{ac}}$ and $\frac{R_{S2}}{\beta_{ac}}$ can be neglected.

$$\therefore \quad (r_e + R_E)i_{e1} + R_E i_{e2} = V_{S1} \qquad \text{...(5)}$$

$$\text{and} \quad (R_E)i_{e1} + (r_e + R_E)i_{e2} = V_{S2} \qquad \text{...(6)}$$

Solve for i_{e1} and i_{e2} by Cramer's rule.

$$D = \begin{vmatrix} r_e + R_E & R_E \\ R_E & r_e + R_E \end{vmatrix} = (r_e + R_E)^2 - R_E^2$$

$$D = \begin{vmatrix} V_{S1} & R_E \\ V_{S2} & r_e + R_E \end{vmatrix} = V_{S1}(r_e + R_E) - V_{S2}R_E$$

$$D = \begin{vmatrix} r_e + R_E & V_{S1} \\ R_E & V_{S2} \end{vmatrix} = V_{S2}(r_e + R_E) - V_{S1}R_E$$

$$\therefore \quad i_{e1} = \frac{D_1}{D} = \frac{V_{S1}(r_e + R_E) - V_{S2}R_E}{(r_e + R_E)^2 - R_E^2} \qquad \text{...(7)}$$

$$\text{and} \quad i_{e2} = \frac{D_2}{D} = \frac{V_{S2}(r_e + R_E) - V_{S1}R_E}{(r_e + R_E)^2 - R_E^2} \qquad \text{...(8)}$$

The output voltage is given by,

$$V_o = V_{C2} - V_{C1} = -R_C i_{C2} - (-R_C i_{C1})$$

$$= R_C i_{C1} - R_C i_{C2} = R_C(i_{e_1} - i_{e_2}) \quad ...(9)$$

This is because $i_c \approx i_e$.

Using expressions of i_{e1} and i_{e2},

$$V_o = R_C\left[\frac{V_{S1}(r_e + R_E) - V_{S2}R_E - [V_{S2}(r_e + r_E) - V_{S1}R_E]}{(r_e + R_E)^2 - R_E^2}\right]$$

$$= R_C\left[\frac{(r_e + R_E)(V_{S1} - V_{S2}) + R_E(V_{S1} - V_{S2})}{(r_E + R_E)^2 - R_E^2}\right]$$

$$= \frac{R_C(V_{S1} - V_{S2})(r_e + 2R_E)}{r_e^2 + 2r_eR_E + R_E^2 - R_E^2} = \frac{R_C(V_{S1} - V_{S2})(r_e + 2R_E)}{r_e(r_e + 2R_E)}$$

$$\therefore \quad V_o = \frac{R_C}{r_e}(V_{S1} - V_{S2}) \quad ...(10)$$

Now $V_d = V_{S1} - V_{S2}$ = Differential input voltage

and $A_d = \frac{V_o}{V_d} = \frac{V_o}{V_{S1} - V_{S2}}$

$$\therefore \quad A_d = \frac{R_C}{r_e} \quad ...(11)$$

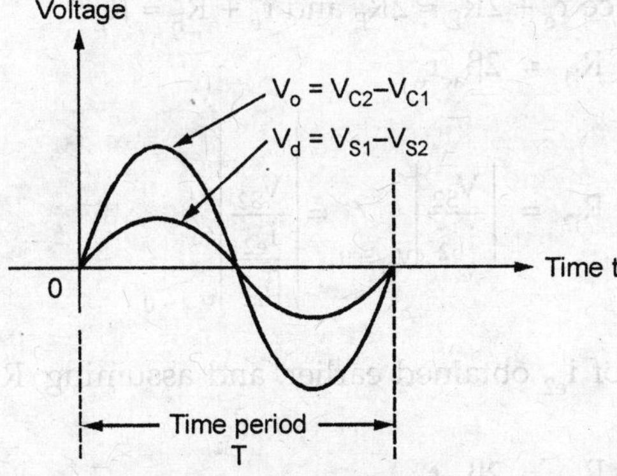

Fig. 2.20 Input and output waveforms

It can be seen that voltage gain is not dependent on R_E. The input and output waveforms are shown in the Fig. 2.20, showing the differential voltage V_d and the output voltage V_o.

2.8.2.2 Differential Input Resistance (R_i)

The differential input resistance is defined as the equivalent resistance between one of the input terminals and the ground when the other input terminal is grounded. Thus R_{i1} is the input resistance seen from the input signal source V_{S1} when the other source V_{S2} is set to zero. While R_{i2} is the input resistance seen from the input signal source V_{S2} when the other source V_{S1} is set to zero.

As the source resistances R_{S1} and R_{S2} are practically very small, these are neglected in the derivation of R_i. Mathematically R_{i1} can be expressed as,

$$R_{i1} = \left|\frac{V_{S1}}{i_{b1}}\right|_{V_{S2}=0} = \left|\frac{V_{S1}}{\frac{i_{e1}}{\beta_{ac}}}\right|_{V_{S2}=0} \quad \text{...(12)}$$

Substituting the value of i_{e1} obtained earlier,

$$R_{i1} = \frac{\beta_{ac} V_{S1}}{\frac{V_{S1}(r_e + R_E) - V_{S2} R_E}{(r_e + R_E)^2 - R_E^2}} \quad \text{but } V_{S2} = 0$$

$$= \frac{\beta_{ac} V_{S1}[(r_e + R_E)^2 - R_E^2]}{V_{S1}(r_e + R_E)} = \frac{\beta_{ac}[r_e^2 + 2r_e R_E + R_E^2 - R_E^2]}{(r_e + R_E)}$$

$$= \frac{\beta_{ac} r_e (r_e + 2R_E)}{(r_e + R_E)}$$

Practically $R_E \gg r_e$ hence $r_e + 2R_E \approx 2R_E$ and $r_e + R_E \approx R_E$

$\therefore$ $$R_{i1} = 2\beta_{ac} r_e \quad \text{...(13)}$$

Similarly, $$R_{i2} = \left|\frac{V_{S2}}{i_{b2}}\right|_{V_{S1}=0} = \left|\frac{V_{S2}}{\frac{i_{e2}}{\beta_{ac}}}\right|_{V_{S1}=0} \quad \text{...(14)}$$

Substituting the value of i_{e2} obtained earlier, and assuming $R_E \gg r_e$ i.e. $r_e + 2R_E \approx 2R_E$ and $r_e + R_E \approx R_E$ we get,

$$R_{i2} = 2\beta_{ac} r_e \quad \text{...(15)}$$

Thus $$\boxed{R_{i1} = R_{i2} = 2\beta_{ac} r_e}$$

2.8.2.3 Output Resistance (R_o)

The output resistance is defined as the equivalent resistance between one of the output terminals and the ground. So the output resistance R_{o1} between collector C_1 and the ground is the collector resistance R_C while the output resistance R_{o2} between collector C_2 and the ground is also the collector resistance R_C.

Thus $$R_{o1} = R_{o2} = R_C \quad ...(17)$$

The current gain of the differential amplifier can not be defined, hence its expression is not derived. The differential amplifier is used as a voltage amplifier and not as a current or power amplifier.

2.8.2.4 Inverting and Non-inverting Inputs

The input voltage V_{S1} produces a positive component in the output voltage i.e, in phase component hence it is called non-inverting input. The input voltage V_{S2} produces a negative component in the output i.e. 180° out of phase inverted component. Hence V_{S2} is called inverting input. The two bases to which V_{S1} and V_{S2} are connected are called non-inverting and inverting input terminals, respectively.

2.8.2.5 CMRR

It has been mentioned that the differential amplifier has ability to reject the unwanted common mode disturbance signals such as 50 Hz noise or hum. Such signals may distort the output. The ability of differential amplifier to suppress undesirable disturbance is measured by a factor called CMRR.

$$\text{CMRR} = \frac{A_d}{A_c}$$

The gain A_d can be calculated using the derived expression. To calculate A_c, apply a known voltage V_c to both the inputs and measure the resultant output V_o which ideally should be zero.

$$\therefore \quad A_c = \frac{V_o}{V_c}$$

2.8.3 A.C. Analysis of Dual Input Unbalanced Output Differential Amplifier

In unbalanced output configuration, the output is taken from one of the two collectors with respect to ground. Let the output is taken from C_2 for the derivation purpose.

The a.c. equivalent circuit for the configuration is shown in the Fig. 2.21.

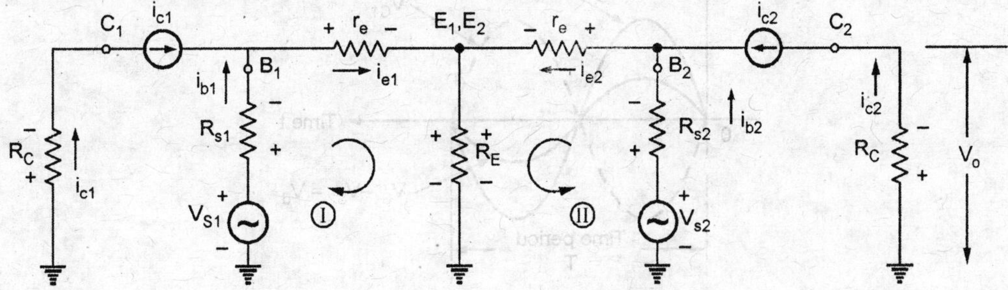

Fig. 2.21 A.C. equivalent circuit.

2.8.3.1 Voltage Gain (A_d)

Applying KVL to the two loops shown in the Fig. 2.21,

Loop I, $-i_{e1}r_e - R_E(i_{e1} + i_{e2}) + V_{S1} - i_{b1}R_{S1} = 0$

Loop II, $-i_{e2}r_e - R_E(i_{e1} + i_{e2}) + V_{S2} - i_{b2}R_{S2} = 0$

The equations are same as equations (1) and (2) in the section 2.9.2.1 obtained earlier hence the expressions of i_{e1} and i_{e2} also remain same.

$$\therefore \quad i_{e1} = \frac{(r_e + R_E)V_{S1} - R_E V_{S2}}{(r_e + R_E)^2 - R_E^2} \qquad \text{...(3)}$$

and

$$i_{e2} = \frac{(r_e + R_E)V_{S2} - R_E V_{S1}}{(r_e + R_E)^2 - R_E^2} \qquad \text{...(3)}$$

The output voltage is now across C_2 and ground only hence,

$$V_o = V_{C2} = -R_C i_{C2} = -R_C i_{e2} \quad \text{as} \quad i_{C2} \approx i_{e2}$$

Substituting value of i_{e2},

$$V_o = \frac{-R_C[(r_e + R_E)V_{S2} - R_E V_{S1}]}{(r_e + R_E)^2 - R_E^2}$$

$$\therefore \quad V_o = +R_C\left[\frac{R_E V_{S1} + (r_e + R_E)V_{S2}}{r_e(r_e + 2R_E)}\right] \qquad \text{...(5)}$$

Now R_E □ r_e hence $r_e + R_E \approx R_E$ and $r_e + 2R_E \approx 2R_E$

$$\therefore \quad V_o = \frac{R_C R_E(V_{S1} - V_{S2})}{2R_E r_e} = \frac{R_C(V_{S1} - V_{S2})}{2r_e} \qquad \text{...(6)}$$

$$\therefore \quad A_d = \frac{V_o}{V_{S1} - V_{S2}} = \frac{R_C}{2r_e} \qquad \text{...(7)}$$

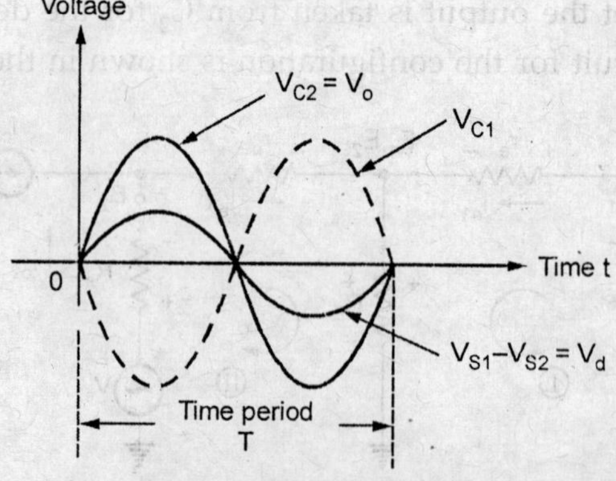

Fig. 2.22 Input and output waveforms.

Thus for dual input unbalanced output, the differential gain is half the gain of dual input balanced output differential amplifier.

The input and output waveforms are shown in the Fig. 2.22.

2.8.3.2 Differential Input Resistance (R_i)

The input resistance does not depend on the way in which output is measured. The input resistance R_i as viewed from either input source of the circuit remains same as obtained for dual input balanced output configuration. This is because, apart from way of measuring output there is no difference between the two configurations.

$$\therefore \quad \boxed{R_{i1} = R_{i2} = 2\beta_{ac} r_e} \quad \text{...(8)}$$

2.8.3.3 Output Resistance (R_o)

The output resistance R_o measured at the collector C_2 with respect to ground is equal to the collector resistance R_C hence,

$$\boxed{R_o = R_C} \quad \text{... (9)}$$

2.8.4 A.C. Analysis of Single Input Balanced Output Differential Amplifier

This configuration uses only single input signal source V_{S1}. The other input terminal is grounded.

The a.c. equivalent circuit is shown in the Fig. 2.23 for this configuration, which uses T equivalent models of transistors Q_1 and Q_2.

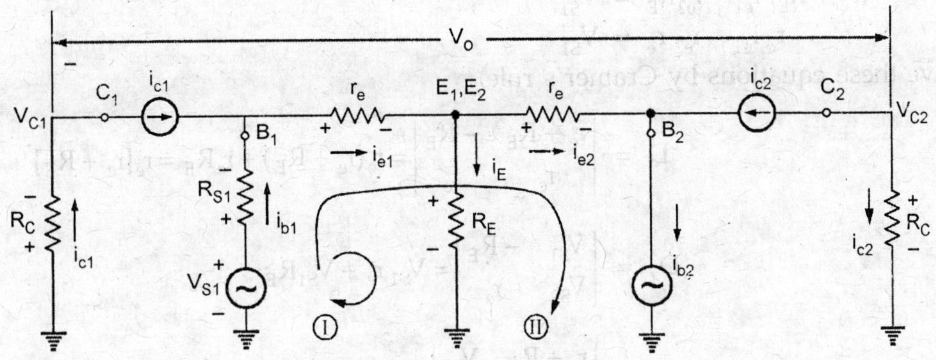

Fig. 2.23 A.C. equivalent circuit.

Note the following important points about the circuit:

1. When V_{S1} is positive, the base emitter voltage of Q_1 is positive, while the base of the transistor Q_2 is negative. Hence the collector current of Q_1 increases and that of Q_2 decreases from the operating point value I_{CQ}. This change in the collector currents is shown in the Fig. 2.23. The two collector currents are shown in the same direction, due to such change, in the positive half cycle of source voltage V_S. During negative half cycle of source voltage V_S, the collector current of Q_2 increases and that of Q_1 decreases. Both i_{c1} and i_{c2} change their directions, during negative half cycle.

2. The polarity of voltage across each collector resistance R_C is consistent with the direction of current sources i_{c1} and i_{c2}.
3. The output voltage V_o is the difference between voltage at collector C_2 and voltage at collector C_1.

$$V_o = V_{C2} - V_{C1}$$

2.8.4.1 Voltage in Gain (A_d)

Applying KVL to the two loops shown,

Loop I, $$-i_{e1}r_e - i_E R_E + V_{S1} - i_{b1}R_S = 0 \qquad ...(1)$$

Loop II, $$-i_{e2}r_e - i_E R_E + V_{S2} - i_{b1}R_S = 0 \qquad ...(2)$$

Now $$i_E = i_{e1} - ie_2, i_{b1} \approx \frac{i_{e1}}{\beta_{ac}} \text{ and } i_{b2} \approx \frac{i_{e2}}{\beta_{ac}}$$

$$\therefore \quad V_{S1} - \frac{i_{e1}R_S}{\beta_{ac}} - i_{e1}r_e - (i_{e1} - i_{e2})R_E = 0 \qquad ...(3)$$

and $$V_{S1} - \frac{i_{e1}R_S}{\beta_{ac}} - i_{e1}r_e - i_{e2}r_e = 0 \qquad ...(4)$$

Now $\frac{R_S}{\beta_{ac}}$ is very low and neglecting it was get,

$$(r_e + R_E)\, i_{e1} - i_{e2}R_E = V_{S1} \qquad ...(5)$$

$$r_e\, i_{e1} - i_{e2}\, r_e = V_{S1} \qquad ...(6)$$

Solve these equations by Cramer's rule,

$$D = \begin{vmatrix} r_e + R_E & -R_E \\ r_e & r_e \end{vmatrix} = r_e(r_e + R_E) + r_e R_E = r_e[r_e + R_E]$$

$$D_1 = \begin{vmatrix} V_{S1} & -R_E \\ V_{S2} & r_e \end{vmatrix} = V_{S1}\, r_e + V_{S1}R_E$$

$$D_2 = \begin{vmatrix} r_e + R_E & V_{S1} \\ R_E & V_{S1} \end{vmatrix} = V_{S1}(r_e + R_E) - V_{S1}r_e = V_{S1}\, R_E$$

$$\therefore \quad i_{e1} = \frac{D_1}{D} = \frac{V_{S1}(r_e + R_E)}{r_e[r_e + 2R_E]} \qquad ...(7)$$

and $$i_{e2} = \frac{D_2}{D} = \frac{V_{S1}R_E}{r_e[r_e + 2R_E]} \qquad ...(8)$$

The output voltage $$V_o = V_{C2} - V_{C1} = i_{c2}R_C - (-i_{c1}R_C)$$

$$= R_C(i_{e2} + i_{e1}) \quad \text{as} \quad i_e \approx i_c$$

Substituting expressions of i_{e1} and i_{e2} we get,

$$V_o = R_C i_{e2} + R_C i_{e1} \quad ...(9)$$

$$\therefore \quad V_o = \frac{R_C V_{S1} R_E}{r_e[r_e + 2R_E]} + \frac{R_C V_{S1}(r_e + R_E)}{r_e[r_e + 2R_E]}$$

$$= \frac{R_C V_{S1}(r_e + 2R_E)}{r_e[r_e + 2R_E]} = \frac{R_C V_{S1}}{r_e} \quad ...(10)$$

$$\therefore \quad \boxed{A_d = \frac{V_o}{V_{S1}} = \frac{R_C}{r_e}} \quad ...(11)$$

This voltage gain of single input, balanced output differential amplifier is equal to that of voltage gain of the dual input balanced output differential amplifier.

The input and output waveforms are shown in the Fig. 2.24.

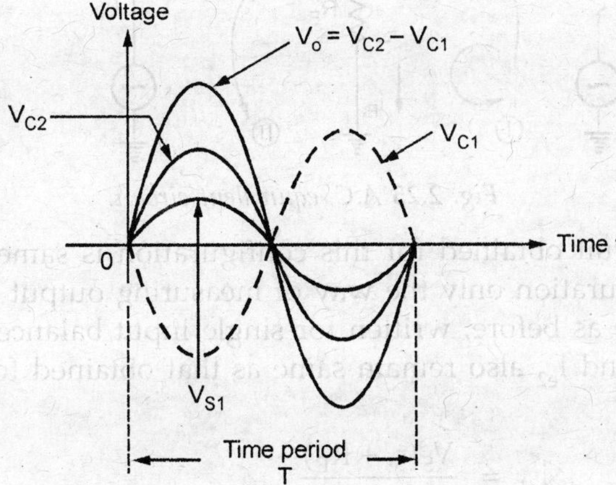

Fig. 2.24 Input and output waveforms.

2.8.4.2 Differential Input Resistance (R_i)

The input resistance R_i as seen from the input signal source is,

$$R_i = \frac{V_{S1}}{i_{b1}} = \frac{V_{S1}}{\frac{i_{e1}}{\beta_{ac}}} = \frac{\beta_{ac} V_{S1}}{i_{e1}} \quad ...(12)$$

Substituting i_{e1},

$$R_i = \frac{\beta_{ac} V_{S1}}{\left[\frac{V_{S1}(r_e + R_E)}{r_e(r_e + 2R_E)}\right]} = \frac{\beta_{ac} r_e (r_e + 2R_E)}{(r_e + R_E)}$$

Now $R_E > r_e$ so $r_e + R_E \approx R_E$ and $r_e + 2R_E \approx 2R_E$

$$\therefore \quad \boxed{R_i = \frac{\beta_{ac} r_e 2R_E}{R_E} = 2\beta_{ac} r_e} \quad ...(13)$$

2.8.4.3 Output Resistance (R_o)

It is the resistance measured at either output terminal with respect to ground. Hence output resistance measured at collector C_1 with respect to ground is same as that measured at collector C_2 with respect to ground given by,

$$\therefore \quad \boxed{R_{o1} = R_{o2} = R_C} \quad ...(14)$$

2.8.5 A.C. Analysis of Single Input Unbalanced Output Differential Amplifier

This configuration used only single input V_{S1} and output is taken between C_2 and ground. The a.c. equivalent circuit is shown in the Fig. 2.25.

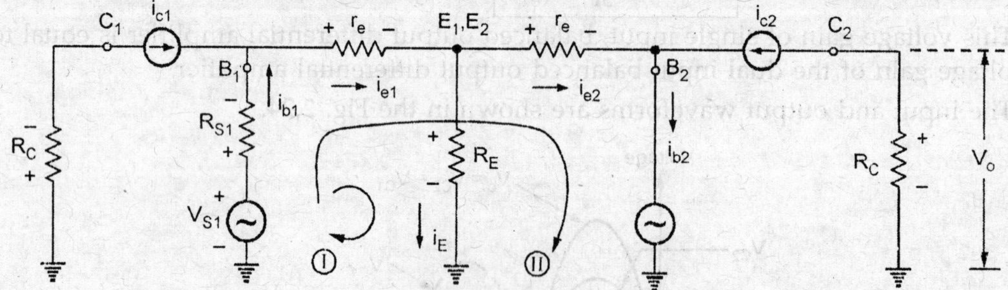

Fig. 2.25 A.C. equivalent circuit.

The equivalent circuit obtained for this configuration is same as that of single input balanced output configuration only the way of measuring output is different. So the loop equations remain same as before, written for single input balanced output circuit. Hence the expressions of i_{e1} and i_{e2} also remain same as that obtained for single input balanced output circuit.

$$i_{e1} = \frac{V_S(r_e + R_E)}{r_e(r_e + 2R_E)} \quad ...(1)$$

$$i_{e2} = \frac{V_S R_E}{r_e(r_e + 2R_E)} \quad ...(2)$$

The output voltage is given by,

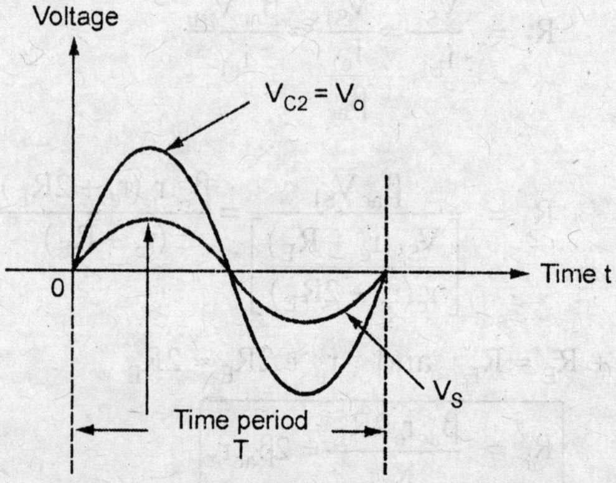

Fig. 2.26 Input and output waveforms.

$$V_o = i_{c2}R_C = i_{e2}R_C \qquad \text{... as } i_{c2} \approx i_{e2}$$

$$\therefore \quad V_o = \frac{V_S R_E R_C}{r_e(r_e + 2R_E)} \qquad \text{...(3)}$$

Now R_E □ r_e hence $r_e + 2R_E \approx 2R_E$

$$\therefore \quad V_o = \frac{V_S R_E R_C}{r_e 2R_E} = \frac{V_S R_C}{2r_e} \qquad \text{...(4)}$$

$$\therefore \quad A_d = \frac{V_o}{V_S} = \frac{V_C}{2r_e} \qquad \text{...(5)}$$

The input and output waveforms are shown in the Fig. 2.26. In this case, it is assumed that the d.c. output voltage is zero.

2.8.5.2 Differential Input Resistance (R_i)

This is same as obtained for single input balanced output configuration i.e.

$$\boxed{R_i = 2\beta_{ac}\, r_e} \qquad \text{...(6)}$$

While deriving this it is assumed that R_E □ r_e.

2.8.5.2 Output Resistance (R_o)

It is measured at collector C_2 with respect to ground and is equal to collector resistance R_C.

$$\therefore \quad \boxed{R_o = R_C} \qquad \text{... (7)}$$

This configuration is rarely used in practice because,

1. It requires more components.
2. The voltage gain is less compared to single input balanced output configuration.
3. There exists a d.c. voltage at the output without any input signal source applied.

Example 2.4: For the differential amplifier shown in Fig. 2.27, calculate:

(*i*) Operating point i.e. I_{CQ} and V_{CEQ}

(*ii*) Voltage gain

(*iii*) Input and output impedance.

Assume $\beta = 100$.

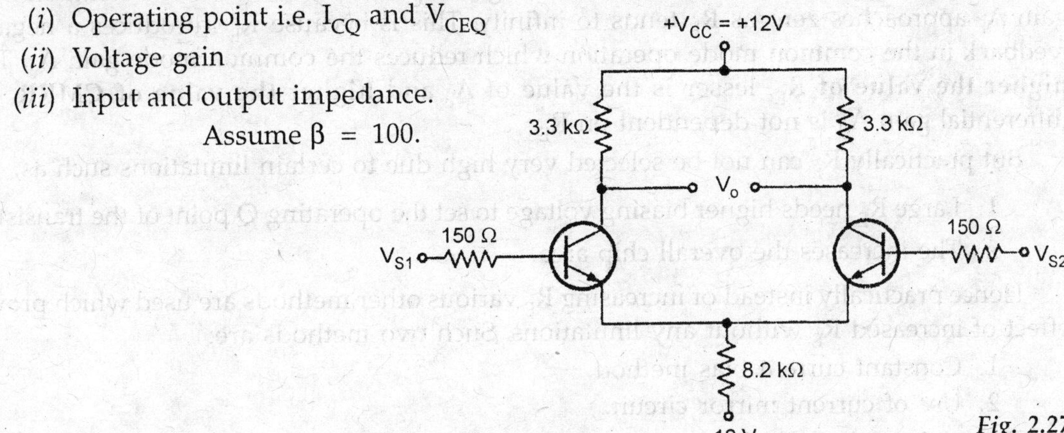

Fig. 2.27

Solution: As h_{ie} is not given, use r parameters to obtain voltage gain.

$R_C = 3.3\ k\Omega$, $R_S = 150\ \Omega$, $R_E = 8.2\ k\Omega$, $V_{CC} = 12\ V = V_{EE}$, $\beta = 100$

(*i*)
$$I_E = \frac{V_{EE} - V_{BE}}{\frac{R_S}{\beta} + 2R_E} = \frac{12 - 0.7}{\frac{150}{100} + 2 \times 8.2 \times 10^3} = 0.688\ mA$$

$$\therefore \quad I_{CQ} = I_E = \mathbf{0.688\ mA}$$

$$V_{CEQ} = V_{CC} + V_{BE} - I_{CQ}R_C = 12 + 0.7 - 0.688 \times 10^{-3} \times 3.3 \times 10^3$$

$$= \mathbf{10.4296\ V}$$

(*ii*) For voltage gain,
$$r_e = \frac{26\ mA}{I_E}$$

$$\therefore \quad r_e = \frac{26 \times 10^{-3}}{0.688 \times 10^{-3}} = 37.7906\ \Omega$$

$$\therefore \quad A_d = \frac{R_C}{r_e} = \frac{3.3 \times 10^3}{37.7906} = \mathbf{87.32}$$

As the configuration is dual input balanced output.

(*iii*)
$$R_i = 2r_e\beta = 2 \times 37.7906 \times 100 = \mathbf{7.558\ k\Omega}$$

(*iv*)
$$R_o = R_C = \mathbf{3.3\ k\Omega}$$

2.9 METHODS OF IMPROVING CMRR

As seen earlier, in the Ex. 2.1, higher the value of CMRR, better is the performance of differential amplifier. Hence in practice the efforts are always to improve the CMRR of the differential amplifier. Alongwith the basic circuit, various other circuits are used to improve the performance of differential amplifier.

2.9.1 Effect of R_E

To improve the CMRR, the common mode gain A_c must be reduced. The common mode gain A_c approaches zero as R_E tends to infinity. This is because R_E introduces a negative feedback in the common mode operation which reduces the common mode gain A_c. Thus **higher the value of** R_E, lesser is the value of A_c and **higher the value of CMRR**. The differential gain A_d is not dependent on R_E.

But practically R_E can not be selected very high due to certain limitations such as,

1. Large R_E needs higher biasing voltage to set the operating Q point of the transistors.
2. The increases the overall chip area.

Hence practically instead of increasing R_E various other methods are used which provide effect of increased R_E without any limitations. Such two methods are:

1. Constant current bias method.
2. Use of current mirror circuit.

The other method used to increase A_d to improve CMRR is called use of an active load.

2.10 DIFFERENTIAL AMPLIFIER WITH CONSTANT CURRENT SOURCE

Without physically increasing the value of R_E, the R_E is replaced by a transistor operated at a constant current. Such a constant current source circuit gives the effect of a very high resistance without affecting the Q point values of the differential amplifier. The differential amplifier using constant current bias circuit instead of R_E is shown in the Fig. 2.28.

The transistor used is Q_3 and the values of R_1, R_2 and R_3 are selected so as to give the same operating point values for the two transistors Q_1 and Q_2.

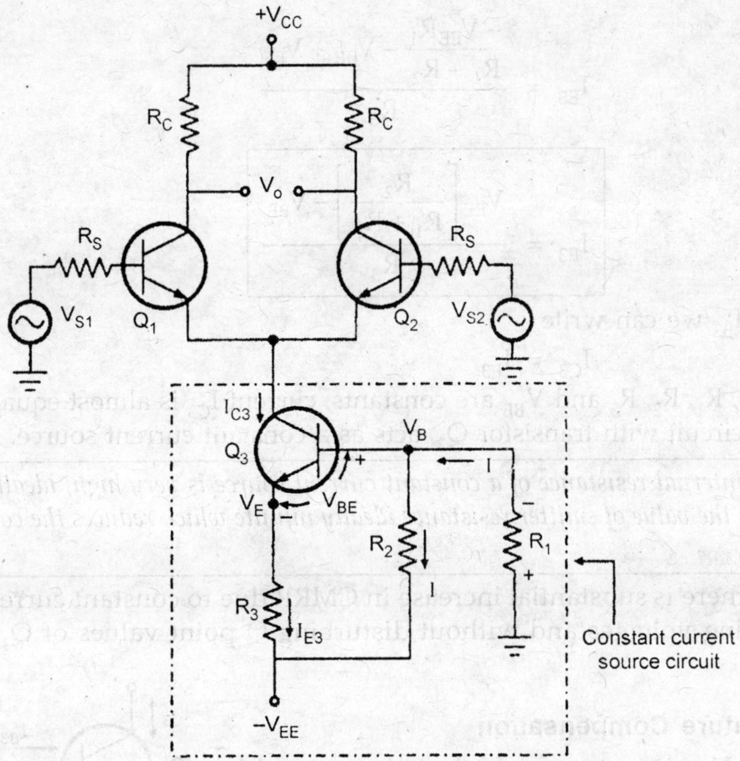

Fig. 2.28 Use of constant current source.

2.10.1 Circuit Analysis

Let current through R_3 be I_{E3} while current through R_1 is I. Neglecting the base current of Q_3 which is very small due to large β_{ac}, we can assume that current through R_2 is also I.

Applying Kirchhoff's law,

$$-IR_1 - IR_2 + V_{EE} = 0$$

$$\therefore \quad \boxed{I = \frac{V_{EE}}{R_1 + R_2}} \quad \ldots(1)$$

Now $\quad V_B = -IR_1$

Negative sign according to the direction of current.

$$\therefore \quad V_B = -\frac{V_{EE}R_1}{R_1 + R_2} \quad \text{...(2)}$$

Now $\quad V_E = V_B - V_{BE}$

and $$I_{E3} = \frac{V_E - (-V_{EE})}{R_3} \quad \text{...(3)}$$

Substituting expressions for V_B and V_E,

$$\therefore \quad I_{E3} = \frac{\frac{-V_{EE}R_1}{R_1 + R_2} - V_{BE} + V_{EE}}{R_3}$$

$$\therefore \quad \boxed{I_{E3} = \frac{V_{EE}\left[\frac{R_2}{R_1 + R_2}\right] - V_{BE}}{R_3}} \quad \text{...(4)}$$

Neglecting I_{B3} we can write

$$I_{C3} = I_{E3}$$

Thus as V_{EE}, R_1, R_2, R_3 and V_{BE} are constants, current I_{C3} is almost equal to I_{E3} and also constant. Thus circuit with transistor Q_3, acts as a constant current source.

***Key Point:** The internal resistance of a constant current source is very high, ideally infinite. Hence this circuit makes the value of emitter resistance ideally infinite which reduces the common mode gain A_c ideally to zero.*

Practically there is substantial increase in CMRR due to constant current bias without increasing biasing voltages and without disturbing Q point values of Q_1 and Q_2.

2.10.2 Temperature Compensation

In practice V_{BE} is not constant but changes with respect to temperature, approximately at the rate of 2.5 mV/°C. This intern may change I_{E3}. Thus there is necessity to provide a thermal compensation.

From junction of emitters of Q_1 and Q_2. Such a thermal compensation is provided by replacing R_2 with the two diodes D_1 and D_2 as shown in the Fig. 2.29.

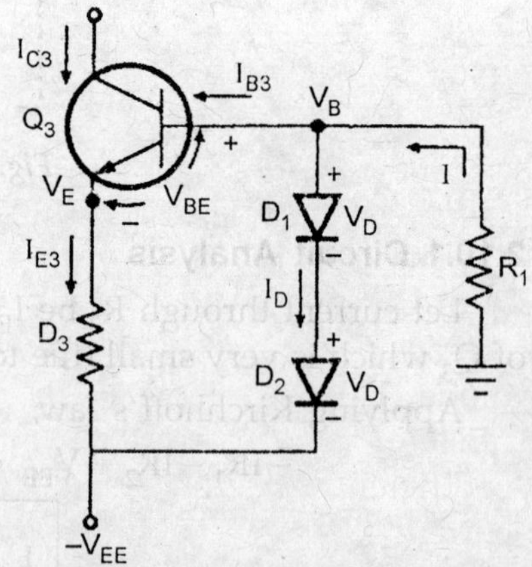

Fig. 2.29 Temperature compensated current source.

As the temperature increases, the transistor current i.e. collector current I_{C3} increases. At the same time the diode current also increases. This is because the diode current is given by

$$I_D = I_o\left[e^{V_D/\eta V_T} - 1\right] \quad \text{...(5)}$$

where V_D is diode bias voltage, V_T is the voltage equivalent of temperature and η is constant.

$$V_T = \frac{T}{11600}$$

$$\eta = 1 \text{ for germanium} = 2 \text{ for silicon}$$

Due to increased diode current I_D, the base current of Q_3 decrease. This compensated for the increase in I_{C3}.

Circuit Analysis

Assume that the voltage drop across the diodes is same as V_D.

$$V_B = -V_{EE} + 2V_D \quad \text{...(6)}$$

Now $$V_E = V_B - V_{BE}$$

$\therefore$ $$V_E = -V_{EE} + 2V_D - V_{BE} \quad \text{...(7)}$$

And $$I_{E3} = \frac{V_E - (-V_{EE})}{R_3} = \frac{-V_{EE} + 2V_D - V_{BE} + V_{EE}}{R_3}$$

$\therefore$ $$\boxed{I_{E3} = \frac{2V_D - V_{BE}}{R_3}} \quad \text{...(8)}$$

If the diode and transistor characteristics are same the V_D is same as V_{BE}.

$\therefore$ $$I_{E3} = \frac{V_D}{R_3} = I_{C3} \quad \text{...(9)}$$

Thus for a given value of R_E, the emitter current I_{E3} depends on the voltage drop across diodes. But drop across the diodes is a function of the current I_D, which is a part of current I, which is determined by the value of R_1. Thus the value of I_{E3} can be varied by changing the value of R_1 or R_E, as per the requirement.

The value of R_1 can be obtained as,

$$\boxed{R_1 = \frac{V_{EE} - 2V_D}{I_{E3}} = \frac{V_{EE} - 1.4\,(V)}{I_{E3}}} \quad \text{...(10)}$$

2.10.3 Constant Current Bias Using Zener

The zener diode is often used in practice in place of the two diodes. Then problem of temperature dependent characteristics of the transistor can be solved by using zener diode. This is because zeners are available over a wide range of voltages and can have a matching temperature coefficient of voltage to those of transistors. Constant current bias circuit using zener is shown in the Fig. 2.30.

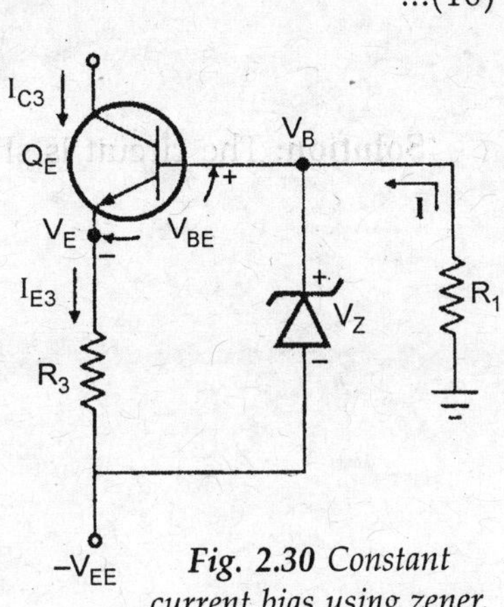

Fig. 2.30 Constant current bias using zener.

Circuit Analysis

The voltage at the base of transistor Q_3 is

$$V_B = -V_{EE} + V_Z \quad ...(11)$$

Now $$V_E = V_B - V_{BE}$$

$\therefore$ $$V_E = -V_{EE} + V_Z - V_{BE} \quad ...(12)$$

And $$I_{E3} = \frac{V_E - (-V_{EE})}{R_3}$$

$$I_{E3} = \frac{-V_{EE} + V_Z - V_{BE} + V_{EE}}{R_3}$$

$\therefore$ $$\boxed{I_{E3} = \frac{V_Z - V_{BE}}{R_3}} \quad ...(13)$$

While the value of R_1 is selected in such a way that zener diode always conducts in reverse region. It can be calculated as

$$\boxed{R_1 = \frac{V_{EE} - V_Z}{I}} \quad ...(14)$$

The common mode gain is almost zero, providing very large value of CMRR, with a constant current bias. To offer extremely large resistance under a.c. conditions, a simple circuit of constant current source, with less number of components is now a day used. This circuit is called **current mirror** or **current repeater circuit.**

Example 2.5: Calculate the constant current I in the circuit shown in the Fig. 2.31.

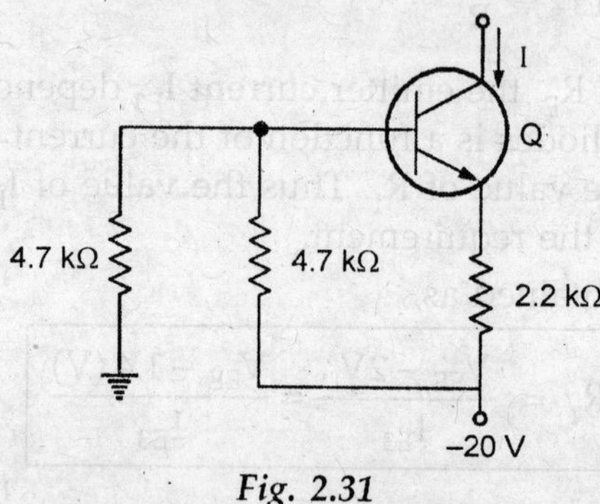

Fig. 2.31

Solution: The circuit is shown in again in the Fig. 2.31 (*a*) with various voltages.

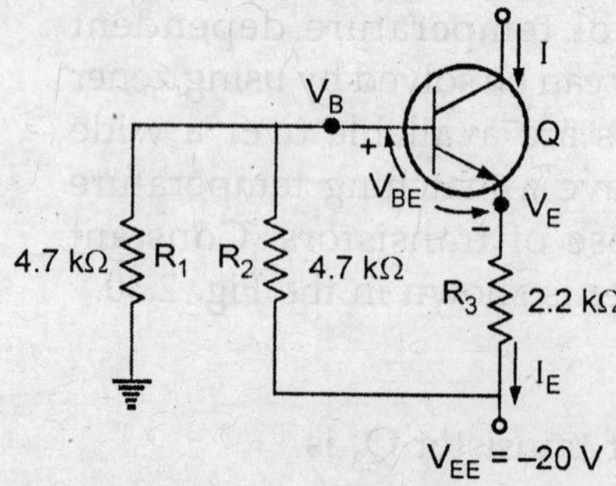

Fig. 2.31 (a)

$$V_B = \frac{-V_{EE}R_1}{R_1 + R_2} = \frac{-20 \times 4.7 \times 10^3}{(4.7 \times 10^3 + 4.7 \times 10^3)}$$

$$= -10 \text{ V} \qquad \text{...(Neglecting transistor base current)}$$

$$V_E = V_B - V_{BE} = -10 - 0.7$$

$$= -10.7 \text{ V}$$

$$I_E = \frac{V_E - (-V_{EE})}{R_3} = \frac{-10.7 + 20}{2.2 \times 10^3}$$

$$= 4.22 \text{ mA}$$

$$I = I_E = \mathbf{4.22\ mA}$$

2.11 CURRENT MIRROR CIRCUIT

The circuit in which the output current is forced to equal the input current is called as **current mirror circuit.** In a current mirror circuit, the output current is the mirror image of input current. The basic block diagram is shown in the Fig. 2.32 (*a*) while the Fig. 2.32 (*b*) shows the circuit diagram.

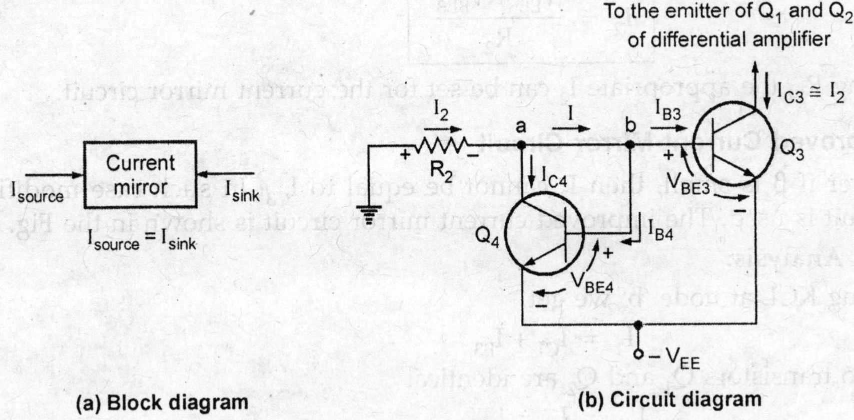

Fig. 2.32 Current mirror

2.11.1 Circuit Analysis

The circuit consists of two matched transistors Q_3 and Q_4. Their base-emitter voltage and base currents are same.

$$\therefore \qquad V_{BES} = V_{BE4} \text{ and } I_{B3} = I_{B4}$$

Similarly their collector currents are also same.

$$\therefore \qquad \boxed{I_{C3} = I_{C4}}$$

Applying KCL at node a,

$$I_2 = I_{C4} + I \qquad \text{...(1)}$$

Now

Applying KCL at node b,

$$I = I_{B3} + I_{B4} = 2I_{B4} = 2I_{B3} \quad ...(2)$$

$$\therefore \quad I_2 = I_{C4} + 2I_{B4} = I_{C3} = 2I_{B3} \quad ...(3)$$

Now $$I_{B3} = \frac{I_{C3}}{\beta}$$

$$\therefore \quad \boxed{I_2 = I_{C3} + I_{C3} + \left(\frac{2}{\beta}\right)} \quad ...(4)$$

Generally β is very large and hence $\left(\frac{2}{\beta}\right)$ is negligibly small.

$$\therefore \quad I_2 \cong I_{C3} \quad ...(5)$$

Thus the collector current of Q_3 is nearly equal to the current I_2. Hence once current mirror circuit is set for current I_2, it provides constant current bias to the differential amplifier. Thus I_2 can be obtained by writing KVL for the base-emitter loop of transistor Q_3.

$$-I_2R_2 - V_{BE3} + V_{EE} = 0$$

$$\therefore \quad \boxed{I_2 = \frac{V_{EE} - V_{BE3}}{R_2}} \quad ...(6)$$

Selecting R_2, the appropriate I_2 can be set for the current mirror circuit.

2.11.2 Improved Current Mirror Circuit

However if β is small, then I_2 cannot be equal to I_{C3}. In such case modified current mirror circuit is used. The improved current mirror circuit is shown in the Fig. 2.33.

Circuit Analysis:

Applying KCL at node 'b' we get,

$$I_1 = I_{C1} + I_{B3} \quad ...(7)$$

The two transistors Q_1 and Q_2 are identical

$$I_{B1} = I_{B2} = I_B \quad ...(8)$$

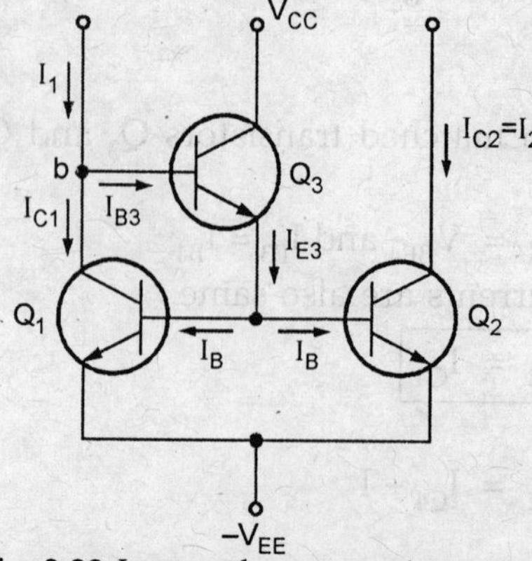

Fig. 2.33 Improved current mirror circuit

Hence the emitter current I_{E3} of transistor Q_3 gets divided equally.

$$\therefore \quad I_{E3} = 2I_B \quad ...(9)$$

Now $\quad I_{E3} = (1+\beta)\, I_{B3}$

$$\therefore \quad 2I_B = (1+\beta)\, I_{B3} \quad ...(10)$$

Now $\quad I_{C1} = I_{E1} = \beta\, I_B \quad ...(11)$

and $\quad I_{C2} = I_{E2} = \beta\, I_B \quad ...(12)$

Substituting (11) and (10) in (7), we get,

$$I_1 = \beta\, I_B + \frac{2I_B}{(1+\beta)}$$

$$\boxed{I_1 = I_B\left[\frac{\beta(1+\beta)+2}{(1+\beta)}\right]} \quad ...(13)$$

Substituting in (12) we get,

$$I_{C2} = \beta\left[\frac{I_1(1+\beta)}{\beta(1+\beta)+2}\right] \quad ...(14)$$

Thus
$$\boxed{I_{C2} = I_{C1} = I_2 = \frac{I_1\,\beta(1+\beta)}{\beta(1+\beta)+2}} \quad ...(15)$$

The ratio $\frac{I_2}{I_1}$ is much less dependent upon β for this modified circuit.

2.11.3 Wilson Current Source Circuit

Another widely used current source using current mirror technique is **Wilson current source.** High output resistance is the feature of Wilson current source. The Wilson current source circuit is shown in the Fig 2.34.

The high output resistance is achieved due to the negative feedback through Q_3. It also provides base currents cancellation making I_2 nearly equal to I_{ref}.

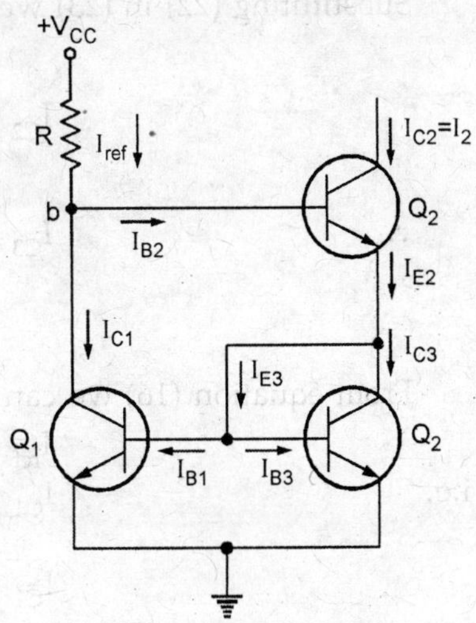

Fig. 2.34 Wilson current source

Circuit Analysis

Applying KCL at node b,

$$I_{ref} = I_{C1} + I_{B2} \quad ...(16)$$

and $\quad I_{E2} = I_{C3} + I_{B1} + I_{B3} \quad ...(17)$

But as all the transistors Q_1, Q_2 and Q_3 are identical hence their base-emitter voltages and base currents are also equal.

$$\therefore \quad I_{B1} = I_{B2} = I_{B3} = I_B$$

$\therefore \quad I_{E2} = I_{C3} + 2I_B$ ...(18)

Also $\quad I_{E2} = I_{C2} + I_{B2}$

i.e. $\quad I_{E2} = I_{C2} + I_B$ ...(19)

Now $\quad I_{C2} = I_{E2} - I_B$

$$= I_{C3} + 2I_B - I_B$$

$$= I_{C3} + I_B \qquad ...(20)$$

But $\quad I_{C3} = I_{C1}$ as transistors are identical.

$\therefore \quad I_{C2} = I_{C1} + I_B$ ...(21)

Comparing equations (21) and (16) we can say that,

$$I_2 = I_{C2} = I_{ref}$$

Now $\quad I_{E2} = I_{C3} + 2\,I_B$

But $\quad I_B = \dfrac{I_{C3}}{\beta}$

$$\therefore \quad I_{E2} = I_{C3} + \frac{2\,I_{C3}}{\beta}$$

$$\therefore \quad I_{E2} = I_{C3}\left[1 + \frac{2}{\beta}\right] \qquad ...(22)$$

Now $\quad I_{C2} = I_{E2}\left[\dfrac{\beta}{1+\beta}\right]$ from (19) ...(23)

Substituting (22) in (23) we get,

$$I_{C2} = I_{C3}\left[1 + \frac{2}{\beta}\right]\left[\frac{\beta}{1+\beta}\right] \qquad ...(24)$$

$$\therefore \quad I_{C3} = I_{C2}\,\frac{1}{\left[1 + \frac{2}{\beta}\right]\left[\frac{\beta}{1+\beta}\right]} \qquad ...(25)$$

From equation (16) we can write,

$$I_{ref} = I_{C1} + I_{B2}$$

i.e. $\quad I_{C1} = I_{ref} - I_B$

$$I_{C1} = I_{ref} - \frac{I_{C2}}{\beta} \qquad ...(26)$$

But as two transistors are identical,

$$I_{C3} = I_{C1}$$

So from the equations (25) and (26)

$$I_{C2}\frac{1}{\left[1+\frac{2}{\beta}\right]\left[\frac{\beta}{1+\beta}\right]} = I_{ref} - \frac{I_{C2}}{\beta}$$

$$\therefore \quad I_{C2}\left[\frac{(1+\beta)\beta}{\beta(2+\beta)}+\frac{1}{\beta}\right] = I_{ref}$$

$$\therefore \quad I_{C2}\left[\frac{\beta^2+\beta+2+\beta}{\beta(2+\beta)}\right] = I_{ref}$$

$$\therefore \quad I_{C2}\left[\frac{\beta^2+2\beta+2}{\beta(2+\beta)}\right] = I_{ref}$$

$$\therefore \quad I_{C2} = I_{ref}\left[\frac{\beta(2+\beta)}{\beta^2+2\beta+2}\right]$$

$$\therefore \quad I_{C2} = I_{ref}\left[\frac{\beta^2+2\beta+2-2}{\beta^2+2\beta+2}\right]$$

$$\therefore \quad \boxed{I_{C2} = I_{ref}\left[1-\frac{2}{\beta^2+2\beta+2}\right]} \quad ...(27)$$

The equation shows that the output current ($I_{C2} = I_2$) and the reference current I_{ref} differ by only a factor which is of the order of $2/\beta^2$. Thus low sensitivity of the circuit towards the transistor base currents is verified.

2.11.4 Widlar Current Source Circuit

In operational amplifier low input current is required. Hence input stage is biased at very low current, typically at a collector current of the order of 5 μA. Currents of such low magnitude can be obtained with a modified circuit called **Widlar current source.** The circuit is shown in the Fig. 2.35.

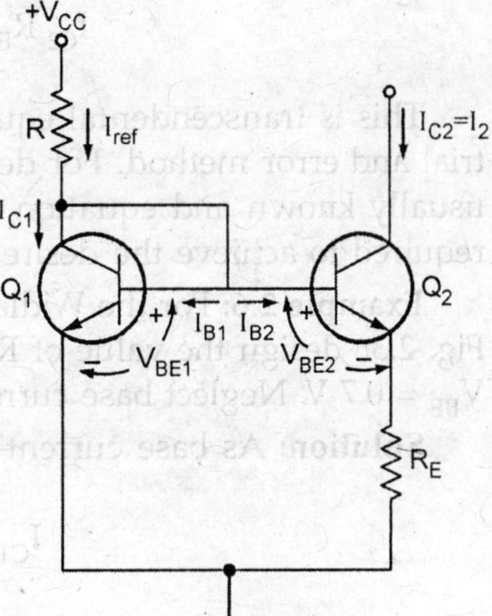

Fig. 2.35 Widlar current source

Circuit Analysis

The two transistors are identical to each other, but due to the emitter resistance R_E, V_{BE1} and V_{BE2} are different. In fact $V_{BE2} < V_{BE1}$ and hence $I_{C2} < I_{C1}$. Due to asymmetric nature of the base emitter loop the circuits is called **'lens'** rather than a **'mirror'**.

Applying KVL to the base emitter loop,

$$V_{BE1} = V_{BE2} + (I_{B2} + I_{C2})\,R_E \qquad ...(28)$$

i.e.

$$V_{BE1} - V_{BE2} = (I_{B2} + I_{C2})\,R_E \qquad ...(29)$$

For a transistor, we can write,

$$I_{C1} = I_S\, e^{V_{BE1}/V_T}$$

And

$$I_{C2} = I_S\, e^{V_{BE2}/V_T}$$

Where I_S is the reverse saturation current, same for the two transitors and V_T is voltage equivalent of temperature. These equations are called **Ebers-Moll** equations for the transistor.

$$\therefore \quad \frac{I_{C1}}{I_{C2}} = e^{(V_{BE1} - V_{BE2})/V_T} \qquad ...(30)$$

Taking natural logarithm of both sides we get,

$$\ln\left(\frac{I_{C1}}{I_{C2}}\right) = \frac{V_{BE1} - V_{BE2}}{V_T}$$

Hence we get,

$$V_{BE1} - V_{BE2} = V_T \ln\left(\frac{I_{C1}}{I_{C2}}\right) \qquad ...(31)$$

Equating equations (29) and (31),

$$(I_{B2} + I_{C2})\,R_E = V_T \ln\left(\frac{I_{C1}}{I_{C2}}\right) \qquad ...(32)$$

Neglecting I_{B2} for large value of β,

$$I_{C2}\,R_E = V_T \ln\left(\frac{I_{C1}}{I_{C2}}\right) \qquad ...(33)$$

This is transcendental equation and must be solved by trial and error method. For design purposes, I_{C1} and I_{C2} are usually known and equation (33) provides the value of R_E required to achieve the desired value of I_{C2}.

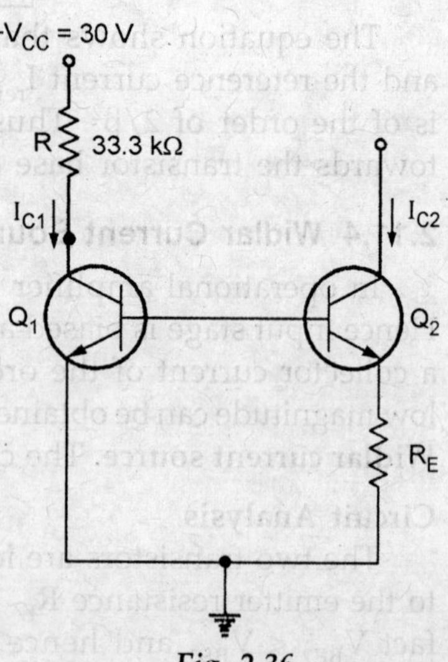

Fig. 2.36

Example 2.6: For the Widlar current source shown in the Fig. 2.36 design the value of R_E to get I_{C2} as 20 μA. Assume V_{BE} = 0.7 V. Neglect base current.

Solution: As base current is neglected

$$I_{C1} = \frac{V_{CC} - V_{BE}}{R}$$

$$= \frac{30 - 0.7}{33.3 \times 10^3} = 0.8798 \text{ mA}$$

Now $\qquad I_{C2} = 20\ \mu A$

$$\therefore \qquad I_{C2} R_E = V_T \ln\left(\frac{I_{C1}}{I_{C2}}\right)$$

Assuming $V_T = 26$ mV at room temperature of 27 °C

$$\therefore \qquad 20 \times 10^{-6} R_E = 20 \times 10^{-3} \ln\left(\frac{0.8798 \times 10^{-3}}{20 \times 10^{-6}}\right)$$

$$\therefore \qquad R_E = \mathbf{4.92\ k\Omega}$$

Example 2.7: In the Widlar current source, reference current is 1 mA and R_E is selected as 5 kΩ.

Determine approximate value of I_{C2} neglecting the base currents.

Solution: As base currents are to be neglected,

$$I_{ref} = I_{C1} = 1\text{ mA}$$

$$R_E = 5\text{ k}\Omega$$

Assuming $\qquad V_T = 26$ m V at 300 °K,

$$I_{C2} R_E = V_T \ln\left(\frac{I_{C1}}{I_{C2}}\right)$$

$$26 \times 10^{-3} \ln\left(\frac{1 \times 10^{-3}}{I_{C2}}\right) - I_{C2} \times 5 \times 10^3 = 0$$

This is to be solved by trial and error method.

Let $\qquad I_{C2} = 12\ \mu A$, L.H.S = 0.05499

At $\qquad I_{C2} = 22\ \mu A$, L.H.S = – 0.01076

So actual value is less than but closer to 22 μA.

At $\qquad I_{C2} = 20\ \mu A$, L.H.S = 0.0017

So L.H.S. is almost zero, hence the value of I_{C2} is 20 μA.

2.11.5 Current Mirror with Magnification

Sometimes it is necessary to have a circuit in which ratio of biasing currents in two transistors is required to be fixed. Such a circuit is called as **current mirror circuit with magnification.**

The circuit is improved version of Widlar circuit. Such a circuit is shown in the Fig. 2.37.

In this circuit, emitter resistances R_{E1} and R_{E2} are added to the transistors Q_1 and Q_2.

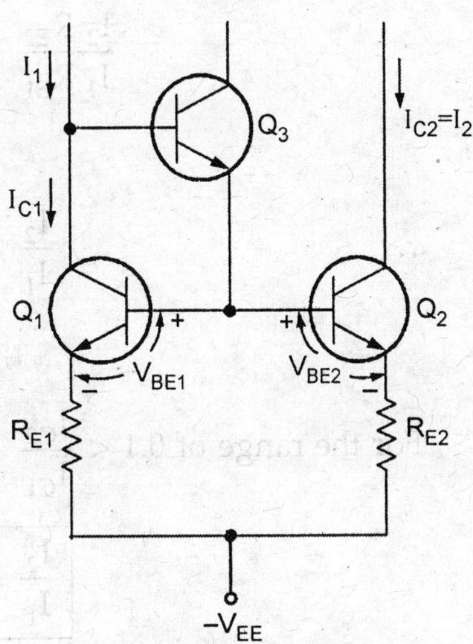

Fig. 2.37 *Current mirror with magnification*

Circuit Analysis

Neglecting the base currents, similar to the equation (30) we can write,

$$\frac{I_{C1}}{I_{C2}} = e^{(V_{BE1} - V_{BE2})/\eta^{V_T}}$$

Assuming $\eta = 1$ and rearranging the terms we can write,

$$V_{BE1} - V_{BE2} = V_T \ln\left(\frac{I_{C1}}{I_{C2}}\right) \quad ...(34)$$

Applying KVL to the loop containing R_{E1} and R_{E2},

$$-V_{BE2} - I_2R_{E2} + I_1R_{E1} + V_{BE1} = 0$$

$$\therefore \quad V_{BE2} - V_{BE1} = I_1 R_{E1} - I_2 R_{E2} \quad ...(35)$$

Equating the two equations (34) and (35)

$$I_1 R_{E1} - I_2 R_{E2} = -\left[V_T \ln\left(\frac{I_{C1}}{I_{C2}}\right)\right]$$

$$\therefore \quad I_1 R_{E1} - I_2 R_{E2} = V_T \ln\left(\frac{I_{C1}}{I_{C2}}\right)$$

$$\therefore \quad 1 - \frac{I_2 R_{E2}}{I_1 R_{E1}} = \frac{V_T \ln\left(\frac{I_{C2}}{I_{C1}}\right)}{I_1 R_{E1}}$$

$$\therefore \quad \frac{I_2 R_{E2}}{I_1 R_{E1}} = 1 - \frac{V_T \ln\left(\frac{I_{C2}}{I_{C1}}\right)}{I_1 R_{E1}}$$

$$\therefore \quad \frac{I_2}{I_1} = \frac{R_{E1}}{R_{E2}}\left[1 - \frac{V_T \ln\left(\frac{I_{C2}}{I_{C1}}\right)}{I_1 R_{E1}}\right] \quad ...(36)$$

For the range of $0.1 < \frac{I_{C2}}{I_{C1}} < 10$, $\ln\left(\frac{I_{C2}}{I_{C1}}\right)$ can be neglected and hence we get

$$\boxed{\frac{I_2}{I_1} = \frac{R_{E1}}{R_{E2}}} \quad ...(37)$$

Hence selecting proper values of R_{E1} and R_{E2}, large ratio of I_2 and I_1 can be obtained by this modified circuit.

2.11.6 Multiple Current Source

Multiple current source circuit can be obtained from basic current mirror circuit. This circuit can be used to supply more than one load. The circuit is shown in the Fig. 2.38.

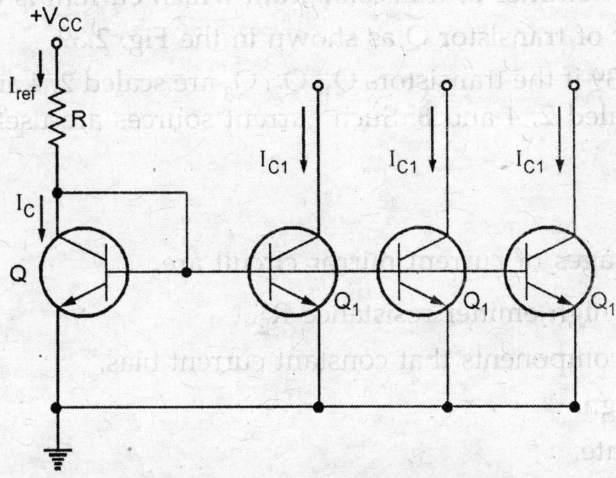

Fig. 2.38 Multiple current source

All the transistors are identical in characteristics. Hence all have same base currents. If n is number of transistors used in multiple current source then

$$I_{ref} = I_B + I_C + n\, I_B \quad ...(38)$$

Where I_B = Base current of Q

I_C = Collector current of Q

$$\therefore \quad I_{ref} = (n+1)\, I_B + \beta\, I_B$$

$$\therefore \quad I_{ref} = (n+1+\beta)\, I_B \quad ...(39)$$

Now $I_C = \beta\, I_B$

$$\therefore \quad \boxed{I_C = I_{ref}\left(\frac{\beta}{n+1+\beta}\right)} \quad ...(40)$$

For large value of β we get

$$\boxed{I_C = I_{ref} = \frac{V_{CC} - V_{BE}}{R}} \quad ...(41)$$

In practice, the transistors like Q_1, Q_2, Q_3 are fabricated as a single multiple-collector merged device. Such a circuit is shown in the Fig. 2.39 where Q_M represents the transistors Q_1, Q_2 and Q_3.

In such multiple current source more general relationship involves the area. The current in each transistor is proportional to the area of the emitter. So

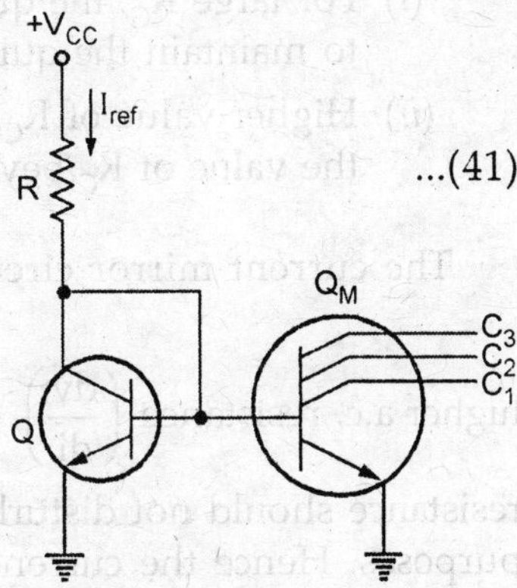

Fig. 2.39 *Multiple collector merged multiple current source*

$$I_C = I_{ref} = \frac{A_Q}{A}$$

Where A_Q = Area of emitter of transistor from which current is to be calculated.

A = Area of emitter of transistor Q as shown in the Fig. 2.39

Hence in the Fig. 2.39 if the transistors Q_1, Q_2, Q_3 are scaled 2, 4 and 8, then the currents I_{C1}, I_{C2}, I_{C3}, are also scaled 2, 4 and 8. Such current sources are useful in digital to analog converters.

2.11.7 Advantages

The various advantages of current mirror circuit are,

1. Provides very high emitter resistance R_E.
2. Requires less components that constant current bias.
3. Simple to design.
4. Easy to fabricate.
5. With properly matched transistors, collector current thermal stability is achieved.

Thus constant current bias can be easily replaced by current mirror circuit to improve CMRR. Due to its advantages, **current mirror circuit is most commonly used in the integrated circuit op-amps.**

2.12 USE OF AN ACTIVE LOAD TO IMPROVE CMRR

To Improve CMRR, it is necessary to Increase A_d.

While, $$A_d = \frac{h_{fe} R_C}{R_S + h_{ie}}$$

Thus to increase A_d, R_C must be high as possible.

But there are limitations to select maximum value of R_C such as:

(*i*) For large R_C, the quiescent drop is more hence higher biasing voltage is necessary to maintain the quiescent collector current.

(*ii*) Higher value of R_C, requires a large chip area. Hence it is not possible to increase the value of R_C beyond a particular limit.

The current mirror circuit discussed earlier has very low d.c. resistance $\left(\frac{dV}{dI}\right)$ and higher a.c. resistance $\left(\frac{dv}{di}\right)$. The requirement to increase the gain is same that the collector resistance should not disturb d.c. conditions while it must provide large resistance for a.c. purposes. Hence the current mirror circuit can be used as a collector load instead of R_C. Such a load is called an **active load.** The quiescent voltage across the current mirror is the fraction of the supply voltage. This eliminates the need of high biasing supply voltage. It basically acts as a current source and provides large a.c. resistance.

The differential amplifier using a current mirror as an active load is shown in the Fig. 2.40.

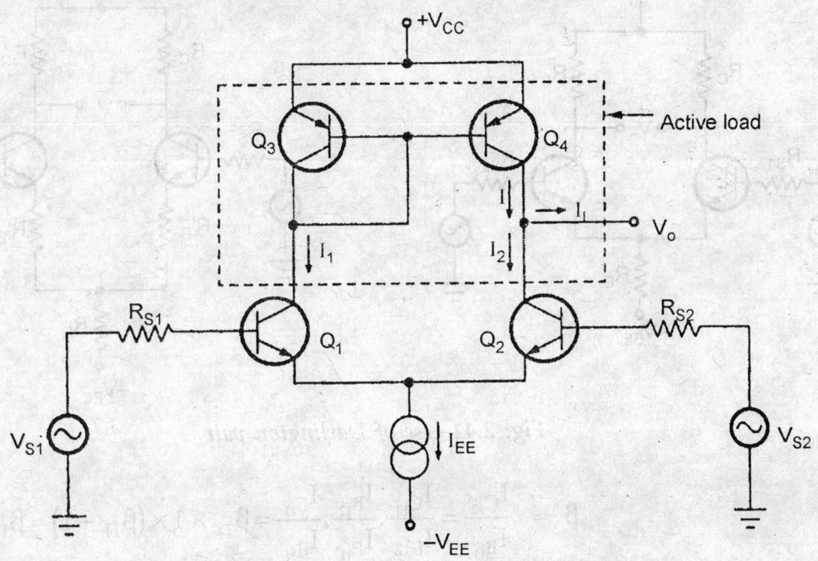

Fig. 2.40 *Differential amplifier with an active load*

Under the d.c. conditions, $V_{S1} = V_{S2} = 0$. As Q_1 and Q_2 are matched transistors hence $I_1 = I_2 = I_{EE}/2$, where base currents of Q_1 and Q_2 are neglected. The transistors Q_3 and Q_4 form a current repeater hence $I = I_1 = I_2$.

The load current I_L entering the next stage is,

$$I_L = I - I_2 = 0 \quad \text{...(1)}$$

But when V_{S1} increases over V_{S2}, the current I_1 increases whereas I_2 decreases as $I_1 + I_2 = I_{EE}$ constant. Also the current I always remains equal to I_1 due to the current mirror action.

Key Point: *Thus the active load provides very high a.c. resistance and hence high differential mode voltage gain. Thus as A_d becomes high, CMRR gets improved.*

2.13 METHODS OF REALIZING HIGH INPUT RESISTANCE

The various methods of realizing the high input resistance for the differential amplifier circuit are,

1. Use of Darlington pair.
2. Use of FET
3. Use of swamping resistors.

2.13.1 Use of Darlington Pair

In this method, instead as using transistor Q_1 and Q_2 in the differential amplifier stage, Darlington pairs are used. This is shown in the Fig. 2.41.

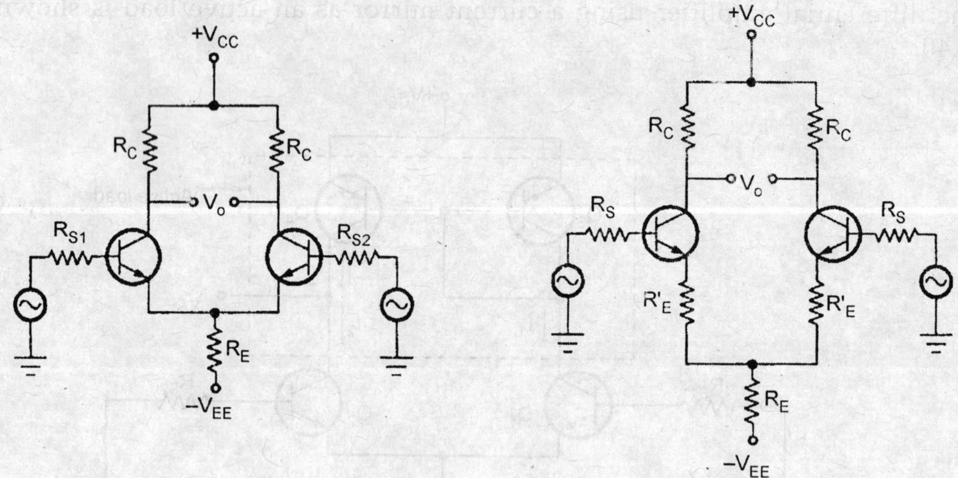

Fig. 2.41 Use of Darlington pair

Now
$$\beta \approx \frac{I_{C_{12}}}{I_{B_{11}}} = \frac{I_{C_{12}}}{I_{B12}} \cdot \frac{I_{B_{12}}}{I_{E_{11}}} \cdot \frac{I_{E_{11}}}{I_{B_{11}}} = \beta_{12} \times 1 \times (\beta_{11}+1) \approx \beta_{11}\ B_{12}$$

Thus the most important advantage of this method is extraordinary high current gain possible due to Darlington pair. Thus if the current gain of individual transistors is 100 then the overall current gain is of the order of 10,000.

The disadvantage of this method is higher offset voltage V_{ios}. Due to cascaded stages, the offset voltage of about twice the ordinary two transistor differential amplifier is possible.

2.13.2 Use of FET

Another method is to use FET differential amplifier with all the remaining stages using BJTs. Due to FET, high input resistance of the order of 10^9 –10^{12} Ω is possible. Another advantage of this method is very low input bias current of the order of 10^{-9} to 10^{-12} A is possible. The disadvantages of using FET differential pair are low voltage gain and higher offset voltage compared to BJT differential pair.

2.13.3 Use of Swamping Resistors

To achieve high input resistance, external resistances R'_E are connected in series with each emitter of each transistor of a differential amplifier. Such resistances are called **swamping resistors.**

The differential amplifier using swamping resistors is shown in the Fig. 2.42

The various effects of swamping resistances on the performance of the differential amplifier are,

1. Increase in the input resistance.
2. Minimization of changes in the transistor parameters due to the factors like temperature.

3. Reduction in the differential gain.
4. Increase in the linearity range of the differential amplifier.

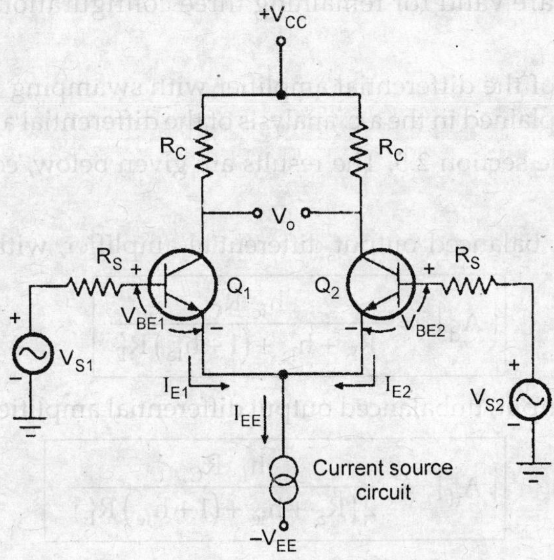

Fig. 2.42 Differential amplifier with swamping resistors

D.C. Analysis

Due to R_E', there are few changes in the d.c. analysis of differential amplifier. The emitter current in each transistor can be determined by applying KVL to base-emitter loop of Q_1. For the d.c. analysis $V_{S1} = V_{S2} = 0$ V. Assuming $R_{S1} = R_{S2} = R_S$, we can write,

$$-I_BR_S - V_{BE} - I_E R_E' - 2\,I_ER_E + V_{EE} = 0 \quad \text{...(1)}$$

The values of I_B, I_C and I_E are same for Q_1 and Q_2 as they are matched.

Now $$I_B = \frac{I_E}{h_{fe}}$$

$$\therefore \quad -\frac{I_E R_S}{h_{fe}} - V_{BE} - I_ER_E' - 2\,I_ER_E + V_{EE} = 0$$

$$\therefore \quad I_E = \frac{V_{EE} - V_{BE}}{2\,R_E + R_E' + \dfrac{R_S}{h_{fe}}} \quad \text{...(2)}$$

As $\frac{R_S}{h_{fe}}$ is very small, as h_{fe} is large, the equation can be modified as,

$$I_E = \frac{V_{EE} - V_{BE}}{2\,R_E + R_E'} \quad \text{...(3)}$$

While the equation for the collector-emitter loop remains same as before which is

$$V_{CE} = V_{CC} + V_{BE} - I_C R_C$$

The expressions are valid for remaining three configurations of differential amplifier.

A.C. Analysis

The a.c. analysis of the differential amplifier with swamping resistors can be carried out similar to the steps explained in the a.c. analysis of the differential amplifier without swamping resistors, earlier in the section 2.5. The results are given below, considering the effect of R'_E in the analysis:

For a dual input, balanced output differential amplifier, with swamping resistors,

$$|A_d| = \frac{h_{fe} R_C}{R_S + h_{ie} + (1 + h_{fe}) R'_E} \quad ...(4)$$

While for a dual input, unbalanced output differential amplifier, with swamping resistors.

$$|A_d| = \frac{h_{fe} R_C}{2[R_S + h_{ie} + (1 + h_{fe}) R'_E]} \quad ...(5)$$

The result for common mode gain for both the configurations is

$$|A_d| = \frac{h_{fe} R_C}{R_S + h_{ie} + (1 + h_{fe})(R'_E + 2R_E)} \quad ...(6)$$

While the differential input impedance gets modified as

$$R_{in} = 2[R_S + h_{ie} + (1 + h_{fe}) R'_E] \quad ...(7)$$

And the output impedance which remains same as before is,

$$R_o = R_C \quad ...(8)$$

Example 2.8: The specifications for the dual input balanced output differential amplifier are given below:

$R_C = 4.7\ k\Omega$, $R_E = 4.7\ k\Omega$, $R_S = 70\ \Omega$,

$h_{ie} = 1.4.7\ k\Omega$, $V_{CC} = 12\ V$, $V_{EE} = -12\ V$

$h_{fe} = 75$, $V_{BE} = 0.7\ V$

(*i*) Determine operating point values, differential voltage gain, common mode gain, input resistance, output resistance.

(*ii*) If the swamping resistances each of 100 Ω is inserted in the circuit, calculate the above parameters, considering the effect of swamping resistance.

Solution: The circuit without swamping resistance is shown in the Fig. 2.43 (*a*).

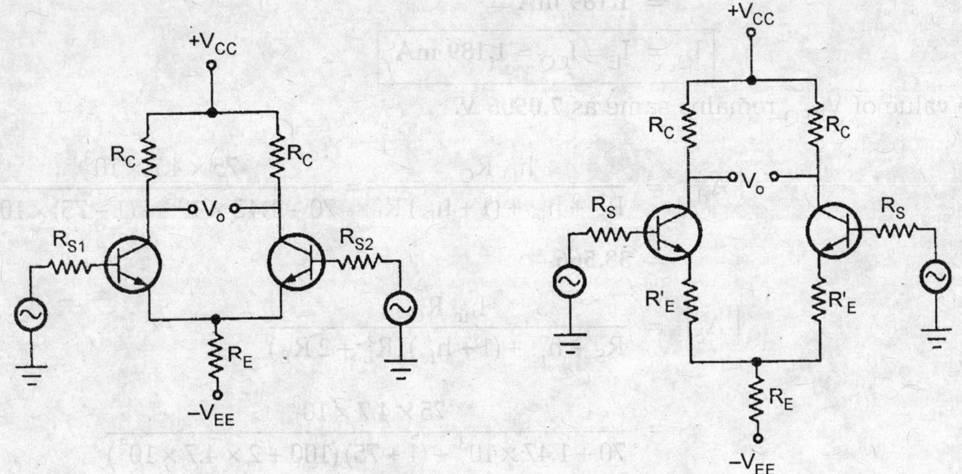

Fig. 2.43

(i) Without swamping resistances

$$I_E = \frac{V_{EE} - V_{BE}}{2\,R_E + \dfrac{R_S}{h_{fe}}} = \frac{12 - 0.7}{2 \times 4.7 \times 10^3 + \dfrac{70}{75}} = 1.202\text{ mA}$$

$$I_C = I_E = I_{CQ}$$

$$V_{CE} = V_{CC} + V_{BE} - I_C R_C = 12 + 0.7 - 1.202 \times 10^{-3} \times 4.7 \times 10^3$$

$$= \mathbf{7.0506\ V}$$

$$|A_d| = \frac{h_{fe}\,R_C}{R_S + h_{ie}} = \frac{75 \times 4.7 \times 10^3}{70 + 1.47 \times 10^3} = \mathbf{228.89}$$

$$|A_c| = \frac{h_{fe}\,R_C}{R_S + h_{ie} + 2R_E\,(1 + h_{fe})}$$

$$= \frac{75 \times 4.7 \times 10^3}{70 + 1.47 \times 10^3 + 2 \times 4.7 \times 10^3\,(1 + 75)} = \mathbf{0.4923}$$

$$R_i = 2\,(R_S + h_{ie}) = 2\,(70 + 1.47 \times 10^3)$$

$$= 3.08\text{ k}\Omega$$

$$\boxed{R_o = R_C = 4.7\text{ k}\Omega}$$

(ii) Now consider the effect of swamping resistance R'_E = 100 W.

The circuit with swamping resistance is shown in the Fig. 2.43 (*b*).

$$I_E = \frac{V_{EE} - V_{BE}}{2R_E + R'_E + \dfrac{R_S}{h_{fe}}} = \frac{12 - 0.7}{2 \times 4.7 \times 10^3 + 100 + \dfrac{70}{75}}$$

$$= \mathbf{1.189\ mA}$$

$$\boxed{I_C = I_E = I_{CQ} = \mathbf{1.189\ mA}}$$

The value of V_{CEQ} remains same as **7.0506 V.**

$$|A_d| = \frac{h_{fe}\, R_C}{R_S + h_{ie} + (1+h_{fe})\, R'_E} = \frac{75 \times 4.7 \times 10^3}{70 + 1.47 \times 10^3 + (1+75) \times 100}$$

$$= \mathbf{38.566}$$

$$|A_c| = \frac{h_{fe}\, R_C}{R_S + h_{ie} + (1+h_{fe})(R'_E + 2\,R_E)}$$

$$= \frac{75 \times 4.7 \times 10^3}{70 + 1.47 \times 10^3 + (1+75)(100 + 2 \times 4.7 \times 10^3)}$$

$$= \mathbf{0.4871}$$

$$R_i = 2\,[R_S + h_{ie} + (1 + h_{fe})\, R'_E\,]$$

$$= 2[70 + 1.47 \times 10^3 + (1 + 75) \times 100]$$

$$= \mathbf{18.28\ k\Omega}$$

$$\boxed{R_o = R_C = 4.7\ k\Omega}$$

Note: It can be observed that the differential gain reduces considerably, the common mode gain almost remains same and the input resistance increases substantially, due to the addition of swamping resistances in the circuit.

2.14 TRANSFER CHARACTERISTICS OF DIFFERENTIAL AMPLIFIER

The transfer characteristics of the differential amplifier is the graph of differential input V_d against the currents I_{C1} and I_{C2}. Such a transfer characteristics is very useful in studying the large signal behaviour of the differential amplifier. This is also useful in identifying the range of the differential input voltage over which the circuit operation is in linear region. The basic dual input, balanced output differential amplifier circuit is shown in the Fig. 2.44.

In practice the resistance R_E is replaced by a constant current source to improve the performance of the differential amplifier. Hence in the Fig. 2.44 the current source circuit is shown instead of the resistance R_E.

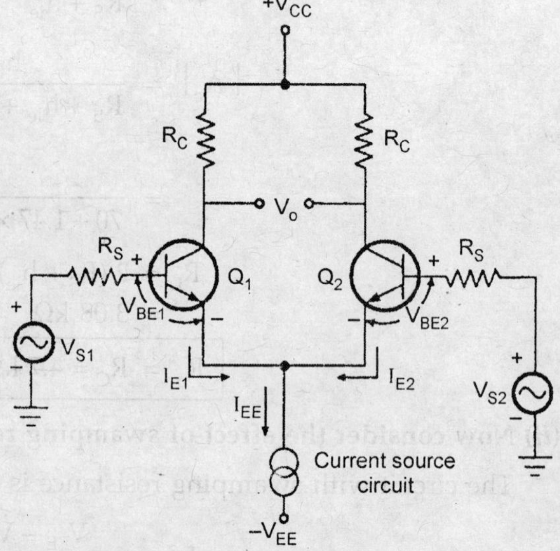

Fig. 2.44 Differential amplifier circuit

To obtain the transfer characteristics, the following assumptions are made:

(i) The current source circuit used with current I_{EE} has infinite output resistance.
(ii) The source resistances R_S in the base of transistors Q_1 and Q_2 are neglected.
(iii) The output resistance of each transistor is infinite.

The assumptions are valid for low frequency, large signals.

For a transistor, we can write the equation for its collector current as

$$I_C = I_S e^{V_{BE}/V_T} \quad ...(1)$$

where I_S = reverse saturation current

V_{BE} = base-emitter voltage

V_T = voltage equivalent of temperature

Thus for two transistors we can write the equations for their collector currents as,

$$I_{C1} = I_S e^{V_{BE1}/V_T} \quad ...(2)$$

$$I_{C2} = I_S e^{V_{BE2}/V_T} \quad ...(3)$$

where $I_{S1} = I_{S2} = I_S$ as transistors are matched.

This equation is called as **Ebers-Moll equation** for the transistor.

$$\therefore \quad \ln \frac{I_{C1}}{I_S} = \frac{V_{BE1}}{V_T}$$

$$\therefore \quad V_{BE1} = V_T \ln\left(\frac{I_{C1}}{I_S}\right) \quad ...(4)$$

and

$$V_{BE2} = V_T \ln\left(\frac{I_{C2}}{I_S}\right) \quad ...(5)$$

Now consider the loop including two inputs and two base emitter junctions, neglecting R_S, as shown in the Fig. 2.45.

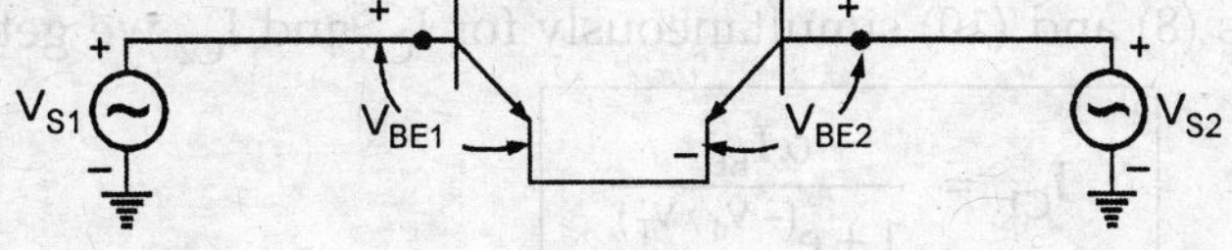

Fig. 2.45

Applying KVL to the loop shown,

$$V_{S1} - V_{BE1} + V_{SE2} - V_{S2} = 0 \quad ...(6)$$

Substituting V_{BE1} and V_{SE2} from equations (4) and (5),

$$V_{S1} - V_T \ln\left(\frac{I_{C1}}{I_S}\right) + V_T \ln\left(\frac{I_{C2}}{I_S}\right) - V_{S2} = 0$$

$$\therefore \quad V_T\left[\ln\left(\frac{I_{C2}}{I_S}\right)-\ln\left(\frac{I_{C1}}{I_S}\right)\right] = V_{S2}-V_{S1}$$

$$\therefore \quad V_T\left[\ln\left(\frac{I_{C1}}{I_S}\right)-\ln\left(\frac{I_{C2}}{I_S}\right)\right] = V_{S1}-V_{S2}$$

$$\therefore \quad V_T\ln\left[\frac{I_{C1}}{I_S}\Big/\frac{I_{C2}}{I_S}\right] = V_{S1}-V_{S2}$$

$$\therefore \quad V_T\ln\left(\frac{I_{C1}}{I_{C2}}\right) = V_{S1}-V_{S2}$$

$$\therefore \quad \ln\left(\frac{I_{C1}}{I_{C2}}\right) = \frac{V_{S1}-V_{S2}}{V_T}$$

$$\therefore \quad \frac{I_{C1}}{I_{C2}} = \frac{V_{S1}-V_{S2}}{V_T} \quad \text{...(7)}$$

Now

$$V_{S1}-V_{S2} = V_d$$

$$= \text{Differential input}$$

$$\therefore \quad \frac{I_{C1}}{I_{C2}} = e^{V_d/V_T} \quad \text{...(8)}$$

Now current through current source I_{EE} is the addition of the two emitter currents I_{E1} and I_{E2}.

$$\therefore \quad I_{EE} = I_{E1}+I_{E2} \quad \text{...(9)}$$

but

$$I_E = \alpha\, I_C$$

$$\therefore \quad I_{EE} = \frac{1}{\alpha}[I_{C1}+I_{C2}] \quad \text{...(10)}$$

Solving equations (8) and (10) simultaneously for I_{C1} and I_{C2} we get,

$$I_{C1} = \frac{\alpha I_{EE}}{1+e^{(-V_d/V_T)}} \quad \text{...(11)}$$

$$I_{C2} = \frac{\alpha I_{EE}}{1+e^{(V_d/V_T)}} \quad \text{...(12)}$$

From these two equations, the transfer characteristics can be obtained.

Let αI_{EE} be constants K, so for various values of V_d we can obtain I_{C1} and I_{C2} as in Table 2.3.

Table 2.3

V_d	I_{C1}	I_{C2}
0	0.5	0.5 K
V_T	0.73	0.27 K
$2V_T$	0.88	0.12 K
$-V_T$	0.27	0.73 K
$-2V_T$	0.12	0.88 K

The corresponding transfer characteristics are shown in the Fig. 2.46.

It can be seen from the characteristics that for large values of V_d, I_{C1} and I_{C2} become independent of V_d. Now V_T is 26 mV at room temperature so for the range of V_T from $-2V_T$ to $+2V_T$ i.e. ± 52 mV, the characteristics are linear in nature.

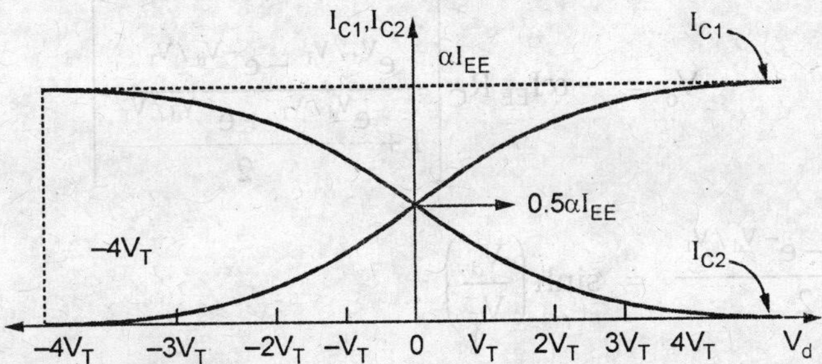

Fig. 2.46 Transfer characteristics of differential amplifier

***Key Point:** Thus for $-2V_T \leq V_d \leq 2V_T$ the circuit behaves linearly and all the quantities I_{C1}, I_{C2} and the output V_o all behave linearly.*

While for the value of $V_d > 4V_T$ i.e. approximately 100 mV, I_{C1} becomes almost equal to (αI_{EE}) while I_{C2} becomes zero. Due to this Q_1 acts as a closed switch and Q_2 acts as open switch. If $V_d < -4V_T$ then Q_2 acts as closed switch while Q_1 as an open switch.

***Key Point:** So the mode of operation for V_d between $\pm 2V_T$ is used for the differential amplifier which is used in analog IC chips.*

Now let us see the effect of change in V_d on the output voltage V_o. The output voltage is balanced output taken across the two collector points.

Let V_{o1} = collector voltage of Q_1

V_{o2} = collector voltage of Q_2

$$\therefore \quad V_o = V_{o1} - V_{o2} \quad ...(13)$$

Now
$$V_{o1} = V_{CC} - I_{C1} R_C \quad ...(14)$$

and
$$V_{o2} = V_{CC} - I_{C2} R_C \quad ...(15)$$

$$\therefore \quad V_o = V_{CC} - I_{C1} R_C - V_{CC} + I_{C2} R_C$$

$$\therefore \quad V_o = -R_C(I_{C1} - I_{C2}) \quad ...(16)$$

Substituting the values of I_{C1} and I_{C2} from the equations (11) and (12),

$$V_o = -R_C\left[\frac{\alpha I_{EE}}{1+e^{-V_d/V_T}} - \frac{\alpha I_{EE}}{1+e^{V_d/V_T}}\right]$$

$$= -\alpha I_{EE} R_C\left[\frac{1+e^{V_d/V_T} - 1 - e^{-V_d/V_T}}{\left(1+e^{-V_d/V_T}\right)\left(1+e^{V_d/V_T}\right)}\right]$$

$$= -\alpha I_{EE} R_C\left[\frac{e^{V_d/V_T} - e^{-V_d/V_T}}{2+e^{V_d/V_T}+e^{-V_d/V_T}}\right]$$

Dividing denominator and numerator by 2 we get,

$$V_o = -\alpha I_{EE} R_C\left[\frac{\dfrac{e^{V_d/V_T} - e^{-V_d/V_T}}{2}}{1+\dfrac{e^{V_d/V_T}+e^{-V_d/V_T}}{2}}\right] \qquad ...(17)$$

But $$\frac{e^{V_d/V_T} - e^{-V_d/V_T}}{2} = \sinh\left(\frac{V_d}{V_T}\right)$$

and $$\frac{e^{V_d/V_T} + e^{-V_d/V_T}}{2} = \cosh\left(\frac{V_d}{V_t}\right)$$

$$\therefore \quad V_o = -\alpha I_{EE}R_C\left[\frac{\sinh\left(\dfrac{V_d}{V_T}\right)}{1+\cosh\left(\dfrac{V_d}{V_T}\right)}\right]$$

$$= -\alpha I_{EE}R_C\left[\frac{2\sinh\left(\dfrac{V_d}{2V_T}\right)\cosh\left(\dfrac{V_d}{2V_T}\right)}{2\cosh^2\left(\dfrac{V_d}{2V_T}\right)}\right]$$

$$\therefore \quad \boxed{V_o = \alpha I_{EE}R_C \tanh\left[-\frac{V_d}{2V_T}\right]} \qquad ...(18)$$

The equation (18) gives the relation between differential input V_d and the output voltage V_o. The relation can be shown graphically as shown in the Fig. 2.47.

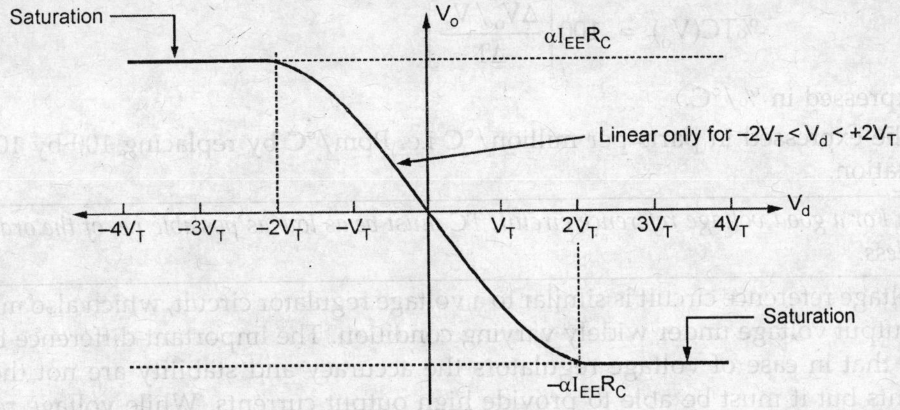

Fig. 2.47 Graph V_o vs V_d

It can be seen that when V_d is zero, the output voltage V_o is also zero. This is very important advantage of the differential amplifier. Due to this, it allows direct coupling of cascade stages without introducing d.c. offsets. The output shows two distinct levels, one positive and other negative, for a change in V_d of about 4 V_T.

***Key Point:** Thus the graph is linear between ± $2V_d$ and such mode of operation is utilized in digital circuits.*

2.15 VOLTAGE REFERENCES

A voltage reference circuit is basically used to provide a constant d.c. voltage which acts as a reference or standard for other circuits and is independent of changes in the parameters like temperature, input line voltage, load current etc.

The basic requirements of any voltage reference circuit are accuracy and stability, with temperature and time. Thus an important characteristics of any voltage reference circuit is its temperature coefficient.

***Key Point:** The temperature coefficient of the output voltage of a voltage reference circuit is the measure of the ability of the circuit to maintain the standard output voltage under varying temperature conditions.*

If V_o is the output voltage of a voltage reference then mathematically its temperature coefficient is defined as,

$$\therefore \quad \boxed{TC(V_o) = \frac{\Delta V_o}{\Delta T}}$$

It is expressed in mV/°C or μV/°C.

It is expressed in percentage form as

$$\boxed{\%TC(V_o) = 100\left[\frac{\Delta V_o / \Delta T}{V_o}\right]}$$

$$\therefore \qquad \%TC(V_o) = 100\left[\frac{\Delta V_o / V_o}{\Delta T}\right]$$

It is expressed in %/°C.

It can be expressed in parts per million/°C i.e. Ppm/°C by replacing 100 by 10° in the above equation.

***Key Point:** For a good voltage reference circuit, TC must be as low as possible i.e. of the order of few ppm/°C or less.*

The voltage reference circuit is similar to a voltage regulator circuit, which also maintains constant output voltage under widely varying condition. The important difference between the two is that in case of voltage regulators the accuracy and stability are not the prime requirements but it must be able to provide high output currents. While voltage reference circuits can provide very low output currents but accuracy and stability are the prime considerations.

2.15.1 Performance Parameters of Voltage Reference Circuits

The various performance parameters of voltage reference circuit, other than temperature coefficient are,

1. Line regulation
2. Load regulation
3. Long term stability
4. Ripple Rejection Ratio (RRR)

1. Line regulation: The effect of the changes in the input line voltage on the output voltage is specified by the parameter line regulation.

It is defined as,

$$\text{Line regulation} = \frac{\Delta V_o}{\Delta V_i}$$

where ΔV_i = Changes in the input line voltage

It is also called input regulation or supply regulation.

In the percentage form it is expressed as,

$$\%\ \text{Line regulation} = 100\left[\frac{\Delta V_o / V_o}{\Delta V_i}\right]$$

It is expressed in mV/V or %/V.

***Key Point:** Its value should be as low as possible and the d.c. output voltage must remain constant irrespective of any changes in the line voltage.*

---

2. Load regulation: Similar to the input voltage, load current I_L also can affect the output voltage. The ability of the circuit to maintain the output voltage under varying load current condition is denoted by the parameter load regulation. For an ideal circuit, output voltage is independent of the load current. It is related to the output impedance of the circuit.

It is defined as,

$$\text{Load regulation} = \frac{\Delta V_o}{\Delta I_L}$$

It is expressed in mV/mA or in mV/A, depending upon output current capability. In the percentage form, it is expressed as,

$$\%\ \text{Load regulation} = 100\left[\frac{\Delta V_o / V_o}{\Delta I_L}\right]$$

It is measured in %/mA or %/A .

***Key Point:** This value should be as low as possible and the output voltage should remain constant for the variation of the load from no load to full load condition.*

3. Long term stability: The circuit characteristics should not get changed with change in time. The ability of the circuit to maintain the output voltage constant with respect to the time is given by the parameter **long-term stability**.

It is usually given in ppm/1000 hours.

Thus if for a particular voltage reference the output voltage is 5 V and long term stability is given as 100 ppm/1000 hours, then it means within the period of 1000 hours i.e. about 42 days the output can change by [V_o × long-term stability]

i.e. $5 \times 100 \times 10^{-6} = 500\ \mu V$ i.e. 0.5 mV

***Key Point:** Ideally output voltage should not change with time.*

4. Ripple Rejection Ratio (RRR): It is denoted as RRR. The input to the voltage reference circuit is usually unregulated which is having some ripples. The output voltage gets affected by such ripples. It produces ripples at the output which is undesirable. The ability of circuit to reject input ripples and an indication of how much ripples are present at the output due to input, is by the factor ripple rejection ratio (RRR).

It is defined as,

$$\text{RRR} = 20 \log_{10}\left[\frac{V_{ri}}{V_{ro}}\right]$$

where V_{ri} = Input ripple magnitude

V_{ro} = Output ripple magnitude

The input and output ripples are expressed usually at 120 Hz. It is generally expressed in dB.

***Key Point:** High the value of RRR, better is the performance of the circuit.*

In practice many time zener diode reference circuits and temperature compensated avalanche diode circuits are used.

2.16 BANDGAP VOLTAGE REFERENCES

The zener diodes and temperature compensated avalanche diodes have breakdown voltage in the range of 6 V to 7 V. Thus the supply voltages required are of the order of 10 V or more. Hence it is not possible to use zener reference circuits in the lower supply voltage circuits using operating supply voltage as 5 V or less. In such cases bandgap voltage reference circuits can be used.

We know that for the semiconductor device like transistor, the base-emitter voltage V_{BE} has negative temperature coefficient while voltage equivalent of the temperature $V_T = kT$ has a positive temperature coefficient. The base emitter voltage is highly predictable and depends on voltage corresponding to energy bandgap of the semiconductor used. The silicon has the energy bandgap voltage of 1.205 V.

Key Point: Thus in bandgap reference circuits, negative TC of V_{BE} is compensated by positive TC of the thermal voltage V_T and small reference voltage of the order of 1.2 V can be obtained.

2.16.1 Basic Bandgap Reference Circuit

The basic bandgap reference circuit is shown in the Fig. 2.48.

It uses a Widlar current source circuit, with transistors Q_1 and Q_2. The Q_1 and Q_2 are matched npn transistors.

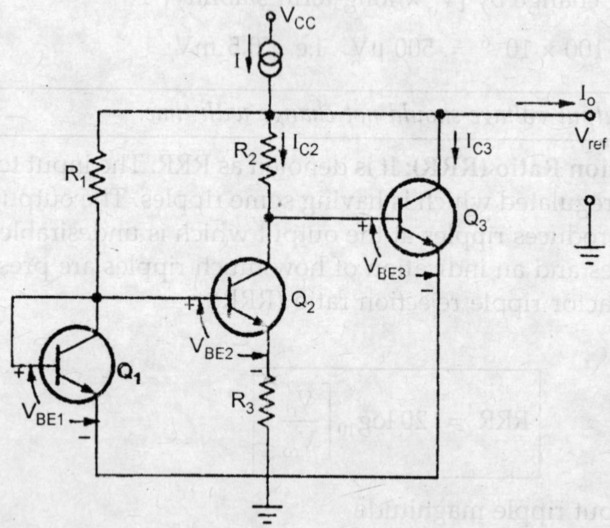

Fig. 2.48 Basic bandgap reference circuit

Key Point: The current density of Q_1 is 10 times higher than Q_2 thus Q_1 is operated at high current density.

Applying KVL to base-emitters of Q_1 and Q_2 through R_3 we get,

$$V_{R3} = \text{Voltage across } R_3 = V_{BE1} - V_{BE2} = \Delta V_{BE} \quad ...(1)$$

According to Ebers Moll equations of transistor,

$$I_{C1} = I_S e^{V_{BE1}/V_T} \quad ...(2)$$

and $$I_{C2} = I_S e^{V_{BE2}/V_T} \quad ...(3)$$

where I_S = collector saturation current

V_T = kT = Thermal voltage

$$\therefore \quad \frac{I_{C1}}{I_{C2}} = e^{(V_{BE1} - V_{BE2})/V_T} \quad ...(4)$$

$$\therefore \quad \ln\left[\frac{I_{C1}}{I_{C2}}\right] = \frac{V_{BE1} - V_{BE2}}{V_T}$$

$$\therefore \quad V_{BE1} - V_{BE2} = V_T \ln\left[\frac{I_{C1}}{I_{C2}}\right]$$

$$\therefore \quad \boxed{V_{R3} = V_T \ln\left[\frac{I_{C1}}{I_{C2}}\right]} \quad ...(5)$$

Neglecting base current of Q_2,

$$I_{C2} = I_{R3} \quad ...(6)$$

and $$I_{R3} = \frac{V_{R3}}{R_3} = \frac{V_T \ln\left[\frac{I_{C1}}{I_{C2}}\right]}{R_3} \quad ...(7)$$

$$\therefore \quad \boxed{I_{C2} = \frac{V_T}{R_3} \ln\left[\frac{I_{C1}}{I_{C2}}\right]} \quad ...(8)$$

Key Point: *Thus I_{C2} is proportional to voltage equivalent of the temperature V_T.*

Applying KVL through R_2, base-emitter of Q_3 and V_{ref} we get,

$$V_{ref} = V_{BE3} + I_{C2} R_2 \quad ...(9)$$

$$\therefore \quad \boxed{V_{ref} = V_{BE3} + \frac{R_2}{R_3} V_T \ln\left[\frac{I_{C1}}{I_{C2}}\right]} \quad ...(10)$$

Now V_{BE3} has negative temperature coefficient (T_C) while V_T has positive temperature coefficient. Hence designing proper values of ratios R_2 to R_3 and I_{C1} to I_{C2}, a **zero temperature coefficient** can be achieved.

We know that V_T = kT where k is Boltzmann's constant given by 8.62×10^{-5} eV/°K. If k is given in J/°K then it is required to be divided by q i.e. charge on one electron which is 1.602×10^{-19} C.

$$\therefore \quad V_{ref} = V_{BE3} + \frac{R_2}{R_3} kT \ln\left[\frac{I_{C1}}{I_{C2}}\right] \quad ...(11)$$

To obtain TC (V_{ref}), differentiate equation of V_{ref} with respect to T,

$$\therefore \quad TC\ (V_{ref}) = \frac{\partial V_{ref}}{\partial T} = \frac{dV_{BE3}}{dT} + \frac{R_2}{R_3} k \ln\left[\frac{I_{C1}}{I_{C2}}\right] \quad ...(12)$$

where $\frac{dV_{BE3}}{dT}$ = TC (V_{BE}) = Temperature coefficient of V_{BE}.

The temperature coefficient of V_{BE} can be obtained similar to the procedure of obtaining temperature dependence of the diode forward voltage. Hence from the p-n junction theory we can write,

$$TC\ (V_{BE}) = \frac{dV_{BE}}{dT} = \frac{V_{BE} - (V_{G0} + 3V_T)}{T}$$

where, V_{G0} = Bandgap voltage at absolute zero temperature
= 1.21 V for Si and 0.785 V for Ge

$$\therefore \quad TC\ (V_{BE}) = -\left[\frac{V_{G0} - V_{BE}}{T} + 3k\right] \quad ...(13)$$

Now to have TC (V_{ref}) = 0 we can write from the equation (11),

$$TC\ (V_{BE3}) = -\frac{R_2}{R_3} k \ln\left[\frac{I_{C1}}{I_{C2}}\right]$$

$$\therefore \quad \frac{V_{G0} - V_{BE}}{T} + 3k = +\frac{R_2}{R_3} k \ln\left[\frac{I_{C1}}{I_{C2}}\right]$$

Multiplying by T on both sides and using $V_T = kT$ we get,

$$V_{BE} = V_{G0} + 3V_T - \frac{R_2}{R_3} V_T \ln\left[\frac{I_{C1}}{I_{C2}}\right] \quad ...(14)$$

Using this in equation (10),

$$V_{ref} = V_{G0} + 3V_T - \frac{R_2}{R_3} V_T \ln\left[\frac{I_{C1}}{I_{C2}}\right] + \frac{R_2}{R_3} V_T \ln\left[\frac{I_{C1}}{I_{C2}}\right]$$

$$\therefore \quad V_{ref} = V_{G0} + 3V_T \quad ...(15)$$

***Key Point:** Thus the reference voltage is constant and determined by the V_{G0} which is bandgap voltage hence the circuit is called bandgap voltage reference circuit.*

It can be seen that at 27 °C we have, $V_T \approx 26$ mV

$$\therefore \quad V_{ref} = 1.205 + 3 \times 26 \times 10^{-3} = 1.283 \text{ V}$$

Key Point: *Thus reference voltage of the order of 1.2 V can be generated with bandgap reference.*

The op-amp can be used at the output to raise the value of V_{ref}. To reduce the output impedance, a high input impedance buffer circuit can be used between the bandgap reference voltage circuit and the load. Such a circuit has good power supply rejection too.

2.16.2 Applications

Due to good temperature and noise characteristics and compatibility with low supply voltages, bandgap reference circuits are used in number of applications. Such circuits are voltage regulators, DA and AD converters. V-F and F-V converters, power supply supervisory circuits and bar graph meters. Nowadays, IC bandgap reference circuits such as LM 385 are also available.

Example 2.9: Design a basic bandgap reference circuit with the parameters as

$$TC\,(V_{ref}) = 0,\ I_{C1} = I_{C3} = 0.2 \text{ mA},\ I_{C2} = \frac{1}{5} I_{C1} \text{ and } I_S = 2 \times 10^{-5} \text{ A for all the transistors.}$$

Assume room temperature of 27°C.

Solution: From Ebers Moll transistor equations.

$$I_{C1} = I_S\, e^{V_{BE1}/V_T}$$

Now
$$V_T = kT = 8.62 \times 10^{-5} \times [273 + 27] = 25.86 \text{ mV}$$

$$\therefore \quad V_{BE1} = V_T \ln\left[\frac{I_{C1}}{I_{C2}}\right] = 25.86 \times 10^{-3} \ln\left[\frac{0.2 \times 10^{-3}}{2 \times 10^{-15}}\right] = 0.655 \text{ V}$$

Applying KVL to Q_1 through R_1 and V_{BE1} we can write, (Refer Fig. 2.39)

$$R_1 = \frac{V_{ref} - V_{BE1}}{I_{C1}}$$

At 27 °C,
$$V_{ref} = V_{G0} + 3V_T = 1.205 + 3 \times 25.86 \times 10^{-3}$$
$$= 1.2825 \text{ V}$$

$$\therefore \quad R_1 = \frac{1.2825 - 0.655}{0.2 \times 10^{-3}} = 3.13 \text{ k}\Omega$$

Now
$$V_{R3} = V_T \ln\left[\frac{I_{C1}}{I_{C2}}\right] = 25.86 \times 10^{-3} \ln[5]$$

$$V = 0.04162 \text{ V}$$

$$\therefore \quad R_3 = \frac{V_{R3}}{I_{C2}} = \frac{0.04162}{\left[\frac{1}{5} \times 0.2 \times 10^{-3}\right]} = 1.04 \text{ k}\Omega$$

$$V_{ref} = V_{BE3} + I_{C2} R_2$$

Hence in the designed bandgap reference circuit,

$$R_1 = \mathbf{3.13\ k\Omega}, \quad R_2 = \mathbf{15.68\ k\Omega} \quad \text{and} \quad R_3 = \mathbf{104\ k\Omega}$$

While $V_{ref} = \mathbf{1.2825\ V}$

Examples with Solutions

Example 2.10: An op-amp has a differential gain of 80 dB and CMRR of 95 dB. If $V_1 = 2\ \mu V$ and $V_2 = 1.6\ \mu V$, then calculate the differential and common mode output values.

Solution: A_d = 80 dB and CMRR = 95 dB

Do not use dB values for the calculation directly.

$$A_d \text{ in dB} = 20 \log A_d$$

$$\therefore \quad 80 = 20 \log A_d$$

$$\therefore \quad A_d = 1 \times 10^4 \text{ (absolute)}$$

$$\text{CMRR in dB} = 20 \log \text{CMRR}$$

$$\therefore \quad 95 = 20 \log \text{CMRR}$$

$$\therefore \quad \text{CMRR} = 5.6234 \times 10^4$$

∴ Differential output can be calculated as

$$V_d = A_d(V_1 - V_2) = 1 \times 10^4 (2 - 1.6) \times 10^{-6}$$

$$= \mathbf{4\ mV}$$

And common mode output can be calculated as,

$$V_c = A_c \frac{(V_1 + V_2)}{2}$$

Now

$$\text{CMRR} = \frac{A_d}{A_c}$$

$$\therefore \quad 5.6234 \times 10^4 = \frac{1 \times 10^4}{A_c}$$

$$\therefore \quad A_c = 0.1778$$

$$\therefore \quad V_c = 0.1778 \times \frac{(2 + 1.6)}{2} \times 10^{-6}$$

$$= \mathbf{0.32\ \mu V}$$

Example 2.11: The common mode input to a certain differential amplifier, having differential gain of 125 is 4 sin 200 π t V. Determine the common mode output if CMRR is 60 dB.

Solution: The CMRR in dB is

$$60 = 20 \log \left| \frac{A_d}{A_c} \right|$$

$$\therefore \quad \log \left| \frac{A_d}{A_c} \right| = 3$$

$$\therefore \qquad \frac{A_d}{A_c} = 1000$$

Now $\qquad A_d = 125$

$$\therefore \qquad A_c = \frac{A_d}{1000} = \frac{125}{1000}$$
$$= 0.125$$

Hence the common mode output is

$$= A_c V_c = 0.125\ (4 \sin 200\ \pi\ t)$$
$$= \mathbf{0.5 \sin (200\ \pi\ t)\ V}$$

Example 2.12: For the differential amplifier shown in the Fig. 2.49, determine

(*i*) The voltages at the collector of each transistor

(*ii*) The differential voltage gain.

Assume $V_{BE} = 0.7$ V, $h_{fe} = 100$, $h_{ie} = 3.9$ kΩ and the source resistance as 1 kΩ.

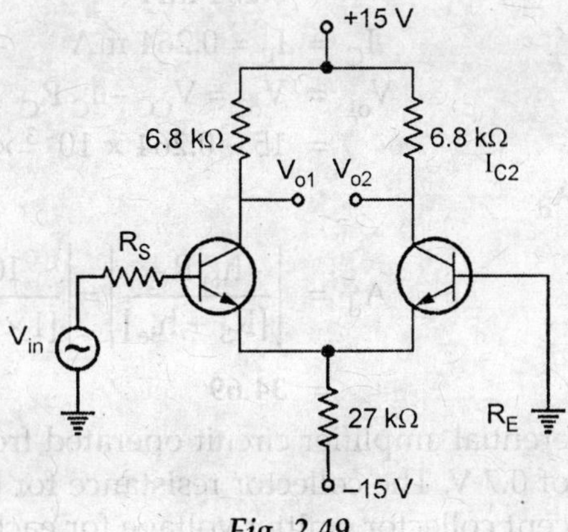

Fig. 2.49

Solution: The circuit is redrawn showing the various voltages and current as shown in the Fig. 2.49 (*a*)

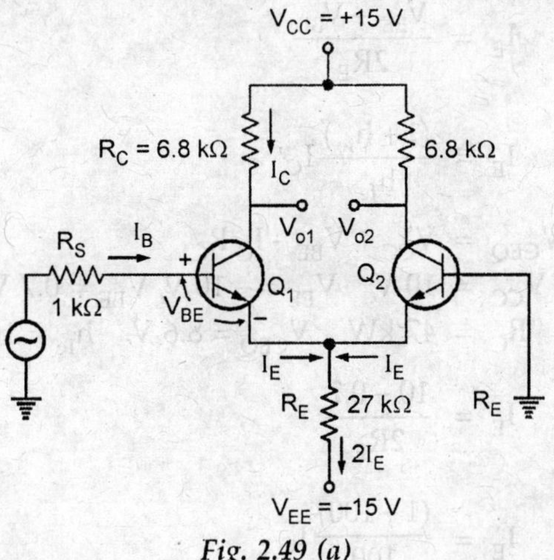

Fig. 2.49 (a)

(*i*) Applying KVL to the base emitter loop of Q_1,

$$-I_B R_S - V_{BE} - 2I_E R_E + V_{EE} = 0$$

Now $\quad I_B = \frac{I_E}{\beta}$ where $\beta = h_{fe}$

$$\therefore \quad \frac{-I_E R_S}{\beta} - V_{BE} - 2I_E R_E + V_{EE} = 0$$

$$\therefore \quad I_E\left[\frac{R_S}{\beta} + 2R_E\right] + V_{BE} - V_{EE} = 0$$

$$\therefore \quad I_E = \frac{V_{EE} - V_{BE}}{\frac{R_S}{\beta} + 2R_E} = \frac{15 - 07}{\frac{1\times10^3}{100} + 1\times27\times10^3}$$

$$= 0.264 \text{ mA}$$

$$\therefore \quad I_C = I_E = 0.264 \text{ mA}$$

$$\therefore \quad V_{o1} = V_{o2} = V_{CC} - I_C R_C$$

$$= 15 - 0.264 \times 10^{-3} \times 6.8 \times 10^3 = 13.2 \text{ V}$$

(*ii*) Differential gain A_d

$$A_d = \left|\frac{h_{fe} R_C}{[R_S + h_{ie}]}\right| = \left|\frac{100\times6.8\times10^3}{[1\times10^3 + 3.9\times10^3]}\right|$$

$$= \mathbf{34.69}$$

Example 2.13: For a differential amplifier circuit operated from ± 10 V, assume matched pair of transistors with V_{BE} of 0.7 V. The collector resistance for each is 47 kΩ. Calculate the value of R_E so that the quiescent collector emitter voltage for each transistor is 8.6 V. Assume h_{fe} = 100.

Solution: From d.c. analysis of differential amplifier,

$$I_E = \frac{V_{EE} - V_{BE}}{2R_E}$$

And $\quad I_E = \frac{(1 + h_{fe})}{h_{fe}} I_C$

And $\quad V_{CEQ} = V_{CC} + V_{BE} - I_C R_C$

$$V_{CC} = 10 \text{ V}, \quad V_{EE} = -10 \text{ V}, V_{BE} = 0.7 \text{ V}$$

$$R_C = 47 \text{ kW}, \quad V_{CEQ} = 8.6 \text{ V}, \quad h_{fe} = 100$$

$$\therefore \quad I_E = \frac{10 - 0.7}{2R_E}$$

But $\quad I_E = \frac{(1 + 100)}{100} I_C$

$$\therefore \quad \frac{101}{100} I_C = \frac{9.3}{2R_E}$$

$$\therefore \quad R_E = \frac{4.6039}{I_C} \quad \text{...(1)}$$

Now $\quad V_{CEQ} = V_{CC} + V_{BE} - I_C R_C$

$$\therefore \quad 8.6 = 10 + 0.7 - I_C \times 47 \times 10^3$$

$$\therefore \quad I_C = \frac{10.7 - 8.6}{47 \times 10^3}$$

$$= 4.468 \times 10^{-5} \text{ A} \quad \text{...(2)}$$

Substituting in equation (1),

$$R_E = \frac{4.6039}{4.468 \times 10^{-5}}$$

$$= \mathbf{103.03\ k\Omega}$$

Example 2.14: Fig. 2.50 shows the dual input differential amplifier.

(*i*) Determine the Q point values for each transistor.

(*ii*) Determine the voltage gain.

(*iii*) Determine input and output resistances.

(*iv*) Determine the output voltage if V_{S1} = 50 mV (p-p) and V_{S2} = 20 mV (p-p).

(*v*) What is the maximum output voltage without clipping?

Assume h_{fe} = 100, V_{BE} = 0.72 V and h_{ie} = 2.5 kΩ.

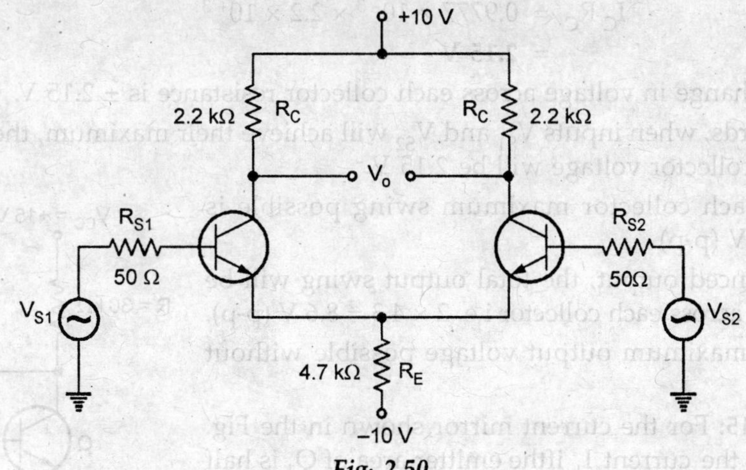

Fig. 2.50

Solution: V_{CC} = 10 V $\quad V_{EE}$ = – 10 V

(*i*) Now $\quad I_E = \frac{V_{EE} - V_{BE}}{2R_E} = \frac{10 - 0.72}{2 \times 4.7 \times 10^3}$

$$= 0.987 \text{ mA}$$

$$I_C = \frac{h_{fe}}{1+h_{fe}} I_E$$

$$I_C = \frac{100}{101} \times 0.987 = 0.9774 \text{ mA}$$

$$\therefore \quad I_{CQ} = I_C = \mathbf{0.9774 \text{ mA}}$$

$$V_{CE} = V_{CC} + V_{BE} - I_C R_C = 10 + 0.72 - 0.9774 \times 10^{-3} \times 2.2 \times 10^3$$

$$= 8.57 \text{ V}$$

$$\therefore \quad V_{CEQ} = V_{CE} = \mathbf{8.57 \text{ V}}$$

(*ii*) For dual input, balanced output

$$A_d = \frac{h_{fe} R_C}{R_S + h_{ie}}$$

$$= \frac{100 \times 2.2 \times 10^3}{50 + 2.5 \times 10^3} = \mathbf{86.274}$$

(*iii*) $R_{in} = 2(R_S + h_{ie}) = 2(50 + 2.5 \times 10^3)$

$= \mathbf{5.1 \text{ k}\Omega}$

$R_o = R_C = \mathbf{2.2 \text{ k}\Omega}$

(*iv*) $V_o = A_d (V_{S1} - V_{S2}) = 86.27 (50 - 20)$

$= \mathbf{2.588 \text{ V (p-p)}}$

(*v*) Now voltage across R_C is

$$I_C R_C = 0.9777 \times 10^{-3} \times 2.2 \times 10^{+3}$$

$$= \mathbf{2.15 \text{ V}}$$

i.e. maximum change in voltage across each collector resistance is ± 2.15 V.

In other words, when inputs V_{S1} and V_{S2} will achieve their maximum, the instantaneous values of each collector voltage will be 2.15 V.

So across each collector maximum swing possible is 2 × 2.15 i.e. 4.3 V (p-p).

Due to balanced output, the total output swing will be twice the swing across each collector i.e. 2 × 4.3 = 8.6 V (p-p).

This is the maximum output voltage possible without clipping.

Example 2.15: For the current mirror shown in the Fig. 2.51, determine the current I_1 ifthe emitter area of Q_1 is half of Q.

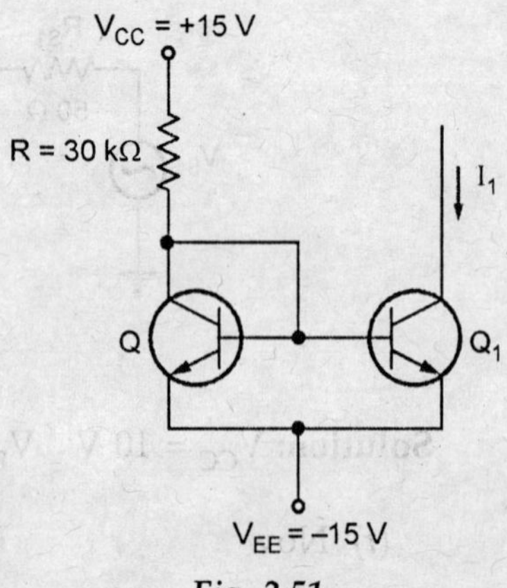

Fig. 2.51

Solution: Current through the collector of Q is say I_C.

Now $$I_C = \frac{V_{CC} + V_{EE} - V_{BE}}{R}$$

$$= \frac{15 + 15 - 0.7}{30 \times 10^3} = 0.976 \text{ mA}$$

Now current I_1 is proportional to the ratio of the areas.

$$\therefore \quad I_1 = I_C \left[\frac{A_Q}{A}\right]$$

Now $\quad A_Q = 0.5 \text{ A}$

$$\therefore \quad I_1 = I_C \times 0.5$$

$$\therefore \quad I_1 = 0.976 \times 0.5 = \mathbf{0.49 \text{ mA}}$$

Example 2.16: Determine the currents I_2, I_3, I_4 if the areas of Q_2, Q_3 and Q_4 are 0.5 A_1, 0.25 A_1 and 0.125 A_1 respectively where A_1 is the area of Q_1. Refer Fig. 2.52.

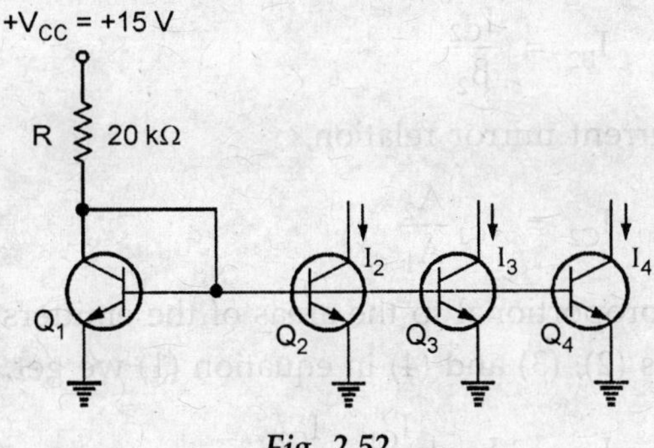

Fig. 2.52

Solution: Let current through Q_1 be I_1. Then

$$I_1 = \frac{V_{CC} - V_{BE}}{R} \quad \text{as emitter of } Q_1 \text{ grounded.}$$

$$= \frac{15 - 0.7}{20 \times 10^3} = 0.715 \text{ mA}$$

Now I_2, I_3 and I_4 are proportional to ratio of the emitter areas. Hence

$$I_2 = I_1\left[\frac{A_2}{A_1}\right] = 0.715\left[\frac{0.5 A_1}{A_1}\right] = \mathbf{0.357 \text{ mA}}$$

$$I_3 = I_1\left[\frac{A_3}{A_1}\right] = 0.715\left[\frac{0.25 A_1}{A_1}\right] = \mathbf{0.1787 \text{ mA}}$$

And

$$I_4 = I_1\left[\frac{A_4}{A_1}\right] = 0.715\left[\frac{0.125 A_1}{A_1}\right] = \mathbf{0.089 \text{ mA}}$$

Example 2.17: For the circuit shown in the Fig. 2.53 show that

$$I_{C2} = I_{ref}\frac{A_2}{A_1}\left[1+\frac{1}{\beta_1}+\frac{A_2}{A_1}.\frac{1}{\beta_2}\right]^{-1}$$

Where the subscripts relate to the transistors Q_1 and Q_2.

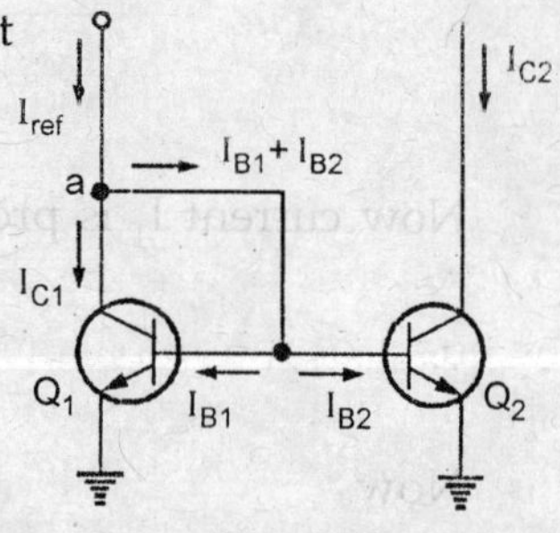

Fig. 2.53

Solution: Applying KCL to the node 'a',

$$I_{ref} = I_{C1} + I_{B1} + I_{B2} \quad ...(1)$$

Now $$I_{B1} = \frac{I_{C1}}{\beta_1} \quad ...(2)$$

And $$I_{B2} = \frac{I_{C2}}{\beta_2} \quad ...(3)$$

Similarly from the current mirror relation,

$$I_{C2} = I_{C1}\frac{A_2}{A_1} \quad ...(4)$$

As the current I_{C2} is proportional to the areas of the emitters of Q_1 and Q_2. Substituting equations (2), (3) and (4) in equation (1) we get,

$$I_{ref} = I_{C1} + \frac{I_{C1}}{\beta_1} + \frac{I_{C2}}{\beta_2}$$

$$= I_{C1} + \frac{I_{C1}}{\beta_1} + \frac{I_{C2}}{\beta_2}.\frac{A_2}{A_1}$$

$$= I_{C1}\left[1+\frac{1}{\beta_1}+\frac{A_2}{A_1}.\frac{1}{\beta_2}\right] \quad ...(5)$$

Now using the relation equation (4) we get,

$$I_{C2} = I_{C1}.\frac{A_2}{A_1}$$

Substituting I_{C1} in terms of I_{ref} from equation (5) above,

$$I_{C2} = \frac{I_{ref}}{\left[1+\frac{1}{\beta_1}+\frac{A_2}{A_1}.\frac{1}{\beta_2}\right]}.\frac{A_2}{A_1}$$

i.e. $$I_{C2} = I_{ref}\frac{A_2}{A_1}\left[1+\frac{1}{\beta_1}+\frac{A_2}{A_1}.\frac{1}{\beta_2}\right]^{-1}$$ ...Hence proved.

Example 2.18: In a constant current source with zener, the zener voltage is 6.2 V. Calculate the current I in the circuit shown in the Fig. 2.54.

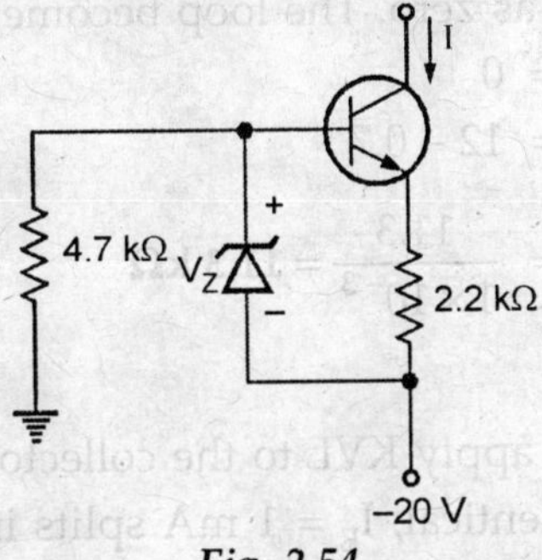

Fig. 2.54

Solution: For the circuit,

$$R_1 = 4.7\text{ k}\Omega,\quad R_3 = 2.2\text{ k}\Omega,\quad V_z = 6.2\text{ V},\quad V_{EE} = -20\text{ V}$$

The base voltage of transistor is

$$V_B = -V_{EE} + V_z = -20 + 6.2 = -13.8\text{ V}$$

Now the emitter voltage is

$$V_E = -V_{EE} + V_z - V_{BE} = -20 + 6.2 - 0.7 = -14.5\text{ V}$$

And the emitter current is

$$I_E = \frac{V_E - (-V_{EE})}{R_3} = \frac{-14.5 - (-20)}{2.2 \times 10^3} = 2.5\text{ mA}$$

$$\therefore \quad I_E = I = \mathbf{2.5\ mA.}$$

Example 2.19: For the differential amplifier shown in the Fig. 2.55 determine

(*i*) The value of R_E to provide an emitter current I_E of 1 mA

(*ii*) The d.c. output voltage V_{o1}.

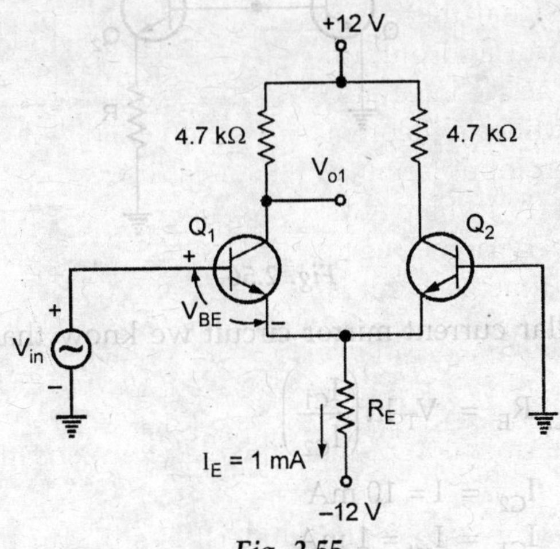

Fig. 2.55

Assume identical transistors and V_{BE} as 0.7 V.

Solution:

(*i*) To determine value of R_E, apply KVL to the base-emitter loop of Q_1. For d.c. analysis assume V_{in} as zero. The loop becomes,

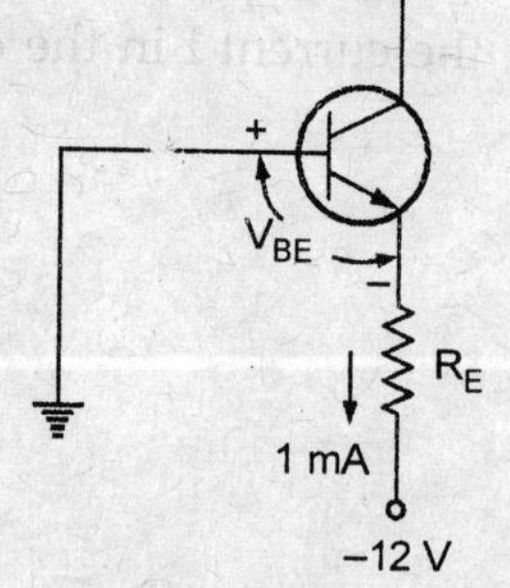

Fig. 2.55 (a)

$$\therefore -R_E(1 \times 10^{-3}) + 12 - V_{BE} = 0$$

$$\therefore \quad R_E(1 \times 10^{-3}) = 12 - 0.7$$

$$\therefore \quad R_E = \frac{11.3}{1 \times 10^{-3}} = \mathbf{11.3\ k\Omega}$$

(*ii*) To obtain the output V_{o1}, apply KVL to the collector circuit of Q_1.

As the two transistors are identical, I_E = 1 mA splits into two equal parts between the two transistors.

$$\therefore \quad I_{E1} = I_{E2} = 0.5\ \text{mA}$$

Now $\quad I_{E1} = I_{C1}$ and $I_{E2} = I_{C2}$

$$\therefore \quad I_{C1} = 0.5\ \text{mA}$$

$$\therefore \quad 0 = +12 - I_{C1}(4.7 \times 10^3) - V_{o1}$$

$$\therefore \quad V_{o1} = 12 - 0.5 \times 10^{-3} \times 4.7 \times 10^3 = \mathbf{9.65\ V.}$$

Example 2.20: Determine the value of the resistance required for the Widlar current mirror shown in the Fig. 2.56, if the reference current is 1 mA and the control current (I) is 10 μA.

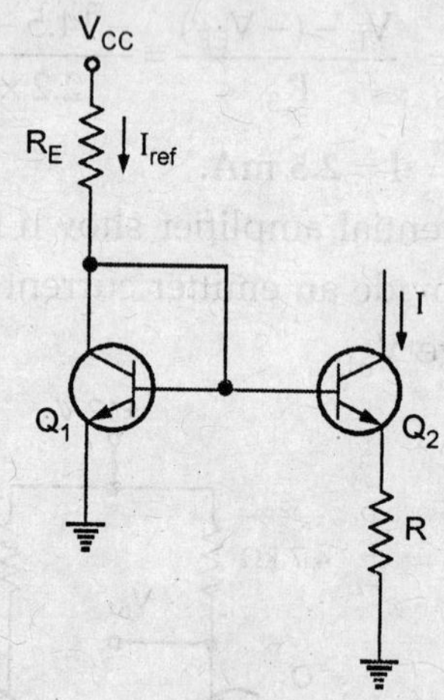

Fig. 2.56

Solution: For the Widlar current mirror circuit we know that

$$I_{C2} R_E = V_T \ln\left(\frac{I_{C1}}{I_{C2}}\right)$$

In this example,

$$I_{C2} = I = 10\ \text{mA}$$

$$I_{C1} = I_{ref} = 1\ \text{mA}$$

$$V_T = 26\ \text{mV}$$

$$\therefore \quad 10 \times 10^{-6} R_E = 26 \times 10^{-3} \ln\left(\frac{1 \times 10^{-3}}{10 \times 10^{-6}}\right)$$

$$\therefore \quad R_E (10 \times 10^{-6}) = 0.1197$$

$$\therefore \quad R_E = \mathbf{11.97\ k\Omega}$$

This is the required value of the resistance.

Example 2.21: For a differential amplifier shown in the Fig. 2.57, select a suitable value of R_E so that the V_{CEQ} for each transistor is 4 V. Assume h_{ie} = 100 and matched silicon transistors.

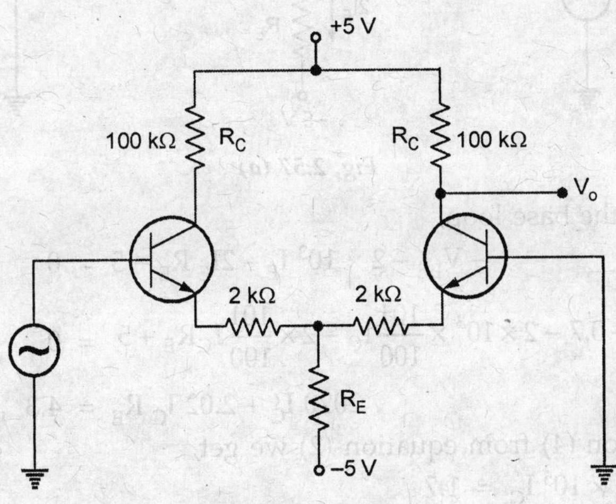

Fig. 2.57

Solution: Note that there is additional resistance of 2 kΩ in the transistor. **Hence the standard equations can not be used in this case.**

Drawing the circuit again with the indication of all d.c. currents as shown in the Fig. 2.57 (*a*).

Applying KVL to the collector to emitter loop,

$$-100 \times 10^3 \times I_C - V_{CE} - 2 \times 10^3 I_E - 2I_E R_E + 5 + 5 = 0 \quad \ldots(1)$$

Now $\quad V_{CE} = V_{CEQ} = 4$ given

And $\quad I_E = \left[\frac{1 + h_{fe}}{h_{fe}}\right] I_C$

$$I_E = \left(\frac{1 + 100}{100}\right) I_C$$

Substituting in equation (1)

$$-100 \times 10^3 I_C - 4 - 2 \times 10^3 \times \frac{101}{100} I_C - 2 \times \frac{101}{100} I_C R_E + 10 = 0$$

$$1.0202 \times 10^5 I_C + 2.02 I_C R_E = +6 \quad \ldots(2)$$

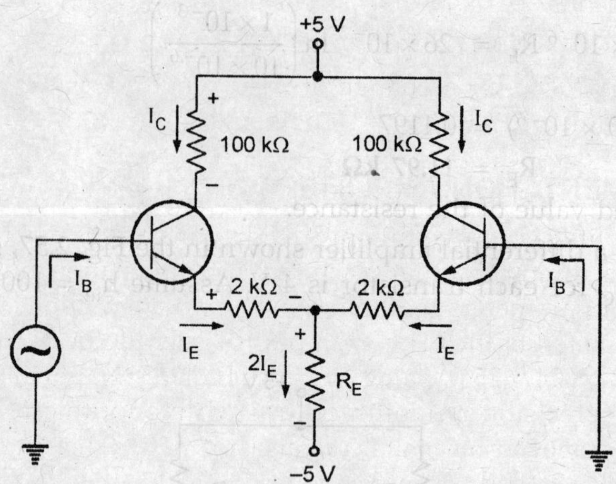

Fig. 2.57 (a)

Applying KVL to the base loop,

$$-V_{BE} - 2 \times 10^3 I_E - 2I_E R_E + 5 = 0 \qquad ...(3)$$

$$\therefore \quad -0.7 - 2\times 10^3 \times \frac{101}{100} I_C - 2\times \frac{101}{100} I_C R_E + 5 = 0$$

$$\therefore \quad 2020\ I_C + 2.02\ I_C\ R_E = 4.3 \qquad ...(4)$$

Subtracting equation (4) from equation (2) we get,

$$1 \times 10^5\ I_C = 1.7$$

$$\therefore \quad I_C = 1.7 \times 10^{-5}$$

Substituting in equation (4), we get

$$2020 \times 1.7 \times 10^{-5} + 2.02 \times 1.7 \times 10^{-5} \times R_E = 4.3$$

$$\therefore \quad R_E = \mathbf{124.21\ k\Omega}$$

Example 2.22: Determine I_{E3}, V_{CE1} and differential mode gain for the circuit shown in the Fig. 2.58. Assume h_{fe} = 250 and h_{ie} = 560 Ω for Q_1 and Q_2.

Solution: For the transistor Q_4, apply KVL to the base emitter loop assuming current flowing through 1.5 kΩ resistor as I,

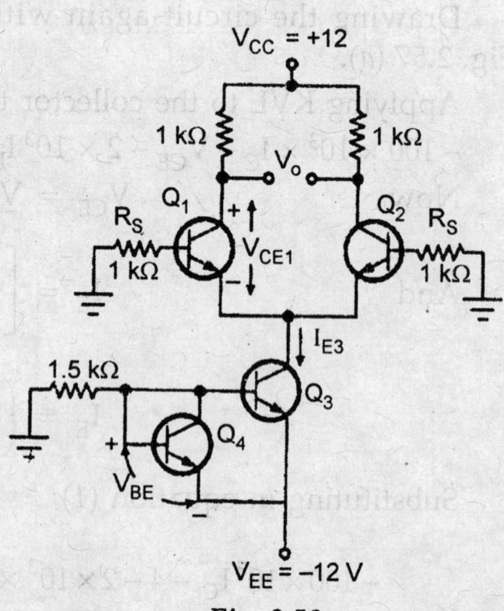

Fig. 2.58

$$-1.5 \times 10^3\ I - V_{BE} + 12 = 0$$

$$\therefore \quad 1.5 \times 10^3\ I = 12. - 0.7$$

$$\therefore \quad I = \frac{11.3}{1.5\times 10^3}$$

$$= 7.533 \text{ mA}$$

As Q_3, Q_4 form a constant current bias,

$$I_{E3} = I = \mathbf{7.533\ mA}$$

This is due to two equal parts of emitter currents I_{E1} and I_{E2}.

$$\therefore \quad I_{E1} = I_{E2} = \frac{I_{E3}}{2} = \frac{7.533}{2}$$

$$= 3.766 \text{ mA}$$

Now $\quad I_{C1} = I_{E1} = 3.766$ mA

Now $\quad V_{CE1} = V_{CC} + V_{BE} - I_{C1} R_C$

$$= 12 + 0.7 - 3.76 \times 10^{-3} \times 1 \times 10^3$$

$$= \mathbf{8.933 \text{ V}}$$

For dual input, balanced output differential amplifier,

$$A_d = \frac{h_{fe}R_c}{(R_S + h_{ie})} = \frac{250 \times 1 \times 10^3}{1 \times 10^3 + 560}$$

$$= \mathbf{160.25}$$

Example 2.23: For the difference amplifier shown in Fig. 2.59. Calculate V_{C1} and V_{C2} if $V_1 = V_2 = 0$ V. Assume $V_{be} = 0$ V.

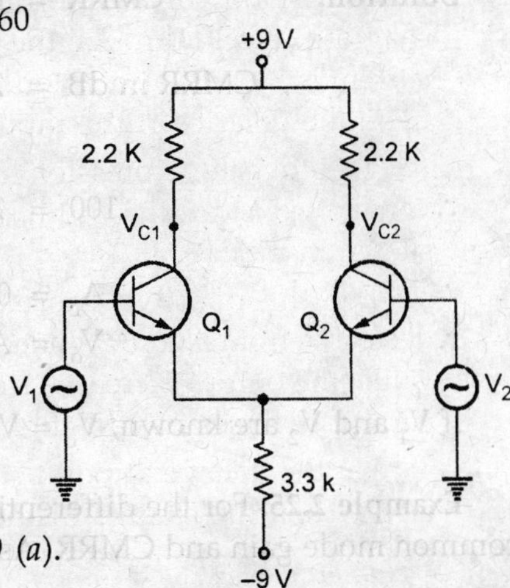

Fig. 2.59

Solution: The currents are as shown in the Fig. 2.59 (*a*).

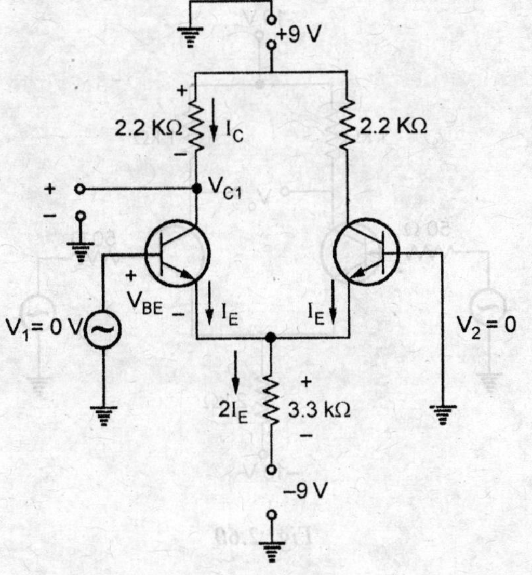

Fig. 2.59 (a)

Applying KVL to B-E loop of Q_1,

$$-V_{BE} - 2I_E \times 3.3 \times 10^3 = 0$$

$$\therefore \quad I_E = \frac{9-0.7}{2\times3.3\times10^3} = 1.2575 \text{ mA} = I_C$$

Applying KVL for output path,

$$-2.2 \times 10^3 I_C - V_{C1} + 9 = 0$$

$$\therefore \quad V_{C1} = 9 - 2.2 \times 10^3 \times 1.2575 \times 10^{-3} = 6.2335 \text{ V}$$

And $\quad V_{C2} = V_{C1} = \mathbf{6.2335\ V}$

Example 2.24: Determine the output voltage of a differential amplifier having differential amplifier gain 2000 and the CMRR 100 dB.

Solution: $\quad \text{CMRR} = 100 \text{ dB}, A_d = 2000$

$$\text{CMRR in dB} = 20 \log \frac{A_d}{A_c}$$

$$\therefore \quad 100 = 20 \log \frac{2000}{A_c}$$

$$\therefore \quad A_c = 0.02$$

$$\therefore \quad V_o = A_d V_d + A_c V_c = \mathbf{2000\ V_d + 0.02\ V_c}$$

If V_1 and V_2 are known, $V_d = V_1 - V_2$ and $V_c = \frac{V_1 + V_2}{2}$ and output can be obtained.

Example 2.25: For the differential amplifier shown in Fig. 2.60 find the differential gain, common mode gain and CMRR. Assume $h_{fe} = 100$, $h_{ie} = 1$ kΩ.

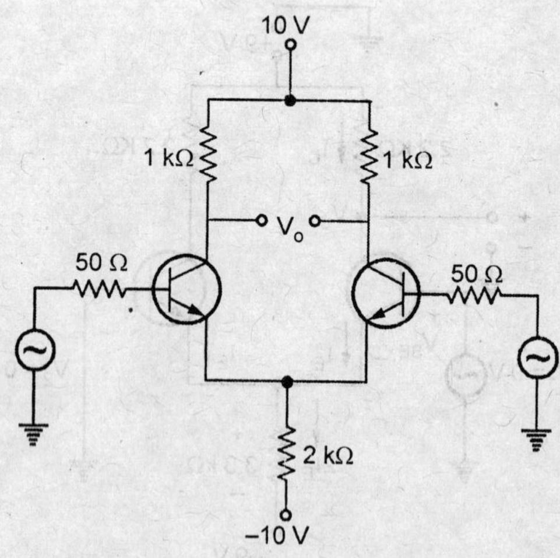

Fig. 2.60

Solution: $R_C = 1\ k\Omega$, $R_E = 2\ k\Omega$, $R_S = 50\ \Omega$, $V_{CC} = +10\ V$

$$V_{EE} = -10\ V,\quad h_{fe} = 100,\ h_{ie} = 1\ k\Omega$$

$$\therefore\quad A_d = \frac{h_{fe}R_C}{(R_S + h_{ie})} = \frac{100(1\times10^3)}{[50 + 1\times10^3]} = \mathbf{95.238}$$

$$A_c = \frac{-h_{fe}R_C}{R_S + h_{ie} + 2R_E(1+h_{fe})} = \frac{-100\times1\times10^3}{50 + 1\times10^3 + 2\times2\times10^3\times101}$$

$$= \mathbf{-0.2468}$$

$$\text{CMRR} = \left|\frac{A_d}{A_c}\right| = \left|\frac{95.238}{-0.2468}\right| = \mathbf{385.7615}$$

$$\text{CMRR in dB} = 20\log 385.7615 = 51.726 = \mathbf{51.726\ dB}$$

Example 2.26: Calculate the output current I of the current source given in Fig. 2.61 Assume h_{FE} of transistor is very high and $V_{BE} = 0.6$ V.

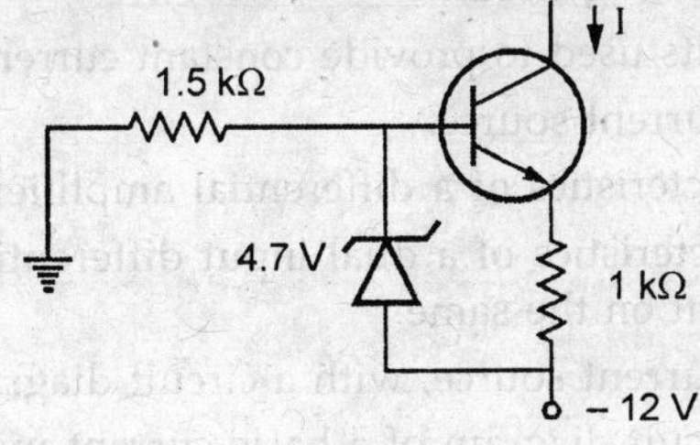

Fig. 2.61

Solution:

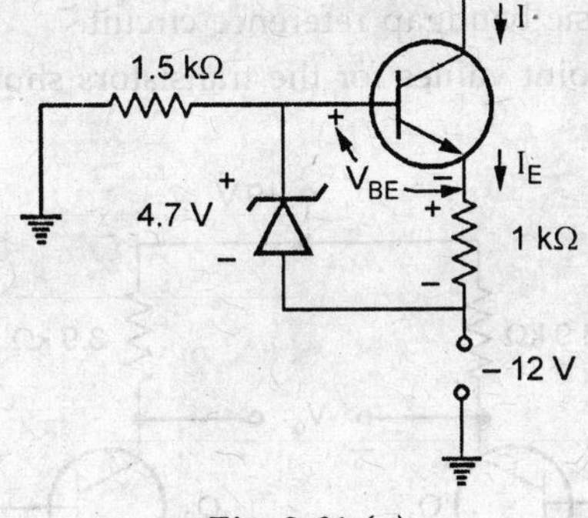

Fig. 2.61 (a)

Applying KVL to the base-emitter loop,

$$-V_{BE} - I_E(1\times10^3) + 4.7 = 0$$

$$I_E = \frac{4.7 - 0.6}{1\times10^3}$$

$$= 4.1\ \text{mA}$$

As h_{FE} is very high,

$$I = I_E = \mathbf{4.1\ mA.}$$

REVIEW QUESTIONS

1. What is meant by differential amplifier?
2. Explain the operation of a basic differential amplifier.
3. Give the four differential amplifier configuration. Which is most commonly used and why?
4. What is the meaning of balanced and unbalanced output?
5. Draw the circuit diagram of emitter coupled differential amplifier and obtain its d.c. analysis.
6. Using the h-parameters, carry out the a.c. analysis of the differential amplifier. Find A_d, A_c, R_i and R_o for dual input balanced output.
7. Define CMRR.
8. Obtain expression for CMRR.
9. Explain the method of improving CMRR.
10. Draw the circuit diagram of a symmetrical emitter coupled difference amplifier and show that a very high CMRR will result if the diff. amp is supplied by a constant current bias.
11. What is the active load ? Where it is used and why ?
12. Explain the operation of differential amplifier with swamping resistors.
13. Write a note on Widlar biasing circuit.
14. Explain the various circuits used to provide constant current bias in a differential amplifier.
15. Write a note on Wilson current source.
16. Derive the transfer characteristics of a differential amplifier.
17. Sketch the transfer characteristics of a dual input differential amplifier showing linear and limiting regions. Comment on the same.
18. Explain the working of current source, with a circuit diagram.
19. Draw and explain the circuit diagram of a basic current mirror.
20. What is voltage reference circuit ?
21. Draw and explain the basic bandgap reference circuit.
22. Calculate the operating point values for the transistors shown in the Fig. 2.62.

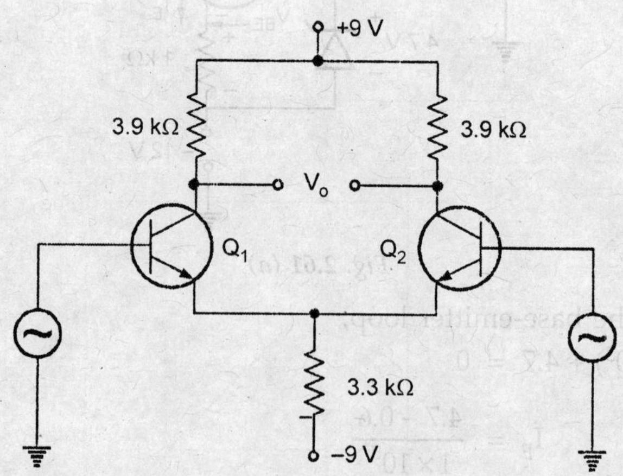

Fig. 2.62

(Ans. 1.25 mA, 4.825 V)

23. For a differential amplifier, two sets of inputs are applied. The first set is V_1 = 50 μV and V_2 = – 50 μV and second set is V_1 = 1050 μV and V_2 = 950 μV. If the CMRR is 100, calculate the percentage difference in the output voltage obtained for the two sets of the input signals.

    If now CMRR is improved to 10000, calculate the percentage difference in the output voltage obtained for the two sets of the input signals. **(Ans. 10%, 0.1%)**

24. Determine the output voltage of an op-amp for the input voltages of V_1 = 150 μV, V_2 = 140 μV. The amplifier has a differential gain of 4000 and the value of CMRR is (*i*) 100 and (*ii*) 10^5. **(Ans. 45.8 mV, 40.006 mV)**

25. For a dual input, balanced output differential amplifier R_C = 47 kΩ, R_{S1} = R_{S2} = 20 kΩ, R_E = 43 kΩ, h_{fe} = 75, h_{ie} = 20 kΩ, V_{CC} = 9 V, V_{EE} = – 9 V and V_{BE} = 0.7 V. Calculate
    (*i*) Operating point values (*ii*) differential gain A_d (*iii*) common mode gain A_c
    (*iv*) CMRR in dB. **(Ans. (95.5 μA, 5.2 V), 87.4, 0.536, 44.24 dB)**

26. For a differential amplifier circuit operated from ± 5 V, assume V_{BE} = 0.7 V and h_{fe} = 100. If R_c for each is 100 kΩ, calculate the value of R_E so that the quiescent collector emitter voltage for each transistor is 4 V. **(Ans.: 125 kΩ)**

27. When a differential input of 1 mV sin ωt is applied to a differential amplifier the differential output is 0.5 V sin ωt. When both inputs are joined together and 1 V sin ωt is applied the output is 200 mV sin ωt. Determine the CMRR. **(Ans. 68 dB)**

28. The parameters for the differential amplifier are given as : R_C = 1 kΩ, R_S = 1 kΩ, h_{fe} = 1 kΩ and R_E = 2 MΩ. Neglecting h_{oe}, calculate the difference mode gain and common mode gain. Hence calculate CMRR in dB. The amplifier is in dual input, balanced output configuration. **(Ans. 100.17 dB)**

29. For the circuit shown in the Fig. 2.63 determine the collector current I_{C2} exactly if A_1 = 2000 μm², A_2 = 200 μm², $β_1$ = 150, $β_2$ = 100 and I_{ref} = 1 mA.

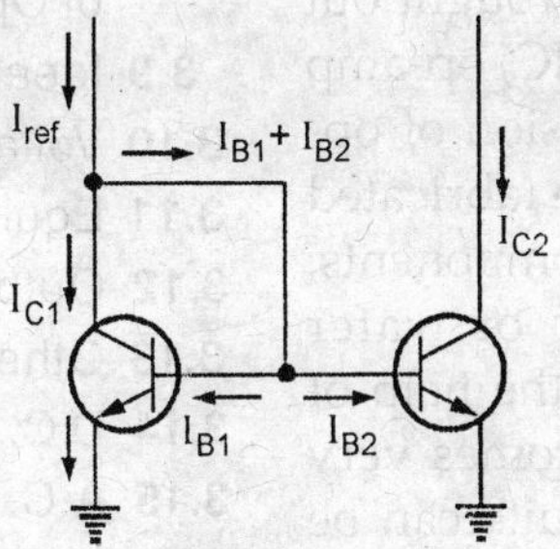

Fig. 2.63

(Ans. 0.099 mA Hint: Refer Ex. 1.18)

❖❖❖

Chapter

3

Operational Amplifier and Its Characteristics

3.1 INTRODUCTION

The operational amplifier, most commonly referred as **'op-amp'** was introduced in 1940s. The first operational amplifier was designed in 1948 using vacuum tubes. In those days, it was used in the analog computers to perform a variety of mathematical operations such as addition, subtraction, multiplication etc. Due to its use in performing mathematical operations it has been given a name operational amplifier. Due to the use of vacuum tubes, the early op-amps were bulky, power consuming and expensive.

Robert J. Widlar at Fairchild brought out the popular 741 integrated circuit (IC) op-amp between 1964 to 1968. The IC version of op-amp uses BJTs and FETs which are fabricated along with the other supporting components, on a single semiconductor chip or wafer which is of a pinhead size. With the help of IC op-amp, the circuit design becomes very simple. The variety of useful circuits can be built without the necessity of knowing about the complex internal circuitry. Moreover, IC op-amps are inexpensive, take up less space and consume less power. The IC op-amp has become an integral part of almost every electronic circuit which uses linear integrated circuit. The modern linear IC op-amp works at lower voltages. It is so low in cost that millions are now in use, annually.

Inside the Chapter

***Key Point:** Because of their low cost, small size, versatility, flexibility and dependability, op-amps are used in the fields of process control, communications, computers, power and signal sources, displays and measuring systems.*

The op-amp is basically an excellent high gain d.c. amplifier.

3.2 OP-AMP SYMBOL AND TERMINALS

The symbol for an op-amp along with its various terminals, is shown in the Fig. 3.1.

The op-amp is indicated basically by a triangle which points in the direction of the signal flow.

All the op-amps have atleast following five terminals:

(*i*) The positive supply voltage terminal V_{CC} or + V.

(*ii*) The negative supply voltage terminal – V_{EE} or – V.

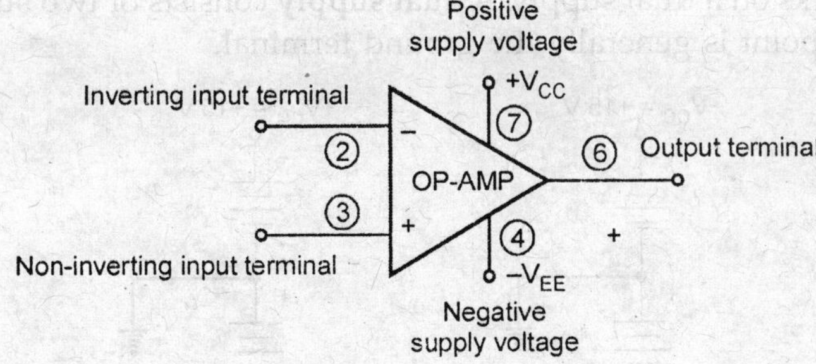

Fig. 3.1 Op-amp symbol.

(*iii*) The output terminal.

(*iv*) The inverting input terminal, marked as negative.

(*v*) The noninverting input terminal, marked as positive.

The input at inverting input terminal results in opposite polarity (antiphase) output. While the input at noninverting input terminal results in the same polarity (phase) output. This is shown in the Fig. 3.2 (*a*) and (*b*). The input and output are in antiphase means having 180° phase difference in between them while inphase input and output means having 0^0 phase difference in between them.

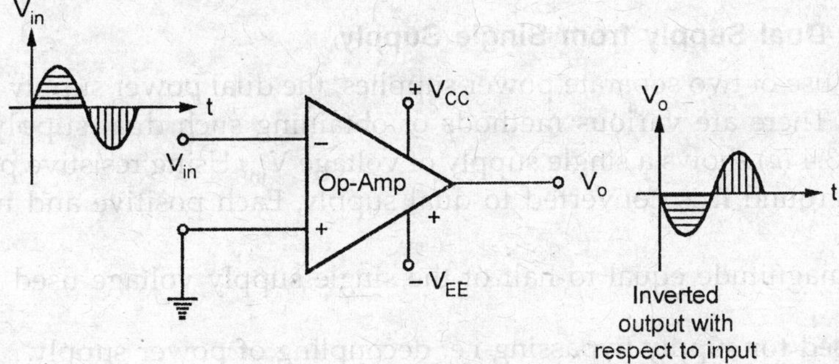

***Fig. 3.2 (a)** Input applied to inverting terminal.*

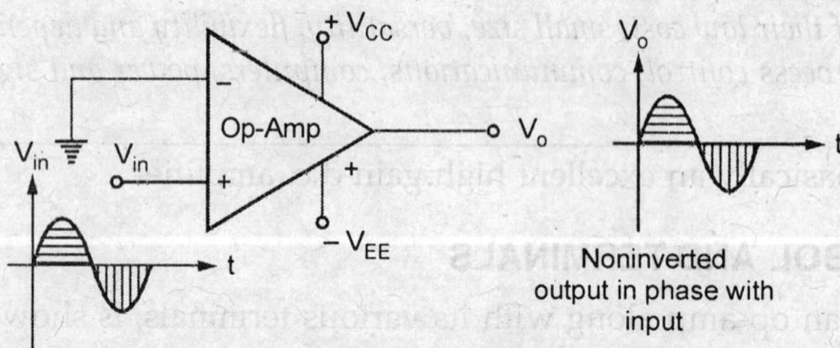

Fig. 3.2 (b) Input applied to noninverting terminal.

The op-amp is fabricated on a tiny silicon chip and packaged in a suitable case. Fine gauge wires are used to connect the chip to the external leads.

3.2.1 Power Supply

The op-amp works on a dual supply. A **dual supply** consists of two supply voltages both d.c., whose middle point is generally the ground terminal.

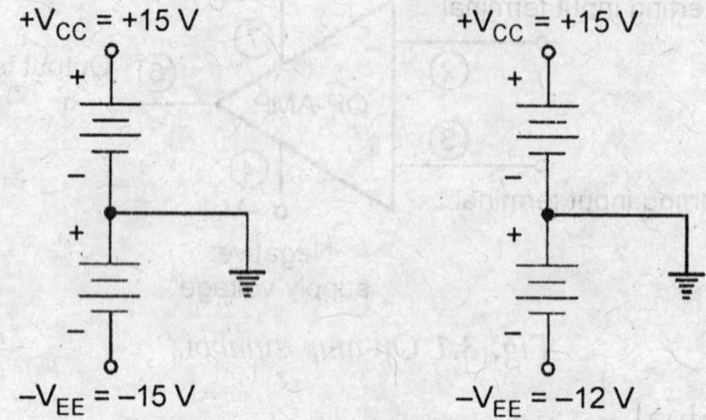

Fig. 3.3 Types of dual supply

The dual supply is generally balanced i.e. the voltages of the positive supply + V_{CC} and that of the negative supply – V_{EE} are same in magnitude. The typical commercially used power supply voltages are ± 15 V. But if the two voltage magnitudes are not same in a dual supply it is called as **unbalanced dual supply**. The balanced and unbalanced types of dual supply are shown in the Fig 3.3 (*a*) and (*b*) respectively.

Practically in most of the op-amp circuits balanced dual supply is used. The other popular balanced dual supply voltages are ± 9 V, ± 12 V, ± 22 V etc.

3.2.2 Obtaining Dual Supply from Single Supply

To avoid the use of two separate power supplies, the dual power supply is derived from a single supply. There are various methods of obtaining such dual supply from a single supply. The Fig. 3.4 (*a*) shows a single supply of voltage V_{in}. Using resistive potential divider network and a ground it is converted to dual supply. Each positive and negative supply obtained has a magnitude equal to half of the single supply voltage used i.e. $\left(\frac{V_{in}}{2}\right)$. The capacitors are used for supply bypassing i.e. decoupling of power supply.

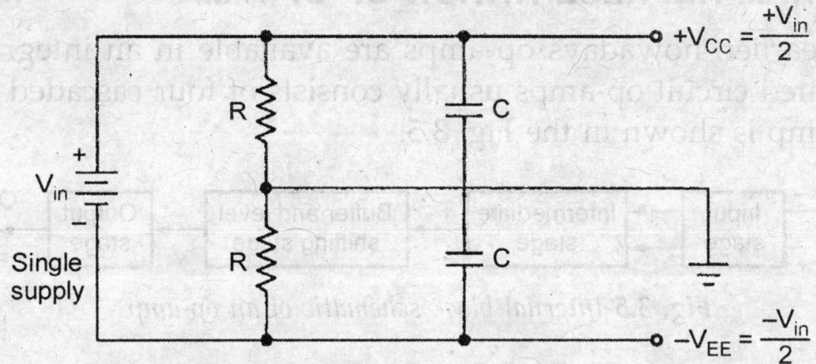

Fig. 3.4 (a) Obtaining dual supply from Single source.

The resistances 'R' should not draw high current from supply hence their values are more than 10 kΩ. The capacitors are in the range 0.01 to 10 μF.

If the voltages required are other than ($V_{in}/2$) then the zener diodes of appropriate voltage rating can be used as shown in the Fig. 3.4 (*b*).

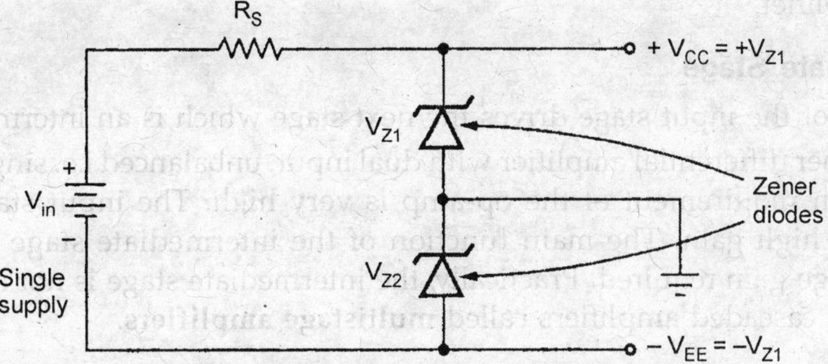

Fig. 3.4 (b) Use of zener diodes.

Many times practically due to mismatch in the devices, equal positive and negative voltages are not available. To adjust them a potentiometer can be used as shown in the Fig. 3.4 (*c*). To avoid the damage due to reversal of polarities connected to IC, the diodes D_1 and D_2 can be used. Such diodes provide reverse polarity protection and can be the part of any of the above discussed circuits.

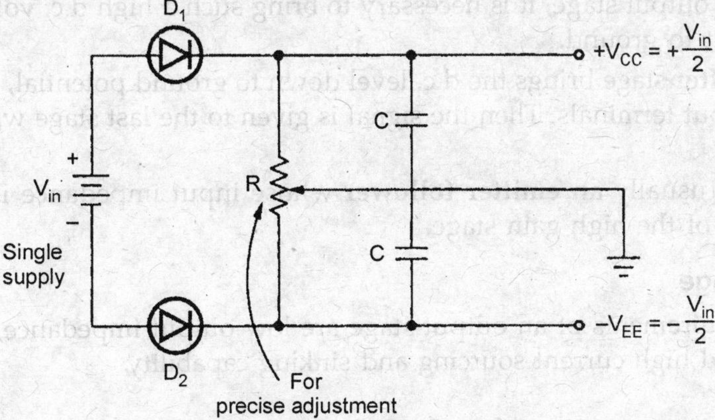

Fig. 3.4 (c) Use of potentiometer.

3.3 BLOCK DIAGRAM REPRESENTATION OF OP-AMP

As mentioned earlier, nowadays op-amps are available in an integrated circuit form. Commercial integrated circuit op-amps usually consists of four cascaded blocks. The block diagram of IC op-amp is shown in the Fig. 3.5.

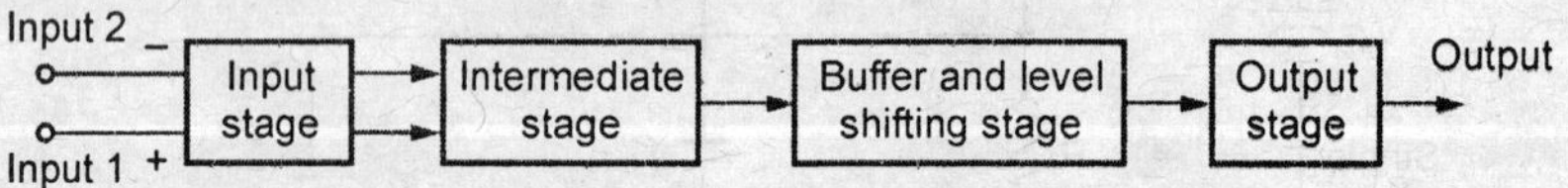

Fig. 3.5 Internal block schematic of an op-amp

3.3.1 Input Stage

The input stage requires high input impedance to avoid loading on the sources. It requires two input terminals. It also requires low output impedance. All such requirements are achieved by using the **dual input, balanced output differential amplifier** as the input stage. The function of a differential amplifier is to amplify the difference between the two input signals. The differential amplifier has high input impedance. This stage provides most of the voltage gain of the amplifier.

3.3.2 Intermediate Stage

The output of the input stage drives the next stage which is an intermediate stage.

This is another differential amplifier with dual input, unbalanced i.e. single ended output. The overall gain requirement of the op-amp is very high. The input stage alone cannot provide such a high gain. The main function of the intermediate stage is to provide an additional voltage gain required. Practically, the intermediate stage is not a single amplifier but the chain of cascaded amplifiers called **multistage amplifiers.**

3.3.3 Level Shifting Stage

All the stages are directly coupled to each other. As the op-amp amplifies d.c. signals also, the coupling capacitors are not used to cascade the stages. Hence the d.c. quiescent voltage level of previous stage gets applied as the input to the next stage. Hence stage by stage d.c. level increases well above ground potential. Such a high d.c. voltage level may drive the transistors into saturation. This further may cause distortion in the output due to clipping. This may limit the maximum a.c. output voltage swing without any distortion. Hence before the output stage, it is necessary to bring such a high d.c. voltage level to zero volts with respect to ground.

The **level shifter** stage brings the d.c. level down to ground potential, when no signal is applied at the input terminals. Then the signal is given to the last stage which is the output stage.

The buffer is usually an **emitter follower** whose input impedance is very high. This prevents loading of the high gain stage.

3.3.4 Output Stage

The basic requirements of an output stage are low output impedance, large a.c. output voltage swing and high current sourcing and sinking capability.

The push-pull complementary amplifier meets all these requirements and hence used as an output stage. This stage increases the output voltage swing and keeps the voltage swing symmetrical with respect to ground. The stage raises the current supplying capability of the op-amp.

In short, the overall block diagram can be shown as in the Fig. 3.6.

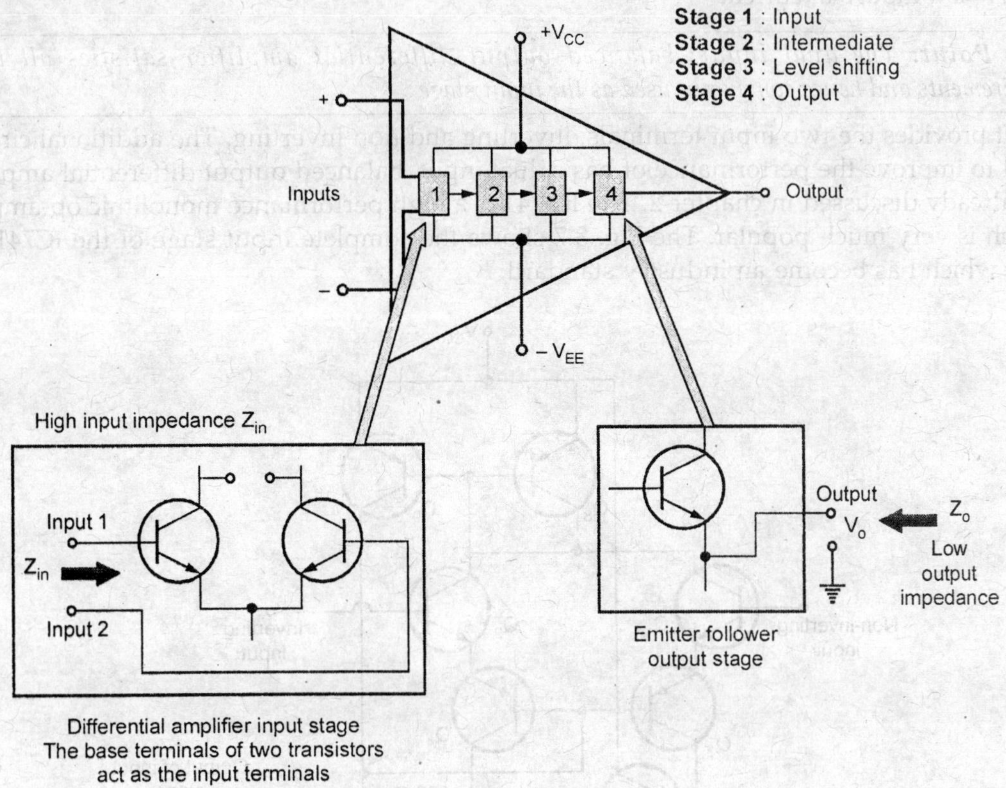

Fig. 3.6 Block diagram of an op-amp.

3.3.5 Currents and Impedances

Practically an input stage is a transistorised differential amplifier stage using the two transistors. The base terminals of these two transistors are the two input terminals of the op-amp. Some small d.c. base current is necessary to maintain the operating state of the transistors. These two base currents are the input bias currents of the op-amp. Typically the d.c. input current is of the order of 500 nA or less.

The input impedance Z_{in} is very high of the order of 1 MΩ or greater. To increase the input impedance field effect transistors (FETs) are used instead of bipolar junction transistors (BJTs). In such a case, gates of the two FETs are the input terminals of the op-amp.

The output stage is an emitter follower having very low output impedance. The output impedance Z_o is usually 75 Ω and the maximum output current is of the order of 25 mA.

3.4 INPUT STAGE OF OP-AMP

The basic requirements of the input stage of op-amp are:

1. high voltage gain
2. high input impedance
3. two input terminals
4. small input offset voltage
5. small input offset current
6. high CMRR
7. low input bias current

***Key Point:** The dual input, balanced output differential amplifier satisfies all these requirements and hence commonly used as the input stage.*

It provides the two input terminals, inverting and non-inverting. The additional circuits used to improve the performance of basic dual input, balanced output differential amplifier are already discussed in chapter 2. The IC 741 is a high performance monolithic op-amp IC, which is very much popular. The Fig. 3.7 shows the complete input stage of the IC741 op-amp which has become an industry standard.

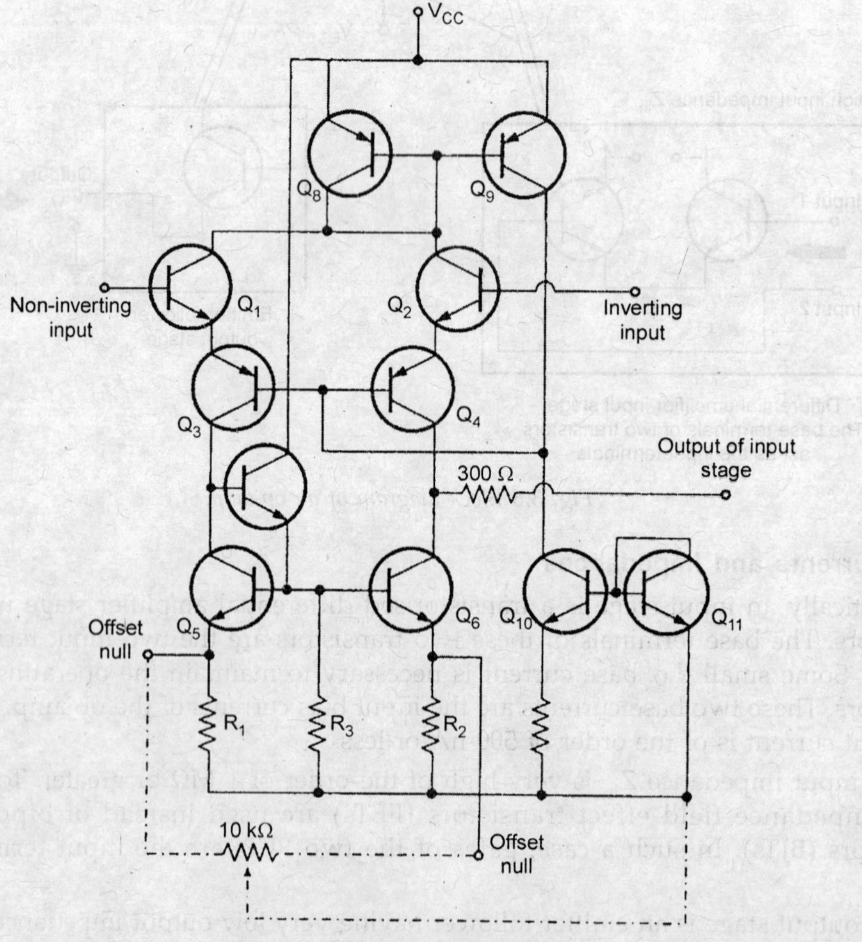

Fig. 3.7 *Input stage of IC 741 op-arnp.*

The input stage is a differential amplifier stage comprising of Q_1 and Q_2 transistors, between which the differential input signal is applied. The transistors Q_3 and Q_4 are also the part of the input differential amplifier used to increase the maximum signal input capacity.

The constant current source consisting of the transistors Q_9 and Q_{10} supplies constant base current for the transistors Q_3 and Q_4. This constant base current in the transistors Q_3 and Q_4 inturn establishes the operating current for the input differential stage. The diode connected transistors Q_8 and Q_{11} stabilize the base potentials of the transistors Q_9 and Q_{10} respectively.

The resistances R_1, R_2 and R_3 along with the transistors Q_5 and Q_6 form a controlled current source. The 10 kΩ potentiometer is connected between offset null terminals and by connecting the wiper to the negative supply. This potentiometer is used to control the emitter currents of the transistors Q_5 and Q_6. This arrangement helps in minimising offset voltage and currents. The transistor Q_7 supplies the base currents to the transistors Q_5 and Q_6. The base currents and hence the collector currents of the transistors Q_5 and Q_6 must be equal at all the times as the transistors Q_5 and Q_6 are identical. The collector voltage of the transistor Q_5 controls the operation of the transistor Q_7. This ensures that any change in Q_5 collector voltage, produces identical changes in the transistors Q_5 and Q_6 collector currents, via the transistor Q_7.

The transistors Q_4 and Q_6 form the complementary symmetry amplifier. The output of the differential stage is taken at the junction of the transistors Q_4 and Q_6. This output at the junction of Q_4 and Q_6 is proportional to the differential input signal.

3.5 INTERMEDIATE STAGE

The output of the input stage drives the next stage which is an intermediate stage. This is another differential amplifier with dual input, unbalanced i.e. single ended output. The overall gain requirement of the op-amp is very high. The input stage alone cannot provide such a high gain. The main function of the intermediate stage is to provide an additional voltage gain required. Practically, the intermediate stage is not a single amplifier but the chain of cascaded amplifiers called as multistage amplifiers.

Key Point: *No coupling capacitor is used to couple the two differential amplifiers.*

The Fig. 3.8 shows the typical input and intermediate stage differential amplifier. This is cascading of differential amplifiers and also called **directly coupled differential amplifiers.**

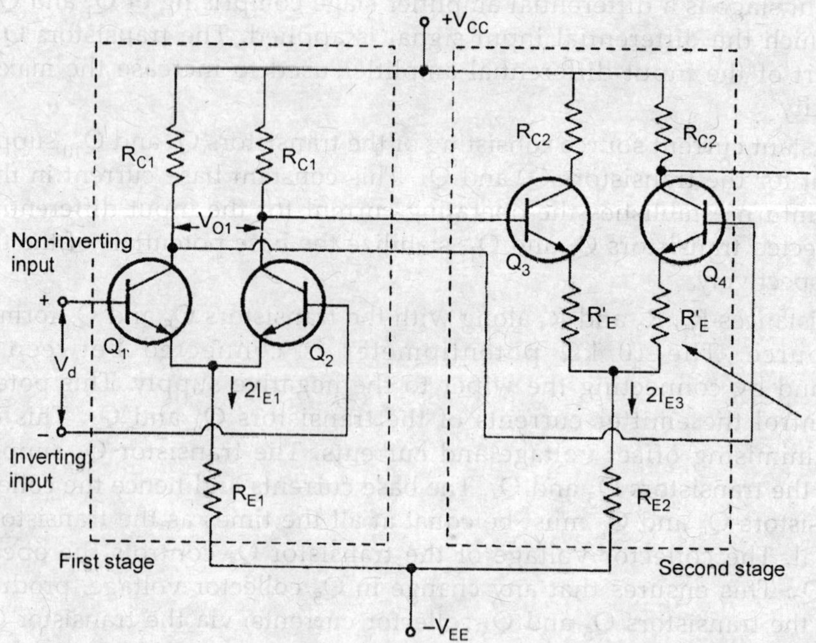

Fig. 3.8 Input and intermediate stage of an Op-arnp.

3.6 LEVEL SHIFT STAGE IN OP-AMP

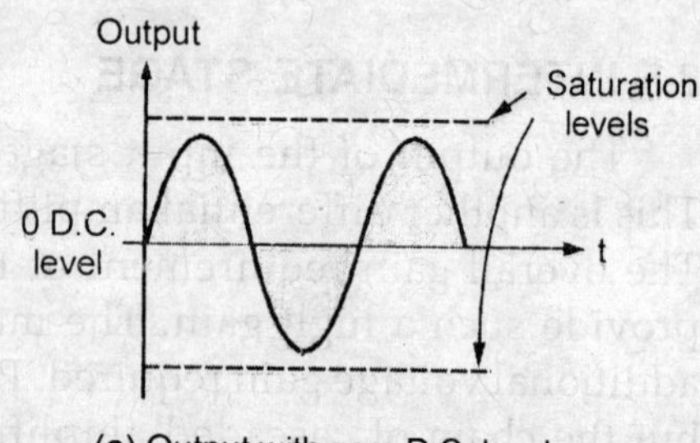

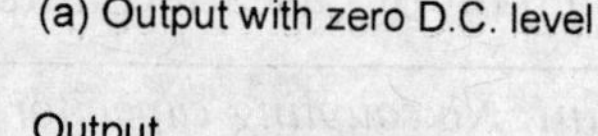

As coupling capacitors are not used to couple the amplifiers in the intermediate stage, the d.c. biasing voltage level propagates through the amplifier chain. This finally appears as a significant d.c. component at the output alongwith a.c. output.

Following are the effects due to such d.c. component at the output,

1. The output gets distorted.
2. It limits the maximum output voltage swing.

This is shown in Fig. 3.9.

The main purpose of the level shifting stage is to shift the output quiescent d.c. level towards the ground, with minimum change in the a.c. signal. This also satisfies the requirement of op-amp that its output should have quiescent voltage level of a v for zero input signal.

Let us study the different level shifting networks used in op-amp.

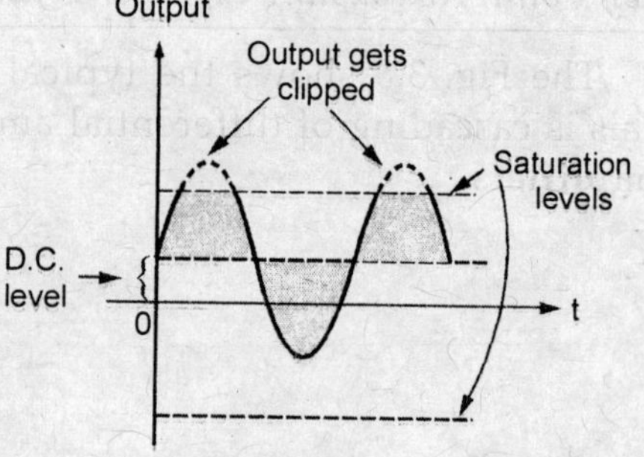

(b) Distorted output due to additional D.C. level

Fig. 3.9

Circuit 1:

The simplest type of level shifting network is shown in the Fig. 3.10.

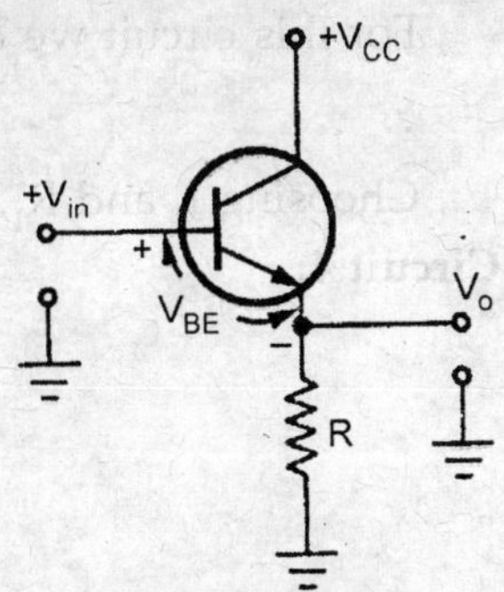

Fig. 3.10 Simplest level shifting network.

It is basically an emitter follower circuit. In this circuit as observed in the Fig. 3.10, the amount of shift obtained is equal to V_{BE} which is almost 0.7 V. So if V_{in} is the increased level signal and V_o is output with reduced level, we can write for the above circuit as,

$$V_o = V_{in} - V_{BE} \quad ...(1)$$

Key Point: *The negative sign indicates downward shift in the level.*

Circuit 2:

But generally a shift of 0.7 V is not sufficient. Hence the circuit is modified with the help of two resistances R_1 and R_2 as shown in the Fig. 3.11.

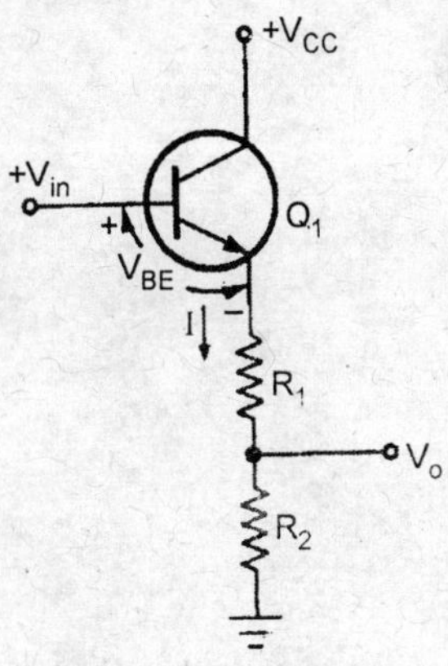

Fig. 3.11 Typical level shifting network.

This is a common collector stage which also acts as a **buffer** to isolate high gain stages from the output stage.

To find the level shift, consider current I flowing through the emitter of the transistor Q_1.

Applying KVL to base emitter loop.

$$V_{in} - V_{BE} - I(R_1 + R_2) = 0$$

$$I = \frac{(V_{in} - V_{BE})}{R_1 + R_2} \quad ...(2)$$

Now $\quad V_o = IR_2$

$$\therefore \quad V_o = \frac{(V_{in} - V_{BE})}{R_1 + R_2}$$

Now $\quad V_o = IR_2$

$$\therefore \quad V_o = \frac{(V_{in} - V_{BE})R_2}{R_1 + R_2} \quad ...(3)$$

By proper selection of R_1 and R_2, level at V_o can be controlled. The main drawback of this circuit is that the signal voltage also gets attenuated by the factor $R_2/(R_1 + R_2)$. As R_2 is decreased to improve the d.c. level shift, the a.c. gain starts decreasing. Another disadvantage of the circuit is that its output impedance is also high.

Circuit 3:

Another improved circuit with constant current source also can be used as a level shifting network. It is shown in the Fig. 3.12. The constant current bias is provided with the help of the transistor Q_2.

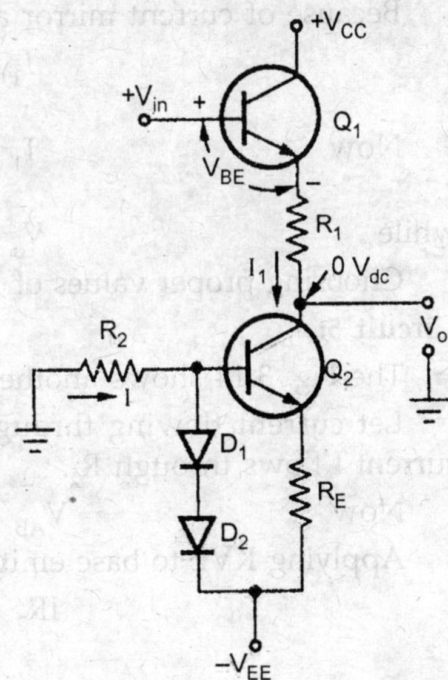

Fig. 3.12 Level shifting with constant current bias.

For this circuit we can write,

$$\therefore \qquad V_o = V_{in} - V_{BE} - I_1R_1 \qquad \dots (4)$$

Choosing I_1 and R_1, V_o can be obtained as almost 0 V.

Circuit 4:

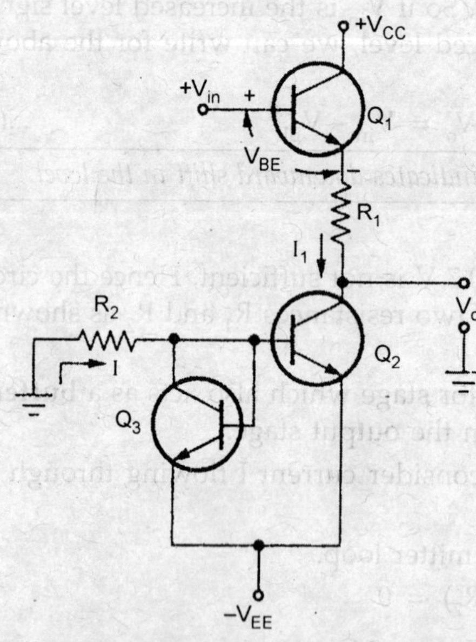

Fig. 3.13 Level shifting network with current mirror.

Another level shifting network circuit uses a current mirror circuit with basic emitter follower circuit. It is shown in the Fig. 3.13.

Because of current mirror action,

$$I_1 = I \qquad \dots(5)$$

Now
$$I_1 = I = \frac{V_{EE} - V_{BE}}{R_2} \qquad \dots(6)$$

while
$$V_o = V_{in} - V_{BE} - I_1R_1 \qquad \dots(7)$$

Choosing proper values of I and R_1, the output V_o can be brought to 0 V level.

Circuit 5:

The Fig. 3.14 shows another level shifting network with **V_{BE} multiplier circuit.**

Let current flowing through emitter of Q_1 is I. Neglecting base current of Q_2, same current I flows through R_2.

Now
$$V_{AB} = I(R_1 + R_2) \qquad \dots (8)$$

Applying KVL to base emitter loop of Q_2,

$$IR_2 = V_{BE}$$

$$\therefore \qquad I = \frac{V_{BE}}{R_2} \qquad \dots (9)$$

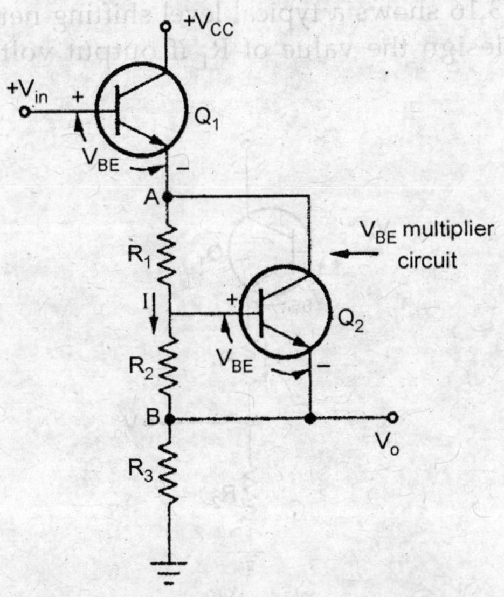

Fig. 3.14 Level shifting with V_{BE} multiplier

$$\therefore \quad V_{AB} = V_{BE}\left[1+\frac{R_1}{R_2}\right] \quad \text{... (10)}$$

Key Point: *Thus the circuit acts as a V_{BE} multiplier circuit.*

While

$$V_o = V_{in} - V_{BE} - V_{AB}$$

$$\therefore \quad V_o = V_{in} - \left[V_{BE} + V_{BE}\left(1+\frac{R_1}{R_2}\right)\right]$$

i.e.

$$V_o = V_{in} - V_{BE}\left[2+\frac{R_1}{R_2}\right] \quad \text{...(11)}$$

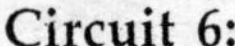

Circuit 6:

A lateral pnp and vertical npn transistor combination can be used in level shifting Fig. 3.15 pnp-npn level shifting network as shown in the Fig. 3.15.

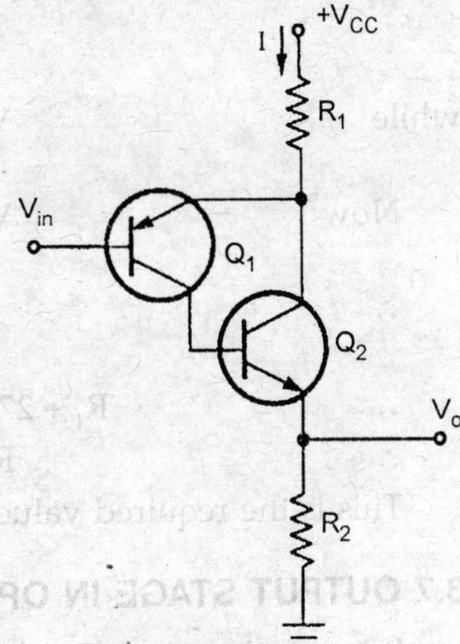

Fig. 3.15 pnp-npn level shifting network

The output voltage V_o for such a circuit is given by the expression

$$V_o = \frac{R_2}{R_1}(V_{CC} - V_{BE} - V_{in}) \quad \text{... (12)}$$

where V_{BE} is the base emitter voltage of the $Q_1 - Q_2$ connection.

Example 3.1: The Fig. 3.16 shows a typical level shifting network. If input d.c. level is 6.84 V and R_2 is 270 Ω, design the value of R_1 if output voltage level required is zero volts.

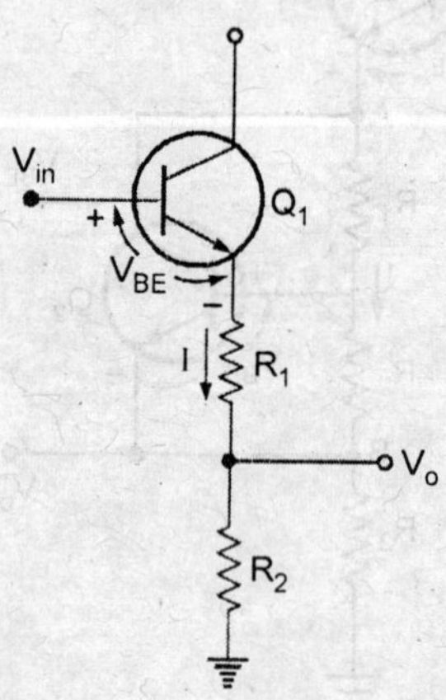

Fig. 3.16

Solution: Let current through emitter branch is I. Applying KVL to base emitter loop,

$$V_{in} - V_{BE} - I(R_1 + R_2) = 0$$

$$\therefore \quad I = \frac{V_{in} - V_{BE}}{R_1 + R_2}$$

while

$$V_o = IR_2 = \frac{(V_{in} - V_{BE})R_2}{R_1 + R_2}$$

Now $V_o = 0$, $V_{in} = 6.84$ V, $V_{BE} = 0.7$ V, $R_2 = 270\ \Omega$.

$$\therefore \quad 0 = \frac{(6.84 - 0.7)\,270}{270 + R_1}$$

$$\therefore \quad R_1 + 270 = 1657.8$$

$$\therefore \quad R_1 = \mathbf{1.387\ k\Omega}$$

This is the required value of R_1 to get 0 V output level.

3.7 OUTPUT STAGE IN OP-AMP

It is mentioned earlier that the output stage of op-amp supplies the load and provides low output impedance. The requirements of good output stage are:

1. large output voltage swing capability
2. large output current swing capability
3. low output impedance
4. low quiescent power dissipation
5. short circuit protection

A push-pull amplifier in class AB or class B operation satisfies all the above requirements and hence commonly used in the output stage of an op-amp.

The Fig. 3.17 shows class B push-pull amplifier which is basically emitter-follower with the complementary transistors.

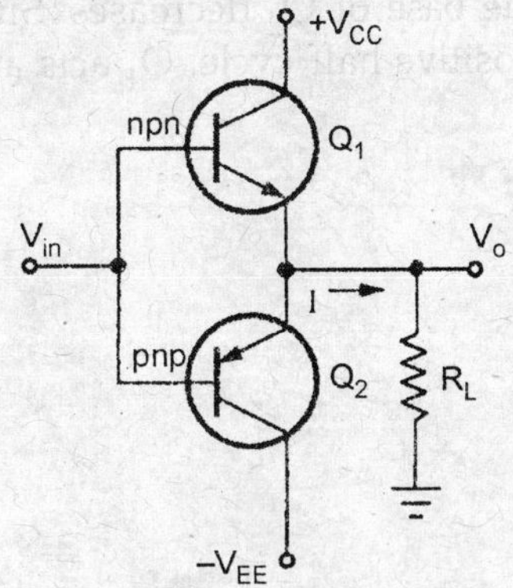

***Fig. 3.17** Output stage of an op-amp.*

The circuit supplies a load of resistance R_L. When V_{in} is positive, the transistor Q_1 supplies the load which becomes ON. While when V_{in} is negative then the transistor Q_1 is cut off and Q_2 acts as sink to remove the current from the load.

One major limitation of this circuit is that the output voltage remains zero as long as V_{in} is less than V_{BE}. This introduces the cross-over distortion in the output. This is shown in the voltage transfer characteristics of the output stage in the Fig. 3.18.

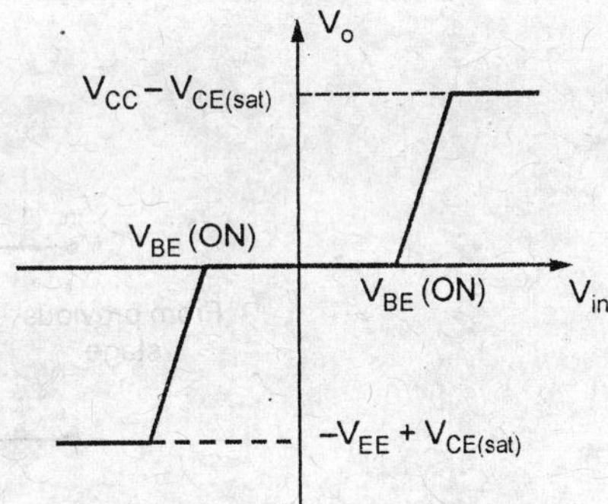

***Fig. 3.18** Cross over distortion in the output.*

Such a cross over distortion can be avoided by applying a bias voltage greater than 2 V_{BE}, between the two bases. This ensures a flow of small current in the transistors even in the quiescent state.

Such a voltage is supplied by employing a pair of series connected diodes as shown in the Fig. 3.19.

The voltage supplied due to D_1, D_2 at the bases of Q_1, Q_2 alongwith the current source of I_{bias} eliminates the cross over distortion completely.

To stabilize the quiescent base current, the small resistance R_E is used between the emitter of Q_1, Q_2 and the output. This resistance provides limited short circuit protection too.

The circuit discussed above is suitable to deliver the output power of the order of several hundred milliwatts or less. But if large amount of power is to be delivered then the class B or AB circuits are not suitable. This is because of the fact that the pnp transistors used in the circuits, have a limited current carrying capability.

The circuit which uses all npn transistors in the output stage is shown in the Fig. 3.20.

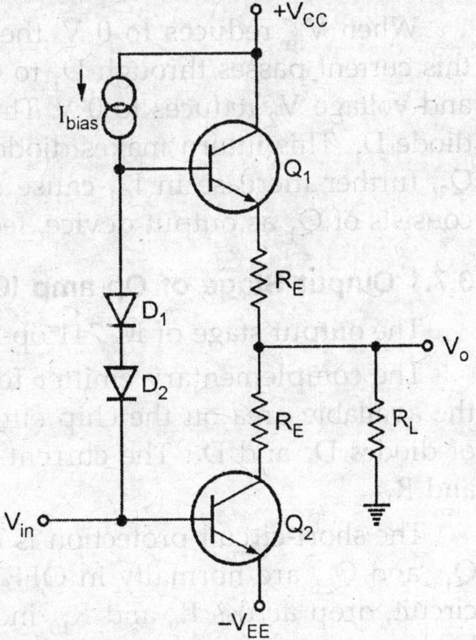

***Fig. 3.19** Output stage with diodes D_1, D_2.*

When V_{in} is positive, the transistor Q_1 gets turn ON to produce finite I_{C1}. As base of Q_2 is more positive than its emitter, the diode D_1 will turn ON in preference to D_2. The current I_{C1} will flow through D_1 and will be drawn from Q_3, which is assumed saturated. As V_{in} becomes more positive, I_{C1} increases and voltage V_2 at the base of Q_2 decreases. Since Q_2 is emitter-follower, V_o also decreases with V_2. Thus for positive half cycle, Q_1 acts as driver and Q_2 as output device.

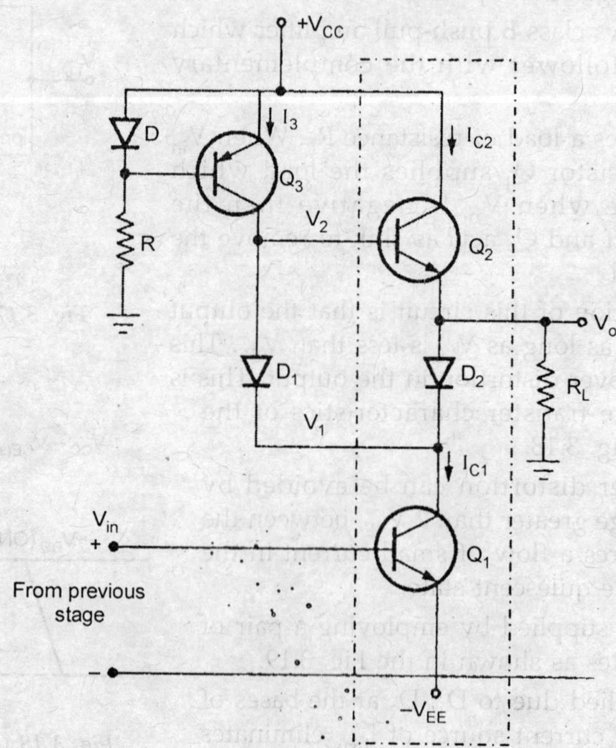

Fig. 3.20 Output stage with npn transistors.

When V_{in} reduces to 0 V, then current I_{C2} is also zero. At this stage $I_{C1} = I_3$ and all this current passes through D_1 to Q_1. If I_{C1} is increased further, V_o remains constant at 0 V and voltage V_2 reduces to 0 V. Thus V_1 is negative by an amount equal to the drop across diode D_1. This inturn makes diode D_2 ON. Since current in D_1 is decided by the transistor Q_3, further increase in I_{C1} cause current to flow through D_2 and the negative half cycle consists of Q_1 as output device, feeding R_L through D_2.

3.7.1 Output Stage of Op-amp IC 741

The output stage of IC 741 op-amp is shown in the Fig. 3.21.

The complementary emitter follower is formed by the transistors Q_{14} and Q_{20}. To use the available area on the chip effectively, the Darlington pair Q_{18} and Q_{19} is used instead of diodes D_1 and D_2. The current limiting at the output is provided by the resistances R_9 and R_{10}.

The short-circuit protection is necessary for the transistors Q_{14} and Q_{20}. The transistors Q_{15} and Q_{21} are normally in OFF condition. When output current increases due to short circuit, drop across R_9 and R_{10} increases. This causes Q_{15} or Q_{21} to turn ON to divert the base current available to Q_{14} or Q_{20}. Hence output current gets limited to V_{BE}/R_9 or

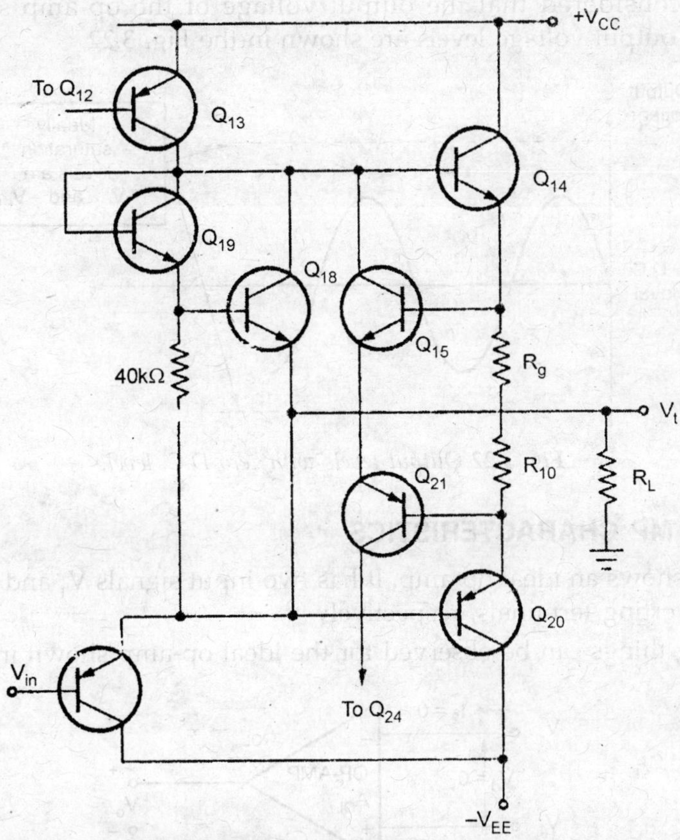

***Fig. 3.21** The output stage of IC 741 op-amp*

V_{BE}/R_{10}. Such a protection is called active current limiting. Using resistances in emitter circuit is passive current limiting which needs large value resistances. This inturn limits the output voltage swing. Hence active current limiting with the help of transistors is preferred which has no effect on the output voltage swing under short circuit condition.

3.8 VOLTAGE LEVELS AND SATURATING PROPERTY OF OP-AMP

The supply voltages of the op-amp are $+ V_{CC}$ and $- V_{EE}$. These supply voltage levels decide the maximum output voltage levels of the op-amp.

***Key Point:** Practically the op-amp output saturates at the voltages slightly less than the supply voltages $+ V_{CC}$ and $- V_{EE}$.*

Thus the output voltage of the op-amp can be driven to within 1 V of $+ V_{CC}$ and $- V_{EE}$ before the output saturation takes place. This is called **saturating property** of the op-amp.

Practically maximum output swing of the op-amp is 0.9 times the supply voltages. Thus the op-amp using ± 15 V supply has maximum output voltage swing as 0.9 × (± 15 V) i.e. ± 13.5 V.

Ideally it is considered that the output voltage of the op-amp saturates at $+ V_{CC}$ $- V_{EE}$ levels. The output voltage levels are shown in the Fig. 3.22.

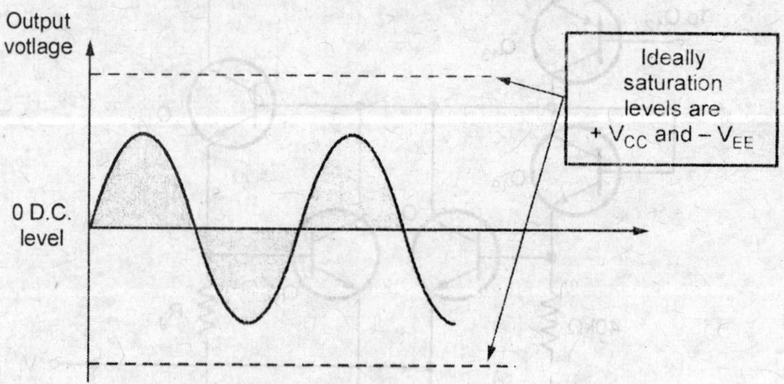

Fig. 3.22 Output levels with zero D.C. level.

3.9 IDEAL OP-AMP CHARACTERISTICS

The Fig. 3.23 shows an ideal op-amp. It has two input signals V_1 and V_2 applied to non-inverting and inverting terminals, respectively.

The following things can be observed for the ideal op-amp shown in the Fig. 3.23.

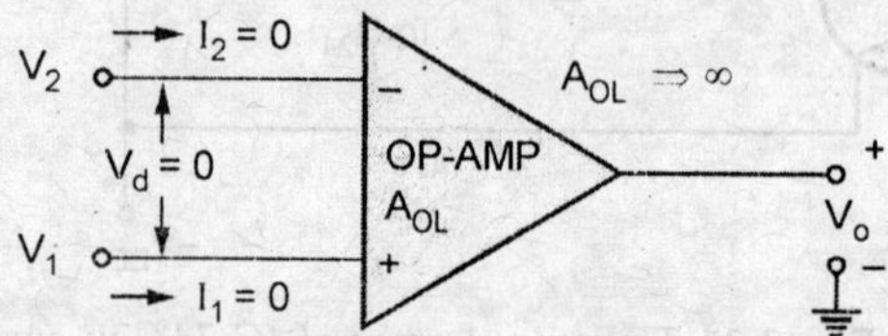

Fig. 3.23 Ideal op-arnp.

1. An ideal op-amp draws no current at both the input terminals i.e. $I_1 = I_2 = 0$. Thus its input impedance is infinite. Any source can drive it and there is no loading on the driver stage.
2. The gain of an ideal op-amp is infinite (∞), hence the differential input $V_d = V_1 - V_2$ is essentially zero for the finite output voltage V_o.
3. The output voltage V_o is independent of the current drawn from the output terminals. Thus its output impedance is zero and hence output can drive an infinite number of other circuits.

These properties are expressed generally as the characteristics of an ideal op-amp. The various characteristics of an ideal op-amp are:

(*a*) Infinite voltage gain: ($A_{OL} = \infty$)

It is denoted as A_{OL}. It is the differential open loop gain and is infinite for an ideal op-amp.

(*b*) Infinite input impedance: (R_{in} = ∞)

The input impedance is denoted as R_{in} and is infinite for an ideal op-amp. This ensures that no current can flow into an ideal op-amp.

(*c*) Zero output impedance: (R_o = 0)

The output impedance is denoted as R_o and is zero for an ideal op-amp. This ensures that the output voltage of the op-amp remains same, irrespective of the value of the load resistance connected.

(*d*) Zero offset voltage: (V_{ios} = 0)

The presence of the small output voltage though $V_1 = V_2 = 0$ is called an offset voltage. It is zero for an ideal op-amp. This ensures zero output for zero input signal voltage.

(*e*) Infinite bandwidth

The range of frequency over which the amplifier performance is satisfactory is called its bandwidth. The bandwidth of an ideal op-amp is infinite. This means the operating frequency range is from a to ∞. This ensures that the gain of the op-amp will be constant over the frequency range from d.c. (zero frequency) to infinite frequency. So op-amp can amplify d.c. as well as a.c. signals.

(*f*) Infinite CMRR: (ρ = ∞)

The ratio of differential gain and common mode gain is defined as CMRR. Thus infinite CMRR of an ideal op-amp ensures zero common mode gain. Due to this common mode noise output voltage is zero for an ideal op-amp.

(*g*) Infinite slew rate: (S = ∞)

This ensures that the changes in the output voltage occur simultaneously with the changes in the input voltage.

The slew rate is important parameter of op-amp. When the input voltage applied is step type which changes instantaneously then the output also must change rapidly as input changes. If output does not change with the same rate as input then there occurs distortion in the output. Such a distortion is not desirable. Infinite slew rate indicates that output changes simultaneously with the changes in the input voltage.

The parameter slew rate is actually defined as the maximum rate of change of output voltage with time and expressed in V/μs.

$$\text{Slew rate} = S = \left.\frac{dV}{dt}\right|_{\text{maximum}}$$

Its ideal value is infinite for the op-amp.

(*h*) No effect of temperature:

The characteristics of op-amp do not change with temperature.

(*i*) Power Supply Rejection Ratio: (PSRR = 0)

The power supply rejection ratio is defined as the ratio of the change in input offset voltage due to the change in supply voltage producing it, keeping other power supply voltage constant. It is also called power supply sensitivity.

So if V_{EE} is constant and due to change in V_{CC}, there is change in input offset voltage then PSRR is expressed as,

$$\text{PSRR} = \left.\frac{\Delta V_{ios}}{\Delta V_{CC}}\right|_{V_{EE}\text{ Constant}}$$

For a fixed V_{CC}, if there is change in V_{EE} causing change in input offset voltage then,

$$\text{PSRR} = \left.\frac{\Delta V_{ios}}{\Delta V_{EE}}\right|_{V_{EE}\text{ Constant}}$$

It is expressed in mV/V or μV/V and its ideal value is zero.

These ideal characteristics of op-amp are summarized in the Table 3.1.

Table 3.1 Ideal op-amp characteristics

Characteristics	Symbol	Values
Open loop voltage gain	A_{OL}	∞
Input impedance	R_{in}	∞
Output impedance	R_o	0
Offset voltage	V_{oo}	0
Bandwidth	B.W.	∞
C.M.R.R	ρ	∞
Slew rate	S	∞
Power supply rejection ratio	PSRR	0

3.10 VOLTAGE TRANSFER CURVE OF OP-AMP

The graph of output voltage V_o plotted against the differential input voltage V_d; assuming gain constant is called voltage transfer curve or characteristics of op-amp.

3.10.1 Ideal Voltage Transfer Curve

Ideally open loop gain of op-amp is ∞.

$$\therefore \quad A_{OL} = \frac{V_o}{V_d} = \infty$$

$$\therefore \quad V_d = \frac{V_o}{\infty} = 0$$

$$V_o = A_{OL} V_d$$

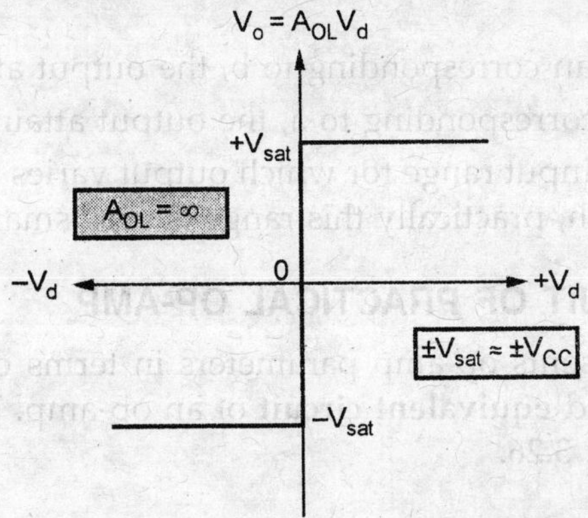

Fig. 3.24 *Ideal voltage transfer curve.*

Thus for zero input, the output of op-amp is always at saturation level $\pm$ V_{sat}, due to infinite gain. Thus voltage transfer curve for ideal op-amp is a vertical line as shown in the Fig. 3.24.

Thus ideally range of input for linear operation of the op-amp is zero.

3.10.2 Practical Voltage Transfer Curve

Practically A_{OL} is finite for the op-amp. For op-amp 741C, it is 2×10^5.

$$\therefore \quad V_o = A_{OL} V_d$$

$$\therefore \quad \pm V_{sat} = 2 \times 10^5 V_d$$

The saturation voltages are almost $\pm$ 15 V.

$$\therefore \quad V_o = \frac{\pm 15}{2 \times 10^5} = \pm 75 \, \mu V$$

Hence practically till V_d is between – 75 μV and + 75 μV, the output will vary linearily with input. But once V_d exceeds $\pm$ 75 μV, the output is saturated.

Thus the practical voltage transfer curve is as shown in the Fig. 3.25.

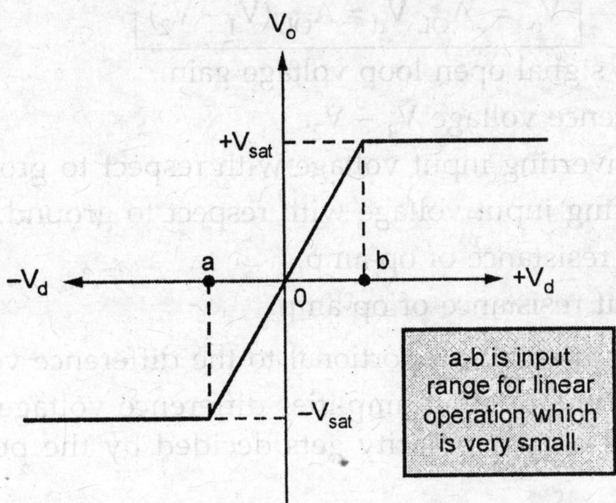

Fig. 3.25 *Practical voltage transfer curve.*

Thus,

(*i*) If V_d is greater than corresponding to b, the output attains + V_{sat}.

(*ii*) If V_d is less than corresponding to a, the output attains – V_{sat}.

(*iii*) Thus range a-b is input range for which output varies linearily with the input. But as A_{OL} is very high, practically this range is very small.

3.11 EQUIVALENT CIRCUIT OF PRACTICAL OP-AMP

The circuit which represents op-amp parameters in terms of physical components, for the analysis purpose is called **equivalent circuit** of an op-amp. The equivalent circuit of an op-amp is shown in the Fig. 3.26.

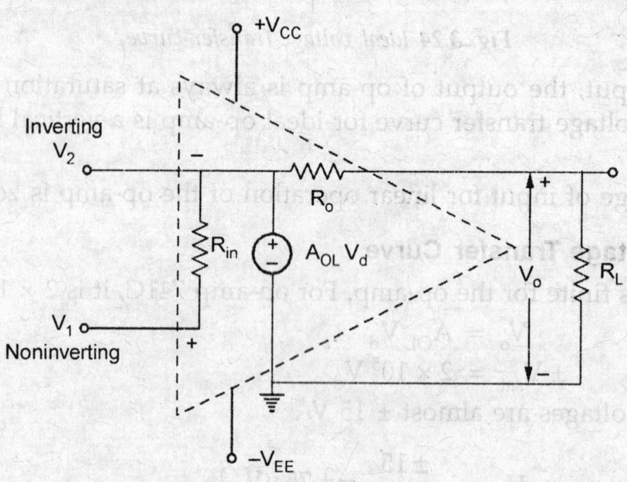

Fig. 3.26 Equivalent circuit of an op-arnp.

The circuit shows the op-amp parameters like input resistance, output resistance, the open loop voltage gain in terms of circuit components like R_{in}, R_o etc. The op-amp amplifies the difference between the two input voltages.

$$V_o = A_{OL} V_d = A_{OL} (V_1 - V_2)$$

where A_{OL} = Large signal open loop voltage gain.

V_d = Difference voltage $V_1 - V_2$.

V_1 = Noninverting input voltage with respect to ground.

V_2 = Inverting input voltage with respect to ground.

R_i = Input resistance of op-amp.

R_o = Output resistance of op-amp.

The output voltage is directly proportional to the difference voltage V_d.

It is to be noted that the op-amp amplifies difference voltage and not the individual input voltages. Thus the output polarity gets decided by the polarity of the difference voltage V_d.

The voltage source $A_{OL}V_d$ is the Thevenin's equivalent voltage source while R_o is the Thevenin's equivalent resistance looking back into the output terminals.

The equivalent circuit plays an important role in analysing various op-amp applications as well as in studying the effects of feedback on the performance of op-amp.

3.11.1 Practical Op-amp Characteristics

The characteristics of an ideal op-amp can be approximated closely enough, for many practical op-amps. But basically the practical op-amp characteristics are little bit different than the ideal op-amp characteristics.

The various characteristics of a practical op-amp can be described as below.

(*a*) Open loop gain: It is the voltage gain of the op-amp when no feedback is applied. Practically it is several thousands.

(*b*) Input impedance: It is finite and typically greater than 1 MΩ. But using FETs for the input stage, it can be increased upto several hundred MΩ.

(*c*) Output impedance: It is typically few hundred ohms. With the help of negative feedback, it can be reduced to a very small value like 1 or 2 ohms.

(*d*) Bandwidth: The bandwidth of practical op-amp in open loop configuration is very small. By application of negative feedback, it can be increased to a desired value.

(*e*) Input offset voltage: Whenever both the input terminals of the op-amp are grounded, ideally, the output voltage should be zero. However, in this condition, the practical op-amp shows a small non zero output voltage. To make this output voltage zero, a small voltage in millivolts is required to be applied to one of the input terminals. Such a voltage makes the output exactly zero. This d.c. voltage, which makes the output voltage zero, when the other terminal is grounded is called input offset voltage denoted as V_{ios}. How much voltage, to which terminal and with what polarity, to be applied, is specified by the manufacturer in the datasheet. The input offset voltage depends on the temperature. The concept of input offset voltage is shown in the Fig. 3.27.

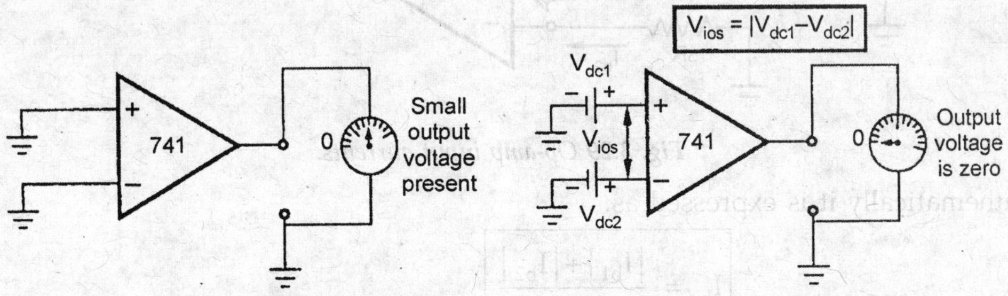

Fig. 3.27 Concept of input offset voltage.

The V_{ios} can be positive or negative hence absolute value of the V_{ios} is mentioned in the data sheet.

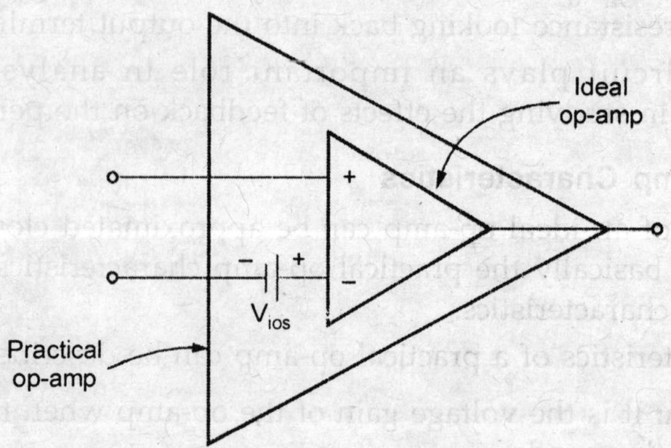

Fig. 3.28 Practical op-amp model with V_{ios}.

The smaller the value of V_{ios}, better is the matching of the input terminals.

The input offset voltage depends on the temperature.

Many time voltage to one of the input terminals is applied with the proper polarity so as to null the output, keeping other input terminal grounded. For ideal op-amp, V_{ios} is zero, hence practical op-amp model is generally shown as in the Fig. 3.28 with the indication of the input offset voltage. For op-amp 741 C the input offset voltage is 6 mV.

***(f)* Input bias Current:** The average value of the two currents flowing into the op-amp input terminals is called input bias current and denoted as I_b. It is shown in the Fig. 3.29.

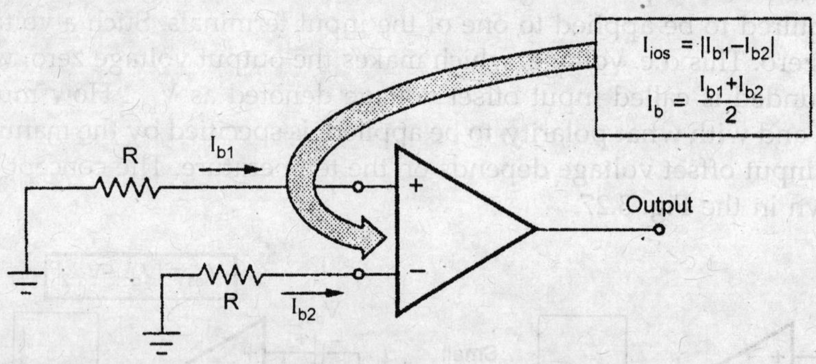

Fig. 3.29 Op-amp input currents.

Mathematically it is expressed as,

$$I_b = \frac{|I_{b1}| + |I_{b2}|}{2}$$

Ideally it should be zero while for op-amp 741C, maximum value of I_b is 500 nA.

***Key Point:** Both I_{ios} and I_b are temperature dependent.*

***(g)* Input offset Current:** It is seen that the input stage of the op-amp is the dual input differential amplifier and the input terminals are the base terminals of the two transistors as shown in the Fig. 3.30. Hence the input currents of op-amp are the base currents of the two transistors Q_1 and Q_2 used in the input stage. Ideally, Q_1 and Q_2 must be perfectly matched and two base currents must be equal. But practically the two input base currents differ by small amount.

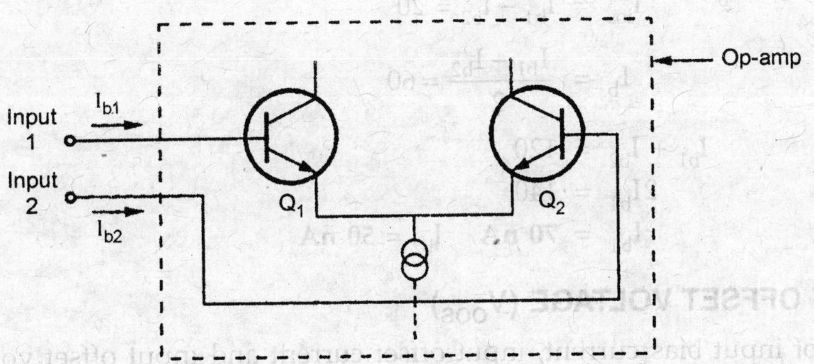

Fig. 3.30 Concept of op-amp input currents.

The algebraic difference between the currents flowing into the two input terminals of the op-amp is called **input offset current** and denoted as I_{ios}. Mathematically it is expressed as,

Where I_{b1} = Current entering into noninverting input terminal

and I_{b2} = Current entering into inverting input terminal

Ideally I_{ios} is zero while for op-amp 741C, maximum value of I_{ios} is 200 nA.

This current is responsible to produce the output though input terminals are grounded.

Example 3.2: If the base currents for the emitter coupled transistors of a differential amplifier are 18 μA and 22 μA, determine

(*i*) Input bias current (*ii*) Input offset current for an op-amp.

Solution: The two input base currents are

$$I_{b1} = 18\ \mu A \quad \text{and} \quad I_{b2} = 22\ \mu A$$

(*i*) The input bias current is

$$I_b = \frac{I_{b1} + I_{b2}}{2}$$

$$= \frac{18 + 22}{2}$$

$$= \mathbf{20\ \mu A}$$

(*ii*) The input offset current is

$$I_{ios} = |I_{b1} - I_{b2}|$$

$$= |18 - 22|$$

$$= \mathbf{4\ \mu A}$$

Example 3.3: For a particular op-amp, the input offset current is 20 nA while input bias current is 60 nA. Calculate the values of two input bias currents.

Solution:

$$I_{ios} = 20 \text{ nA}, \quad I_b = 60 \text{ nA}$$

Now $I_{ios} = I_{b1} - I_{b2} = 20$

$$I_b = \frac{I_{b1} + I_{b2}}{2} = 60$$

$\therefore \quad I_{b1} + I_{b2} = 120$

$\therefore \quad 2I_{b1} = 140$

$\therefore \quad I_{b1} = \mathbf{70\ nA}, \quad I_{b2} = \mathbf{50\ nA}$

3.12 OUTPUT OFFSET VOLTAGE (V_{OOS})

The effect of input bias current, input offset current and input offset voltage is to add the error to the expected d.c. output voltage. These parameters produce the output voltage when both the input terminals are grounded.

Thus the d.c. voltage present at the output terminals when both the input terminals are grounded is called output offset voltage denoted as V_{oos}.

The output voltage caused by input offset voltage V_{ios} can be either positive or negative with respect to ground. Similarly, the output voltage due to the input bias current I_{b2} can be also positive or negative with respect to ground. If these output voltages are of different polarity then the total output offset voltage is very small. But if these voltages are of same polarity then output offset voltage is large. Thus the output offset voltage is given by,

$$\boxed{V_{oos} = V_o \text{ due to } V_{ios} + V_o \text{ due to } I_b}$$

To compensate the output offset voltage due to the input bias current a compensating resistance is used as shown in the Fig. 3.31. It is denoted as R_{comp}. This is called **offset minimizing resistor.**

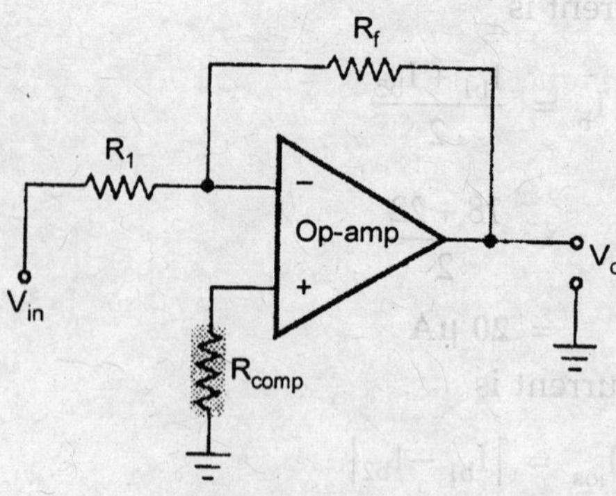

Fig. 3.31 Use of R_{comp}.

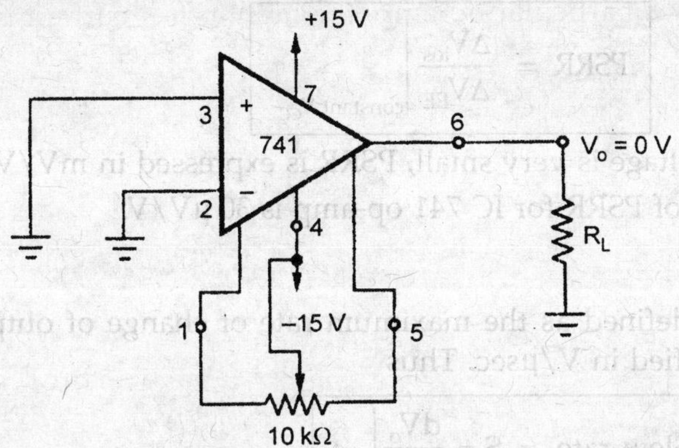

Fig. 3.32 Offset voltage compensation for IC 741 op-amp.

While the effect of offset voltage is nullified using voltage compensating network.

For many op-amps, the offset compensation pins are provided to nullify the offset voltage. The manufacturer's specifications must be referred to use these compensation pins. For example, in case of IC 741 op-amp, the pins 1 and 5 are provided as compensation pins. The manufacturer recommends that a 10 kΩ potentiometer must be connected between the pins 1 and 5. A wiper must be connected to the negative supply pin which is pin number 4. By adjusting this potentiometer wiper, the output can be nullified.

The arrangement for IC 741 op-amp is shown in the Fig. 3.32.

If the op-amp does not have the offset compensation pins, then the separate external compensating network is required to be designed.

3.13 OTHER IMPORTANT OP-AMP PARAMETERS

Two other important op-amp parameters are the slew rate and the power supply rejection ratio (PSRR). Let us discuss and obtain the expressions for these two parameters.

3.13.1 Power Supply Rejection Ratio

The power supply rejection ratio (PSRR) is defined as the ratio of the change in input offset voltage due to the change in supply voltage producing it, keeping other power supply voltage constant. It is also called **power supply sensitivity (PSV).**

Fig. 3.33

Now if V_{EE} is constant and due to certain change in V_{CC}, there is change in input offset voltage then PSRR is defined as,

$$\text{PSRR} = \left.\frac{\Delta V_{ios}}{\Delta V_{CC}}\right|_{\text{constant } V_{EE}} \quad ...(1)$$

For a fixed V_{CC}, if there is a change in V_{EE} then

$$PSRR = \left.\frac{\Delta V_{ios}}{\Delta V_{EE}}\right|_{constant\ V_{CC}} \quad ...(2)$$

As input offset voltage is very small, PSRR is expressed in mV/V or μV/V.

The typical value of PSRR for IC 741 op-amp is 30 μV/V.

3.13.2 Slew Rate

The **slew rate** is defined as the maximum rate of change of output voltage with time. The slew rate is specified in V/μsec. Thus

$$\text{Slew rate} = S = \left.\frac{dV_o}{dt}\right|_{max} \quad ...(3)$$

The slew rate is caused due to limited charging rate of the compensating capacitor and current limiting and saturation of the internal stages of an op-amp, when a high frequency, large amplitude signal is applied. The internal capacitor voltage cannot change instantaneously. It is given by $\frac{dV_c}{dt} = \frac{I}{C}$. For large charging rate, the capacitor should be small or charging current should be large.

Hence the slew rate for the op-amp whose maximum internal capacitor charging current is known, can be obtained as

$$S = \frac{I_{max}}{C}$$

For example, for IC 741 the charging current is 15 μA and the internal capacitor is 30 pF, hence its slew rate is

$$S = \frac{15 \times 10^{-6}}{30 \times 10^{-12}} = \frac{0.5}{10^{-6}}\ \text{V/sec}$$

$$= 0.5\ \text{V/μsec}$$

3.13.2.1 Effect of Slew Rate

Consider a circuit using op-amp having **unity gain**. Thus **output is same as input**. If the input is square wave, output has to be square wave. But this is observed for certain frequency of input. Due to slew rate of an op-amp, for a particular input frequency, output gets distorted as shown in the Fig. 3.34.

Then observing such a distorted waveform on CRO the slew rate can be obtained as,

$$S = \frac{\Delta V_o}{\Delta t}\ \text{V/sec}$$

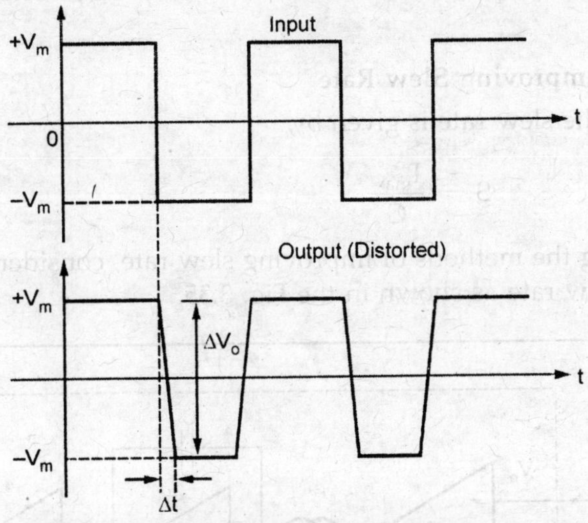

Fig. 3.34 Effect of slew rate

The typical value of S for IC 741 op-amp is 0.5×10^6 V/sec i.e. 0.5 V /μsec. Ideally it should be infinite.

***Key Point:** Higher the value of S, better is the performance of op-amp.*

3.13.2.2 Slew Rate Equation

Consider unity gain op-amp circuit with purely sinusoidal input. The output must be same as input.

$$V_s = V_m \sin \omega t \quad \text{...(5)}$$

$$V_o = V_m \sin \omega t \quad \text{...(6)}$$

$$\therefore \quad \frac{dV_o}{dt} = V_m (\omega \cos \omega t) \quad \text{...(7)}$$

But $$S = \text{slew rate} = \left.\frac{dV_o}{dt}\right|_{max} \quad \text{...(8)}$$

The equation (7) has maximum value when cos ωt = 1.

$$\therefore \quad S = V_m \omega = 2\pi f V_m$$

$$\therefore \quad \boxed{S = 2\pi fr V_m \text{ V/sec}} \quad \text{...(9)}$$

This is the required slew rate equation.

For distortion free output, the maximum allowable input frequency f_m can be obtained as,

$$\therefore \quad \boxed{f_m = \frac{S}{2\pi V_m} \text{ Hz}} \quad \text{...(10)}$$

This is also called full power bandwidth of the op-amp. The V_m is peak of output waveform.

3.13.2.3 Methods of Improving Slew Rate

It is known that the slew rate is given by,

$$S = \frac{I_{max}}{C}$$

For understanding the methods of improving slew rate, consider the op-amp model for the analysis of the slew rate as shown in the Fig. 3.35.

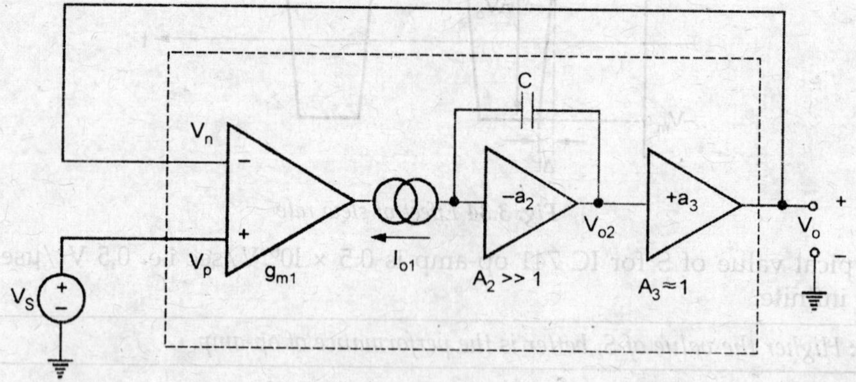

Fig. 3.35 Op-amp model for slew rate analysis.

The op-amp used is in voltage follower mode, in which the output $V_o = V_i$. The circuit is similar to that used earlier to derive slew rate equation, only the op-amp is replaced by its model.

When input overdrives the input stage then $I_{max} = \pm I_{01}$ (sat) which are saturation current levels of the input stage. Under this condition, op-amp is said to be operating under large-signal conditions.

The saturation of the input stage limits the slew rate because under saturation condition, the rate at which capacitor C can charge or discharge, according to the input overdrive is at its maximum.

From the Fig. 3.35 we can write,

$$I_{o1}(\text{sat}) = C\frac{dV_o}{dt}$$

$$\therefore \quad \frac{dV_{o2}}{dt} = \frac{I_{o1}(\text{sat})}{C} \qquad \text{...(11)}$$

This rate of change of V_{o2} is maximum, due to the saturation effect. Now the gain of third stage $a_3 \approx 1$ hence $V_o = V_{o2}$.

$$\therefore \quad \left.\frac{dV_o}{dt}\right|_{max} = \frac{dV_{o2}}{dt} = \frac{I_{o1}(\text{sat})}{C} \qquad \text{...(12)}$$

But maximum rate of change of output voltage is the slew rate.

$$\therefore \quad S = \frac{I_{o1}(sat)}{C} \quad \text{...slew rate}$$

Analysing the op-amp model used we can write,

$$V_{o2} = \text{drop across } C = Z_C\, I_{o1}$$

The input stage is a **transconductance amplifier** i.e. voltage input, current output amplifier. For sufficiently small differential input voltage, the relation between input voltage and output current for such an amplifier is,

$$\text{Output current} = g_m \text{ (differential input)}$$

$$\therefore \quad I_{o1} = g_{m1}(V_p - V_n) \quad \text{...For input stage}$$

$$\therefore \quad V_{o2} = Z_C\, g_m (V_p - V_n)$$

$$\text{But} \quad V_{o2} \approx V_o \quad \text{as} \quad a_3 \approx 1$$

$$\therefore \quad V_o = Z_C\, g_{m1}(V_p - V_n)$$

$$= \left[\frac{1}{j\omega C}\right] g_{m1}(V_p - V_n)$$

$$\text{As} \quad Z_C = X_C = -\frac{j}{\omega C} = \frac{1}{j\omega C}$$

$$\therefore \quad \text{Op-amp gain} = \frac{|V_o|}{|V_p - V_n|}$$

$$\therefore \quad |a| = \frac{|V_o|}{|V_p - V_n|} = \frac{g_{m1}}{\omega C} = \frac{g_{m1}}{2\pi f C}$$

$$\therefore \quad |a| f = \frac{g_{m1}}{2\pi C} \quad \text{... (13)}$$

Now the gain-bandwidth product of op-amp is denoted as f_t given by the product of gain $|a|$ and bandwidth f as,

$$f_t = |a| f \quad \text{... (14)}$$

$$\therefore \quad f_t = \frac{g_{m1}}{2\pi C}$$

$$\therefore \quad \boxed{C = \frac{g_{m1}}{2\pi f_t}} \quad \text{... (14)}$$

The gain bandwidth product is also called **unity gain bandwidth** of op-amp.

Substituting value of C in the slew rate,

$$\therefore \quad \boxed{S = 2\pi \frac{I_{o1(sat)}}{g_{m1}} f_t}$$

From this equation, we can get an idea about improvement in slew rate.

Hence methods of improving slew rate are,

1. Increasing f_t: Higher the value of gain-bandwidth product of op-amp, higher is its slew rate. To increase f_t, it can be seen from the equation (15), the internal capacitor value must be reduced. Hence for uncompensated op-amps, user can use the compensating network in such a way to reduce C and increase f_t to improve the slew rate. The frequency compensation schemes like feed-forward compensation, discussed earlier, can be used to achieve high f_t and hence higher slew rate.

2. Increasing I_{o1}(sat): This method is difficult because I_{o1}(sat) for the op-amp can not be controlled externally by the user. But to increase I_{o1} without affecting the value of g_{m1} and f_t is to provide alternative path to the rapid charging and discharging of C, when large signal condition exists. This is possible by using additional input transistor pair which can be designed such that it will go into conduction region when sufficiently large amplitude input signal is applied. This can increase I_{o1} (sat) to improve the slew rate.

But such adjustments are possible only in programmable op-amps where operating point of the device can be set by the user with external current I_{set}. The separate pin is provided for such op-amps to implement such current setting. Examples of such programmable op-amps are μA776 by Fairchild, LM346 by National Semiconductor, TL066 by Texas Instruments etc.

3. Reducing g_{m1}: The gain of transistorised differential input stage can be reduced by using suitable resistances in series with the emitters of the differential input transistors. This technique is called emitter degeneration. Reducing input stage transconductance by this method, slew rate can be improved.

Another method of reducing transconductance of the input stage is to use FET differential input pair instead of BJT differential input pair. The g_m values for the FETs are less than those for the BJTs. In addition to reducing g_m and improving slew rate, the added advantages of using FETs are very low input bias currents and very low input offset currents.

Thus by using proper frequency compensation and other compensating networks and referring to the data sheets, higher slew rate op-amps can be selected for the high speed applications.

Example 3.4: An op-amp operates as a unity gain buffer with 3 V (peak to peak) square wave input. If op-amp is ideal with slew rate 0.5 V/μsec, find the maximum frequency of operation.

Solution: Given circuit is unity gain.

Peak to peak = 3 V of square wave

$$\therefore \quad V_m = \frac{V_{p-p}}{2} = \frac{3}{2} = 1.5$$

From slew rate equation,

$$f_m = \frac{S}{2\pi V_m} = \frac{\left(\frac{0.5}{10^{-6}}\right)}{2\pi \times 1.5} \quad \text{...Use S in V/sec}$$

$$= \mathbf{53.051\ kHz}$$

This is the maximum frequency of operation.

Example 3.5: For a typical op-amp, I_{CQ} = 15 μA and C = 35 pF. The peak value of input is 12 V. Determine slew rate and maximum possible frequency of input voltage that can be applied to get undistorted output.

Solution: $I_{max} = I_{CQ} = 15\ \mu A,\ C = 35\ pF,\ V_m = 12\ V.$

$$\therefore \quad S = \frac{I_{max}}{C}$$

$$= \frac{15 \times 10^{-6}}{35 \times 10^{-12}} = 0.4285 \times 10^6\ V/sec = \mathbf{0.4285\ V/\mu sec}$$

$$= \mathbf{5.684\ kHz}$$

Upto this frequency, output will be undistorted.

3.13.3 Transient Response Rise Time

When the output of the op-amp is suddenly changing like pulse type, then the rise time of the response depends on the cut-off frequency f_H of the op-amp. Such a rise time is called **cut-off frequency limited rise time or transient response rise time.** It is inversely proportional to the cut-off frequency and given by,

$$\boxed{t_r = \frac{0.35}{f_H}}$$

where t_r = rise time and f_H = cut-off frequency.

Example 3.6: The open loop gain of a certain op-amp falls to 0 dB at a frequency of 10 MHz. Find the transient response rise time, if it is used as unity gain amplifier.

Solution: For unity gain amplifier, UGB = f_H = 10 MHz given.

$$\therefore \quad t_r = \frac{0.35}{f_H} = \frac{0.35}{10 \times 10^6}$$

$$= \mathbf{25\ nsec} \quad \text{... Rise time}$$

3.14 D.C. CHARACTERISTICS OF OP-AMP

The important d.c. characteristics of op-amp are,

1. Input bias current (I_b)
2. Input offset current (I_{ios})

3. Input offset voltage (V_{ios})
4. Thermal drift

The input bias current, input offset current and input offset voltage are already discussed. The effect of these parameters is to add the error to the expected d.c. output voltage. These parameters produce **output offset voltage** V_{oos}.

3.14.1 Thermal Drift

The op-amp parameters input offset voltage V_{ios}, input bias current I_b and input offset current I_{ios} are not constants but vary with the factors:

(*i*) Temperature (*ii*) Supply voltage changes and (*iii*) Time

The effect of change in temperature on the parameters is most severe. Let us discuss the effect of change in temperature on these parameters.

3.14.1.1 Effect on Input Offset Voltage

The effect of change in temperature on the input offset voltage is defined by a factor called **thermal voltage drift**. It is also called as input offset voltage drift.

The thermal voltage drift is defined as average rate of change of input offset voltage per unit change in temperature. Mathematically it is given by,

$$\text{Input offset voltage drif} = \frac{\Delta V_{ios}}{\Delta T}$$

where ΔV_{ios} = change in input voltage

ΔT = change in temperature

It is expressed in µV/°C. The drift is not constant and it is not uniform over specified operating temperature range. The value of the input offset voltage may increase or decrease with the increasing temperature.

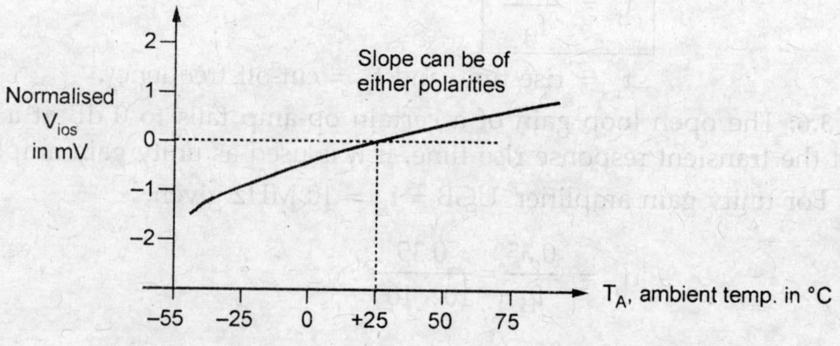

Fig. 3.36 Input offset voltage drift.

The Fig. 3.36 shows the graph of normalized values of input offset voltage versus temperature, for MC1741 op-amp.

The input offset voltage is zero at room temperature of 25 °C. Referring to the graph shown in the Fig. 3.36, the thermal voltage drift values can be obtained.

3.14.1.2 Effect on Input Offset and Bias Currents

Similar to the input offset voltage, input bias current and input offset current are not constants but vary with temperature. The effect of temperature on input bias current is defined by a factor called **input bias current drift** while effect on input offset current is defined by a factor called **input offset current drift.**

The average rate of change of input bias current per unit change in temperature is called input bias current drift.

The average rate of input offset current per unit change in temperature is called input offset current drift.

Mathematically these drifts are given by,

$$\text{Thermal drift in input bias current} = \frac{\Delta I_b}{\Delta T} \quad \text{... (2)}$$

$$\text{Thermal drift in input offset current} = \frac{\Delta I_{ios}}{\Delta T} \quad \text{... (3)}$$

Both the drifts are measured in nA/°C or pA/°C. These parameters vary randomly with temperature i.e. they may be positive in one temperature range and negative in another.

The Fig. 3.37 (*a*) and 3.37 (*b*) show the graphs of normalized values of input bias current and input offset current versus temperature, for MC1741 op-amp. These curves are different for different types of op-amps and are generally provided by the manufacturers.

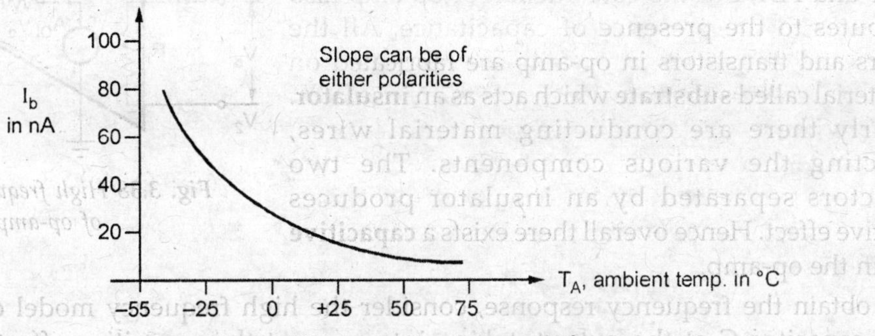

Fig. 3.37 (a) Input bias current drift.

The input offset current is assumed to be zero at room temperature of 25°C. Practically no information is available about the change in input bias current versus temperature. Infact when compensating resistance R_{comp} is used, there is no need to consider the change in input bias current as a function of change in temperature.

3.15 A.C. CHARACTERISTICS OF OP-AMP

The important a.c. characteristics of op-amp are,

1. Slew rate 2. Frequency response

The slew rate is already discussed. It indicates the ability of op-amp with which it can change its output according to changes in the input.

3.15.1 Frequency Response of Op-amp

Ideally, an op-amp should have an infinite bandwidth. This means the gain of op-amp must remain same for all the frequencies from zero to infinite. Uptill now we have assumed gain of the op-amp as constant but practically op-amp gain decreases at higher frequencies. Such a gain reduction with respect to frequency is called roll off.

This happens because gain of the op-amp depends on the frequency and hence mathematically it is a complex number. Its magnitude and the phase angle changes with the frequency. The plot showing the variations in magnitude and phase angle of the gain due to the change in frequency is called **frequency response** of the op-amp. In such plots, to accommodate large range of frequency, it is plotted on a logarithmic scale. The gain magnitude can be plotted as a numerical value or may be expressed in decibels. When the gain in decibels, phase angle in degrees are plotted against logarithmic scale of frequency, the plot is called **Bode plot**. The manner in which the gain of the op-amp changes with variation in frequency is known as the **magnitude plot** and the manner in which the phase shift changes with variation in frequency is known as **the phase angle plot**. Generally magnitude plot is supplied by the manufacturers.

The dependence of gain of the op-amp on frequency is basically because of presence of capacitive component in the equivalent circuit of the op-amp. As op-amp uses BJT and FET, which have the junction capacitances which is very small. But at high frequency, these offer decreased reactance. Not only the BJT and FET, but the construction of op-amp also contributes to the presence of capacitance. All the resistors and transistors in op-amp are fabricated on the material called **substrate** which acts as an **insulator**. Similarly there are conducting material wires, connecting the various components. The two conductors separated by an insulator produces capacitive effect. Hence overall there exists a **capacitive effect** in the op-amp.

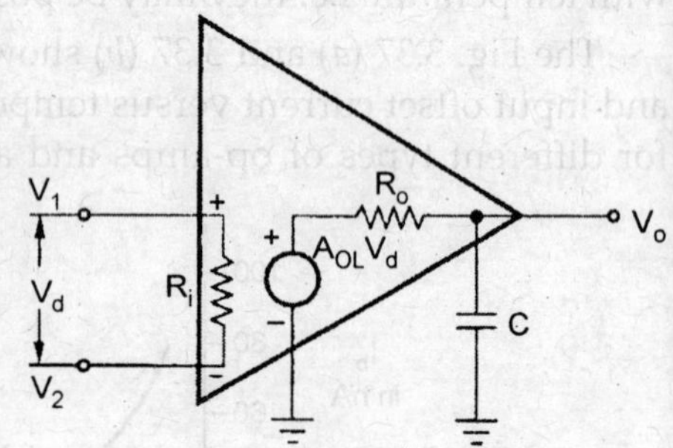

Fig. 3.38 High frequency model of op-amp.

To obtain the frequency response, consider the high frequency model of the op-amp with a capacitor C at the output, taking into account the capacitive effect present. It is shown in the Fig. 3.38.

Let $-jX_C$ be the capacitive reactance due to the capacitor C. From the Fig. 3.38, using voltage divider rule,

$$V_o = -jX_C\left[\frac{A_{OL} V_d}{R_o - jX_C}\right] \quad \text{... (1)}$$

Now $$-j = \frac{1}{j}$$

and $$X_C = \frac{1}{2\pi f C}$$

$$\therefore \quad V_o = \frac{\dfrac{1}{j2\pi fC}}{R_o + \dfrac{1}{j2\pi fC}}(A_{OL}\ V_d) \quad \text{...(2)}$$

$$\therefore \quad V_o = \frac{A_{OL} V_d}{1 + j2\pi fCR_o} \quad \text{...(3)}$$

Hence the open loop voltage gain as a function of frequency is

$$A_{OL} = \frac{V_o}{V_d}$$

$$= \frac{A_{OL}}{1 + j2\pi fR_o C} \quad \text{... (4)}$$

Let
$$f_o = \frac{1}{2\pi R_o C}$$

$$\therefore \quad \boxed{A_{OL}(f) = \frac{A_{OL}}{1 + j\left(\dfrac{f}{f_o}\right)}} \quad \text{...(5)}$$

where $A_{OL}(f)$ = Open loop voltage gain as a function of frequency
A_{OL} = Gain of op-amp at a Hz i.e. d.c.
f = Operating frequency
f_o = Break frequency or cut off frequency of op-amp

For a given op-amp and selected value of C, the frequency f_o is constant. The equation (5) can be written in the polar form as

$$\boxed{|A_{OL}(f)| = \frac{A_{OL}}{\sqrt{1 + \left(\dfrac{f}{f_o}\right)^2}}} \quad \text{...(6)}$$

$$\boxed{\angle A_{OL}(f) = \phi(f) = -\tan^{-1}\left(\frac{f}{f_o}\right)} \quad \text{...(7)}$$

At f = 0 Hz, the magnitude is AOL, while $\phi(f) = 0°$.

For IC 741 op-amp, f_o = 5 Hz and the open loop gain 200,000, we can calculate gain and phase shifts at various frequencies as shown in Table 3.2.

Table 3.2

Frequency f in Hz	$\lvert A_{OL}(f)\rvert$ in dB =	$\phi(f) = -\tan^{-1}\left(\frac{f}{f_o}\right)$
$20\log\frac{A_{OL}}{\sqrt{1+\left(\frac{f}{f_o}\right)^2}}$ in dB	in degrees	
0	106.02 dB	0°
5	103.01 dB	– 45°
10	99.03 dB	– 63.43°
100	79.98 dB	– 87.13°
1000	60.00 dB	– 89.71°
100×10^3	20.00 dB	– 89.99°
1×10^6	0 dB	– 89.999°

As the frequency increases till f_o, the gain is almost constant but after f_o, the gain reduces with a rate of – 20 dB/decade. The maximum possible phase shift is – 90°. Hence the frequency response is shown as in the Fig. 3.39 (See Fig. on page 3-40).

The following observations can be made from the frequency response of an op-amp:

(*i*) The open loop gain A_{OL} is almost constant from 0 Hz to the break frequency f_o.

(*ii*) At $f = f_o$, the gain is 3 dB down from its value at 0 Hz. Hence the frequency f_o is also called as **– 3 dB frequency**. It is also known as **corner frequency**.

(*iii*) After $f = f_o$, the gain $A_{OL}(f)$ decreases at a rate of 20 dB/decade or 6 dB/octave. A decade is 10 times change in frequency while octave is 2 times change in frequency. As gain decreases, slope of the magnitude plot is – 20 dB/decade or – 6 dB/octave, after $f = f_o$.

(*iv*) At a certain frequency, the gain reduces to 0 dB. This means $20\log\lvert A_{OL}(f)\rvert$ is 0 dB i.e. $\lvert A_{OL}(f)\rvert = 1$. Such a frequency is called **gain cross-over frequency** or **unity gain bandwidth (UGB).** It is also called **closed loop bandwidth**. UGB is the **gain bandwidth product** only if an op-amp has a single break frequency, before $A_{OL}(f)$ dB is zero. IC 741 op-amp has a single break frequency and its UGB is approximately 1 MHz. So for an op-amp with single break frequency f_o, after f_o the gain bandwidth product is constant equal to UGB.

$$\therefore \quad \boxed{\text{USB} = A_{OL}\, f_o} \qquad \text{...(8)}$$

UGB is also called gain bandwidth product and denoted as f_t. Thus f_t is product of gain of op-amp and bandwidth.

The break frequency is nothing but a corner frequency f_o. At this frequency, slope of the magnitude plot changes. The op-amp for which there is only once change in the slope of the magnitude plot, is called as **single break frequency** op-amp. The IC 741 op-amp, is single break frequency i.e. slope of the plot changes only once from 0 to – 20 dB/decade at f_o, as shown in the Fig. 3.39.

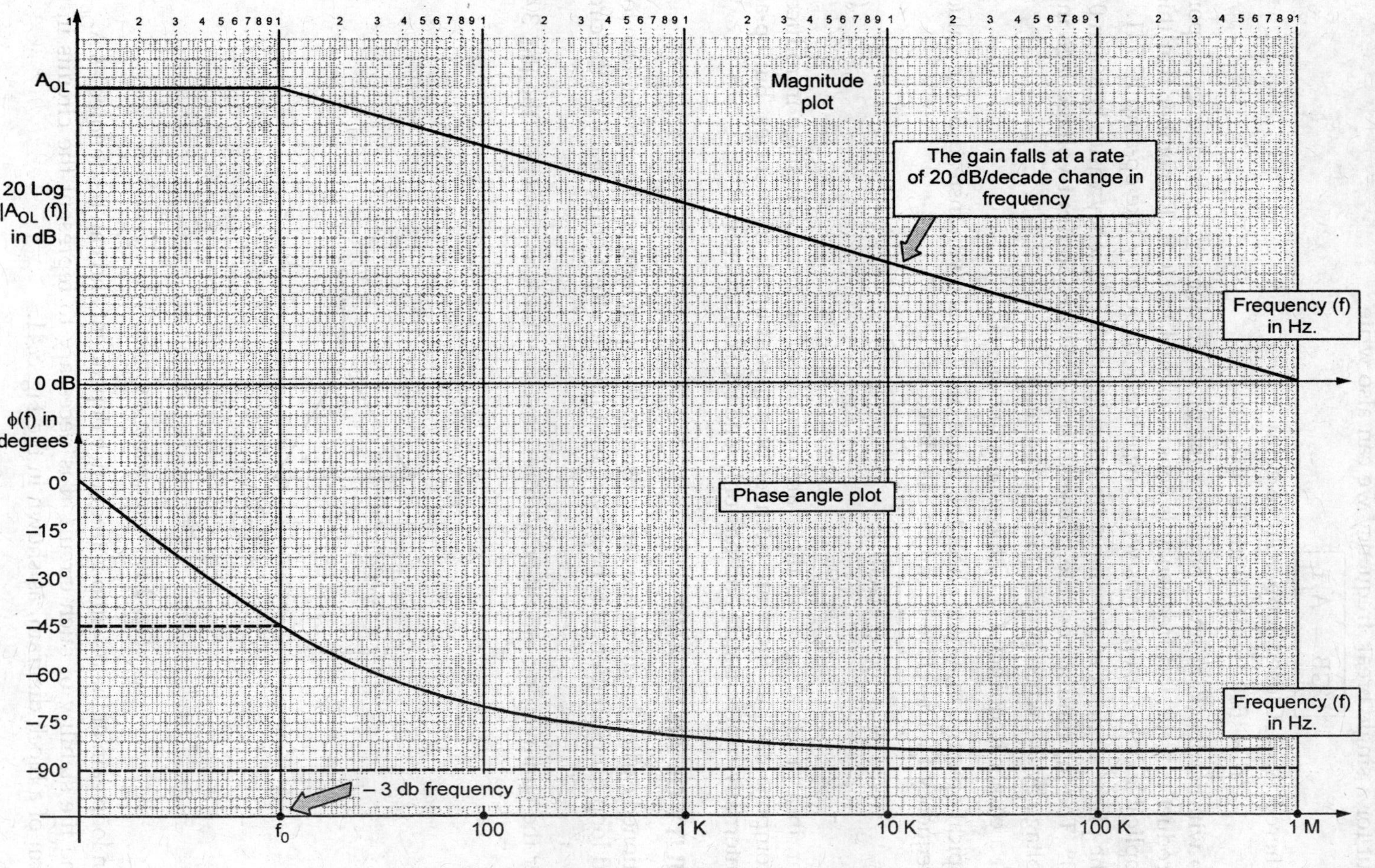

Fig. 3.39 Frequency response of an op-amp

But for a single break frequency we can also write

$$\boxed{UGB = A_f f_f} \quad ...(9)$$

where A_f = Closed loop voltage gain

f_f = Bandwidth with feedback

So with feedback, at any point after the break frequency on frequency response, product of gain and frequency is constant equal to UGB. Remember that this is applicable only for op-amps with single break frequency like op-amp IC 741.

(*v*) The phase angle of an op-amp with a single break frequency varies between 0° to 90°. The maximum possible phase shift is – 90°, i.e. output voltage lags input voltage by 90° when phase shift is maximum.

(*vi*) At a corner frequency $f = f_o$, the phase shift is – 45°.

For a typical op-amp, in a data sheet the value of UGB is given instead of the value of the break frequency f_o. Therefore f_o can be calculated as

$$\boxed{f_o = \frac{UGB}{A_{OL}}} \quad ...(10)$$

However, in a practical op-amp there are number of stages. Each stage introduces a capacitive component. Thus there are number of different break frequencies. But op-amp like 741 is internally compensated and has only one break frequency.

3.16 STABILITY OF AN OP-AMP

As mentioned earlier, op-amps are rarely used in the open loop configuration. Due to its use in closed loop configuration, it is necessary to study the effect of feedback on the circuit stability.

Consider the noninverting amplifier with resistive feedback as shown in the Fig. 3.40.

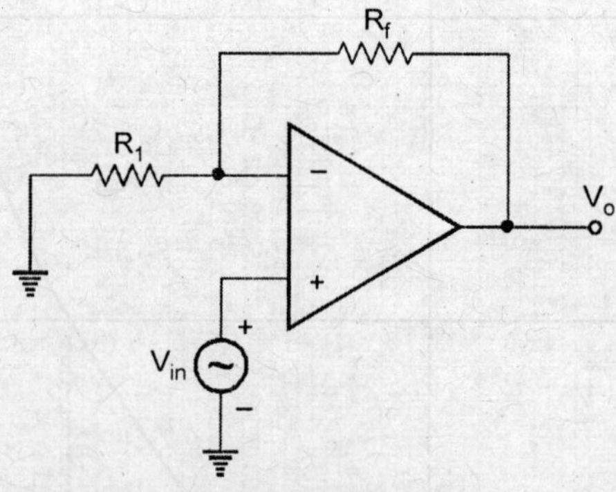

Fig. 3.40 Noninverting amplifier.

The open loop gain of the op-amp is frequency dependent and can be denoted as A_{OL} (f). To determine the stability of such circuit, it is necessary to represent the circuits in the standard form of a block diagram, as shown in the Fig. 3.41.

The block between output and input is called forward block while the block between output signal and the feedback is called feedback block. The transfer function or gain of forward block is $A_{OL}(f)$ while the transfer function or gain of feedback block is β. Let us investigate the stability of such a system with the help of frequency response approach.

***Key Point:** The feedback block uses only resistive elements and hence its transfer function β is independent of frequency and is constant.*

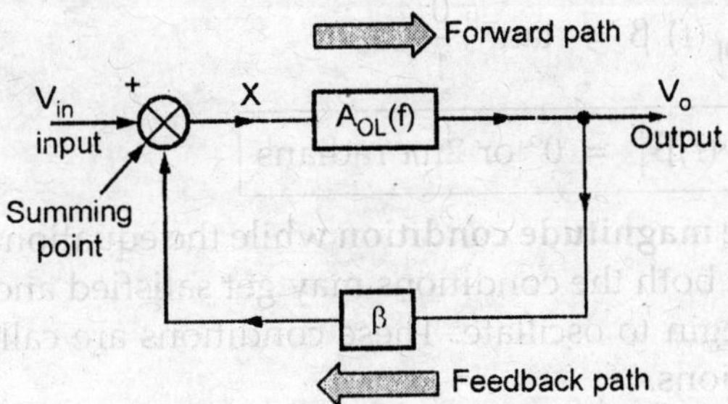

Fig. 3.41 Block diagram of noninverting amplifier.

From the Fig. 3.41 we can write,

$$x = V_{in} - \beta V_o \quad \text{...(1)}$$

$$V_o = x[A_{OL}(f)] \quad \text{...(2)}$$

$$\therefore \quad \boxed{x = \frac{V_o}{A_{OL}(f)}} \quad \text{...(3)}$$

Substituting in (1),

$$\frac{V_o}{A_{OL}(f)} = V_{in} - \beta V_o$$

$$\therefore \quad V_o\left[\frac{1}{A_{OL}(f)} + \beta\right] = V_{in}$$

$$\therefore \quad \boxed{\frac{V_o}{V_{in}} = \frac{A_{OL}(f)}{1 + A_{OL}(f)\beta}} \quad \text{... (4)}$$

The ratio V_o/V_{in} with feedback is called **closed loop gain** of the amplifier while the product $A_{OL}(f)$ β is called **loop gain** which is generally used to determine the stability.

The stability of the circuit entirely depends on the behaviour of the roots of the characteristic equation of the system, which is

$$1 + A_{OL}(f)\,\beta = 0 \quad \text{...(5)}$$

i.e. $$\boxed{A_{OL}(f)\,\beta = -1} \quad \text{...(6)}$$

This can be expressed in the complex form as

$$A_{OL}(f)\,\beta = -1 + j0 \quad \text{...(7)}$$

From this, the conditions for the sustained oscillations can be obtained.

Equating angles and magnitudes of both sides of the equation (7) we get,

$$|A_{OL}(f)\beta| = \sqrt{1^2 + 0^2}$$

i.e. $$\boxed{|A_{OL}(f)\beta| = 1} \quad ...(8)$$

While $$\angle A_{OL}(f)\,\beta = \tan^{-1}\frac{0}{1}$$

i.e. $$\boxed{|A_{OL}(f)\beta| = 0° \text{ or } 2n\pi \text{ radians}} \quad ...(9)$$

The equation (8) is the **magnitude condition** while the equation (9) is the **angle condition.** At a particular frequency, both the conditions may get satisfied and there is every possibility that the amplifier may begin to oscillate. These conditions are called **Barkhausen Criterion** for the sustained oscillations.

Now oscillations is the verge of instability. The system is said to be **stable** if its output reaches a fixed value within a finite time. While if output increases with time indefinitely or there is output without input, the system is said to be **unstable.**

Key Point: *So determining stability of the circuit means actually deciding how close the circuit is to the conditions of sustained oscillations which indicates the starting point of instability.*

This is possible from the frequency response of the circuit by defining certain specifications.

3.16.1 Stability Specifications from Frequency Response

The loop gain $A_{OL}(f)\,\beta$ is used to obtain the frequency response of the system over a desired range of frequency, as mentioned earlier. Once the magnitude plot and phase angle plot (Bode plot) is obtained, the following specifications are determined, from which system stability can be decided.

(*i*) Gain Cross Over Frequency (ω_{gc}):

The frequency at which the loop gain magnitude $\left(|A_{OL}(f)\,\beta|\right)$ is unity i.e.

$20 \log |A_{OL}(f)\,\beta| = 0$ dB is called **gain cross over frequency** and denoted as ω_{gc}.

(*ii*) Phase Cross Over Frequency (ω_{pc}):

The frequency at which the phase shift introduced by the loop gain is $-180°$ or $n\pi$ radians is called phase cross over frequency and denoted as ω_{pc}.

(*iii*) Gain Margin (G.M.):

The critical values of loop gain are 0 dB and $-180°$ from the stability point of view. How far away the circuit is from this point of instability can be decided by determining loop gain at phase cross over frequency. This gain measured at the frequency $\omega = \omega_{pc}$ is called gain margin. So loop gain $|A_{OL}(f)\beta|$ measured at $\omega = \omega_{pc}$ is the gain margin of the circuit. Mathematically it can be expressed as

$$\boxed{G.M = -20\log|A_{OL}(f)\beta|_{\omega=\omega_{pc}} \ dB} \quad ...(10)$$

(*iv*) Phase Margin (P.M.):

Similar to the gain at phase cross over frequency, one more specification can be defined from the phase shift of loop gain existing at gain cross over frequency. This specification is called phase margin.

The amount by which angle of the loop gain ($\angle A_{OL}(f)\ \beta$) differ from the $-180°$ phase shift, at $\omega = \omega_{gc}$ is called phase margin. Mathematically it is expressed as

$$\boxed{P.M = 180° + \angle A_{OL}(f)\beta|_{\omega=\omega_{gc}}} \quad ...(11)$$

3.16.2 Stability Criterion

When $\omega = \omega_{gc} = \omega_{pc}$, both angle and magnitude conditions as per Barkhausen criterion get satisfied and circuit oscillates. The system in such a case is called **marginally stable.**

If G.M. is positive i.e. gain $|A_{OL}(f)\beta|$ in dB is negative at $\omega = \omega_{pc}$, the system is said to be stable. Similarly if P.M. is positive i.e. $\angle A_{OL}(f)\ \beta$ is greater than $-180°$, the system is said to be stable.

The **negative values of C.M. and P.M.,** indicate **instability** of the system.

Key Point: *More positive are the values of G.M. and P.M., more stable is the system.*

The Fig. 3.42 shows the positive G.M. and P.M. for the stable system on the frequency response.

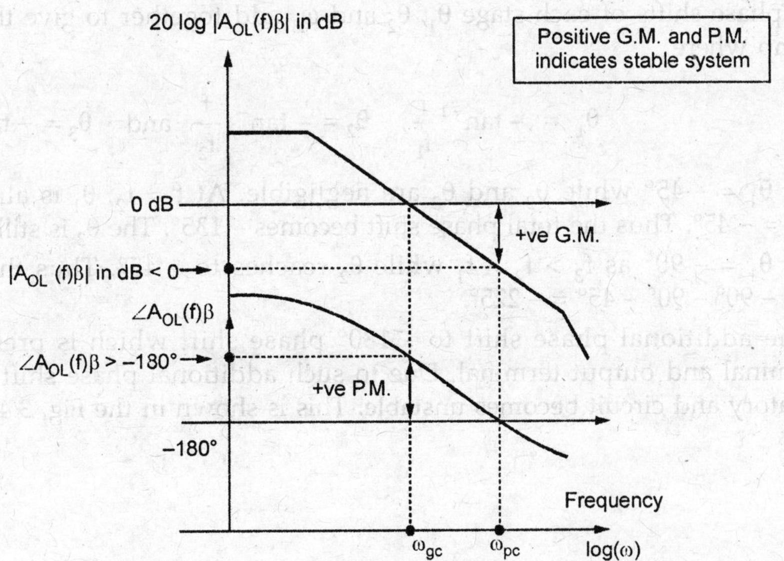

Fig. 3.42 Typical frequency response for stable ***op-amp***

3.17 CONCEPT OF FREQUENCY COMPENSATION

It is seen earlier that the op-amp circuit with single break or comer frequency is inherently stable. Now consider the system with three break or comer frequencies. This is possible in practice due to the capacitive component produced by the various number of stages. The open loop frequency dependent transfer function of such a system can be assumed as

$$A_{OL}(f) = \frac{A_{OL}(\text{at } f=0)}{\left(1+j\frac{f}{f_1}\right)\left(1+j\frac{f}{f_2}\right)\left(1+j\frac{f}{f_3}\right)} \quad ...(1)$$

where $0 < f_1 < f_2 < f_3$

In the Laplace domain it can be expressed as

$$A_{OL}(s) = \frac{A_{OL}}{(s+\omega_1)(s+\omega_2)(s+\omega_3)} \quad ...(2)$$

where A_{OL} = gain at a frequency or d.c. gain and ω_1, ω_2, ω_3 are the comer frequency values such that $0 < \omega_1 < \omega_2 < \omega_3$.

The phase shift introduced by such loop gain is

$$\phi(f) = -\tan^{-1}\left(\frac{f}{f_1}\right) - \tan^{-1}\left(\frac{f}{f_2}\right) - \tan^{-1}\left(\frac{f}{f_3}\right) \quad ...(3)$$

As $f \to \infty$, $\tan^{-1}\left(\frac{f}{f_1}\right), -\tan^{-1}\left(\frac{f}{f_2}\right)$ and $\tan^{-1}\left(\frac{f}{f_3}\right)$ all contributes a phase shift of 90°.

Hence at high frequencies $\phi(f) \to -270°$.

Thus the phase shifts of each stage θ_1, θ_2 and ω_3 add together to give the total phase shift of op-amp where,

$$\theta_1 = -\tan^{-1}\frac{f}{f_1}, \quad \theta_2 = -\tan^{-1}\frac{f}{f_2} \quad \text{and} \quad \theta_3 = -\tan^{-1}\frac{f}{f_3}$$

At $f = f_1$, $\theta_1 = -45°$ while θ_3 and θ_3 are negligible. At $f = f_2$, θ_1 is almost $-90°$ as $f_2 > f_1$ and $\theta_{12} = -45°$. Thus the total phase shift becomes $-135°$. The θ_3 is still small.

At $f = f_3$, $\theta_1 = -90°$ as $f_3 > f_2 > f_1$ while θ_2 reaches to $-45°$. Thus the total phase shift becomes $-90° - 90° - 45° = -225°$.

This is the additional phase shift to $-180°$ phase shift which is present between inverting terminal and output terminal. Due to such additional phase shift op-amp may behave oscillatory and circuit becomes unstable. This is shown in the Fig. 3.43.

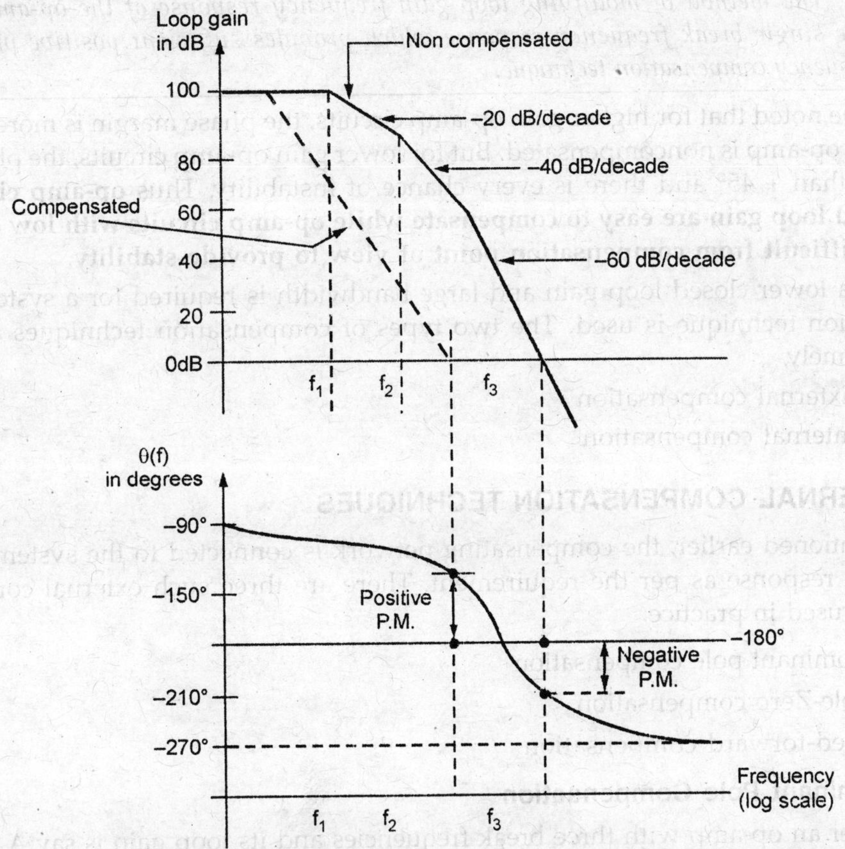

Fig. 3.43 Concept of frequency compensation.

Thus as seen from the frequency response of noncompensated op-amp, due to additional phase shift, at $\omega = \omega_{gc}$, there is chance that total phase shift becomes almost – 360° and circuit becomes unstable. The phase margin is negative in such a case. To build the oscillations the total phase shift need not be exactly – 360° but if it is more than – 315° and $A_{OL}\,\beta > 1$ then oscillations may start.

To avoid the oscillations, the total phase shift should not be greater than – 315° i.e. due to the op-amp, excluding – 180° between inverting terminal and output, the phase shift should not be greater than – 315° + 180° = – 135°.

∴ Minimum phase margin = 180° – 135° = + 45°

Thus for op-amp circuit stability, it is necessary to have phase margin of + 45°.

This is possible by providing necessary compensation so that only one breakover frequency exists. Due to this, phase shift of op-amp can not increase beyond – 90°. Thus there is no chance that op-amp phase shift becomes – 135° and phase margin is automatically more than +45°. This technique of compensation is called frequency compensation technique. Using this technique, the op-amp is converted to single break frequency op-amp. The frequency response of such op-amp has only one roll off rate of – 20 dB/dec and is as shown dotted in the Fig. 3.44.

***Key Point:** The method of modifying loop gain frequency response of the op-amp so that it behaves like single break frequency response which provides sufficient positive phase margin is called frequency compensation technique.*

It can be noted that for higher gain op-amp circuits, the phase margin is more than + 45°, though the op-amp is noncompensated. But for lower gain op-amp circuits, the phase margin is smaller than + 45° and there is every chance of instability. Thus **op-amp circuits with high closed loop gain are easy to compensate while op-amp circuits with low closed loop gain are difficult from compensation point of view to provide stability.**

When a lower closed loop gain and large bandwidth is required for a system, suitable compensation technique is used. The two types of compensation techniques are used in practice namely,

(*i*) External compensation

(*ii*) Internal compensation

3.18 EXTERNAL COMPENSATION TECHNIQUES

As mentioned earlier, the compensating network is connected to the system externally to alter the response as per the requirement. There are three such external compensation techniques used in practice.

1. Dominant pole compensation
2. Pole-Zero compensation
3. Feed-forward compensation

3.18.1 Dominant Pole Compensation

Consider an op-amp with three break frequencies and its loop gain is say A.

$$A = \frac{A_{OL}}{\left(1+j\frac{f}{f_1}\right)\left(1+j\frac{f}{f_2}\right)\left(1+j\frac{f}{f_3}\right)}$$

The equation is same as (1) of section 3.17 and reproduced for the convenience.

In this loop gain, the dominant pole is introduced by adding a compensating network. Such a network is nothing but a simple R-C network as shown in the Fig. 3 44.

The dominant pole means the pole with magnitude much smaller than the existing poles. And hence the break frequency of the compensating network is the smallest compared to the existing break frequencies.

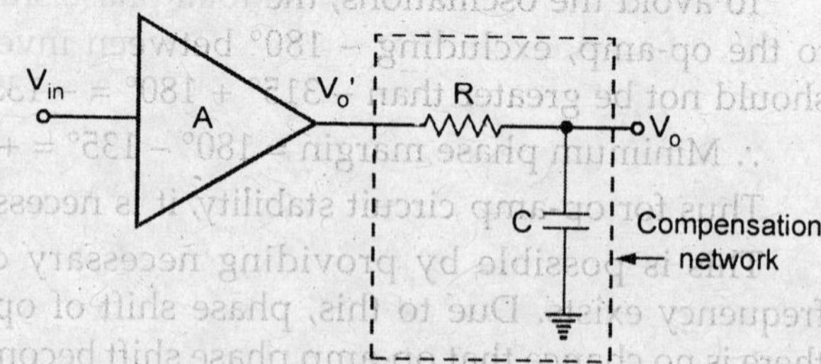

Fig. 3.44 Dominant pole compensation.

The transfer function of the compensating network can be obtained as:

$$A_1 = \text{Transfer function of compensating network}$$

$$= \frac{V_o}{V_o'}$$

By the voltage divider rule applied to the network,

$$A_1 = \frac{V_o}{V_o'} = \frac{-jX_C}{R - jX_C} \quad ...(1)$$

$$= \frac{-\dfrac{j}{2\pi fC}}{R - \dfrac{j}{2\pi fC}} = \frac{1}{\dfrac{R}{\left(-\dfrac{j}{2\pi fC}\right)} + 1} \quad ...(2)$$

As $\frac{1}{-j} = +j$, we can write

$$\therefore \quad A_1 = \frac{1}{1 + j2\pi fRC} \quad ...(3)$$

$$\text{Let} \quad f_d = \frac{1}{2\pi fC} \quad ...(4)$$

$$A_1 = \frac{1}{1 + j\left(\dfrac{f}{f_d}\right)} \quad ...(5)$$

Where f_d = break frequency of the compensating network

Hence the compensated transfer function becomes

$$A' = AA_1 \quad ...(6)$$

$$\therefore \quad A' = \frac{A_{OL}}{\left(1 + j\dfrac{f}{f_d}\right)\left(1 + j\dfrac{f}{f_1}\right)\left(1 + j\dfrac{f}{f_2}\right)\left(1 + j\dfrac{f}{f_3}\right)} \quad ...(7)$$

The values of R and C are selected in such a way that the loop gain drops to 0 dB with a slope of – 20 dB/decade and at a frequency where the poles of uncompensated system contributes very small phase shift. This ensures that at $\omega = \omega_{gc}$, the phase shift $\phi(f)$ is greater than – 180° and hence positive phase margin. Generally f_d is selected so that magnitude plot for A′ passes through 0 dB at the pole f_1 of A i.e. $\omega_{gc} = f_1$. The uncompensated and compensated magnitude plots are shown in the Fig. 3.45.

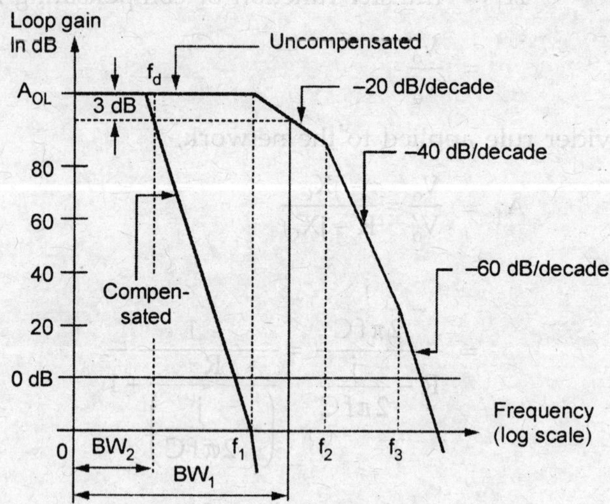

Fig. 3.45 Dominant pole compensation.

It can be observed from the plot that 3 dB down bandwidth for noncompensated system is BW_1 while for compensated it becomes BW_2. There is drastic reduction in the bandwidth.

Advantages:

(*i*) As the noise frequency components are outside the smaller bandwidth, the noise immunity of the system improves.

(*ii*) Adjusting value of f_d, adequate phase margin and the stability of the system is assured.

Disadvantage:

(*i*) The only disadvantage of the method is that the bandwidth reduces drastically, as mentioned earlier.

3.18.2 Pole-Zero Compensation

Consider the same op-amp described by the open loop gain A with three break frequencies as

$$A = \frac{A_{OL}}{\left(1+j\frac{f}{f_1}\right)\left(1+j\frac{f}{f_2}\right)\left(1+j\frac{f}{f_3}\right)}$$

In this method, the transfer function A is modified by adding a pole and a zero with the help of compensating network. The zero added is at higher frequency while a pole is at lower frequency. Such a compensating network is shown in the Fig. 3.46.

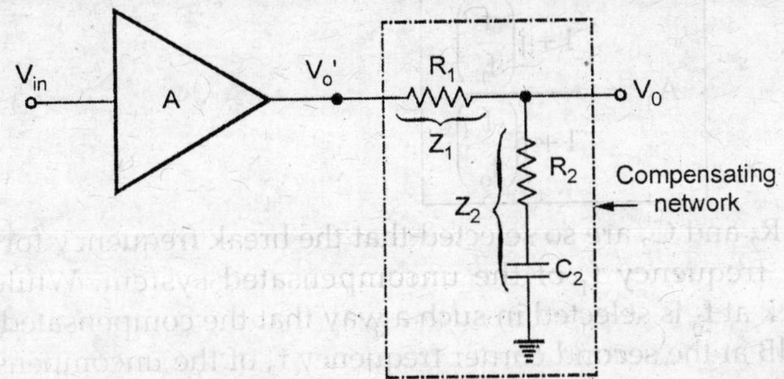

Fig. 3.46 Pole-Zero compensation.

The transfer function of the compensating network is say A_1.

$$A_1 = \frac{V_o}{V_o'}$$

By the voltage divider rule,

$$A_1 = \frac{Z_1}{Z_1 + Z_2} \quad ...(8)$$

Now $$Z_2 = R_2 - jX_{C2} \quad ...(9)$$

And $$Z_1 = R_1 \quad ...(10)$$

$$\therefore \quad A_1 = \frac{R_2 - jX_{C2}}{R_1 + R_2 - jX_{C2}} = \frac{R_2 - \dfrac{j}{2\pi f C_2}}{R_1 + R_2 - \dfrac{j}{2\pi f C_2}} = \frac{\dfrac{R_2}{\left[\dfrac{-j}{2\pi f C_2}\right]} + 1}{\dfrac{R_1 + R_2}{\left[\dfrac{-j}{2\pi f C_2}\right]} + 1}$$

Now $$-\frac{1}{j} = +j$$

$$\therefore \quad \boxed{A_1 = \frac{1 + j2\pi f R_2 C_2}{1 + j2\pi f (R_1 + R_2) C_2}} \quad ...(11)$$

Now let $$f_1 = \frac{1}{2\pi R_2 C_2} \quad ...(12)$$

And $$f_o = \frac{1}{2\pi (R_1 + R_2) C_2} \quad ...(13)$$

$$\therefore \qquad \boxed{A_1 = \frac{1+j\left(\dfrac{f}{f_1}\right)}{1+j\left(\dfrac{f}{f_o}\right)}} \qquad \text{...(14)}$$

The values of R_1, R_2 and C_2 are so selected that the break frequency for the zero matches with the first corner frequency f_1 of the uncompensated system. While the pole of the compensating network at f_o is selected in such a way that the compensated transfer function A′ passes through 0 dB at the second corner frequency f_2 of the uncompensated system. The resultant loop gain becomes

$$A' = AA_1$$

$$\boxed{A_1 = \frac{A_{OL}\left(1+j\dfrac{f}{f_1}\right)}{\left(1+j\dfrac{f}{f_o}\right)\left(1+j\dfrac{f}{f_1}\right)\left(1+j\dfrac{f}{f_2}\right)\left(1+j\dfrac{f}{f_3}\right)}} \qquad \text{... (15)}$$

where $0 < f_o < f_1 < f_2 < f_3$

The first corner frequency is now f_0 and the gain starts rolling off at – 20 dB/decade at f_o. At $f = f_1$ there is pole-zero cancellation and rolling rate continues as – 20 dB/decade. The values of R_1, R_2 and C_2 are selected such that plot passes through 0 dB at f_2 i.e., $\omega_{gc} = f_2$. The response for uncompensated and compensated system is shown in the Fig. 3.47.

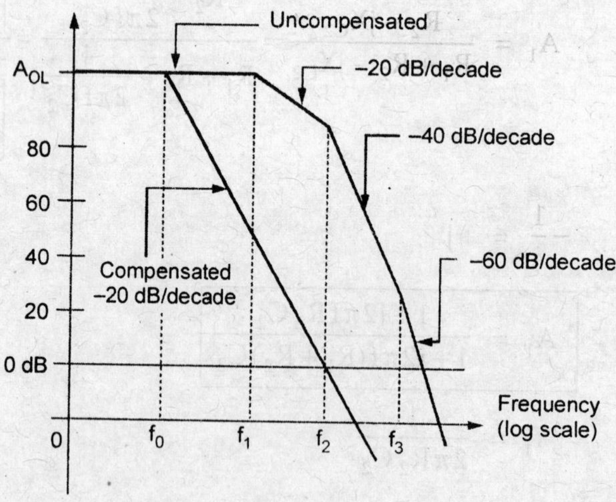

Fig. 3.47 Pole-zero compensation.

As compared to the dominant pole compensation there is improvement in the bandwidth, equal to $f_2 - f_1$. This is the additional advantage of pole-zero compensation technique.

The value of compensation capacitor is generally very large. Hence it is not possible to build such capacitor into standard integrated circuits. Generally the connections are brought out from IC to connect the compensation elements externally.

3.18.3 Feed-Forward Compensation

In multistage amplifiers it is possible that at gain cross-over frequency ω_{gc} the phase shift may become less than – 180° due to individual contributions by the various stages. This is undesirable from the amplifier stability point of view.

Usually there is one stage which is very dominant in providing at least one pole which overrides the effects of other stages hence it acts as bandwidth bottleneck for the over all device. The feed-forward compensation is used to provide a high frequency bypass around such dominant stage to avoid its phase lag. Thus this stage becomes ineffective due to the compensation and the remaining stages provide sufficient phase log in order to obtain a safe value of phase margin.

Consider the multistage amplifier with dominant bottleneck stage with response $A_1(jf)$ and all the remaining stages are combined together to get response $A_2(jf)$. The Fig. 3.48 shows the two stages while Fig. 3.49 shows the individual as well as total frequency response without any compensation used.

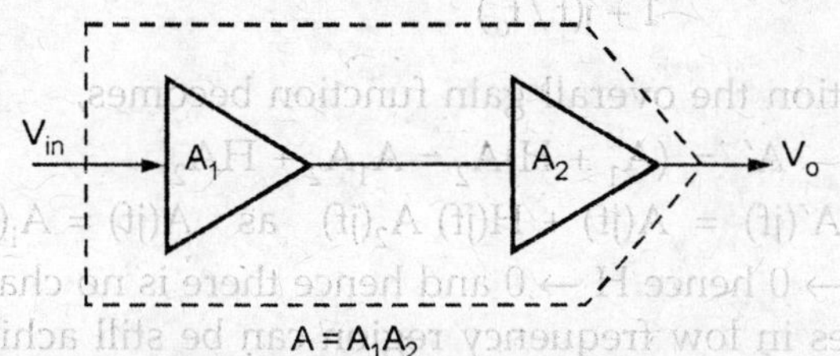

Fig. 3.48

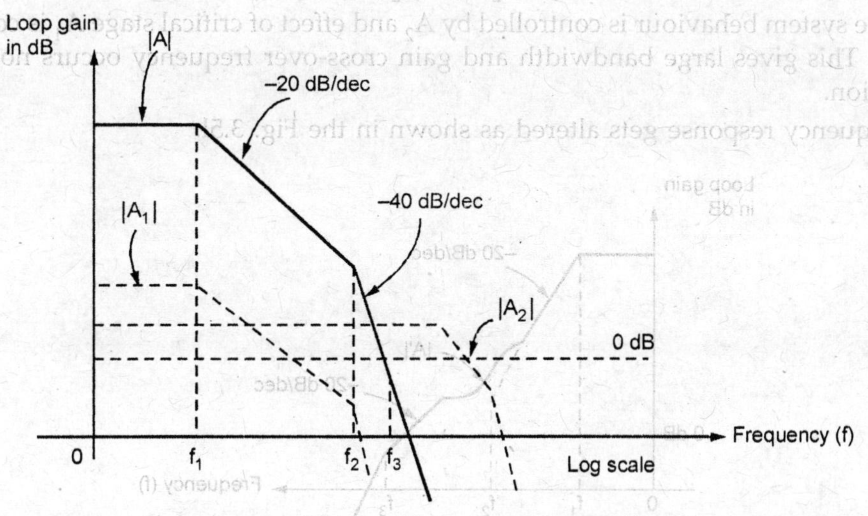

Fig. 3.49

As seen from the frequency response the gain cross-over frequency occurs in the region of roll off − 40 dB/dec which indicates instability of the overall system.

In feed-forward compensation an ordinary high-pass function with transfer function H(jω) is used in the system as shown in the Fig. 3.50.

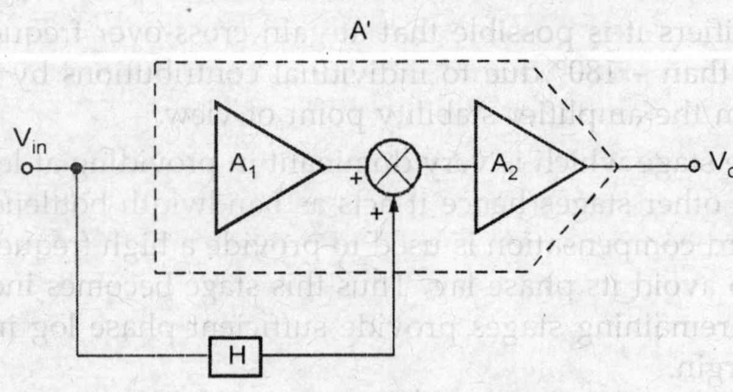

Fig. 3.50

The transfer function of high pass section is,

$$H(jf) = \frac{j(f/f_o)}{1+j(f/f_o)}$$

Now due to this function the overall gain function becomes,

$$A' = (A_1 + H)A_2 = A_1A_2 + HA_2$$

i.e. $$A'(jf) = A(jf) + H(jf)\,A_2(jf) \quad \text{as} \quad A(jf) = A_i(jf)\,A_2(jf)$$

At low frequencies, $f \to 0$ hence $H \to 0$ and hence there is no change in uncompensated response. Thus advantages in low frequency region can be still achieved.

At high frequencies, $f \to \infty$ hence $H \to 1$ and $A \to 0$

$\therefore$ $$A' = H(jf)\,A_2(if) = A_2(jf) \quad \text{for high frequencies}$$

Thus the system behaviour is controlled by A_2 and effect of critical stage A_1 is completely eliminated. This gives large bandwidth and gain cross-over frequency occurs now in low roll off region.

The frequency response gets altered as shown in the Fig. 3.51.

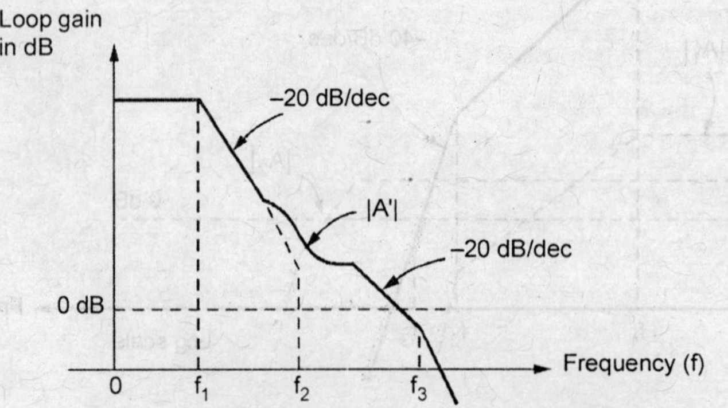

Fig. 3.51 Effect of feed forward compensation.

The disadvantages of this method are:

1. At the time of transition from A to A_2, the phase of A may approach to – 180° before transition and this may produce excessive ringing effect.
2. If phase lag – 180° occurs at transition, signal cancellation results which produces a notch in the compensated response.

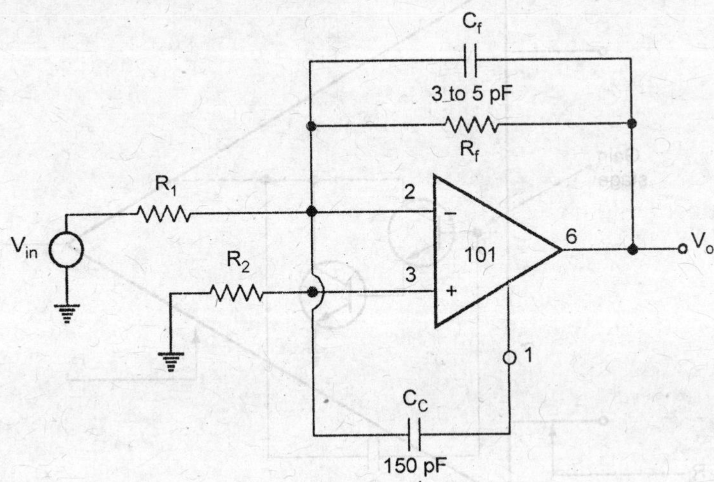

Fig. 3.52 Feed-forward compensation.

In practice feed-forward compensation is provided with a feed-forward capacitor C_c. This is shown for the popular op-amp 101 in the Fig. 3.52. The capacitor C_c is connected between the negative terminal (2) and the compensating terminal (1). A small capacitor C_f is needed across R_f to ensure frequency stability.

The expression for the comer frequency of the bypass function is,

$$f_o = \frac{1}{2\pi R_{ic} C_c}$$

where R_{ic} is the equivalent resistance seen by C_c. This type of compensation gives high slew rate of about 10 V/μs and full-power bandwidth over 200 kHz but it is effective only for inverting type applications.

3.19 INTERNAL COMPENSATION TECHNIQUE

Recently in op-amps like IC 741, the compensation is provided internally, which is generally built in lag compensation.

A capacitor ranging from 10 to 30 pF is fabricated between input and output stage to achieve the necessary compensation. This type of compensation is called **Miller effect compensation**. Such op-amps are called **compensated op-amps.**

3.19.1 Miller Effect Compensation

The main drawback of dominant pole compensation is the reduction in the bandwidth. Similarly the value of capacitance required in this method is very large and fabricating such large capacitance on the chip is not practicable.

These drawbacks are avoided by using Miller effect compensation, using the Miller effect. In dominant pole compensation, a capacitor is connected to ground from the output terminal but in Miller effect compensation it is connected in the feedback path of the Darlington pair used in the output stage of op-amp. This is shown in the Fig. 3.53.

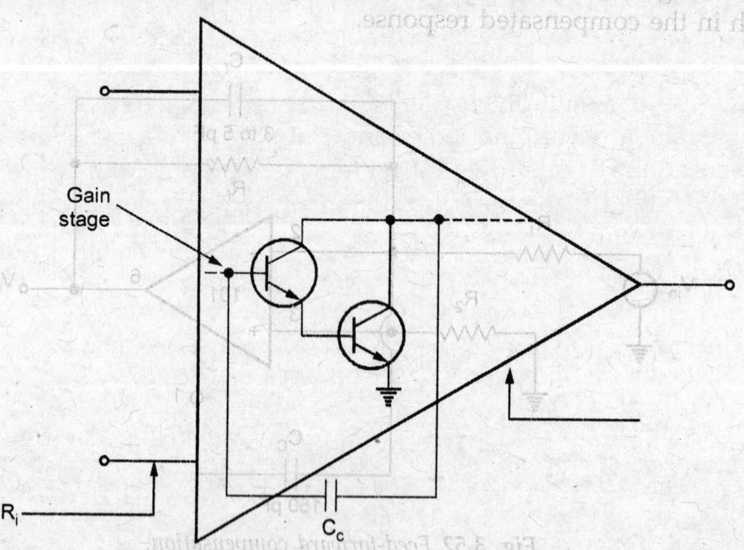

Fig. 3.53 Miller effect compensation.

The C_c is the compensating capacitor, R_i is the input resistance and R_o is the output resistance of the Darlington stage. The gain of the Darlington stage is given by,

$$a_2 = -G_{mc}R_o$$

Where G_m = Transconductance of the stage

Looking through the input terminal C, appears as the Miller capacitance C_M and from the results of Miller effect we can write,

$$Z_{C_M} = \frac{Z_{C_c}}{1+a_2}$$

Where $Z_{C_M} = \frac{1}{j\omega C_M}$ and $Z_{C_c} = \frac{1}{j\omega C_c}$

$$\therefore \quad \frac{1}{j\omega C_M} = \frac{\frac{1}{j\omega C_c}}{(1+a_2)}$$

$$\therefore \quad C_M = (1+a_2)\,C_c$$

Thus effectively C_c gets multiplied by $(1 + a_2)$ where a_2 is the gain of the stage which is large, as viewed through the input terminals. Thus practically small C_c values can be used, which is helpful from monolithic fabrication point of view.

This Miller equivalent capacitance C_M forms a low pass RC section with input resistance R_i whose corner frequency is given by,

$$f_d = \frac{1}{2\pi C_M R_i} \quad ...(16)$$

The uncompensated and compensated magnitude plots are shown in the Fig. 3.54.

In addition to the multiplying the capacitance, Miller effect has another advantage. It causes rearrangement of original poles and cause **Pole splitting**. This means due to Miller effect compensation, f_1 gets lowered while f_2 gets raised. Thus poles get diverged. This increases the bandwidth compared to dominant pole compensation.

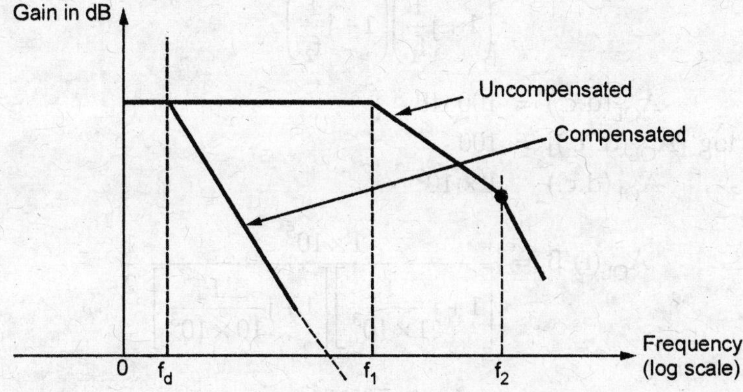

Fig. 3.54 Miller effect compensation.

***Key Point:** The Miller effect compensation is simplest and most commonly used for internally compensated op-amps.*

Such compensated op-amps usually have single break frequency and are inherently stable in nature irrespective of value of closed loop gain. External compensating network is not required for such op-amps.

Some internally compensated op-amps are Fairchild's μA 741, National semiconductor's LM 107, LM 741, LM 112 and Motorola's MC 1858.

Example 3.7: An op-amp circuit is shown in the Fig. 3.55. For this op-amp A_{OL} (d.c.) is 100 dB. The open loop frequency response exhibits two breaks at 1 kHz and 10 kHz. Draw the Bode plot and determine gain margin and phase margin of the op-amp circuit.

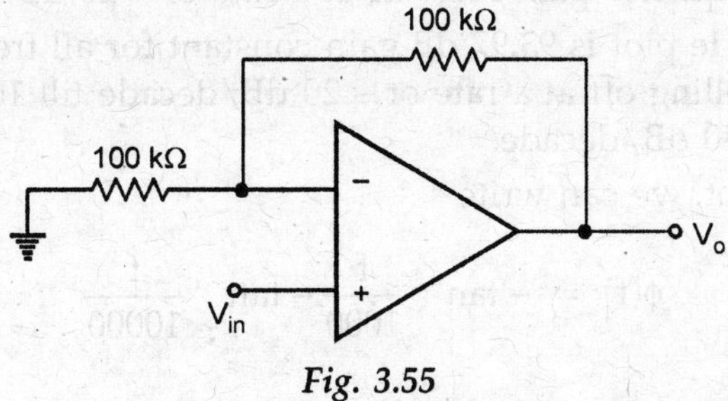

Fig. 3.55

Solution: The circuit is non-inverting amplifier. The feedback factor of the circuit (β) is

$$\beta = \frac{1}{\left[1+\frac{R_f}{R_1}\right]} = \frac{1}{1+\left[\frac{100}{100}\right]} = \frac{1}{2}$$

It has two break frequencies as

$$f_1 = 1 \text{ kHz} \quad \text{and} \quad f_2 = 10 \text{ kHz}$$

Hence the loop gain of the circuit is

$$A_{OL}(f)\,\beta = \frac{A_{OL}(d.c.)}{\left(1+j\frac{f}{f_1}\right)\left(1+j\frac{f}{f_2}\right)}.\beta \quad f = \text{input frequency}$$

Now $A_{OL}(d.c.) = 100 \text{ dB}$

$\therefore$ $20 \log [A_{OL}(d.\ c.)] = 100$

$\therefore$ $A_{OL}(d.c.) \quad 1 \times 10^5$

$$\therefore \quad A_{OL}(f)\,\beta = \frac{1\times 10^5}{\left[1+j\frac{f}{1\times 10^3}\right]\left[1+j\frac{f}{10\times 10^3}\right]}.\frac{1}{2}$$

$$= \frac{5\times 10^4}{\left[1+j\frac{f}{1000}\right]\left[1+j\frac{f}{10000}\right]}$$

$$|A_{OL}(f)\beta| = \frac{5\times 10^4}{\sqrt{1+\left(\frac{f}{1000}\right)^2}\cdot\sqrt{1+\left(\frac{f}{10000}\right)^2}}$$

$$\therefore \quad \text{Magnitude in dB} = 20 \log \frac{5\times 10^4}{\sqrt{1+\left(\frac{f}{1000}\right)^2}\cdot\sqrt{1+\left(\frac{f}{10000}\right)^2}}$$

At every corner frequency gain rolls off at a rate of – 20 dB/dec.

So overall magnitude plot is 93.97 dB gain constant for all frequencies less than 1 kHz. From 1 kHz it starts rolling off at a rate of – 20 dB/decade till 10 kHz. From 10 kHz, gain rolls off at a rate of – 40 dB/decade.

For phase angle plot, we can write

$$\phi(f) = -\tan^{-1}\frac{f}{1000} - \tan^{-1}\frac{f}{10000}$$

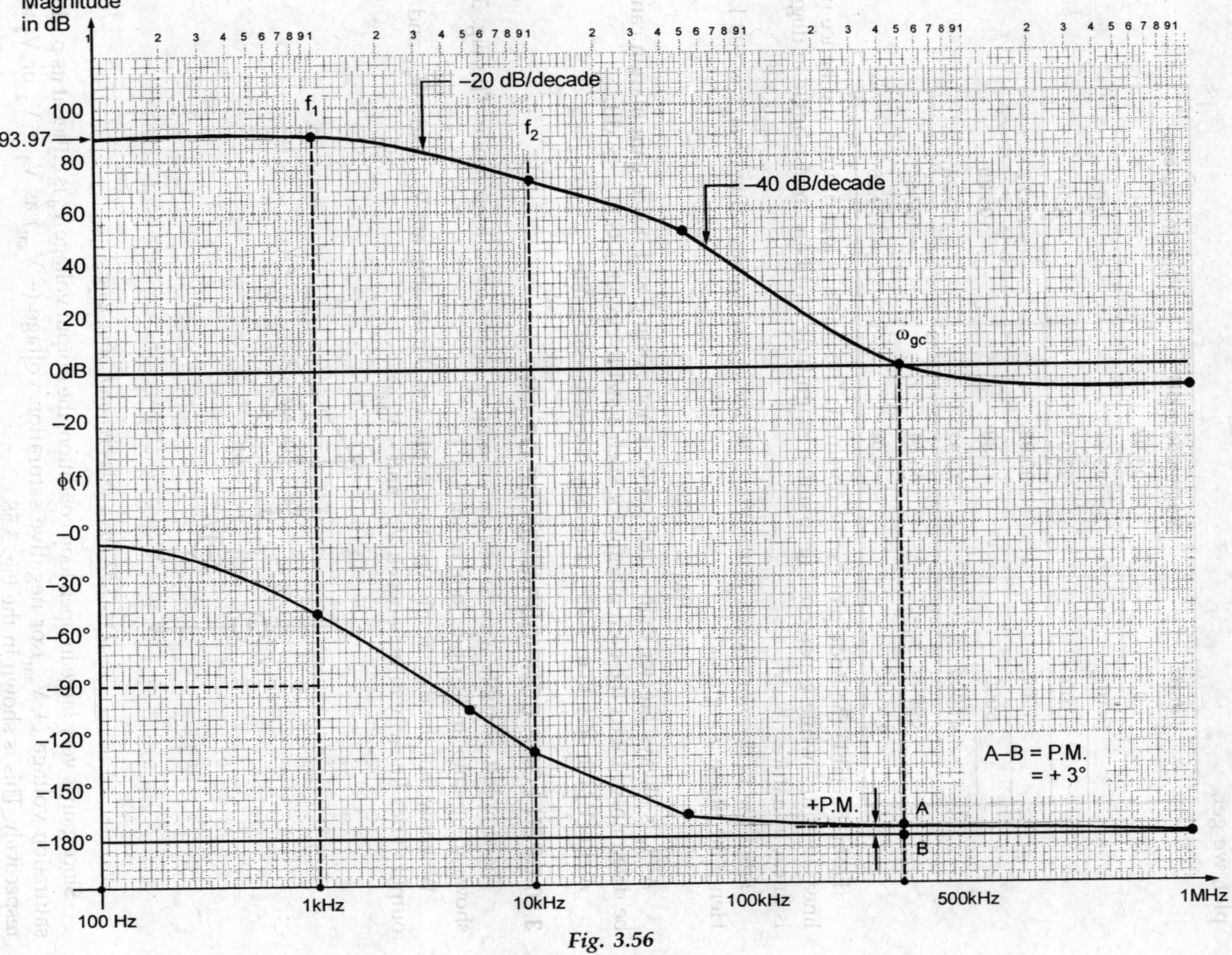
Magnitude in dB
100
93.97
80
60
40
20
0dB
–20
φ(f)
–0°
–30°
–60°
–90°
–120°
–150°
–180°
f1
f2
–20 dB/decade
–40 dB/decade
ωgc
+P.M.
A
B
A–B = P.M.
= + 3°
100 Hz
1kHz
10kHz
100kHz
500kHz
1MHz

Fig. 3.56

Now tabulating magnitude and phase angles for various frequencies to sketch Bode plot, we get,

f in Hz	Magnitude in dB	ϕ (f) in degrees
0	93.97	0°
10	93.97	– 0.63
100	93.93	– 6.28
1000	90.92	– 50.71
5000	77.89	– 105.25
10000	70.92	– 129.28
50000	45.84	– 167.54
1×10^6	– 6.02	– 179.36

From this table the Bode plot can be obtained. The x-axis is logarithmic and hence non-linear in nature while y-axis used for plotting magnitude in dB and phase angle in degrees is linear in nature.

From the Bode plot shown in the Fig. 3.56 the approximate value of ω_{gc}. is 500 kHz. Hence the phase margin is

$$P.M = +3°$$

As ϕ_f is always greater than – 180°, cure is not existing and hence gain margin cannot be determined. Infact the gain margin for such a stable circuit is G.M. = ∞ dB.

As the P.M. is less than + 45°, partically circuit is not stable.

3.20 OPEN LOOP CONFIGURATION OF OP-AMP

The simplest possible way to use an op-amp is in the open loop mode. The Fig. 3.57 shows an op-amp in the open loop condition.

We know that the d.c. supply voltages applied to the op-amp are V_{CC} and – V_{EE} and the output varies linearily only between V_{CC} and – V_{EE}.

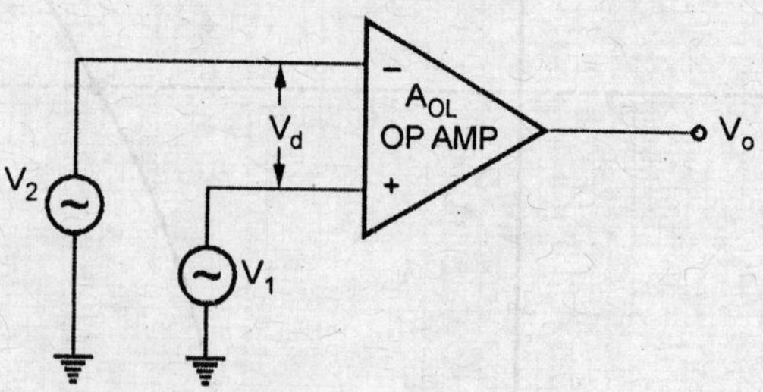

Fig. 3.57 Open loop operation of op-amp.

Since gain is very large in open loop condition, the output voltage V_o is either at its positive saturation voltage (+ V_{sat}) or negative saturation voltage (– V_{sat}) as $V_1 > V_2$ or $V_2 > V_1$ respectively. This is shown in the Fig. 3.58.

Thus very small noise voltage present at the input also gets amplified due to its high open loop gain and op-amp gets saturated.

It can be seen from the Fig. 3.58, only for small range of input signal (from point a to b), it behaves linearily. This range is very small and practically due to high open loop gain, op-amp either shows $+V_{sat}$ or $-V_{sat}$ level.

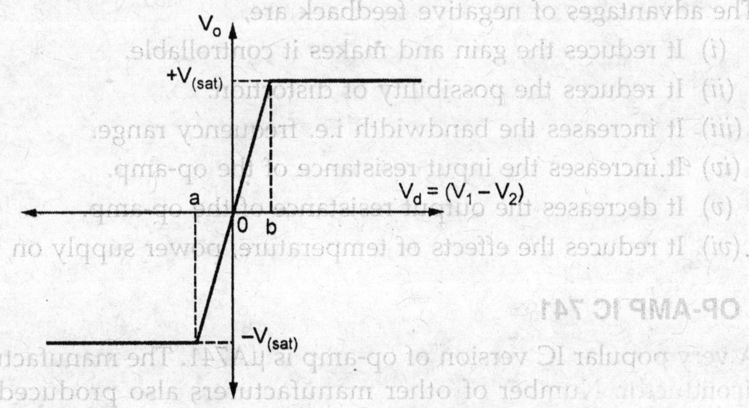

Fig. 3.58 Voltage transfer characteristics.

This indicates the inability of op-amp to work as a linear small signal amplifier in the open loop mode.

Hence, the op-amp is generally not used in the open loop configuration.

Key Point: Such an open loop behaviour of the op-amp finds some rare applications like voltage comparator, zero crossing detector etc.

3.21 CLOSED LOOP CONFIGURATION OF OP-AMP

The utility of op-amp increases considerably if it is used in a closed loop mode. The closed loop mode is possible using feedback. The feedback allows to feed some part of the output back to the input. In linear applications the op-amp is always used with negative feedback. The feedback helps to control gain which otherwise drives op-amp into saturation.

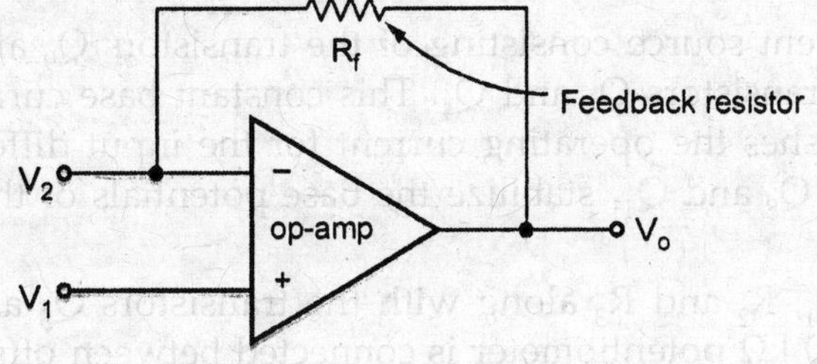

Fig. 3.59 Op-amp with negative feedback.

The negative feedback is possible by adding a resistor as shown in the Fig. 3.59, called **feedback resistor.** The feedback is said to be negative as the feedback resistor connects the output to the inverting input terminal.

The gain resulting with feedback is called **closed loop** gain of the op-amp. Due to feedback resistance there is reduction in the gain. The closed loop gain is much less than the open loop gain and is independent of it.

Most of the linear circuits use op-amp in a closed loop mode with negative feedback with R_f. This is because, due to reduced gain, the output is not driven into the saturation and the circuit behaves in a linear manner.

The advantages of negative feedback are,

(*i*) It reduces the gain and makes it controllable.

(*ii*) It reduces the possibility of distortion.

(*iii*) It increases the bandwidth i.e. frequency range.

(*iv*) It increases the input resistance of the op-amp.

(*v*) It decreases the output resistance of the op-amp.

(*vi*) It reduces the effects of temperature, power supply on the gain of the circuit.

3.22 OP-AMP IC 741

A very popular IC version of op-amp is μA741. The manufacturer of μA741 is Fairchild Semiconductor. Number of other manufacturers also produced 741 op-amp IC version namely MC1741 by Motorola, LM741 by National Semiconductor etc. For convenience, this widely used op-amp IC is called simply IC 741 op-amp dropping the prefixes.

3.22.1 Internal Schematic of IC 741

The Fig. 3.60 shows the schematic diagram of the IC 741 op-amp that has become an industry standard. This equivalent circuit illustrates the internal structure of the 741 IC op-amp. The diagram helps to understand all the internal stages along with their capabilities and the limitation. The most important point about this equivalent circuit is that all the 741 models have this same equivalent circuit. The data sheet of IC includes absolute maximum ratings, electrical specifications, the performance curves and typical applications. The data sheets for IC μA 741 are given in the appendix for the further information.

The input stage is a differential amplifier stage comprising of Q_1 and Q_2 transistors, between which the differential input signal is applied. The transistors Q_3 and Q_4 are also the part of the input differential amplifier used to increase the maximum signal input capacity.

The constant current source consisting of the transistors Q_9 and Q_{10} supplies constant base current for the transistors Q_3 and Q_4. This constant base current in the transistors Q_3 and Q_4 inturn establishes the operating current for the input differential stage. The diode connected transistors Q_8 and Q_{11} stabilize the base potentials of the transistors Q_9 and Q_{10} respectively.

The resistances R_1, R_2 and R_3 along with the transistors Q_5 and Q_6 form a controlled current source. The 10 kΩ potentiometer is connected between offset null terminals and by connecting the wiper to the negative supply. This potentiometer is used to control the emitter currents of the transistors Q_5 and Q_6. This arrangement helps in minimising offset voltage and currents. The transistor Q_7 supplies the base currents to the transistors Q_5 and

Q_6. The base currents and hence the collector currents of the transistors Q_5 and Q_6 must be equal at all the times as the transistors Q_5 and Q_6 are identical. The collector voltage of the transistor Q_5 controls the operation of the transistor Q_7. This ensures that any change in Q_5 collector voltage, produces identical changes in the transistors Q_5 and Q_6 collector currents, via the transistor Q_7.

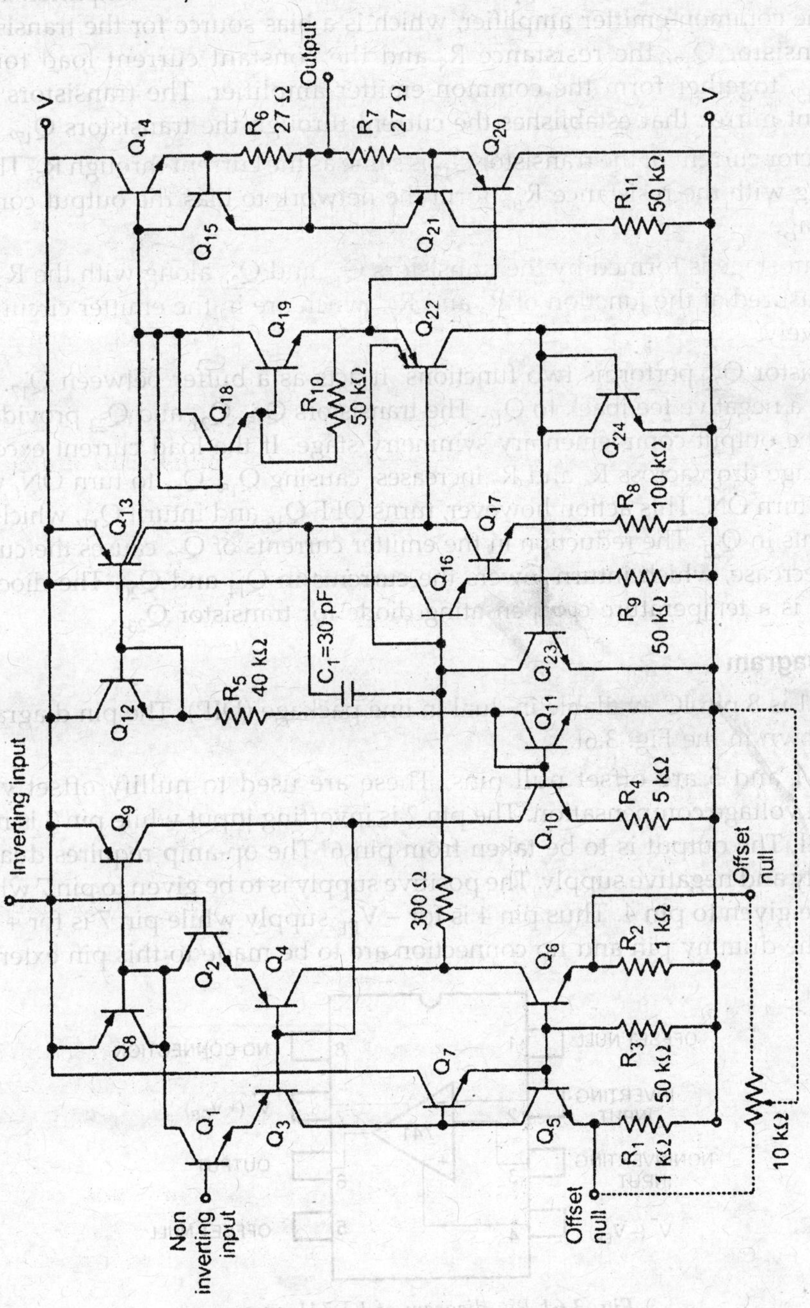

Fig. 3.60 Internal schematic of IC 741.

The transistors Q_4 and Q_6 form the complementary symmetry amplifier. The output of the differential stage is taken at the junction of the transistors Q_4 and Q_6. This output at the junction of Q_4 and Q_6 is proportional to the differential input signal.

The common collector amplifier is formed by the transistor Q_{16} and resistance R_9. The output of the differential stage is applied to this common collector amplifier. The output of Q_{16} drives the common-emitter amplifier, which is a bias source for the transistors Q_{18} and Q_{19}. The transistor Q_{17}, the resistance R_8 and the constant current load formed by the transistor Q_{13}, together form the common emitter amplifier. The transistors Q_{12} and Q_{13} form a current mirror that establishes the current through the transistors Q_{17}, Q_{18} and Q_{19}.

The collector current of the transistor Q_{13} is same as the current through R_5. The transistors Q_{18}, Q_{19} along with the resistance R_{10}, form the network to bias the output complementary symmetry stage.

The output stage is formed by the transistors Q_{14} and Q_{20} along with the R_6 and R_7. The output is measured at the junction of R_6 and R_7, which are in the emitter circuits of Q_{14} and Q_{20}, respectively.

The transistor Q_{22} performs two functions. It acts as a buffer between Q_{17} and Q_{20} and also provides a negative feedback to Q_{16}. The transistors Q_{15}, Q_{21} and Q_{23} provide the current limiting for the output complementary symmetry stage. If the load current exceeds the safe limit, the voltage drop across R_6 and R_7 increases, causing Q_{15}, Q_{21} to turn ON, which inturn causes Q_{23} to turn ON. This action however, turns OFF Q_{16} and inturn Q_{17}, which reduces the emitter currents in Q_{22}. The reduction in the emitter currents of Q_{22} causes the currents in Q_{18} and Q_{19} to decrease, which inturn lowers the currents in Q_{14} and Q_{20}. The diode connected transistor Q_{24} is a temperature compensating diode for transistor Q_{23}.

3.22.2 Pin Diagram

The IC 741 is 8 pin IC available in dual in line package (DIP). The pin diagram of IC 741 op-amp is shown in the Fig. 3.61.

The pins 1 and 5 are offset null pins. These are used to nullify offset voltages and provide offset voltage compensation. The pin 2 is inverting input while pin 3 is noninverting input terminal. The output is to be taken from pin 6. The op-amp requires dual supply i.e. positive supply and negative supply. The positive supply is to be given to pin 7 while negative supply is to be given to pin 4. Thus pin 4 is for $-V_{EE}$ supply while pin 7 is for $+V_{CC}$ supply. The pin 8 is the dummy pin and no connection are to be made to this pin externally.

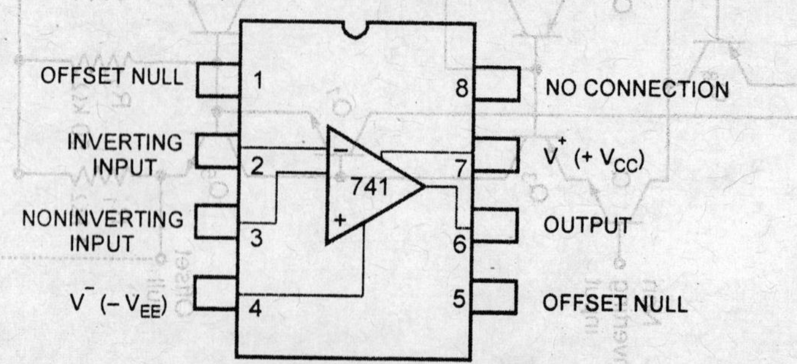

Fig. 3.61 Pin diagram of IC 741 op-amp.

3.22.3 Ideal Vs Practical Characteristics of IC 741 Op-amp

Let us compare the ideal op-amp characteristics with practical characteristics of IC 741 op-amp.

The Table 3.3 lists the ideal op-amp characteristics and the typical characteristics for 741 IC, a popular general purpose op-amp IC.

Table 3.3

Sr. No	Parameter	Symbol	Ideal	Typical for 741 rc
1.	Open loop voltage gain	A_{OL}	∞	2×10^5
2.	Output impedance	Z_{out}	0	75 Ω
3.	Input impedance	Z_{in}	∞	2 MΩ
4.	Input offset current	I_{ios}	0	20 nA
5.	Input offset voltage	V_{ios}	0	2 mV
6.	Bandwidth	B.W.	∞	1 MHz
7.	CMRR	ρ	∞	90 dB
8.	Slew rate	S	∞	0.5 V/μsec
9.	Input bias current	I_b	0	80 nA
10.	Power supply rejection ratio	PSRR	0	30 μV/V

3.22.4 Features of IC 741

(*i*) No frequency compensation required.

(*ii*) Short circuit protection provided.

(*iii*) Offset voltage null capability.

(*iv*) Large common mode and differential voltage range.

(*v*) No latch up.

3.22.5 Packages of Op-amp

The op-amp ICs are available in various packages. The three popular packages available for op-amp are,

1. The metal can package denoted as TO.
2. The dual in line package denoted as DIP.
3. The flat package or flat pack.

The metal can packages are available with 3, 5, 8, 10 and 12 pins. The metal sealing plane is at the bottom, used over which the silicon chip bonded. This plane is effective for the dissipation of heat. The Fig. 3.62 shows the 8 pin metal can package and the connection diagram. The tab is used to identify pin 8 and the pins are numbered counterclockwise when metal can is viewed from the top.

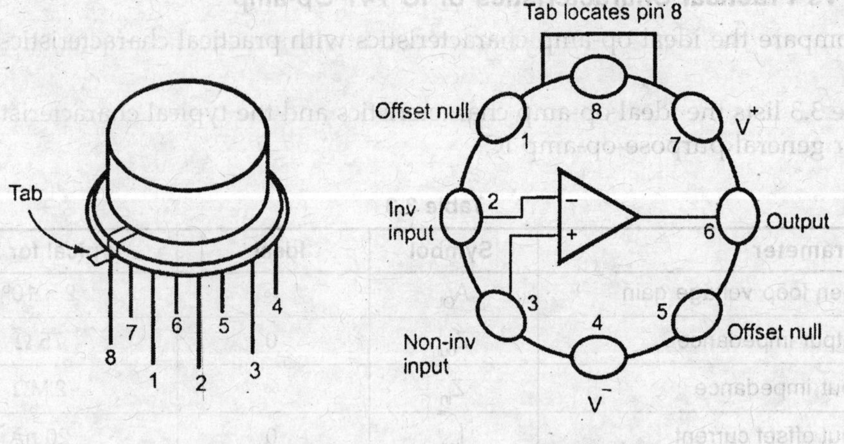

Fig. 3.62 Metal can package.

The dual in line pack (DIP) is popular for commercial applications. It is easy to handle, fits standard mounting hardware and is inexpensive when molded in plastic. Ceramic DIPs are used for high temperature, high performance (usually military) equipment. The mini DIP provides a low cost/space efficient package.

The Fig. 3.63 shows 8 pin and 14 pin DIPs and their connection diagrams. For DIPs either plastic or ceramic cases are available. The pin 1 is indicated by a notch or dot, as viewed from the top and other terminals are numbered counterclockwise.

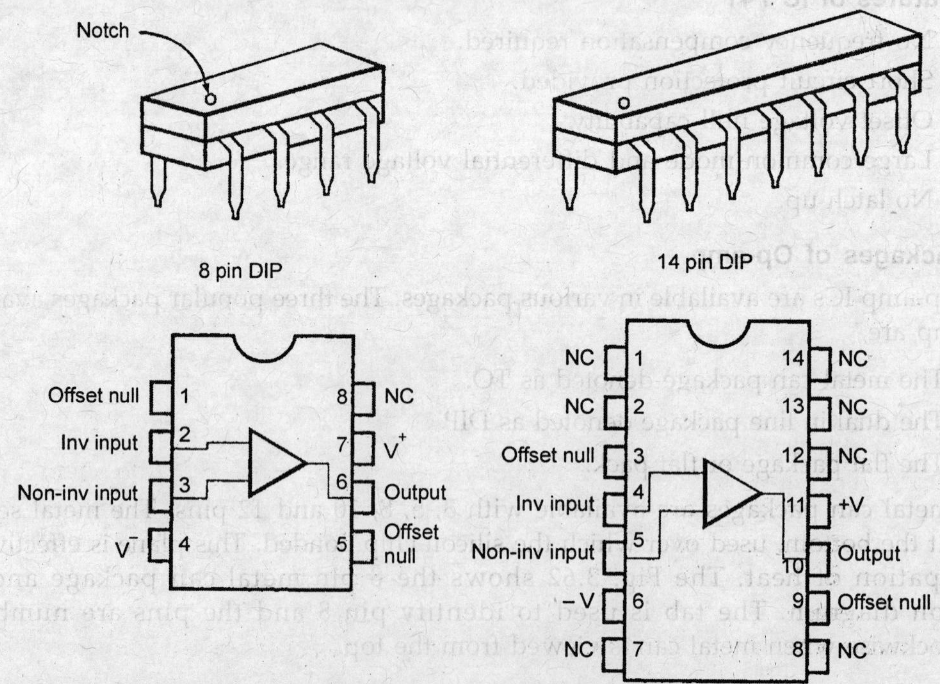

Fig. 3.63 Dual in line packages.

For circuits where space is critical, the flat pack gives a compact package. However, flat packs are much more difficult to handle than DIPs and often do not dissipate power as well. The metal can packages allow easy connection to a heat sink, and are chosen when heat dissipation is the main consideration.

The Fig. 3.64 shows 10 pin flat package where the chip is enclosed in a rectangular ceramic case. The terminals are taken out through the sides and ends.

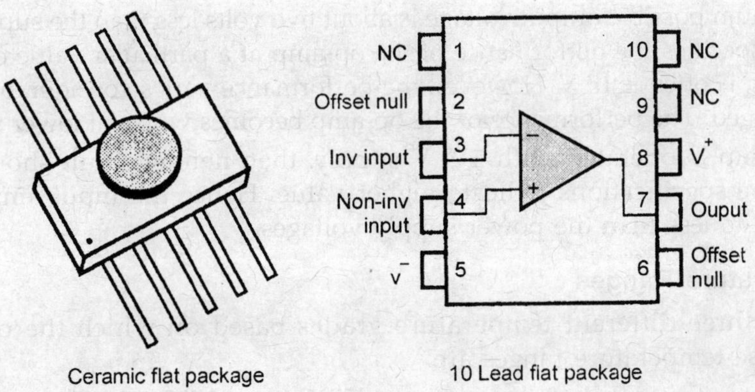

Fig. 3.64 Flat package.

Nowadays, the surface mounted technology (SMI) is used where higher circuit density for a package of a given size is required. This is complex circuit in which many op-amps, other res and many large chips are placed in a single package. Such packages have lower noise and improved frequency response characteristics.

3.22.6 Commercial Op-amp Parameters

Before selecting an op-amp for any application, it is necessary to study its absolute maximum ratings included in the datasheet. The absolute maximum ratings of the commercial op-amp include the ratings such as:

(i) Maximum power dissipation:

This is the maximum power which can be dissipated, In the internal stages of the op-amp in the form of heat.

(ii) Operating temperature range:

As specified in the data sheet, op-amp can work satisfactorily, over the operating temperature range, as required for the given application.

(iii) Maximum supply voltage:

This is the maximum d.c. supply voltage which can be applied to the op-amp.

(iv) Maximum differential input voltage:

This rating gives the maximum value of difference between the two input voltages, applied to the two input terminals of the op-amp.

(v) Maximum common mode input voltage:

This is the maximum value of the input voltage which can be simultaneously applied to the two input terminals.

(vi) Storage temperature range:

This gives the temperature range over which the op-amp can be stored safely. If these maximum ratings are exceeded, op-amp can be damaged.

The maximum possible output voltage is about two volts less than the supply voltage. The manufacturer specifies the performance of the op-amp at a particular value of power supply voltages, which is often ± 15 V. However, the performance varies considerably as the supply voltage is changed. The performance of the op-amp becomes worse at lower supply voltages.

If an op-amp is operated with ± 9 V supply, then neither input should exceed 9 V, eventhough the specifications indicate higher value. Hence the input amplitude is to be always limited to less than the power supply voltages.

3.22.7 Temperature Ranges

There are three different temperature grades based on which the op-amp ICs are classified. These temperature ranges are:

(*i*) Military temperature range : – 55°C to + 125°C (or – 55°C to + 85°C)

(*ii*) Industrial temperature range : – 20°C to + 85°C (or – 40°C to + 85 °C)

(*iii*) Commercial temperature range : 0°C to 70°C (or 0°C to + 75°C)

Individual data sheet of each IC include the various IC parameters and the temperature conditions under which the parameters are determined. The military grade devices are always superior and with very precisely controlled parameters. The commercial grade ICs have the worst tolerance and hence are the cheapest.

3.22.8 Classes of Op-amp IC 741

The op-amp IC 741 is available in different classes such as 741, 741 A, 741 C, 741 E, 741 S and 741 SC. The classes are basically based on military grade or commercial grade of op-amp IC and depending upon improved electrical specifications.

Table 3.4 Classes of op-amp IC 741

Class	Type	Remark
741	Military grade	Temperature range – 55°C to + 125°C
741 A	Improved version of 741	Better electrical characteristics
741 C	Commercial grade	Temperature range 0° to 70°C
741 E	Improved version of 741C	Better electrical characteristics
741 S	Military grade	Higher slew rate
741 SC	Commercial grade	Higher slew rate

The following table gives comparison of electrical characteristics of the classes of op-amp IC 741:

Table 3.5 Comparison of classes of op-amp IC 741

No.	Parameter	741	741 A	741 C	741 E
1.	Input offset voltage	1 mV	0.8 mV	2 mV	0.8 mV
2.	Input offset current	20 nA	3 nA	20 nA	3 nA
3.	Input bias current	80 nA	30 nA	80 nA	30 nA
4.	PSRR	30 μV/V	15 μV/V	30 μV/V	15 μV/V
5.	Input impedance	2 MΩ	6 MΩ	2 MΩ	6 MΩ
6.	Bandwidth	1 MHz	1.5 MHz	1 MHz	1.5 MHz
7.	Slew rate	0.5 V/μs	0.7 V/μs	0.5 V/μs	0.7 V/μs
8.	CMRR	90 dB	95 dB	90 dB	95 dB
9.	Power consumption	60 mW	80 mW	50 mW	80 mW

3.23 APPLICATIONS OPEN LOOP OP-AMP CONFIGURATIONS

The configuration in which output depends on input, but output has no effect on the input is called open loop configuration. No feedback from output to input is used in such configuration. The op-amp works as high gain amplifier in such open loop configuration. The op-amp can be used in three modes while using it in open loop configuration. These three open loop configurations are,

1. Differential amplifier
2. Inverting amplifier
3. Noninverting amplifier

Let us discuss these in detail.

3.23.1 Differential Amplifier

The amplifier which amplifies the difference between the two input voltages is called differential amplifier. The Fig. 3.65 shows the open loop differential amplifier.

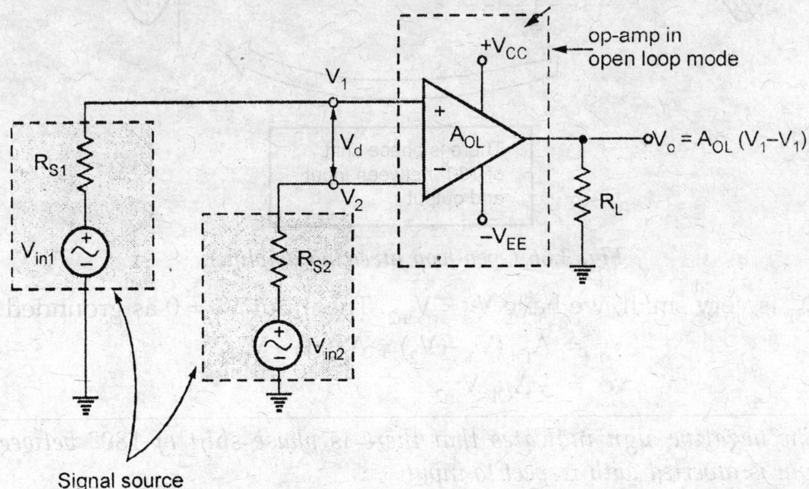

Fig. 3.65 Open loop differential amplifier.

The V_{in1} and V_{in2} are the signal sources. R_{S1} and R_{S2} are the internal source resistances. The V_1 and V_2 can be a.c. or d.c. because the op-amp can amplify both a.c. and d.c. voltages. The values of R_{S1} and R_{S2} are practically very small. Hence the drop across these resistances can be neglected which gives us $V_1 = V_{in1}$ and $V_2 = V_{in2}$.

As op-amp amplifies the difference voltage we can write,

$$V_o = A_{OL} V_d = A_{OL}(V_1 - V_2) = A_{OL}(V_{in1} - V_{in2})$$

Thus it works as the differential amplifier.

The output polarity depends on the polarity of the difference voltage V_d. The load resistance R_L is generally selected more than 2 kΩ.

***Key Point:** For very small V_d, output gets driven into saturation due to high A_{OL}, hence this application is applicable for very small range of differential input voltage.*

3.23.2 Inverting Amplifier

The amplifier in which the output is inverted i.e. having 180° phase shift with respect to the input is called an inverting amplifier.

In such configuration the input is applied to the inverting input only and the non-inverting terminal is grounded. The configuration is shown in the Fig. 3.66.

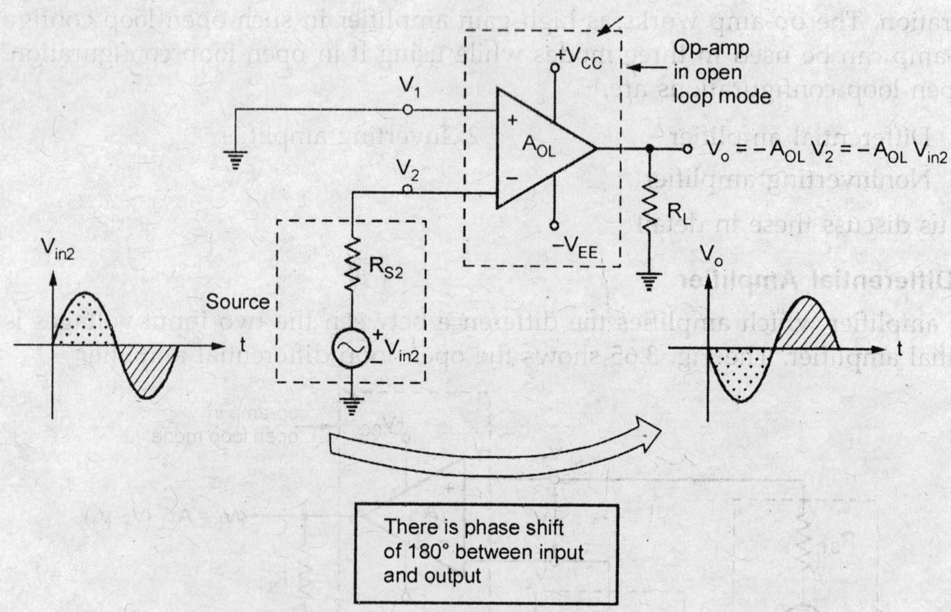

Fig. 3.66 Open-loop inverting amplifier.

Now as R_{S2} is very small, we have $V_2 = V_{in2}$. The input $V_1 = 0$ as grounded.

So
$$V_o = A_{OL}(V_1 - V_2) = A_{OL}(0 - V_{in2})$$
$$V_o = -A_{OL} V_{in2}$$

***Key Point:** The negative sign indicates that there is phase shift of 180° between input and output i.e. output is inverted with respect to input.*

The input V_{in2} is amplified A_{OL} times, so configuration works as an inverting amplifier.

3.23.3 Non-inverting Amplifier

The amplifier in which the output is amplified input without any phase shift in between input and output is called noninverting amplifier. The output and input are in phase.

In such configuration the input is applied to the noninverting input only and the inverting terminal is grounded. The configuration is shown in the Fig. 3.68.

As R_{S1} is very small, we have $V_1 = V_{in1}$. The input $V_2 = 0$ as grounded.

So $$V_o = A_{OL}(V_1 - V_2)$$

$\therefore$ $$V_o = A_{OL}(V_{in1} - 0) = A_{OL}\, V_{in1}$$

***Key Point:** The positive output shows that input and output are in phase and input is amplified A_{OL} times to get the output. So configuration works as noninverting amplifier.*

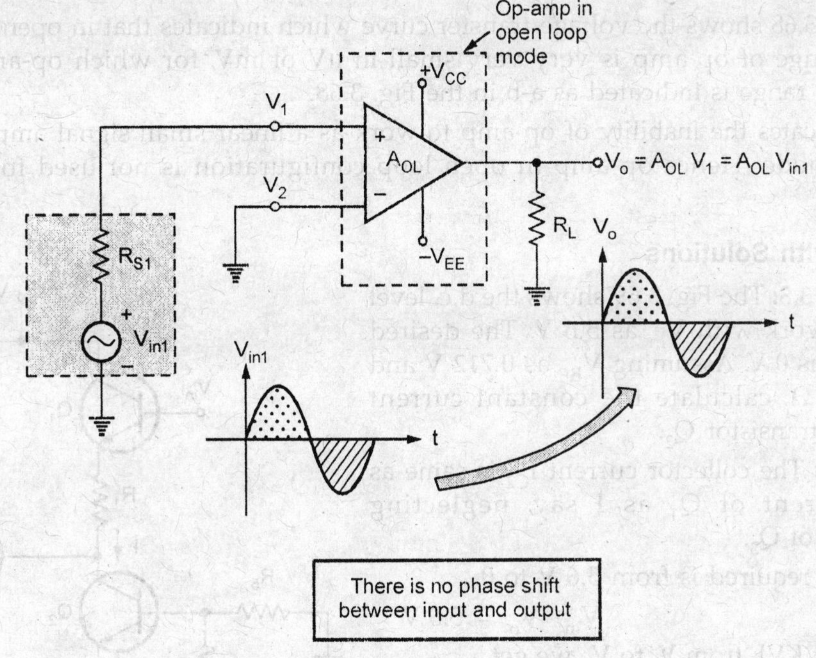

Fig. 3.67 Open-loop non-inverting amplifier.

Why Op-Amp is generally not used in open loop mode?

As open loop gain of op-amp is very large, very small input voltage drives the op-amp voltage to the saturation level. Thus in open loop configuration, the output is at its positive saturation voltage ($+V_{sat}$) or negative saturation voltage ($-V_{sat}$) depending on which input V_1 or V_2 is more than the other. For a.c. input voltages, output may switch between positive and negative saturation voltages.

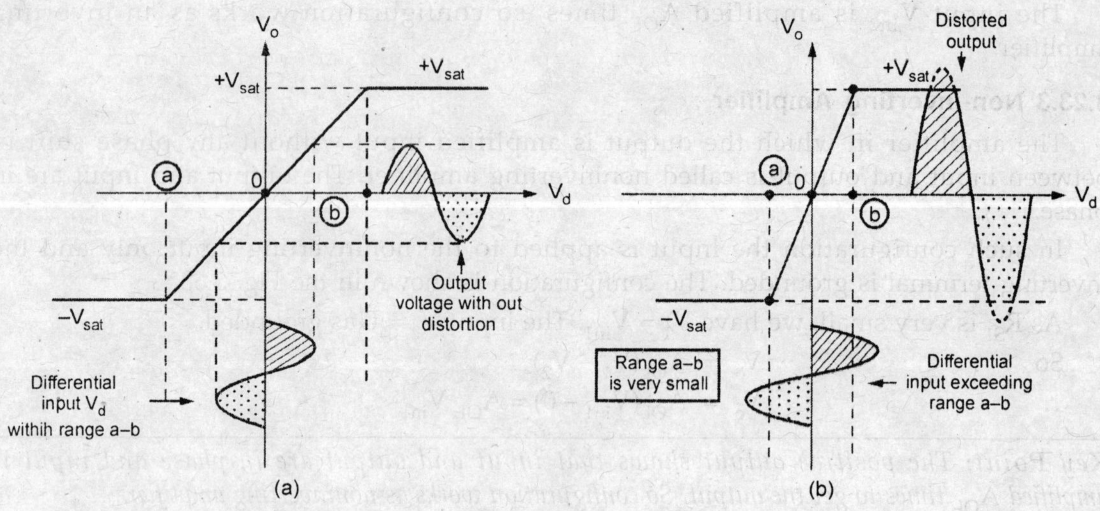

Fig. 3.68

The Fig. 3.68 shows the voltage transfer curve which indicates that in open loop mode, the input range of op-amp is very very small in μV or mV, for which op-amp behaves linearly. This range is indicated as a-b in the Fig. 3.68.

This indicates the inability of op-amp to work as a linear small signal amplifier in the open loop mode. Hence op-amp in open loop configuration is not used for the linear applications.

Examples with Solutions

Example 3.8: The Fig. 3.69 shows the d.c. level shifting network with V_{in} as 5.6 V. The desired output level is 0 V. Assuming V_{BE} as 0.712 V and R_1 as 1.47 kΩ, calculate the constant current supplied by transistor Q_2.

Solution: The collector current I_{C2} is same as emitter current of Q_1 as I say, neglecting base current of Q_3.

The shift required is from 5.6 V to 0.

$$V_{in} - V_o = 5.6 \text{ V}$$

Applying KVL from V_i to V_o we get,

$$V_{in} - I_{R1} - V_{BE1} - V_{BE2} = 0$$

$$V_{BE1} = V_{BE2} = 0.712 \text{ V}$$

$$\therefore \quad 5.6 - 1 \times 1.47 \times 10^3 - 2 \times 0.712 = 0$$

$$\therefore \quad I = \mathbf{2.84 \text{ mA}}$$

This is the constant current source value required to get 0 V d.c. output level.

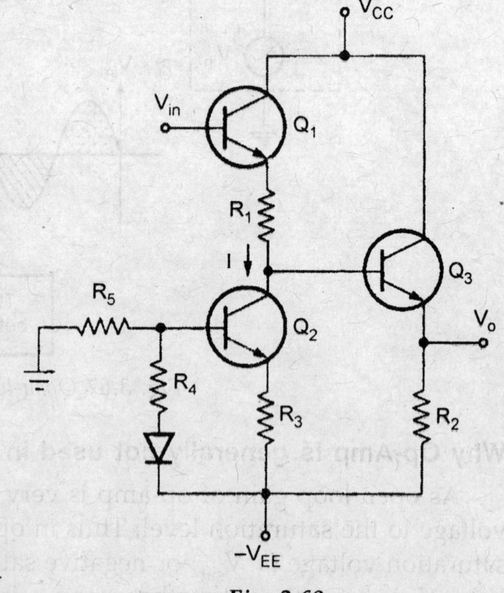

Fig. 3.69

Example 3.9: Derive the expression for the d.c. shift between input and output provided by the circuit shown in the Fig. 3.70.

Assume the base currents of transistors as negligibly small.

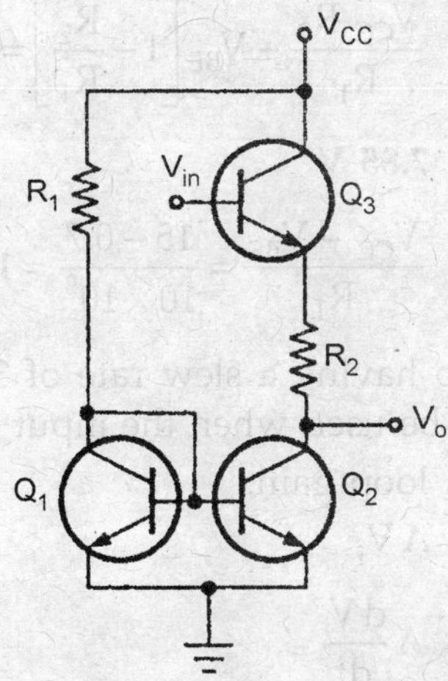

Fig. 3.70

Solution: The various currents and voltages in the circuit are shown in the Fig. 3.71. Transistors Q_1 and Q_2 forms a current mirror. Neglecting base currents of Q_1 and Q_2,

$$I_{C1} = I = I_{C2}$$

Applying KVL from V_{CC} to ground through R_1 and Q_1 we get,

$$V_{CC} - R_1 - V_{BE} = 0$$

$$\therefore \quad I = I_{C1} = I_{C2} = \frac{V_{CC} - V_{BE}}{R_1}$$

Applying KVL from V_{in} to V_o,

$$V_{in} - V_{BE} - I_{C2} R_2 = V_o$$

But the shift is $V_{in} - V_o$,

$$\therefore \quad V_{in} - V_o = V_{BE} + I_{C2} R_2$$

Substituting I_{C2},

$$V_{in} - V_o = V_{BE} + \frac{(V_{CC} - V_{BE})R_2}{R_1}$$

$$\therefore \quad \boxed{V_{in} - V_o = \frac{V_{CC}R_2}{R_1} + V_{BE}\left[1 - \frac{R_2}{R_1}\right]}$$

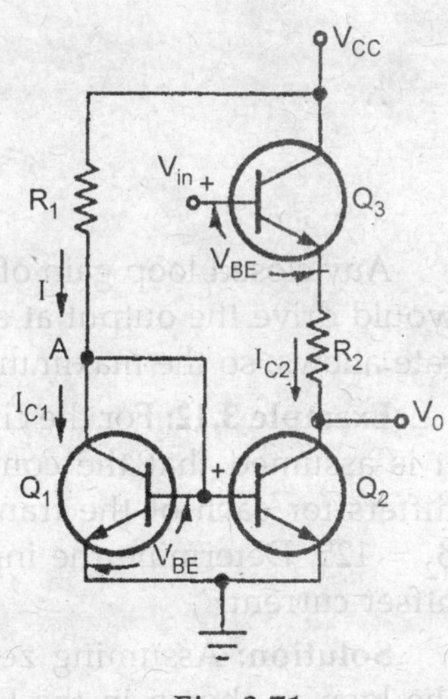

Fig. 3.71

This is the expression for the d.c. shift.

Example 3.10: In the example 3.8 if V_{CC} is 15 V, R_1 is 10 kΩ, R_2 is 5 kΩ, calculate the shift assuming V_{BE} as 0.7 V. Also determine the current I.

Solution: The shift is,

$$V_{in} - V_o = \frac{V_{CC} R_2}{R_1} + V_{BE}\left[1 - \frac{R_2}{R_1}\right] = \frac{15 \times 5 \times 10^3}{10 \times 10^3} + 0.7\left[1 - \frac{5 \times 10^3}{10 \times 10^3}\right]$$

$$= \mathbf{7.85\ V}$$

And $$I = \frac{V_{CC} - V_{BE}}{R_1} = \frac{15 - 0.7}{10 \times 10^3} = \mathbf{1.43\ mA.}$$

Example 3.11: For an op-amp having a slew rate of 3 V/µsec. What is the maximum closed loop voltage gain that can be used when the input signal varies by 0.4 V in 12 µsec?

Solution: Let A be the closed loop gain.

$$\therefore \quad V_o = A V_i$$

$$\therefore \quad \frac{dV_o}{dt} = A \frac{dV_i}{dt}$$

but $$\frac{dV_o}{dt} = \text{slew rare} = S$$

$$\therefore \quad S = A \frac{dV_i}{dt}$$

$$\therefore \quad A = \frac{S}{\left(\frac{dV_i}{dt}\right)} = \frac{\left(\frac{3}{10^{-6}}\right)}{\left(\frac{0.4}{12 \times 10^6}\right)} = \mathbf{90}$$

Any closed loop gain of magnitude greater than 90 would drive the output at a rate greater than the slew rate allows, so the maximum closed loop gain is 90.

Example 3.12: For the circuit shown in the Fig. 3.72, it is assumed that the common-emitter current gain differs for each of the transistors, with $\beta_1 = 100$ and $\beta_2 = 125$. Determine the input bias current and input offset current.

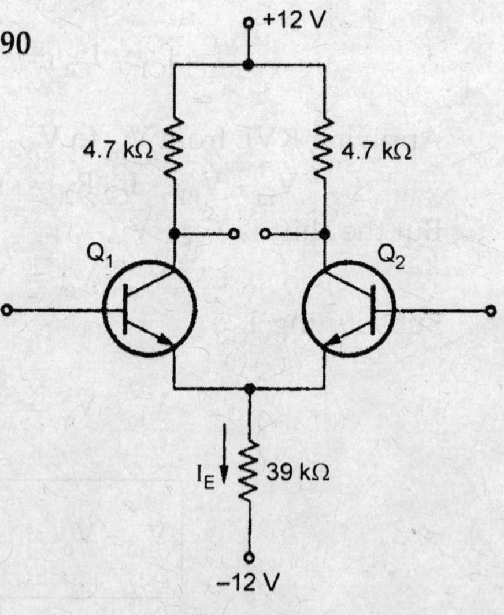

Fig. 3.72

Solution: Assuming zero input voltage, consider the loop as shown in the Fig. 3.73 and applying KVL we get,

$$-V_{BE1} - 39 \times 10^3 I_E + 12 = 0$$

$$\therefore \quad I_E = \frac{12 - V_{BE1}}{39 \times 10^3}$$

$$= \frac{12 - 0.7}{39 \times 10^3} = 289.743\ \mu A$$

Fig. 3.73

$$\therefore \quad I_{E1} = I_{E2} = \frac{I_E}{2} = \frac{289.743}{2}$$

$$= 144.871\ \mu A$$

$$I_{b1} = \frac{I_{E1}}{\beta_1} = \frac{144.871}{100}$$

$$= 1.44871\ \mu A$$

$$I_{b2} = \frac{I_{E2}}{\beta_2} = \frac{1.44871}{125}$$

$$= 1.15897\ \mu A$$

$$\therefore \quad I_b = \frac{I_{b1} + I_{b2}}{2} = \frac{1.44871 + 1.15897}{2}$$

$$= \mathbf{1.3038\ \mu A}$$

and

$$I_{ios} = |I_{b1} - I_{b2}| = |1.44871 - 1.15897|$$

$$= \mathbf{0.28974\ \mu A}$$

Example 3.13: An op-amp has 7 kHz sine wave input signal. Find the largest amplitude that the output of the amplifier can have without distortion with I_{CQ} of 8 μA and C_C of 27 pF.

Solution: The slew rate is

$$S = \frac{1}{C} = \frac{8 \times 10^{-6}}{27 \times 10^{-12}} = 0.2962\ V/\mu\ sec$$

Now

$$f_m = \frac{S}{2\pi V_m} = 7\ kHz$$

$$\therefore \quad V_m = \frac{S}{2\pi f_m} = \frac{\left(\frac{0.296}{10^{-6}}\right)}{2\pi \times 7 \times 10^3} = \mathbf{6.736\ volts}$$

Example 3.14: A square wave of peak to peak amplitude of 750 mV has to be amplified to a peak to peak amplitude of 3.8 V, with a rise time of 4.5 μsec or less. Can IC 741 op-amp be used?

Solution: The IC 741 has a slew rate of 0.5 V/μsec. According to definition,

$$S = \frac{\Delta V}{\Delta t}$$

Now rise time is the time required by the output to rise from 10% to 90% of its final value.

$$\therefore \quad \Delta V = (0.9 - 0.1) \times 3.8 = 3.04 \text{ V}$$

This happens in 4.5 μsec, i.e. $\Delta t = 4.5$ μsec

$$\therefore \quad S = \frac{3.04}{4.5} = \mathbf{0.675\ V/\mu sec}$$

***Key Point:** The slew rate of IC 741 is 0.5 V/μsec which is too low compared to what is required. Hence IC 741 op-amp cannot be used.*

Example 3.15: The input signal V_i to an op-amp is $0.04 \sin 1.13 \times 10^5$ t is to be amplified to the maximum extent. How much maximum gain can be had by using op-amp with slew rate of 0.4 V/μsec?

Solution: The given input can be compared with

$$V_i = V_m \sin wt$$

and $$V_i = 0.04 \sin 1.13 \times 10^5 t$$

$$\therefore \quad V_{m(input)} = 0.04$$

and $$\omega = 1.13 \times 10^5$$

Now $$f_m = \frac{S}{2\pi V_m}$$

and $$\omega = 2 p f_m = 1.13 \times 10^5$$

$$\therefore \quad f_m = \frac{1.13 \times 10^5}{2\pi}$$

Substituting in (1), with $S = 0.4$ V/μsec $= \frac{0.4}{10^{-6}}$ V/sec

$$\frac{1.13 \times 10^5}{2\pi} = \frac{(0.4/10^{-6})}{2\pi V_m}$$

$$\therefore \quad V_m = 3.5398 \text{ V}$$ This is the magnitude of the output voltage.

$$\therefore \quad \text{Gain} = \frac{V_m(\text{output})}{V_m(\text{input})} = \frac{3.5398}{0.04} = \mathbf{88.5}$$

Example 3.16: In response to a square wave input, the output of an op-amp changed from – 3 V to + 3 V over a time interval of 0.25 μs. Determine the slew rate of the op-amp.

Solution: Change in output voltage = – 3 V to + 3 V

$$\therefore \quad dV_o = 3 - (-3) = 6 \text{ V}$$

and $$dt = 0.25 \ \mu s$$

$$\therefore \quad S = \frac{dV_o}{dt} = \frac{6}{0.25\ \mu s} = 24 \text{ V}/\mu s$$

Example 3.17: An op-amp has slew rate of 2 V/μsec. Find the rise time for an output voltage of 10 V amplitude resulting from a rectangular pulse I/P if op-amp is slew rate limited.

Solution: The op-amp is slew rate limited so it allows output to change by 2 V in 1 μsec as the slew rate is 2 V/μsec.

Now the output is 10 V peak and due to slew rate it will not be exactly rectangular. The output will be as shown in the Fig. 3.74. The rise time is the time taken by the output to rise form 10% to 90% of its maximum value. So T, is time required by the output to rise from (10% of 10 V) i.e. 1 V to (90% of 10 V) i.e. 9 V. It takes 1 usee to rise by 2 V so to rise from 1 V to 9 V i.e. total rise by 8 V it will take,

$$T_r = \frac{8}{2} = 4\ \mu\text{sec}$$

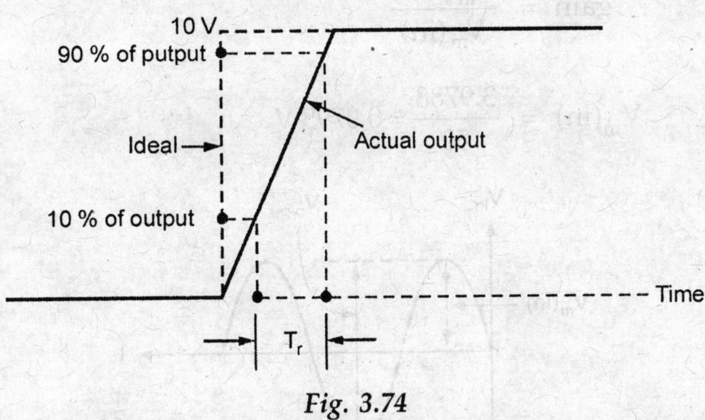

Fig. 3.74

Example 3.18: Determine the slew rate of the op-amp following are the input and output waveforms of a voltage follower.

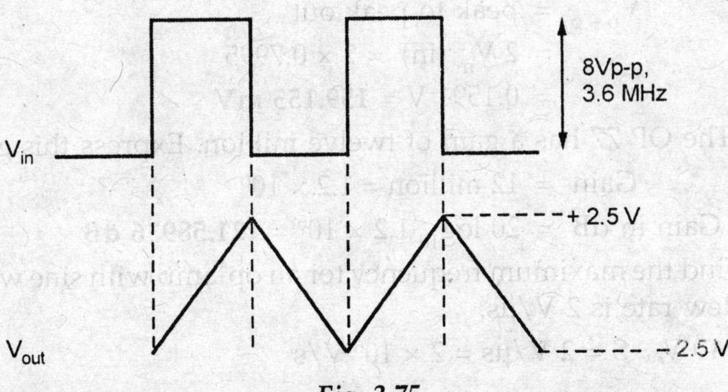

Fig. 3.75

Solution:

$$V_m = \text{Peak value of the output} = 2.5\ \text{V}$$

$$f = 3.6\ \text{MHz}$$

$$\therefore \quad S = 2\pi f V_m = 2\pi \times 3.6 \times 10^6 \times 2.5$$

$$= 56.548 \times 10^6\ \text{V/sec}$$

$$= \mathbf{56.548\ V/\mu sec}$$

Example 3.19: A 741 Cop-amp whose slew rate is 0.5 V/μs is used as an inverting amplifier with a gain of 50. The voltage gain against frequency curve of 741 C is flat upto 20 kHz. What maximum peak to peak input signal can be applied without distorting the output?

Solution:

$$S = 0.5\text{ V}/\mu\text{s} = 0.5 \times 10^6\text{ V/s, gain} = 50,\ f_m = 20\text{ kHz}$$

$$S = 2\pi f_m V_m \text{ where } V_m = \text{magnitude of output voltage}$$

$$\therefore \quad 0.5 \times 10^6 = 2p \times 20 \times 10^3 \times V_m$$

$$\therefore \quad V_m = \frac{0.5 \times 10^6}{2\pi \times 20 \times 10^3} = 3.9788\text{ V} = V_m(\text{out})$$

Now

$$\text{gain} = \frac{V_m(\text{out})}{V_m(\text{in})}.$$

$$\therefore \quad V_m(\text{in}) = \frac{3.9788}{50} = 0.0795\text{ V}$$

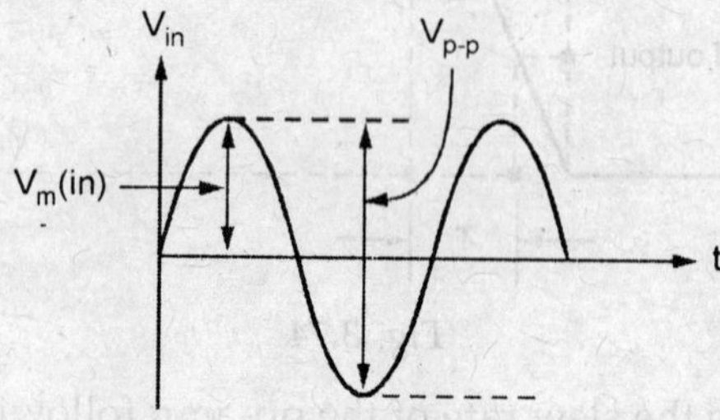

Fig. 3.76

$$\therefore \quad V_{p-p} = \text{peak to peak out}$$
$$= 2V_m(\text{in}) = 2 \times 0.7995$$
$$= 0.1591\text{ V} = \mathbf{159.155\text{ mV}}$$

Example 3.20: The OP 77 has a gain of twelve million. Express this gain in dB.

Solution: $\text{Gain} = 12 \text{ million} = 1.2 \times 10^6$

$$\therefore \quad \text{Gain in dB} = 20 \log_{10} 1.2 \times 10^6 = \mathbf{121.58936\text{ dB}}$$

Example 3.21: Find the maximum frequency for an op-amp with sine wave output voltage of 10 V peak and slew rate is 2 V/μs.

Solution: $V_m = 10$ V, $S = 2\text{ V}/\mu\text{s} = 2 \times 10^6\text{ V/s}$

$$f_m = \frac{S}{2\pi V_m} = \frac{2 \times 10^6}{2\pi \times 10} = \mathbf{31.8309\text{ kHz}}$$

REVIEW QUESTIONS

1. Draw the op-amp symbol and explain the various op-amp terminals.
2. Mention any four important characteristic of ideal op-amp.
3. Write down the ideal characteristics of ideal operational amplifier.
4. Draw and explain the ideal voltage transfer curve of an op-amp.
5. Which are the various practical characteristics of an op-amp ? What is the difference between ideal and practical characteristics of an op-amp ?
6. Draw the block schematic of an op-amp and briefly explain each block ?
7. Define following parameters as applied to an op-amp:
   (*i*) Input bias current
   (*ii*) Input offset current
   (*iii*) Input offset current
   (*iv*) C.M.R.R.
   (*v*) P.S.R.R
   (*vi*) Slew rate
8. Briefly explain the necessity and function of different stages of op-amp with respect to its block schematic.
9. Explain in detail frequency compensation applied to operational amplifier.
10. List the various requirements of input and output stage of a typical op-amp.
11. Explain the various level shifting networks which are used in a typical op-amp.
12. Draw and explain the output stage of IC 741 op-amp.
13. Write a note on frequency response and explain UGB.
14. Draw a neat sketch showing frequency response of μA 741 op-amp.
15. Define the absolute maximum ratings of a commercial op-amp.
16. Write a note on packages of op-amp.
17. Why open loop op-amp configurations is not used in linear applications?
18. Why differential amplifier is used as an input stage of op-amp ?
19. Name the important terminals of an integrated op-amp.
20. Derive the slew rate equation for an op-amp.
21. What is the stability of an op-amp ? Explain the various stability specifications.
22. Explain the different external compensation techniques.
23. Write a note on methods of improving the slew rate.
24. Explain the internal schematic of IC 741.
25. Write a note on ideal op-amp characteristics.
26. Write a note on characteristics of ideal op-amp 741.
27. Which are the various of IC 741 ? Explain them.
28. Explain the applications of open loop op-amp configuration.
29. List the a.c. and d.c. characteristics of op-amp.
30. How input offset voltage is compensated ?
31. What is output offset voltage ? On which factors it depends ?
32. For the circuit shown in the Fig. 3.77, determine the maximum frequency that may be used with a slew rate of 0.5 V/μsec.

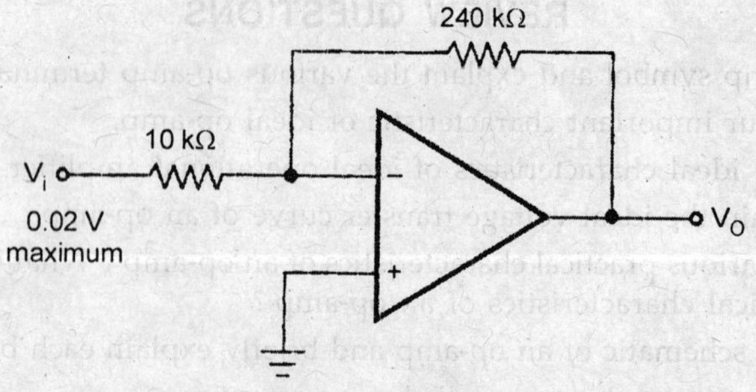

Fig. 3.77

33. For the circuit shown in the Fig. 3.78, it is assumed that the common-emitter current gain differs for each of the transistors, as $\beta_1 = 100$ and $\beta_2 = 120$. Determine

    (*i*) Input bias current (*ii*) Input offset current

[Ans.: 1.682 μA, 0.306 μA]

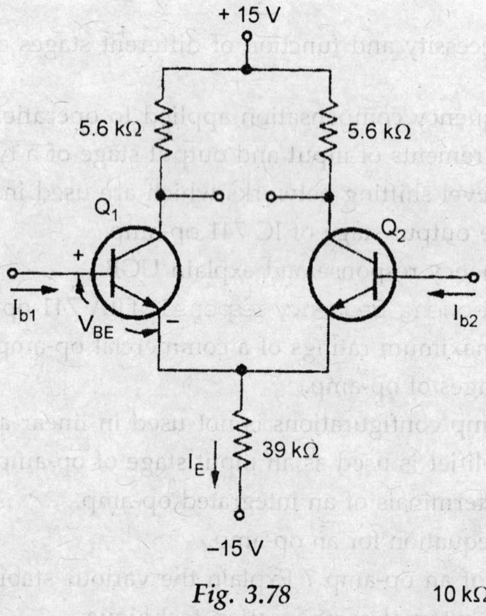

Fig. 3.78

34. Calculate $V_1 - V_2$ for the level shifter shown in the Fig. 3.79. Assume identical silicon transistors with $V_{BE} = 0.7$ V and very large values of β.

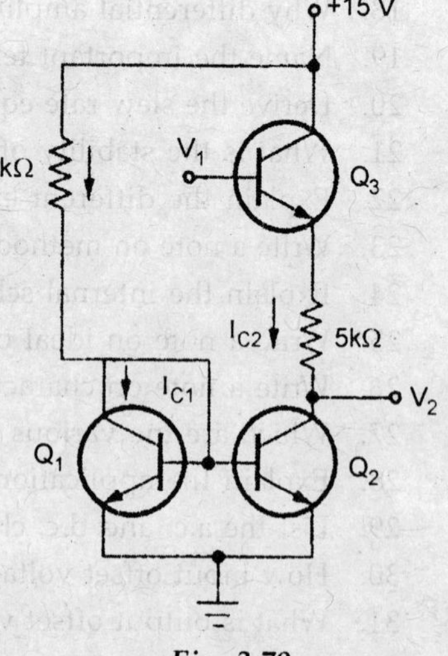

Fig. 3.79

35. The output of an op-amp voltage follower is a triangular wave as shown in the Fig. 3.80 for a square wave input of frequency 2 MHz and 8 V peak to peak amplitude. What is the slew rate of the op-amp ?

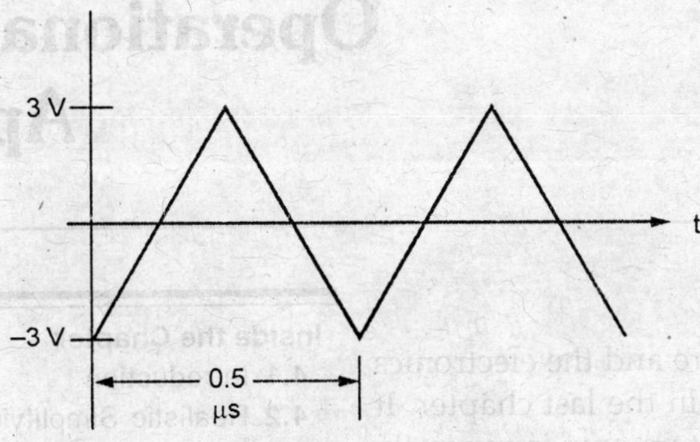

Fig. 3.80

36. An op-amp 741 with a slew rate of 0.5 V/μ sec is used in a noninverting amplifier with gain of 2. Determine the highest frequency that may be processed that will produce output of 20 V peak to peak without exceeding slew rate. **[Ans.: 7.95 kHz]**
37. In what way is 741 S better than 741 ?
38. What is the main difference between the OP AMPs 741, 741 C, 741 S and 741 SC ?

❋❋❋

Chapter

4

Operational Amplifier Applications

4.1 INTRODUCTION

The internal structure and the electronics of op-amp is discussed in the last chapter. It has been mentioned that op-amp is generally used in the closed loop configuration, using feedback.

The countless simple circuits using one or more operational amplifiers, along with some external resistors and the capacitors can be constructed. Such applications of op-amp are classified as linear and non-linear type of applications.

In the **linear applications**, output voltage varies linearly with respect to the input voltage. In such linear applications, a negative feedback is used from the amplifier output to the inverting input terminal.

Key Point: *Some of the linear applications are voltage follower, adder, subtractor, instrumentation amplifier, integrators, differentiators etc.*

In the non-linear applications, a feedback is provided from the output to the non-inverting input terminal. The feedback may be provided to the inverting input terminal using non-linear elements like diodes, transistors etc. The non-linear circuits have highly non-linear input to output characteristics.

Inside the Chapter

Key Point : *The typical non-linear applications are precision rectifier, log amplifiers, antilog amplifiers, comparators, Schmitt trigger circuits etc.*

The various linear and non-linear op-amp applications are discussed in this chapter.

The concept of virtual ground plays an important role in studying various application circuits.

4.2 REALISTIC SIMPLIFYING ASSUMPTIONS

We can make two assumptions which are realistic and simplify the analysis of op-amp circuits to a great extent.

4.2.1 Zero Input Current

The current drawn by either of the input terminals (inverting and non-inverting) is zero.

In practice, the current drawn by the input terminals is very small, of the order μA or nA. Hence the assumption of zero input current is realistic.

4.2.2 Virtual Ground

This means the differential input voltage V_d between the non-inverting and inverting input terminals is essentially zero.

This is obvious because even if output voltage is few volts, due to large open loop gain of op-amp, the difference voltage V_d at the input terminals is almost zero.

e.g. If output voltage is 10 V and the A_{OL} i.e. open loop gain is 10^4 then

$$V_o = V_d A_{OL}$$

$$\therefore \quad V_d = \frac{V_o}{A_{OL}} = \frac{10}{10^4} = 1 \text{ mV}$$

Hence V_d is very small. As $A_{OL} \to \infty$, the difference voltage $V_d \to 0$ and realistically assumed to be zero for analyzing the circuits.

$$\therefore \quad V_d = \frac{V_o}{A_{OL}}$$

$$\therefore \quad (V_1 - V_2) = \frac{V_o}{\infty} = 0$$

$$\therefore \quad V_1 = V_2$$

Thus we can say that under linear range of operation there is virtually short circuit between the two input terminals, in the sense that their voltages are same. No current flows from the input terminals to the ground. The Fig. 4.1 shows the concept of the virtual ground. The thick line indicates the virtual short circuit between the input terminals.

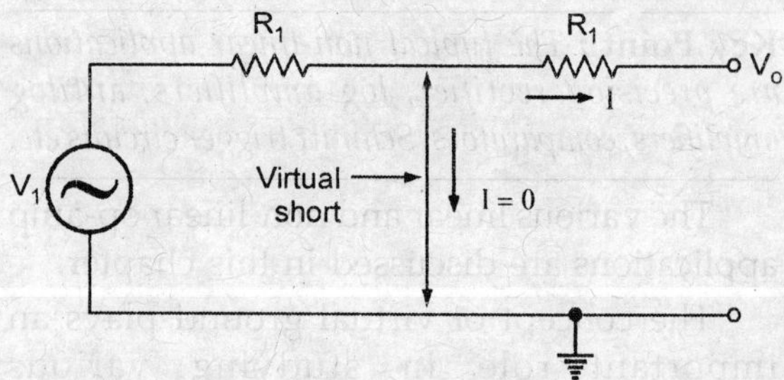

Fig. 4.1 Concept of virtual ground in an op-amp.

Now if the non-inverting terminal is grounded, by the concept of virtual short, the inverting terminal is also at ground potential, though there is no physical connection between the inverting terminal and the ground. This is the principle of virtual ground.

Key Point: *Thus from the equation* (1), *the voltage at the non-inverting input terminal of an op-amp can be realistically assumed to be equal to the voltage at the inverting input terminal.*

The realistic simplifying assumptions are always used to analyze the practical op-amp applications. The steps of analysis based on the assumptions are,

Step 1 : Input current of the ideal op-amp is always zero. Using this, the current distribution in the circuit is obtained.

Step 2: The input terminals of the op-amp are always at the same potential. Thus if one is grounded, the other can be treated to be virtually grounded. From this, the expressions for various branch currents can be obtained.

Step 3 : Analyzing the various expressions obtained, eliminating unwanted variables, the output expression interms of input and circuit parameters can be obtained.

4.3 IDEAL INVERTING AMPLIFIER

As the name suggests the output of such an amplifier is inverted as compared to the input signal. The inverted output signal means having a phase shift of 180° as compared to the input signal.

So, an amplifier which provides a phase shift of 180° between input and output is called inverting amplifier. The basic circuit diagram of an inverting amplifier using op-amp is shown in the Fig. 4.2 (*a*).

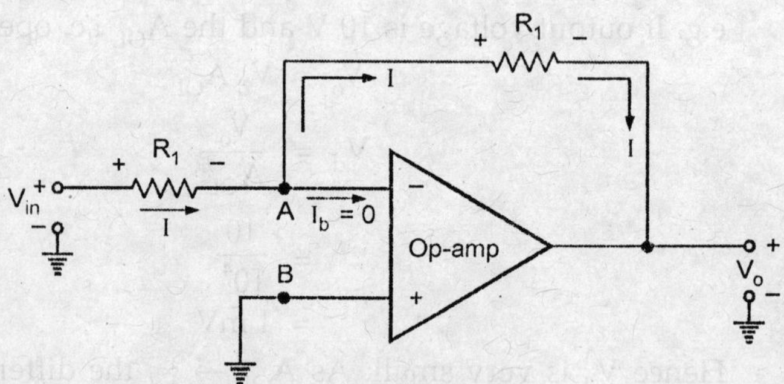

Fig. 4.2 (a) Inverting amplifier.

Derivation of closed loop gain

As node B is grounded, node A is also at ground potential, from the concept of virtual ground, so $V_A = 0$.

$$\therefore \qquad I = \frac{V_{in} - V_A}{R_1}$$

$$\therefore \qquad I = \frac{V_{in}}{R_1} \qquad \text{...(1)}$$

Now from the output side, considering the direction of current I we can write,

$$I = \frac{V_A - V_o}{R_f}$$

$$\therefore \qquad I = \frac{-V_o}{R_f} \qquad \text{... (2)}$$

Entire current I passes through R_f as op-amp input current is zero.

Equating equations (1) and (2) we get,

$$\frac{V_{in}}{R_1} = \frac{-V_o}{R_f}$$

$$\therefore \qquad A_{VF} = \frac{V_o}{V_{in}} = -\frac{R_f}{R_1} \text{ (Gain with feedback)} \qquad \text{...(3)}$$

The $\frac{R_f}{R_1}$ is the gain of the amplifier while negative sign indicates that the polarity of output is opposite to that of input. Hence it is called inverting amplifier.

The input and output waveforms are shown in the Fig. 4.2 (*b*).

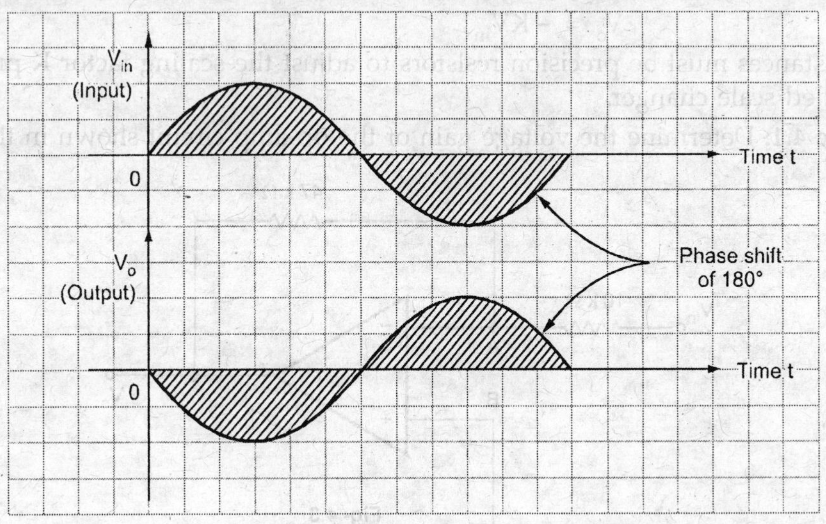

Fig. 4.2 (b) Waveforms of inverting amplifier.

Observations :

1. The output is inverted with respect to input, which is indicated by minus sign.
2. The voltage gain is independent of open loop gain of the op-amp, which is assumed to be large.
3. The voltage gain depends on the ratio of the two resistances. Hence selecting R_f and R_1, the required value of gain can be easily obtained.
4. If $R_f > R_1$, the gain is greater than 1.
   If $R_f < R_1$, the gain is less than 1.
   If $R_f = R_1$, the gain is unity.
   Thus the output voltage can be greater than, less than or equal to the input voltage, in magnitude.
5. If the ratio of R_f and R_1 is K which is other than one, the circuit is called **scale changer** while for $R_f / R_1 = 1$ it is called **phase inverter**.
6. The closed loop gain is denoted as A_{VF} or A_{VCL} i.e. gain with feedback.

4.3.1 Sign Changer

In the ideal inverting amplifier if $R_f = R_1$ then the gain is $A_{CL} = -1$. Thus the magnitude of output is same as that of the input but its sign is opposite to that of the input.

$$\therefore \quad V_o = -V_{in} \quad \text{for} \quad R_f \neq R_1$$

This circuit is called **sign changer** or **phase inverter.**

4.3.2 Scale Changer

In the ideal inverting amplifier if $R_f \neq R_1$ then the gain is $A_{CL} = -K$ where $K = R_f / R_1$. Thus the circuit is used to multiply input by a constant K called scaling factor.

$$\therefore \quad V_o = -KV_{in}$$

The resistances must be precision resistors to adjust the scaling factor K precisely. This circuit is called scale changer.

Example 4.1: Determine the voltage gain of the op-amp circuit shown in the Fig. 4.3.

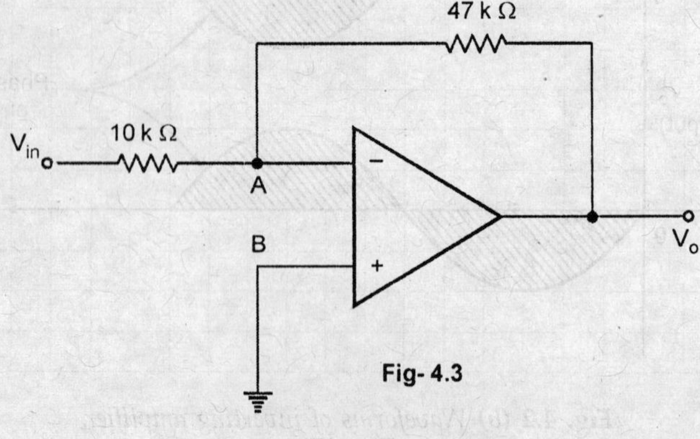

Fig. 4.3

Solution: From the Fig. 4.3, $R_1 = 10\ k\Omega$, $R_f = 47\ k\Omega$

The circuit is inverting amplifier.

$$\therefore \quad \text{Gain} = \frac{V_o}{V_{in}} = -\frac{R_f}{R_1}$$

$$= -\frac{47 \times 10^3}{10 \times 10^3} = -\mathbf{4.7}$$

The gain is 4.7 and negative sign indicates phase shift i.e. inverting mode.

Example 4.2: A sine wave of 0.5 V peak voltage is applied to an inverting amplifier using $R_1 = 10\ k\Omega$ and $R_f = 50\ k\Omega$. It uses the supply voltages of ± 12 V. Determine the output and sketch the waveform.

If now the amplitude of input sine wave is increased to 5 V what will be the output ? Is it practically possible ? Sketch the waveform.

Solution: For an inverting amplifier.

$$\text{Gain} = \frac{V_o}{V_{in}} = -\frac{R_f}{R_1} = \frac{-50}{10} = -5$$

Now $V_m = 0.5$ V for the input hence,

$$(V_o)_m = (V_{in})_m \times \text{Gain} = 0.5 \times 5 = 2.5 \text{ V peak}$$

The input and output waveforms are inverted with respect to each other and are shown in the Fig. 4.4 (*a*).

Now $V_m = 5$ V for the input hence,

$$(V_o)_m = (V_{in})_m \times \text{Gain} = 5 \times 5 = \mathbf{25\ V\ peak}$$

But op-amp output saturates at ± 12 V i.e. at supply voltages used. So portion above ± 12 V and below – 12 V will be clipped off from the output. So 25 V peak output is not practically possible. The input and output waveforms are shown in the Fig. 4.4 (*b*).

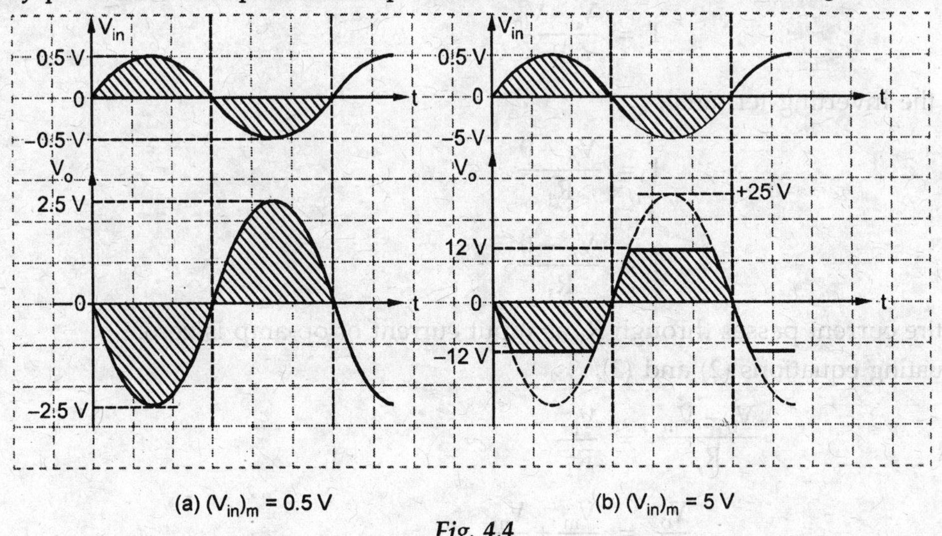

Fig. 4.4

In both cases, there exists a phase shift of 180° between input and output.

4.4 IDEAL NON-INVERTING AMPLIFIER

An amplifier which amplifies the input without producing any phase shift between input and output is called non-inverting **amplifier.** The basic circuit diagram of a non-inverting amplifier using op-amp is shown in the Fig. 4.5 (*a*). The input is applied to the non-inverting input terminal of the op-amp.

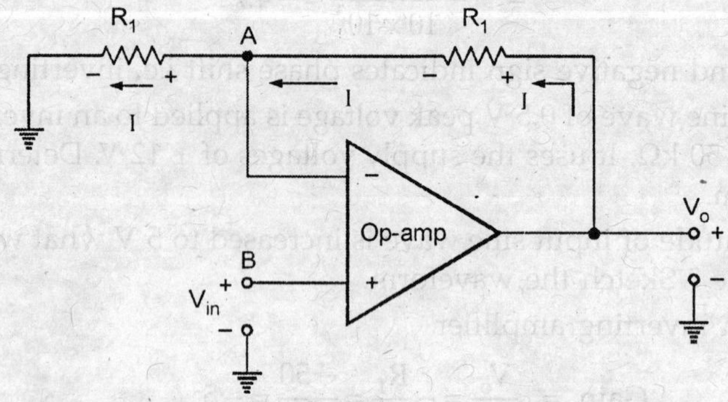

Fig. 4.5 (a) Non-inverting amplifier.

Derivation of closed loop gain

The node B is at potential V_{in}, hence the potential of point A is same as B which is V_{in}, from the concept of virtual share.

$$\therefore \quad V_A = V_B = V_{in} \qquad \text{...(1)}$$

From the output side we can write,

$$I = \frac{V_o - V_A}{R_f}$$

$$\therefore \quad I = \frac{V_o - V_{in}}{R_f} \qquad \text{...(2)}$$

At the inverting terminal,

$$I = \frac{V_A - 0}{R_1}$$

$$\therefore \quad I = \frac{V_A - 0}{R_1} \qquad \text{...(3)}$$

Entire current passes through R_1 as input current of op-amp is zero.

Equating equations (2) and (3),

$$\therefore \quad \frac{V_o - V_{in}}{R_f} = \frac{V_{in}}{R_1}$$

$$\therefore \quad \frac{V_o}{R_f} = \frac{V_{in}}{R_f} + \frac{V_{in}}{R_1}$$

$$\therefore \quad \frac{V_o}{R_f} = V_{in}\left[\frac{(R_1 + R_f)}{R_1 R_f}\right]$$

$$\therefore \quad \frac{V_o}{V_{in}} = \frac{(R_1 + R_f)R_f}{R_1 R_f} = \frac{R_1 + R_f}{R_1}$$

$$\therefore \quad A_{VF} = \frac{V_o}{V_{in}} = 1 + \frac{R_f}{R_1} \qquad ...(4)$$

The positive sign indicates that there is no phase shift between input and output.

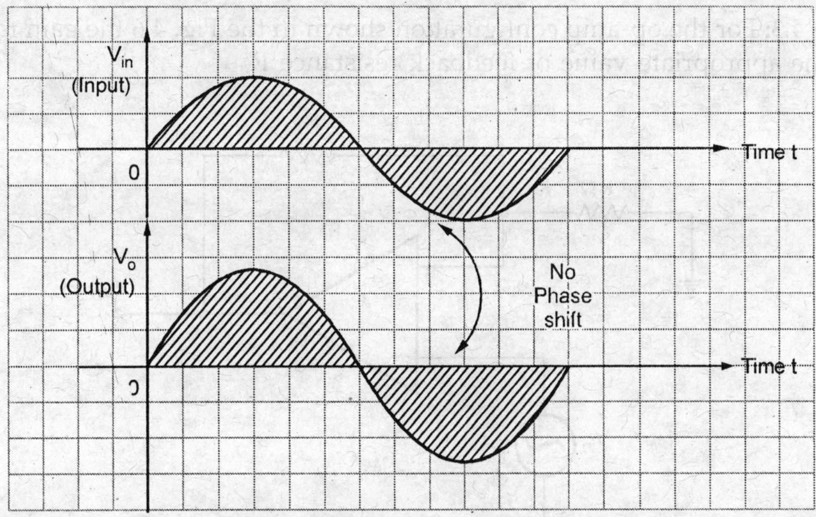

Fig. 4.5 (b) Waveforms of non-inverting amplifier.

The input and output waveforms are shown in the Fig. 4.5 (*b*).

The Table 4.1 provides the comparison of the ideal inverting and non-inverting amplifier op-amp circuits.

Table 4.1

Sr. No.	Ideal inverting amplifier	Ideal non-inverting amplifier
1.	Voltage gain = $-R_f / R_1$	Voltage gain = $1 + (R_f / R_1)$
2.	The output is inverted with respect to input.	No phase shift between input and output.
3.	The voltage gain can be adjusted as greater than, equal to or less than one.	The voltage gain is always greater than one.
4.	The input impedance is R_1.	The input impedance is extremely large.

Observations

1. The voltage gain is always greater than one.
2. The voltage gain is positive indicating that for a.c. input, the output and input are in phase while for d.c. input, the output polarity is same as that of input.
3. The voltage gain is independent of open loop gain of op-amp, but depends only on the two resistance values.
4. The desired voltage gain can be obtained by selecting proper values of R_f and R_1.

Example 4.3: For the op-amp configuration shown in the Fig. 4.6 the gain required is 61. Determine the appropriate value of feedback resistance R_f.

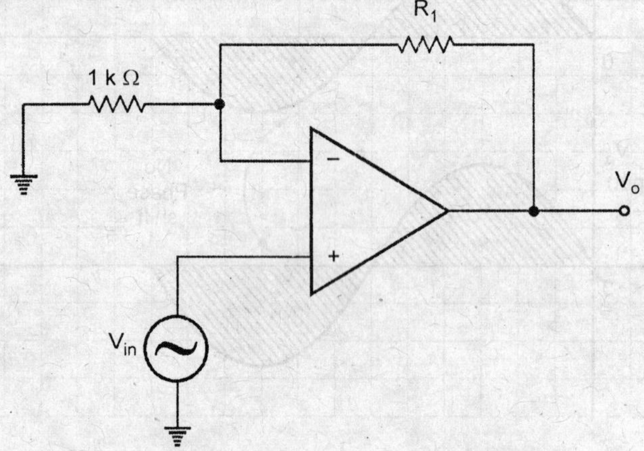

Fig. 4.6

Solution: From Fig. 4.6, R_1 = 1 kΩ and gain required is 61.
The configuration is non-inverting amplifier.

$$\therefore \quad \text{Gain} = \frac{V_o}{V_{in}} = 1 + \frac{R_f}{R_1}$$

$$\therefore \quad 61 = 1 + \frac{R_f}{R_1}$$

$$\therefore \quad \frac{R_f}{1} = 61 - 1 = 60$$

$$\therefore \quad R_f = 60 \text{ k}\Omega$$

4.5 IMPORTANT POINT ABOUT NON-INVERTING AMPLIFIER

Consider a non-inverting amplifier shown in the Fig. 4.7 (*a*). In this case,

$$V_o = \left(1+\frac{R_f}{R_1}\right)V_{in}$$

But consider a non-inverting amplifier shown in the Fig. 4.7 (*b*). In this case V_{in} is not directly applied to the non-inverting terminal.

Key Point: *Remember that the non-inverting amplifier always amplifies voltage at its non-inverting terminal i.e.* V_B *by* $(1 + R_f / R_1)$ *times.*

In second case the output voltage V_o is given by,

$$V_o = \left(1+\frac{R_f}{R_1}\right)V_B \quad \text{and} \quad V_o \neq \left(1+\frac{R_f}{R_1}\right)V_{in}$$

Then analyzing circuit at non-inverting terminal, V_B can be obtained interms of V_{in}. In the circuit shown in the Fig. 4.7 (*b*), $R_2 - R_3$ forms potential divider and as op-amp input current is zero,

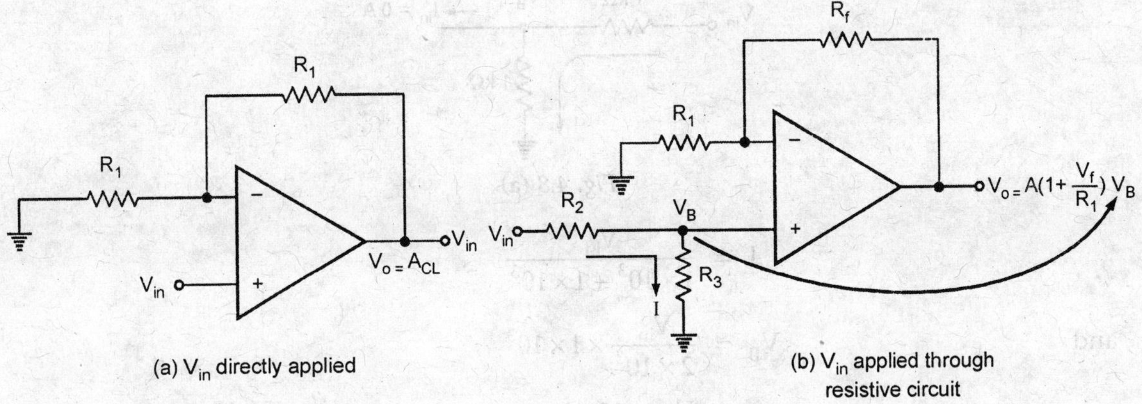

Fig. 4.7

$$V_B = \frac{V_{in}}{R_2+R_3} \times R_3$$

$$\boxed{V_o = \left(1+\frac{R_f}{R_1}\right)V_B = \left(1+\frac{R_f}{R_1}\right)\left(\frac{R_3}{R_2+R_3}\right)V_{in}}$$

This fact helps to analyze complex op-amp circuits in which non-inverting amplifier is a subpart.

Example 4.4: Find V_o for the circuit shown in the Fig. 4.8.

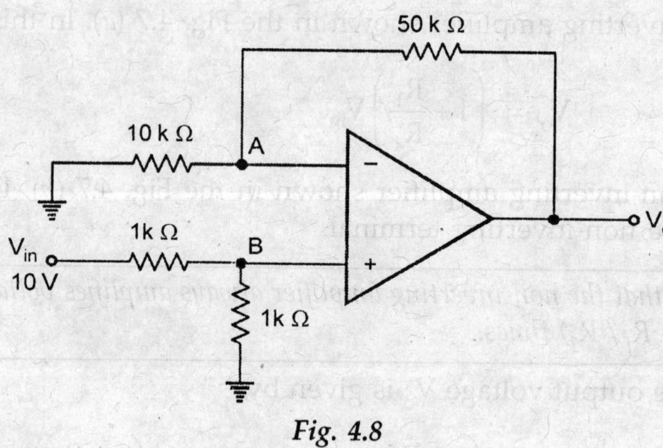

Fig. 4.8

Solution : This is non-inverting amplifier but it will amplify V_B by $1+\frac{R_f}{R_1}$ If and not V_{in} = 10 V. The **op-amp input current** is zero hence part of the circuit is as shown in the Fig. 4.8 (*a*).

Fig. 4.8 (a)

$$\therefore \quad I = \frac{V_{in}}{1\times 10^3 + 1\times 10^3}$$

and

$$V_B = \frac{V_{in}}{2\times 10^3}\times 1\times 10^3$$

$$\therefore \quad V_B = \frac{V_{in}}{2} = \frac{10}{2} = 5\text{ V}$$

$$= \frac{V_{in}}{2} = \frac{10}{2} = 5\text{ V}$$

$$\therefore \quad V_B = \left(1+\frac{R_f}{R_1}\right)V_B$$

$$= \left(1+\frac{50\times 10^3}{10\times 10^3}\right)5 = \mathbf{30\ V}$$

Example 4.5: In the Fig. 4.9, an inverting amplifier is shown with R_1 = 20 kΩ and R_f = 100 kΩ. A load of 10 kΩ is connected to the output with input voltage of 0.7 V. Calculate (*i*) I_1 (*ii*) V_o (*iii*) I_L (*iv*) Total current I_o.

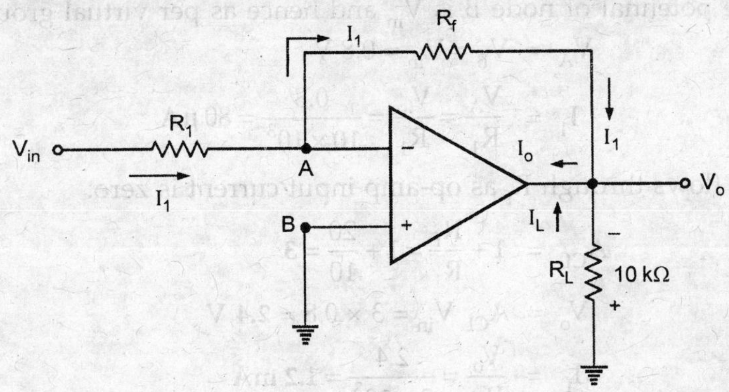

Fig. 4.9

Solution: From the given circuit, node A is at virtual ground hence $V_B = V_A = 0$.

(*i*) $$I_1 = \frac{V_{in} - V_A}{R_1} = \frac{V_{in}}{R_1} = \frac{0.7}{20 \times 10^3} = \mathbf{35\ \mu A}$$

(*ii*) $$V_o = V_{in} A_{CL} = -\frac{R_f}{R_1} \times V_{in} = -\frac{100}{20} \times 0.7 = \mathbf{-3.5\ V}$$

(*iii*) $$I_L = \frac{V_o}{R_L} = \frac{-3.5}{10 \times 10^3} = \mathbf{-0.35\ mA}$$

The negative sign indicates that it flows from ground towards output terminal.

(*iv*) As op-amp input current is zero, I_1 flows through R_f as well. Applying KCL at output node,

$$I_o = I_L + I_1 = 0.35 \times 10^{-3} + 35 \times 10^{-6}$$
$$= 0.385 \text{ mA into the op-amp output.}$$

Note that while calculating I_o, negative sign is not considered as it indicates direction. **Once direction is fixed from ground towards output, negative sign need not be considered.**

Example 4.6: For a non-inverting amplifier shown in the Fig. 4.10, calculate (*i*) A_{CL} (*ii*) V_o (*iii*) I_L and (*iv*) I_o.

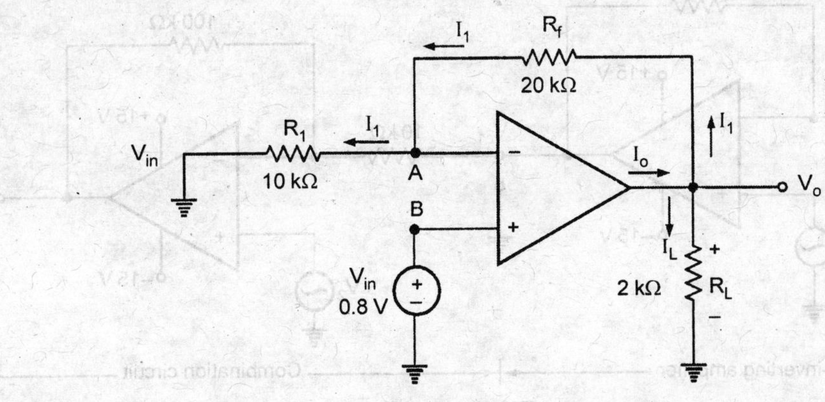

Fig. 4.10

Solution: The potential of node B is V_{in} and hence as per virtual ground,

$$V_A = V_B = V_{in} = 0.8 \text{ V}$$

$$\therefore \quad I_1 = \frac{V_A}{R_1} = \frac{V_{in}}{R_1} = \frac{0.8}{10 \times 10^3} = 80 \ \mu\text{A}$$

Same current flows through R_f as op-amp input current is zero.

(*i*) $$A_{CL} = 1 + \frac{R_f}{R_1} = 1 + \frac{20}{10} = 3$$

(*ii*) $$V_o = A_{CL} V_{in} = 3 \times 0.8 = \mathbf{2.4 \ V}$$

(*iii*) $$I_L = \frac{V_o}{R_L} = \frac{2.4}{2 \times 10^3} = \mathbf{1.2 \ mA}$$

(*iv*) Applying KCL at the output node,

$$I_o = I_1 + I_L = 80 \times 10^{-6} + 1.2 \times 10^{-3} = \mathbf{1.28 \ mA}$$

The I_o flows from op-amp towards output terminal.

Example 4.7: Find output voltage V_o interms of V_1 and V_2 for the op-amp circuit shown in the Fig. 4.11. Assume ideal op-amps.

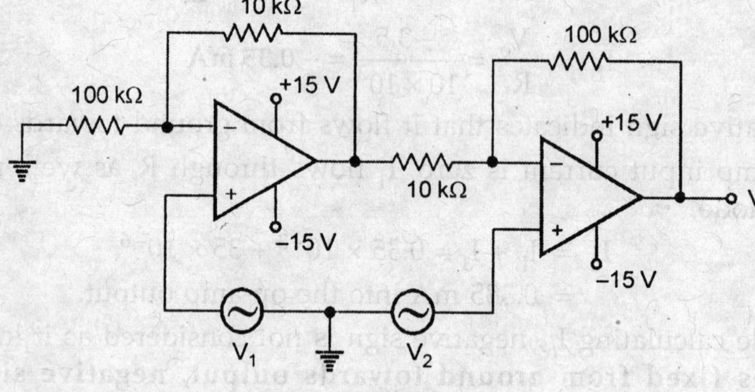

Fig. 4.11

Solution: Split the given circuit as shown in the Fig. 4.11 (*a*).

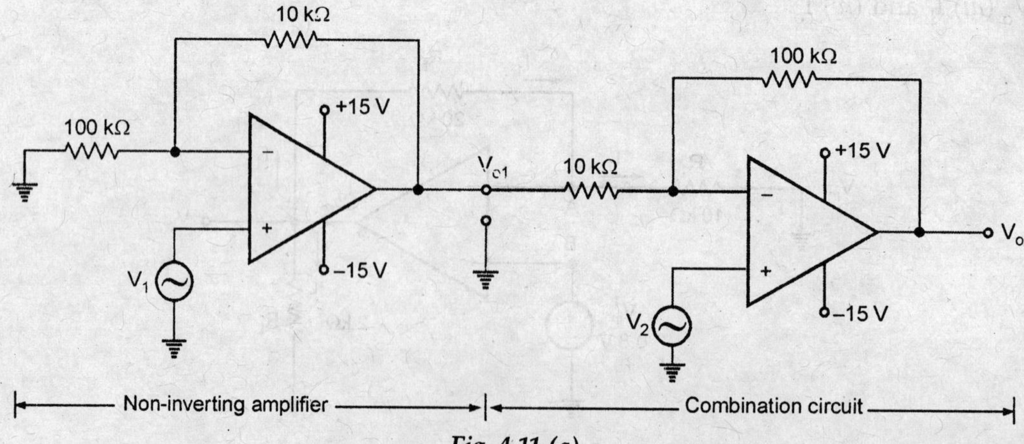

Fig. 4.11 (a)

For first stage, $$V_{o1} = \left(1+\frac{R_f}{R_1}\right)V_{in} = \left(1+\frac{10\times10^3}{100\times10^3}\right)V_1 = 1.1\,V_1$$

For the second stage use Superposition principle. Use each input one at a time, assuming other input zero.

Assume V_{o1} active and $V_2 = 0$ V i.e grounded. This is **inverting amplifier,** as shown in the Fig. 4.11 (*b*).

$$\therefore \quad V_o' = V_{o1}\left(-\frac{R_f}{R_1}\right)$$

$$= \frac{-100\times10^3}{10\times10^3}[1.1\,V_1] = -11\,V_1 \qquad \text{...due to } V_{o1} \text{ alone}$$

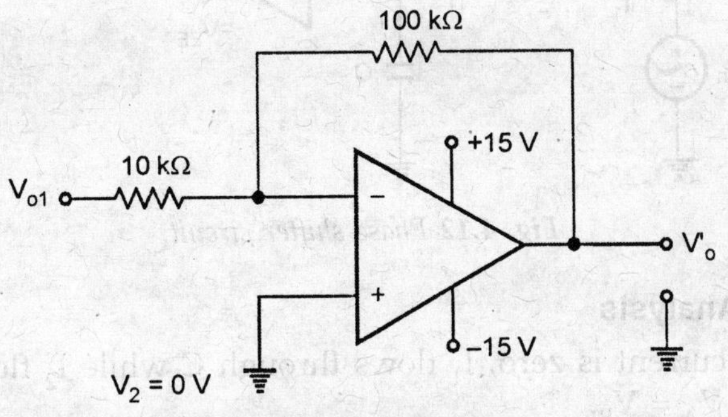

Fig. 4.11 (b)

Assume V_2 active and $V_{o1} = 0$ V i.e grounded

This is **non-inverting amplifier.**

$$\therefore \quad V_o'' = V_2\left(1+\frac{R_f}{R_1}\right)$$

$$= V_2\left(1+\frac{100\times10^3}{10\times10^3}\right)$$

$$= 11\,V_2 \qquad \text{...due to } V_2 \text{ alone}$$

Fig. 4.11 (c)

Hence the net output V_o is sum of V_o' and V_{o1}''.

$$\therefore \quad V_o = V_o' + V_o'' = -11\,V_1 + 11\,V_2 = 11(V_2 - V_1)$$

4.6 PHASE SHIFT CIRCUIT

Using the concept of inverting and non-inverting amplifier circuits, a phase shifter circuit can be designed.

A phase shifter circuit is one in which all signals are transmitted from input to output without change in amplitude but the circuit introduces a phase shift as signal transmits from input to output.

The Fig. 4.12 shows a phase shifter circuit.

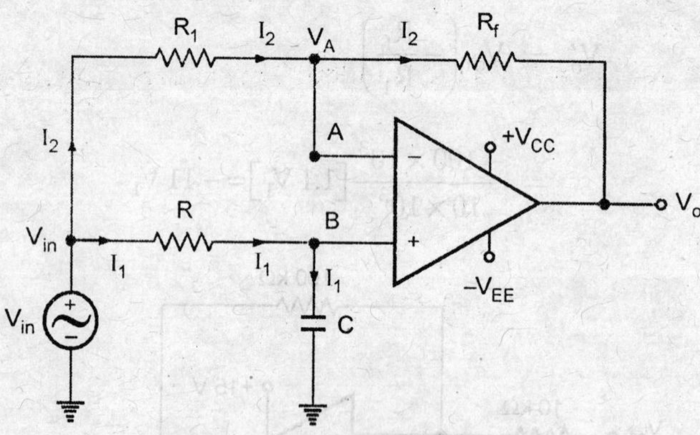

Fig. 4.12 Phase shifter circuit.

4.6.1 Mathematical Analysis

As op-amp input current is zero, I_1 flows through C while I_2 flows through R_f. While due to virtual ground $V_A = V_B$.

From the branch 'C', $V_B = \frac{1}{C}\int I_1 \, dt$

Using Laplace transform for ease of analysis,

$$V_B(s) = \frac{1}{sC} I_1(s) \qquad \text{...(1)}$$

While, $$I_1(s) = \frac{V_{in}(s) - V_B(s)}{R} \qquad \text{...From branch R}$$

$$\therefore \quad sCV_B(s) = \frac{V_{in}(s) - V_B(s)}{R} \qquad \text{...From equation (1)}$$

$$\therefore \quad V_B(s) = \frac{V_{in}(s)}{1 + sCR} \qquad \text{...(2)}$$

From branch R_1, $$I_2(s) = \frac{V_{in}(s) - V_A(s)}{R_1} = \frac{V_{in}(s) - V_B(s)}{R_1} \qquad \text{...}V_A = V_B$$

From branch R_f, $I_2(s) = \dfrac{V_A(s) - V_o(s)}{R_f} = \dfrac{V_B(s) - V_o(s)}{R_f}$

$$\therefore \quad \frac{V_{in}(s) - V_B(s)}{R_1} = \frac{V_B(s) - V_o(s)}{R_f}$$

$$\therefore \quad \frac{V_{in}(s)}{R_1} - V_B(s)\left[\frac{1}{R_1} + \frac{1}{R_f}\right] = \frac{-V_o(s)}{R_f}$$

$$\therefore \quad -\frac{R_f}{R_1}V_{in}(s) + V_B(s)\left[\frac{R_1 + R_f}{R_1}\right] = V_o(s) \qquad \text{...(3)}$$

Using equation (2) in equation (3),

$$\frac{-R_f}{R_1}V_{in}(s) + \frac{V_{in}(s)}{1+sCR}\left(1 + \frac{R_f}{R_1}\right) = V_o(s)$$

$$\therefore \; V_{in}(s)\left[\frac{\frac{-R_f}{R_1}(1+sCR) + \left(1 + \frac{R_f}{R_1}\right)}{1+sCR}\right] = V_o(s)$$

$$\therefore \quad \boxed{\frac{V_o(s)}{V_{in}(s)} = \frac{-\frac{R_f}{R_1}sCR + 1}{1+sCR}} \qquad \text{...(4)}$$

Select the values as $R_f = R_1$

$$\therefore \quad \boxed{\frac{V_o(s)}{V_{in}(s)} = \frac{1-sCR}{1+sCR}} \qquad \text{...(5)}$$

Putting $s = j\omega$ where ω is the variable frequency of the input signal we get,

$$\boxed{\frac{V_o(j\omega)}{V_{in}(j\omega)} = \frac{1-j\omega CR}{1+j\omega CR}} \qquad \text{...(6)}$$

$$\therefore \quad \begin{bmatrix}\text{Magnitude} \\ \text{of the ratio}\end{bmatrix} = \frac{\sqrt{1+\omega^2C^2R^2}}{\sqrt{1+\omega^2C^2R^2}} = 1 \quad \text{hence } |V_o| = |V_{in}|$$

Key Point: *Thus there is no change in the amplitude from input to the output.*

$$\phi = \angle\left(\frac{V_o(j\omega)}{V_{in}(j\omega)}\right) = -\tan^{-1}\left(\frac{\omega CR}{1}\right) - \tan^{-1}\left(\frac{\omega CR}{1}\right) \quad ...(7)$$

The angle ϕ is the phase shift introduced by the circuit as signal passes from input to output.

As ω varies from 0 to $\neq$, the phase shift ϕ varies from 0° to – 180°. Due to negative sign it is called **phase lag.** This variable phase lag is shown in the Fig. 4.13.

Thus the circuit acts as a **phase shifter**.

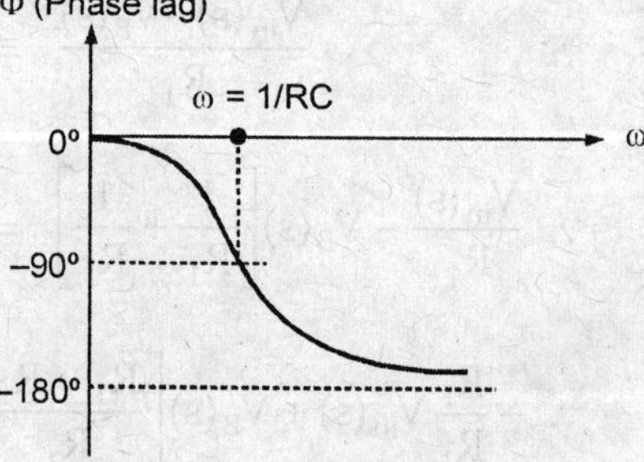

Fig. 4.13 Variation of phase shift against frequency.

4.7 VOLTAGE FOLLOWER

A circuit in which the output voltage follows the input voltage is called voltage follower circuit. The voltage follower circuit using op-amp is shown in the Fig. 4.14.

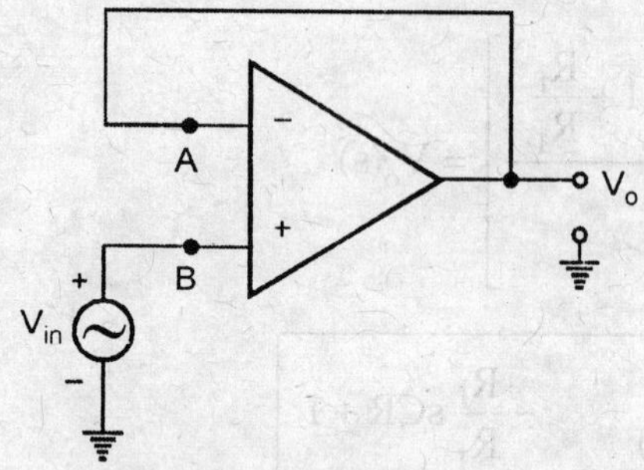

Fig. 4.14 Voltage follower.

The node B is at potential V_{in}. Now node A is also at the same potential as B i.e. V_{in}.

$$\therefore \quad V_A = V_B = V_m \quad ... (1)$$

Now node A is directly connected to the output. Hence we can write,

$$V_o = V_A \quad ...(2)$$

Equating the equations (1) and (2),

$$V_o = V_{in} \quad ...(3)$$

For this circuit, the voltage gain is unity.

Thus the output voltage V_o is equal to the input voltage V_{in}. If V_{in} increases, V_o also increases. If V_{in} decreases, then V_o also decreases. Thus output follows the input hence the circuit is called voltage follower circuit. It is also called source follower, unity gain amplifier, buffer amplifier or isolation amplifier. The input and output waveforms are shown in the Fig. 4.15.

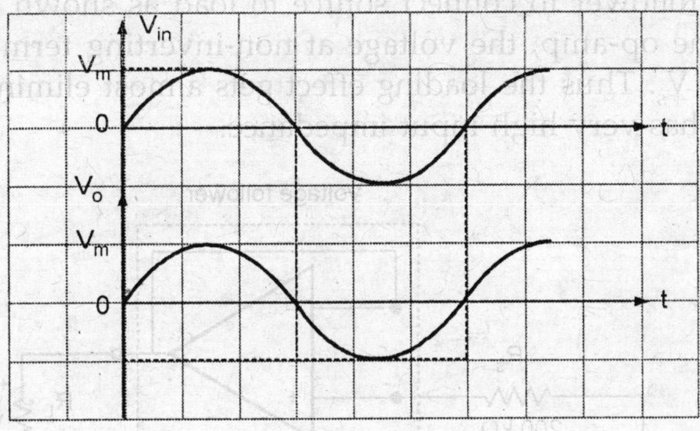

Fig. 4.15

4.7.1 Advantages of Voltage Follower

The advantages of such voltage follower circuit are,

1. Very large input resistance, of the order of MΩ.
2. Low output impedance, almost zero. Hence it can be used to connect high impedance source to a low impedance load, as a buffer.
3. It has large bandwidth.
4. The output follows the input exactly without a phase shift.

Because the input resistance of the voltage follower is extremely large, it draws negligible current from the signal source. Therefore the signal source is not loaded. This property is most suitable in the instrumentation amplifier, which is used to amplify the signal output from a transducer. If the output of a transducer is directly applied to the input of an amplifier which does not have high input impedance, then there is every possibility that the transducer may get loaded. Loading alters the characteristics of the transducer. To prevent this, the voltage follower circuit can be used in between the transducer and the amplifier. Such an use of the voltage follower is called an isolator or a buffer.

4.7.2 Practical Use of Voltage Follower

Consider a circuit consisting of a voltage source driving a load. The source resistance is 200 kΩ and the load resistance is 2 kΩ. It is shown in the Fig. 4.16. The source with R_s represents an equivalent circuit of many practical situations like transducer driving a load. According to potential divider,

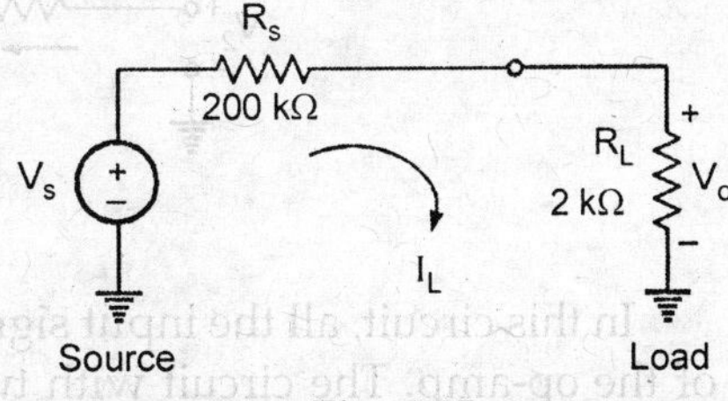

Fig. 4.15

$$V_o = \frac{V_s R_L}{R_s + R_L} = 0.0099\ V_s \approx 0.01\ V_s$$

Fig. 4.16 Source driving load.

Thus only 0.01 times Vs is available to drive the load and there is severe attenuation. This is nothing but loading effect.

Now use voltage follower to connect source to load as shown in the Fig. 4.17. As no current can go into the op-amp, the voltage at non-inverting terminal is V_s. And due to voltage follower $V_o = V_s$. Thus the loading effect gets almost eliminated due to the use of voltage follower as it has very high input impedance.

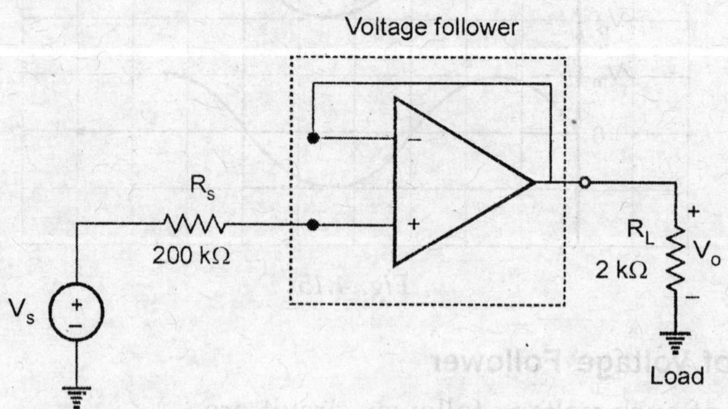

Fig. 4.17 Reducing loading effect.

4.8 SUMMER OR ADDER CIRCUIT

As the input impedance of an op-amp is extremely large, more than one input signal can be applied to the inverting amplifier. Such circuit gives the addition of the applied signals at the output. Hence it is called **summer** or **adder** circuit. Depending upon the sign of the output, the summer circuits are classified as inverting summer and non-inverting summer.

4.8.1 Inverting Summer

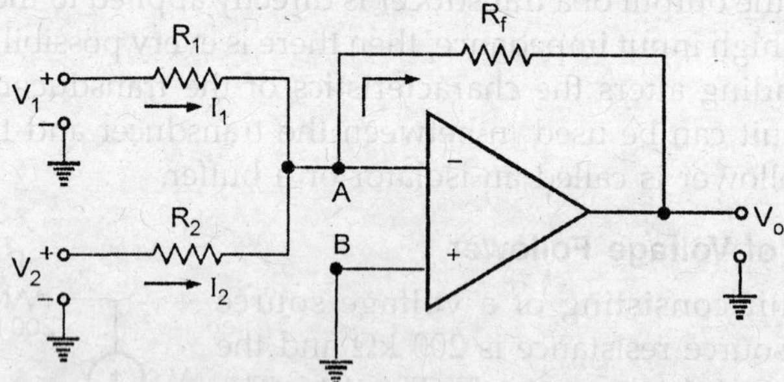

Fig. 4.18 Inverting summer.

In this circuit, all the input signals to be added are applied to the inverting input terminal of the op-amp. The circuit with two input signals is shown in the Fig. 4.18.

As point B is grounded, due to virtual ground concept the node A is also at virtual ground potential.

$$\therefore \quad V_A = 0 \qquad \text{...(1)}$$

Now from the input side,

$$I_1 = \frac{V_1 - V_A}{R_1} = \frac{V_1}{R_1} \qquad \text{...(2)}$$

$$I_2 = \frac{V_2 - V_A}{R_2} = \frac{V_2}{R_2} \qquad ...(3)$$

Applying KCL at node A and as input op-amp current is zero,

$$I = I_1 + I_2 \qquad ...(4)$$

From the output side,

$$I = \frac{V_A - V_o}{R_f} = \frac{V_o}{R_f} \qquad ...(5)$$

Substituting equations (5), (2) and (3) in equation (4),

$$-\frac{V_o}{R_f} = \frac{V_1}{R_1} + \frac{V_2}{R_2}$$

$$\therefore \quad \boxed{V_o = -\left[\frac{R_f}{R_1}V_1 + \frac{R_f}{R_2}V_2\right]} \qquad ...(6)$$

If the three resistances are equal, $R_1 = R_2 = R_f$

$$\boxed{V_o = -(V_1 + V_2)} \qquad ...(7)$$

By properly selecting R_f, R_1 and R_2, we can have weighted addition of the input signals like $aV_1 + bV_2$, as indicated by the equation (6).

Infact in such a way, n input voltages can be added.

Key Point: *Thus the magnitude of the output voltage is the sum of the input voltages and hence circuit is called summer or adder circuit.*

4.8.2 Non-inverting Summing Amplifier

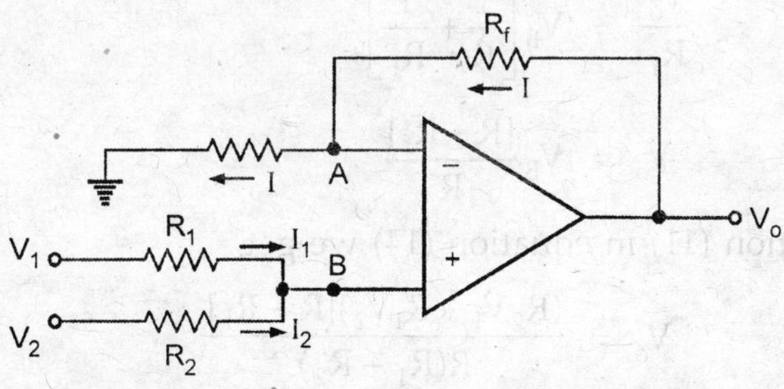

Fig. 4.19 Non-inverting summing amplifier.

The circuit discussed above is inverting summing amplifier, which can be noticed from the negative sign in the equation (6). But a summer that gives **non-inverted** sum of the input signals is called **non-inverting summing amplifier.** The circuit is shown in the Fig. 4.19.

Let the voltage of node B is V_B. Now the node A is at the same potential as that of B, due to virtual ground.

$$\therefore \quad V_A = V_B \quad \text{...(8)}$$

From the input side,

$$I_1 = \frac{V_1 - V_B}{R_1} \quad \text{and} \quad I_2 = \frac{V_2 - V_B}{R_2} \quad \text{...(9)}$$

But as the input current of op-amp is zero,

$$I_1 + I_2 = 0 \quad \text{...(10)}$$

$$\therefore \quad \frac{V_1 - V_B}{R_1} + \frac{V_2 - V_B}{R_2} = 0$$

$$\therefore \quad \frac{V_1}{R_1} + \frac{V_2}{R_2} = V_B\left[\frac{1}{R_1} + \frac{1}{R_2}\right]$$

$$V_o = \frac{(R_2 V_1 + R_1 V_2)}{(R_1 + R_2)} \quad \text{...(11)}$$

Now at node A,

$$I = \frac{V_A}{R} = \frac{V_B}{R} \quad \text{as} \quad V_B = V_A \quad \text{...(12)}$$

and

$$I = \frac{V_o - V_A}{R_f} = \frac{V_o - V_B}{R_f}$$

Equating the two equations (12) and (13),

$$\therefore \quad \frac{V_B}{R} = \frac{V_o - V_B}{R_f}$$

$$\therefore \quad \frac{V_o}{R_f} = V_B\left[\frac{1}{R} + \frac{1}{R_f}\right]$$

$$\therefore \quad V_o = V_B \frac{[R + R_f]}{R} \quad \text{...(14)}$$

Substituting equation (11) in equation (14) we get,

$$V_o = \frac{(R_2 V_1 + R_1 V_2)[R + R_f]}{R(R_1 + R_2)}$$

i.e.

$$\boxed{V_o = \frac{R_2(R + R_f)}{R(R_1 + R_2)} V_1 + \frac{R_1(R + R_f)}{R(R_1 + R_2)} V_2}$$

The equation (15) shows that the output is weighted sum of the inputs.

If $R_1 = R_2 = R = R_f$, we get

$$\boxed{V_o = V_1 + V_2} \quad \text{...(16)}$$

Key Point: *As there is no phase difference between input and output, it is called non-inverting summer amplifier.*

4.8.3 Average Circuit

If in the inverting summer circuit, the values of resistance are selected as,

$$R_1 = R_2 = R$$

and $$R_f = \frac{R}{2}$$

Then from the equation (6) we get,

$$V_o = -\left[\frac{R/2}{R}V_1 + \frac{R/2}{R}V_2\right]$$

$$\therefore \quad \boxed{V_o = \frac{(V_1 + V_2)}{2}} \qquad \dots (17)$$

Key Point: *Thus the magnitude of the output voltage is the average of the two input voltages. So circuit acts like an average.*

Similarly average of *n* inputs can be calculated by selecting,

$$R_1 = R_2 = R_3 = \dots = R_n = R \text{ and } R_f = \frac{R}{n}$$

Example 4.8: *Determine the output voltage for the configuration shown in the Fig. 4.20.*

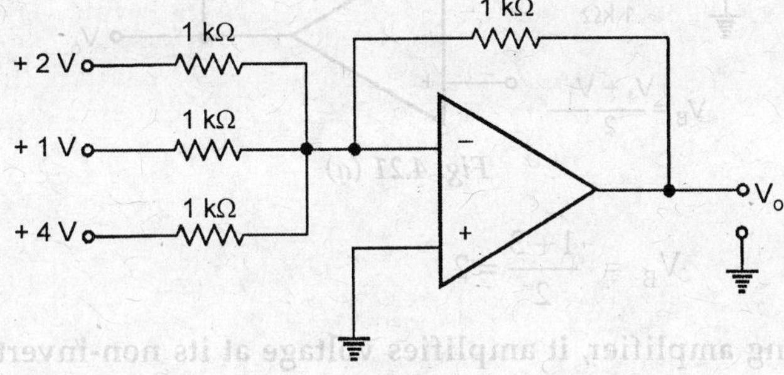

Fig. 4.20

Solution: The circuit is inverting adder circuit.

$$R_1 = R_2 = R_3 = 1\ k\Omega$$

and $$R_f = 1\ k\Omega$$

$$\therefore \quad V_o = -\left[\frac{R_f}{R_1}V_1 + \frac{R_f}{R_2}V_2 + \frac{R_f}{R_3}V_3\right]$$

$$= -[V_1 + V_2 + V_3] = -[2 + 1 + 4]$$

$$= -7\ V$$

Example 4.9: Determine the output voltage for the circuit shown in the Fig. 4.21, for $V_1 = 1$ V and $V_2 = 3$ V.

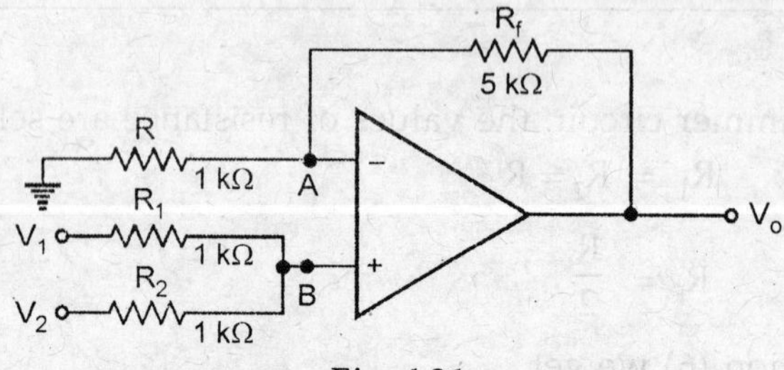

Fig. 4.21

Solution: In this circuit $R_1 = R_2 = 1$ kΩ

The voltage at non-inverting terminal is given by the equation (11),

$$\therefore \quad V_B = \frac{R_2 V_1 + R_1 V_2}{R_1 + R_2}$$

$$= \frac{V_1 + V_2}{2} \qquad \ldots R_1 = R_2 = R'$$

The circuit reduces to a non-inverting amplifier with $V_B = V_1 + V_2/2$ as shown in the Fig. 4.21 (*a*).

Fig. 4.21 (a)

$$\therefore \quad V_B = \frac{1+3}{2} = 2$$

For a non-inverting amplifier, it amplifies voltage at its non-inverting input terminal by the gain $1 + \frac{R_f}{R}$.

$$\therefore \quad V_o = V_B\left(1 + \frac{R_f}{R}\right) = 2\left(1 + \frac{5}{1}\right)$$

$$= \textbf{12 V}$$

Alternatively equation (15) can be used to find V_o,

$$V_o = \frac{R_2(R + R_f)}{R(R_1 + R_2)} V_1 + \frac{R_1(R + R_f)}{R(R_1 + R_2)} V_2$$

$$= \left[\frac{1(6)}{1(1+1)}\right]\times 1+\left[\frac{1\times 6}{1\times(1+1)}\right]\times 3$$

$$= 3 + 9 = \mathbf{12\ V}$$

Example 4.10: Calculate the value of R_s in the Fig. 4.22 shown, such that V_o becomes zero.

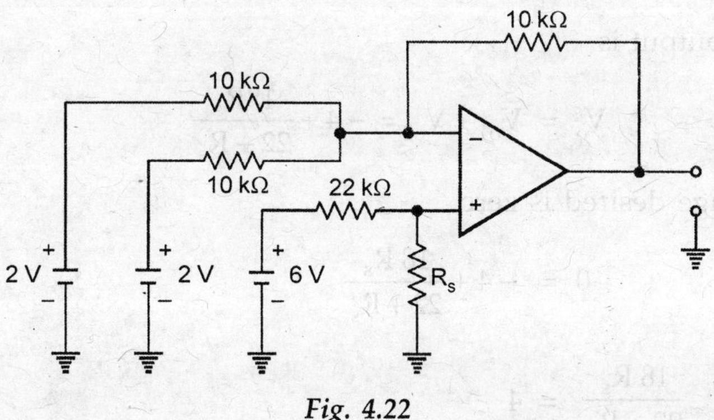

Fig. 4.22

Solution: Using the Superposition principle, suppress 6 V battery and hence non-inverting terminal, becomes grounded. The circuit acts as an inverting summer,

$$\therefore \quad V_{o1} = -\left[\frac{10\ K}{10\ K}\times 2+\frac{10\ K}{10\ K}\right] = -4\ V$$

Now ground both 2 V batteries hence the circuit becomes as shown in the Fig. 4.22 (*a*).

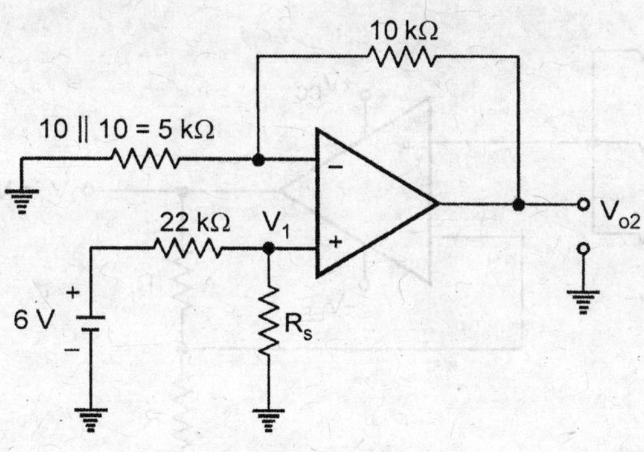

Fig. 4.22 (a)

Using the result of non-inverting amplifier,

$$V_{o2} = \left[1+\frac{R_f}{R_1}\right]V_1 = \left[1+\frac{10\ K}{5\ K}\right]\times V_1 = 3V_1$$

Using potential divider rule,

$$V_1 = \left[\frac{6}{22+R_s}\right]R_s$$

$$\therefore \quad V_{o2} = 3\times\frac{6}{22+R_s}\times R_s = \frac{18\,R_s}{22+R_s}$$

Hence the total output is

$$V_o = V_{o1}+V_{o2} = -4+\frac{18R_s}{22+R_s}$$

The output voltage desired is zero.

$$\therefore \quad 0 = -4+\frac{18\,R_s}{22+R_s}$$

$$\therefore \quad \frac{18\,R_s}{22+R_s} = 4$$

$$\therefore \quad 18\,R_s = 88+4\,R_s$$

$$\therefore \quad 14\,R_s = 88$$

$$\therefore \quad R_s = \mathbf{6.285\ k\Omega}$$

Example 4.11: For non-inverting amplifier shown in the Fig. 4.23 find the relation between R_f and R such that $V_o = V_1 + V_2 + V_3$.

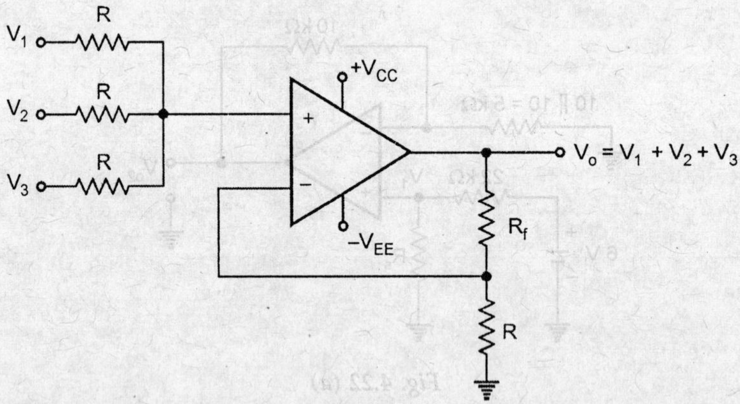

Fig. 4.23

Solution: The circuit is redrawn as shown in the Fig. 4.23(*a*).

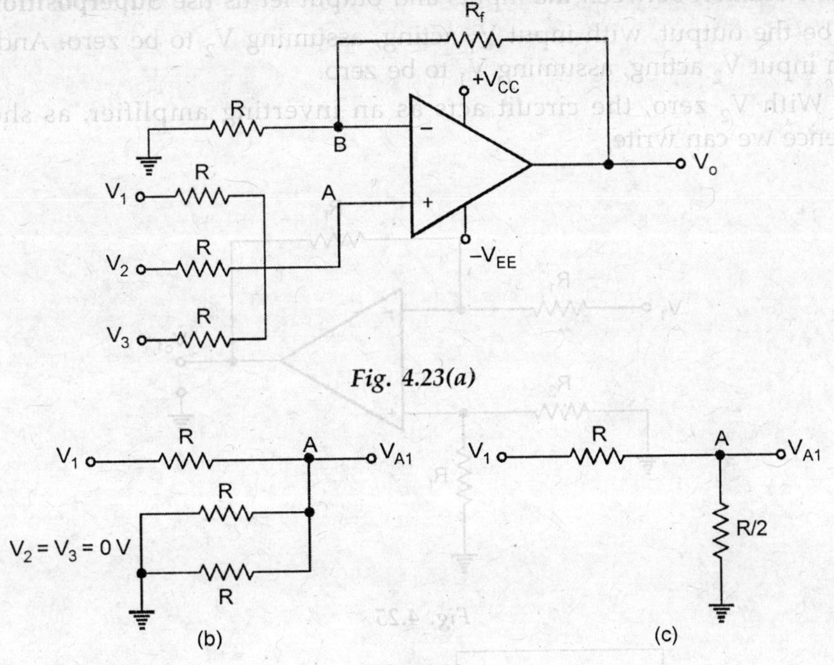

Fig. 4.23(a)

Fig. 4.23

Similarly for V_2 acting, with $V_1 = V_3 = 0$, $V_{A2} = V_2/3$.

Similarly for V_3 acting, with $V_1 = V_2 = 0$, $V_{A3} = V_3/3$.

$$\therefore \quad V_A = V_{A1} + V_{A2} + V_{A3} = \frac{V_1 + V_2 + V_3}{3}$$

$$\therefore \quad V_o = \left[1 + \frac{R_f}{R}\right] V_A = \left[1 + \frac{R_f}{R}\right] \times \frac{V_1 + V_2 + V_3}{3}$$

But $\quad V_o = V_1 + V_2 + V_3 \quad$ (given)

$$\therefore \quad 1 + \frac{R_f}{R} = 3$$

$$\therefore \quad R_f = 2R$$

4.9 SUBTRACTOR OR DIFFERENCE AMPLIFIER

Similar to the summer circuit, the subtraction of two input voltages is possible with the help of op-amp circuit, called subtractor or difference amplifier circuit.

The circuit diagram is Fig. 4.24.

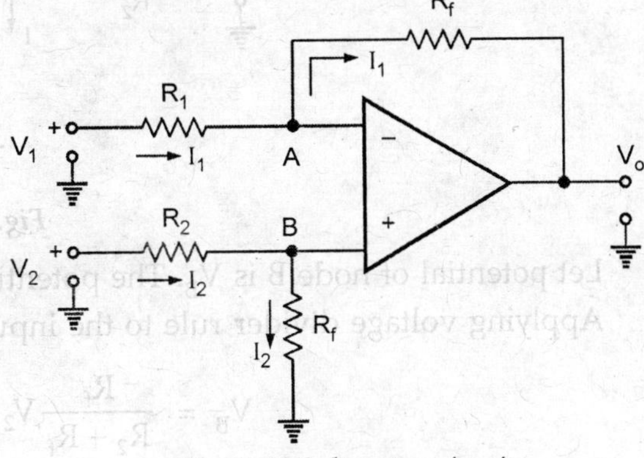

Fig. 4.24 Subtractor circuit

To find the relation between the inputs and output let us use **Superposition principle.**

Let V_{o1} be the output, with input V_1 acting, assuming V_2 to be zero. And V_{o2} be the output, with input V_2 acting, assuming V_1 to be zero.

Case 1: With V_2 zero, the circuit acts as an inverting amplifier, as shown in the Fig. 4.25. Hence we can write,

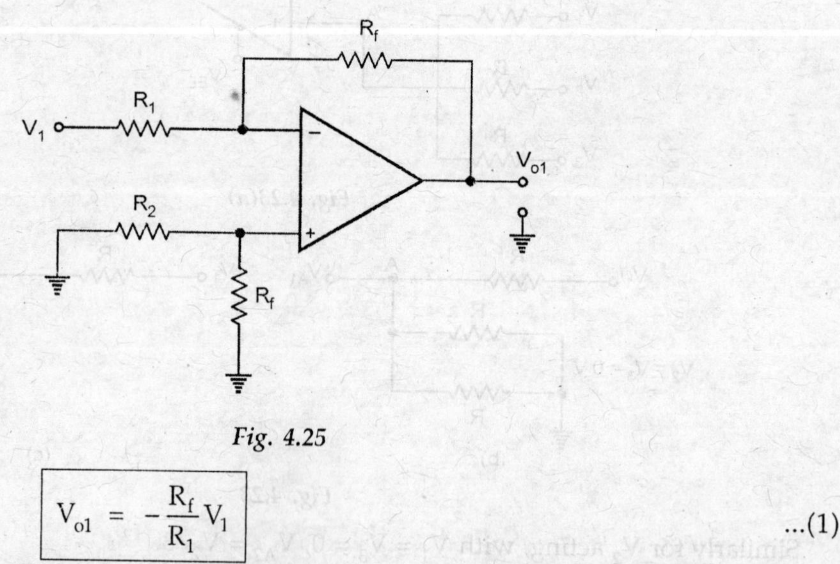

Fig. 4.25

$$\boxed{V_{o1} = -\frac{R_f}{R_1} V_1} \qquad \text{...(1)}$$

Case 2: While with V_1 as zero, the circuit reduces to as shown in the Fig. 4.26.

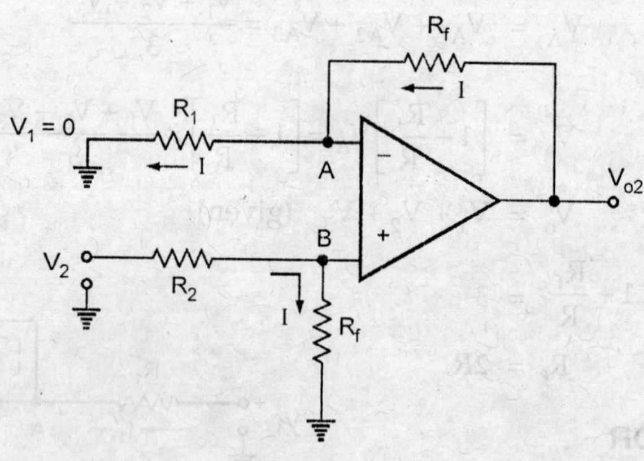

Fig. 4.26

Let potential of node B is V_B. The potential of node A is same as B i.e. $V_A = V_B$. Applying voltage divider rule to the input V_2 loop,

$$V_B = \frac{R_f}{R_2 + R_f} V_2 \qquad \text{...(2)}$$

Now $$I = \frac{V_A}{R_1} = \frac{V_B}{R_1} \quad ...(3)$$

And $$I = \frac{V_{o2} - V_A}{R_f} = \frac{V_{o2} - V_B}{R_f} \quad ...(4)$$

Equating the equations (3) and (4),

$$\frac{V_B}{R_1} = \frac{V_{o2} - V_B}{R_f}$$

$$\therefore \quad V_{o2} = \frac{R_1 + R_f}{R_1} V_B$$

$$\therefore \quad V_{o2} = \left[1 + \frac{R_f}{R_1}\right] V_B \quad ...(5)$$

Substituting V_B from equation (2) in equation (5) we get,

$$V_{o2} = \left[1 + \frac{R_f}{R_1}\right]\left[\frac{R_f}{R_2 + R_f}\right] V_2 \quad ...(6)$$

Hence using Superposition principle,

$$V_o = V_{o1} + V_{o2}$$

$$= -\frac{R_f}{R_1} V_1 + \left[1 + \frac{R_f}{R_1}\right]\left[\frac{R_f}{R_2 + R_f}\right] V_2 \quad ...(7)$$

Now if the resistances are selected as $R_1 = R_2$,

$$V_o = -\frac{R_f}{R_1} V_1 + \left[1 + \frac{R_f}{R_1}\right]\left[\frac{R_f}{R_1 + R_f}\right] V_2 = \frac{R_f}{R_1} V_1 + \frac{R_f}{R_1} V_2$$

$$\therefore \quad V_o = +\frac{R_f}{R_1}(V_2 - V_1) \quad ...(8)$$

Key Point: *Thus the output voltage is proportional to the difference between the two input voltages. Thus it acts as a subtractor or difference amplifier.*

If $R_1 = R_2 = R_f$ is selected,

$$V_o = V_2 - V_1 \quad ...(9)$$

But by selecting proper values of R_1, R_2 and R_f, we can have the subtraction of two inputs with appropriate strengths like

$$V_o = aV_2 - bV_1$$

Thus using adders and sub tractors, various mathematical equations like

$V_o = (aV_1 + bV_2) - (cV_3 + dV_4)$ can be solved. In such equation, two adder circuits are used and the outputs of these circuits are used as input to a subtractor to obtain the required equation.

Example 4.12: Design the op-amp circuit which can give the output as

$V_o = 2V_1 - 3V_2 + 4V_3 - 5V_4$.

Solution: The positive and negative terms can be added separately using two adders and then subtractor can be used.

$V_o = 2V_1 + 4V_3$, let $R_{f1} = 100\ k\Omega$

$$V_{o1} = -\left(\frac{R_{f1}}{R_1}V_1 + \frac{R_{f1}}{R_3}V_3\right)$$

$$\therefore \qquad \frac{R_{f1}}{R_1} = 2 \text{ hence } R_1 = 50\ k\Omega$$

and
$$\frac{R_{f1}}{R_3} = 4 \text{ hence } R_3 = 25\ k\Omega$$

For $(3V_2 + 5V_4)$, let $R_{f2} = 120\ k\Omega$

$$V_{o2} = -\left(\frac{R_{f1}}{R_2}V_1 + \frac{R_{f1}}{R_4}V_3\right)$$

$$\therefore \qquad \frac{R_{f2}}{R_2} = 3 \text{ hence } R_2 = 40\ k\Omega$$

and
$$\frac{R_{f2}}{R_4} = 5 \text{ hence } R_4 = 24\ k\Omega$$

Use the subtractor with all the resistances of same value of $R = 100\ k\Omega$.

Hence the output of the subtractor is $V_o = V_{o2} - V_{o1}$ where V_{o2} and V_{o1} are the two inputs of the subtractor, derived from the previous adders.

The output voltage is,

$$V_o = V_{o2} - V_{o1} = -3V_2 - 5V_4 - (-2V_1 - 4V_3)$$
$$= 2V_1 - 3V_2 + 4V_3 - 5V_4$$

This is the required output.

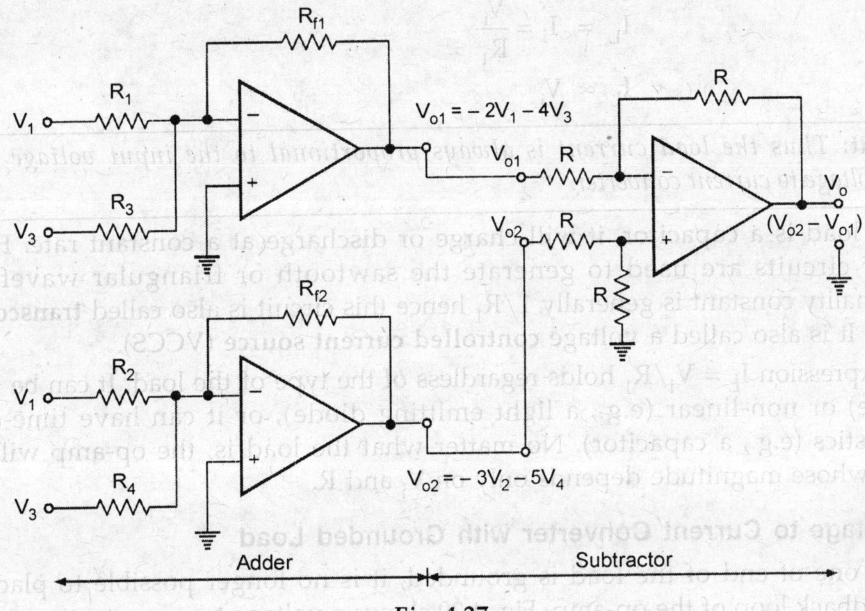

Fig. 4.27

4.10 VOLTAGE TO CURRENT CONVERTER

In a voltage to current converter, the output load current is proportional to the input voltage. According to the connection of load there are two types of V to I converters : Floating type and Grounded type. In floating type V to I converter R_L is not connected to the ground whereas in grounded type, one end of R_L is connected to the ground.

4.10.1 Voltage to Current Converter with Floating Load

The Fig. 4.28 shows a voltage to current converter in which load resistor R_L is floating.

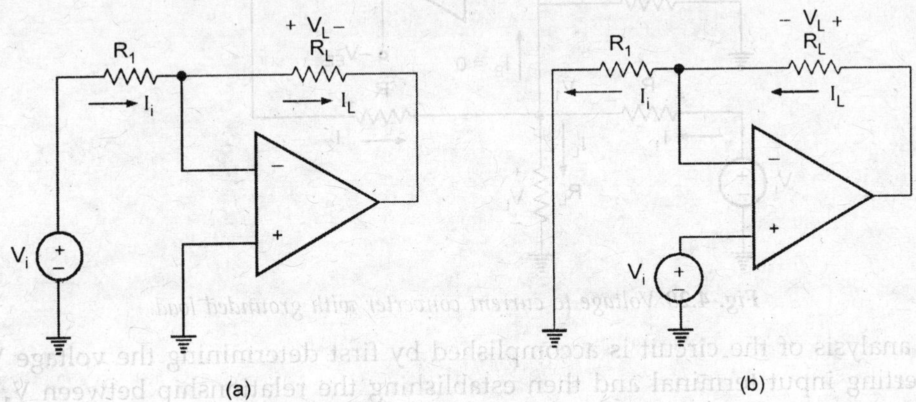

Fig. 4.28 Floating load V-I converters.

As input current of op-amp is zero.

$$I_L = I_1 = \frac{V_i}{R_1} \quad ...(1)$$

$$\therefore \quad I_L \propto V_i \quad ...(2)$$

Key Point: *Thus the load current is always proportional to the input voltage and circuit works as voltage to current converter.*

If the load is a capacitor, it will charge or discharge at a constant rate. Hence such converter circuits are used to generate the sawtooth or triangular waveforms. The proportionality constant is generally $1/R_1$ hence this circuit is also called **transconductance amplifier.** It is also called a voltage **controlled current source** (VCCS).

The expression $I_L = V_i/R_1$ holds regardless of the type of the load. It can be linear (e.g., a resistive) or non-linear (e.g., a light emitting diode), or it can have time-dependent characteristics (e.g., a capacitor). No matter what the load is, the op-amp will draw the current I_i whose magnitude depends only on V_i and R.

4.10.2 Voltage to Current Converter with Grounded Load

When one of end of the load is grounded, it is no longer possible to place the load within feedback loop of the op-amp. Fig. 4.29 shows a voltage to current converter in which one end of load resistor R_L is grounded. It is also known as **'Howland Current Converter'** from the name of its inventor.

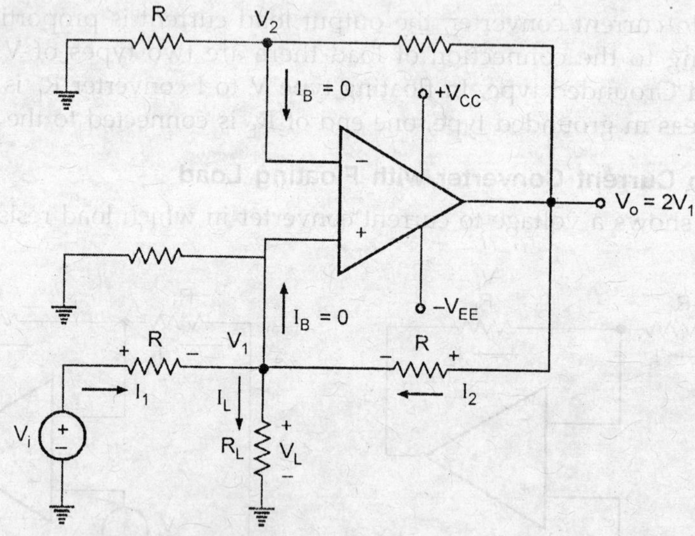

Fig. 4.29 Voltage to current converter with grounded load.

The analysis of the circuit is accomplished by first determining the voltage V_1 at the non-inverting input terminal and then establishing the relationship between V_1 and the load current.

Applying KCL at node V_1 we get,

$$I_1 + I_2 = I_L$$

$$\frac{V_i - V_1}{R} + \frac{V_o - V_1}{R} = I_L$$

$$V_i + V_o - 2V_1 = I_L R$$

$$\therefore \quad V_1 = \frac{V_i + V_o - I_L R}{2}$$

The gain of op-amp in non-inverting mode is given as A = 1 + R_f / R_1. For this circuit it is 1 + R/R = 2. Hence, output voltage can be written as,

$$V_o = 2V_1 = V_i + V_o - I_L R \quad ...(4)$$

$$0 = V_i + I_L R$$

$$\therefore \quad V_i = I_L R$$

$$\text{or} \quad \boxed{I_L = \frac{V_i}{R}} \quad ...(5)$$

From the above equation we can say that the load current depends on the input voltage V_i and resistor R.

4.10.3 Applications of V-I Converter

4.10.3.1 Low Voltage D.C. Voltmeter

The V-I converter circuit with floating load can be modified as a low voltage d.c. voltmeter circuit. The resistance R_L is to be replaced by D'Arsonval meter movement. This is shown in the Fig. 4.30.

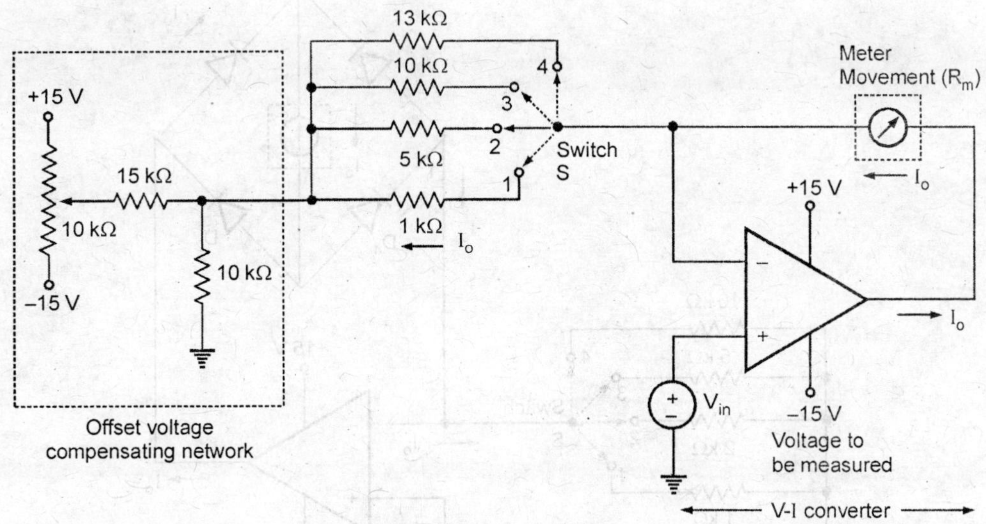

Fig. 4.30 D.C. voltmeter.

For accurate readings it is necessary to make the nulling of op-amp. This is done by offset voltage compensating network. Let R_m be the meter resistance. The effective Thevenin's equivalent of compensating network is 10 Ω. Hence when switch S is in position 1, the effective R_1 is 1 kΩ + 10 Ω ≈ 1 kΩ.

Let V_{in} = 1 V then,

$$I_o = \frac{V_{in}}{R_1} \approx \frac{1\,V}{1\,k\Omega} = 1\,mA$$

So current of 1 mA i.e. input of 1 V will result full scale deflection, if meter has full scale deflection current of 1 mA.

When switch is in position 2 then R_1 = 5 kΩ. Thus for full scale deflection,

$$1 \times 10^{-3} = \frac{V_{in}}{5 \times 10^3}$$

$$\therefore \quad V_{in} = 5\,V$$

Thus 5 V are required to cause full scale deflection. Thus range of voltmeter increases by 5. Similarly at positions 3 and 4, ranges of X10 and X13 are also obtained.

Remember that for ± 15 V supply, maximum input voltage range is ± 14 V hence with ± 15 V supply, maximum range possible is X13.

It is important to note that meter resistance R_m does not affect the value of I_o. Thus meter can be calibrated properly to measure d.c. voltages.

4.10.3.2 Low Voltage A.C. Voltmeter

To measure a.c. voltages, the full wave bridge rectifier is used and D'Arsonval meter movement is used. The combination of meter and rectifier is used in the feedback loop of a.c. voltmeter. The circuit is shown in the Fig. 4.31.

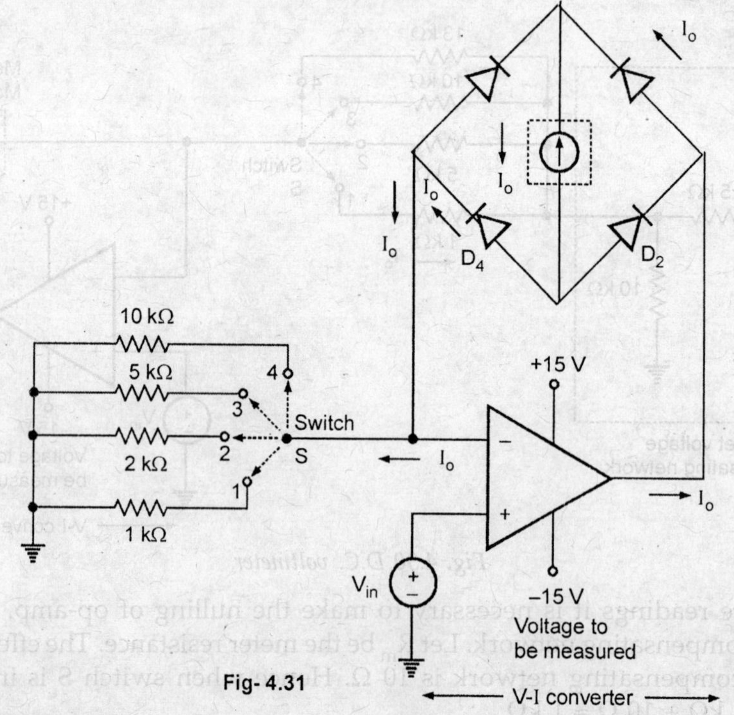

Fig. 4.31 A.C. voltmeter.

The alternating current I_o is converted to d.c. During positive half cycle of V_{in} the diodes D_3 and D_4 conduct while for negative half cycle of V_{in} the diodes D_1 and D_2 conduct. The direction of current through meter is always same for both the half cycles of V_{in}. Thus meter gives the average (d.c.) value of the rectified current. If it is necessary to calibrate the meter interms of r.m.s. values of input voltage then the factor 1.11 must be considered which relates r.m.s. and average values for purely sinusoidal waveforms. The range can be extended using the selector switch S and various resistances as shown in the Fig. 3.55. At position 2, X2 can be obtained. At position 3, X5 range can be obtained. At position 4, X10 range can be obtained. It must be noted that due to low slew rate of 741 op-amp, the input a.c. voltage frequency should not be greater than 4 kHz. Similarly PIV ratings of the diodes must be large than the saturation voltages of op-amp.

4.10.3.3 Diode Tester and Match Finder

Another application of V-I converter is a diode tester and diode match finder. The resistance R_L is replaced by a diode as shown in the Fig. 4.32.

The current 10 is set by V_{in} and resistance R_1.

$$I_o = \frac{V_{in}}{R_1}$$

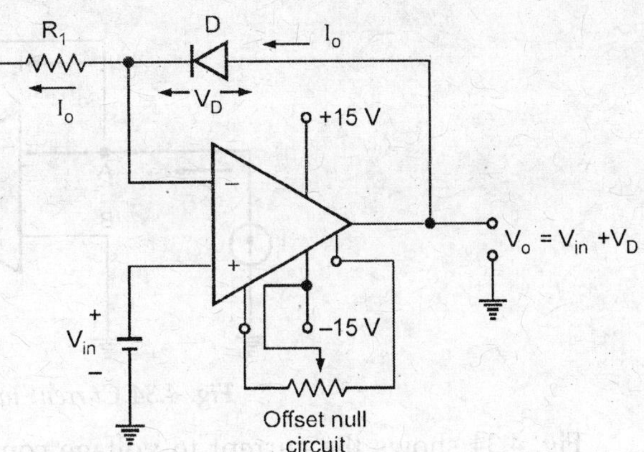

Fig. 4.32 Diode tester and match finder.

So as long as V_{in} and R_1 are constants, the current I_o is also constant. The drop across diode can be directly measured by the voltmeter or the output voltage will be $V_o = V_{in} + V_D$. From the value of V_D, the diode can be judged for its performance. Note that to avoid error in the output voltage, op-amp should be initially nulled using offset null circuit.

Similarly replacing diodes one after the other, the perfectly matched diodes can be found out. R_1 can be selected according to V_{in} and testing current 10 required in the circuit.

4.10.3.4 Zener Diode Tester

The V-I converter circuit can also be used to test zener diode. The zener diode is connected in reverse biased condition in the feedback path as shown in the Fig. 4.33.

The current I_o is set by V_{in} and R_1. If this current is more than knee current of the zener then voltage across zener is V_Z for which it is rated. Similarly matched zeners also can be obtained using same circuit.

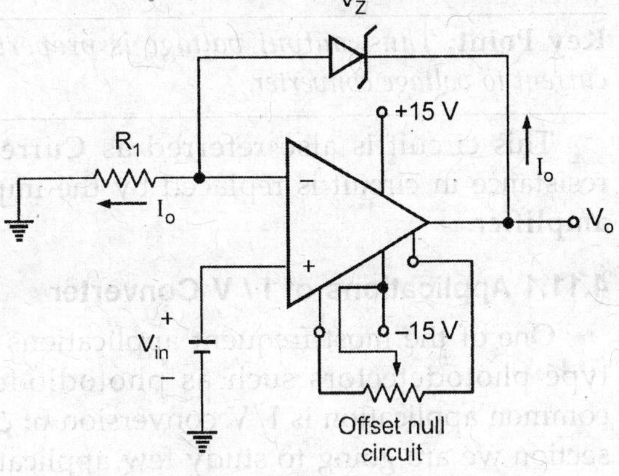

Fig. 4.33 Zener diode tester.

The same circuit can be used for LED testing and match finding by replacing the feedback path by LED under test. The brightness of LEDS can be tested by setting specific current using V_{in} and R_1.

In all the applications, as 741 is used it is necessary to note that short circuit current rating of op-amp i.e. 25 mA should not be exceeded by the current I_o.

4.11 CURRENT TO VOLTAGE CONVERTER

In such a converter, the output voltage is proportional to the input current. It accepts an input current I_i and yields an output voltage V_o such that $V_o = A\,I_i$, where A is the gain of the circuit. Since A is measured in ohms, it is more appropriate to denote gain by the symbol R. Because of this, I-V converters are also called transresistance amplifiers.

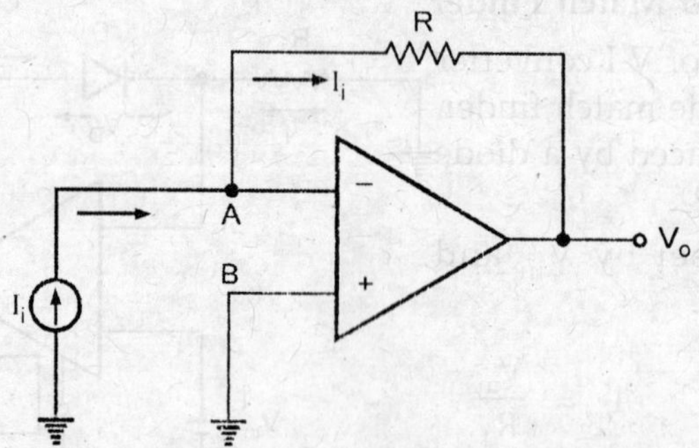

Fig. 4.34 Current to voltage converter.

Fig. 4.34 shows the current to voltage converter.

The node A is a virtual ground as node B is grounded. Hence $V_A = 0$.

$$I_i = \frac{V_A - V_o}{R} = -\frac{V_o}{R} \quad ...(1)$$

$$V_o = -I_i R \quad ...(2)$$

$$\therefore \quad V_o \propto I_i$$

Key Point: *Thus, output voltage is proportional to the input current and circuit works as a current to voltage converter.*

This circuit is also referred as **Current Controlled Voltage Source (CCVS).** If the resistance in circuit is replaced by the impedance Z, the circuit is called **transimpedance amplifier.**

4.11.1 Applications of I / V Converter

One of the most frequent applications of I/V converters is in connection with current type photodetectors such as photodiodes, photoFETs and photomultipliers. Another common application is I/V conversion of current output digital to analog converter. In this section we are going to study few applications of I/V converters.

4.11.1.1 Photodiode Detector

Fig. 4.35 shows the connection diagram of one of the most widely used photodetectors, photodiode.

The photodiode produces electrical current in response to incident light. This current flows through R. The voltage across R, the output voltage is proportional to the diode current.

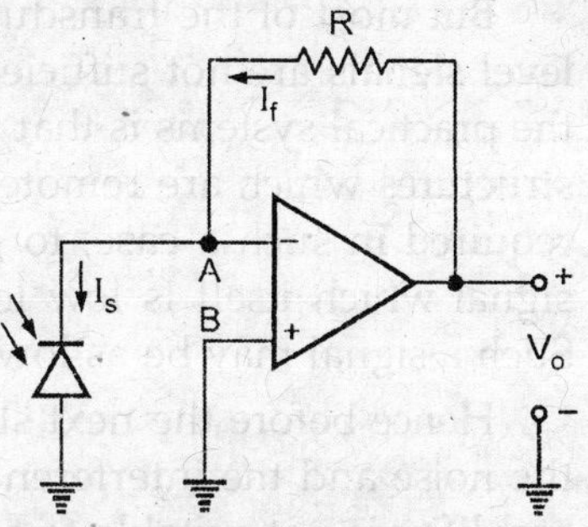

Fig. 4.35 Photodiode as a photodetector.

4.11.1.2 PhotoFET Detector

Fig. 4.36 shows another photodetector circuit with photoFET.

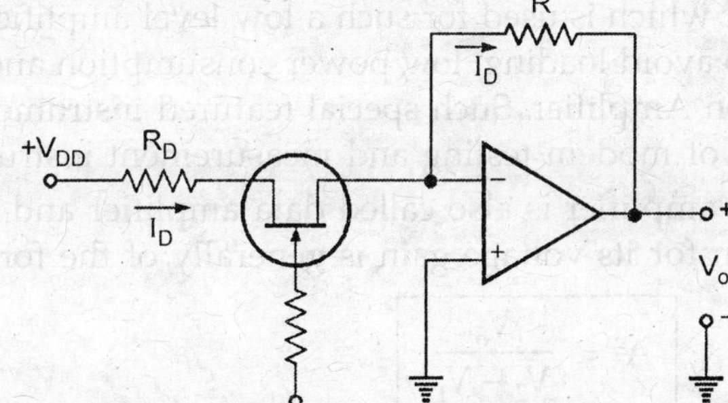

Fig. 4.36 Photodetector circuit with photoFET.

The photoFET is similar to a conventional junction FET, with the exception of a lens for focusing light onto the gate junction. On application of light, photons enter the gate area and excite valence electrons into the conduction band. The photon-excited current carriers cause a small current ΔI_G resulting in large current change ΔI_D. The change in 10 results change the voltage drop across R and hence the output voltage. Therefore the op-amp circuit acts as an I/V converter and gives the indication of light interms of voltage.

4.12 INTRODUCTION TO INSTRUMENTATION AMPLIFIERS

Many industrial systems, consumer systems and process control systems require a precise measurement of the physical quantities like temperature, pressure, humidity, weight etc. A sugar factory requires a measurement of flow, level and temperature of juice. The plastic furnaces require precise measurement of the temperature. The dairy plant requires a precise measurement of temperature and the humidity. Such measurements help the industries to have the production of quality products.

The measurement of the physical quantities is generally carried out with the help of a device called as **transducer**. A transducer is a device which converts one form of energy into another. For example a thermocouple converts the heat energy into an electrical energy, microphone converts the sound energy into an electrical energy, strain gauge converts pressure, force like mechanical energy into an electrical energy. Such a proportional electrical signal output from a transducer can be used further to control or operate the other parts of the system or can be used to get the display or recording of the measured physical quantity.

But most of the transducer outputs are generally of very low level signals. Such a low level signals are not sufficient to drive the next stage of the system. One more difficulty in the practical systems is that the transducer used may be mounted on pieces of equipment or structures which are remote from the control location. Long connecting wires or cables are required in such a case, to get the transducer output to the control room. Due to this, the signal which itself is low level, gets subjected to the noise and atmospheric interference. Such a signal may be as low as few mV or even μV.

Hence before the next stage, it is necessary to amplify the level of such signal, rejecting the noise and the interference. Hence general single ended amplifier like high gain emitter amplifier is not suitable to amplify such signals. For rejection of noise, such amplifiers must have high common mode rejection ratio. Hence a special amplifier is used to amplify such signals.

The special amplifier which is used for such a low level amplification with high CMRR, high input impedance to avoid loading, low power consumption and some other features is called an Instrumentation Amplifier. Such special featured instrumentation amplifiers have become an integral part of modem testing and measurement instrumentation.

The instrumentation amplifier is also called data amplifier and is basically a difference amplifier. The expression for its voltage gain is generally of the form,

$$\boxed{A = \frac{V_o}{V_2 - V_1}} \qquad \text{...(1)}$$

where V_o = Output of the amplifier

$V_2 - V_1$ = Differential input which is to be amplified

Let us list down the requirements of an instrumentation amplifier.

4.13 REQUIREMENTS OF A GOOD INSTRUMENTATION AMPLIFIER

As mentioned above, the instrumentation amplifiers are used to amplify the low level differential signals very precisely, in presence of the large common mode noise and interference signals. Hence a good instrumentation amplifier has to meet the following specifications:

1. **Finite, accurate and stable gain:** As very low level signals are required to be amplified by the instrumentation amplifiers, high and finite gain is the basic requirement. It is usually in the range of 1 to 1000. The gain has to be accurate and closed loop gain must be stable in nature.
2. **Easier gain adjustment:** Not only finite and stable gain is required but a variable gain over the prescribed range is also required. The gain adjustment must be easier and precise. Generally such gain adjustment is done continuously using a potentiometer or is done digitally with the help of switches, which are JFET or MOSFET switches.
3. **High input impedance:** To avoid the loading of input sources, input impedance of the instrumentation amplifier must be very high (ideally infinite). The differential mode input impedance Z_{id} is the equivalent impedance between the two input

terminals. The common mode input impedance Z_{ic} is the equivalent impedance between each input terminal and ground.

4. **Low output impedance:** Extremely low output impedance (ideally zero) to avoid the loading on the immediate stage.
5. **High CMRR:** The output of transducer, when transmitted with long transmission lines has presence of large common mode noise voltages. The instrumentation amplifier must amplify only the differential input, completely rejecting the common mode input component. Thus it must have ideally infinite CMRR.
6. **Low power consumption:** The power consumption of an instrumentation amplifier should be as low as possible.
7. **Low thermal and time drifts:** The parameters of the instrumentation amplifier, should not drift with temperature or time.
8. **High slew rate:** The slew rate of the instrumentation amplifier must be as high as possible to provide maximum undistorted output voltage swing.
9. The amplifier must have differential input so that it can be amplified.

Let us see the various circuits which can be used as the instrumentation amplifier and their advantages and disadvantages.

4.14 DIFFERENCE AMPLIFIER USING ONE OP-AMP

The basic difference amplifier using op-amp amplifies the differential input and hence can be considered as the instrumentation amplifier. The basic difference amplifier using one op-amp is shown in the Fig. 4.37.

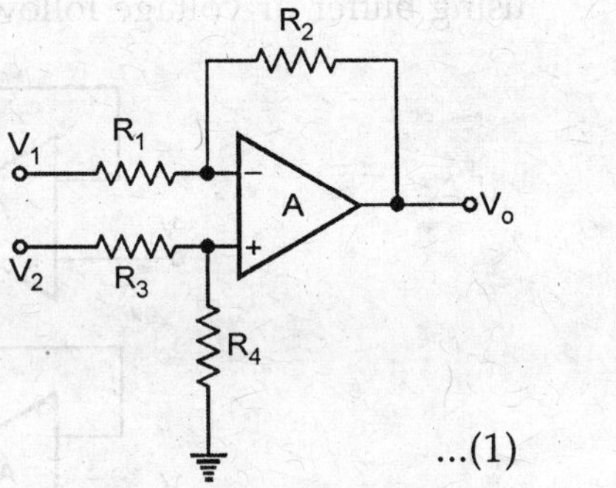

Fig. 4.37 Difference amplifier.

The voltage gain of such a difference amplifier is given by,

$$V_o = -\frac{R_2}{R_1}V_1 + \left(1+\frac{R_2}{R_1}\right)\left(\frac{R_4}{R_3+R_4}\right)V_2 \quad ...(1)$$

Selecting the resistance values as,

$$R_1 = R_3 \quad \text{and} \quad R_2 = R_4$$

$$V_o = -\frac{R_2}{R_1}V_1 + \left(1+\frac{R_2}{R_1}\right)\left(\frac{R_4}{R_3+R_4}\right)V_2 = -\frac{R_2}{R_1}V_1 + \frac{R_2}{R_1}V_2$$

$$\therefore \quad V_o = (V_2 - V_1)\frac{R_2}{R_1}$$

$$\therefore \quad \boxed{\frac{V_o}{V_2 - V_1} = \frac{R_2}{R_1}}$$

Thus the differential input $V_2 - V_1$ gets amplified by the ratio R_2/R_1. This gain is same as that of the inverting amplifier.

The main limitations of this circuit to be used as an instrumentation amplifier is its input impedance. The input impedance of the circuit is very low. Not even this, but when $V_2 = 0$, the input impedance is R_1 while when $V_1 = 0$, the input impedance is $R_1 + R_4$ i.e. $R_1 + R_2$. Thus the input impedances seen by the signal sources V_1 and V_2 are not the same. Such inequality in the input impedances is also the major limitation, as an instrumentation amplifier required balanced and high input impedance.

Another limitation is that the internal circuit components are not ideal and matched and external resistance values are not precise. Hence imbalance may result as ideal equations can not be satisfied by the resistance values.

Hence the basic difference amplifier is generally not used as an instrumentation amplifier. But it can be seen in some applications like output from a Wheatstone bridge, which could be a strain gauge transducer.

4.15 MODIFIED DIFFERENCE AMPLIFIER

The difference amplifier using op-amp can be used as instrumentation amplifier, but the main problem in using it as an instrumentation amplifier is its input impedance. The input impedance of difference amplifier is low while the instrumentation amplifier need very high input impedance.

To get a very high input impedance, the difference amplifier circuit can be modified by using buffer or voltage follower circuits at the inputs. This is shown in the Fig. 4.38.

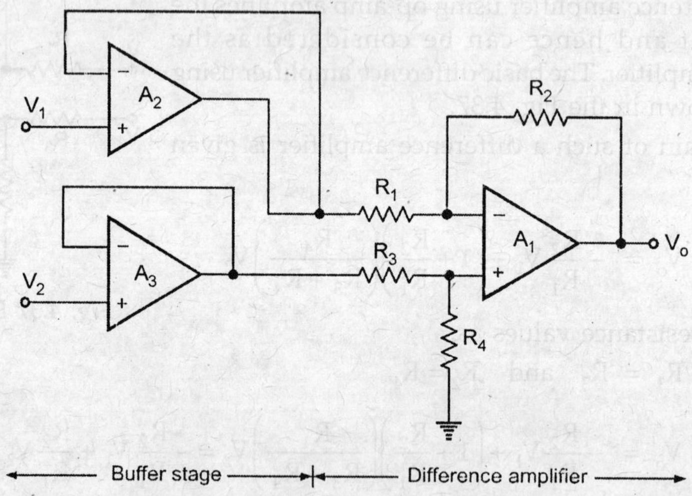

Fig. 4.38 Modified difference amplifier.

The gain of the voltage follower circuit is unity while its input impedance is very high. Hence the circuit provides same voltage gain as provided by the op-amp difference amplifier.

Hence the gain expression for the circuit is

$$V_o = -\frac{R_2}{R_1}V_1 + \left(1+\frac{R_2}{R_1}\right)\left(\frac{R_4}{R_3+R_4}\right)V_2 \quad ...(1)$$

Now $\quad R_3 = R_1 = R_1$

And $\quad R_4 = R_2 = R_2$

$$\therefore \quad V_o = \frac{R_2}{R_1}(V_2 - V_1) \quad \ldots(2)$$

Due to unity gain voltage follower circuits, the gain remains same as that of basic difference amplifier. But its input impedance is very high as input impedance of voltage follower circuit is very very high.

Such a circuit requires three op-amps like μA 741. The IC LM324 contains four op-amps inside and hence can be conveniently used to design this circuit, which can be used as an instrumentation amplifier.

4.16 INSTRUMENTATION AMPLIFIER WITH TWO OP-AMPS

As seen earlier, the high input impedance is achieved by using voltage follower circuit. Such a high input impedance along with variable voltage gain can be achieved by using a two op-amp instrumentation amplifier circuit. Such a circuit is shown in the Fig. 4.39.

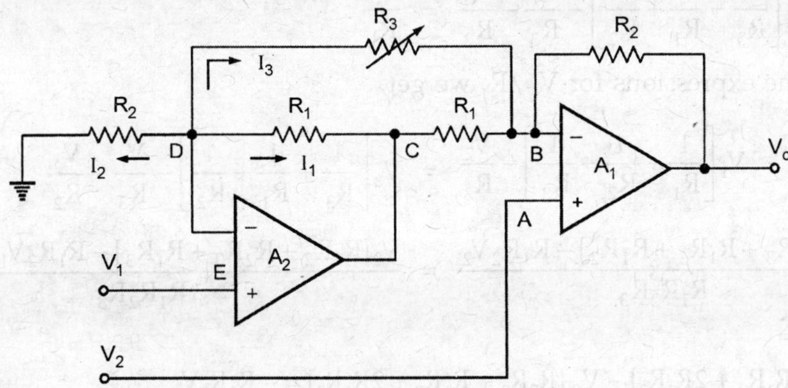

Fig. 4.39 Instrumentation amplifier using two op-amps.

For amplifier A_1, we can say that potential of node A is same as that of node B.

$$\therefore \quad V_A = V_B = V_2 \quad \ldots(1)$$

Similarly for amplifier A_2, the potential of node E is same as that of node D.

$$V_E = V_D = V_1 \quad \ldots(2)$$

Now

$$I_2 = \frac{V_D}{R_2} = \frac{V_1}{R_2} \quad \ldots(3)$$

$$I_1 = \frac{V_D - V_C}{R_1} = \frac{V_1 - V_C}{R_1} \quad \ldots(4)$$

$$I_3 = \frac{V_D - V_B}{R_3} = \frac{V_1 - V_2}{R_3} \quad \ldots(5)$$

Applying KCL at node D,

$$I_1 + I_2 + I_3 = 0 \qquad ...(6)$$

$$\therefore \quad \frac{V_1 - V_C}{R_1} + \frac{V_1}{R_2} + \frac{V_1 - V_2}{R_3} = 0 \qquad ...(7)$$

$$\therefore \quad V_1\left[\frac{1}{R_1} + \frac{1}{R_2} + \frac{1}{R_3}\right] - \frac{V_2}{R_3} = \frac{V_C}{R_1} \qquad ...(8)$$

Similarly applying KCL at node B and neglecting input current of op-amps we can write,

$$\frac{V_B - V_D}{R_3} + \frac{V_B - V_C}{R_1} + \frac{V_B - V_o}{R_2} = 0 \qquad ...(9)$$

$$\therefore \quad \frac{V_2 - V_1}{R_3} + \frac{V_2 - V_C}{R_1} + \frac{V_B - V_o}{R_2} = 0 \qquad ...(10)$$

$$\therefore \quad V_2\left[\frac{1}{R_3} + \frac{1}{R_1} + \frac{1}{R_2}\right] - \frac{V_1}{R_3} - \frac{V_o}{R_2} = \frac{V_C}{R_1} \qquad ...(11)$$

Equating the expressions for V_C/R_1 we get,

$$V_1\left[\frac{1}{R_1} + \frac{1}{R_2} + \frac{1}{R_3}\right] - \frac{V_2}{R_3} = V_2\left[\frac{1}{R_3} + \frac{1}{R_1} + \frac{1}{R_2}\right] - \frac{V_1}{R_3} - \frac{V_o}{R_2} \qquad ...(12)$$

$$\frac{V_1[R_2R_3 + R_1R_3 + R_1R_2] - R_1R_2V_2}{R_1R_2R_3} = \frac{V_2[R_1R_2 + R_2R_3 + R_1R_3] - R_1R_2V_1 - R_1R_3V_o}{R_1R_2R_3} \qquad ...(13)$$

$$V_1[R_2R_3 + R_1R_3 + 2R_1R_2] - V_2[R_2R_3 + R_1R_3 + 2R_1R_2] = -R_1R_3V_o \qquad ...(14)$$

$$R_1R_3V_o = V_2 - V_1[R_2R_3 + R_1R_3 + 2R_1R_2] \qquad ...(15)$$

$$\therefore \quad \frac{V_o}{(V_2 - V_1)} = 1 + \frac{R_2}{R_1} + \frac{2R_2}{R_3} \qquad ...(16)$$

The $V_2 - V_1$ is the differential input while V_o is the output voltage. Thus the voltage gain of this circuit is

$$\boxed{A_V = 11 + \frac{R_2}{R_1} + \frac{2R_2}{R_3}} \qquad ...(17)$$

As the resistance R_3 is variable, the circuit permits gain variation, precisely.

4.16.1 Advantages

The **advantages** of this circuit are,

1. The gain variation is easy and precise.

2. The CMRR value is completely independent of the setting of resistance R_3. Hence with the precision ratios for R_2/R_1, the gain can be changed without degrading the performance of the amplifier.
3. The resistance R_3 is separate from the accurately matched resistances R_1 and R_2, which are required for symmetric arrangement.

The **main problem** is that the resistances R_1 and R_2 must be accurately matched. Another limitation of the circuit is that it treats the input asymmetrically. The input V_1 has to propagate through A_2 before reaching to A_1. Due to this additional delay, common mode components of the two signals will no longer cancel out with each other, at high frequencies. This decreases CMRR with frequency. The typical set of values of R_1, R_2 and R_3 is

$R_1 = 9\ k\Omega \quad R_2 = 3\ k\Omega \quad R_3 = 0.5\ k\Omega$ to $9\ k\Omega$.

4.17 THREE OP-AMP INSTRUMENTATION AMPLIFIER

Another commonly used instrumentation amplifier circuit is one using three op-amps. This circuit provides high input resistance for accurate measurement of signals from transducers. In this circuit, a non-inverting amplifier is added to each of the basic difference amplifier inputs. The circuit is shown in the Fig. 4.40.

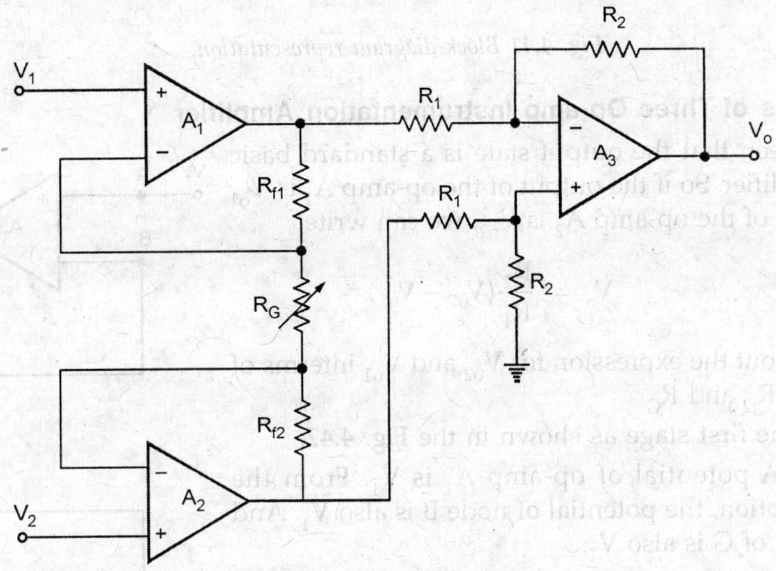

Fig. 4.40 Three op-amp instrumentation amplifier.

The op-amps A_1 and A_2 are the non-inverting amplifiers forming the input or first stage of the instrumentation amplifier. The op-amp A_3 is the normal difference amplifier forming an output stage of the amplifier.

The block diagram representation of the three op-amp instrumentation amplifier is shown in the Fig. 4.41.

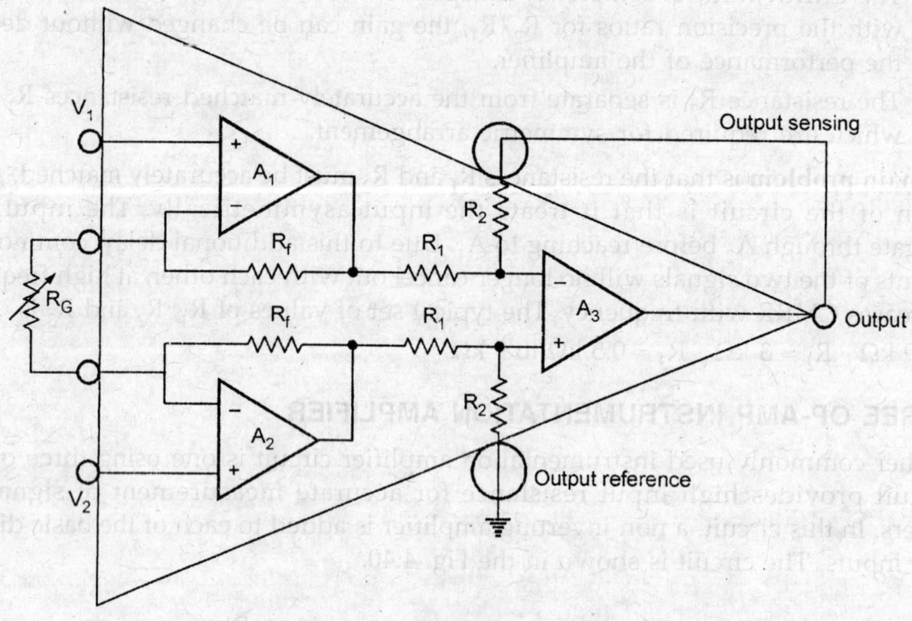

Fig. 4.41 Block diagram representation.

4.17.1 Analysis of Three Op-amp Instrumentation Amplifier

It can be seen that the output state is a standard basic difference amplifier. So if the output of the op-amp A_1 is V_{o1} and the output of the op-amp A_2 is V_{o2}, we can write,

$$V_o = \frac{R_2}{R_1}(V_{o2} - V_{o1})$$

Let us find out the expression for V_{o2} and V_{o1} interms of V_1, V_2, R_{f1} and R_{f2} and R_G.

Consider the first stage as shown in the Fig. 4.42.

The node A potential of op-amp A_1 is V_1. From the realistic assumption, the potential of node B is also V_1. And hence potential of G is also V_1.

The node D potential of op-amp A_2 is V_2. From the realistic assumption, the potential of node C is also V_2. And hence potential of H is also V_2.

The input current of op-amp A_1 and A_2 both are zero. Hence current I remains same through R_{f1} R_G and R_{f2}.

Applying Ohm's law between the nodes E and F we get,

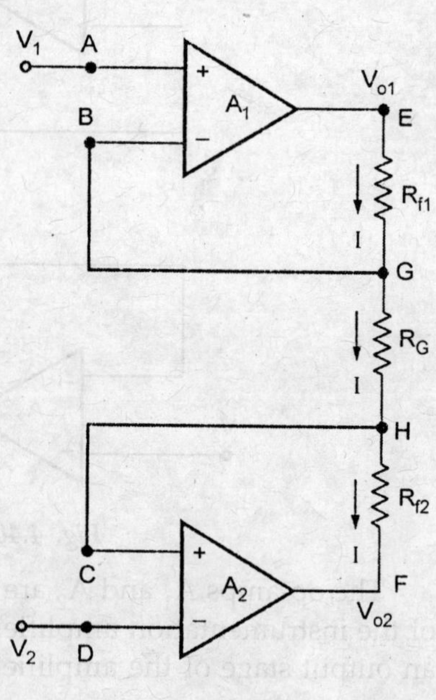

Fig. 4.42

$$I = \frac{V_{o1} - V_{o2}}{R_{f1} + R_G + R_{f2}} \qquad ...(2)$$

Let $$R_{f1} = R_{f2} = R_f \qquad ...(3)$$

$\therefore$ $$I = \frac{V_{o1} + V_{o2}}{2R_f + R_G} \qquad ...(4)$$

Now from the observation of nodes G and H,

$$I = \frac{V_G - V_H}{R_H} = \frac{V_1 - V_2}{R_G} \qquad ...(5)$$

Equating the two equations (4) and (5),

$$\frac{V_{o1} - V_{o2}}{2R_f + R_G} = \frac{V_2 - V_1}{R_G} \qquad ...(6)$$

$\therefore$ $$\frac{V_{o2} - V_{o1}}{2R_f + R_G} = \frac{V_1 - V_2}{R_G} \qquad ...(7)$$

$\therefore$ $$V_{o2} - V_{o1} = \frac{(2R_f + R_G)(V_2 - V_1)}{R_G} \qquad ...(8)$$

Substituting the $V_{o2} - V_{o1}$, in the equation (1),

$$V_o = \frac{R_2}{R_1} \cdot \left[\frac{2R_f + R_G}{R_G}\right](V_2 - V_1) \qquad ...(9)$$

$\therefore$ $$\boxed{V_o = \frac{R_2}{R_1} \cdot \left[1 + \frac{2R_f}{R_G}\right](V_2 - V_1)} \qquad ...(10)$$

This is the overall gain of the Circuit.

4.17.2 Advantages

Following are the advantages of three op-amp instrumentation amplifier circuit:

(*i*) With the help of variable resistance R_G, the gain can be easily varied, without disturbing the symmetry of the circuit.

(*ii*) Gain depends on external resistances and hence can be adjusted accurately and made stable by selecting high quality resistances.

(*iii*) The input Impedance depends on the input impedance of non-inverting amplifiers which is extremely high.

(*iv*) The output impedance is the output impedance of the op-amp A_3 which is very very low. This is as required by any instrumentation amplifier.

(v) The CMRR of the op-amp A_1 is very high and most of the common mode signal will be rejected.

(vi) By trimming one of the resistances of the output stage, CMRR can be made extremely high, as required by a good instrumentation amplifier.

Thus the circuit satisfies all the requirements of a good instrumentation amplifier and hence very commonly used in practical applications.

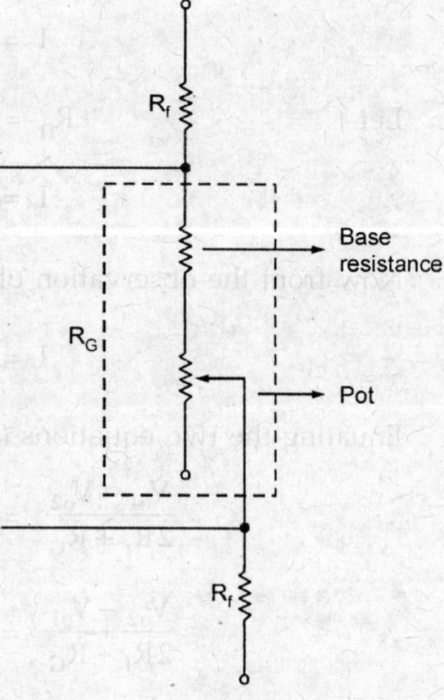

Fig. 4.43

Key Point: *The resistance RG is generally implemented as the series combination of the suitable base resistance and the pot. This is shown in the Fig. 4.43.*

Example 4.13: The circuit shown in the Fig. 4.44 is an instrumentation amplifier. Determine the range over which its gain can be varied if potentiometer is varied over its entire range.

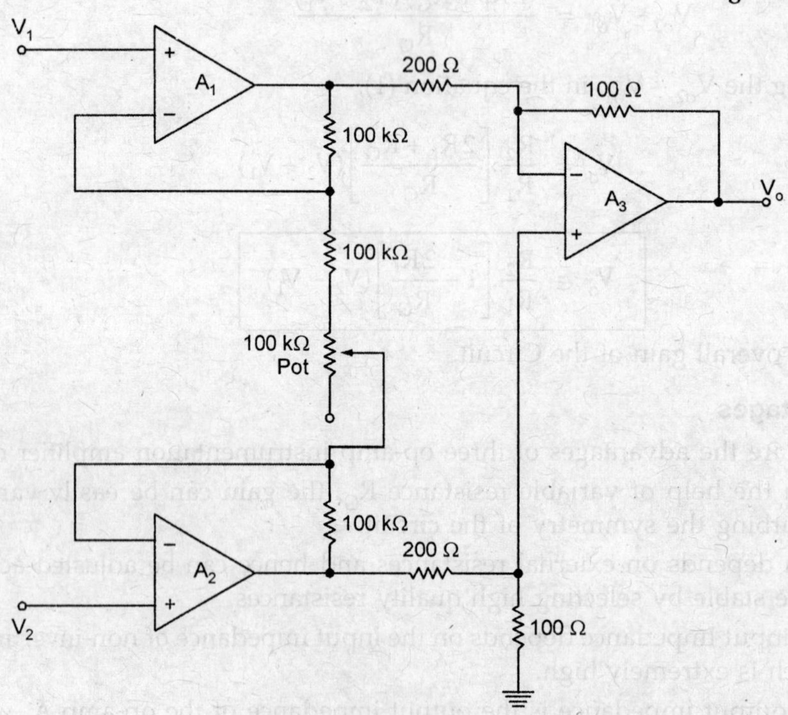

Fig. 4.44

Solution: The gain of the instrumentation amplifier is given by,

$$\frac{V_o}{V_2 - V_1} = \left(1 + \frac{2R_f}{R_G}\right)\left(\frac{R_2}{R_1}\right)$$

In the circuit given,

$$R_2 = 100\ \Omega, \quad R_1 = 200\ \Omega, \quad R_f = 100\ k\Omega$$

And R_G is the series combination of 100 Ω and potentiometer of 100 kΩ. Let potentiometer resistance is 0 Ω at start.

Hence, $$R_G = 100 + 0 = 100\ \Omega$$

$$\therefore \quad \text{Gain} = \left(1 + \frac{2 \times 100 \times 10^3}{100}\right)\left(\frac{100}{200}\right) = \mathbf{1005}$$

When the potentiometer resistance is at its maximum value,

$$R_G = 100 + 100\ k\Omega = 100.1\ k\Omega$$

$$\therefore \quad \text{Gain} = \left(1 + \frac{2 \times 100 \times 10^3}{100.1 \times 10^3}\right)\left(\frac{100}{200}\right)$$

$$= \mathbf{1.499}$$

For all practical purposes, the gain can be varied from **1.5** to **1005.**

Example 4.14: For the instrumentation amplifier shown in the Fig. 4.45, determine the value of R_G if the gain required is 1000.

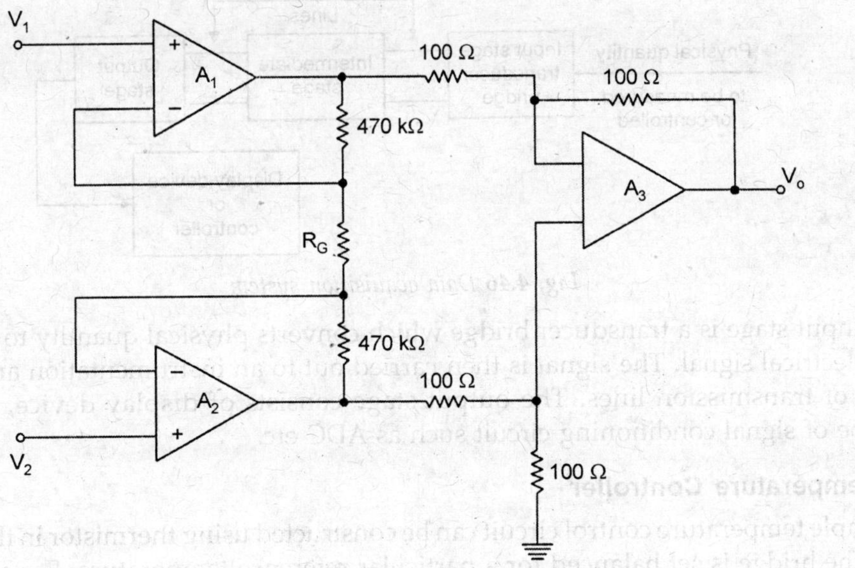

Fig. 4.45

Solution: The values of various resistances are

$$R_1 = 100\ k\Omega,\quad R_2 = 100\ k\Omega,\quad R_f = 470\ k\Omega$$

The R_G is to be selected for gain of 1000.

$$\text{Gain} = \left(1+\frac{2R_f}{R_G}\right)\left(\frac{R_2}{R_1}\right)$$

$$\therefore \quad 100 = \left(1\times\frac{2\times 470}{R_G}\right)\left(\frac{100}{100}\right)$$

$$\therefore \quad 100\times 1 = 1+\frac{940}{R_G}$$

$$\therefore \quad \frac{940}{R_G} = 99$$

$$\therefore \quad R_G = 9.4949\ \Omega$$

4.18 APPLICATIONS OF INSTRUMENTATION AMPLIFIER

As mentioned earlier, the instrumentation amplifier along with the transducer bridge can be used in many practical applications. Let us study some of such practical applications. The general form of such systems can be called as 'data acquisition system' and can be represented in the block diagram form as shown in the Fig. 4.46.

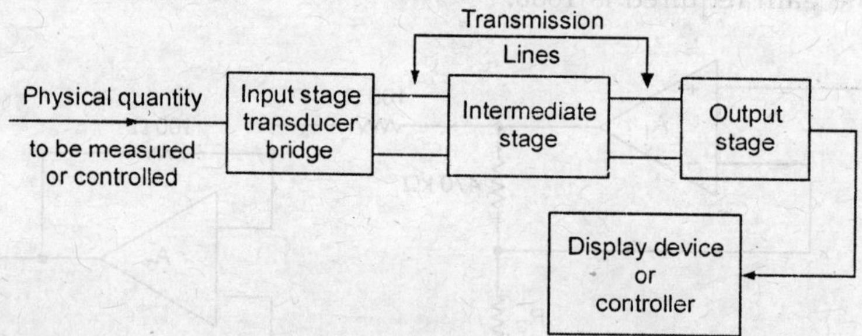

Fig. 4.46 Data acquisition system.

The input stage is a transducer bridge which converts physical quantity to be measured into an electrical signal. The signal is then carried out to an instrumentation amplifier, with the help of transmission lines. The output stage consists of display device, controller or some type of signal conditioning circuit such as ADC etc.

4.18.1 Temperature Controller

A simple temperature control circuit can be constructed using thermistor in the transducer bridge. The bridge is set balanced for a particular reference temperature. For any change in this temperature, the instrumentation amplifier produces the output voltage. Now this voltage can be used to drive the relays which inturn controls the ON-OFF of the heating unit, to control the temperature.

The system is shown in the Fig. 4.47.

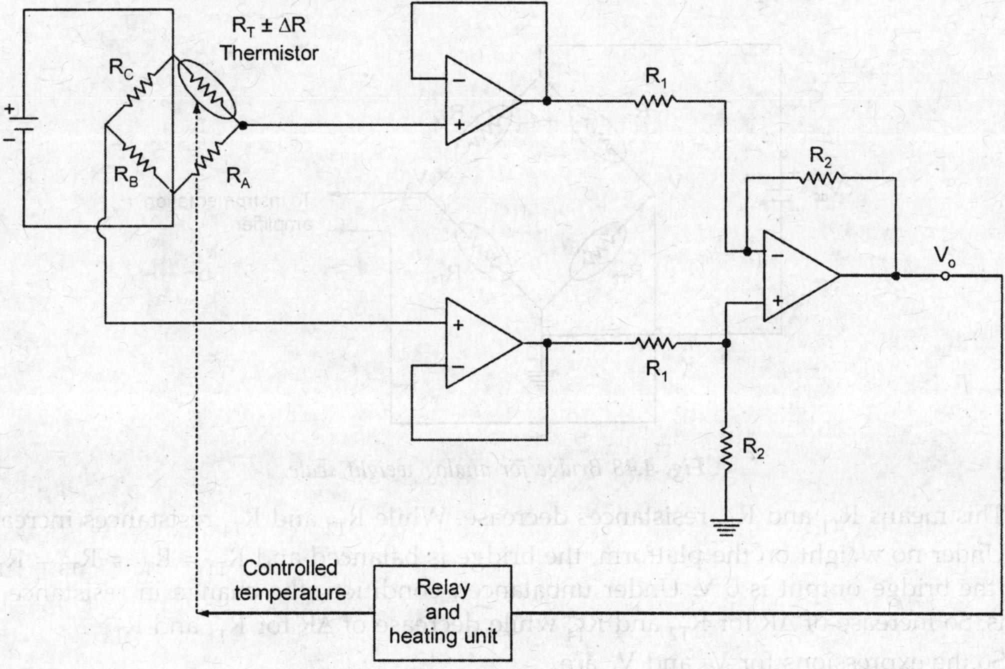

Fig. 4.47 Simple temperature controller.

4.18.2 Temperature Indicator

The circuit shown above in the Fig. 4.47 can be used as a temperature indicator. As explained earlier, bridge is kept balanced at some reference temperature when $V_o = 0$ V. The meter connected at the output is calibrated to reference temperature, corresponding to this reference condition. As temperature changes, amplifier output also changes. The meter can be calibrated to indicate the desired temperature range by selecting the appropriate gain of the amplifier.

4.18.3 Light Intensity Meter

The same circuit, replacing thermistor with a photocell can be used as a simple light intensity meter. The bridge is made balanced for the darkness condition. When light falls on the photocell its resistance changes and produces unbalanced bridge condition. This produces the output, which inturn produces the meter deflection. The meter can be calibrated in terms of lux to measure the light intensity. Such a light intensity meter is very accurate and stable.

4.18.4 Analog Weight Scale

The similar circuit, but using strain gauges in all the four arms of the bridge can be used as a simple analog weight scale. The bridge is shown in the Fig. 4.48.

The elements are mounted on the base of the weight platform. When the weight is placed on the platform the strain gauges in the opposite arm elongate while the strain gauges in other two opposite arms get compressed.

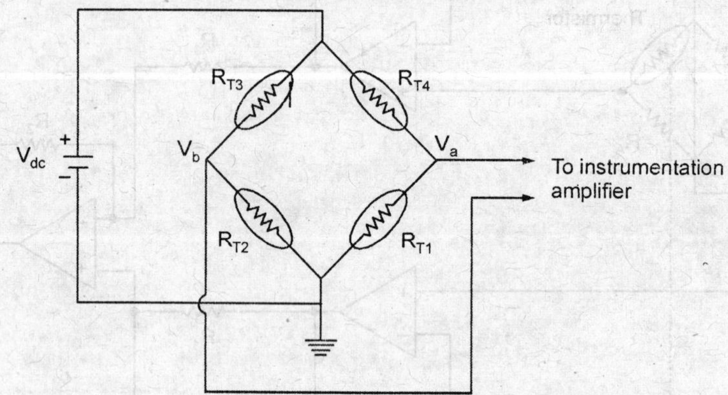

Fig. 4.48 Bridge for analog weight scale.

This means R_{T1} and R_{T3} resistances decrease. While R_{T2} and R_{T4} resistances increase.

Under no weight on the platform, the bridge is balanced and $R_{T1} = R_{T2} = R_{T3} = R_{T4} = R$ and the bridge output is 0 V. Under unbalanced condition, the change in resistance is ΔR ohms. So increase of ΔR for R_{T2} and R_{T4} while decrease of ΔR for R_{T1} and R_{T3}.

So the expressions for V_a and V_b are,

$$V_b = \frac{(R+\Delta R)V_{dc}}{(R+\Delta R+R+\Delta R)} = \frac{(R+\Delta R)V_{dc}}{2R} \qquad ...(1)$$

$$V_a = \frac{(R+\Delta R)V_{dc}}{(R-\Delta R+R+\Delta R)} = \frac{(R-\Delta R)V_{dc}}{2R} \qquad ...(2)$$

$$\therefore \quad V_{ab} = V_b - V_a = \frac{(R+\Delta R)V_{dc}}{2R} - \frac{(R-\Delta R)V_{dc}}{2R} \qquad ...(3)$$

$$= \frac{2\Delta R\, V_{dc}}{2R} = \frac{\Delta R\, V_{dc}}{R} \qquad ...(4)$$

Hence the net output voltage is,

$$V_o = \frac{A\,\Delta R\, V_{dc}}{R} \qquad ...(5)$$

The output voltage is proportional to the change in the resistance, which inturn depends upon the applied weight. The meter connected at the output can be calibrated in kilograms.

Example 4.15: A strain gauge force transducer in full bridge configuration gives the output 2 mV/V for 10 kV force. Design instrumentation amplifier to obtain 10 V output for full scale value when bridge excitation is 5 V d.c. What should be CMRR of this instrumentation amplifier ?

Solution: $V_{dc} = 5\text{ V}, \quad V_o = 10\text{ V}$

For the bridge, $$V_{ab} = \frac{\Delta R\, V_{dc}}{2(2R + \Delta R)} = 2\text{ mV/V}$$

i.e. $$\frac{V_{ab}}{V_{dc}} = \frac{\Delta R}{2(2R + \Delta R)} = 2\text{ mV/V}$$

For $V_{dc} = 5\text{ V}, \quad V_{ab} = 2 \times 5 = 10\text{ mV}$

$= \text{Bridge output}$

For three op-amp instrumentation amplifier,

$$V_o = A\, V_{ab}$$

$$\therefore \quad 10 = A \times 10 \times 10^{-3}$$

$$\therefore \quad A = \mathbf{1000}$$

It is necessary to design instrumentation amplifier of gain 1000.

Now $$A = \frac{R_2}{R_1}\left(1 + \frac{2R_f}{R_G}\right) = A_1 A_2 = 1000$$

Let $$A_1 = 10 = \frac{R_2}{R_1}$$

Choose $R_1 = 10\text{ kW}$ hence $R_2 = R_1 = 100\text{ k}\Omega$

Let $$A_2 = 1000 = 1 + \frac{2R_f}{R_G}$$

Choose $R_G = 100\ \Omega$

$$\therefore \quad 1000 = 1 + \frac{2R_f}{R_G}$$

$$\therefore \quad R_f = 50\text{ k}\Omega$$

The designed circuit is shown in the Fig. 4.49.

The CMRR should be as high as possible and enough as provided by three op-amp instrumentation amplifier. The resistances must be matched to 0.1 % to keep error less than 0.1%.

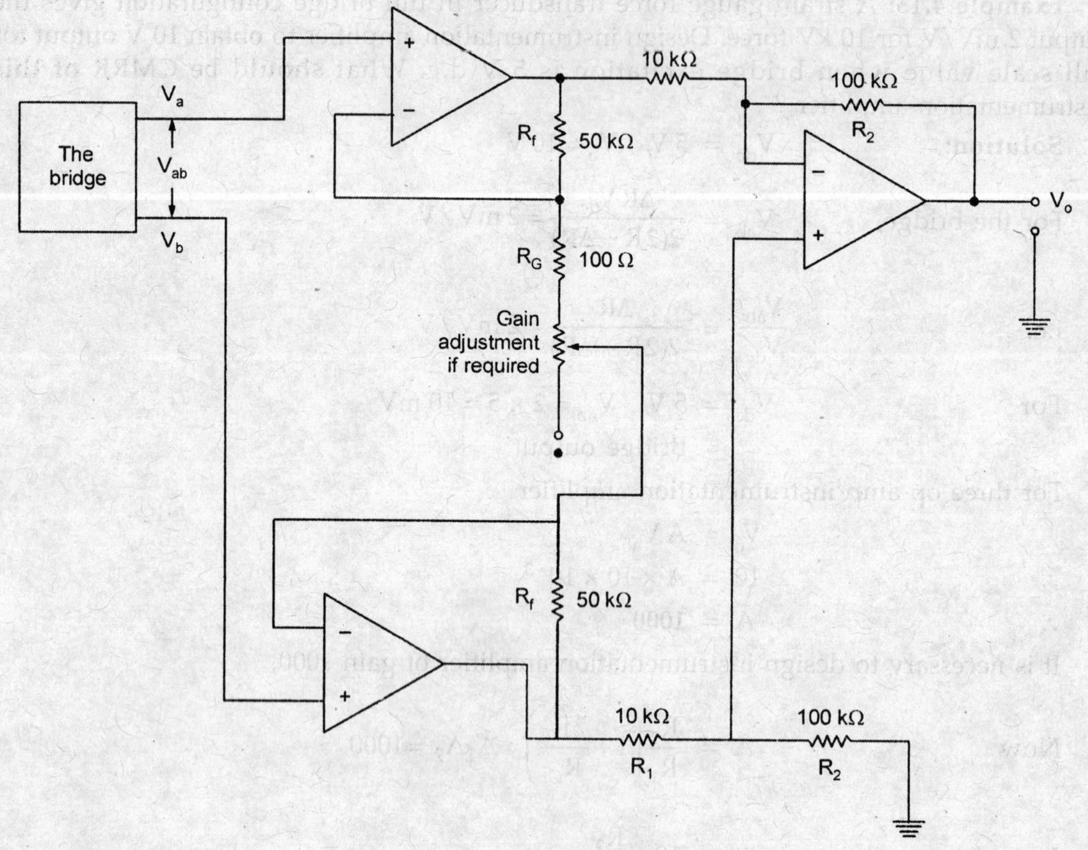

Fig. 4.49

Example 4.16: In the circuit shown in the Fig. 4.50 (See Fig. 4.50 on next page) the transducer used in thermistor with a reference resistance of 80 kΩ at 25°C. The temperature coefficient of thermistor is – 1.2 kΩ/°C. Determine the output voltage at 80°C and 10°C.

Solution: From the bridge we can write,

$$V_b = \frac{V_{dc}R_B}{R_B + R_C} = \frac{12 \times 80 \times 10^3}{80 \times 10^3 + 80 \times 10^3}$$

$$= 6 \text{ V Constant}$$

The voltage V_a changes with temperature as it depends on transducer resistance.

Now R_T = 80 kΩ at 25 °C

$$\Delta R = \alpha \Delta T = -1.2 \text{ k}\Omega/°\text{C} \times (80 - 25) = -66 \text{ k}\Omega$$

Now the gain of the instrumentation amplifier is

$$A = -\frac{R_2}{R_1} = \frac{-4.7 \times 10^3}{1.2 \times 10^3} = -3.92$$

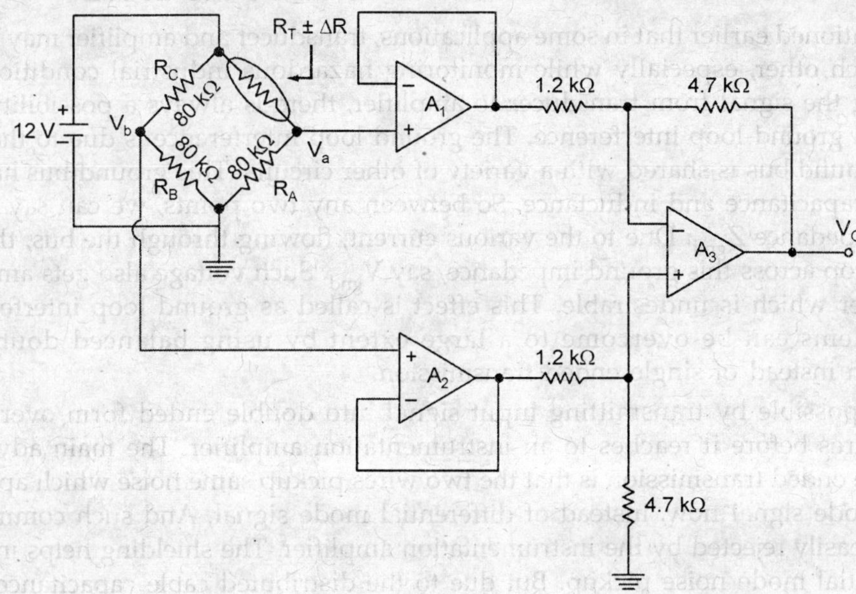

Fig. 4.50

$$\therefore \quad V_o = \frac{A\Delta R}{4R} V_{dc} \quad \text{where } R = R_T$$

$$= \frac{-3.92 \times -66 \times 10^3}{4 \times 80 \times 10^{23}} \times 12 = \mathbf{9.702\ V}$$

...This is the output at 80°C

Now at $T = 10°C$, we have

$$\Delta R = a\ DT = -1.2\ kW/°C \times (10 - 25)$$

$$= 18\ kW$$

The gain A remains same hence the output voltage is

$$V_o = \frac{A\Delta R}{4R} V_{dc} = \frac{-3.92 \times 18 \times 10^3}{4 \times 80 \times 103} \times 12$$

$$= \mathbf{-2.646\ V}$$

This shows that meter at the output can be calibrated with centre zero, at T = 25°C.

The making corresponding to V_o = 9.702 V is 80°C while for V_o = – 2.646 V it can be marked as 10°C. Calculating V_o at T = 0°C, the 0°C can be marked and thus meter can be calculated from 0°C to 80°C. Note that V_o = 0 at T = 25°C.

4.19 INSTRUMENTATION AMPLIFIER WITH ACTIVE GUARD DRIVE

It is mentioned earlier that in some applications, transducer and amplifier may be located far from each other, especially while monitoring hazardous industrial conditions. While transmitting the signal from transducer to amplifier, there is always a possibility of noise pick up and ground loop interference. The ground loop interference is due to the fact that common ground bus is shared with a variety of other circuits. This ground bus has its own resistance, capacitance and inductance. So between any two points, we can say that there exists an impedance Z_{gnd}. Due to the various current, flowing through the bus, there exists a voltage drop across this ground impedance, say V_{gnd}. Such voltage also gets amplified by the amplifier which is undesirable. This effect is called as ground loop interference. All these problems can be overcome to a large extent by using balanced double ended transmission instead of single ended transmission.

This is possible by transmitting input signal into double ended form over a pair of shielded wires before it reaches to an instrumentation amplifier. The main advantage of such double ended transmission is that the two wires pickup same noise which appears as a common mode signal now, instead of differential mode signal. And such common mode signal gets easily rejected by the instrumentation amplifier. The shielding helps in reducing the differential mode noise pickup. But due to the distributed cable capacitances another problem arises called **degradation of CMRR**.

Due to the cable capacitances and source resistances imbalance, shown in the Fig. 4.51, common mode signal gets converted to the differential mode which reduces the CMRR value substantially.

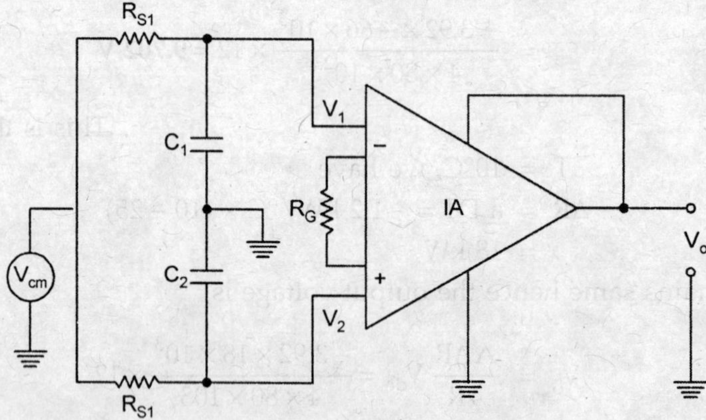

Fig. 4.51 Effect of source resistances and cable capacitances.

The value of CMRR due to the RC imbalance is given by the expression,

$$\text{CMRR} = 20 \log_{10}\left[\frac{1}{2\pi f R_u C_{cm}}\right] \quad ...(1)$$

where

$$R_u = |R_{s1} - R_{s2}| \quad ...(2)$$

$$= \text{Source resistance imbalance}$$

$$\boxed{C_m = \frac{1}{2}(C_1 + C_2)} \quad \text{...(3)}$$

$$= \text{Common mode capacitance}$$

While C_1 and C_2 are the capacitances of each wire with respect to the ground.

The effect of the imbalance in the cable capacitances is to convert the common mode signal to the differential mode signal. Hence such a signal gets amplified by the amplifier which is totally undesirable. Due to all these things, the value of CMRR of the system deteriorates significantly which affects the performance of the instrumentation amplifier badly.

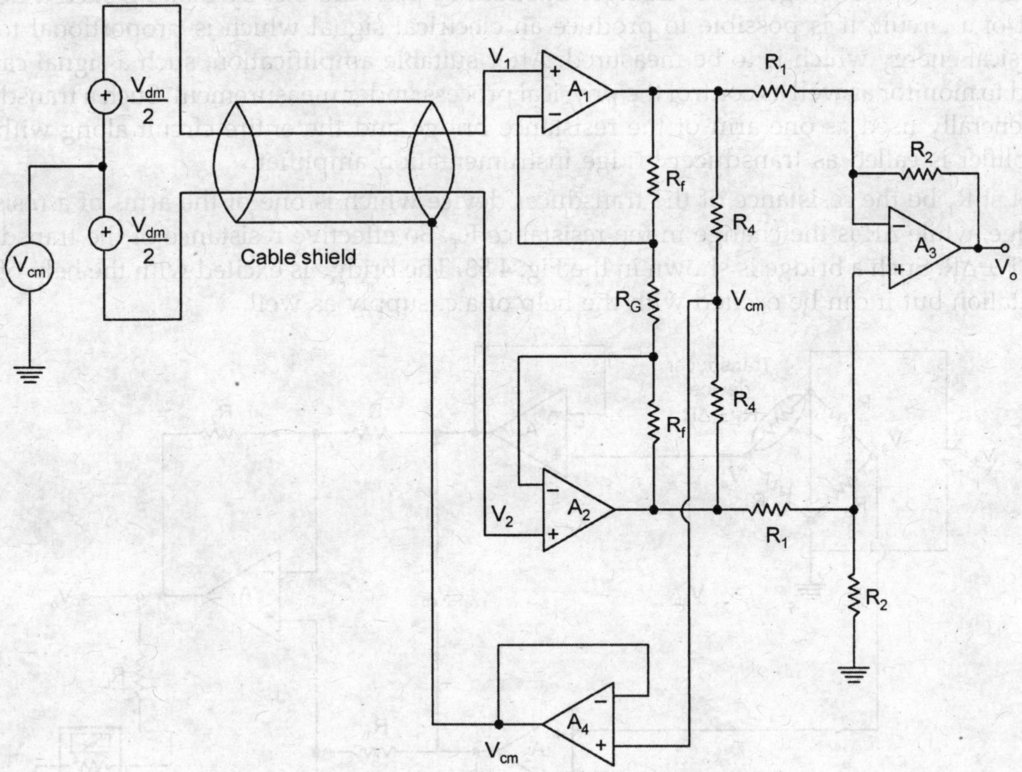

Fig. 4.52 Instrumentation amplifier with active guard drive.

The effect of cable capacitances can be eliminated by driving the shield with the common mode voltage itself. This reduces the common mode swing across cable capacitance to zero.

The Fig. 4.52 shows that the common mode voltage is obtained at the junction of the two R_4 resistances. Then this voltage is subsequently buffered by the voltage follower A_4 to the cable shield. This maintains the cable shield at the common mode voltage level.

Such an arrangement is called as **'active guard drive'**. The active guard drive reduces the effective common mode input capacitance and also reduces the input leakage currents.

Hence this active guard drive feature acts to increase the CMRR of the system. Examples of this type of instrumentation amplifiers are INA 104 (Burr-Brown), 3630 (Burr-Brown) and LM 0036 (National Semiconductor). The INA 104 has a CMRR of 106 dB minimum at 60 Hz.

4.20 INSTRUMENTATION AMPLIFIER USING TRANSDUCER BRIDGE

The resistive transducer bridge is a device whose resistance varies due to changes in some physical condition. The examples of such resistive transducers are such as thermistors or resistance temperature detectors (RTDs) whose resistance changes with temperature, photoresistors whose resistance changes with changes in light sensitivity, strain gauges whose resistance changes with changes in strain or pressure etc. By making such a device part of a circuit, it is possible to produce an electrical signal which is proportional to the physical energy which is to be measured. After suitable amplification, such a signal can be used to monitor as well as control the physical process under measurement. Such a transducer is generally used as one arm of the resistance bridge and the entire circuit along with the amplifier is called as transducer bridge instrumentation amplifier.

Let R_T be the resistance of the transducer device which is one of the arms of a resistive bridge, while ΔR is the change in the resistance R_T. So effective resistance of the transducer is RT ± ΔR. Such a bridge is shown in the Fig. 4.53. The bridge is excited with the help of d.c. excitation but it can be excited with the help of a.c. supply as well.

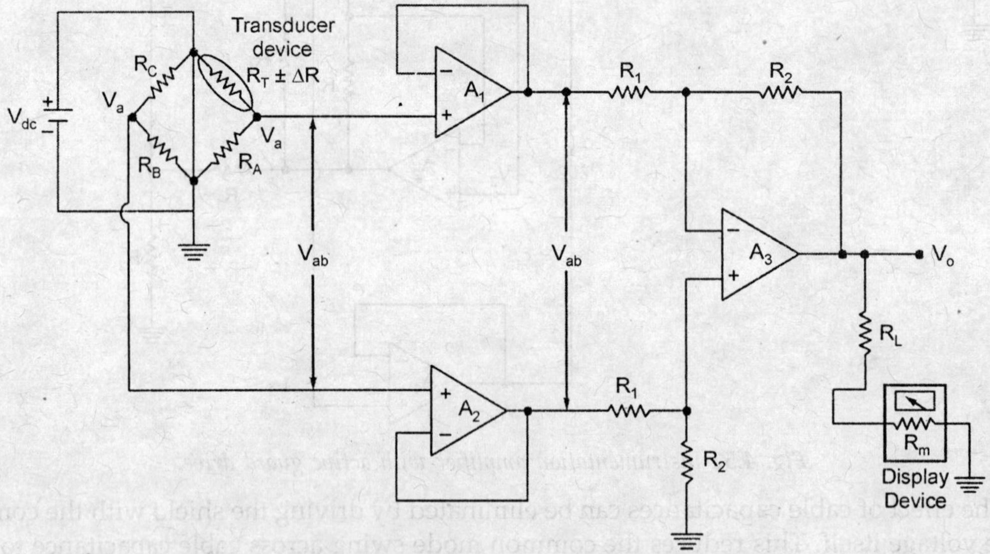

Fig. 4.53 Instrumentation amplifier using transducer bridge.

At some reference condition of the physical energy, the bridge is balanced and we get,

$$V_b = V_a \quad \text{...(1)}$$

i.e.

$$\boxed{\frac{R_B(V_{dc})}{R_B + R_C} = \frac{R_A(V_{dc})}{R_A + R_T}} \quad \text{...(2)}$$

where R_T = Resistance of the device at the reference condition

For example if the device is thermistor then the reference is selected as 0° and R_T is the resistance of the thermistor at 0°, for which bridge is balanced.

Under this condition, the differential input to the instrumentation amplifier is $V_d = V_b - V_a = 0$ hence output of the amplifier is also zero. Thus the meter connected at the output displays the reference value of the physical energy to be measured.

From the equation (2), the bridge balancing condition can be written as,

$$\frac{R_C}{R_B} = \frac{R_T}{R_A} \quad ...(3)$$

The reference condition is generally established by the designer and depends on the transducer device characteristics, the type of the physical energy to be measured and the desired application.

When there is a change in the physical quantity to be measured, the voltage V_a no longer remains same as V_b. This is due to the fact that the resistance of transducer device changes from R_T to $R_T + \Delta R$ and hence balancing of the bridge gets disturbed. This produces a differential input for the instrumentation amplifier. Let us derive the expression for this differential input.

The resistances R_B and R_C are constant and voltage V_b will remain same as before, which is

$$V_b = \frac{R_B(V_{dc})}{R_B + R_C} \quad ...(4)$$

Due to the change in the resistance of the transducer device, the voltage V_a becomes,

$$V_a = \frac{R_A(V_{dc})}{(R_T + \Delta R)} \quad ...(5)$$

$$\therefore \quad V_{ab} = V_b - V_a \quad ...(6)$$

$$= \frac{R_B(V_{dc})}{R_B + R_C} - \frac{R_A(V_{dc})}{(R_T + \Delta R)} \quad ...(7)$$

If the values of all the resistances are selected of same value then,

Then $$V_{ab} = \frac{RV_{dc}}{2R} - \frac{\Delta R\, V_{dc}}{2R + \Delta R} \quad ...(9)$$

$$= \frac{RV_{dc}(2R + \Delta R) - RV_{dc}.RC}{2R(2R + \Delta R)} \quad ...(10)$$

$$\therefore \quad V_{ab} = \frac{2R^2V_{dc} + R\Delta RV_{dc} - 2R^2V_{dc}}{2R(2R + \Delta R)}$$

$$\therefore \quad \boxed{V_{ab} = \frac{\Delta R\, V_{dc}}{2(2R + \Delta R)}} \quad \text{...(11)}$$

The positive V_{ab} i.e. $V_b - V_a$ indicates that $V_b > V_a$.

Now the gain of the first stage of instrumentation amplifier is unity as it is voltage follower circuit while the gain of the second stage is $-R_2/R_1$.

$$\therefore \quad V_o = V_{ab} \times \left(-\frac{R_2}{R_1}\right) \quad \text{...(12)}$$

$$\therefore \quad \boxed{V_o = \frac{R_2}{R_1} \cdot \frac{\Delta R V_{dc}}{2(2R + \Delta R)}} \quad \text{...(13)}$$

As change in resistance $\Delta R \ll 2R$, neglecting it

we get, $$\boxed{V_o = -\frac{R_2}{R_1} \cdot \frac{\Delta R}{4R} \cdot V_{dc}} \quad \text{...(14)}$$

Thus the output voltage is proportional to the change in the resistance ΔR. Thus the meter connected at the output can be calibrated interms of the units of physical quantity to be measured.

In general we can say that if A is the gain of instrumentation amplifier whether it is any instrumentation amplifier circuit, then

$$\boxed{V_o = \frac{A\,\Delta R}{4R} \cdot V_{dc}} \quad \text{...(15)}$$

where A = Gain of the instrumentation amplifier

Such bridge circuits are very common in various applications. So transducers play an important role in the measurement and the control instrumentation.

Example 4.17: A transducer resistive bridge uses a platinum resistance temperature detector (Pt RTD) as a transducer. Its resistance is 100 Ω at T = 0 °C. Its resistance temperature coefficient at 0 °C is 0.00392/°C.

(*i*) Write the expression for the resistance as a function of temperature T.

(*ii*) Calculate the transducer resistance at 25°C and 100 °C.

Solution: (*i*) The reference resistance of the transducer is R = 100 Ω and T = 0 °C. The expression for the resistance at T °C is

$$\boxed{R_T = R_0\,[1 + \alpha T]}$$

where R_T = R at temperature T

R_0 = R at 0 °C

(*ii*) at T = 25 °C,

$$R\,(25\,°C) = R\,(0\,°C)\,[1 + 0.00392 \times T]$$

$$\therefore \quad R\,(25\,°C) = 100\,[1 + 0.00392 \times 25] = \mathbf{109.8\ \Omega}$$

Similarly at T = 100°C,

$$R\ (25\ ^\circ C) = 100\ [1 + 0.00392 \times 100] = \mathbf{139.2\ \Omega}$$

It can be seen that ΔR is 9.8 Ω for a change of 25°C while ΔR is 39.2 Ω for a change of 100°C, in the resistance value of Pt RTD.

4.21 STUDY OF INSTRUMENTATION AMPLIFIER IC

The instrumentation amplifier is very commonly required in various practical applications, as seen earlier. Many manufacturers have manufactured instrumentation amplifier ICs which allow higher optimization of parameters required for the particular application like CMRR, gain control, gain linearity, stability etc. The examples of such dedicated instrumentation amplifier ICs are AD 521/524/624/625 (Analog devices), LM 363 (National Semiconductors) and Amp-01 and Amp-05 (Precision monolithic). In such monolithic ICs the first stage differential amplification is achieved by matched transistor pair rather than op-amps. This is because transistor pair is faster than op-amps and less sensitive to common mode signals. Hence the restriction to get matched resistances gets relaxed. Let us study the internal circuit details of instrumentation amplifier IC, Amp-01.

4.21.1 Circuit Diagram of AMP-01

The Fig. 4.54 shows the simplified internal circuit diagram of instrumentation amplifier IC AMP-01.

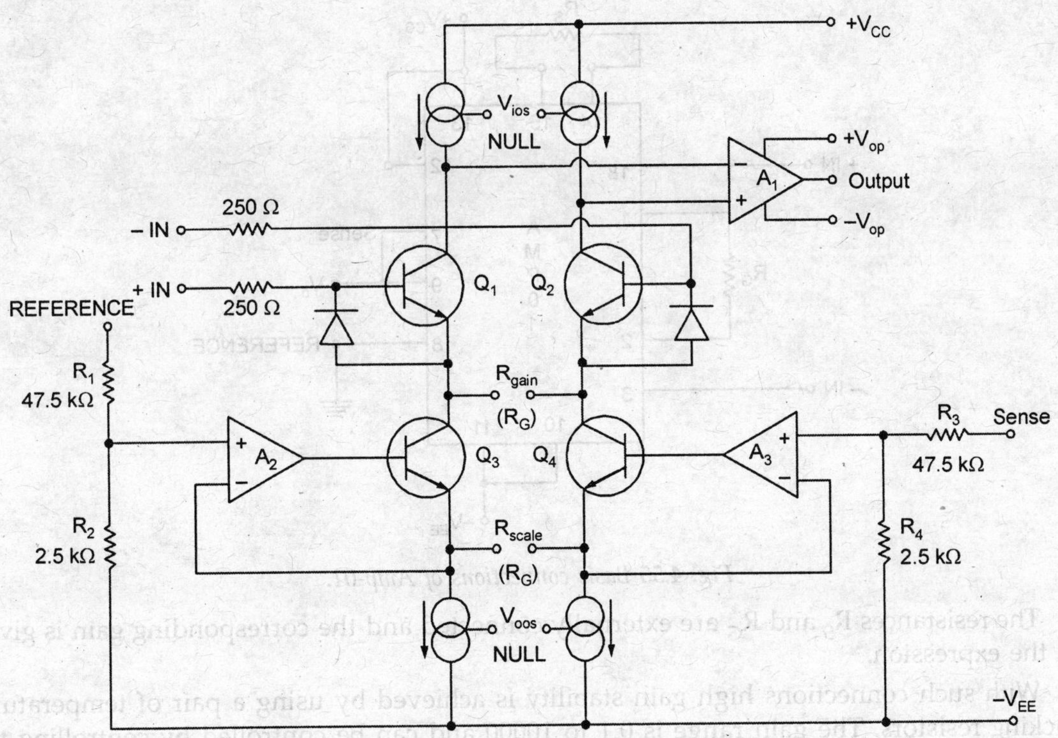

Fig. 4.54 Internal circuit diagram of IC Amp-01.

The circuit shows that the resistance R_G and R_S are to be connected externally which in turn control the gain of the instrumentation amplifier. The V_{ios} i.e., input offset voltage and V_{ios} i.e. output offset voltage, NULL provisions are also available. The pair of transistors Q_1 and Q_2 forms a differential amplifier and plays an important role in differential input amplification, required at the first stage. + IN and – IN are the input terminals where differential input is to be applied.

When a differential signal is applied between the input terminals available from bases of the transistors Q_1 and Q_2, the currents through Q_1 and Q_2 get unbalanced. Due to this, there exists a potential difference between the inverting and non-inverting terminals of op-amp A_1. For op-amp the voltage of non-inverting terminals V_p and inverting terminal V_n must be same and potential difference between them must be zero. So for unbalanced currents in Q_1 and Q_2, the op-amp A_1 responds in the opposite direction in order to restore the balanced condition at its input which is $V_n = V_p$.

To achieve this, A_1 applies a suitable drive to the bottom transistor pair Q_3 and Q_4 through the op-amp A_3. The amount of drive needed depends on two things, one the ratio of the externally connected resistances R_S and R_G and the second is the magnitude of the differential input. This drive forms the output of the instrumentation amplifier.

The pin diagram of IC-Amp-01 is shown in the Fig. 4.55. The diagram also shows the basic connections required to work with gain ranging from the values 0.1 to 10000.

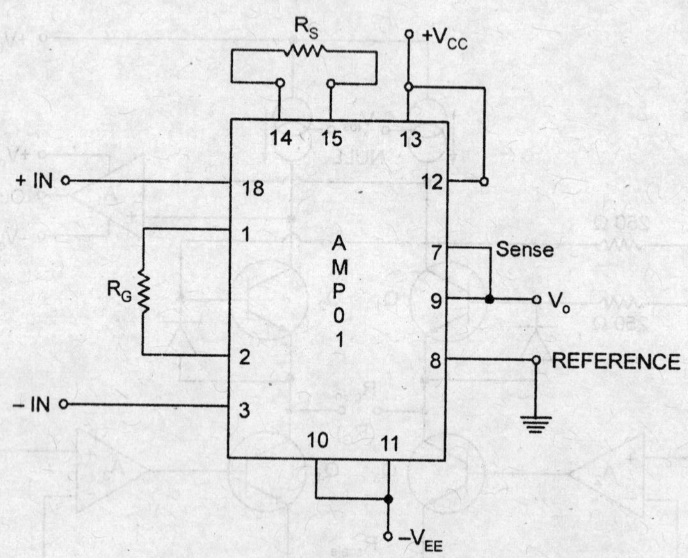

Fig. 4.55 Basic connections of Amp-01.

The resistances R_S and R_G are externally connected and the corresponding gain is given by the expression.

With such connections high gain stability is achieved by using a pair of temperature-tracking resistors. The gain range is 0.1 to 10000 and can be controlled by controlling the ratio R_S/R_G.

The Table 4.2 gives the characteristics of instrumentation amplifier IC Amp-01.

Table 4.2 : Characteristics of AMP-01

Parameter	Value
Offset voltage	15 μV
Offset voltage drift	0.1 μV/°C
Noise	0.2 μV p-p (0.1 Hz to 10 Hz)
Output drive	± 10 V at ± 50 mA
Capacitive load stability	upto 1 μF
Gain range	0.1 to 10000
Linearity	16 bit at gain 1000
CMRR	140 dB at gain 1000
Bias current	1 nA
Provides output stage thermal shutdown	

4.22 INTEGRATOR

In an integrator circuit, the output voltage is the integration of the input voltage. The integrator circuit can be obtained without using active devices like op-amp, transistors etc. In such a case an integrator is called passive integrator. While an integrator using an active devices like op-amp is called active integrator. In this section, we will discuss the operation of active op-amp integrator circuit.

4.22.1 Ideal Active Op-amp Integrator

Consider the op-amp integrator circuit as shown in the Fig. 4.56.

The node B is grounded. The node A is also at the ground potential from the concept of virtual ground.

$$\therefore \qquad V_A = 0 = V_B$$

As input current of op-amp is zero, the entire current I flowing through R_1, also flows through C_f, as shown in the Fig. 4.56.

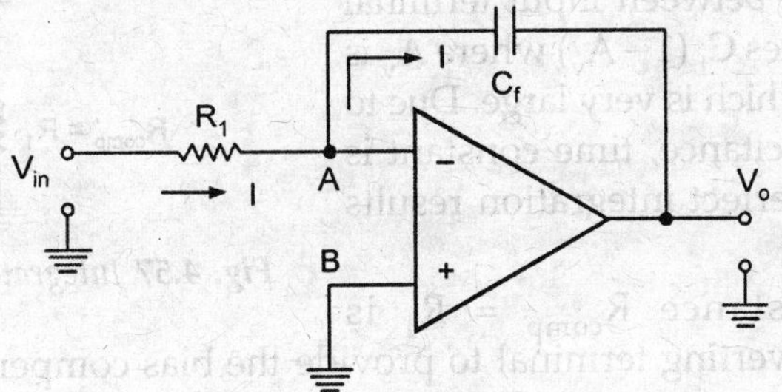

Fig. 4.56 Op-amp integrator.

From input side we can write,

$$I = \frac{V_{in} - V_A}{R_1} = \frac{V_{in}}{R_1} \quad ...(1)$$

From output side we can write,

$$I = C_f \frac{d(V_A - V_o)}{dt}$$

$$\therefore \quad I = -C_f \frac{dV_o}{dt} \quad ...(2)$$

Equating the two equations (1) and (2),

$$\frac{V_{in}}{R_1} = -C_f \frac{dV_o}{dt} \quad ...(3)$$

Integrating both sides,

$$\int_0^t \frac{V_{in}}{R_1} dt = -C_f \int \frac{dV_o}{dt} dt$$

i.e. $$\int_0^t \frac{V_{in}}{R_1} dt = -C_f V_o \quad ...(4)$$

$$\therefore \quad V_o = -\frac{1}{R_1 C_f} \int_0^t V_{in}\, dt + V_o(0)$$

where V_o (0) is the constant of integration, indicating the initial output voltage.

The equation (5) shows that the output is $-1/R_1C_f$ times the integral of input and $R_1 C_f$ is called **time constant** of the integrator.

The negative sign indicates that there is a phase shift of 180° between input and output. The main advantage of such an active integrator is the large time constant. By Miller's theorem the effective capacitance between input terminal A and the ground becomes $C_f (1 - A_v)$ where A_v is the gain of the op-amp which is very large. Due to such large effective capacitance, time constant is very large and thus a perfect integration results due to such circuit.

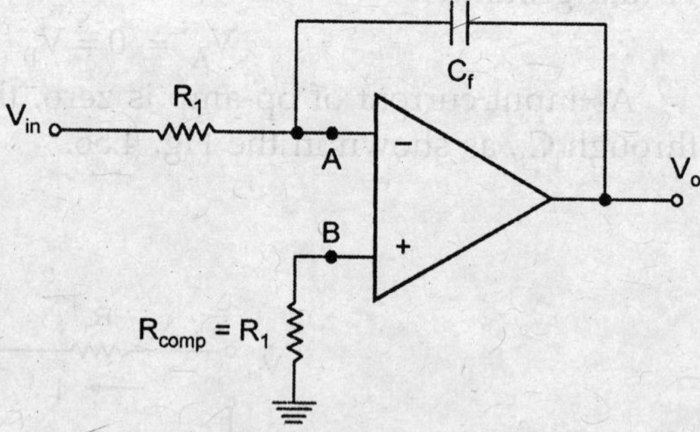

Fig. 4.57 Integrator with bias compensation.

Sometimes a resistance $R_{comp} = R_1$ is connected to the non-inverting terminal to provide the bias compensation. This is shown in the Fig. 4.57.

As the input current of op-amp is zero, the node B is still can be treated at ground potential in this circuit.

Key Point: *Hence the above analysis is equally applicable to the integrator circuit with bias compensation. And the output is the perfect integration of the input.*

4.22.2 Input and Output Waveforms

Let us see the output waveforms, for various input signals. For simplicity of understanding, assume that the time constant $R_1 C_f = 1$ and the initial voltage $V_o(0) = 0$ V.

(*i*) Step input signal

Let the input waveform is of step type, with a magnitude of A units as shown in the Fig. 4.58.

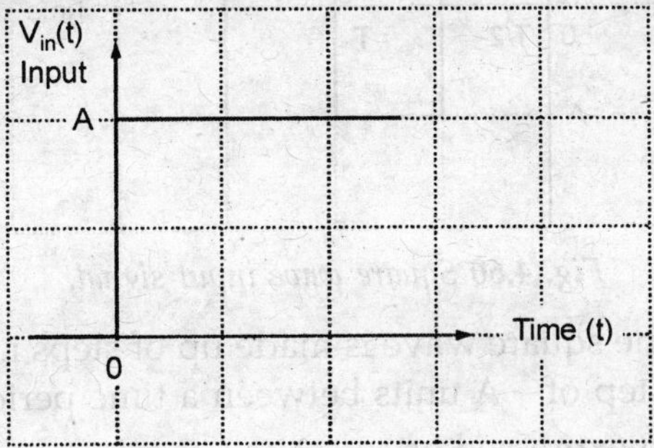

Fig. 4.58 Step input signal.

Mathematically the step input can be expressed as,

$$\therefore \quad V_{in}(t) = A \quad \text{for} \quad t \geq 0 \qquad \text{...(6)}$$

$$\text{And} \quad = 0 \quad \text{for} \quad t < 0$$

From equation (5), with $R_1C_f = 1$ and $V_o(0) = 0$,

We can write,

$$\therefore \quad V_o(t) = -\int_0^t V_{in}(t)\,dt = -\int_0^t A\,dt = -A\int_0^t dt = -A[t]_0^t$$

$$\therefore \quad \boxed{V_o(t) = -At} \qquad \text{...(7)}$$

Fig. 4.59 Output waveform for step input.

Thus output waveform is a straight line with a slope of – A where A is magnitude of the step input. The output waveform is shown in the Fig. 4.59.

(*ii*) Square wave input signal

Let the input waveform is a square wave as shown in the Fig. 4.60.

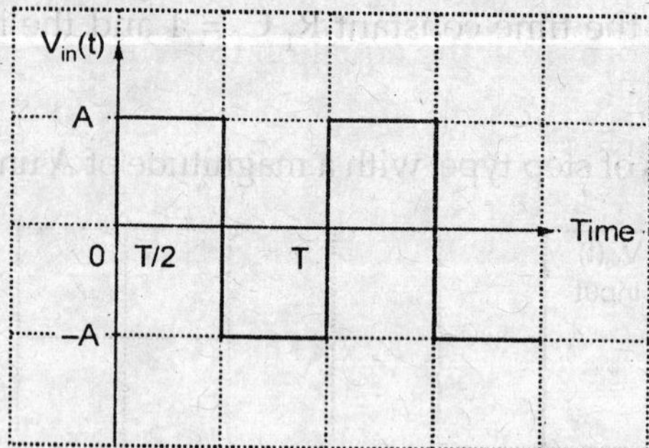

Fig. 4.60 *Square wave input signal.*

It can be observed that the square wave is made up of steps i.e. a step of A between time period of 0 to T/2 while a step of – A units between a time period of T/2 to T and so on.

Mathematically it can be expressed as,

$$\begin{aligned} V_{in}(t) &= A, \quad 0 < t < T/2 \\ &= -A, \quad T/2 < t < T \end{aligned} \qquad ...(8)$$

This is the expression for the input signal for one period.

As discussed earlier, the output for step input is a straight line with a slope of – A. So for the period 0 to T/2 output will be straight line with slope – A. From t = T/2 till t = T, the slope of the straight line will become – (– A) i.e. + A.

So the output can be expressed mathematically for one period as,

$$\begin{aligned} V_o(t) &= -At \quad 0 < t < T/2 \\ &= +At, \quad T/2 < t < T \end{aligned} \qquad ...(9)$$

The output waveform is shown in the Fig. 4.61.

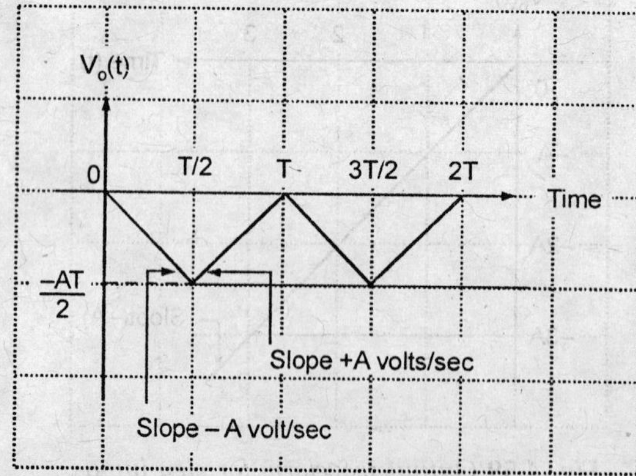

Fig. 4.61 *Output waveform for square wave input.*

(*iii*) Sine wave input signal

Let the input waveform is purely sinusoidal with a frequency of W rad/sec. Mathematically it can be expressed as,

$$\boxed{V_{in}(t) = V_m \sin \omega t} \qquad ...(10)$$

where V_m is the amplitude of the sine wave and T be the period of the waveform.

To find the output waveform, use the equation (5) with $R_1C_f = 1$ and $V_o(0) = 0$ V.

$$\therefore \quad V_o(t) = -\int V_{in}\, dt$$

$$= -\int V_m \sin dt = -V_m\left[\frac{1}{\omega}(-\cos \omega t)\right]$$

$$\therefore \quad V_o(t) = -\frac{V_m}{\omega}(-\cos \omega t) \qquad ...(11)$$

Thus it can be seen that the output of an integrator is a cosine waveform for a input. Due to inverting integrator, the output waveform is as shown in the Fig. 4.62.

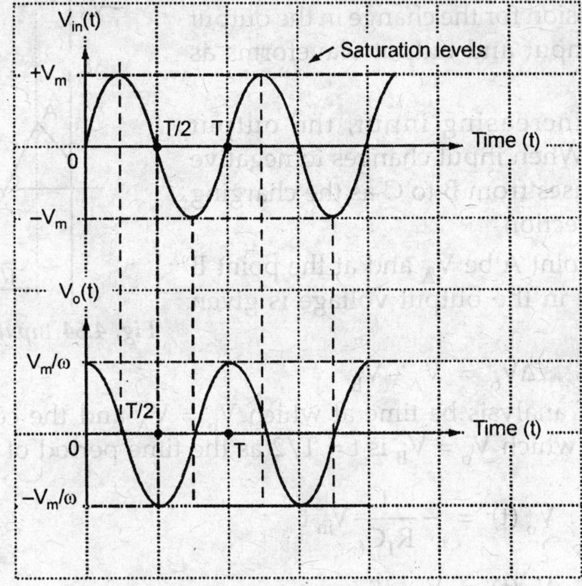

Fig. 4.62 Sine wave input and cosine output.

Table 4.3

	$V_{in}(t)$	$V_o(t)$
Integrator	Step	Ramp
	Square wave	Triangular wave
	Sine wave	Cosine wave

4.22.3 Expression for Change in Output Voltage

The Fig. 4.63 shows an ideal integrator, which gives a triangular wave output for the square wave input.

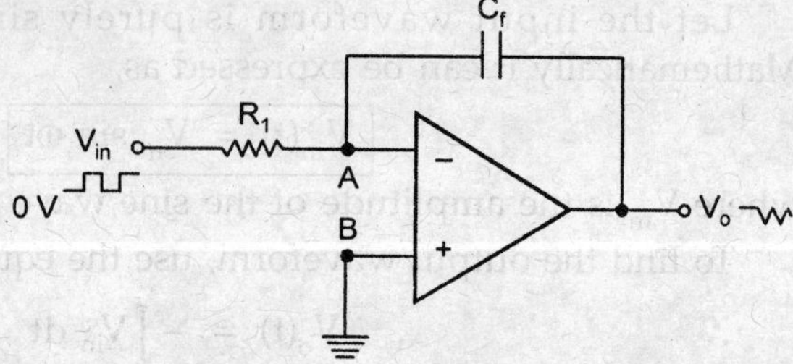

Fig. 4.63 Relationship between input and output.

The input is initially 0 V, for which output is also 0 V. When input increases, then due to the inverting integrator, output ramps down. While when the input decreases towards 0 V, it produces increasing ramp at the output, provided that output does not reach the negative saturation level. Thus the square wave input produces triangular wave, at the output. The behaviour of the output is basically due to the charging current through the capacitor C_f. The charging current is V_{in} / R_1 also reverses its direction, depending upon the polarity of the input.

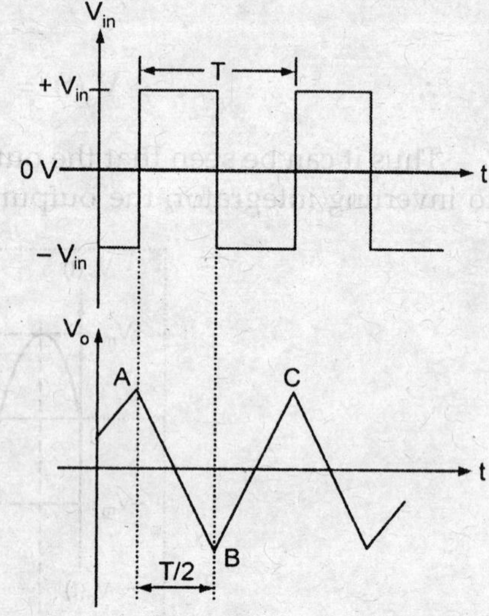

Fig. 4.64 Input and output waveforms.

To derive the expression for the change in the output voltage, consider the input and output waveforms as shown in the Fig. 4.64.

For the positive increasing input, the output decreases from A to B. When input changes to negative value, the output increases from B to C as the charging current reverses its direction.

Let the voltage at point A be V_A and at the point B is V_B. Then the change in the output voltage is given by,

$$\Delta V_o = V_A - V_B \qquad ...(12)$$

Let the reference of analysis be time at which $V_o = V_A$ and the corresponding time be t = 0. While the time at which $V_o = V_B$ is t = T/2 as the time period of the input is T.

Now $$V_o(t) = -\frac{1}{R_1 C_f} V_{in}\, t \qquad ...(13)$$

At t = 0, $$V_o(t) = V_A = 0 \qquad ...(14)$$

At t = T/2, $$V_o(t) = V_B = -\frac{V_{in}\, T}{2 R_1 C_f} \qquad ...(15)$$

$$\therefore \quad \Delta V_o = 0 - \left(\frac{V_{in}\, T}{2 R_1 C_f}\right)$$

$$\therefore \quad \boxed{\Delta V_o = \frac{V_{in}\, T}{2 R_1 C_f}} \qquad ...(16)$$

This is the required expression for the change in the output voltage.

4.22.4 Frequency Response of Ideal Integrator

Consider the output equation of an ideal integrator,

$$\therefore \quad V_o(t) = -\frac{1}{R_1 C_f}\int_0^t V_{in}\, dt$$

Assume the initial voltage $V_o(0)$ as zero.

Taking Laplace transform of this equation, we get,

$$V_o(s) = -\frac{1}{sR_1C_f} V_{in}(t) \qquad ...(17)$$

To get the frequency response replaces by jω.

$$\therefore \quad V_o(j\omega) = -\frac{1}{j\omega R_1 C_f} V_{in}(j\omega)$$

Hence the gain of the integrator is,

$$\frac{V_o(j\omega)}{V_{in}(j\omega)} = -\frac{1}{j\omega R_1 C_f}$$

To get the frequency response, obtain the magnitude of the gain, which is

$$A = \left|\frac{V_o(j\omega)}{V_{in}(j\omega)}\right| = \left|-\frac{1}{j\omega R_1 C_f}\right|$$

$$\therefore \quad A = \frac{1}{\omega R_1 C_f} = \frac{1}{2\pi f R_1 C_f} \qquad ...(20)$$

At very low frequencies such as d.c. (f = 0), the gain becomes infinite. This is because the capacitive reactance becomes very high for low frequencies. Hence it acts as an open circuit and thus op-amp works in open loop configuration. And the open loop gain of the op-amp is almost infinite. As the frequency increases, the gain drops. The response is similar to that of low pass filter.

Let $f = f_b$ be the frequency at which gain of the op-amp becomes one i.e. 0 dB.

$$\text{At } f = f_b, \quad 20 \log A = 20 \log \frac{1}{2\pi f_b R_1 C_f} = 0 \text{ dB}$$

$$\text{i.e.} \quad A = \frac{1}{2\pi f_b R_1 C_f} = 1 \qquad ...(21)$$

$$\therefore \quad f_b = \frac{1}{2\pi R_1 C_f} \qquad ...(22)$$

The gain drops to 0 dB at a frequency $f = f_b$, from its very high value at low frequencies. The frequency response of such an ideal integrator is shown in the Fig. 4.65.

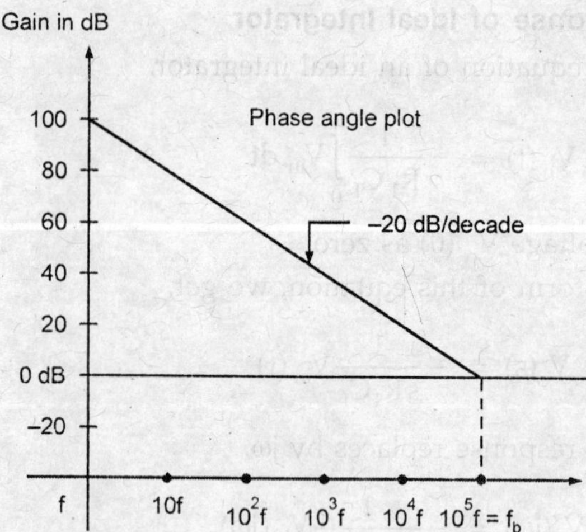

Fig. 4.65 Frequency response of ideal integrator.

The gain rolls off at a rate of 20 dB/decade. It can be seen from the frequency response that 3 dB down bandwidth of an ideal integrator is very small.

4.22.5 Errors in an Ideal Integrator

The operation amplifier has input offset voltage (V_{ios}) and the input bias current (I_b). In the absence of input voltage or at zero frequency (d.c.), op-amp gain is very high. The input offset voltage gets amplified and appears at the output as an error voltage. The bias current also results in a capacitor charging current and adds its effect in an output error voltage. The two components, due to high d.c. gain of op-amp cause output to ramp up or down, depending upon the polarities of offset voltage and/or bias current. After some time, output of op-amp may achieve its saturation level. Hence there is a possibility of op-amp saturation due to such an error voltage and it is very difficult to pull op-amp out of saturation. Thus the output of an ideal integrator in the absence of input signal is likely to be offset towards the positive or negative saturation levels.

In the presence of the input signal also, the two components namely offset voltage and bias current, contribute an error voltage at the output. Thus it is not possible to get a true integration of the input signal at the output. Output waveform may be distorted due to such an error voltage.

Another limitation of an ideal integrator is its bandwidth, which is very small. Hence an ideal integrator can be used for a very small frequency range of the input only.

Due to all these limitations, an ideal integrator is not used in practice. Some additional components are used along with the basic integrator circuit to reduce the effect of an error voltage, in practice. Such an integrator is called Practical Integrator Circuit.

4.23 PRACTICAL INTEGRATOR

The limitations of an ideal integrator can be minimized in the practical integrator circuit, which uses a resistance R_f in parallel with the capacitor C_f.

The practical integrator circuit is shown in the Fig. 4.66.

The resistance R_{comp} is also used to overcome the errors due to the bias current.

The resistance R_f reduces the low frequency gain of the op-amp.

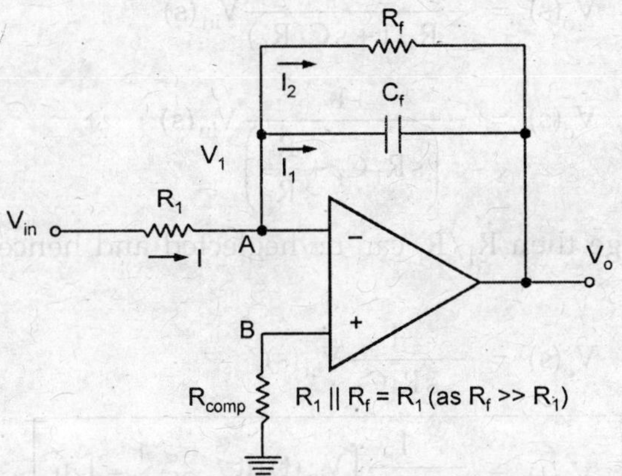

Fig. 4.66 Practical integrator circuit.

4.23.1 The Analysis of Practical Integrator

As the input current of op-amp is zero, the node B is still at ground potential. Hence the node A is also at the ground potential from the concept of virtual ground. So $V_A = 0$.

Referring to Fig. 4.66,

$$I = \frac{V_{in} - V_A}{R_1} = \frac{V_{in}}{R_1} \quad \text{...(1)}$$

Similarly $$I_1 = C_f \frac{d(V_A - V_o)}{dt} = -C_f \frac{dV_o}{dt} \quad \text{...(2)}$$

And $$I_2 = \frac{V_A - V_o}{R_f} = \frac{-V_o}{R_f} \quad \text{...(3)}$$

At node A, applying KCL,

$$I = I_1 + I_2 \quad \text{...(4)}$$

$$\therefore \quad \frac{V_{in}}{R_1} = -C_f \frac{dV_o}{dt} - \frac{V_o}{R_f} \quad \text{...(5)}$$

Taking Laplace of this equation,

$$\frac{V_{in}(s)}{R_1} = -s\,C_f\,V_o(s) - \frac{V_o(s)}{R_f} \quad \text{...(6)}$$

$$\therefore \quad \frac{V_{in}(s)}{R_1} = -V_o(s)\left[s\,C_f + \frac{1}{R_f}\right]$$

$$\therefore \quad \frac{V_{in}(s)}{R_1} = \frac{-V_o(s)[1+sC_f R_f]}{R_f}$$

$$\therefore \quad V_o(s) = -\frac{R_f}{R_1(1+sC_f R_f)} V_{in}(s) \qquad \text{...(7)}$$

$$\therefore \quad V_o(s) = -\frac{1}{\left(sR_1 C_f + \frac{R_1}{R_f}\right)} V_{in}(s) \qquad \text{...(8)}$$

When R_f is very large then R_1/R_f can be neglected and hence circuit behaves like an ideal integrator as,

$$V_o(s) = -\frac{1}{sR_1C_f} V_{in}(s) \qquad \text{...(9)}$$

i.e.

$$\boxed{V_o(t) = -\frac{1}{R_1C_f}\int V_{in}(t)\,dt \quad \text{as } \frac{1}{s} = \int dt} \qquad \text{...(10)}$$

4.23.2 Frequency Response of Practical Integrator

To determine the frequency response, let us obtain the expression for the gain of the practical integrator interms of the frequency.

From the equation (7) we can write,

$$\frac{V_o(s)}{V_{in}(s)} = -\frac{R_f/R_1}{1+sC_f R_f} \qquad \text{...(11)}$$

Replacing s by jω in the steady state,

$$\frac{V_o(j\omega)}{V_{in}(j\omega)} = -\frac{R_f/R_1}{1+j\omega C_f R_f} \qquad \text{...(12)}$$

$$\therefore \quad A = -\frac{R_f/R_1}{1+j2\pi f C_f R_f}$$

$$= \text{Gain as a function of frequency}$$

$$\therefore \quad A = \frac{R_f/R_1}{1+j\frac{f}{f_a}} \qquad \text{...(13)}$$

where

$$f_a = \frac{1}{2\pi C_f R_f} \qquad \text{...(14)}$$

This is the break frequency or the corner frequency of the practical integrator. Thus in the frequency response, d.c. gain remains constant for all frequencies less than f_a and from the frequency f_a onwards, as frequency increases, gain reduces at a rate of 20 dB/decade.

The magnitude of the gain A is,

$$|A| = \frac{R_f / R_1}{\sqrt{1+\left(\frac{f}{f_a}\right)^2}} \qquad \text{...(15)}$$

At f = 0 i.e. d.c. condition,

$$|A| = R_f / R_1$$

i.e.
$$= 20 \log \left(\frac{R_f}{R_1}\right) \text{dB} \qquad \text{...(16)}$$

Thus an infinite d.c. gain of op-amp in case of an ideal integrator, gets limited to R_f / R_1 in the practical integrator.

Similarly at $f = f_a$ we get,

$$|A| = \frac{R_f / R_1}{\sqrt{1+1}}$$

$$= 0.707 \frac{R_f}{R_1} \qquad \text{...(17)}$$

i.e.
$$20 \log |A| = -3 \text{ dB} + \text{d.c. gain} \qquad \text{...(18)}$$

Thus the magnitude of gain drops by 3 dB at the frequency $f = f_a$ which is the break frequency. Now for the integration, the frequency response must be straight line of slope – 20 dB/decade, which is possible for the frequencies greater than f_a and less than f_b. Thus in between f_a and f_b practical integrator acts as an integrator. Below f_a, integration does not take place. The frequency response is shown in the Fig. 4.67.

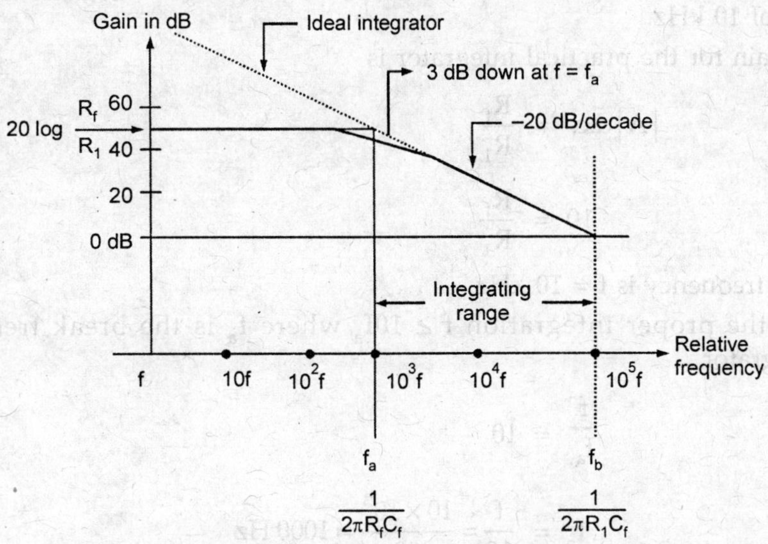

Fig. 4.67 Frequency response of the practical integrator.

Key Point: *It can be seen from the frequency response that the bandwidth of practical integrator is f_a which is much higher than an ideal integrator.*

$$\therefore \quad \boxed{\text{B.W. of Practical Integrator} = f_a} \qquad ...(19)$$

For any input bias current or current due to offset voltage, the path of current is through R, rather than through the capacitor C_f. Thus the output voltage is decided by the resistance ratio (R_f/R_1) which is typically selected as ≥ 10. For (R_f/R_1) of 10, the frequency f_a becomes $(f_b/10)$, this ensures the true integration of the input signal.

For proper integration, the time period T of the input signal has to be larger than or qual to $R_f C_f$, so

$$\boxed{T \geq R_f C_f}$$

where $$R_f C_f = \frac{1}{2\pi f_a}$$

The practical integrator circuit is also called **lossy integrator** as it behaves as integrator only over the upper frequency range.

4.23.3 Applications of Practical Integrator

The integrator circuits are most commonly used in the following applications:

(*i*) In the analog computers.

(*ii*) In solving the differential equations.

(*iii*) In analog to digital converters.

(*iv*) Various signal wave shaping circuits.

(*v*) In ramp generators.

Example 4.18: Design a practical integrator circuit with a d.c. gain of 10, to integrate a square wave of 10 kHz.

The d.c. gain for the practical integrator is

$$|A| \text{d.c.} = \frac{R_f}{R_1}$$

$$\therefore \quad 10 = \frac{R_f}{R_1}$$

The input frequency is f = 10 kHz.

Now for the proper integration f ≥ $10f_a$ where f_a is the break frequency of the practical integrator.

$$\therefore \quad \frac{f}{f_a} = 10$$

$$\therefore \quad f_a = \frac{f}{10} = \frac{10 \times 10^3}{10} = 1000 \text{ Hz}$$

Now for the practical integrator,

$$f_a = \frac{1}{2\pi R_f C_f}$$

$$\therefore \quad 1000 = \frac{1}{2\pi R_f C_f}$$

$$\therefore \quad R_f C_f = 1.5915 \times 10^{-4}$$

Selecting $\quad R_1 = 10 \text{ k}\Omega$, from equation (1)

$$R_f = 10\, R_1 = 10 \times 10 = 100 \text{ k}\Omega$$

$$\therefore \quad C_f = \frac{1.5915 \times 10^{-4}}{100 \times 10^3} = 1.5915 \times 10^{-9}\,\text{F} \approx 16\,\text{nF}$$

And $\quad R_{comp} = R_1 \| R_f$

$$\therefore \quad R_{comp} = \frac{10 \times 100}{10 + 100} = 9.09\,\text{k}\Omega$$

Hence the practical integrator circuit is as shown in the Fig. 4.68.

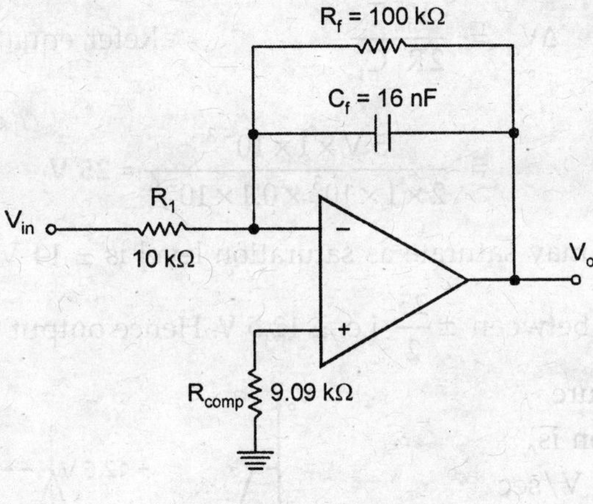

Fig. 4.68

Example 4.19: A 1 kHz square wave is applied to the integrator shown in the Fig. 4.69. The amplifier uses ± 15 V supply and the output saturates at ± 14 V. If the input alternates between ± 5 V,

(*i*) Determine the maximum change in the output voltage.

(*ii*) Determine the minimum slew rate required.

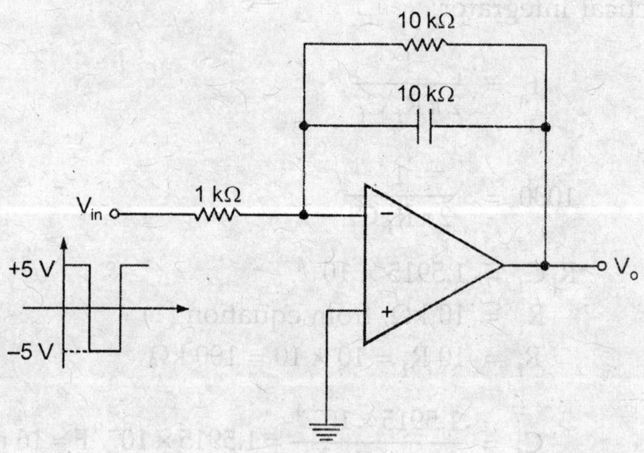

Fig. 4.69

Solution: If output does not reach saturation, for square wave input, output is triangular.

(*i*) For f = 1 kHz, the time period is

$$T = \frac{1}{f} = \frac{1}{1 \times 10^3} = 1 \times 10^{-3} \text{ sec}$$

The change in the output is,

$$\Delta V_o = \frac{V_{in} T}{2 R_1 C_f} \quad \text{Refer equation (16) of section 4.22.3}$$

$$= \frac{5 \text{ V} \times 1 \times 10^{-3}}{2 \times 1 \times 10^3 \times 0.1 \times 10^{-6}} = \mathbf{25 \ V}$$

It seems that output may saturate as saturation level is ± 14 V. But 25 V change means the output will alternate between $\pm \frac{25}{2}$ i.e. ± 12.5 V. Hence output will not be saturated and will be triangular in nature.

(*ii*) Slew rate equation is,

$$S = 2 \pi f_m V_m \text{ V/sec}$$

$$\therefore \quad S = 2 \pi \times 1 \times 10^3 \times 5 \text{ V/sec}$$

$$= 0.03141 \text{ V/sec.}$$

This is the minimum slew rate required.

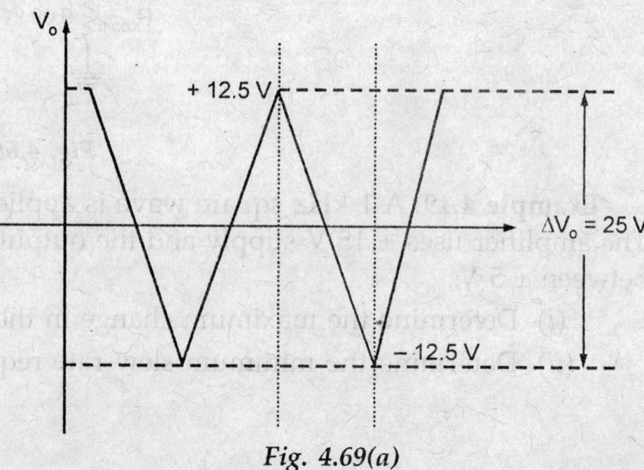

Fig. 4.69(a)

4.24 NON-INVERTING INTEGRATOR

The Fig. 4.70 shows the non-inverting integrator circuit.

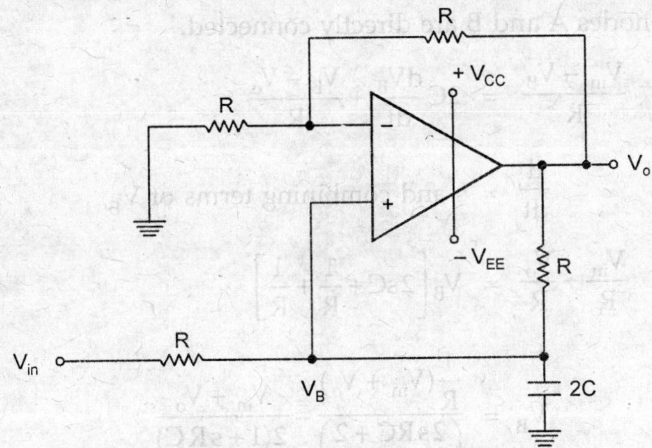

Fig. 4.70 Non-inverting integrator.

The output voltage V_o is,

$$V_o = \left(1+\frac{R}{R}\right)V_B = 2\,V_B \qquad \text{...(1)}$$

To find V_B interms of V_{in}, assume currents as shown in the Fig. 4.71.

No current flows into the op-amp hence current I flows upto the node A.

At node A, $\qquad I = I_1 + I_2 \qquad$ ...(2)

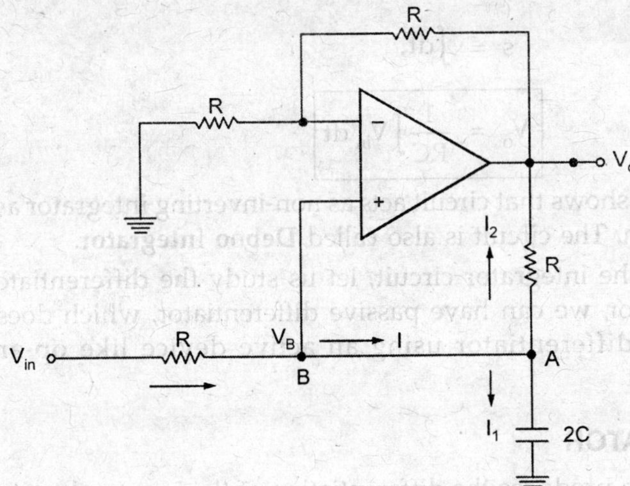

Fig. 4.71 Current distribution.

Now $$I = \frac{V_{in} - V_B}{R}, \quad I_1 = 2C\frac{dV_A}{dt}, \quad I_2 = \frac{V_A - V_o}{R}$$

But $V_B = V_A$ as nodes A and B are directly connected.

$$\therefore \quad \frac{V_{in} - V_B}{R} = 2C\frac{dV_B}{dt} + \frac{V_B - V_o}{R} \qquad ...(3)$$

Using $\frac{d}{dt} = s$ and combining terms of V_B,

$$\therefore \quad \frac{V_{in}}{R} + \frac{V_o}{R} = V_B\left[2sC + \frac{1}{R} + \frac{1}{R}\right] \qquad ...(4)$$

$$\therefore \quad V_B = \frac{\frac{1}{R}(V_{in} + V_o)}{\left(\frac{2sRC + 2}{R}\right)} = \frac{V_{in} + V_o}{2(1 + sRC)} \qquad ...(5)$$

Using equation (5) in equation (1), $V_o = 2\frac{(V_{in} + V_o)}{2(1 + sRC)}$

$$\therefore \quad V_o\left[1 - \frac{1}{1 + sRC}\right] = \frac{V_{in}}{1 + sRC}$$

$$\therefore \quad V_o = \frac{1}{sRC}V_{in} \qquad ...(6)$$

Replacing $$s = \int dt,$$

$$\boxed{V_o = \frac{1}{RC}\int V_{in}\, dt} \qquad ...(7)$$

The equation (7) shows that circuit acts as non-inverting integrator as there is no negative sign in that equation. The circuit is also called **Deboo integrator**.

After studying the integrator circuit, let us study the differentiator circuit. Similar to the passive integrator, we can have passive differentiator, which does not use any active device. While the differentiator using an active device like op-amp is called active differentiator circuit.

4.25 DIFFERENTIATOR

The circuit which produces the differentiation of the input voltage at its output is called differentiator. The differentiator circuit which does not use any active device is called passive differentiator. While the differentiator using an active device like op-amp is called an active differentiator. Let us discuss first the operation of ideal active op-amp differentiator circuit.

4.25.1 Ideal Active Op-amp Differentiator

The active differentiator circuit can be obtained by exchanging the positions of R and C in the basic active integrator circuit. The op-amp differentiator circuit is shown in the Fig. 4.72.

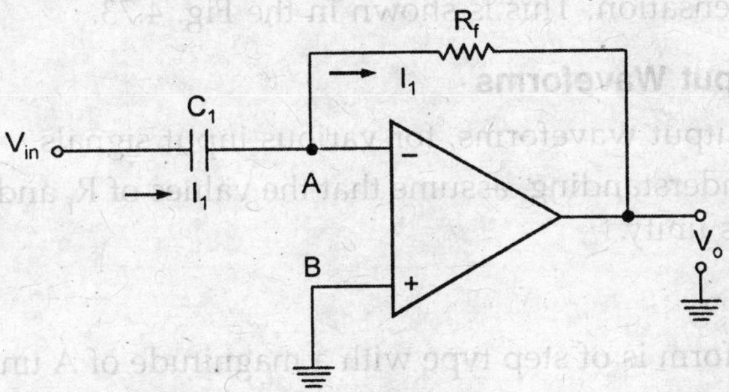

Fig. 4.72 Op-amp differentiator.

The node B is grounded. The node A is also at the ground potential hence $V_A = 0$. As input current of op-amp is zero, entire current I_1 flows through the resistance R_f. From the input side we can write,

$$I_1 = C_1 \frac{d(V_{in} - V_A)}{dt} = C_1 \frac{dV_{in}}{dt} \quad ...(1)$$

From the output side we can write,

$$I = \frac{(V_A - V_o)}{R_f} = \frac{V_o}{R_f} \quad ...(2)$$

Equating the two equations,

$$C_1 \frac{dV_{in}}{dt} = -\frac{V_o}{R_f} \quad ...(3)$$

$$V_o = -C_1 R_f \frac{dV_{in}}{dt} \quad ...(4)$$

The equation shows that the output is C_1 R_f times the differentiation of the input and product C_1 R_f is called **time constant** of the differentiator.

The negative sign indicates that there is a phase shift of 180^0 between input and output. The main advantage of such an active differentiator is the small time constant required for differentiation.

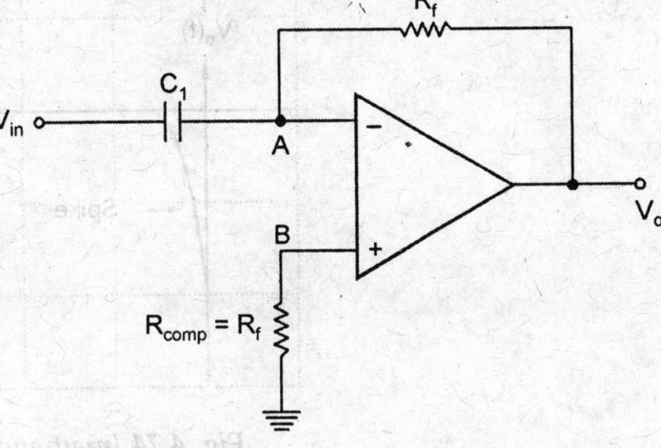

Fig. 4.73 Differentiator with bias compensation.

By Miller's theorem, the effective resistance between input node A and ground $\frac{R_f}{1+A_v} \approx \frac{R_f}{A_v}$ where A_v is the gain of the op-amp which is very large. Hence effective R_f becomes very very small and hence the condition $R_f C_1 \ll T$ gets satisfied at all the frequencies.

In practice a resistance $R_{comp} = R_f$ is connected to the non-inverting terminal to provide the bias compensation. This is shown in the Fig. 4.73.

4.25.2 Input and Output Waveforms

Let us study the output waveforms, for various input signals.

For simplicity of understanding, assume that the values of R_f and C_1 are selected to have time constant ($R_f C_1$) as unity.

(i) Step input signal

Let the input waveform is of step type with a magnitude of A units. Mathematically it is expressed as,

$$V_{in}(t) = A \quad \text{for} \quad t \geq 0 \qquad \text{...(5)}$$

Now mathematically, the output of the differentiator must be,

$$V_o(t) = \frac{dV_{in}}{dt} = \frac{dV_{in}}{dt} = \frac{d(A)}{dt} = 0 \qquad \text{...(6)}$$

This is because A is constant.

Actually the step input takes a finite time to rise from 0 to A volts.

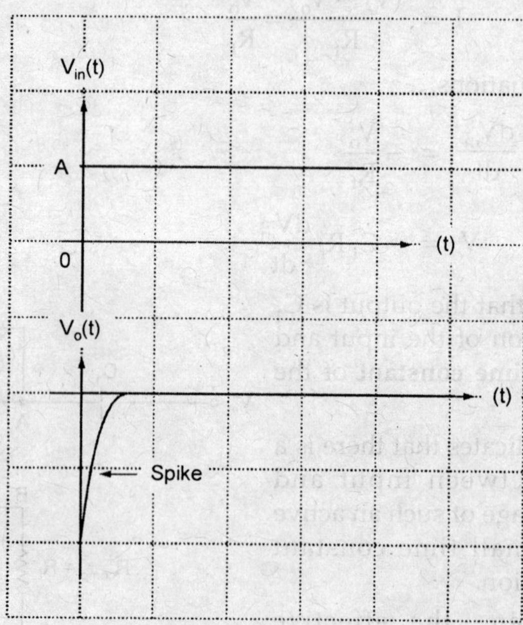

Fig. 4.74 Input and output for step input.

Due to thisfinite time, the differentiator output is not zero but appears in the form of a spike at $t = 0$.

As the circuit acts as an inverting differentiator, the negative going spike or impulse appears at $t = 0$ and after that output remains zero.

Both input and output waveforms of the differentiator with a step input, are shown in the Fig. 4.74.

(ii) Square wave input signal

Input and output for square wave input

The square wave is made of steps i.e. step of A volts from $t = 0$ to $t = T/2$, while a step of $-$ A volts from $t = T/2$ to $t = T$ and so on.

Mathematically it can be expressed as,

$$V_{in}(t) = A \quad 0 < t < T/2 \qquad \text{...(7)}$$
$$= -A \quad T/2 < t < T$$

The differentiator behaves similar to its behaviour to step input.

For positive going impulse, the output shows negative going impulse and for negative going input, the output shows positive going impulse.

Hence the total output for the square wave input is in the form of train of impulses or spikes.

The input and output waveforms are shown in the Fig. 4.75.

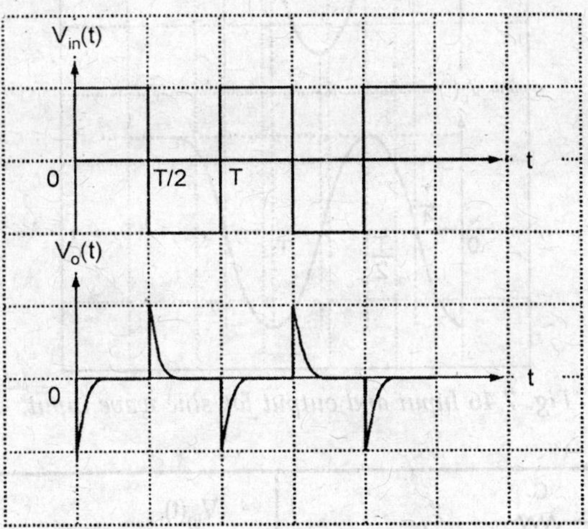

Fig. 4.75 Input and output for square wave input.

(iii) Sine wave input

Let the input waveform be purely sinusoidal with a frequency of m rad/sec.

Mathematically it can be expressed as,

$$V_{in}(t) = V_m \sin \omega t \qquad \text{...(8)}$$

where V_m is the amplitude of the sine wave and T is the period of the waveform. Let us find out the expression for the output.

$$V_o(t) = \frac{-d\,V_{in}(t)}{dt} \quad \text{for} \quad R_f C_1 = 1 = -\frac{d}{dt}(V_m \sin \omega t)$$

$$\therefore \quad V_o(t) = -V_m\, \omega \cos \omega t$$

So at

$$t = 0, \quad V_o(t) = -V_m\, \omega$$

$$t = \frac{T}{2}, \quad V_o(t) = 0$$

$$t = \frac{T}{2}, \quad V_o(t) = +V_m \omega \text{ and so on.}$$

Thus the output of the differentiator is a cosine waveform, for a sine wave input. The input and output waveform is shown in the Fig. 4.76.

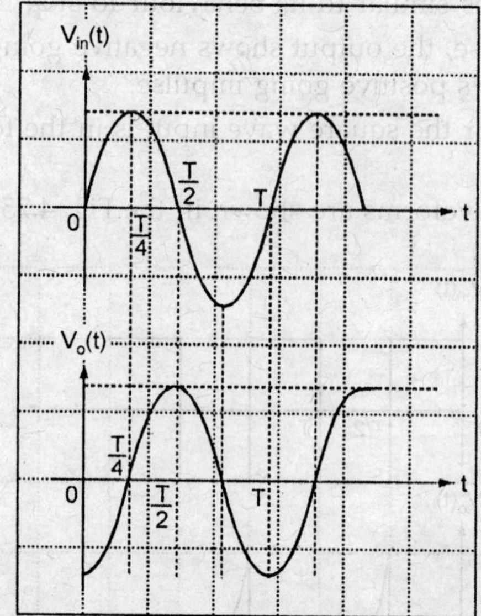

Fig. 7.46 Input and output for sine wave input.

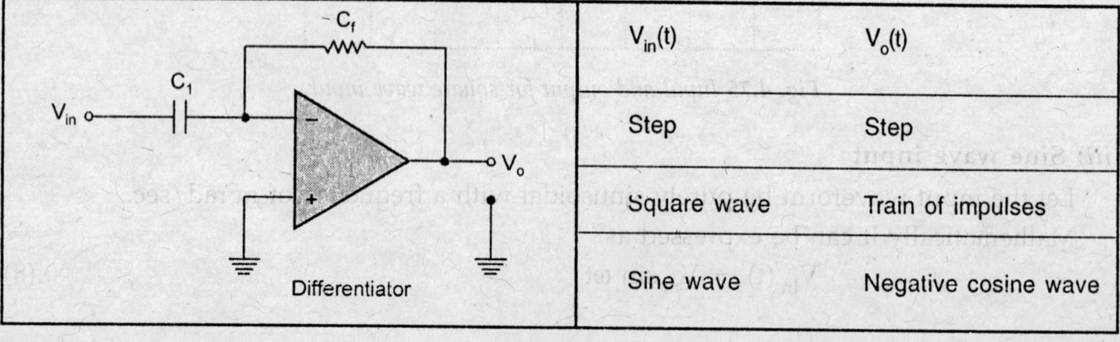

$V_{in}(t)$	$V_o(t)$
Step	Step
Square wave	Train of impulses
Sine wave	Negative cosine wave

4.25.3 Frequency Response of Ideal Differentiator

Consider the output equation of an ideal differentiator which is,

$$V_o(t) = -R_f C_1 \frac{dV_{in}}{dt}$$

Taking Laplace transform of this equation, we get,

$$V_o(s) = -s R_f C_1 V_{in}(s) \qquad ...(10)$$

To get the frequency response, replace s by (jω)

$$\therefore \quad V_o(j\omega) = -j\omega R_f C_1 V_{in}(j\omega) \qquad ...(11)$$

Hence the gain of the differentiator, is

$$\frac{V_o(j\omega)}{V_{in}(j\omega)} = -j\omega R_f C_1 \qquad ...(13)$$

To get the frequency response obtained the magnitude of the gain which is,

$$A = \left|\frac{V_o(j\omega)}{V_{in}(j\omega)}\right| = \left|-j\omega R_f C_1\right|$$

$$\therefore \quad A = \omega R_f C_1 = 2\pi f R_f C_1 \qquad ...(13)$$

So at very low frequency such as d.c. (f = 0) the gain is zero. And as frequency increases, gain also increases.

The expression of the gain can be written as,

$$A = \frac{f}{f_a} \qquad ...(14)$$

where

$$f_a = \frac{1}{2\pi R_f C_1} \qquad ...(15)$$

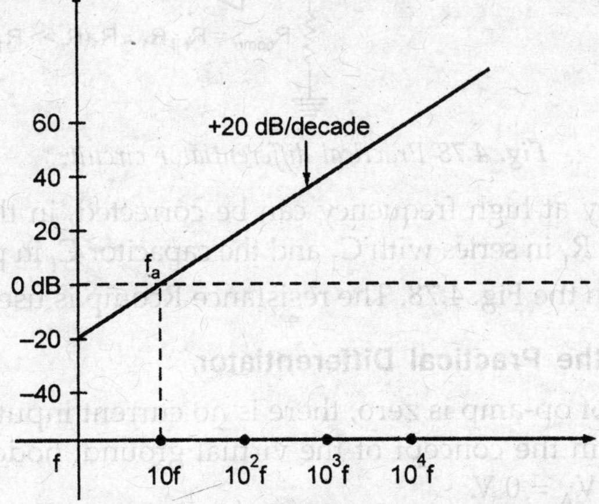

Fig. 4.77 Frequency response of an ideal differentiator.

The frequency f_a is the frequency at which gain becomes 1 or 20 log 1 i.e. 0 dB. The increase in the gain is at a rate of 20 dB/decade. The frequency response of such an ideal differentiator is shown in the Fig. 4.77.

It can be seen from the response that for $f < f_a$, the ratio f/f_a is less than unity. Hence $20 \log (f/f_a)$ is negative. While at $f = f_a$, the gain becomes 0 dB and for frequencies f_a onwards, gain increases with a rate of 20 dB/decade.

4.25.4 Disadvantages of an Ideal Differentiator

The gain of the differentiator increases as frequency increases. Thus at some high frequency, the differentiator may become unstable and break into the oscillations. There is possibility that op-amp may go into the saturation.

Also the input impedance $X_{C1} = (1/2\pi f C_1)$ decreases as frequency increases. This makes the circuit very much sensitive to the noise. Thus when such noise gets amplified due to high gain at high frequency, noise may completely override the differentiated output.

Hence the differentiator circuit suffers from the limitations on its stability and noise problems, at high frequencies. These problems can be corrected using some additional parameters in the basic differentiator circuit. Such a differentiator circuit is called **practical differentiator circuit.**

4.26 PRACTICAL DIFFERENTIATOR

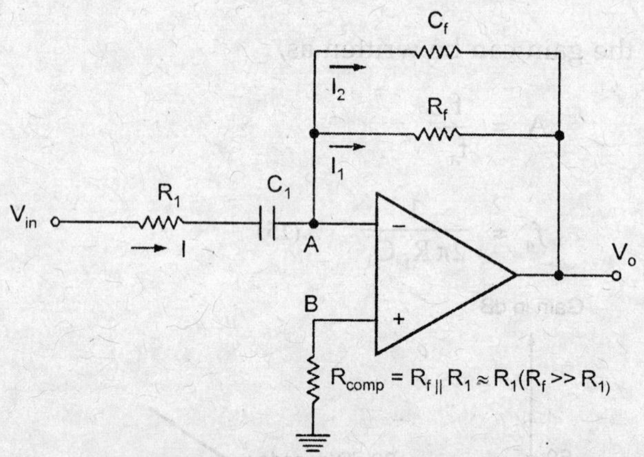

Fig. 4.78 Practical differentiator circuit.

The noise and stability at high frequency can be corrected, in the practical differentiator circuit using the resistance R_1 in series with C_1 and the capacitor C_f in parallel with resistance R_f.

The circuit is shown in the Fig. 4.78. The resistance Rcomp is used for bias compensation.

4.26.1 The Analysis of the Practical Differentiator

As the input current of op-amp is zero, there is no current input at node B. Hence it is at the ground potential. From the concept of the virtual ground, node A is also at the ground potential and hence $V_B = V_A = 0$ V.

For the current I, we can write

$$I = \frac{V_{in} - V_A}{Z_1} = \frac{V_{in}}{Z_1} \quad ...(1)$$

where $Z_1 = R_1$ in series with C_1

So in Laplace domain we can write,

$$Z_1 = R_1 + \frac{1}{sC_1} = \frac{1+sR_1C_1}{sC_1} \quad ...(2)$$

$$\therefore \quad I = \frac{sR_1 V_{in}(s)}{(1+sR_1 C_1)} \quad ...(3)$$

Now the current I_1 is,

$$I_1 = \frac{V_A - V_o}{R_f} = -\frac{V_o}{R_f} \quad ...(4)$$

In Laplace,

$$I_1 = -\frac{V_o(s)}{R_f} \quad ...(5)$$

And

$$I_2 = C_f \frac{d(V_A - V_o)}{dt} = -C_f \frac{dV_o}{dt} \quad ...(6)$$

Taking Laplace transform we get,

$$I_2 = -s\, C_f\, V_o(s) \quad ...(7)$$

Applying KCL at node A,

$$I = I_1 + I_2 \quad ...(8)$$

$$\therefore \quad \frac{sC_1V_{in}(s)}{(1+sR_1C_1)} = -\frac{V_o(s)}{R_f} - sC_f\, V_o(s)$$

$$\therefore \quad \frac{sC_1V_{in}(s)}{(1+sR_1C_1)} = -V_o(s)\left[\frac{1+sR_f\,C_f}{R_f}\right] \quad ...(9)$$

$$\therefore \quad V_o(s) = \frac{-sR_f\,C_1\,V_{in}(s)}{(1+sR_f\,C_1)\,(1+sR_1\,C_1)} \quad ...(10)$$

If $R_f\,C_f = R_1C_1$ then

$$\therefore \quad V_o(s) = \frac{-sR_f\,C_1\,V_{in}(s)}{(1+sR_1\,C_1)^2} \quad ...(11)$$

The time constant $R_f\,C_1$ is much greater than $R_1\,C_1$ or $R_f\,C_f$ and hence the equation (11) reduces to,

$$V_o(s) = -s\,R_f\,C_1\,V_{in}\,(s)$$

i.e.
$$V_o(t) = -R_f C_1 \frac{dV_{in}(t)}{dt} \quad \text{as } s = \frac{d}{dt} \qquad ...(12)$$

Thus the output voltage is the $R_f\,C_1$ times the differentiation of the input.

It may be noted that though $R_f\,C_1$ is much larger than $R_f\,C_f$ or $R_1\,C_1$, it is less than or equal to the time period T of the input, for the true differentiation.

$$R_f C_1 \le T \qquad ...(13)$$

4.26.2 Frequency Response of Practical Differentiator

To determine the frequency response, let us obtain the expression for the gain of the practical differentiator interms of the frequency.

From the equation (11) we can write,

$$\frac{V_o(s)}{V_{in}(s)} = \frac{-sR_f C_1}{(1+sR_1C_1)^2} \quad \text{where } R_f\,C_f = R_1\,C_1$$

Replacing s by jω in the steady state,

$$\frac{V_o(j\omega)}{V_{in}(j\omega)} = \frac{-j\omega R_f C_1}{(1+j\omega R_1C_1)^2} \qquad ...(14)$$

$$\therefore \quad \frac{V_o(j\omega)}{V_{in}(j\omega)} = \frac{-j2\pi f R_f C_1}{(1+j2\pi R_1C_1)^2} \qquad ...(15)$$

Now let,
$$f_b = \frac{1}{2\pi R_1 C_1} \qquad ...(16)$$

And
$$f_a = \frac{1}{2\pi R_f C_1} \qquad ...(17)$$

$$\therefore \quad \frac{V_o(j\omega)}{V_{in}(j\omega)} = \frac{-j\left(\frac{f}{f_a}\right)}{\left(1+j\frac{f}{f_b}\right)^2} \qquad ...(18)$$

The frequencies f_a and f_b are two break frequencies.

$$\therefore \quad A = \frac{V_o(j\omega)}{V_{in}(j\omega)} = \frac{(f/f_a)}{\left[\sqrt{1+\left(\frac{f}{f_o}\right)^2}\right]^2}$$

$$\therefore \quad A = \frac{f/f_a}{1+\left(\frac{f}{f_b}\right)^2} \qquad ...(19)$$

Now as $R_f C_1$ is much larger than $R_1 C_1$, we can write

$$f_a < f_b \qquad \text{... (20)}$$

Hence as frequency increases, the gain increases till $f = f_b$ at a rate of + 20 dB/decade. However after $f = f_b$ the gain decreases at a rate of 20 dB/decade. This 40 dB/decade change at $f = f_b$ occurs due to the combination of R_1C_1 and $R_f C_f$.

So for $R_fC_1 \ll T$, the true differentiation results.

The frequency response is shown in the Fig. 4.79. Refer Fig. 4.79 on next page.

Key Point: *It can be observed from the frequency response that the gain reduces as frequency increases greater than f_b. Hence the problem of instability at high frequency gets eliminated.*

Also the combination of $R_1 C_1$ and $R_f C_f$ help to reduce effectively the impact of high frequency noise and offsets.

It is important to remember that if f_c is the unity gain bandwidth (UGB) then the values of f_a and f_b must be selected in such a way that,

$$\boxed{f_a < f_b < f_c} \qquad \text{...(21)}$$

where f_c is UGB of op-amp in the open loop configuration.

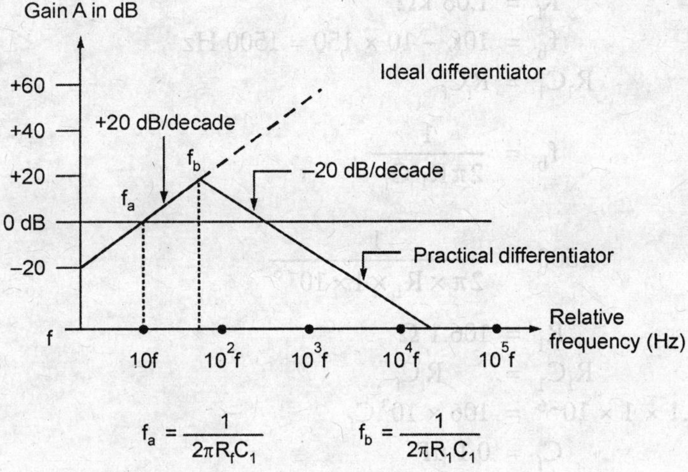

Fig. 4.79 Frequency response of the practical differentiator.

4.26.3 Steps to Design Practical Differentiator

By using following steps, a good practical differentiator can be designed:

(*i*) Choose f_a as the highest frequency of the input signal.

(*ii*) Choose C_1 to be less than 1 μF and calculate the value of R_f.

(*iii*) Choose f_b as 10 times f_a which ensures that $f_a < f_b$.

(*iv*) Finally calculate the values of R_1 and C_f from the expression $R_1C_1 = R_fC_f$.

(*v*) The R_{comp} can be selected as $R_1 || R_f$ but practically it is almost equal to R_1.

4.26.4 Applications of Practical Differentiator

The practical differentiator circuits are most commonly used in :

(*i*) In the wave shaping circuits to detect the high frequency components in the input signal.

(*ii*) As a rate-of-change detector in the FM demodulators.

The differentiator circuit is avoided in the analog computers.

Example 4.20: Design a practical differentiator circuit that will differentiate an input signal with the f_{max} = 150 Hz.

Solution: Refer to the steps for design.

(*i*) $f_a = f_{max} = 150$ Hz

(*ii*) Choose $C_1 = 1\ \mu F$

$$\text{Now} \quad f_a = \frac{1}{2\pi R_f C_1}$$

$$\therefore \quad 150 = \frac{1}{2\pi R_f \times 1 \times 10^{-6}}$$

$$\therefore \quad R_f \doteq 1.06\ k\Omega$$

(*iii*) Now $f_b = 10 f_a - 10 \times 150 = 1500$ Hz

And $R_1 C_1 = R_f C_f$

$$\text{Now} \quad f_b = \frac{1}{2\pi R_1 C_1}$$

$$\therefore \quad 1500 = \frac{1}{2\pi \times R_1 \times 1 \times 10^{-6}}$$

$$\therefore \quad R_1 = 106.1\ \Omega$$

And $R_1 C_1 = R_f C_f$

$$\therefore \quad 106.1 \times 1 \times 10^{-6} = 106 \times 10^3\ C_f$$

$$\therefore \quad C_f = 0.1\ \mu F$$

(*iv*)

$$R_{comp} = R_1 || R_f = \frac{R_1 R_f}{R_1 + R_f}$$

$$= \frac{106.1 \times 1.06 \times 10^3}{1061 + 1.06 \times 10^3} = 96.44\ m.$$

But generally R_{comp} is selected equal to R_1. The designed differentiator circuit is shown in Fig. 4.80.

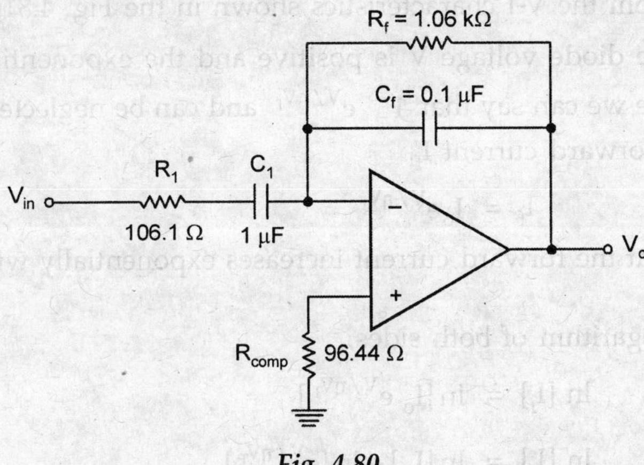

Fig. 4.80

4.27 FUNDAMENTALS OF LOG AMPLIFIERS

Consider a diode as shown in the Fig. 4.81(*a*) and the corresponding volt-ampere (V-I) characteristics in the Fig. 4.81(*b*).

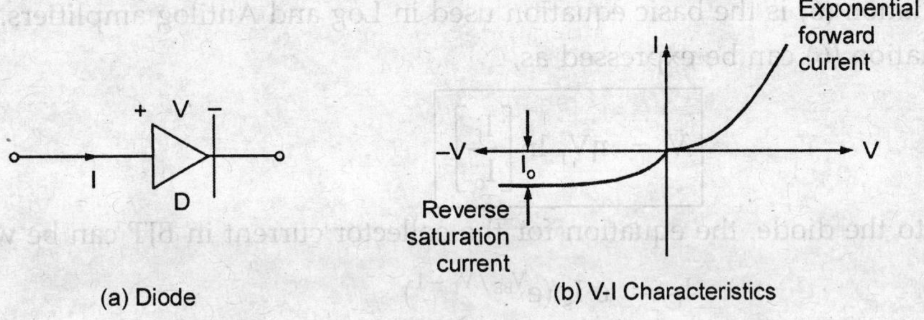

Fig. 4.81

The basic volt-ampere relationship for a diode is given as,

$$I = I_o(e^{V/\eta V_T} - 1)$$

where I = Diode current

I_o = Reverse saturation current

V = Diode voltage

η = 1 for Ge diode, 2 for Si diode

V_T = kT = Voltage equivalent of temperature

and k = Boltzmann's constant = 8.62×10^{-5} eV/°K

T = Temperature in °K

It can be observed that at room temperature of 27°C i.e. T = 300°K, the voltage equivalent of temperature is,

$$V_T = 0.02586 \text{ V} = 26 \text{ mV at } 300 \text{ °K}$$

This value is frequently used in the diode circuits.

It can be seen from the V-I characteristics shown in the Fig. 4.81(*b*) that for a forward biased condition, the diode voltage V is positive and the exponential term $e^{V/\eta V_T}$ has a positive index. Hence we can say that $1 \ll e^{V/\eta V_T}$ and can be neglected. The corresponding diode current is its forward current I_f.

$$\therefore \quad I_f = I_o e^{V/\eta V_T} \qquad ...(2)$$

This indicates that the forward current increases exponentially with respect to the bias voltage V.

Taking natural logarithm of both sides,

$$\ln [I_f] = \ln [I_o \, e^{V/\eta V_T}] \qquad ...(3)$$

$$\therefore \quad \ln [I_f] = \ln [I_o] + \ln [e^{V/\eta V_T}] \qquad ...(4)$$

$$\therefore \quad \ln [I_f] = \ln [I_o] + \frac{V}{\eta V_T} \qquad ...(5)$$

$$\therefore \quad \boxed{V = \eta V_T[\ln (I_f) - \ln(I_o)]} \qquad ...(6)$$

This equation (6) is the basic equation used in Log and Antilog amplifiers.

The equation (6) can be expressed as,

$$\boxed{V = \eta V_T \ln \left[\frac{I_f}{I_o}\right]} \qquad ...(7)$$

Similar to the diode, the equation for the collector current in BJT can be written as,

$$I_C = \alpha I_s (e^{V_{BE}/V_T - 1}) \qquad ...(8)$$

where

I_C = collector current

I_s = emitter saturation current

V_{BE} = base emitter voltage

$\alpha \approx$ 1 and constant

$$\therefore \quad \frac{I_C}{I_s} = e^{V_{BE}/V_T} - 1 \qquad ...(9)$$

Now emitter saturation current is very small of the order of 10^{-13} A.

$$\therefore \quad e^{V_{BE}/V_T} = \frac{I_C}{I_s} + 1 = \frac{I_C}{I_s} \qquad ...(10)$$

As I_s is very small, $1 \ll (I_C/I_s)$.

$$\therefore \quad \ln (e^{V_{BE}/V_T}) = \ln \left(\frac{I_C}{I_s}\right) \qquad ...(11)$$

$$\frac{V_{BE}}{V_T} = \ln\left(\frac{I_C}{I_s}\right) \qquad ...(12)$$

$$\therefore \qquad \boxed{V_{BE} = V_T \ln\left(\frac{I_C}{I_s}\right)} \qquad ...(13)$$

This equation forms the basis of Log and Antilog amplifier circuits where BJT is used instead of a diode.

In the equation (8), V_{BE} represents voltage across transistor, similar to the V in the equation (1), representing voltage across diode.

Now voltage across BJT is actually $V_{CB} + V_{BE}$. But in op-amp circuits when BJT is used in the feedback path as a diode, V_{CB} voltage becomes zero. Hence V is replaced by V_{BE} while writing the equation for the BJT.

For BJT used as a diode, $V_{CB} = 0$.

4.28 BASIC LOGARITHMIC AMPLIFIER

With the background of diode and current equations, let us study the basic logarithmic amplifier circuit using op-amp. The fundamental log amplifier is formed by placing a diode or a transistor in the negative feedback path of the op-amp.

4.28.1 Basic Log Amplifier using Diode

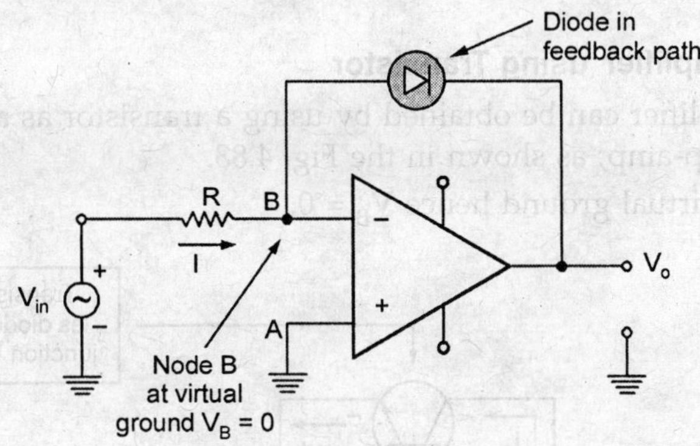

Fig. 4.82 Basic log amplifier.

The circuit diagram of basic log amplifier using a diode is shown in the Fig. 4.82.

The diode D is used in the negative feedback path. The node A is grounded hence node B is at virtual ground. Hence $V_B = 0$.

$$\therefore \qquad I = \frac{V_{in} - V_B}{R} = \frac{V_{in}}{R} \qquad ...(1)$$

As the op-amp input current is zero,

$$I = I_f = \text{Diode current} \qquad ...(2)$$

Now If is the current through diode and voltage across diode is $V_B - V_o$ i.e. $- V_o$.

Hence using equation (7) of section 4.27 we can write,

$$-V_o = \eta V_T \ln\left[\frac{I_f}{I_o}\right] \quad ...(3)$$

Substituting $I_f = I = \frac{V_{in}}{R}$

$$\therefore \quad V_o = -\eta V_T \ln\left[\frac{V_{in}}{RI_o}\right] \quad ...(4)$$

As I_o R is constant d.c. voltage, let us denote it as V_{ref}.

$$\therefore \quad V_o = -\eta V_T \ln\left[\frac{V_{in}}{V_{ref}}\right] \quad ...(5)$$

where $V_{ref} = I_o R$

Thus the output voltage V_o is a function of logarithm of the input voltage V_{in}. The circuit gives the logarithm to base e i.e., natural logarithm. But the same circuit can be used to find out logarithm values to base 10 by the proper scaling as,

$$\log_{10} X = 0.4343 \ln (X) \quad ...(6)$$

The factor 0.4343 can be obtained by connecting attenuator at the output of the basic circuit.

4.28.2 Basic Log Amplifier using Transistor

The basic log amplifier can be obtained by using a transistor as a diode in the negative feedback path of an op-amp, as shown in the Fig. 4.83.

The node B is at virtual ground hence $V_B = 0$.

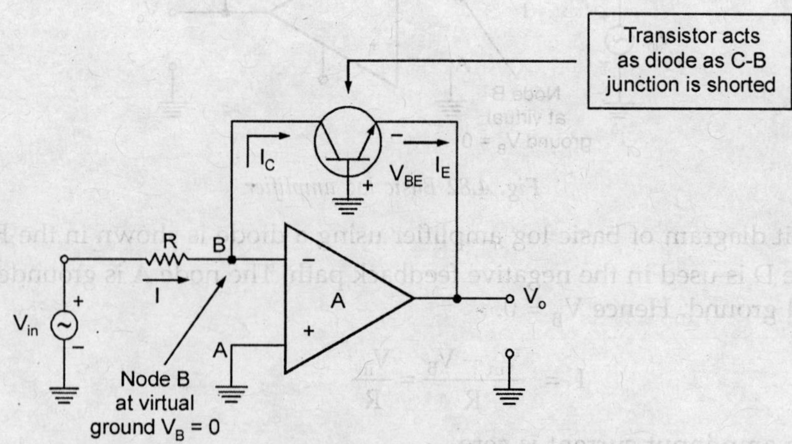

Fig. 4.83 Basic log amplifier.

$$\therefore \quad I = \frac{V_{in} - V_B}{R} = \frac{V_{in}}{R} \qquad ...(7)$$

As the op-amp input current is zero

$$I = I_C = \text{Collector current} \qquad ...(8)$$

The voltage $V_{CB} = 0$ as the collector is at virtual ground and base is grounded. Hence we can write the equation of I_C as,

$$V_{BE} = V_T \ln\left(\frac{I_C}{I_s}\right) \qquad ...(9)$$

Applying to the output side we get,

$$V_o + V_{BE} = 0 \qquad ...(10)$$

$$\therefore \quad V_{BE} = -V_o \qquad ...(11)$$

and $\quad I_C = I = \frac{V_{in}}{R}$

Substituting in the equation (9),

$$-V_o = V_T \ln\left(\frac{V_{in}}{R I_s}\right) \qquad ...(12)$$

Let $\quad V_{ref} = R I_s,$

$$\therefore \quad \boxed{V_o = -V_T \ln\left[\frac{V_{in}}{V_{ref}}\right]} \qquad ...(13)$$

The equation is similar to the equation (5), which gives the output, proportional to the logarithm of the input voltage V_{in}.

4.28.3 Disadvantages of Basic Circuit

1. The reverse saturation current I_o for the diode changes with temperature. In fact it doubles for every ten degree celcius rise in the temperature.
2. Similarly the emitter saturation current varies significantly from one transistor to other and also with the temperature.
3. Hence it is very difficult to set the term V_{ref} for the circuit.
4. The term V_T which is kT also changes with temperature, which appears in both the equations.
5. Thus temperature affects the performance and accuracy of the basic logarithmic amplifier circuit. Hence it is must to provide some sort of temperature compensation to reduce the errors.

4.29 TEMPERATURE COMPENSATED LOG AMPLIFIER

The techniques used for temperature compensation of diode used log amplifier and transistor used log amplifier, are different.

For diode used log amplifier, another matched diode is used in the circuit. The overall circuit with the required temperature compensation is shown in the Fig. 4.84.

The two diodes D_1 and D_2 are perfectly matched and V_T, I_o values for both are same. A constant reference voltage is generated using a constant current source I_{ref}. The voltage across D_1 is V_{f1} while voltage across D_2 is V_{f2}.

The input current of op-amp A_2 is zero, hence the entire current I_{ref} passes through the diode D_2.

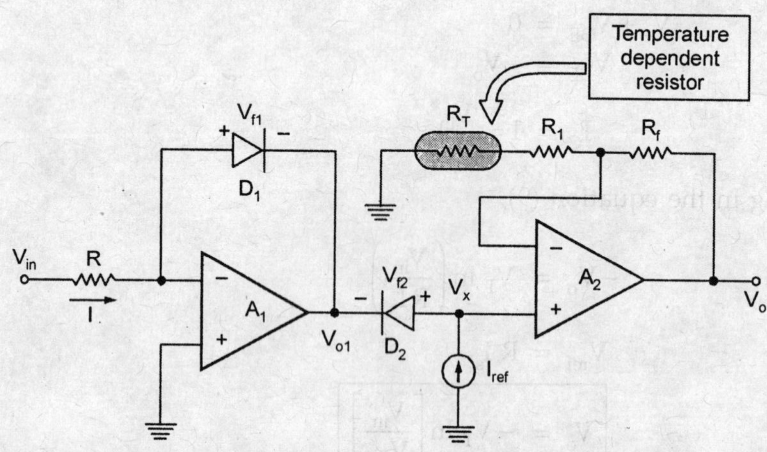

Fig. 4.84 Temperature compensated log amplifier.

From the Fig. 4.84, we can write

$$V_x = V_{o1} + V_{f2} \quad ...(1)$$

Now the voltage V_{o1} is the output of the op-amp A_1 which is the basic log amplifier.

$$\therefore \quad V_{o1} = -\eta V_T \ln\left[\frac{V_{in}}{RI_o}\right] \quad ...(2)$$

While V_{f2} is voltage across diode hence can be expressed from the diode equation as,

$$V_{f2} = \eta V_T \ln\left[\frac{I_{f2}}{I_o}\right] \quad ...(3)$$

where I_{f2} = Current through diode D_2

Here $I_{f2} = I_{ref}$ ...(4)

$$\therefore \quad V_{f2} = \eta V_T \ln\left[\frac{I_{ref}}{I_o}\right] \quad ...(5)$$

Substituting from equations (2) and (5) in equation (1) we get,

$$V_x = -\eta V_T \ln\left[\frac{V_{in}}{R I_o}\right] + \eta V_T \ln\left[\frac{I_{ref}}{I_o}\right] \quad ...(6)$$

$$\therefore \quad V_x = -\eta V_T\left[\ln\left(\frac{V_{in}}{R}\right) - \ln(I_o) - \ln(I_{ref}) + \ln(I_o)\right] \quad ...(7)$$

$$\therefore \quad V_x = -\eta V_T\left[\ln\left(\frac{V_{in}}{R}\right) - \ln(I_{ref})\right] \quad ...(8)$$

$$\therefore \quad V_x = -\eta V_T \ln\left[\frac{V_{in}}{R.I_{ref}}\right] \quad ...(9)$$

Thus the term I_o gets cancelled. But the term V_T is still temperature dependent. It can be compensated by making gain of the op-amp A_2, temperature dependent.

Now op-amp A_2 acts as non-inverting amplifier and hence,

$$V_o = V_x\left[1 + \frac{R_f}{R_1 + R_T}\right] \quad ...(10)$$

$$\therefore \quad V_o = -\eta V_T \frac{(R_1 + R_T + R_f)}{(R_1 + R_T)} \ln\left[\frac{V_{in}}{R I_{ref}}\right] \quad ...(11)$$

The resistance R_T is positive temperature coefficient thermistor.

Thus when temperature increases, R_T increases. Hence the overall term $[\eta V_T(R_1 + R_T + R_f)/(R_1 + R_T)]$ remains constant with respect to temperature. Assuming this term as constant K we can write,

$$V_o = -K \ln\left[\frac{V_{in}}{R I_{ref}}\right] \quad ...(12)$$

Thus the output voltage is proportional to the logarithm of the input voltage V_{in}. The value of I_{ref} is decided by the current source and not dependent on the temperature. Thus the circuit is completely temperature independent and errors due to temperature are completely eliminated.

Let us study now the temperature compensation technique used for the transistor used basic log amplifier circuit.

In this technique two log amplifiers are used. One of the log amplifiers is given as a reference voltage V_{ref} while the other is given an input voltage V_{in}. The two transistors used in the two log amplifier circuits are integrated close together in the same silicon wafer. This ensures a close match of the saturation currents and provides good thermal tracking. The circuit is shown in the Fig. 4.85.

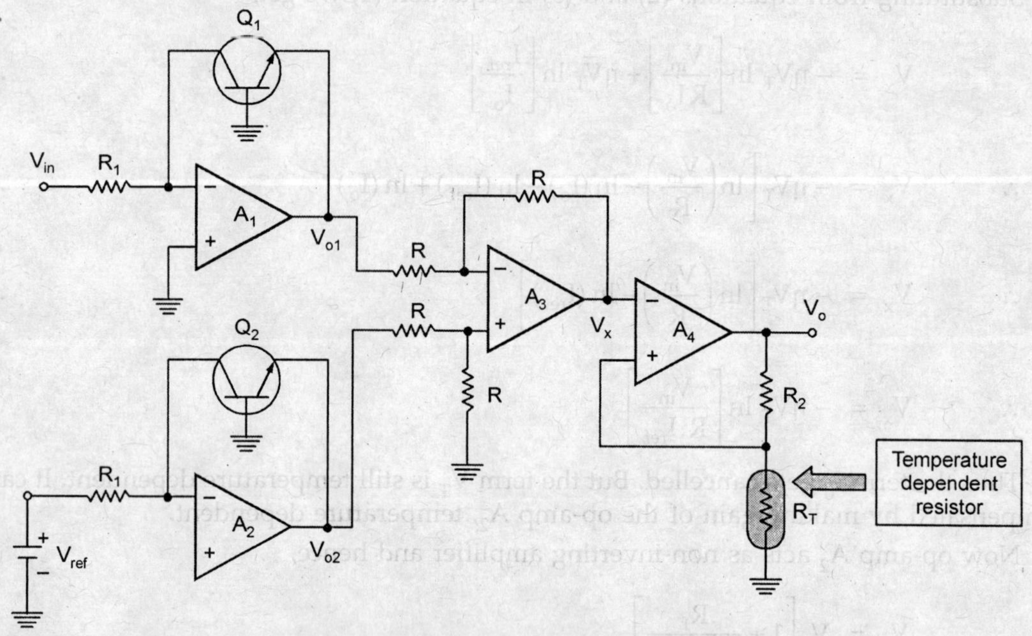

Fig. 4.85 Temperature compensated log amplifier.

The transistors Q_1, Q_2 are fabricated on same chip. The op-amp A_1 and A_2 work as the basic transistor logarithmic amplifiers. For Q_1, Q_2 we can write,

$$I_{s1} = I_{s2} = I_s \quad \text{...(13)}$$

While for the voltages V_{o1} and V_{o2} we can write,

$$V_{o1} = -V_T \ln\left[\frac{V_{in}}{R_1 I_s}\right] \quad \text{...(14)}$$

and

$$V_{o2} = -V_T \ln\left[\frac{V_{ref}}{R_1 I_s}\right] \quad \text{...(15)}$$

The op-amp A_3 acts as a difference amplifier with the unity gain.

$$\therefore \quad V_x = V_{o1}\left[-\frac{R}{R}\right] + V_{o2}\left[1+\frac{R}{R}\right]\left[\frac{R}{R+R}\right]$$

$$\therefore \quad V_x = V_{o2} - V_{o1}$$

$$\therefore \quad V_x = -V_T \ln\left[\frac{V_{ref}}{R_1 I_s}\right] - \left[-V_T \ln\left(\frac{V_{in}}{R_1 I_s}\right)\right]$$

$$= -V_T\left[\ln\left(\frac{V_{ref}}{R_1}\right) - \ln(I_s) - \ln\left(\frac{V_{in}}{R_1}\right) + \ln(I_s)\right] \quad \text{...(17)}$$

$$\therefore \qquad V_x = -V_T \ln\left(\frac{V_{ref}}{V_{in}}\right) \qquad \text{...(18)}$$

Absorbing negative sign,

$$V_x = V_T \ln\left(\frac{V_{in}}{V_{ref}}\right) \qquad \text{...(19)}$$

The V_{ref} is decided by external battery and it is not temperature dependent.

The temperature dependent component I_s gets cancelled out.

Still the term V_T is present in V_x which depends on the temperature. This is compensated by the last stage of the op-amp A_4. It acts as a non-inverting amplifier with gain of $\left(1+\frac{R_2}{R_T}\right)$ where R_T is the thermistor having positive temperature coefficient.

$$\therefore \qquad V_o = V_x\left(1+\frac{R_2}{R_T}\right)$$

$$\therefore \qquad V_o = \frac{V_T(R_2+R_1)}{R_T} \ln\left(\frac{V_{in}}{V_{ref}}\right) \qquad \text{...(20)}$$

As temperature changes, V_T changes but at the same time R_T also changes, so that the ratio $V_T\ (R_2 + R_T)/R_T$ remains constant.

Hence assuming this term as K, we can write,

$$V_o = K \ln\left(\frac{V_{in}}{V_{ref}}\right) \qquad \text{...(21)}$$

Thus properly matched coefficients and the resistance values provide a perfect temperature compensation for the circuit. The output is proportional to the logarithm of the input and temperature independent.

The main disadvantage of the above circuit is that it requires four op-amps. The same effect can be obtained by a circuit which uses only two op-amps. Thus errors due to bias currents and the offset voltages get reduced.

The circuit with two op-amps, providing the temperature compensation is shown in the Fig. 4.86.

The op-amp A_1 acts as a basic log amplifier and hence we can write,

$$V_x = -V_T \ln\left[\frac{V_{in}}{R_1 I_s}\right] \qquad \text{...(22)}$$

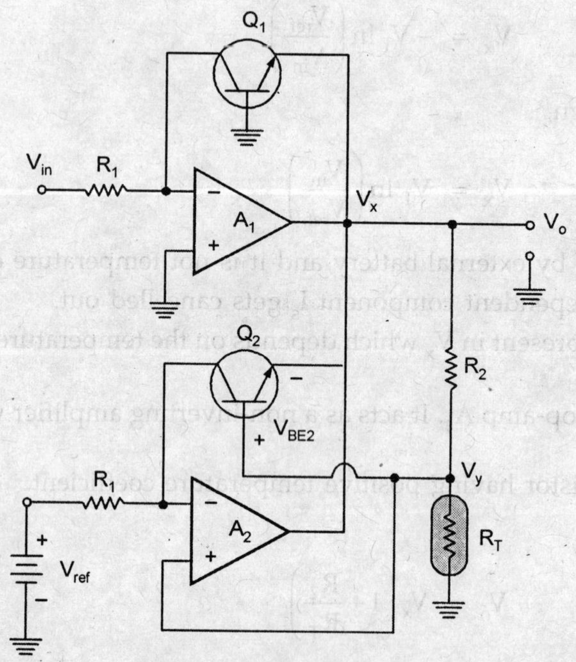

Fig. 4.86 Modified temperature compensation circuit

From the equation (9) of section 4.28.2 we can write an equation for V_{BE2} as,

$$V_{BE2} = V_T \ln\left(\frac{V_{ref}}{R_1 I_s}\right) \qquad ...(23)$$

For the transistor Q_2 we can write,

$$V_{B2} = V_{BE2} + V_{E2} \qquad ...(24)$$

where V_{B2} = Base voltage of Q_2 and V_{E2} = Emitter voltage of Q_2

From the Fig. 4.86 we can say that,

$$V_{B2} = V_y$$

and $$V_{E2} = V_x$$

$$\therefore \quad V_y = V_{BE2} + V_x \qquad ...(25)$$

Substituting V_{BE2} and V_x from equations (23) and (22) in the equation (25), we get

$$\therefore \quad V_y = V_T \ln\left(\frac{V_{ref}}{I_s R_1}\right) + \left(-V_T \ln\left[\frac{V_{in}}{R_1 I_s}\right]\right)$$

$$\therefore \quad V_y = V_T\left[\ln\left(\frac{V_{ref}}{R_1}\right) - \ln(I_s) - \ln\left(\frac{V_{in}}{R_1}\right) + \ln(I_s)\right]$$

$$\therefore \quad V_y = V_T \ln\left(\frac{V_{ref}}{V_{in}}\right)$$

$$\therefore \quad V_y = -V_T \ln\left(\frac{V_{in}}{V_{ref}}\right) \quad ...(26)$$

At the output side, there is a voltage divider of the resistances R_2 and R_T, where R_T is the thermistor.

Now from voltage divider rule,

$$V_y = \frac{V_o}{(R_2 + R_T)}.R_T \quad ...(27)$$

$$\therefore \quad V_o = \frac{V_y(R_2 + R_T)}{R_T} = \frac{V_T(T_2 + R_T)}{R_T} \ln\left(\frac{V_{in}}{V_{ref}}\right) \quad ...(28)$$

So the ratio, $-V_T(R_2 + R_T)/R_T$ is constant, as temperature changes. Hence we get,

$$V_o = K \ln\left(\frac{V_{in}}{V_{ref}}\right) \quad ...(29)$$

The output voltage is temperature independent, hence circuit provides temperature compensation. The circuit uses only two op-amps but must be used of very high quality.

The errors due to bias currents and the offset voltages are still present, affecting the output. Such errors can be by carefully selecting the specifications of both the op-amps. These errors also can be minimized by using FET input op-amps.

4.29.1 Stability Considerations

Due to the presence of an active element in the feedback path, the circuit shows the tendency to oscillate. The circuit is compensated by means of an emitter resistance R_E and a capacitor C_f across the negative feedback element. This arrangement lowers the gain as the frequency rises and avoids the possibility of the oscillations. Hence such a compensation is called as frequency compensation.

The typical frequency compensation network is shown in the Fig. 4.87.

The values of C_f and R_E are determined experimentally as parasitic effects play an important role. Generally C_f values range begins with 100 pF and R_E with 1 kΩ.

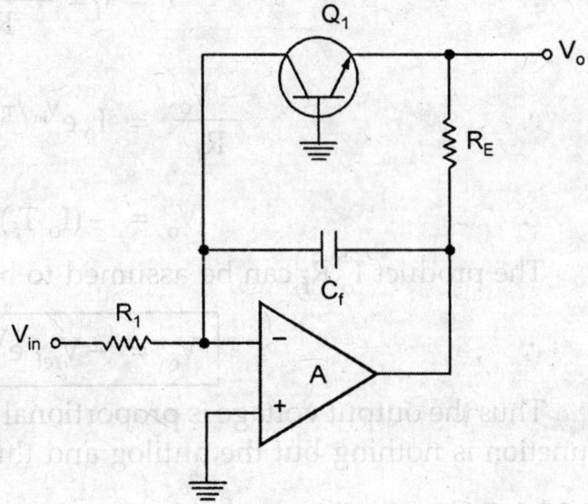

Fig. 4.87 Log amplifier with frequency compensation.

4.30 ANTILOG AMPLIFIER

The log amplifier can easily be turned around to provide the antilog or exponential function which is called antilog amplifier. The basic antilog amplifier can be obtained by using a diode or a transistor.

4.30.1 Basic Antilog Amplifier using Diode

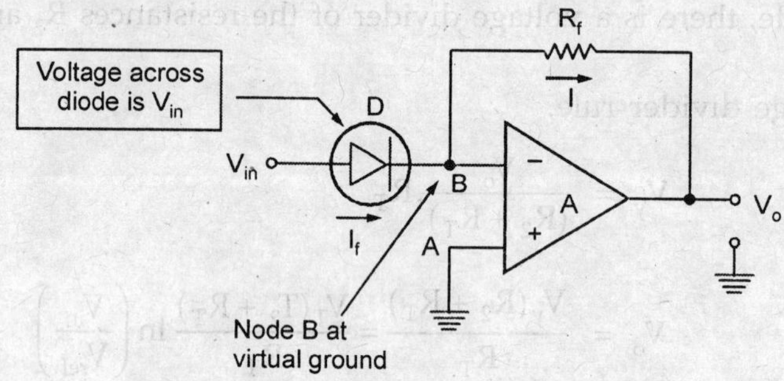

Fig. 4.88 Basic antilog amplifier.

The circuit diagram of basic antilog amplifier using diode is shown in the Fig. 4.88.

The positions of diode and resistance are exchanged as compared to log amplifier circuit. The node A is grounded and hence node B is at virtual ground. Hence $V_B = 0$.

Now the current flowing through the diode is I_f and the voltage across diode is V_{in} itself, as B is at virtual ground. Hence from the diode current equation we can write,

$$I_f = V_o e^{V_{in}/\eta V_T} \quad \text{...(1)}$$

As op-amp input current is zero, the current 1 must be same as I_f.

$$\therefore \quad I = I_f = \frac{V_B - V_o}{R_f} = \frac{-V_o}{R_f} \quad \text{...(2)}$$

$$\therefore \quad \frac{-V_o}{R_f} = I_o e^{V_{in}/\eta V_T}$$

$$\therefore \quad V_o = -(I_o T_f) e^{V_{in}/\eta V_T} \quad \text{...(3)}$$

The product $I_o R_f$ can be assumed to be V_{ref}, so we get

$$\therefore \quad \boxed{V_o = -V_{ref} e^{V_{in}/\eta V_T}} \quad \text{...(4)}$$

Thus the output voltage is proportional to the exponential function of V_{in}. The exponential function is nothing but the antilog and thus circuit works as an antilog amplifier.

4.30.2 Basic Antilog Amplifier using Transistor

The same circuit providing antilog of the input can be obtained by using a transistor instead of a diode. This is shown in the Fig. 4.89.

The node B is at virtual ground hence $V_B = 0$. Thus both collector and base of the transistor are at ground potential and $V_B = 0$. Hence the voltage across the transistor is V_{BE} and we can write the expression for its collector current as,

$$I_C = I_s e^{V_{BE}/V_T} \qquad ...(5)$$

But as seen from the Fig. 4.89, $V_{BE} = V_{in}$

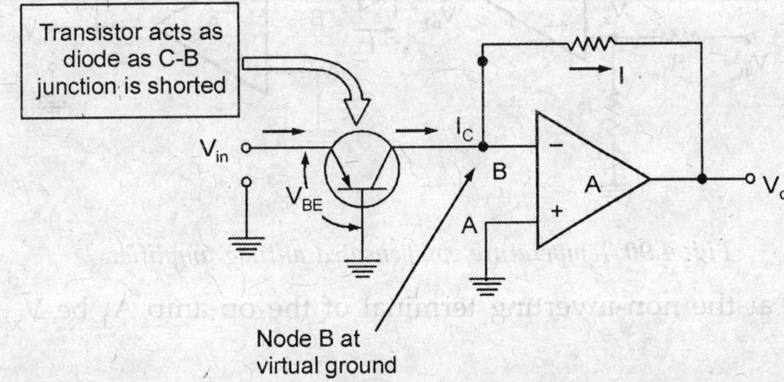

Fig. 4.89 Basic antilog amplifierl.

$$\therefore \quad I_C = I_s e^{V_{in}/V_T} \qquad ...(6)$$

Now the current I_C and current I are same as op-amp input current is zero.

$$\therefore \quad I = I_C = \frac{V_B - V_o}{R_f} = \frac{-V_o}{R_f} \qquad ...(7)$$

$$\therefore \quad \frac{-V_o}{R_f} = I_s e^{V_{in}/V_T}$$

$$\therefore \quad V_o = -I_s R_f e^{V_{in}/V_T} \qquad ...(8)$$

Assuming $I_s R_f$ as V_{ref}, we can write

$$\therefore \quad \boxed{V_o = -V_{ref} e^{V_{in}/\eta V_T}} \qquad ...(9)$$

Thus the output voltage is proportional to the exponential of V_{in} i.e, antilog of V_{in}. Thus circuit works as basic antilog amplifier.

In both the above circuits, it can be seen that the terms I_o, I_s and V_T are present in the output equation. All these are the function of temperature. Hence as temperature changes, these parameters also change and cause serious errors at the output. So the basic antilog circuits also face the same limitations as that of basic log amplifier circuits. And hence temperature compensation is must for the antilog amplifier circuits as well.

4.31 TEMPERATURE COMPENSATED ANTILOG AMPLIFIER

By using, another matched diode in the circuit, the temperature compensation can be provided to the antilog amplifier using diode. The circuit diagram is shown in the Fig. 4.90.

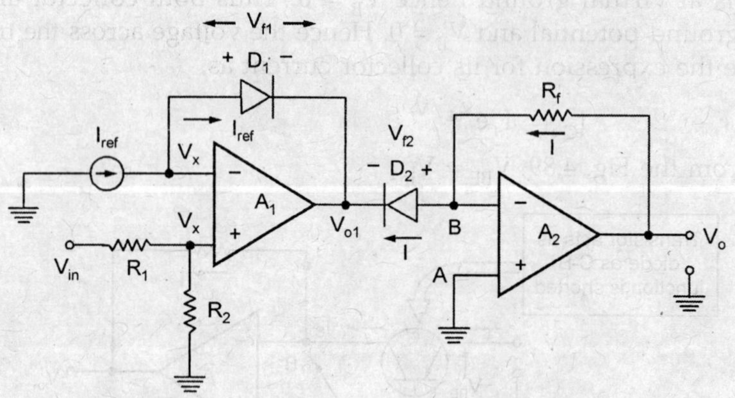

Fig. 4.90 Temperature compensated antilog amplifier.

Let the voltage at the non-inverting terminal of the op-amp A_1 be V_x. By the voltage divider rule,

$$V_x = V_{in}\frac{R_2}{(R_1+R_2)} \qquad ...(1)$$

The current source I_{ref} is used at the inverting terminal of op-amp A_1. The same current flows through diode D_1 as op-amp current zero. Let the drop across the diode D_1 be V_{f1}. And the voltage at the inverting terminal of the op-amp A_1 is V_x.

$$\therefore \quad V_{o1} = V_x - V_{f1} \qquad ...(2)$$

From the diode equation (7) of section 4.27, we can write,

$$V_{f1} = \eta V_T \ln\left(\frac{I_f}{I_o}\right) \qquad .(3)$$

Substituting in equation (2) we get,

$$V_{o1} = V_{in}\left(\frac{R_2}{R_1+R_2}\right) - \eta V_T \ln\left(\frac{I_f}{I_o}\right) \qquad ...(4)$$

For op-amp A_2, the node B is at virtual ground.

$$\therefore \quad V_{o1} = -V_{f2} \qquad ...(5)$$

where V_{f2} is the voltage drop across diode D_2. Also from the diode equation we can write,

$$V_{f2} = \eta V_T \ln\left(\frac{I_{f2}}{I_o}\right) \qquad ...(6)$$

Now $$I_{f2} = I = \frac{V_o - V_B}{R_f} = \frac{V_o}{R_f} \qquad ...(7)$$

While $I_{f1} = I_{ref}$ ...(8)

Substituting equations (6) and (7) in equation (5),

$$\therefore \quad V_{o1} = -\eta V_T \ln\left(\frac{V_o}{I_o R_f}\right) \quad ...(9)$$

Equation equations (4) and (9),

$$V_{in}\left(\frac{R_2}{R_1+R_2}\right) - \eta V_T \ln\left(\frac{I_{ref}}{I_o}\right) = -\eta V_T \ln\left(\frac{V_o}{I_o R_f}\right) \quad ...(10)$$

$$\therefore \quad V_{in}\left(\frac{R_2}{R_1+R_2}\right) = \eta V_T\left[\ln(I_{ref}) - \ln(I_o) - \ln\left(\frac{V_o}{R_f}\right) + \ln(I_o)\right]$$

$$\therefore \quad V_{in}\left(\frac{R_2}{R_1+R_2}\right) = \eta V_T \ln\left(\frac{I_{ref} R_f}{V_o}\right)$$

$$\therefore \quad V_{in}\left(\frac{R_2}{R_1+R_2}\right) = -\eta V_T \ln\left(\frac{V_o}{I_{ref} R_f}\right) \quad ...(11)$$

i.e.

$$\ln\left(\frac{V_o}{I_{ref} R_f}\right) = \frac{-V_{in} R_2}{\eta V_T (R_1+R_2)}$$

$$\therefore \quad \boxed{V_o = I_{ref} R_f \ln^{-1}\left[\frac{-V_{in} R_2}{\eta V_T (R_1+R_2)}\right]} \quad ...(12)$$

Thus output is proportional to the antilog ($\ln^{-1}$) of the input. As component I_o gets cancelled, the circuit provides temperature compensation. Still V_T term, which is temperature dependent exists in the expression. If the resistive divider $R_1 - R_2$ is made temperature sensitive using thermistor, the effect of change in V_T due to the temperature can be eliminated. As V-I characteristics of diode is not accurate and due to problem of resistance associated with the bulk semiconductor material in diode, the diode based antilog amplifiers are generally not used in practice.

Let us see now, the transistor based antilog amplifier, with the facility of temperature compensation. The circuit of transistor based antilog amplifier with temperature compensation is shown in the Fig. 4.91.

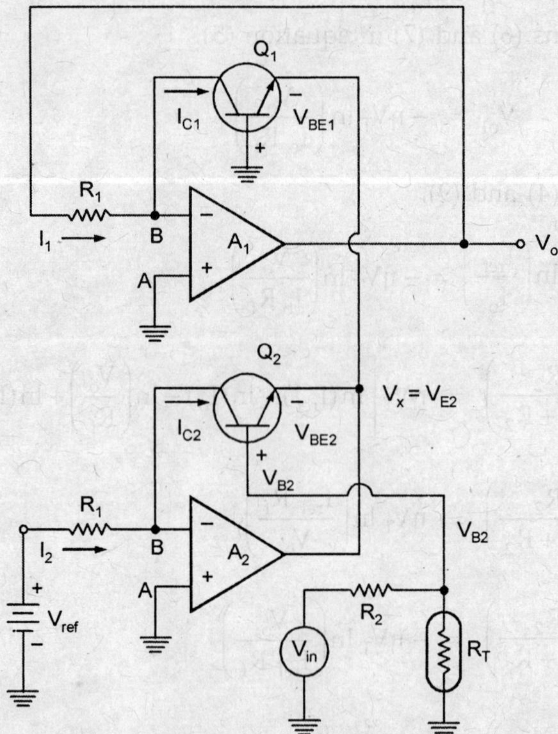

Fig. 4.91 Temperature compensated antilog amplifier.

The two matched transistors Q_1 and Q_2 are used. The external voltage V_{ref} is connected to inverting terminal of op-amp A_2 through resistance R_1. The voltage divider receives the input voltage V_{in}. The thermistor R_T is used in the voltage divider.

As shown in the Fig. 4.91,

$$V_{B2} = \text{Voltage of base of } Q_2$$

$$V_{E2} = \text{Voltage of emitter of } Q_2 = V_x$$

Let us derive the expression for the output voltage V_o.

For both the op-amps, as node A is grounded, node B is at virtual ground and hence V_B for both the op-amps is zero. Referring to the equation (13) of section 4.27, we can write the equations for base-emitter voltages of Q_1 and Q_2 as,

$$V_{BE1} = V_T \ln\left(\frac{I_{C1}}{I_s}\right) \quad ...(13)$$

and

$$V_{BE2} = V_T \ln\left(\frac{I_{C2}}{I_s}\right) \quad ...(14)$$

As op-amp currents are zero,

$$I_{C1} = I_1$$

$$= \frac{V_o - V_B}{R_1} = \frac{V_o}{R_1} \quad \text{... (15)}$$

and $$I_{C2} = I_2$$

$$= \frac{V_{ref} - V_B}{R_1} = \frac{V_{ref}}{R_1} \quad \text{... (16)}$$

$$\therefore \quad V_{BE1} = V_T \ln\left(\frac{V_o}{I_s R_1}\right) \quad \text{...(17)}$$

and $$V_{BE2} = V_T \ln\left(\frac{V_{ref}}{I_s R_1}\right) \quad \text{...(18)}$$

Now the voltage V_x shown in the Fig. 4.91 is,

$$V_x = -V_{BE1} = -V_T \ln\left(\frac{V_o}{I_s R_1}\right) \quad \text{...(19)}$$

While V_{B2} is the voltage output from potential divider which is,

$$V_{B2} = V_{in}\left(\frac{R_T}{R_2 + R_T}\right) \quad \text{...(20)}$$

Now for the transistor O_2, we can write,

$$V_{B2} = V_{E2} + V_{BE2}$$

$$\therefore \quad V_{in}\left(\frac{R_T}{R_2 + R_T}\right) = V_x + V_T \ln\left(\frac{V_{ref}}{I_s R_1}\right)$$

$$\therefore \quad V_{in}\left(\frac{R_T}{R_2 + R_T}\right) = -V_T \ln\left(\frac{V_o}{I_s R_1}\right) + V_T \ln\left(\frac{V_{ref}}{I_s R_1}\right)$$

$$V_{in}\left(\frac{R_T}{R_2 + R_T}\right) = -V_T\left[\ln\left(\frac{V_o}{R_1}\right) - \ln(I_s) - \ln\left(\frac{V_{ref}}{R_1}\right) + \ln(I_s)\right]$$

$$\therefore \quad V_{in}\left(\frac{R_T}{R_2 + R_T}\right) = -V_T \ln\left(\frac{V_o}{V_{ref}}\right) \quad \text{...(21)}$$

$$\therefore \quad \ln\left(\frac{V_o}{V_{ref}}\right) = \frac{-V_{in} R_T}{V_T (R_2 + R_T)}$$

$$\therefore \quad \boxed{V_o = V_{ref} \ln^{-1}\left[\frac{-V_{in} R_T}{V_T (R_2 + R_T)}\right]} \quad \text{...(22)}$$

Thus the output is proportional to the antilog ($\ln^{-1}$) of the input. And the term $R_T/V_T (R_2 + R_T)$ is constant though temperature changes as R_T also changes proportional to the changes in V_T. Thus a complete temperature compensation is achieved with this circuit.

The antilog amplifier is also subjected to the various errors like noise, bias currents, offset voltages, drifts and also faces the frequency stability problem, similar to the log amplifier. These problems can be solved for the antilog amplifier as they were for the log amplifier.

Both log amplifiers and antilog amplifiers are practically obtained by using transistors as compared to diodes because the transistors give high accuracy, reduced bulk resistance effect and high range of operations.

4.32 OP-AMPS IN SWITCHING CIRCUITS

It is seen that op-amp in open loop configuration produces the output voltage which is either at its positive saturation or at its negative saturation level. This is because open loop gain of op-amp is very large. The output saturation voltages are denoted as $+V_{sat}$ and $-V_{sat}$ are usually given by,

$$+V_{sat} = +V_{CC} - 1\text{ V} \quad ...(1)$$

$$-V_{sat} = -V_{EE} + 1\text{ V} \quad ...(2)$$

The 1 V difference may change from op-amp to op-amp. For very small change in input voltage, the op-amp output switches from $+V_{sat}$ to $-V_{sat}$ and vice-versa. So though this property of op-amp is not suitable for linear applications, it is suitable for various non-linear switching circuits such as comparators, crossing detectors, Schmitt triggers etc. Such non-linear circuits using op-amp are discussed in this chapter. In these circuits many times positive feedback is used and op-amp generally does not require any frequency compensation.

4.33 OP-AMP AS A COMPARATOR

The op-amp in open loop configuration can be used as a basic comparator. When two inputs are applied to the open loop op-amp then it compares the two inputs. Depending upon the comparison, it produces output voltage which "is either positive saturation voltage ($+V_{sat}$) or negative saturation voltage ($-V_{sat}$).

A **comparator** is a circuit which compares a signal voltage applied at one input of an op-amp with a known reference voltage at the other input and produce either a high or a low output voltage, depending on which input is higher. As comparator output has two voltage levels, either high or low, it is not linearly proportional to input voltage.

There are two types of comparator circuits which are,

1. Non-inverting comparator
2. Inverting comparator

4.33.1 Basic Non-inverting Comparator

In this comparator, the input voltage is applied to the non-inverting terminal and no reference voltage is applied to other terminal. So inverting terminal is grounded. The input voltage is denoted as V_{in} while the voltage applied to other terminal with which V_{in} is compared is denoted as V_{ref}. In the basic comparator, $V_{ref} = 0$ V. The **basic non-inverting comparator** is shown in the Fig. 4.92.

In the non-inverting comparator, if V_{in} is greater than V_{ref} then output is $+ V_{sat}$ i.e. almost equal to $+ V_{CC}$. While if V_{in} is less than V_{ref} then output is $- V_{sat}$ i.e. almost equal to $- V_{EE}$.

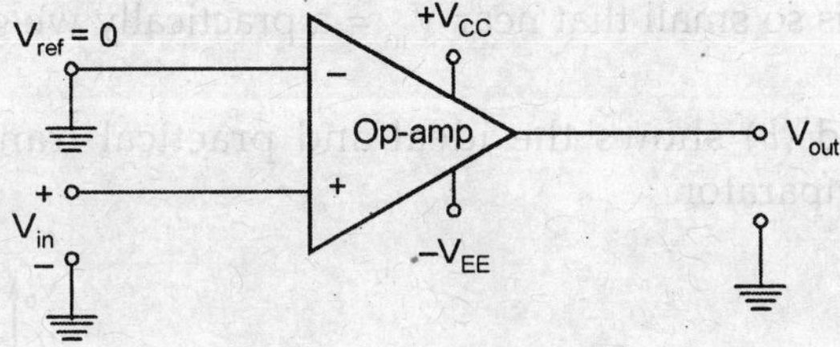

Fig. 4.92 Basic non-inverting comparator.

Thus for Fig. 4.92, as $V_{ref} = 0$ V when V_{in} is positive then $V_o = + V_{sat} = + V_{CC}$ while when V_{in} is negative then $V_o = - V_{sat} = - V_{EE}$. This is because, as open loop gain op-amp (A_{OL}) is very very high even for very small V_{in} the op-amp output saturates.

Thus the two possible output levels of the comparator are $+ V_{sat}$ and $- V_{sat}$, indicating whether the input voltage is greater than or less than the reference voltage. Such type of the comparator, in which the operation is at saturation level is known as **saturating** type of comparator. Assuming symmetrical conditions, the two possible output levels of the saturating type comparator are $+ V_{sat}$ and $- V_{sat}$.

Note that no feedback is applied to the op-amp and it is operated in open loop conditions, because of which the op-amp is operating in saturating conditions.

The input and output waveforms for a basic non-inverting comparator, for sinusoidal input are shown in the Fig. 4.93.

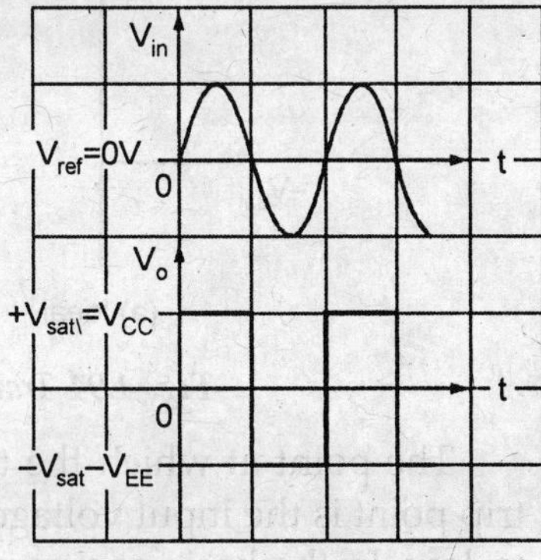

4.93 Op-amp as a Comparator.

The op-amp differential voltage gain A_{OL} is very large. So when inverting input is grounded, very small input voltage V_{in} in the range of microvolt is enough to saturate the op-amp. The $\pm V_{sat}$, the saturation voltage levels of op-amp are mentioned in the data sheet. Hence knowing V_{sat} and differential voltage gain A, we can determine the minimum input voltage level required to saturate op-amp as,

$$V_{in\,(min)} = \frac{V_{sat}}{A_{OL}} \quad \text{for saturation}$$

Now the transfer characteristics is the graph of V in and V_{out}. As A_{OL} is very large hence for very very small positive or negative V_{in}, the output saturates. Hence at $V_{in} = 0$, the transfer characteristics is almost a straight line as shown in the Fig. 4.94(*a*). For example, for 741 C op-amp, A_{OL} is 100,000 while $\pm V_{sat}$ levels are $\pm$ 13.5 V for supply of $\pm$ 15 V.

$$\therefore \quad V_{in} = \frac{V_{sat}}{A_{OL}} = \frac{\pm 13.5}{100000} = \pm 135\,\mu V$$

Thus for + 135 μV of V_{in}, output saturates to + V_{sat} while – 135 μV of V_{in} is enough to saturate output at – V_{sat}. Hence region – 135 μV to + 135 μV of V_{in} the graph of V_{in} and V_{out} is linear. But this range is so small that near V_{in} = a practically we get a straight line transfer characteristics.

The Fig. 4.94(*a*) and (*b*) shows the ideal and practical transfer characteristics of a basic non-inverting comparator.

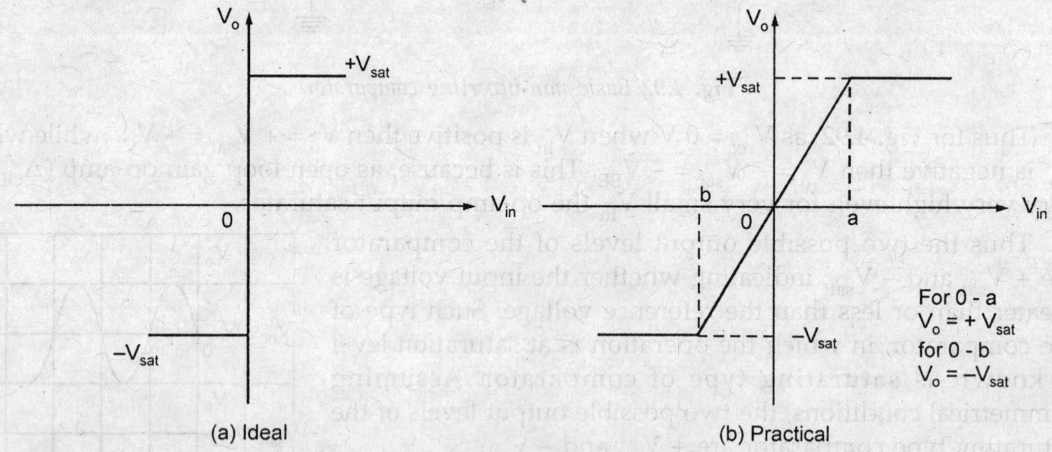

Fig. 4.94 Transfer characteristics of basic non-inverting comparator.

The point at which the transfer characteristics is straight line is called a trip point. The trip point is the input voltage at which the output changes its states from low to high or high to low. In the basic comparator this trip point is zero as at V_{in} = 0, the output changes its states.

Key Point: *So we can say that when V_{in} is greater than trip point, the output is high while if V_{in} is less than the trip point the output is low.*

As this change over occurs at V_{in} = 0, the basic comparator can be used to detect occurrence of zero in the input voltage. Hence this circuit is called zero crossing detector. But in practice it is possible to change the trip point from zero to other voltage. This is achieved by some modifications in the basic comparator circuit.

Moving a Trip Point

By application of a reference voltage to the inverting input rather than grounding it, the trip point can be moved.

The Fig. 4.95 shows the application of reference voltage to the inverting input of a basic comparator using a potential divider consisting of resistors R_1 and R_2.

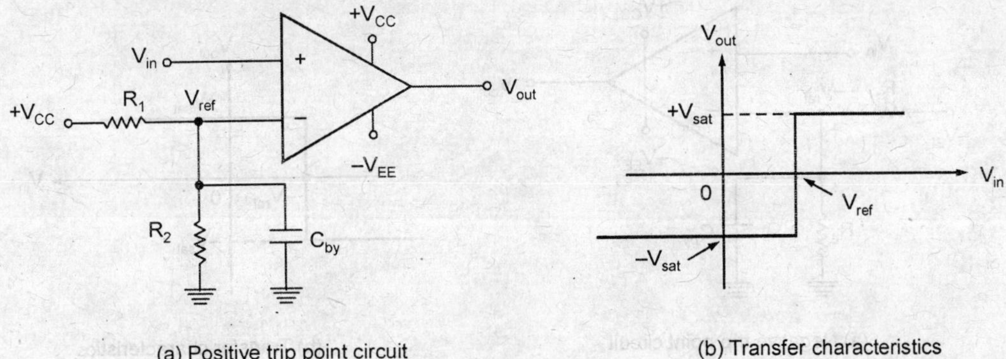

(a) Positive trip point circuit (b) Transfer characteristics

Fig. 4.95

The reference voltage V_{ref} is derived using supply $+ V_{CC}$ and potential divider R_1 and R_2. Mathematically V_{ref} is expressed as,

$$V_{ref} = \frac{+ V_{CC}}{R_1 + R_2} . R_2$$

Now as long as input voltage V_{in} is less than V_{ref}, the output is low i.e. $- V_{sat}$. When V_{in} becomes slightly greater than V_{ref}, the op-amp output becomes high i.e. $+ V_{sat}$. Thus the trip point is moved from $V_{in} = 0$ to $V_{in} = V_{ref}$ due to reference voltage applied to the inverting input terminal.

A bypass capacitor is used on the inverting input to reduce the amount of power supply ripple and noise appearing at the inverting input of op-amp. For effective bypassing of ripple and noise, the critical frequency of the bypass circuit must be much lower than the ripple frequency of power supply.

The transfer characteristics of such a comparator is shown in the Fig. 4.95(*b*) which indicates positive trip point. Such a comparator is also called a **limit detector** as it detects the particular positive level of input beyond which output goes high. The resistances R_1 and R_2 can be used to set the trip point anywhere between 0 and $+ V_{CC}$.

It is possible to obtain a negative trip point by providing a negative reference voltage to the inverting input. This is achieved by using a supply $- V_{EE}$ to the potential divider of R_1 and R_2.

$$\therefore \quad V_{ref} = \frac{- R_2}{R_1 + R_2} V_{EE}$$

This is shown in the Fig. 4.96(*a*). When V_{in} is positive than $- V_{ref}$, error voltage is positive which drives op-amp into positive saturation. When V_{in} is more negative than $- V_{ref}$, error voltage is negative which drives op-amp into negative saturation producing low output. The transfer characteristics is shown in the Fig. 4.96(*b*).

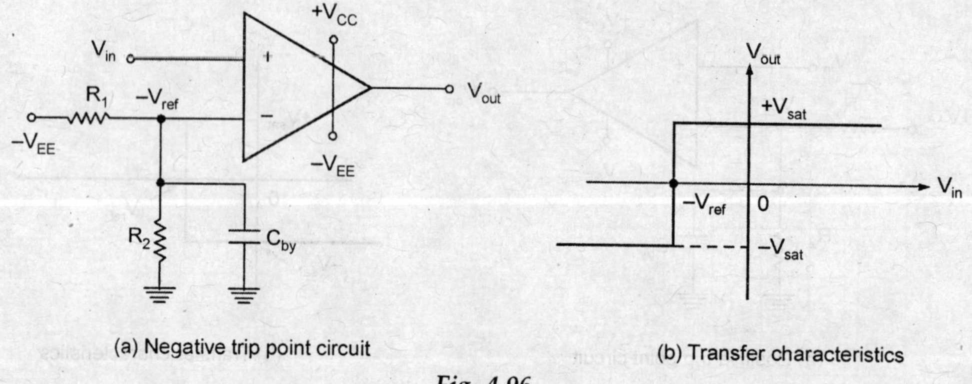

(a) Negative trip point circuit (b) Transfer characteristics

Fig. 4.96

The Fig. 4.97(a) shows input and output waveforms with positive reference voltage while the Fig. 4.97(b) shows input and output waveforms with negative reference voltage.

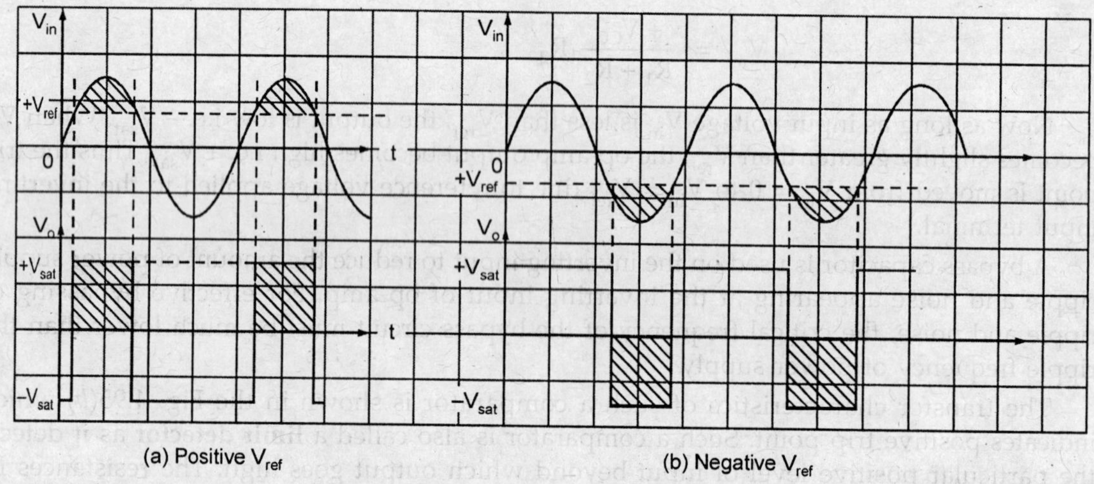

(a) Positive V$_{ref}$ (b) Negative V$_{ref}$

Fig. 4.97 Input and output waveforms.

4.33.2 Inverting Comparator

The Fig. 4.98 shows inverting comparator in which the reference voltage V_{ref} is applied to the non-inverting (+) input and signal voltage (V_{in}) is applied to the inverting (–) input of the op-amp. The V_{ref} can be set using a battery and potential divider as discussed earlier for non-inverting comparator.

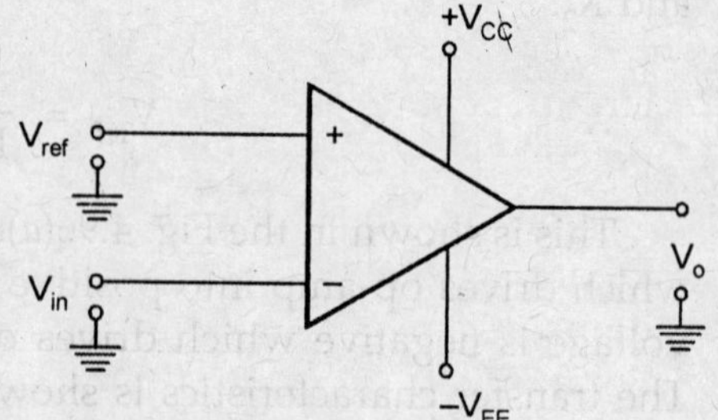

Fig. 4.98 Basic inverting comparator with $V_{ref} = 0$ V.

When V_{in} is less than V_{ref}, the output voltage V_o is at + V_{sat} ($\equiv$ + V_{CC}) because the voltage at the inverting input (–) is less than that at the non-inverting (+) input. On the other hand, when V_{in} is greater than V_{ref}, the non-inverting (+) input becomes negative with respect to the inverting (–) input and V_o goes to $- V_{sat}$ ($= - V_{EE}$). The Fig. 4.99 shows the input and output waveforms for inverting comparator.

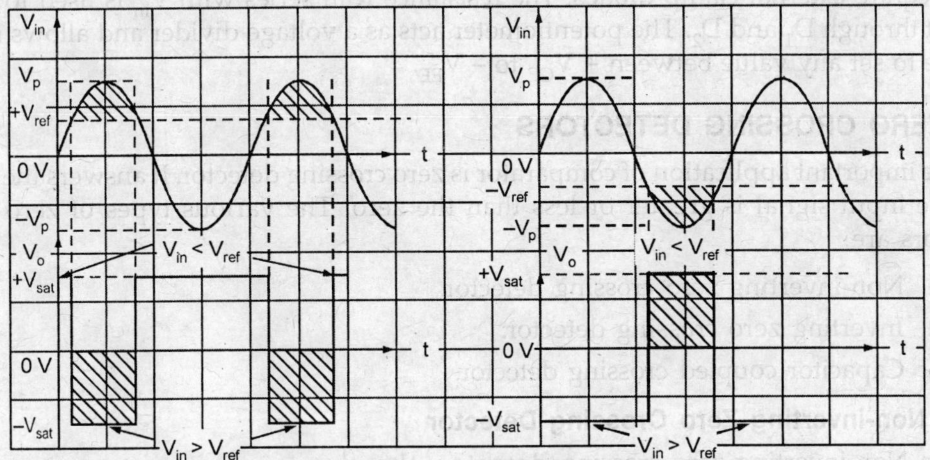

Fig. 4.99 Input and output waveforms for inverting comparator.

Transfer characteristics for inverting comparator with $+ V_{ref}$ is shown in the Fig. 4.100.

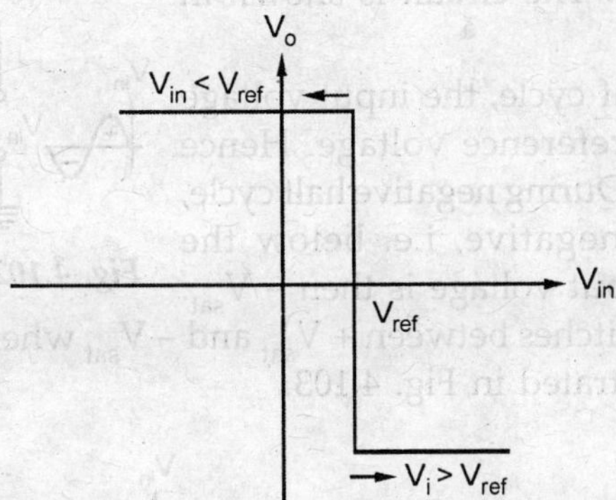

Fig. 4.100 Transfer characteristics for inverting comparator.

4.33.3 Practical Comparator

The Fig. 4.101 shows practical comparator circuit. It consists of protective diodes and potentiometer to adjust the reference voltage.

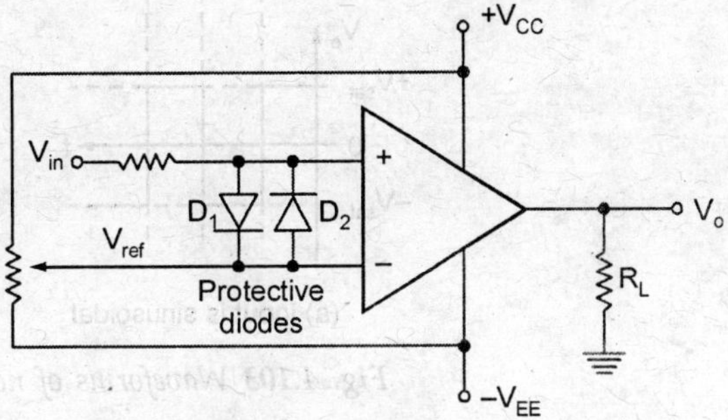

Fig. 4.101 Practical comparator circuit.

As shown in Fig. 4.101 diode D_1 and diode D_2 are connected to protect the op-amp from damage due to excessive input voltage V_{in}. Because of these diodes, the difference input voltage V_{id} is always less than 0.7 V or – 0.7 V. In case of excess input voltage, the difference input voltage V_{id} of the op-amp is clamped to either 0.7 V or – 0.7 V due to the forward biasing of one of the diodes. Hence, these diodes are also called as **clamp diodes**. Some op-amps have built-in input protection circuitry; they don't require external clamp diodes. The resistance R in series with V_{in} is used to limit the current through D_1 and D_2. The potentiometer acts as a voltage divider and allows reference voltage to set any value between + V_{CC} to – V_{EE}.

4.34 ZERO CROSSING DETECTORS

The important application of comparator is zero crossing detector. It answers the question that the input signal is greater or less than the zero. The various types of zero crossing detectors are,

1. Non-inverting zero crossing detector.
2. Inverting zero crossing detector.
3. Capacitor coupled crossing detector.

4.34.1 Non-inverting Zero Crossing Detector

In a Non-inverting zero crossing detector, the op-amp is used in open loop made. Inverting terminal of the op-amp is grounded and input is applied to the non-inverting terminal. The circuit is shown in the Fig. 4.102.

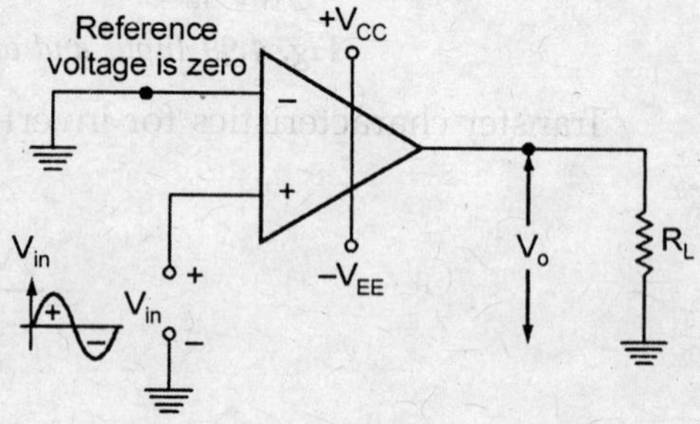

Fig. 4.102 Inverting zero crossing detector

During the positive half cycle, the input voltage is positive i.e. above the reference voltage. Hence the output voltage is + V_{sat}. During negative half cycle, the input voltage V_{in} is negative, i.e. below the reference voltage. The output voltage is then – V_{sat}. Thus the output voltage switches between + V_{sat} and – V_{sat} whenever the input signal crosses the zero level. This is illustrated in Fig. 4.103.

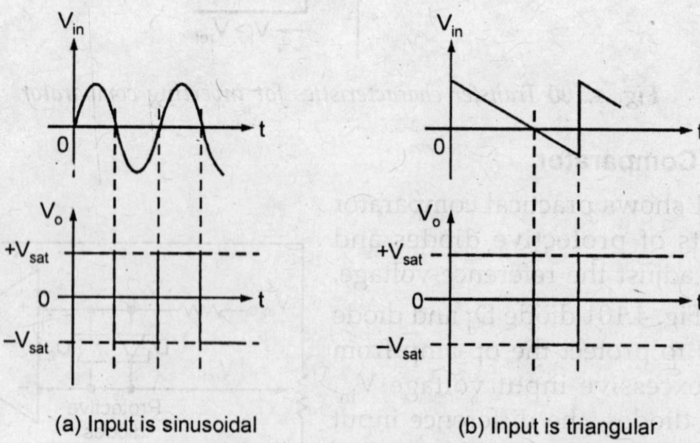

Fig. 4.103 Waveforms of non-inverting zero crossing detector.

Key Point: *From the waveforms of non-inverting zero crossing detector it can be seen that the circuit can be used as a square wave generator.*

4.34.2 Inverting Zero Crossing Detector

In the inverting zero crossing detector, input is directly applied to the inverting input terminal while the non-inverting terminal is grounded. The circuit is shown in the Fig. 4.104.

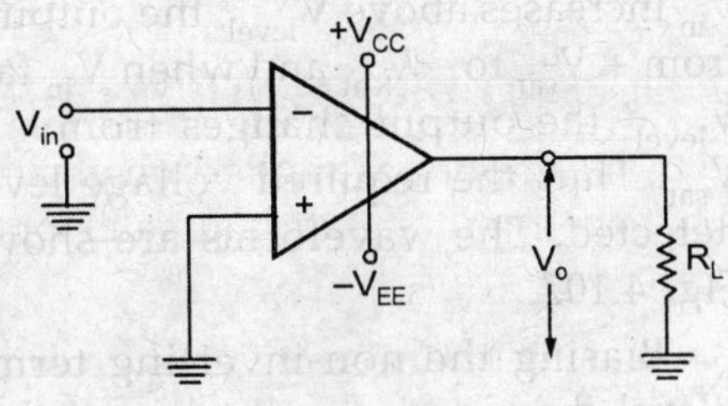

Fig. 4.104 Inverting zero crossing detector.

During positive half cycle, inverting terminal is more positive than non-inverting terminal is more positive than non-inverting terminal. So output is $-V_{sat}$. While for negative half cycle, the output is $+V_{sat}$. This circuit is called an inverter. The waveforms are shown in the Fig. 4.105.

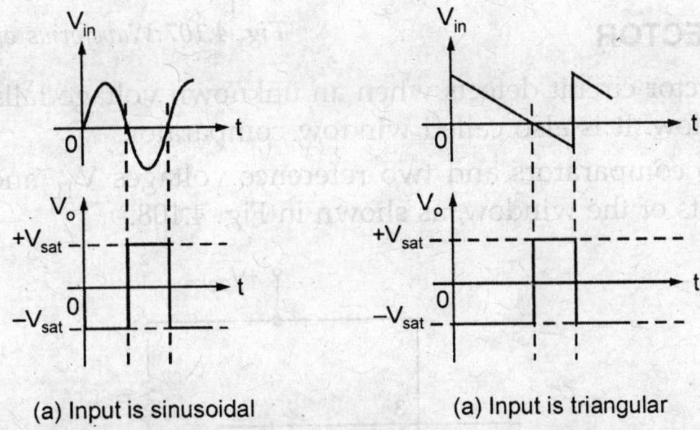

Fig. 4.105 Waveforms of inverting zero crossing detector.

4.35 VOLTAGE LEVEL DETECTOR

The Fig. 4.106 shows the voltage level detector. One of the op-amp input terminals is biased to the input voltage level which is to be detected.

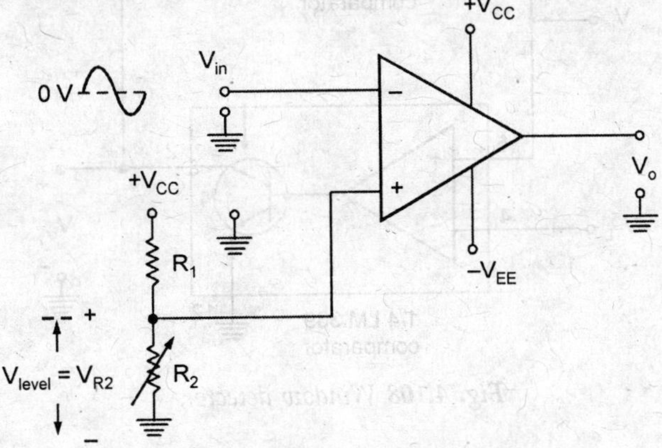

Fig. 4.106 Voltage level detector.

The non-inverting terminal is biased to V_{level} voltage which is drop across R_2 and can be set to a level to be detected.

As basically it is inverting mode, when V_{in} increases above V_{level}, the output changes from $+ V_{sat}$ to $- V_{sat}$ and when V_{in} falls below V_{level}' the output changes from $- V_{sat}$ to $+ V_{sat}$. Thus the required voltage level can be detected. The waveforms are shown in the Fig. 4.107.

Biasing the non-inverting terminal to -Vlevel the negative voltage level detection is also possible using the same circuit.

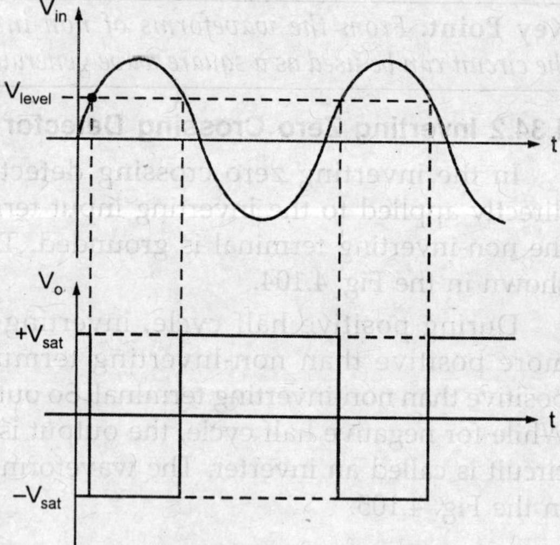

Fig. 4.107 Waveforms of voltage level detector.

4.36 WINDOW DETECTOR

The window detector circuit detects when an unknown voltage falls within a specified voltage band or window. It is also called window comparator.

It consists of two comparators and two reference voltages V_{TL} and V_{TH} defining the lower and upper limits of the window, as shown in Fig. 4.108.

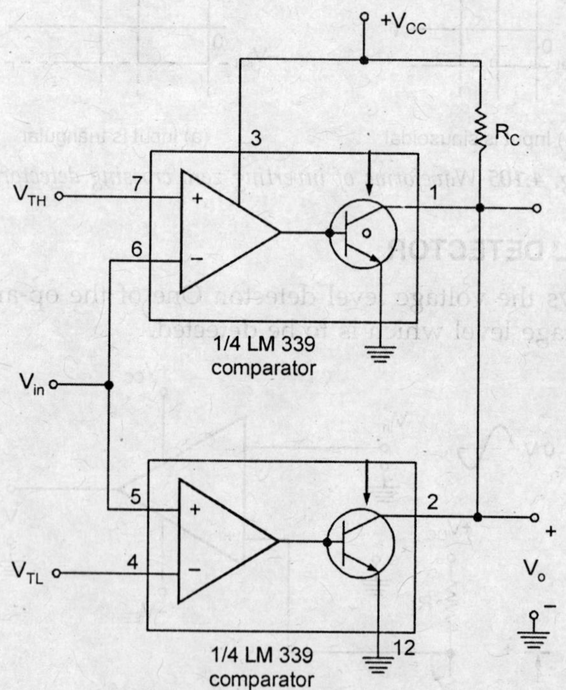

Fig. 4.108 Window detector.

When V_i is within the range $V_{TL} < V_i < V_{TH}$, the output transistors of both compactors are off and RC pulls V_o to V_{CC} to yield a high output. However, when V_i falls outside the range, the output transistor of one of the comparators will saturate (upper comparator for $V_i > V_{TH}$, lower comparator for $V_i < V_{TL}$). This brings output voltage equal to V_{CE} (sat).

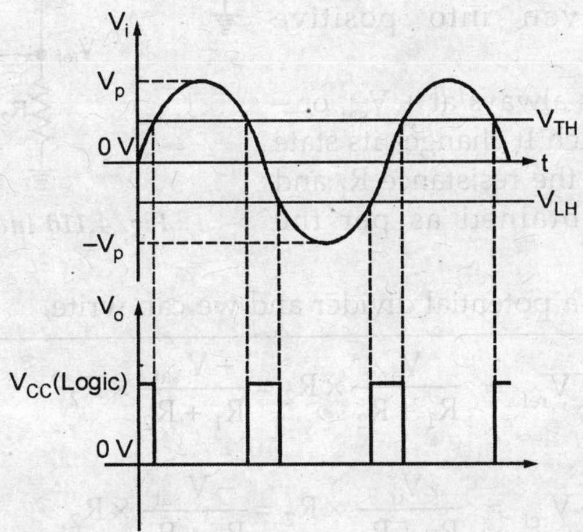

Fig. 4.109(a) Input and Output waveforms of window detector.

As we know that, V_{CE} (sat) is of the order of fraction of volts, by neglecting V_{CE} (sat), output voltage is 0 volt. This is illustrated in Fig. 4.109(*a*). The Fig. 4.109(*b*) shows the transfer characteristics for window detector.

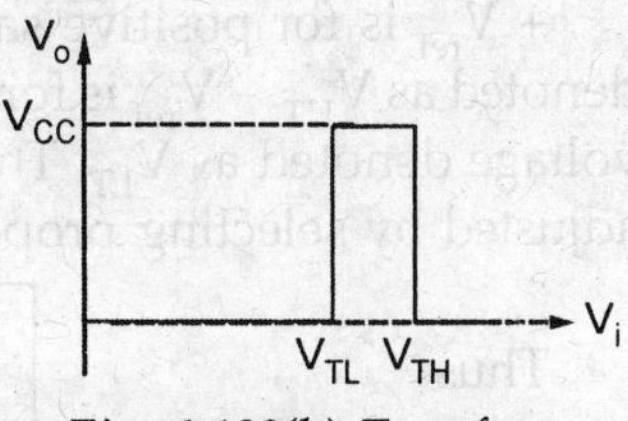

Fig. 4.109(b) Transfer characteristics for window detector.

4.37 SCHMITT TRIGGER [REGENERATIVE COMPARATOR]

We have seen that in a basic comparator, a feedback is not used and the op-amp is used in the open loop mode. As open loop gain of op-amp is large, very small noise voltages also can cause triggering of the comparator, to change its state. Such a false triggering may cause lot of problems in the applications of comparator as zero crossing detector. This may give a wrong indication of zero crossing due to zero crossing of noise voltage rather than zero crossing of input wanted signal. Such unwanted noise causes the output to jump between high and low states. The comparator circuit used to avoid such unwanted triggering is called regenerative **comparator** or **Schmitt** trigger, which basically uses a positive feedback.

4.37.1 Inverting Schmitt Trigger

The Fig. 4.110 shows the basic Schmitt trigger circuit. As the input is applied to the inverting terminal, it is also called inverting Schmitt trigger circuit. The inverting mode produces opposite polarity output. This is fed back to the non-inverting input which is of same polarity as that of output. This ensures positive feedback.

When V_{in} is slightly positive than V_{ref}, the output gets driven into negative saturation at $-V_{sat}$ level.

When V_{in} becomes more negative than $-V_{ref}$, then output gets driven into positive saturation at $+V_{sat}$ level.

Thus output voltage is always at $+V_{sat}$ or $-V_{sat}$ but the voltage at which It changes its state now can be controlled by the resistance R_1 and R_2. Thus V_{ref} can be obtained as per the requirement.

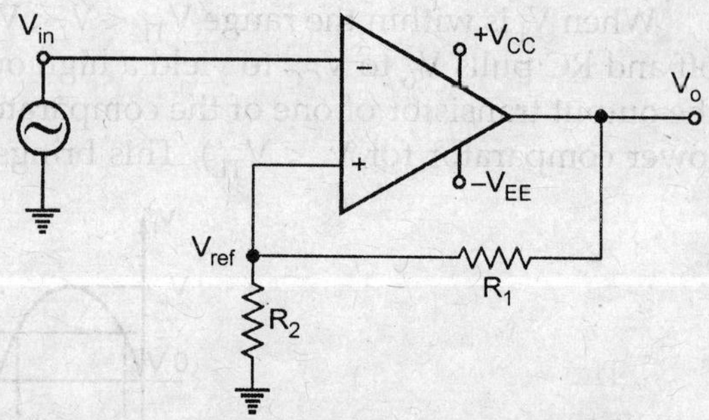

Fig. 4.110 Inverting Schmitt trigger.

Now R_1 and R_2 forms a potential divider and we can write,

$$+V_{ref} = \frac{V_o}{R_1 + R_2} \times R_2 = \frac{+V_{sat}}{R_1 + R_2} \times R_2 \quad \text{...positive saturation}$$

$$-V_{ref} = \frac{V_o}{R_1 + R_2} \times R_2 = \frac{-V_{sat}}{R_1 + R_2} \times R_2 \quad \text{...negative saturation}$$

$+V_{ref}$ is for positive saturation when $V_o = +V_{sat}$ and is called upper threshold voltage denoted as V_{UT}. $-V_{ref}$ is for negative saturation when $V_o = -V_{sat}$ and is called lower threshold voltage denoted as V_{LT}. The values of these threshold voltage levels can be determined and adjusted by selecting proper values of R_1 and R_2.

Thus $$V_{UT} = \frac{+V_{sat} R_2}{(R_1 + R_2)}$$

and $$V_{LT} = \frac{-V_{sat} R_2}{(R_1 + R_2)}$$

The output voltage remains in a given state until the input voltage exceeds the threshold voltage level either positive or negative.

The Fig. 4.111 shows the graph of output voltage against input voltage. This is called transfer characteristics of Schmitt trigger.

The graph indicates that once the output changes its state, it remains there indefinitely until the input voltage crosses any of the threshold voltage levels. This is called hysteresis of Schmitt trigger. The hysteresis is also called **dead band** or **dead zone.**

The difference between V_{UT} and V_{LT} is called width of the hysteresis denoted as H.

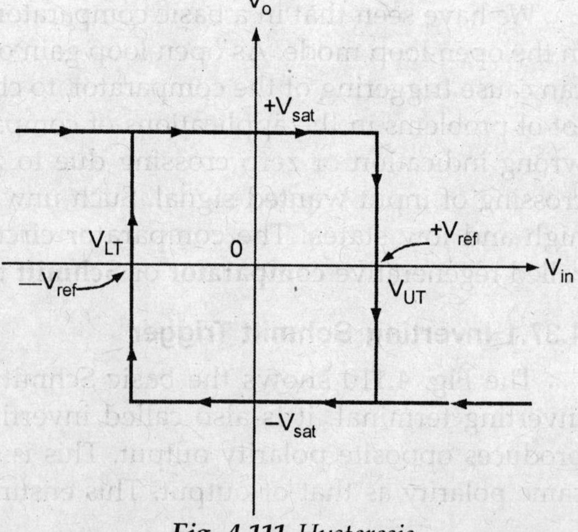

Fig. 4.111 Hysteresis.

$$H = V_{UT} - V_{LT}$$

$$= \frac{2V_{sat} R_2}{R_1 + R_2} - \left[\frac{-V_{sat} R_2}{R_1 + R_2}\right]$$

$$= \frac{2V_{sat} R_2}{R_1 + R_2}$$

The Schmitt trigger eliminates the effect of noise voltage present the noise voltages less than the hysteresis H, cannot cause triggering. As for positive V_{in} greater than V_{UT}, the output becomes - V_{sat} and for negative Vin less than VLT' the output becomes + V_{sat} this is called **inverting Schmitt trigger.**

In short,

$V_{in} < V_{LT'}$	$V_o = +V_{sat}$
$V_{in} < V_{UT'}$	$V_o = -V_{sat}$
$V_{LT} < V_{in} < V_{UT'}$	V_o = Previous state achieved

If input applied is purely sinusoidal, the input and output waveforms for inverting Schmitt trigger can be shown as in the Fig. 4.112.

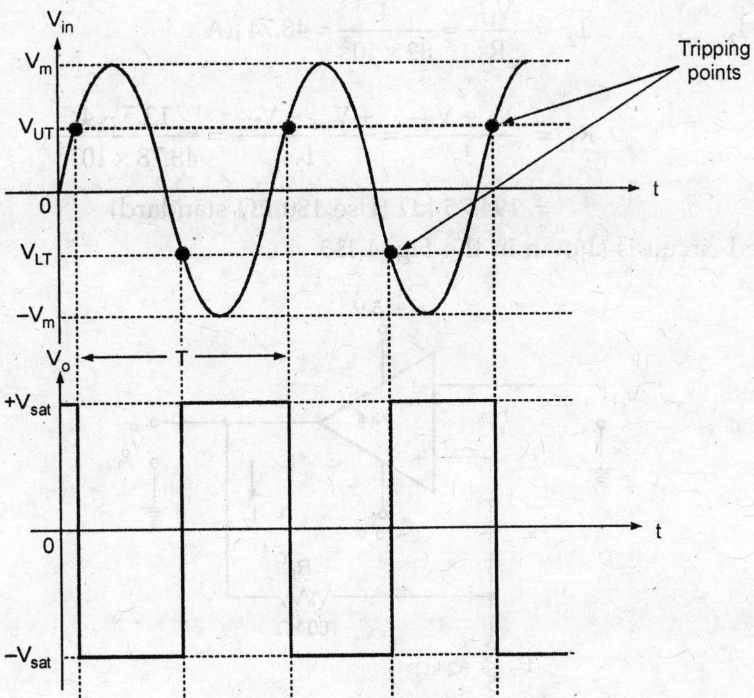

Fig. 4.112 Input and output waveforms.

Design steps: Let the current through R_1 and R_2 is I_2. The op-amp input current is assumed zero.

$$\boxed{R_2 = \frac{\text{Trigger voltage}}{I_2}}$$

where I_2 is selected much higher than op-amp input bias current.

While $$\boxed{R_1 = \frac{V_o - (\text{Trigger voltage})}{I_2}} \qquad ...V_o = V_{sat}$$

For BiFET op-amps, select largest of the two resistors as 1 MΩ and then calculate other.

Example 4.21: Design an inverting Schmitt trigger to have trigger voltages of ± 4 V. Use op-amp 741 with supply of ± 15 V.

Solution: $V_{UT} = +4\text{ V}, \quad V_{LT} = -4\text{ V}, \quad \text{Supply} = \pm 15\text{ V}$

$$\pm V_{sat} = 0.9 \times [\text{Supply}] = \pm 13.5\text{ V} = V_o$$

For op-amp 741, $I_{B(max)} = 500\text{ nA}$

$$\therefore \quad I_2 = 100\, I_{B(max)} = 50\ \mu\text{A}$$

$$\therefore \quad R_2 = \frac{V_{UT}}{I_2} = \frac{4}{50 \times 10^{-6}} = 80\text{ k}\Omega \quad \text{(Use 82 k}\Omega\text{ standard)}$$

Recalculate I_2, $$I_2 = \frac{V_{UT}}{R_2} = \frac{4}{82 \times 10^3} = 48.74\ \mu\text{A}$$

$$\therefore \quad R_1 = \frac{V_o - V_{UT}}{I_2} = \frac{+V_{sat} - V_{UT}}{I_2} = \frac{13.5 - 4}{48.78 \times 10^{-6}}$$

$$= 194.75\text{ k}\Omega \text{ (Use 180 k}\Omega\text{ standard)}$$

The designed circuit is shown in the Fig. 4.113.

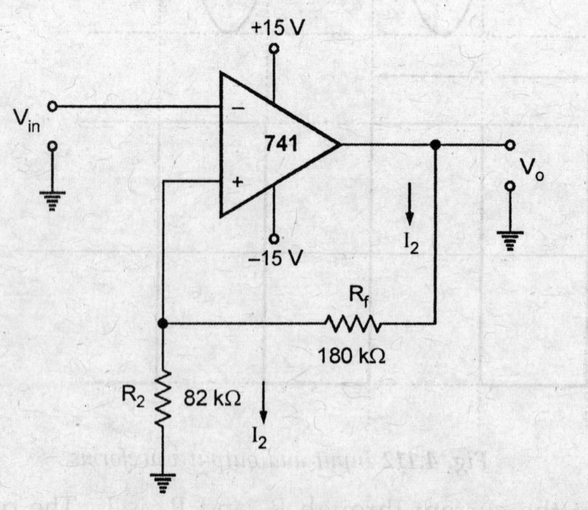

Fig. 4.113

4.37.2 Non-inverting Schmitt Trigger

The Fig. 4.114 shows the non-inverting Schmitt trigger circuit. The input is applied to the non-inverting input terminal of the op-amp.

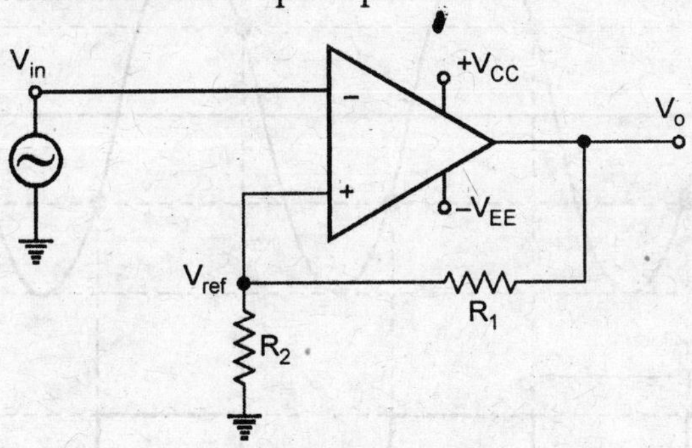

Fig. 4.114 Non-inverting Schmitt trigger.

To understand the working of the circuit, let us assume that the output is positively saturated i.e. at + V_{sat}. This is fed back to the non-inverting input through R_1. This is a positive feedback.

Now though V_{in} is decreased, the output continues its positive saturation level unless and until the input becomes more negative than V_{LT}. At lower threshold, the output changes its state from positive saturation + V_{sat} to negative saturation – V_{sat}. It remains in negative saturation till Vin increases beyond its upper threshold level V_{UT}. The transfer characteristics is shown in the Fig. 4.115.

Fig. 4.115 Hysteresis.

Now $\quad V_A$ = Voltage at point

$$A = I_{in} R_2 = V_{UT}$$

As op-amp input current is zero, I_{in} entirely passes through R_1.

$$\therefore \quad I_{in} = \frac{V_o}{R_1} = \frac{+V_{sat}}{R_1}$$

$$\therefore \quad V_{UT} = I_{in} R_2 = \frac{R_2}{R_1}(+V_{sat}) = V_{sat}\frac{R_2}{R_1}$$

and

$$V_{LT} = \frac{R_2}{R_1}(-V_{sat}) = -V_{sat}\frac{R_2}{R_1}$$

and

$$H = V_{UT} - V_{LT} = 2V_{sat}\frac{R_2}{R_1}$$

If sinusoidal input is applied to the non-inverting Schmitt trigger, the input and output waveforms can be shown as in the Fig. 4.116.

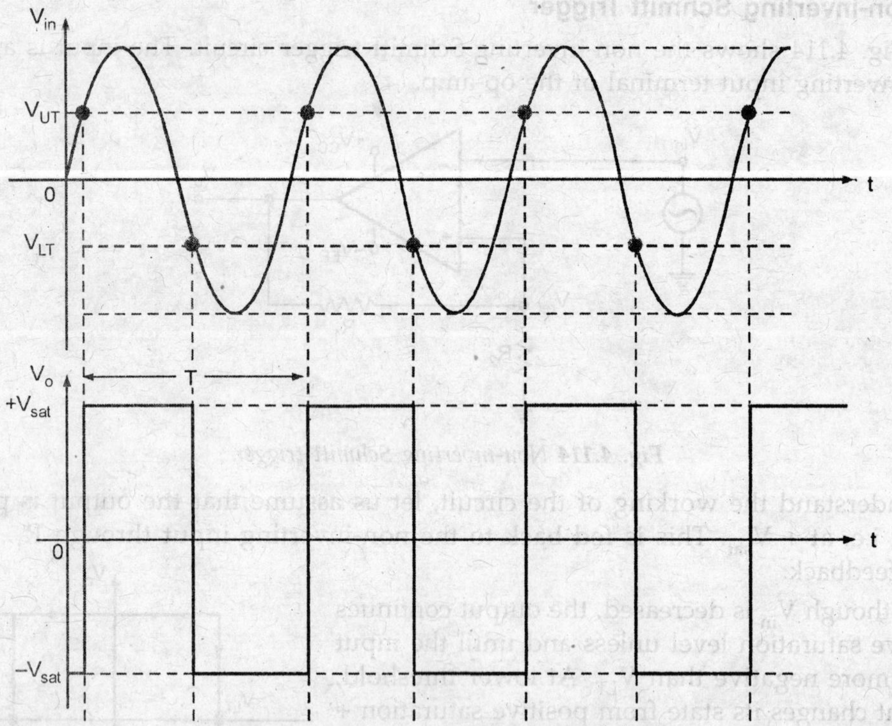

Fig. 4.116 Input and output waveforms.

Example 4.22: For a Schmitt trigger shown in the Fig. 4.117, calculate threshold voltage levels and hysteresis. Assume $V_{sat} = 0.9\ V_{CC}$.

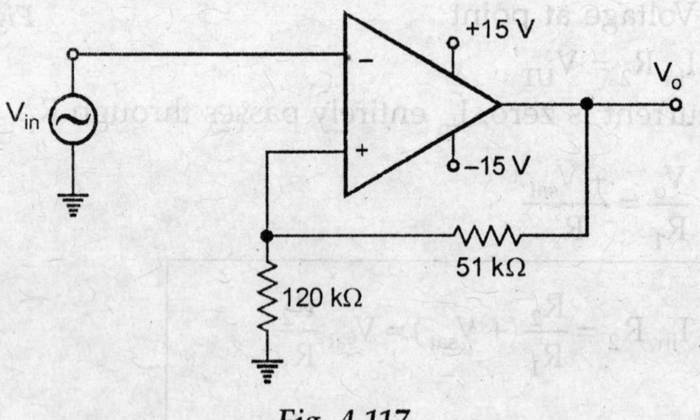

Fig. 4.117

Solution:

$$V_{CC} = +15\ V$$

$$\therefore \quad V_{sat} = 0.9\ V_{CC} = 0.9 \times 15 = 13.5\ V$$

$$R_1 = 51\ k\Omega, \quad R_2 = 120\ \Omega$$

$$V_{UT} = \frac{+V_{sat}\ R_2}{R_1 + R_2} = \frac{13.5 \times 120}{51 \times 10^3 + 120} = \mathbf{0.03169\ V}$$

$$V_{LT} = \frac{-V_{sat}\ R_2}{R_1 + R_2} = \frac{-13.5 \times 120}{51 \times 10^3 + 120} = \mathbf{-0.03169\ V}$$

$$H = V_{UT} - V_{LT} = 0.03169 - (-0.03169)$$
$$= 0.06338\ V = \mathbf{63.38\ mV}$$

Example 4.23: For a non-inverting regenerative comparator shown in the Fig. 4.118, calculate tripping voltages. Assume $V_{sat} = \pm 13.5$ V.

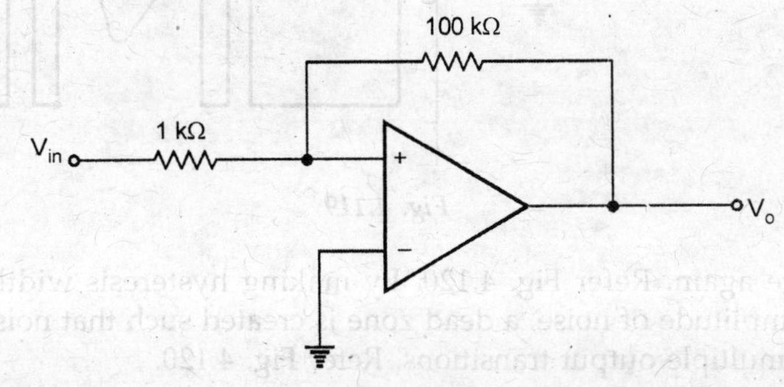

Fig. 4.118

Solution: As input is applied to the non-inverting terminal, the circuit is non-inverting Schmitt trigger.

$R_1 = 100\ k\Omega$, $R_2 = 1\ k\Omega$

$$\therefore \quad V_{UT} = +V_{sat}\frac{R_2}{R_1} = 13.5 \times \frac{1}{100} = \mathbf{0.135\ V}$$

$$V_{LT} = -V_{sat}\frac{R_2}{R_1} = \frac{-13.5 \times 1}{100} = \mathbf{-0.135\ V}$$

4.37.3 Schmitt Trigger Applications

We have already seen one important application of Schmitt trigger as sine to square wave converter. It can be used to eliminate comparator chatter in signal shaping and in ON/OFF control. It is a building block of relaxation oscillators.

4.37.3.1 Schmitt Triggers for Eliminating Comparator Chatter

Chattering can be defined as production of multiple output transition as the input signal swings through the threshold region of a comparator. This happens due to the fact that a.c. noise is present in the practical circuits. Fig. 4.119 shows input signal with a.c. noise and how comparator output chatters. Even if noise is very less, it takes a very small noise spike to cause chatter due to high comparator gains.

This problem can be solved by using Schmitt trigger circuits. This is because they exhibit hysteresis. In case of hysteresis, as soon as the input signal crosses the present threshold level once, the output changes its state and activates the other threshold level, so that the input signal must swing back to the new threshold in order to make the output of the circuit

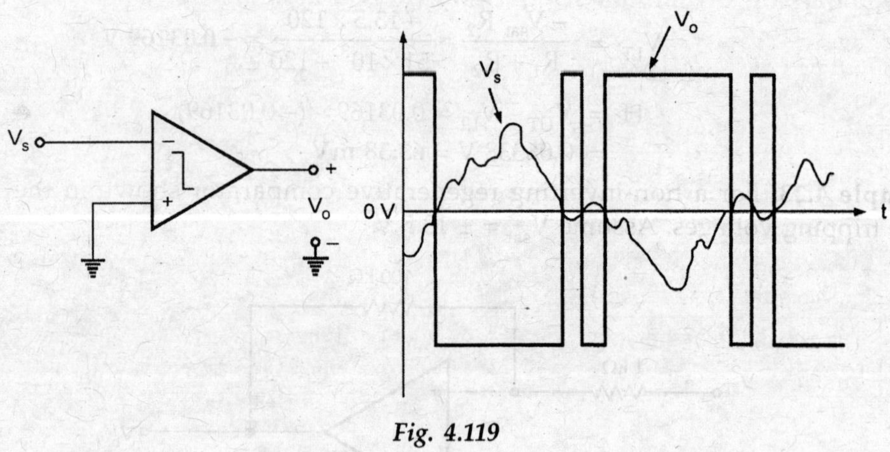

Fig. 4.119

to change its state again. Refer Fig. 4.120. By making hysteresis width greater than the maximum peak amplitude of noise, a dead zone is created such that noise within this zone no longer causes multiple output transitions. Refer Fig. 4.120.

4.37.3.2 Schmitt Triggers in ON/OFF Controllers

In ON/OFF controller, such as temperature controller, when temperature is below setpoint heater is made ON and when it is above setpoint heater is made OFF. Now consider that temperature is just below the setpoint and heater is ON. As soon as the temperature reaches the setpoint, the comparator which is controlling the heater makes it OFF. The smallest temperature drop following the heater OFF will make comparator to go in its activate state and switch on the heater. As a result, the heater will be cycled ON and OFF by the comparator at a rapid pace. This is not desirable. Temperature usually need not be regulated to such a sharp degree. By allowing a dead band of a few degrees we can ensure a comfortable environment and reduce heater cycling significantly. This is possible by introducing hysteresis in the comparator.

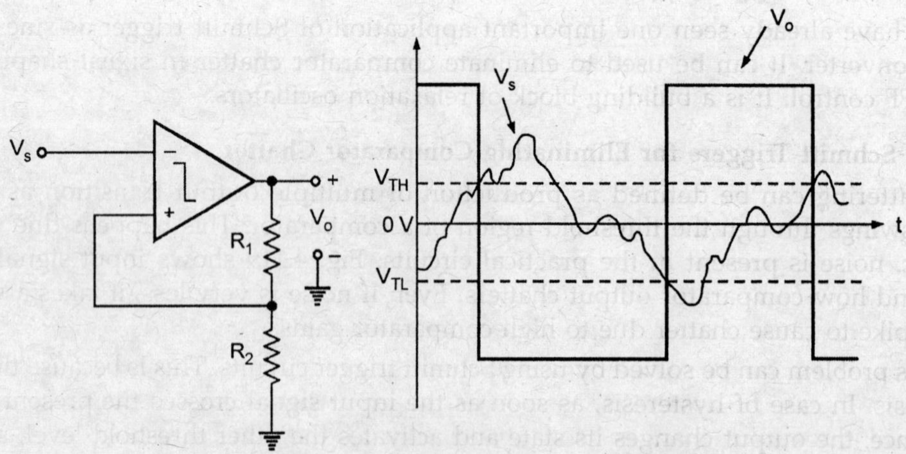

Fig. 4.120

4.37.4 Comparison of Schmitt Trigger and Comparator

The various differences and similarities between Schmitt trigger and comparator are given in the following table.

Sr. No.	Schmitt trigger	Comparator
1.	The feedback is used.	The feedback is not used.
2.	The op-amp used is in closed loop mode.	The op-amp used is in open loop mode.
3.	False triggering due to noise voltages is not possible due to feedback.	False triggering due to noise voltages is possible.
4.	The two different thershold voltages exist as V_{UT} and V_{LT}.	A single reference voltage exists which acts as triggering voltage. i.e. V_{ref} or $-V_{ref}$.
5.	Hysteresis exists with a width $H = V_{UT} - V_{LT}$.	The hysteresis does not exists.
6.	The output voltage is either $+V_{sat}$ or $-V_{sat}$.	The output voltage is either $+V_{sat}$ or $-V_{sat}$.
7.	It can be used as inverting Schmitt trigger or non-inverting Schmitt trigger.	It can be used as inverting comparator or non-inverting comparator.
8.	The various applications are square wave generators, in ON-OFF controllers, eliminating comparator chatter and to convert any waveform to square wave.	The various applications are zero crossing detector, window detector etc.

4.38 SCHMITT TRIGGER WITH DIFFERENT UTP AND LTP LEVELS

Uptill now Schmitt trigger with equal positive and negative values of UTP and LTP values is studied. In practice, it is required to have different UTP and LTP levels for the Schmitt trigger.

In such a case, the basic Schmitt trigger circuit is modified as shown in the Fig. 4.121.

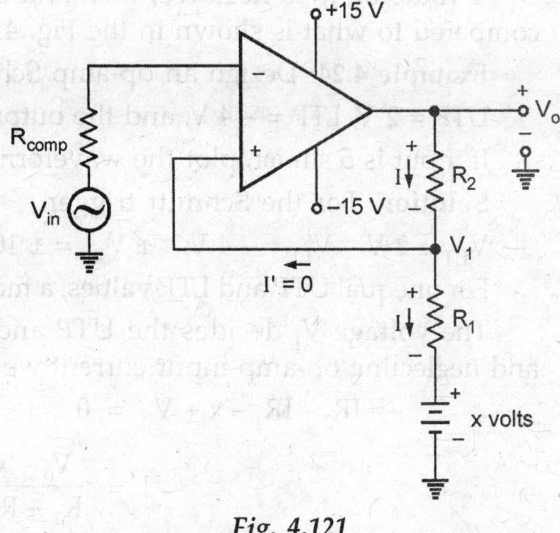

Fig. 4.121

An additional voltage source of 'x' volts is connected in series between R_1 and ground. The voltage V_1 now depends on 'x' and decides the UTP and LTP levels. For $\pm V_{sat}$, two different values of V_1 can be obtained. Thus the circuit acts as a Schmitt trigger with different UTP and LTP levels. For designing such circuit, apply KVL to the output circuit, neglecting op-amp input current.

$$-IR_2 - IR_1 - x + V_o = 0$$

$$\therefore \quad I = \frac{V_o - x}{R_1 + R_2} \quad \text{...(1)}$$

And $$V_1 = IR_1 + x \quad ...(2)$$

$$\therefore \quad V_1 = \frac{(V_o - x)R_1}{R_1 + R_2} + x \quad ...(3)$$

For $V_1 = V_{UT}$, $V_o = +V_{sat}$

$$\therefore \quad \boxed{V_{UT} = \frac{(+V_{sat} - x)\,R_1}{R_1 + R_2} + x} \quad ...(4)$$

For $V_1 = V_{LT}$, $V_o = -V_{sat}$

$$\therefore \quad \boxed{V_{LT} = \frac{(+V_{sat} - x)\,R_1}{R_1 + R_2} + x} \quad ...(5)$$

Subtracting equation (5) from equation (4),

$$\boxed{(V_{UT} - V_{LT}) = \frac{2V_{sat}R_1}{R_1 + R_2}} \quad ...(6)$$

This gives the relation between R_1 and R_2. Substituting $(R_I + R_2)$ interms of R_1 in equation (4), the value of x can be obtained. Then by choosing R_1, the value of R_2 can be obtained.

Finally $$R_{comp} = R_1 || R_2$$

If value of 'x' is negative, it should be connected with opposite polarity in the circuit, compared to what is shown in the Fig. 4.121.

Example 4.24: Design an op-amp Schmitt trigger with the following specifications: UTP = 2 V, LTP = – 4 V, and the output swings between ± 10 V.

If input is 5 sin ωt, plot the waveforms of input and output.

Solution: For the Schmitt trigger,

$V_{UT} = 2$ V, $V_{LT} = -4$ V, $\pm V_{sat} = \pm 10$ V

For unequal UTP and LTP values, a modified circuit is required as shown in the Fig. 4.122.

The voltage V_I decides the UTP and LTP levels. Applying KVL to the output circuit and neglecting op-amp input current we can write,

$$-IR_2 - IR_1 - x + V_o = 0$$

$$\therefore \quad I = \frac{V_o - x}{R_1 + R_2} \quad ...(7)$$

And $$V_1 = IR_1 + x \quad ...(8)$$

$$\therefore \quad V_1 = \frac{V_o - x}{R_1 + R_2}.R_1 + x \quad ...(9)$$

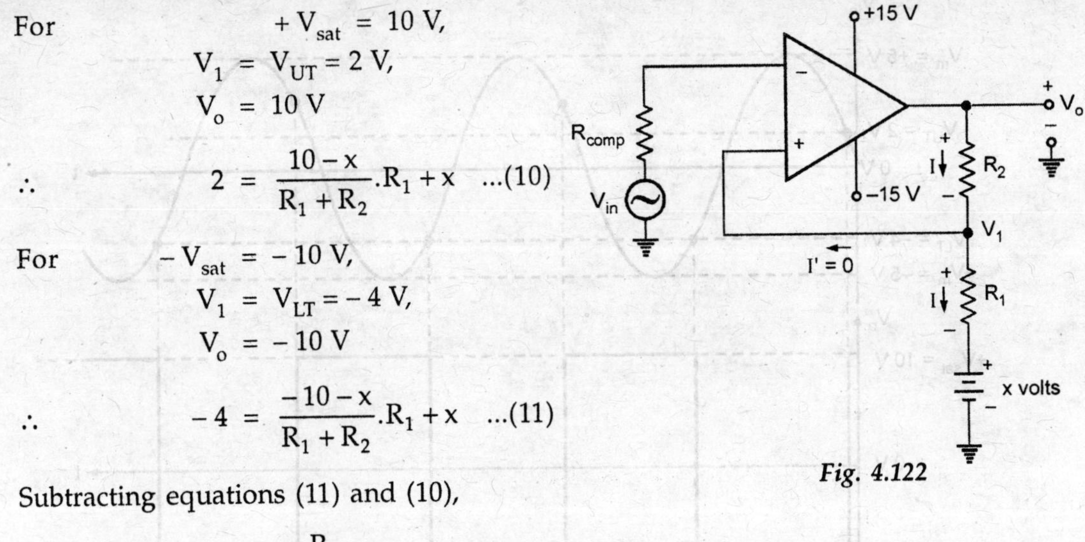

Fig. 4.122

For $+V_{sat} = 10$ V,

$V_1 = V_{UT} = 2$ V,

$V_o = 10$ V

$$\therefore \quad 2 = \frac{10-x}{R_1+R_2}.R_1 + x \quad ...(10)$$

For $-V_{sat} = -10$ V,

$V_1 = V_{LT} = -4$ V,

$V_o = -10$ V

$$\therefore \quad -4 = \frac{-10-x}{R_1+R_2}.R_1 + x \quad ...(11)$$

Subtracting equations (11) and (10),

$$[2-(-4)] = \frac{R_1}{R_1+R_2}[10-x+10+x]$$

$$\therefore \quad 6 = \frac{20R_1}{R_1+R_2}$$

$$\therefore \quad R_1 + R_2 = 3.333\,R_1 \quad ...(12)$$

$$\therefore \quad R_2 = 2.333\,R_1 \quad ...(13)$$

Substituting equation (13) in equation (10),

$$2 = \frac{(10-x)R_1}{3.333R_1} + x$$

$$\therefore \quad 2 \times 3.333 = 10 - x + 3.333\,x$$

$$\therefore \quad 2.333\,x = -3.3334$$

$$\therefore \quad x = -1.4286$$

So actually polarity of the voltage source 'x' must be opposite to what is assumed earlier as shown in the Fig. 4.122.

Choose $R_1 = 1\ k\Omega$ hence $R_2 = 2.333\ k\Omega$

$$\therefore \quad R_{comp} = R_1 || R_2 = 0.7\ k\Omega$$

Now as long as V_{in} is less than V_{UT}, the output is at $+V_{sat} = 10$ V and when $V_{in} > V_{UT}$, the output switches from $+V_{sat}$ to $-V_{sat}$. While as long as $V_{in} > V_{LT}$, the output is at $-V_{sat} = -10$ V and when $V_{in} < V_{LT}$, the output switches from $-V_{sat}$ to $+V_{sat}$.

For the given input, the input and output waveforms are as shown in the Fig. 4.123.

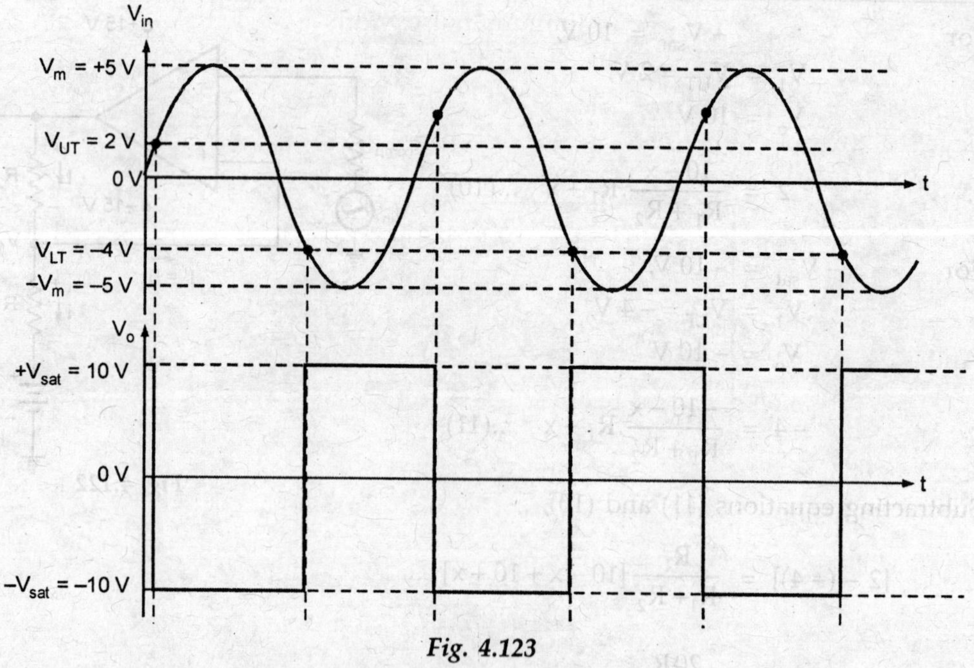

Fig. 4.123

4.38.1 Another Method of Obtaining Different Trigger Levels

The Fig. 4.124 shows another method of obtaining different trigger levels for an inverting Schmitt trigger circuit.

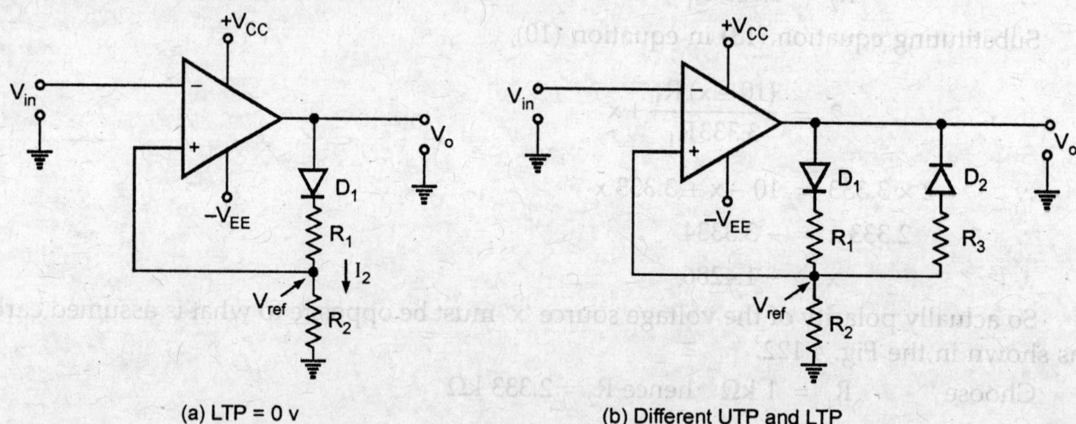

Fig. 4.124 Modified inverting Schmitt trigger circuits.

When the output is positive, D_1 gets forward biased and V_{UT} is the drop across R_2.

When the output is negative, D_1 is reverse biased and only op-amp input current flows through R_2. Hence V_{LT} is almost zero.

The diode D_1 must be selected to have reverse breakdown voltage more than the supply voltage. Its reverse recovery time must be smaller than the minimum pulse width of the input signal.

$$t_{rr} \text{ (for diode)} \leq \frac{\text{Minimum pulse width}}{10} \qquad ...(10)$$

Using two diodes and two resistors as shown in the Fig. 4.124 (*b*) different UTP and LTP levels can be selected. When the output is positive D_1 is forward biased while when the output is negative D_2 is forward biased.

$$\therefore \qquad \text{UTP} = V_{UT} = \frac{[|V_o| - V_F] \times R_2}{R_1 + R_2} \qquad ...(15)$$

And

$$\text{LTP} = V_{LT} = \frac{[|V_o| - V_F] \times R_2}{R_2 + R_3} \qquad ...(16)$$

where V_F is the forward voltage drop of D_1.

Example 4.25: The op-amp comparator circuit is shown in the Fig. 4.125.

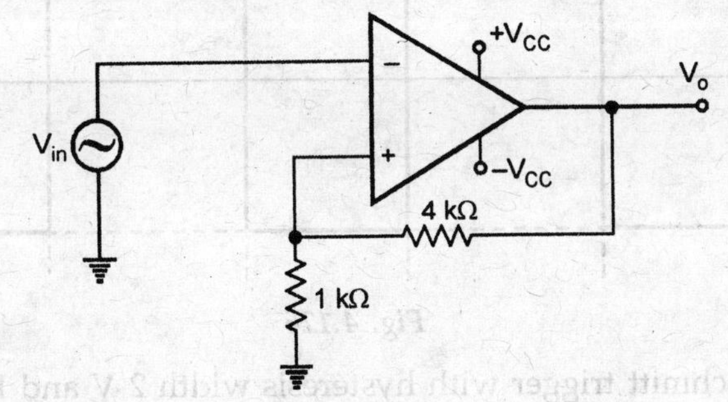

Fig. 4.125

Assume $\pm V_{CC} = \pm 12$ V, $V_{sat} = 0.9\ V_{CC}$

If a sine wave of 10 V is applied, calculate threshold levels and plot input output waveforms.

Solution: The given comparator is inverting Schmitt trigger with $R_1 = 4$ kΩ and $R_2 = 1$ kΩ.

$$V_{sat} = 0.9\ V_{CC} = 0.9 \times 12 = 10.8 \text{ V}$$

The threshold levels are,

$$V_{UT} = \frac{+V_{sat}\ R_2}{R_1 + R_2}$$

$$= \frac{10.8 \times 1}{(1+4)} = \mathbf{2.16\ V}$$

$$V_{LT} = \frac{-V_{sat}\ R_2}{R_1 + R_2}$$

$$= \frac{-10.8 \times 1}{(1+4)} = \mathbf{-2.16\ V}$$

$$V_{in} = 10 \text{ V sine wave}$$

So input and output waveforms are,

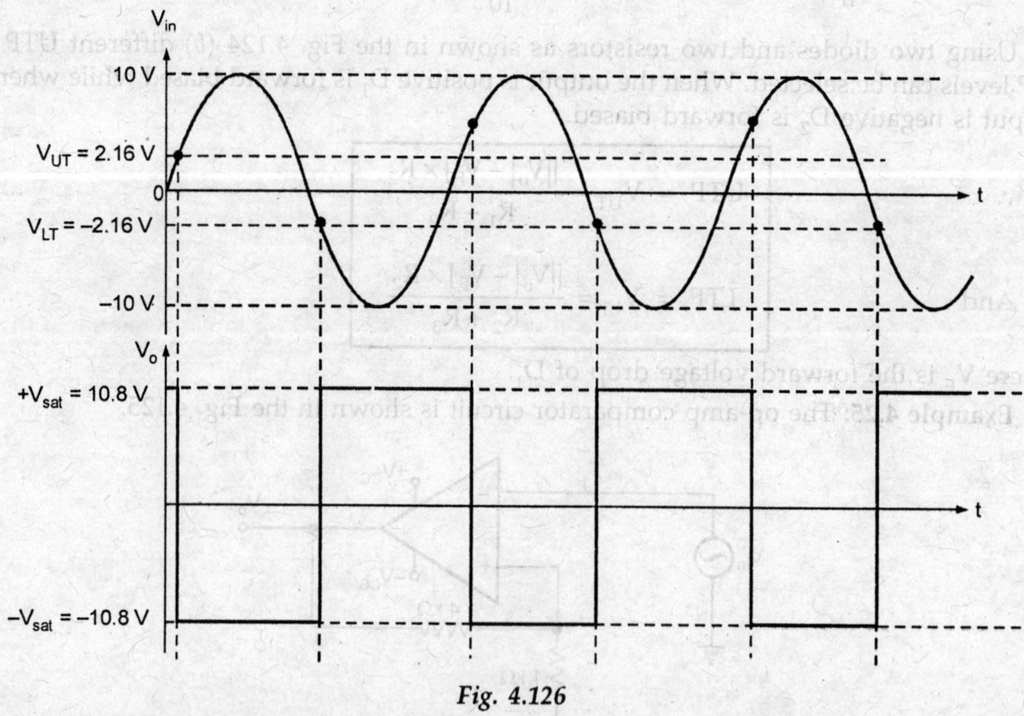

Fig. 4.126

Example 4.26: A Schmitt trigger with hysteresis width 2 V and LTP of – 1.5 V converts a 1 kHz sine wave of 5 V peak into a square wave. Calculate the time duration of the negative and positive portion of the output waveform.

Solution: Let us draw the input and output waveforms as shown in the Fig. 4.127.

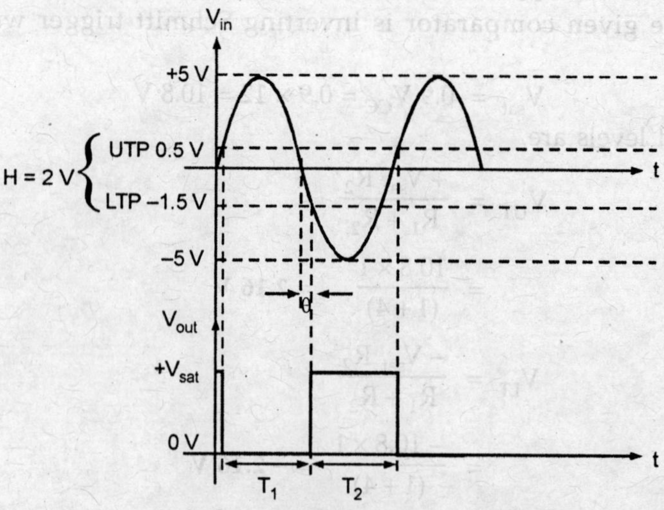

Fig. 4.127

$$LTP = -1.5 \text{ V and } H = 2 \text{ V}$$

Now $\quad H = UPT - LTP$

$\therefore \quad 2 = UTP - (-1.5)$

$\therefore \quad UTP = 0.5 \text{ V}$

$$V_p = 5 \text{ V}$$

In the Fig. 4.127, the angle θ can be obtained from equation of sine wave. Sine wave is represented as,

$$V_{in} = V_p \sin(\pi + \theta) \text{ when } \pi < \omega t < 2\pi$$

At LTP, $\quad -1.5 = 5 \sin(\pi + \theta)$

$$= -5 \sin\theta$$

$\therefore \quad \sin\theta = 0.3$

$\therefore \quad \theta = 17.45°$

The time period of sine wave is,

$$T = \frac{1}{f} = \frac{1}{1 \times 10^3} = 1 \text{ ms}$$

At UTP, $\quad 0.5 = V_p \sin\theta$

$\therefore \quad 0.5 = 5 \sin\theta$

$\therefore \quad \sin\theta = 0.1$

$\therefore \quad \theta = 5.739°$

The time T_1 for output is from 5.739° to (180° + 17.45°)

$\therefore \quad T_1 = 197.45° - 5.739° = 191.71°$

i.e. $\quad T_1 = \frac{191.71° \times 1 \text{ ms}}{360°} = \mathbf{0.5325 \text{ ms}} = 0.5325 \text{ ms}$

and $\quad T_2 = T - T_1 = 1 - 0.5325$

$$= \mathbf{0.4674 \text{ ms}}$$

Example 4.27: For the circuit shown in the Fig. 4.128 calculate the values of R_1 and R_2 if saturation voltages are + 12 V and – 12 V. Assume hysteresis width = 6 V.

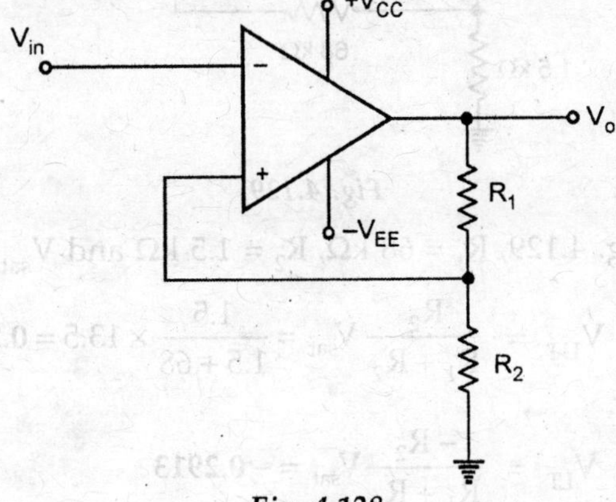

Fig. 4.128

Solution: Given $+V_{sat} = 12$ V, $-V_{sat} = -12$ V, $V_H = 6$ V

We know that hysteresis width is given as,

$$V_H = \frac{R_2}{R_1 + R_2}[+V_{sat} - V_{sat}]$$

$$\therefore \quad \frac{R_2}{R_1 + R_2} = \frac{V_H}{[+V_{sat} - V_{sat}]}$$

$$\therefore \quad \frac{R_2}{R_1 + R_2} = \frac{6}{[12 - (-12)]}$$

$$\therefore \quad \frac{R_2}{R_1 + R_2} = 0.25$$

$$\therefore \quad R_2 = 0.25\ R_1 + 0.25\ R_2$$

$$\therefore \quad 0.75\ R_2 = 0.25\ R_1$$

$$\therefore \quad \frac{R_2}{R_1} = 0.3333$$

Assuming $R_2 = 10\ k\Omega$

$$R_1 = \frac{10000}{0.3333} = \mathbf{30\ k\Omega}$$

Example 4.28: For the Schmitt trigger shown in the Fig. 4.129, calculate the trip points and hysteresis, if $V_{sat} = \pm 13.5$ V. If the resistances have a tolerance of ± 5 %, what is the minimum hysteresis?

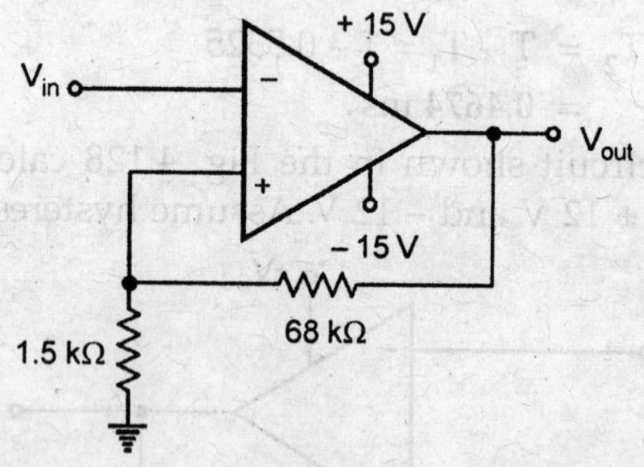

Fig. 4.129

Solution: From the Fig. 4.129, $R_1 = 68$ kΩ, $R_2 = 1.5$ kΩ and $V_{sat} = 13.5$ V

$$V_{UT} = \frac{R_2}{R_1 + R_2} V_{sat} = \frac{1.5}{1.5 + 68} \times 13.5 = \mathbf{0.2913}$$

$$V_{LT} = \frac{-R_2}{R_1 + R_2} V_{sat} = \mathbf{-0.2913}$$

$\therefore \quad H = V_{UT} - V_{LT} = 0.2913 - (-0.2913) = \mathbf{+ 0.5828\ V}$

Now $\quad H = \dfrac{2R_2}{R_1 + R_2} . V_{sat}$

For minimum H, R_2 must be minimum and R_1 must be maximum.

$\therefore \quad R_2 \min = R_2 - 5\%\ R_2 = 1.5 \times 10^3 - 0.05 \times 1.5 \times 10^3 = 1.425\ k\Omega$

and $\quad R_1 \min = R_1 + 5\%\ R_1 = 6.8 \times 10^3 + 0.05 \times 68 \times 10^3 = 71.4\ k\Omega$

$$\therefore \quad H_{min} = \left[\frac{2 \times 1.425 \times 10^3}{71.4 \times 10^3 + 1.525 \times 10^3}\right] \times 13.5 = 0.528\ V$$

Example 4.29: A system uses an ON-OFF temperature controller. Temperature is to be maintained between 25°C to 30°C. The temperature transducer generates a voltage of + 0.5 V at 25°C and + 1.5 V at 30°C. Heater is operated through a relay of 12 V, 100 mA.

Design suitable circuit using op-amp and draw detailed circuit diagram.

Solution: The UTP and LTP levels corresponding to 25 DC and 30 DC are 0.5 V and 1.5 V.

$\therefore \quad V_{UT} = 1.5\ V, \quad V_{LT} = 0.5\ V$

As the levels are different, modified circuit is required as shown in the Fig. 4.130.

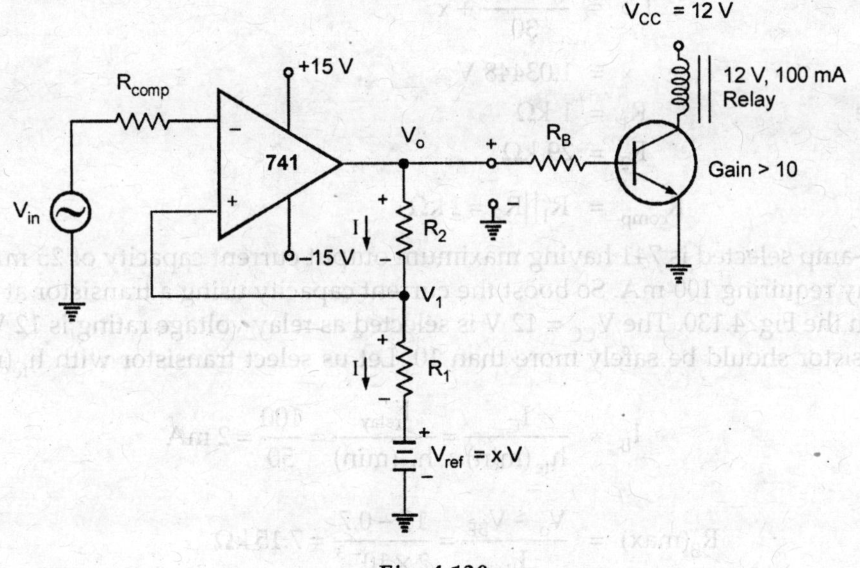

Fig. 4.130

V_1 decides the tripping voltage. Supply used is ± 15 V. So $\pm V_{sat} = \pm 15$ V.

Applying KVL to the output loop,

$- IR_2 - IR_1 - x + V_o = 0$

$$\therefore \quad I = \frac{V_o - x}{R_1 + R_2}$$

And $$V_1 = IR_1 + x = \frac{(V_o - x)}{R_1 + R_2}R_1 + x$$

At $V_o = +V_{sat} = +15\text{ V}, \quad V_1 = V_{UT} = 1.5\text{ V}$

$$\therefore \quad 1.5 = \frac{(15 - x)R_1}{R_1 + R_2} + x \qquad \text{...(1)}$$

At $V_o = -V_{sat} = -15\text{ V}, \quad V_1 = V_{LT} = 0.5\text{ V}$

$$\therefore \quad 1.5 = \frac{(-15 - x)R_1}{R_1 + R_2} + x \qquad \text{...(2)}$$

Subtract equation (2) from equation (1),

$$\therefore \quad 1 = \frac{2 \times 15 R_1}{R_1 + R_2}$$

$$\therefore \quad R_1 + R_2 = 30 R_1 \qquad \text{...(3)}$$

Substitute equation (3) in equation (1),

$$1.5 = \frac{(15 - x)}{30} + x$$

$$x = 1.03448\text{ V}$$

Choose $R_1 = 1\text{ k}\Omega$

$$\therefore \quad R_2 = 29\text{ k}\Omega$$

$$R_{comp} = R_1 || R_2 = 1\text{ k}\Omega$$

The op-amp selected is 741 having maximum output current capacity of 25 mA. It has to drive a relay requiring 100 mA. So boost the current capacity using a transistor at the output as shown in the Fig. 4.130. The V_{CC} = 12 V is selected as relay voltage rating is 12 V. The gain of the transistor should be safely more than 10. Let us select transistor with h_{fe}(min) = 50.

$$\therefore \quad I_B = \frac{I_C}{h_{fe}(\text{min})} = \frac{I_{relay}}{h_{fe}(\text{min})} = \frac{100}{50} = 2\text{ mA}$$

Now $$R_B(\text{max}) = \frac{V_o - V_{BE}}{I_B} = \frac{15 - 0.7}{2 \times 10^{-3}} = 7.15\text{ k}\Omega$$

Use $R_B = 6.8\text{ k}\Omega$

Example 4.30: A Schmitt trigger with the upper threshold level V_{UT} = 0 V and hysteresis width V_H = 0.2 V converts a 1 kHz sine wave of amplitude 4 V_{pp} into a square wave. Calculate the time duration of the negative and positive portion of the output waveform.

Solution: $V_H = 0.2$ V, $f = 1$ kHz, $V_{peak} = \frac{V_{p\text{-}p}}{2} = \frac{4}{2} = 2$ V

$V_{UT} = 0$ V, $V_{LT} = V_{UT} - V_H = 0 - 0.2 = -0.2$ V

The waveforms are shown in the Fig. 4.131.

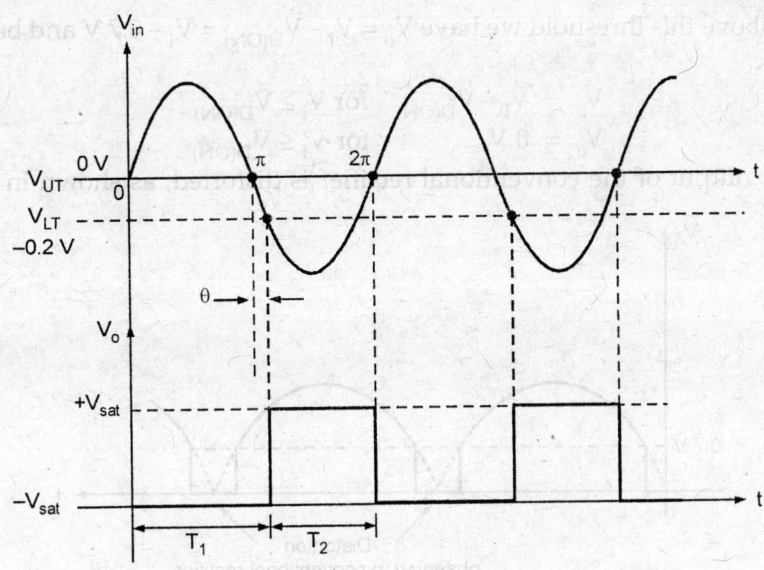

Fig. 4.131

The angle 8 is to be calculated. The sine wave is represented by,

$$V_{in} = V_p \sin(\theta')$$

For $\pi < \omega t < 2\pi$

$$\theta' = \pi + \theta$$

$\therefore$ $V_{in} = V_p \sin(\pi + \theta)$

At LTP, $V_{in} = -0.2$ V

$\therefore$ $-0.2 = 2\sin(\pi + \theta) = -2\sin\theta$ ...As $\sin(\pi + \theta) = -\sin\theta$

$\therefore$ $\sin\theta = 0.1$

$\therefore$ $\theta = 5.739°$

$\therefore$ T_1 is time corresponds to angle (180° + 5.739°).

Now $T = \frac{1}{f} = \frac{1}{1 \times 10^3} = 1$ msec

So T corresponds to 360° i.e. 1 complete cycle.

$\therefore$ $T_1 = \frac{180° + 5.739°}{360°} =$ **0.5159 msec**

$\therefore$ $T_1 = T - T_1 =$ **0.484 msec**

4.39 PRECISION RECTIFIERS

Recall from basic circuit principles that a rectifier circuits can be implemented with a diode/diodes (half wave rectifier or full wave rectifier). The major limitations of these circuits is that they cannot rectify voltages below $V_{D(ON)} = 0.7$ V, the cut-in voltage of the diode. In these circuits V_i has to rise to a threshold of the order of V_D (ON) before any appreciable change can be seen at the output.

Therefore, above this threshold we have $V_o = V_i - V_{D(ON)} = V_i - 0.7$ V and below threshold $V_o = 0$ V i.e.

$$V_o = V_i - V_{D(ON)} \quad \text{for } V_i \geq V_{D(ON)} \quad \text{...(1a)}$$

$$V_o = 0 \text{ V} \quad \text{for } V_i \leq V_{D(ON)} \quad \text{...(1b)}$$

Due to this, output of the conventional rectifier is distorted, as shown in the Fig. 4.132.

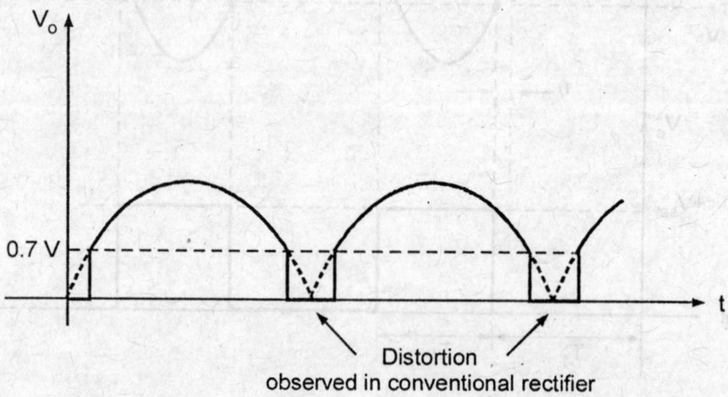

Fig. 4.132 Output waveforms of conventional rectifier.

To achieve precision rectification we need a circuit that keeps V_o equal to V_i for $V_1 > 0$ V. This can be achieved by using op-amp along with the diodes and these circuits are called precision rectifiers. These are used to precisely rectify voltages having amplitudes less than 0.7 V. Hence these circuits are called small signal precision rectifiers.

4.40 PRECISION HALF WAVE RECTIFIERS

The precision half wave rectifiers are classified as,

1. Positive half wave rectifier 2. Negative half wave rectifier

4.40.1 Positive Precision Half Wave Rectifier

The Fig. 4.133 shows the positive small signal precision half wave rectifier.

The diode D_1 is used in the feedback loop of the circuit. The analysis can be done for $V_i > 0$ V and $V_i < 0$ V.

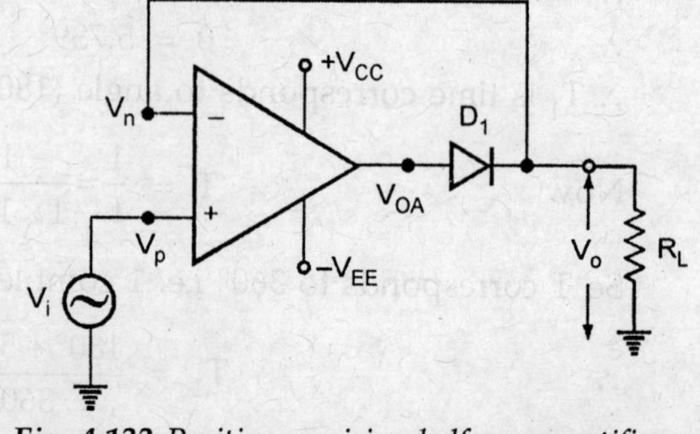

Fig. 4.133 Positive precision half wave rectifier.

Case 1 : $V_i > 0$ V

Consider that input voltage is positive going. Due to high open loop gain of op-amp it produces high voltage V_{OA}. This provides enough drive to the diode D_1 to make it forward biased. So it acts as a ideal diode and in forward biased condition behaves as a switch as shown in the Fig. 4.134.

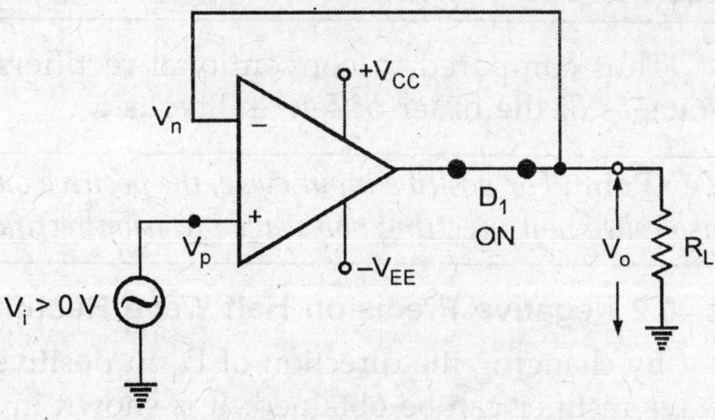

Fig. 4.134 $V_i > 0$ V

The cut-in voltage of diode of 0.7 V gets divided by A_{OL} which is very high. Hence immediately when V_i starts increasing, D_1 becomes ON.

Then circuit works as a voltage follower. From virtual ground $V_n = V_p = V_i$ and $V_o = V_n$ due to the feedback path. Hence entire positive half cycle is available across the load.

Case 2 : $V_i < 0$ V

When V_i goes negative, immediately V_{OA} produced attains $-V_{sat}$ value making diode D_1 reverse biased. Thus diode D_1 acts as an open circuit as shown in the Fig. 4.135.

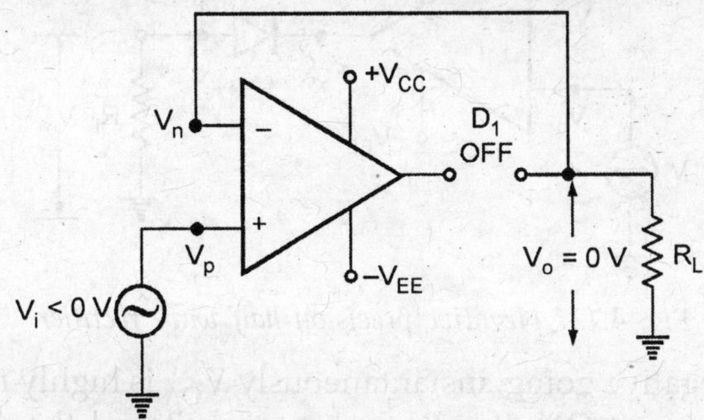

Fig. 4.135 $V_i < 0$ V

Due to OFF diode D_1, feedback path gets opened and no current can flow through R_L. The voltage V_o is 0 V and negative half cycle of V_i gets clipped from the output.

The input-output waveforms are shown in the Fig. 4.136.

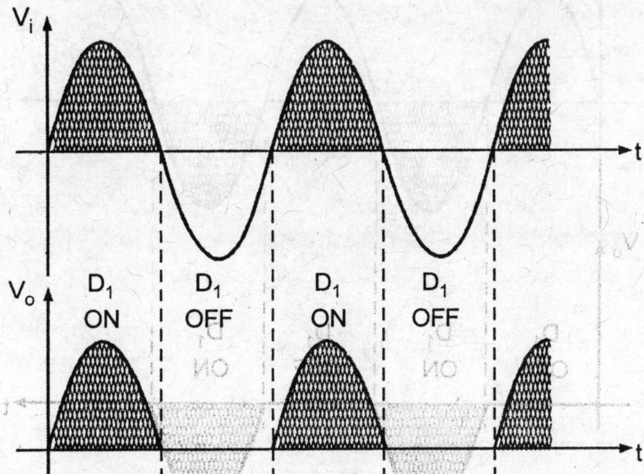

Fig. 4.136 Waveforms of half wave rectifier.

Thus compared to conventional rectifiers, precision rectifiers can rectify very small voltages of the order of few millivolts.

Key Point: *For positive input cycle, the positive output cycle exists at the output hence the circuit is also called non-inverting half wave precision rectifier.*

4.40.2 Negative Precision Half Wave Rectifier

By changing the direction of D_1 in positive half wave rectifier circuit, the negative half wave rectifier can be obtained. It is shown in the Fig. 4.137.

When $V_i > 0$ V i.e. positive going, immediately V_{OA} is very high due to high open loop gain. This reverse biases the diode, making it open. Thus $V_o = 0$ V as no current can pass through R_L.

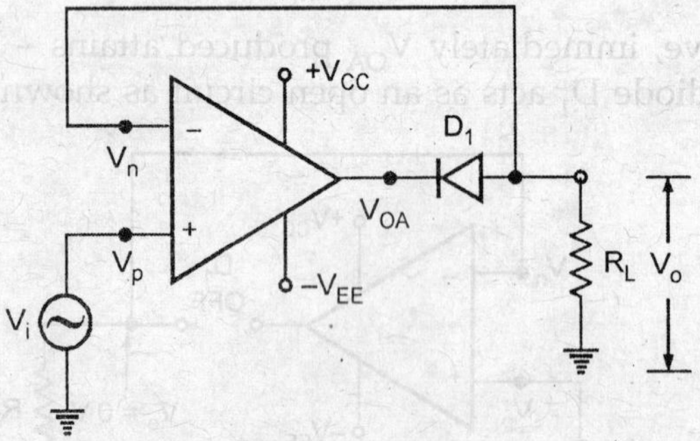

Fig. 4.137 Negative precision half wave rectifier.

When $V_i < 0$ V i.e. negative going, instantaneously V_{OA} is highly negative which forward biases the diode D_1 making it ON. It acts as short circuit and the circuit acts as a voltage follower. Hence the output voltage is same as input voltage. Thus the entire negative half cycle is available across the load. Hence the circuit is called negative half wave rectifier. The waveforms are as shown in the Fig. 4.138.

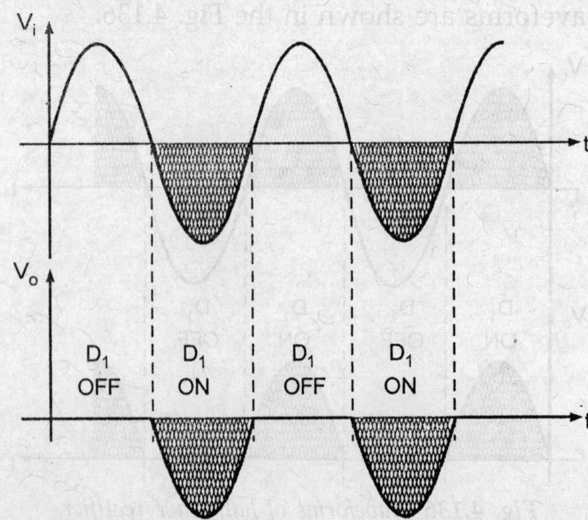

Fig. 4.138 Waveforms of negative half wave rectifier.

Another type of negative half wave rectifier which acts as an **inverting precision half wave rectifier** is shown in the Fig. 4.139.

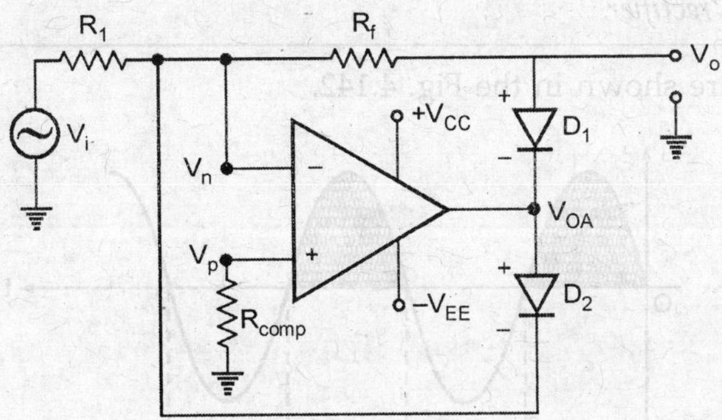

Fig. 4.139 Inverting half wave rectifier.

The basic op-amp circuit is used in inverting mode and the input is applied to inverting input terminal. When $V_i > 0$ V i.e. positive going, due to high open loop gain V_{OA} is highly negative. This forward biases the diode D_1 and reverse biases D_2 and circuit becomes as shown in the Fig. 4.140.

Thus it acts as an inverting amplifier as feedback path gets established through ON diode D_1.

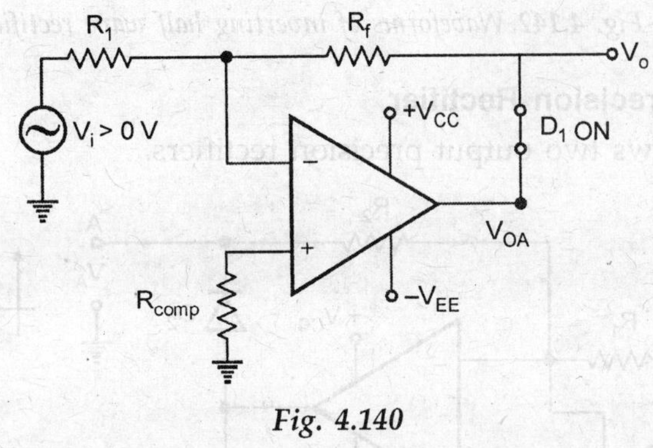

Fig. 4.140

$$\therefore \quad V_o = -\frac{R_f}{R_1} V_i$$

Using $R_f = R_1$, $V_o = -V_i$

Thus for positive half cycle of the input, negative half cycle is available at the output. When $V_i < 0$ V, immediately V_{OA} becomes highly positive due to which D_2 becomes forward biased and D_1 reverse biased. As D_1 is OFF; $V_o = 0$ V while diode D_2 prevents op-amp to go into the positive saturation. The circuit is shown in the Fig. 4.141.

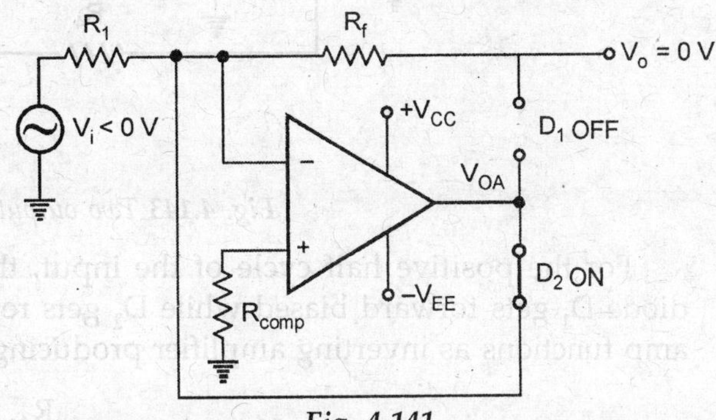

Fig. 4.141

Key Point: *As negative half cycle is produced for positive half cycle of the input, the circuit is called the inverting half wave rectifier.*

The waveforms are shown in the Fig. 4.142.

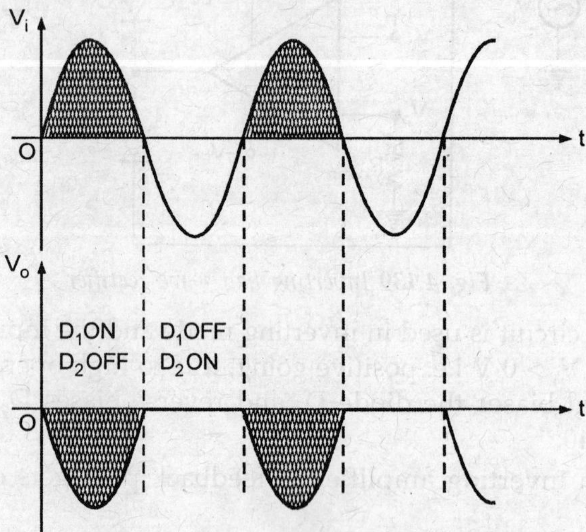

Fig. 4.142 Waveforms of inverting half wave rectifier.

4.40.3 Two Output Precision Rectifier

The Fig. 4.143 shows two output precision rectifiers.

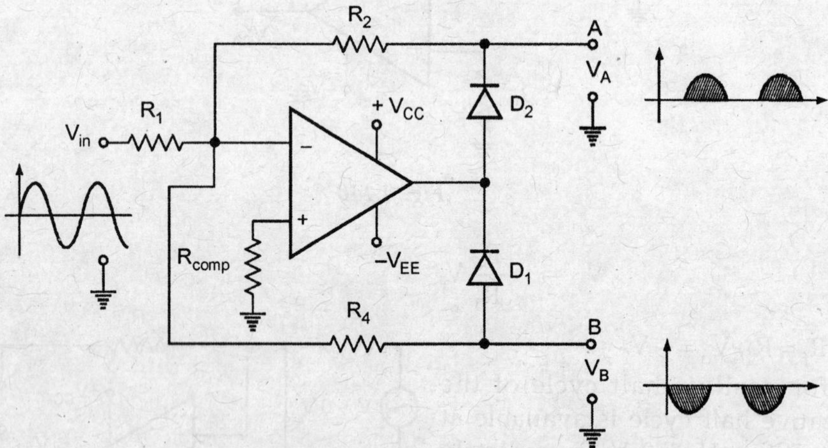

Fig. 4.143 Two output precision rectifier.

For the positive half cycle of the input, the op-amp output goes negative hence the diode D_1 gets forward biased while D_2 gets reverse biased. Hence with R_4 and R_1 the op-amp functions as inverting amplifier producing negative half cycle at the terminal B.

$$\therefore \qquad V_B = -(+V_m)\frac{R_4}{R_1} \qquad \text{...(1)}$$

The inverting terminal is at virtual ground and no current flows through R_2. Hence terminal A remains grounded as long as D_2 is reverse biased.

For the negative half cycle of the input, the op-amp output goes positive hence the diode D_1 gets reverse biased and the diode D_2 gets forward biased. With R_2 and R_1, op-amp behaves as inverting amplifier producing positive half cycle at the terminal A.

$$\therefore \qquad V_A = -(V_{in})\frac{R_2}{R_1} = +V_{in}\frac{R_2}{R_1} \qquad \text{...(2)}$$

The terminal B remains grounded till the diode D_2 is forward biased.

Thus precisely positive half cycles are available at the terminal A while negative half cycles are available at the terminal B.

4.41 PRECISION FULL WAVE RECTIFIERS

The full wave rectifier circuits accept an a.c. signal at the input, inverts either the negative or the positive half, and delivers both the inverted and non-inverted halves at the output, as shown in the Fig. 4.144.

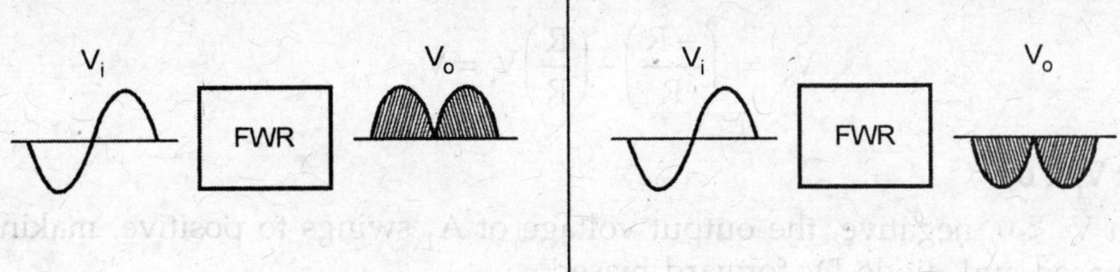

Fig. 4.144 Positive and negative full wave rectifiers.

The operation of the positive full wave rectifier is expressed as,

$$V_o = |V_i| \qquad \text{...(1)}$$

And that of the negative rectifier as

$$V_o = -|V_i| \qquad \text{...(2)}$$

Looking at equations (1) and (2) we can say that precision full wave rectifier circuits are precision absolute value circuits. Fig. 4.145 shows a full wave rectifier or absolute value circuit.

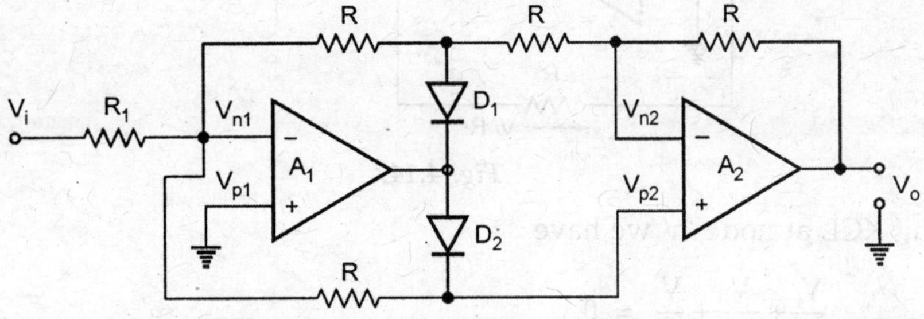

Fig. 4.145 Full wave rectifier.

CASE 1 : $V_i > 0$

When $V_i > 0$, inverting side of A_1 will force its output to swing negative, thus forward biasing D_1 and reverse biasing D_2. Since no current flows through resistance R connected between V_{n1} and V_{p2}, both are equipotential.

i.e. $$V_{n1} = V_{p2} = 0 \text{ V}.$$

The Fig. 4.146 shows the equivalent circuit.

From equivalent circuit, the output voltage can be given as,

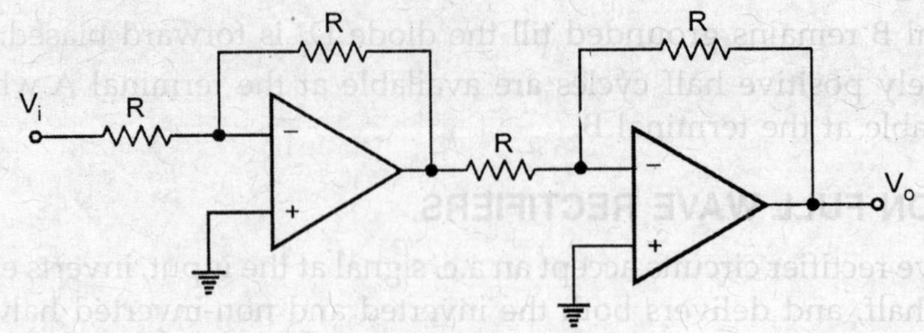

Fig. 4.146

$$V_o = \left(\frac{-R}{R}\right) - \left(\frac{R}{R}\right) V_i = V_i \quad \text{... (3)}$$

CASE 2 : $V_i < 0$

When $V_i < 0$, negative, the output voltage of A_1 swings to positive, making diode D_1 reverse biased and diode D_2 forward biased.

The Fig. 4.147 shows the equivalent circuit.

Let the output voltage of op-amp A_1 be V. Since the differential input to A_2 is zero, the inverting input terminal is also at voltage V, as shown in the Fig. 4.147.

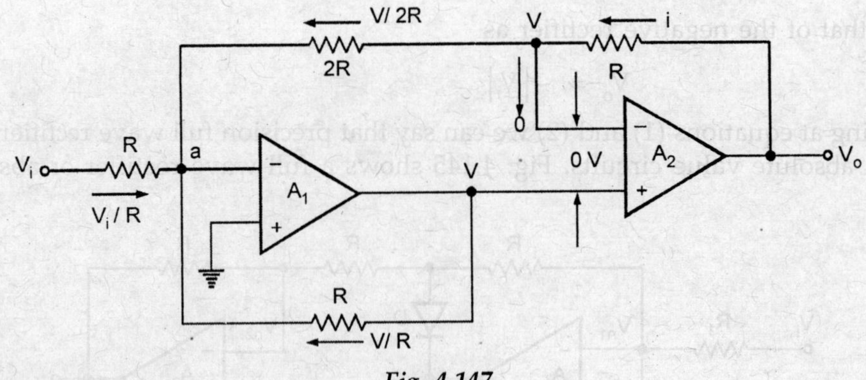

Fig. 4.147

Applying KCL at node 'a' we have

$$\frac{V_i}{R} + \frac{V}{2R} + \frac{V}{R} = 0$$

$$\frac{3V}{2R} = \frac{-V_i}{R}$$

$$\therefore \quad V = \frac{-2}{3} V_i \quad \text{...(4)}$$

To find V_o interms of V we concentrate on the equivalent circuit of A_2, as shown in the Fig. 4.148.

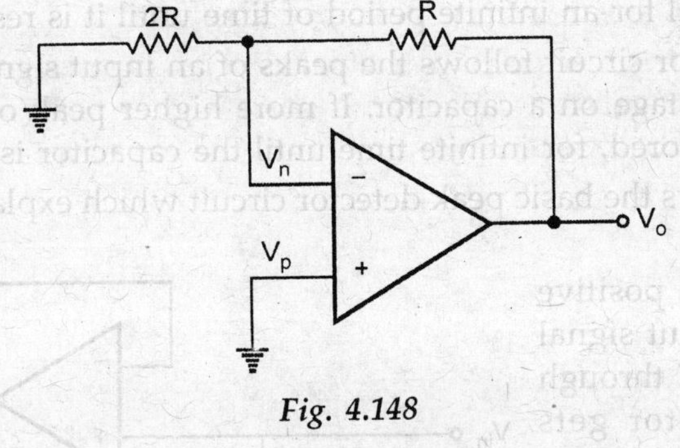

Fig. 4.148

$$\therefore \quad V_o = \left(1 + \frac{R}{2R}\right) V = \left(\frac{2R + R}{2R}\right) V = \frac{3}{2} V \quad \text{... (5)}$$

Substituting value of V in above equation we get,

$$V_o = \frac{3}{2}\left(\frac{-2}{3} V_i\right) = -V_i \quad \text{... (6)}$$

Hence for $V_i < 0$ the output is positive. This is illustrated in Fig. 4.149.

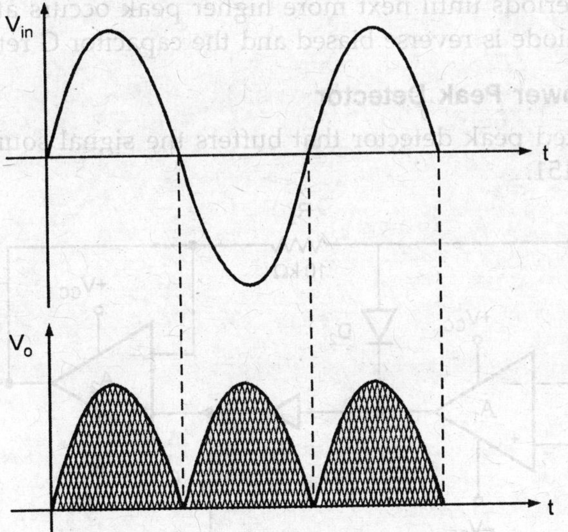

Fig. 4.149 Input and output waveforms for full wave rectifier

4.42 ACTIVE PEAK DETECTORS

In practice number of non-sinusoidal input voltages may exist as the input to the various applications. Such non-sinusoidal waveforms are square, triangular, sawtooth or any other random waveform. A conventional a.c. voltmeter cannot be used to measure the peak of such waveforms as it measures r.m.s. value of purely sinusoidal waveform. Hence to measure the peak of non-sinusoidal waveforms a peak detector circuit is used.

A peak detector is a circuit which notes and remember the peak positive or negative value of an input signal for an infinite period of time until it is reset.

Such a peak detector circuit follows the peaks of an input signal and stores the highest value, interms of a voltage on a capacitor. If more higher peak occurs in an input signal, new peak value gets stored, for infinite time until the capacitor is discharged.

The Fig. 4.150 shows the basic peak detector circuit which explains the basic principle of such circuit.

The circuit is basic positive peak detector. The input signal charges the capacitor C through diode D. The capacitor gets charged equal to the highest input voltage, V_o (Peak of input) neglecting the small diode drop. The capacitor remains charged to the peak value of input unless and until discharged with the help of MOSFET reset switch.

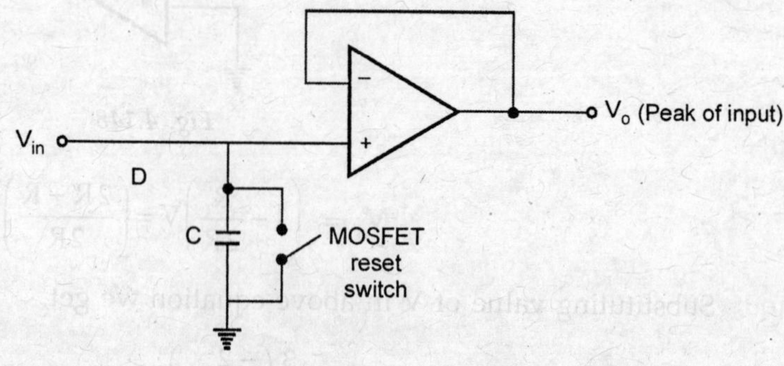

Fig. 4.150 Basic peak detector.

The op-amp is connected as voltage follower and its output voltage will be equal to the drop across capacitor which is positive peak of value of the applied voltage and will remain that way for long periods until next more higher peak occurs at the input. For negative cycles of input the diode is reverse biased and the capacitor C retains its voltage.

4.42.1 Voltage Follower Peak Detector

More sophisticated peak detector that buffers the signal source from the capacitor is shown in the Fig. 4.151.

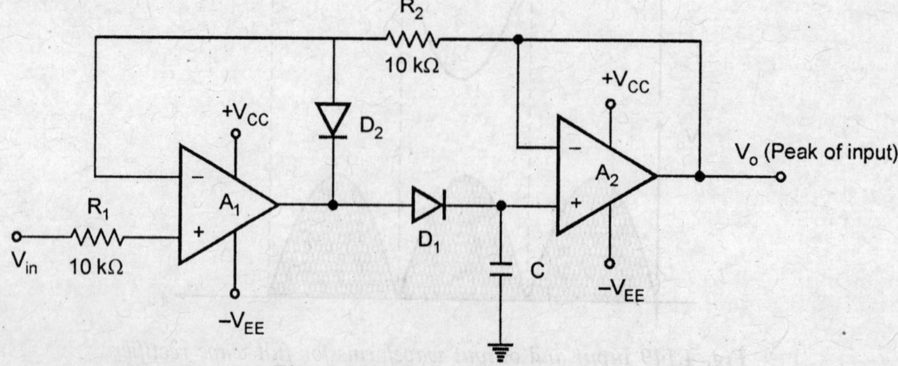

Fig. 4.151 Improved peak detector.

The op-amp A_1 offers a high impedance load to the source. The op-amp A_2 acts as a buffer between the capacitor and the load.

With the same basic principle, the output voltage V_o at any given time is equal to the voltage on the capacitor, which is nothing but the peak of the input voltage occurred upto that time.

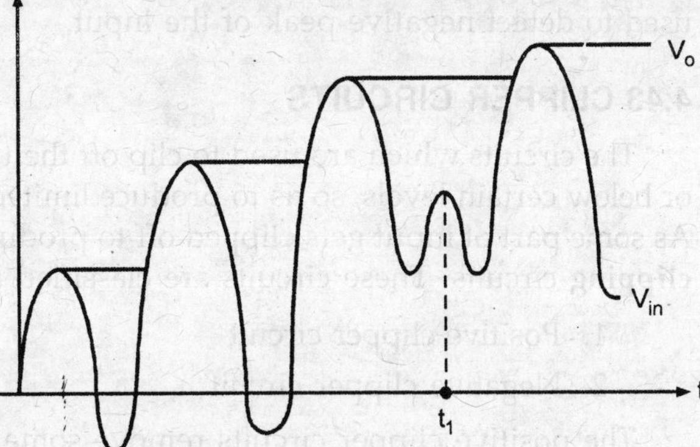

Fig. 4.152 Waveforms for peak detector.

Whenever the input signal has more higher peak than the current one, the capacitor charges upto the new higher input level. But if the input level gets dropped then the capacitor retains the peak of input voltage as diode D_1 gets reverse biased and diode D_2 prevents the output of A_1 from going into the negative saturation. This then serves to improve the recovery time of A_1 when the input attains more positive value. Resistance R_2 provides the path for input bias current to A_1. The resistance R_1 is selected equal to R_2 so as to minimize the effect of offset voltage. To provide the stability against the oscillations, the required frequency compensation must be provided to the op-amp A_1.

The Fig. 4.152 shows the waveforms for the positive peak detector.

The peak at t_1 cannot be recognized as it is less than the previously occurred peak in the input signal.

The circuit can be modified to hold the negative peak of a input signal by reversing the diode connections.

Peak detectors are used for amplitude modulation in communication and in test and measurement instrumentation applications.

4.42.2 Peak Detector using Integrator

Another arrangement for the peak detection is to use op-amp A_2 as an integrator. This avoids the necessity to drive a grounded capacitive load. Because of this, it is easier to maintain the closed loop stability than the previous circuits. The arrangement is shown in the Fig. 4.153.

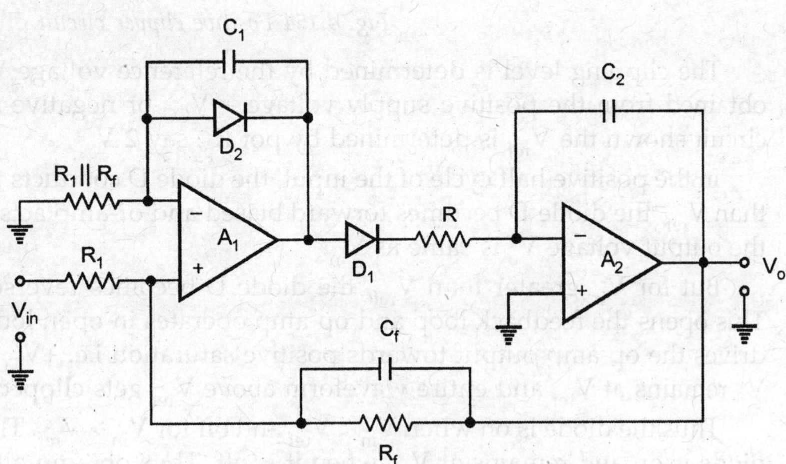

Fig. 153 Peak detector using integrator.

The capacitors C_1 and C_f are required to stabilize the loop. The circuit detects the positive peak of the input V_{in}.

By reversing the connections of the diodes D_1 and D_2, the same configuration can be used to detect negative peak of the input.

4.43 CLIPPER CIRCUITS

The circuits which are used to clip off the unwanted portions of the input voltage above or below certain levels, so as to produce limited required output are called **limiting** circuits. As some part of input gets clipped off to produce output, these circuits are commonly called **clipping** circuits. These circuits are classified as,

1. Positive clipper circuit
2. Negative clipper circuit

The positive clipper circuits remove some positive part from the input to produce the output. The negative clipper circuits remove some negative part from the input to produce the output.

4.43.1 Positive Clipper Circuit

A positive clipper circuit using op-amp is shown in the Fig. 4.154.

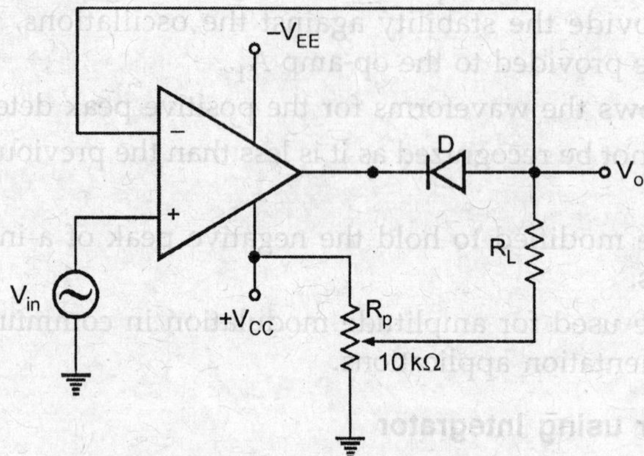

Fig. 4.154 Positive clipper circuit.

The clipping level is determined by the reference voltage V_{ref}. This reference voltage is obtained from the positive supply voltage $+ V_{CC}$ or negative supply voltage $- V_{EE}$. In the circuit shown the V_{ref} is determined by pot R_p, say 2 V.

In the positive half cycle of the input, the diode D conducts till $V_{in} = V_{ref}$. When V_{in} is less than V_{ref}, the diode D becomes forward biased and or-amp acts as a voltage follower. Hence the output voltage V_o is same as V_{in}.

But for V_{in} greater than V_{ref}, the diode D becomes reverse biased and becomes open. This opens the feedback loop and op-amp operates in open loop. This open loop operation, drives the op-amp output towards positive saturation i.e. $+V_{CC}$. Due to this, output voltage V_o remains at V_{ref} and entire waveform above V_{ref} gets clipped off.

Thus the diode is on when $V_{in} < V_{ref}$ and off for $V_{in} > V_{ref}$. The output follows input when diode is on and remains at V_{ref} when it is off. Thus op-amp alternates between closed loop and open loop operation and hence op-amp used must be high speed and compensated for unity gain. The high speed op-amp like HA 2500, LM 310 and μA 318 can be used for such applications. The input and output waveforms are shown in the Fig. 4.155.

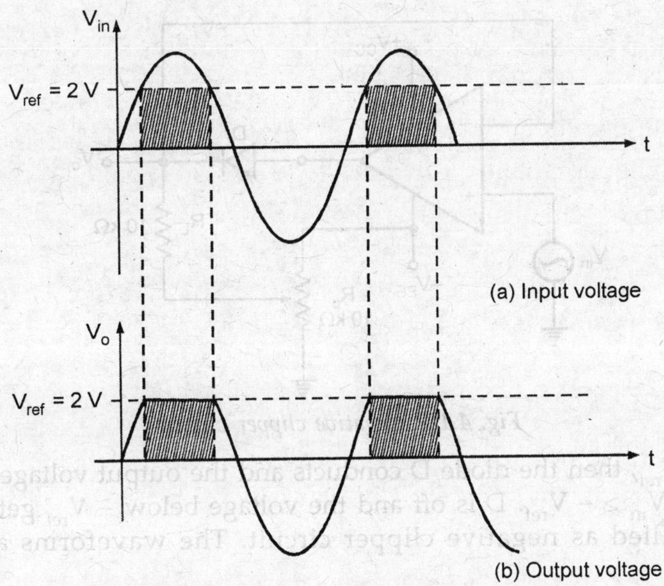

Fig. 4.155 Waveforms of positive clipper circuit.

In the same circuit if the pot R_p is used with $-V_{EE}$ to generate negative V_{ref}, instead of V_{CC} then the entire waveforms above $-V_{ref}$ gets clipped off. Let $V_{ref} = -2$ V in the same circuit, the output follows input only when $V_{in} < V_{ref}$ i.e. $V_{in} < -2$ V. The waveforms with $V_{ref} = -2$ V are shown in the Fig. 4.156.

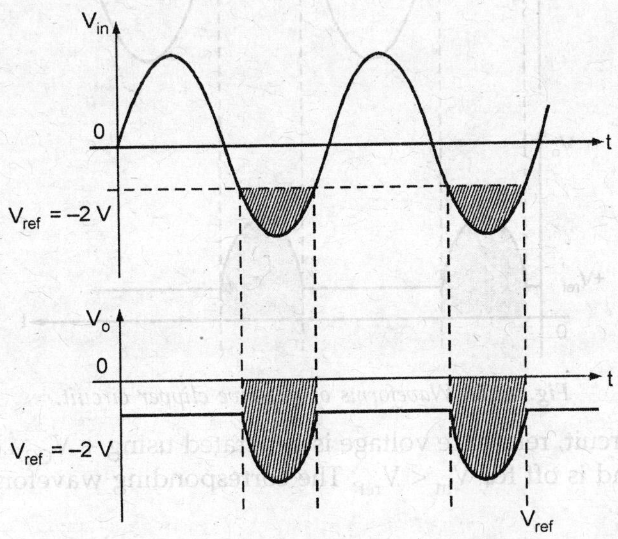

Fig. 4.156 Waveforms with negative.

4.43.2 Negative Clipper Circuit

The negative clipper circuit can be obtained by reversing the connections of diode D and using pot R_p to generate negative reference voltage V_{ref}. This circuit is shown in the Fig. 4.157.

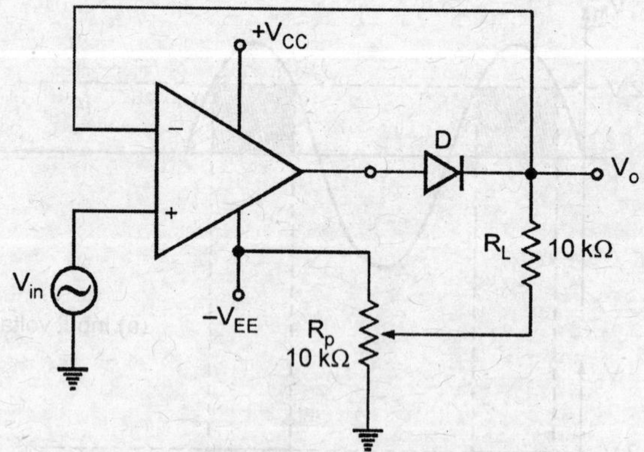

Fig. 4.157 Negative clipper circuit

When $V_{in} > - V_{ref}$, then the diode D conducts and the output voltage follows the input voltage. But when $V_{in} > - V_{ref}$. D is off and the voltage below $- V_{ref}$ gets clipped off. The circuit hence is called as negative clipper circuit. The waveforms are shown in the Fig. 4.158.

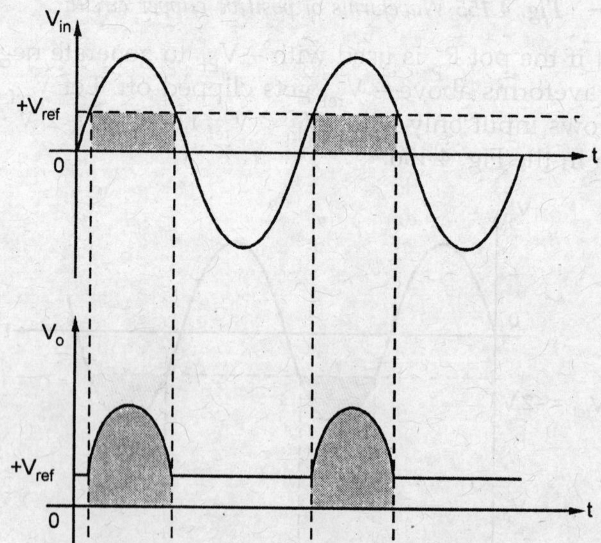

Fig. 4.158 Waveforms of negative clipper circuit.

If for the same circuit, reference voltage is generated using $+ V_{CC}$ i.e. $+ V_{ref}$ the diode D is on for $V_{in} > V_{ref}$ and is off for $V_{in} < V_{ref}$. The corresponding waveforms are shown in the Fig. 4.159.

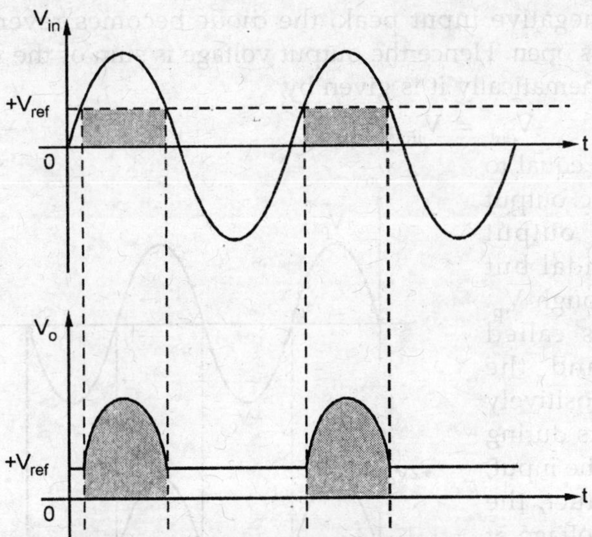

Fig. 4.159 Waveform with positive V_{ref}

4.44 CLAMPER CIRCUITS

Sometimes it is necessary to add a d.c. level to the a.c. output signal. The circuits which are used to add a d.c. level as per the requirement, to the a.c. output are called as clamper circuits. Sometimes such circuits are also called as **d.c. restorer circuits**. If the clamped d.c. level is positive, the circuits are called as **positive clamper circuits** and if clamped d.c. level is negative, the circuits are called as **negative clamper circuits**.

4.44.1 Positive Clamper Circuit

The Fig. 4.160 shows the positive clamper circuit using op-amp.

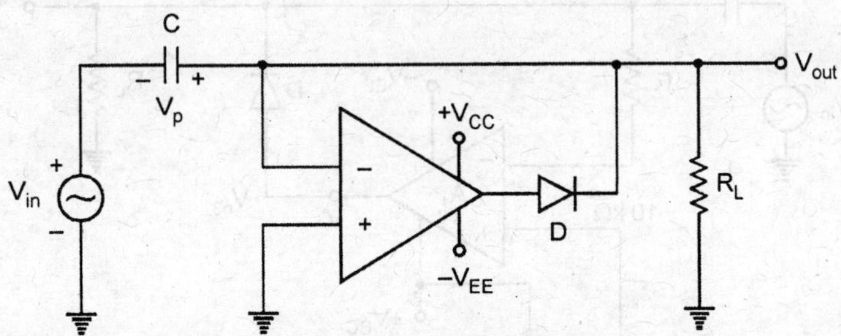

Fig. 4.160 Positive clamper circuit.

When the input voltage is first time negative, due to inverting mode of op-amp, the op-amp output is positive. This turns on the diode making it forward biased. Thus the capacitor charges to the peak value of the negative cycle of input with the polarities as shown in the Fig. 4.160.

Just beyond the negative input peak, the diode becomes reverse biased and stops conducting. It becomes open. Hence the output voltage is sum of the input voltage and the capacitor voltage mathematically it is given by,

$$V_{out} = V_{in} + V_p$$

Thus the d.c. level equal to V_p gets added in the a.c. output signal. The final output waveform is sinusoidal but shifted positively through V_p. Hence the circuit is called positive clamper and the waveform is called positively clamped waveform. As during positive half cycle of the input, diode does not conduct, the capacitor retains its voltage at V_p all the time after the first negative peak of the input.

The waveforms are shown in the Fig. 4.161.

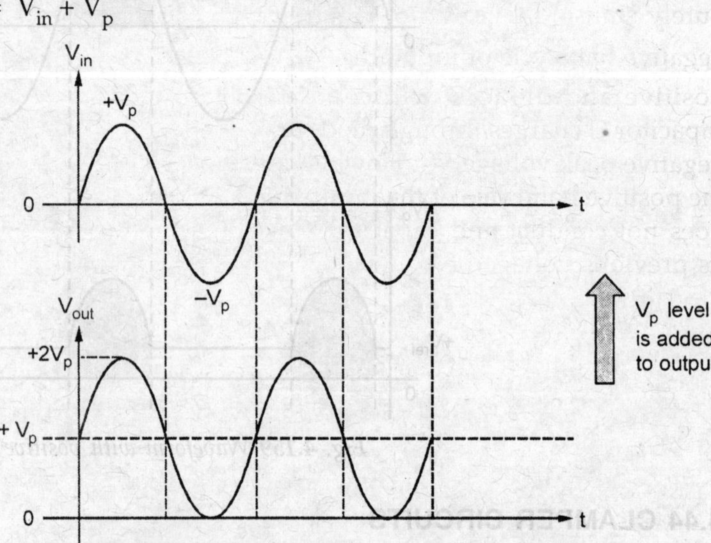

Fig. 4.161 Waveforms of positive clam per circuit.

It can be observed that output waveform swings from a to + 2 V_p i.e. peak to peak voltage of output waveform is 2 V_p which remains same as the input voltage. Thus the circuit shows that the total swing of the output voltage remains same as the total swing of the input voltage in clamper circuit.

In this circuit, the d.c. level added is equal to peak value of the input used. Another circuit in which, variable positive d.c. level can be added is shown in the Fig. 4.162.

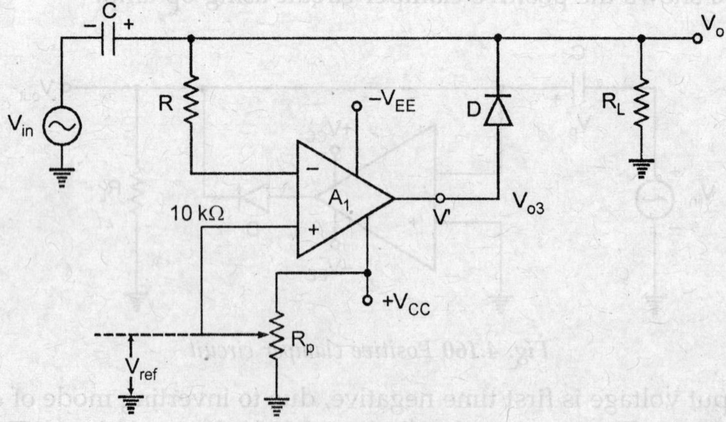

Fig. 4.162 Positive clamper circuit.

The input voltage is applied to the inverting terminal of the op-amp A_1 while variable positive d.c. voltage is applied to the non-inverting input terminal of the op-amp A_1. The circuit can be analyzed using the Superposition theorem considering only one input active at a time.

Let V_{ref} is acting alone and input V_{in} is zero. For positive V_{ref}, the output voltage V' is also positive. Due to this the diode D becomes forward biased. Hence the circuit acts as a voltage follower. Hence the net output voltage V_o is same as positive V_{ref}.

Now let input at inverting terminal be purely sinusoidal i.e. $V_{in} = V_m \sin \omega t$. For negative half cycle of input, the V_0' will be positive and diode D will conduct. The capacitor C charges through diode D to the negative peak voltage V_m. However during the positive half cycle of the input, diode D does not conduct and capacitor C retains its previous voltage of V_m.

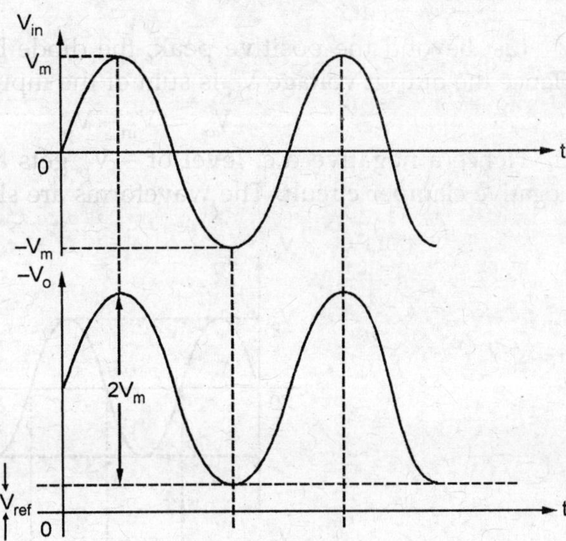

Fig. 4.163 Waveforms for positive clamper circuit.

This voltage V_m is in series with the a.c. input voltage, the output voltage becomes $V_{in} + V_m$.

Hence the net output voltage due to the effect of both the inputs becomes $V_{in} + V_m + V_{ref}$.

The resistance R is used to protect the op-amp against excessive discharge currents from the capacitor C, especially when the d.c. supply voltages are switched off.

The waveforms are shown in the Fig. 4.163.

As the circuit clamps the peaks of the input waveforms hence the circuit is also called as a **peak clamper circuit.**

4.44.2 Negative Clamper Circuit

The Fig. 4.164 shows the negative clamper circuit obtained by reversing the diode connections in positive clamper circuit.

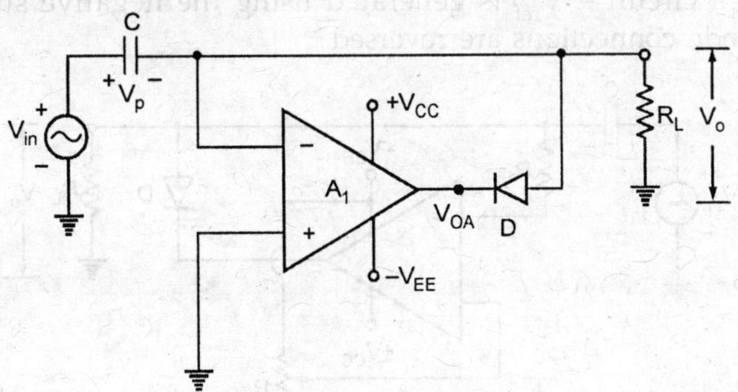

Fig. 4.164 Negative clamper circuit.

When V_{in} is first positive going, due to inverting mode of op-amp, the voltage V_{OA} goes negative. Thus diode D becomes forward biased and capacitor charges to peak value with polarities as shown.

Just beyond the positive peak, the diode becomes reverse biased and becomes open. Hence the output voltage V_o is sum of the input and capacitor voltages.

$$\therefore \quad V_o = V_{in} - V_p$$

Hence a negative d.c. level of $- V_p$ gets added to the output hence circuit is called negative clamper circuit. The waveforms are shown in the Fig. 4.165.

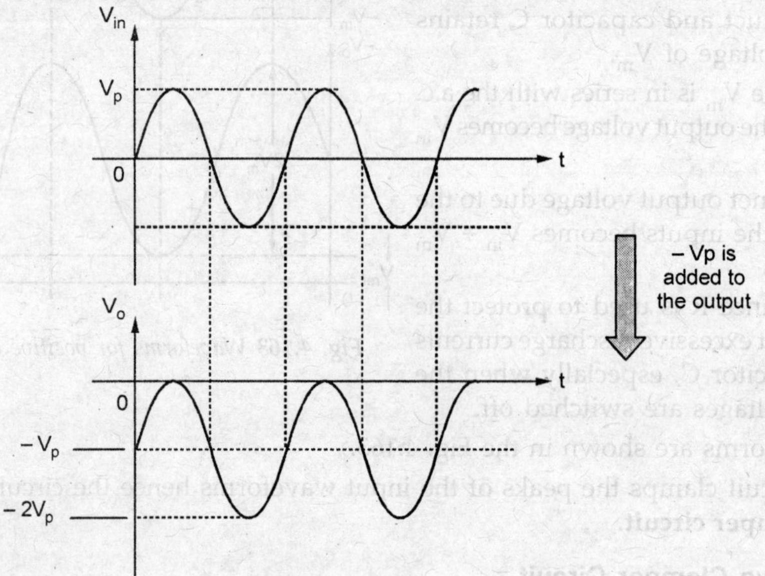

Fig. 4.165 Negative clam per circuit.

Another circuit in which variable negative d.c. level can be added is shown ill the Fig. 4.166. In such a circuit $- V_{ref}$ is generated using the negative supply $- V_{EE}$ of the op-amp and the diode connections are reversed.

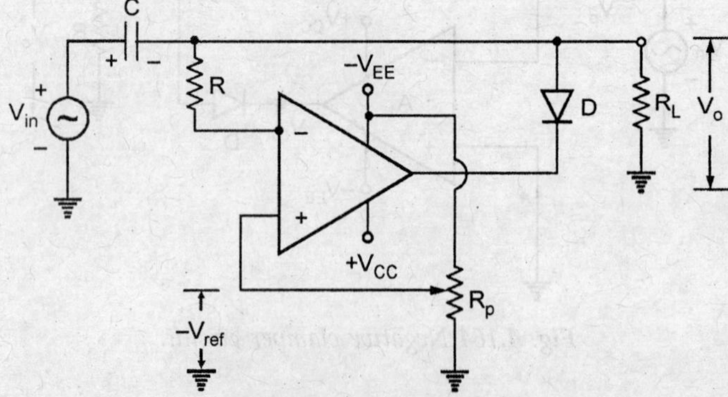

Fig. 4.166 Negative clam per circuit.

Due to opposite connection of diode D, the capacitor charges in reverse direction and hence negative clamper is obtained. The waveforms are shown in the Fig. 4.167.

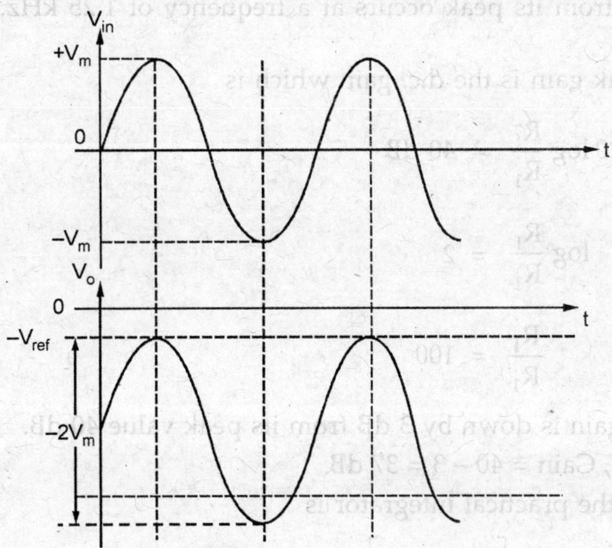

Fig. 4.167 Waveforms for negative clam per circuit.

4.44.3 Clamper Applications

The clamper circuits are often used in the television receivers as d.c. restorer. The video signal in television is processed through capacitively coupled amplifiers hence the signal losts its d.c. component which effectively loses black and white reference levels and the blanking level. Hence it is necessary to restore these levels back before applying signal to the picture tube. This is done by a clamper circuit acting as a d.c. restorer. This is shown in the Fig. 4.168.

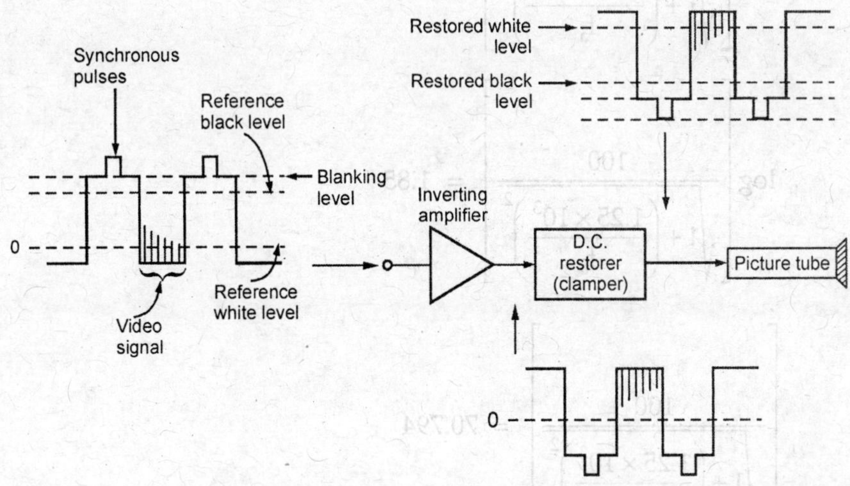

Fig. 4.168 Use of clamper circuit in television receiver.

Examples with Solutions

Example 4.31: Find R_1 and R_f in the loosy integrator so that the peak gain is 40 dB and the gain 3 dB down from its peak occurs at a frequency of 1.25 kHz. Use capacitance of 0.01 μF.

Solution: The peak gain is the d.c. gain which is

$$20 \log \frac{R_f}{R_1} = 40 \text{ dB}$$

$$\therefore \quad \log \frac{R_f}{R_1} = 2$$

$$\therefore \quad \frac{R_f}{R_1} = 100 \qquad \text{...(1)}$$

At f = 1.25 kHz, gain is down by 3 dB from its peak value 40 dB.

So at f = 1.25 kHz, Gain = 40 – 3 = 37 dB.

Now the gain of the practical integrator is

$$20 \log \left[\frac{R_f / R_1}{\sqrt{1+\left(\frac{f}{f_a}\right)}} \right] = 37$$

$$\therefore \quad 20 \log \left[\frac{100}{\sqrt{1+\left(\frac{1.25\times 10^3}{f_a}\right)^2}} \right] = 37$$

$$\therefore \quad \log \left[\frac{100}{\sqrt{1+\left(\frac{1.25\times 10^3}{f_a}\right)^2}} \right] = 1.85$$

$$\therefore \quad \left[\frac{100}{\sqrt{1+\left(\frac{1.25\times 10^3}{f_a}\right)^2}} \right] = 70.794$$

$$\therefore \qquad \sqrt{1+\left(\frac{1.25\times10^3}{f_a}\right)^2} = 1.4125$$

$$\therefore \qquad 1+\left(\frac{1.25\times10^3}{f_a}\right)^2 = 1.9952$$

$$\left(\frac{1.25\times10^3}{f_a}\right)^2 = 0.9952$$

$$\therefore \qquad \frac{1.25\times10^3}{f_a} = 0.9976$$

$$\therefore \qquad f_a = 1.253 \text{ kHz} \qquad \text{...(2)}$$

Now $\qquad f_a = \dfrac{1}{2\pi C_f R_f} \quad$ and $\quad C_f = 0.01\ \mu F$

$$\therefore \qquad 1.253\times10^3 = \frac{1}{2\pi\times0.01\times10^{-6}\times R_f}$$

$$\therefore \qquad R_f = \mathbf{12.70\ k\Omega}$$

From equation (1), $\qquad R_1 = \dfrac{R_f}{100}$

$$= \frac{1270\times10^3}{100}$$

$$\therefore \qquad R_1 = 127\ \Omega$$

Example 4.32: For the op-amp circuit shown in the Fig. 4.169, find the output voltage V_o.

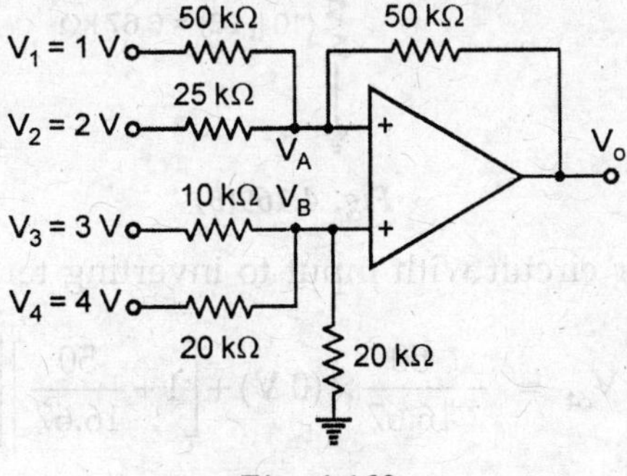

Fig. 4.169

Solution: Use the principle of Superposition.

Ground the inputs V_3 and V_4. Hence as the op-amp input current is zero, the non-inverting terminal becomes grounded. Hence the circuit acts as an inverting summer.

$$\therefore \quad V_{o12} = -\left[\frac{50}{50}V_1 + \frac{50}{25}V_2\right] = -[V_1 + 2V_2] = -[1+4] = -5\text{ V}$$

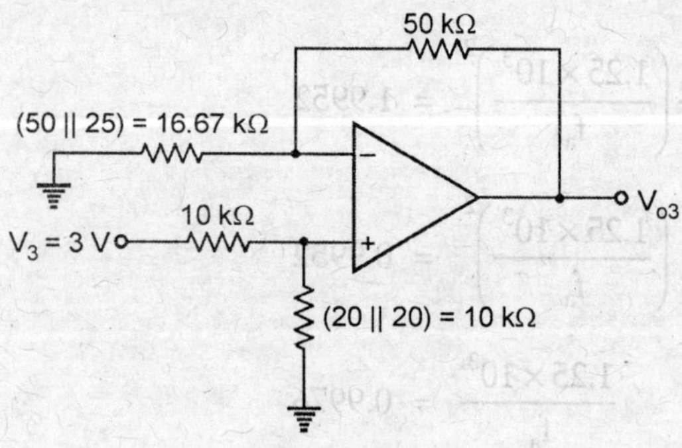

Fig. 4.169(a)

Now ground V_1, V_2 and V_4. Due to this circuit reduces to as shown in Fig. 4.169(*a*). This is subtractor circuit with input to inverting terminal zero.

$$V_{o3} = -\frac{50}{16.67}\times(0\text{ V}) + \left[1+\frac{50}{16.67}\right]\left[\frac{10}{10+10}\right](3\text{ V})$$

$$= 6\text{ V}$$

Similarly ground V_1, V_2 and V_3. Due to this circuit reduces to

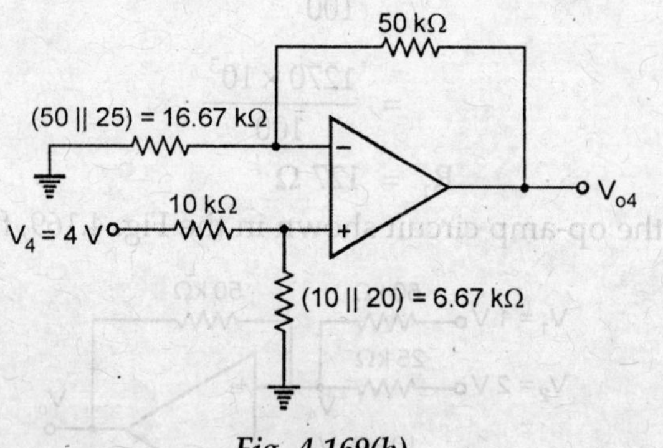

Fig. 4.169(b)

This is also a subtractor circuit with input to inverting terminal zero.

$$\therefore \quad V_{o4} = -\frac{50}{16.67}\times(0\text{ V}) + \left[1+\frac{50}{16.67}\right]\left[\frac{6.67}{20+6.67}\right](4\text{ V}) = 4\text{ V}$$

$\therefore$ Total output voltage is the algebraic sum of all the voltages,

$$V_o = V_{o12} + V_{o3} + V_{o4} = -5 + 6 + 4 = \mathbf{5\text{ V}}$$

This is the output voltage.

Example 4.33: An input d.c. voltage as shown in Fig. 4.170 is fed to an operational amplifier integrator with RC = 1 second. Find the output and sketch. Operational amplifier is nulled initially.

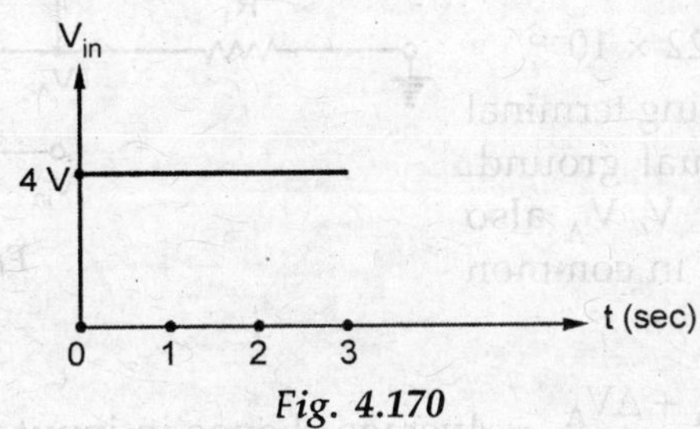

Fig. 4.170

Solution:

RC = 1 see, V_{in} = Step of 4 V starting at t = 0

$$V_{in} = 4\text{ V} \quad \text{for} \quad t \geq 0$$

Then $$V_o(t) = -\frac{1}{RC}\int_0^t V_{in}\, dt$$

$$= -\int_0^t 4\, dt = -4t$$

Initial conditions are zero.

Thus output waveform is a ramp of slope – 4, starting at t = 0. The waveform is shown in the Fig. 4.170(*a*).

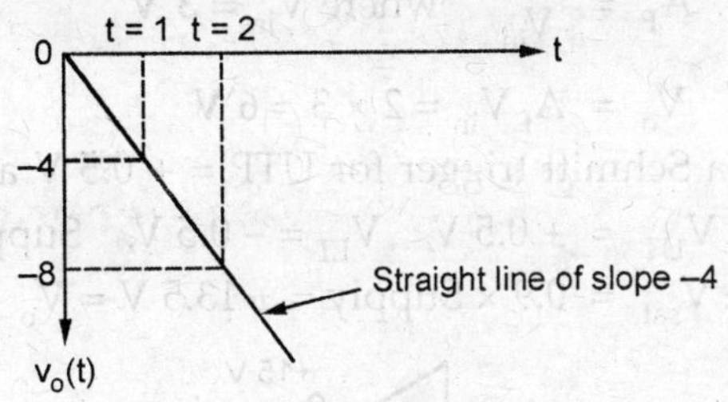

Fig. 4.170(a)

Example 4.34: For the non-inverting operational amplifier with input resistance 100 kΩ and the feedback resistance of 900 kΩ, find the effect on output voltage due to common mode voltage when the input voltage changes by 1 V. Assume CMRR as 70 dB.

Solution: R_1 = 100 kΩ, R_f = 900 kΩ, CMRR = 70 dB

Now $$\text{CMRR in dB} = 20 \log \frac{A_d}{A_c}$$

where $$A_d = 1 + \frac{R_f}{R_1} = 1 + \frac{900}{100} = 10$$ ...Non inverting amplifier

$$\therefore \quad 70 = 20 \log\left(\frac{10}{A_c}\right)$$

$$\therefore \quad A_c = 3.1622 \times 10^{-3}$$

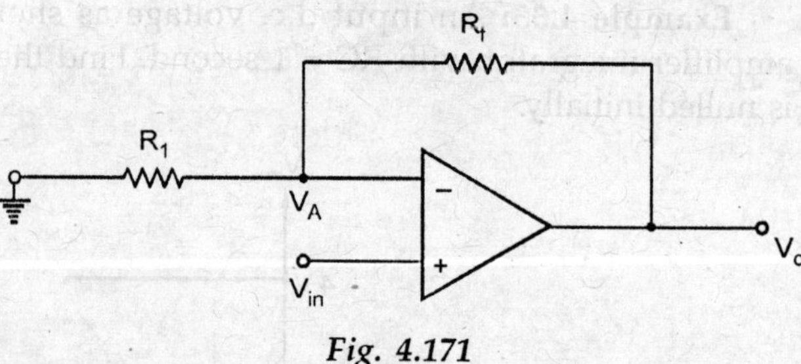

Fig. 4.171

The voltage V_A at inverting terminal is same as V_{in} due to virtual ground. Thus if V_{in} increases by 1 V, V_A also increases by 1 V. So change in common mode voltage is,

$$\Delta V_c = \frac{\Delta V_{in} + \Delta V_A}{2} = \text{Average change in input voltages}$$

$$= \frac{1+1}{2} = 1\,V$$

Thus output voltage will change due to change in common mode input voltage by,

$$\Delta V_o = A_c\,\Delta V_c = 3.1622 \times 10^{-3} \times 1 = 3.1622 \text{ mV}$$

Example 4.35: An input of 3 V is fed to the non-inverting terminal of an operational amplifier. The amplifier has R_i of 10 kΩ and R_f of 10 kΩ. Find the output voltage.

Solution: For the non-inverting amplifier,

$$A_F = 1 + \frac{R_f}{R_i} = 1 + \frac{10}{10} = 2$$

And $$A_F = \frac{V_o}{V_{in}} \quad \text{where } V_{in} = \mathbf{3\ V}$$

$$\therefore \quad V_o = A_f\,V_{in} = 2 \times 3 = \mathbf{6\ V}$$

Example 4.36: Design a Schmitt trigger for UTP = + 0.5 V and LTP = – 0.5 V.

Solution: $V_{UT} = +0.5\text{ V}, \quad V_{LT} = -0.5\text{ V}, \quad \text{Supply} = +15\text{ V}$

$$\pm V_{sat} = 0.9 \times \text{Supply} = \pm 13.5\text{ V} = V_o$$

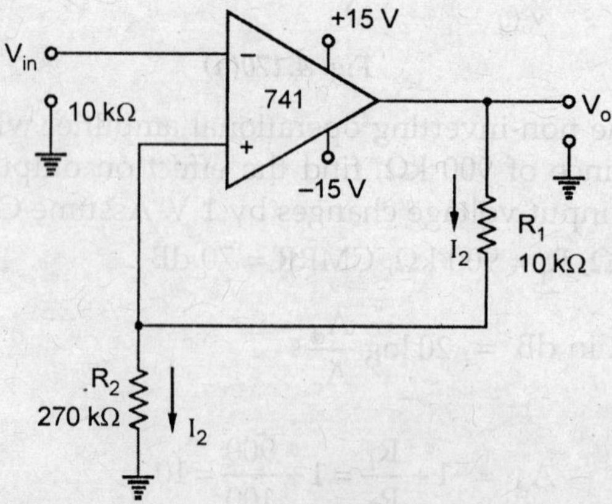

Fig. 4.172

Use op-amp 741 with $I_{B(max)} = 500$ nA

Let I_2 be much higher than $I_{B\,(max)}$.

$$\therefore \quad I_2 = 100\, I_{B(max)} = 50\ \mu A$$

$$\therefore \quad R_2 = \frac{V_{UT}}{I_2} = \frac{0.5}{50 \times 10^{-6}}$$

$$= 10\ k\Omega$$

$$\therefore \quad R_1 = \frac{V_o - V_{UT}}{I_2} = \frac{13.5 - 0.5}{50 \times 10^{-6}}$$

$$= \mathbf{260\ k\Omega} \quad \text{(Use 270 k}\Omega\text{)}$$

The designed circuit is shown in the Fig. 4.172.

Example 4.37: Calculate the output of the circuit shown in Fig. 4.173.

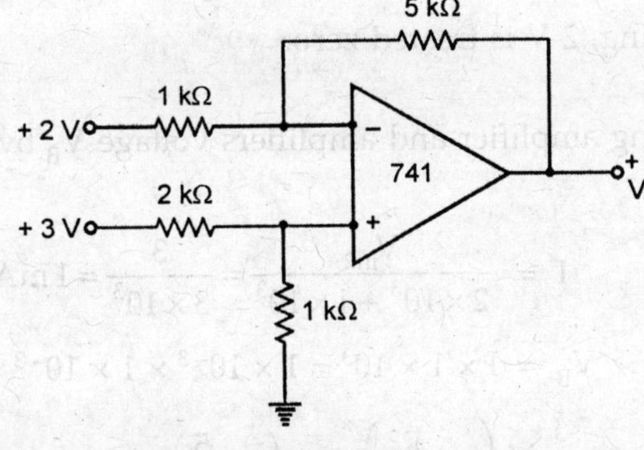

Fig. 4.173

Solution: Use Superposition theorem.

Case 1: Only 2 V acting, 3 V shorted and treated zero.

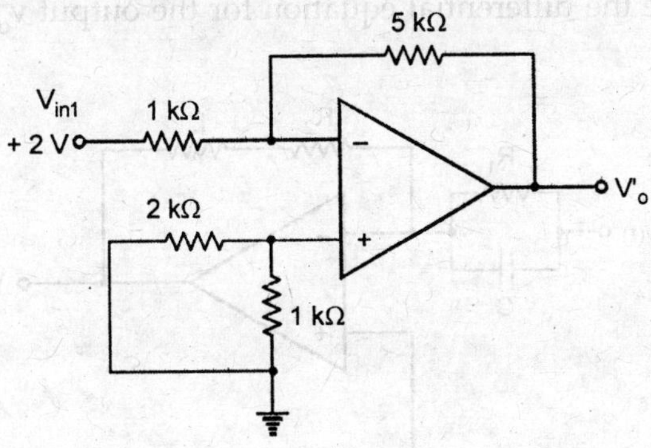

Fig. 4.173(a)

The circuit acts as an inverting amplifier with $R_f = 5$ kΩ and $R_1 = 1$ kΩ.

$$\therefore \quad V_o' = -\frac{R_f}{R_1} V_{in1} = -\frac{5}{1} \times 2 = -\mathbf{10\ V} \qquad \text{...Due to 2 V}$$

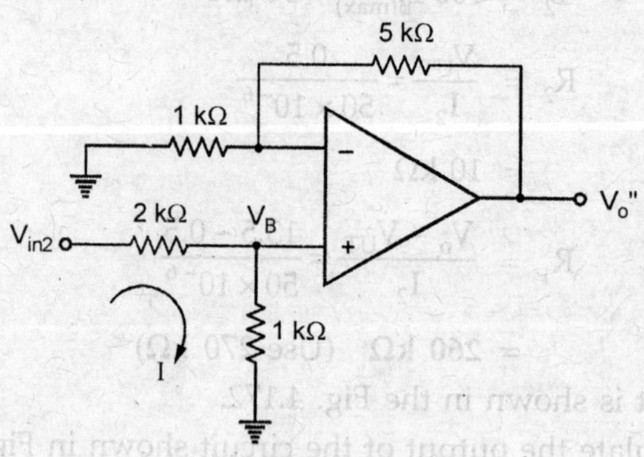

Fig. 4.173(b)

Case 2: Only 3 V acting, 2 V is treated zero.

It acts as non-inverting amplifier and amplifiers voltage V_B by $\left(1+\frac{R_f}{R_1}\right)$ times.

Now
$$I = \frac{V_{in2}}{2\times 10^3 + 1\times 10^3} = \frac{3}{3\times 10^3} = 1\,\text{mA}$$

$$\therefore \quad V_B = I \times 1 \times 10^3 = 1 \times 10^{-3} \times 1 \times 10^{-3} = 1\ \text{V}$$

$$\therefore \quad V_o'' = \left(1+\frac{R_f}{R_1}\right) V_B = \left(1+\frac{5}{1}\right) \times 1 = \mathbf{6\ V} \qquad \text{...Due to 3 V}$$

$\therefore$ Total output voltage $V_o = V_o' + V_o'' = -10 + 6 = \mathbf{-4\ V}$

Example 4.38: Write the differential equation for the output $v_o(t)$ of the circuit shown in Fig. 4.174.

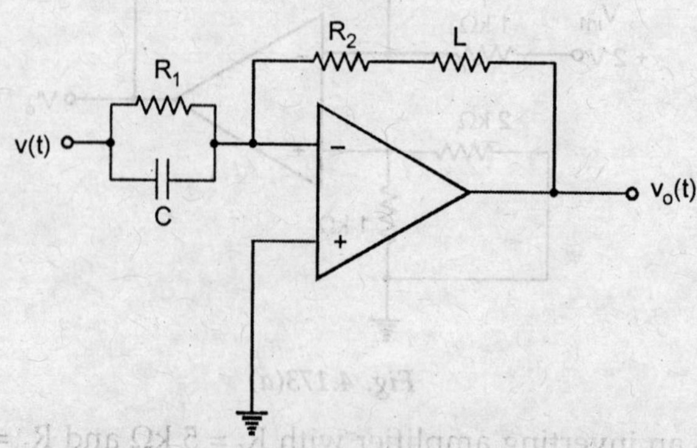

Fig. 4.174

Solution: Taking Laplace transform of the circuit,

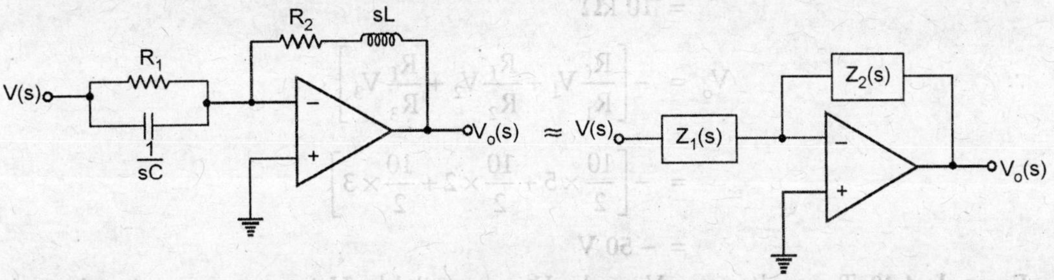

Fig. 4.174(a)

$$Z_1(s) = R_1 || \frac{1}{sC} = \frac{R_1}{1+sCR_1} \quad \text{and} \quad Z_2(s) = R_2 + sL$$

The circuit acts as an inverting amplifier.

$$\therefore \quad \frac{V_o(s)}{V(s)} = -\frac{Z_2(s)}{Z_1(s)} = -\frac{R_2 + sL}{\left(\frac{R_1}{1+sCR_1}\right)} = -\frac{(R_2 + sL)(1+sCR_1)}{R_1}$$

$$\therefore \quad R_1 V_o(s) = -V(s)\left\{+R_2 + s\left[L + CR_1R_2\right] + s^2LCR_1\right\}$$

Putting $\quad s = \frac{d}{dt} \quad \text{and} \quad s^2 = \frac{d^2}{dt^2}$

$$\therefore \quad R_1 V_o(t) = -R_2 v(t) - (L + CR_1C_2)\frac{dv(t)}{dt} - LCR_1\frac{d^2v(t)}{dt^2}$$

$$\therefore \quad v_o(t) = -LC\frac{d^2v(t)}{dt^2} - \left[\frac{L + CR_1CR_2}{R_1}\right]\frac{dv(t)}{dt} - \frac{R_2}{R_1}v(t)$$

This is the expression for $v_o(t)$ in the differential equation form.

Example 4.39: Determine the output V_o for the following circuit.

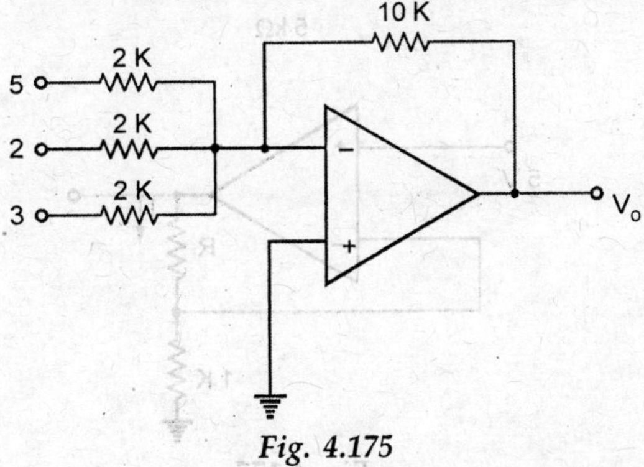

Fig. 4.175

Solution: $R_1 = R_2 = R_3 = 2\text{ kW}$

$$= 10\text{ k}\Omega$$

$$\therefore \quad V_o = -\left[\frac{R_f}{R_1}V_1 + \frac{R_f}{R_2}V_2 + \frac{R_f}{R_3}V_3\right]$$

$$= -\left[\frac{10}{2}\times 5 + \frac{10}{2}\times 2 + \frac{10}{2}\times 3\right]$$

$$= \mathbf{-\,50\ V}$$

Example 4.40: Two voltages + V_1 and + V_2 are available. Using two operational amplifiers draw a circuit to compute 5 V_1 – 3 V_2.

Solution: Use two op-amps.

Design an inverting amplifier with gain – 5 to obtain – 5 V_1 and then use inverting adder.

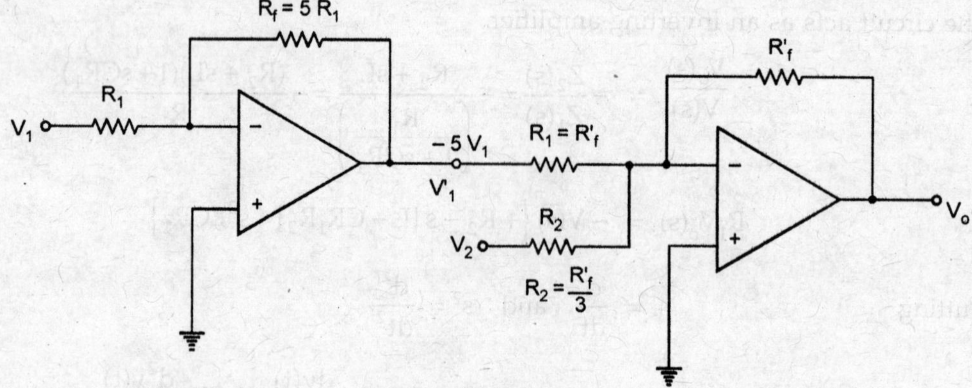

Fig. 4.176

$$\therefore \quad V_o = -\left[\frac{R_f'}{R_1}V_1' + \frac{R_f'}{R_2}V_2\right] = -\left[V_1' + \frac{R_f'}{(R_f'/3)}V_2\right]$$

$$= -[-5\,V_1 + 3\,V_2] = \mathbf{5\,V_1 - 3\,V_2}$$

Example 4.41: Calculate the current I in the circuit shown in Fig. 4.177.

5 kΩ

5 V

I

R

1 K

Fig. 4.177

Solution:

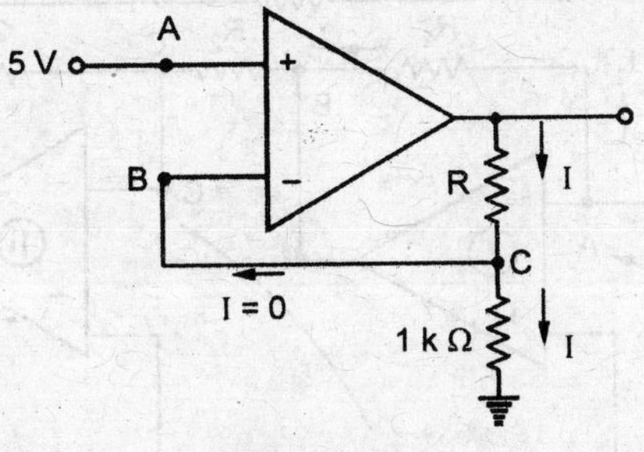

Fig. 4.177(a)

Due to virtual ground,

$$V_B = V_A = 5\text{ V}$$

and

$$V_C = V_B = 5\text{ V}$$

As op-amp input current is zero, current through 1 kΩ is I.

$$\therefore \quad I = \frac{V_C}{1\text{ k}\Omega} = \mathbf{5\text{ mA}}$$

Example 4.42: In the circuit give in Fig. 4.178, show that the input resistance

$$R_i = \frac{V_i}{I_i} = \frac{R_1 R_3}{R_3 - R_1}.$$

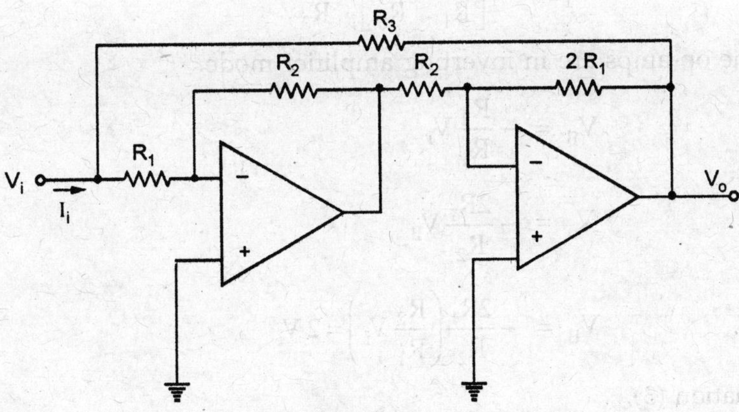

Fig. 4.178

Solution: The current distribution with zero op-amp input current is shown in the Fig. 4.178(*a*).

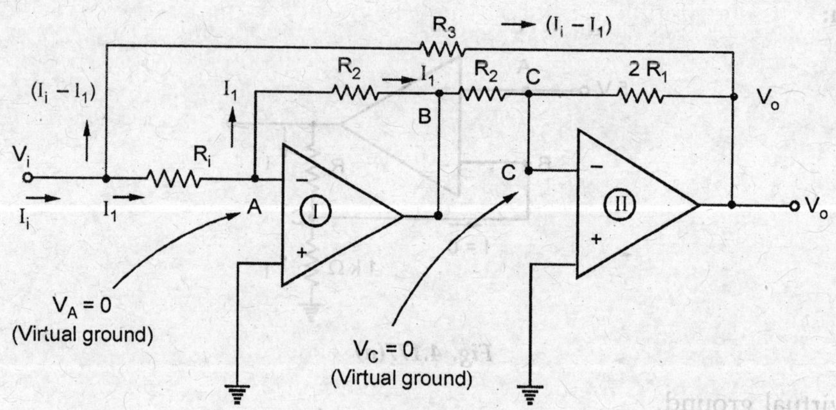

Fig. 4.178(a)

$$I_i - I_1 = \frac{V_i - V_o}{R_3} \quad \text{(branch } R_3\text{)} \qquad ...(1)$$

$$I_1 = \frac{V_i - V_A}{R_1} = \frac{V_i}{R_1} \quad \text{(branch } R_1\text{)} \qquad ...(2)$$

$$\therefore \quad I_1 - \frac{V_i}{R_1} = \frac{V_i}{R_3} - \frac{V_o}{R_3}$$

$$\therefore \quad I_i = V_i\left[\frac{1}{R_1} + \frac{1}{R_3}\right] - \frac{V_o}{R_3} \qquad ...(3)$$

Now both the op-amps are in inverting amplifier mode.

$$\therefore \quad V_B = -\frac{R_2}{R_1} V_1 \qquad ...(4)$$

$$V_o = -\frac{2R_1}{R_2} V_B \qquad ...(5)$$

$$\therefore \quad V_B = -\frac{2R_1}{R_2}\left(\frac{R_2}{R_1} V_i\right) = 2\,V_i \qquad ...(6)$$

Using in equation (3),

$$I_i = V_i\left[\frac{1}{R_1} + \frac{1}{R_3}\right] - \frac{2V_i}{R_3}$$

$$\therefore \quad I_i = V_i\left[\frac{1}{R_1} - \frac{1}{R_3}\right] = V_i\left[\frac{R_3 - R_1}{R_1 R_3}\right]$$

$$\therefore \quad R_i = \frac{V_i}{I_i} = \frac{R_1 R_3}{R3 - R_1} \qquad \text{...proved}$$

Example 4.43: Calculate the current through R_L in the circuit shown in Fig. 4.179.

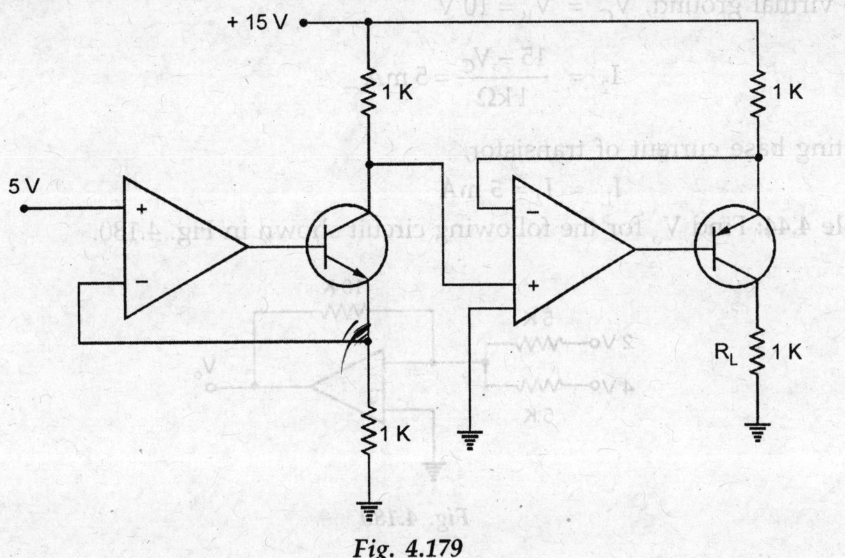

Fig. 4.179

Solution: The voltages and currents are shown in the Fig. 4.179(*a*).

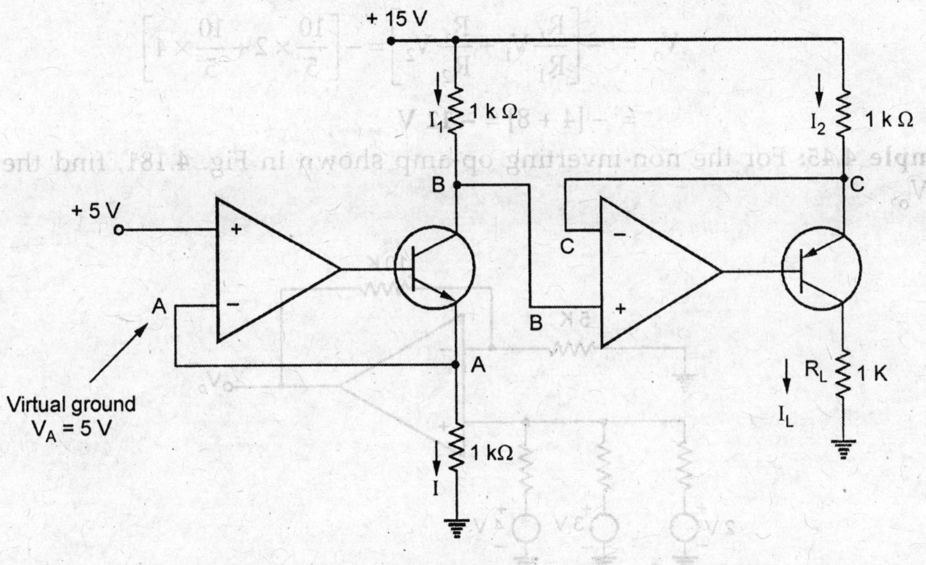

Fig. 4.179(a)

Due to virtual ground, $V_A = 5$ V

$$\therefore \quad I = \frac{V_A}{1\,k\Omega} = 5\,mA$$

Neglecting base current of both the transistors and op-amp input currents,

$$I_1 = I = 5\,mA$$

$$\therefore \quad V_B = 15 - I_1 \times 1\,k\Omega = 10\text{ V}$$

Due to virtual ground, $V_C = V_B = 10\text{ V}$

$$\therefore \quad I_2 = \frac{15 - V_C}{1\,k\Omega} = 5\text{ mA}$$

Neglecting base current of transistor,

$$I_L = I_2 = \mathbf{5\ mA}$$

Example 4.44: Find V_o for the following circuit shown in Fig. 4.180.

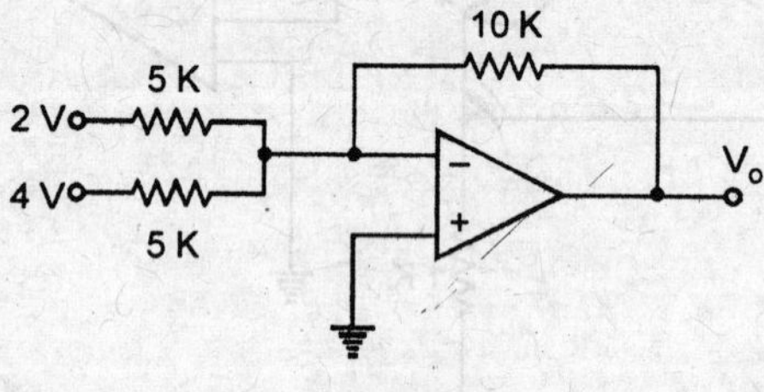

Fig. 4.180

It is an inverting summer.

$R_f = 10\text{ k}\Omega$, $R_1 = 5\text{ k}\Omega$, $R_2 = 5\text{ k}\Omega$, $V_1 = 2\text{ V}$, $V_2 = 4\text{ V}$

$$\therefore \quad V_o = -\left[\frac{R_f}{R_1}V_1 + \frac{R_f}{R_2}V_2\right] = -\left[\frac{10}{5}\times 2 + \frac{10}{5}\times 4\right]$$

$$= -[4 + 8] = \mathbf{-12\ V}$$

Example 4.45: For the non-inverting op-amp shown in Fig. 4.181, find the output voltage V_o.

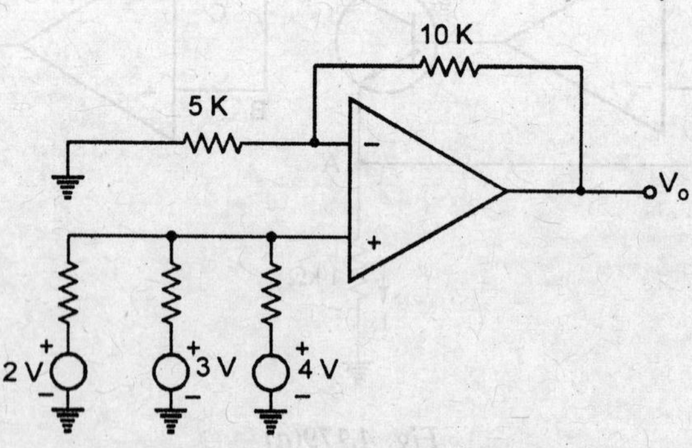

Fig. 4.181

Solution: As shown in the Fig. 4.181, it is a non-inverting summing amplifier. Use the Superposition principle.

Step 1: 4 V acting, 2 V and 3 V are grounded.

$$\therefore \quad V_B = \left[\frac{4}{R+\frac{R}{2}} \times \frac{R}{2}\right] = \frac{4}{3}\,V$$

$$\therefore \quad V_{o2} = \left[1+\frac{R_t}{R_1}\right] V_B = \left[1+\frac{10}{5}\right] \times \frac{4}{5}$$

$$= \mathbf{4\ V}$$

Fig. 4.181 (a)

Step 2: V acting, 2 V and 4 V are grounded.

$$\therefore \quad V_B = \left[\frac{3}{R+\frac{R}{2}} \times \frac{R}{2}\right] = 1\,V$$

$$\therefore \quad R_B = \left[1+\frac{10}{5}\right] \times 1 = \mathbf{3\ V}$$

Fig. 4.181 (b)

Step 3: 2 V acting, 3 V and 4 V are grounded.

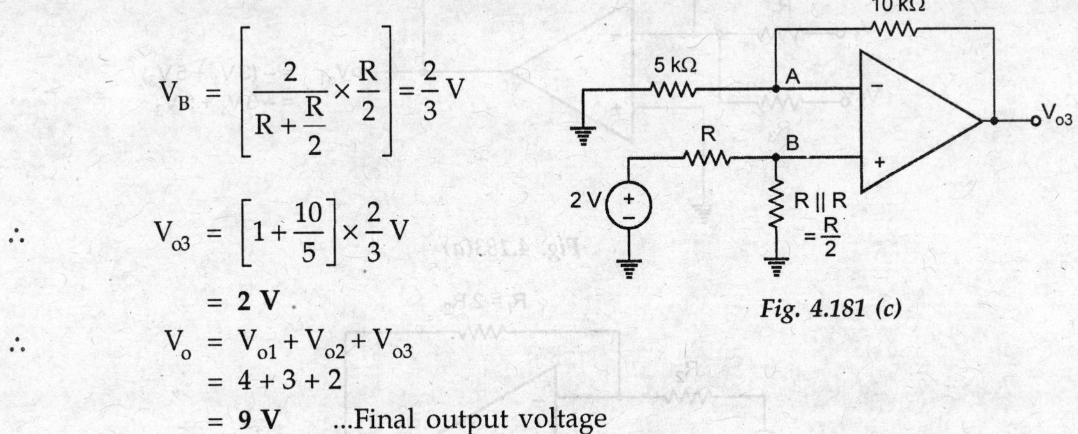

$$V_B = \left[\frac{2}{R+\frac{R}{2}} \times \frac{R}{2}\right] = \frac{2}{3}\,V$$

$$\therefore \quad V_{o3} = \left[1+\frac{10}{5}\right] \times \frac{2}{3}\,V$$

$$= \mathbf{2\ V}$$

Fig. 4.181 (c)

$$\therefore \quad V_o = V_{o1} + V_{o2} + V_{o3}$$

$$= 4 + 3 + 2$$

$$= \mathbf{9\ V} \quad \text{...Final output voltage}$$

Example 4.46: Determine the output voltage V_o for the following circuit shown in Fig. 4.182.

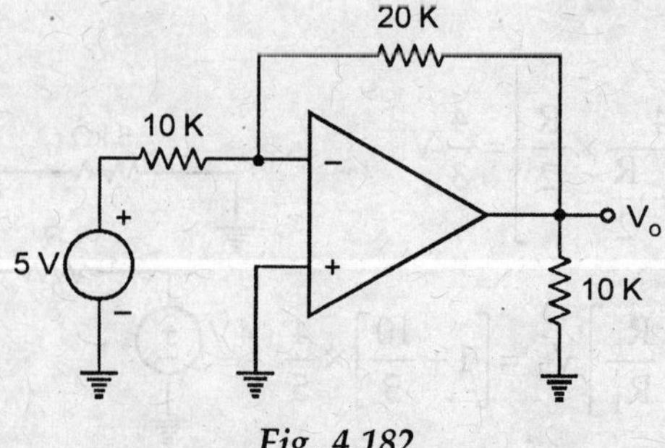

Fig. 4.182

Solution: It is inverting amplifier with R_f = 20 kΩ, R_1 = 10 kΩ.

$$\therefore \quad V_o = -\frac{R_f}{R_1} V_{in} = -\frac{20}{20} \times 5 = \mathbf{-10\ V}$$

Example 4.47: Design an op-amp circuit to give an output voltage $V_o = 3V_1 - 2V_2 + 5V_3$ where V_1, V_2 and V_3 are inputs.

Solution: In the first step, design an adder to get $3V_1 + 5V_3$.

$$V_{o1} = -\left[\frac{R_{f1}}{R_1} V_1 + \frac{R_{f1}}{R_3} V_3\right]$$

$$\therefore \quad \frac{R_{f1}}{R_1} = 3 \quad \text{and} \quad \frac{R_{f1}}{R_3} = 5$$

Choose R_f = 100 kΩ, R_1 = 33.33 kΩ, R_3 = 20 kΩ

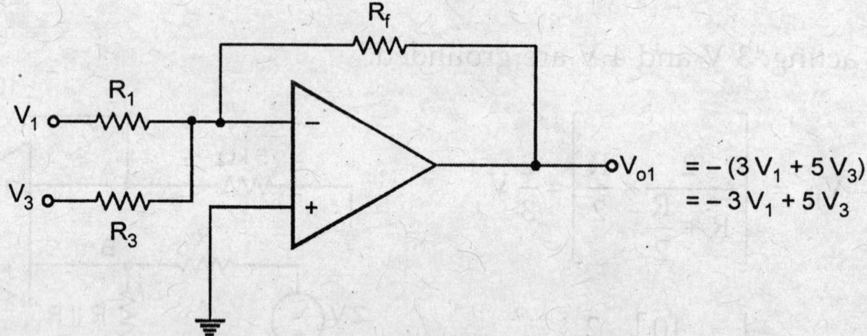

Fig. 4.183(a)

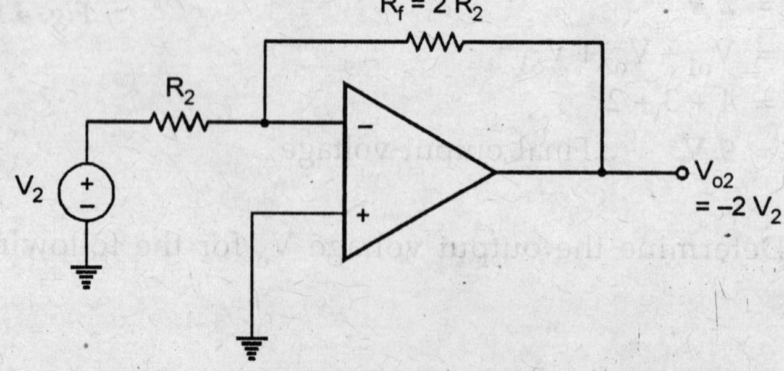

Fig. 4.183(b)

Now generate 2 V_2 by using a inverting amplifier of gain 2.

$$V_{o2} = -\frac{R_f}{R_1} V_2 = -2V_2$$

$\therefore$ $R_f = 2 R_2$

Choose $R_f = 10\ k\Omega$

$R_2 = 5\ k\Omega$

Now use subtractor for getting $V_{o2} - V_{o1}$.

$\therefore$ $V_o = V_{o2} - V_{o1} = -2 V_2 - (-3 V_1 - 5 V_3) = \mathbf{3\ V_1 + 5\ V_3 - 2\ V_2}$

Thus the overall circuit is shown in the Fig. 4.183(*c*).

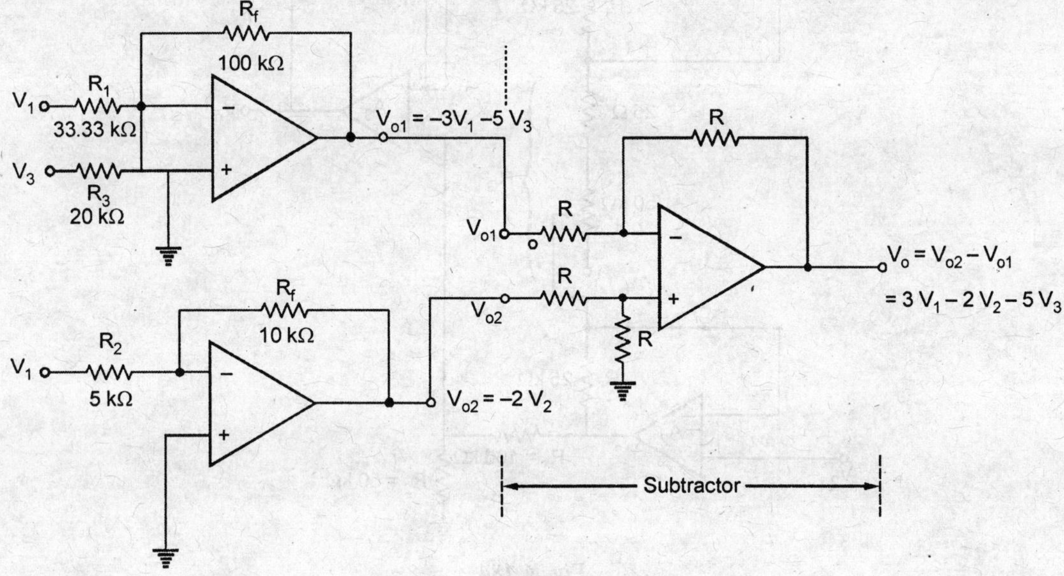

Fig. 4.183(c)

Example 4.48: Design an instrumentation amplifier whose gain can be varied continuously over the range $1 \leq A \leq 1000$. Assume all other relevant details.

Solution: The gain of the instrumentation amplifier is,

$$A = \left(1 + \frac{2R_f}{R_G}\right)\left(\frac{R_2}{R_1}\right) = A_1 A_2$$

As A_1 is greater than 1, A_2 must be less than 1, to get gain A is unity.

Let $A_2 = \frac{1}{2}$ i.e. $\frac{R_2}{R_1} = \frac{1}{2}$

Choose $R_1 = 100\ k\Omega$, $R_2 = 50\ k\Omega$

Let R_G be a pot of 50 kΩ and a base resistance of R_{GB} in series with it. With pot on zero resistance position we get,

$$\left(1+\frac{2R_f}{R_{GB}}\right)\left(\frac{1}{2}\right) = 1000 \quad \text{i.e.} \quad \frac{R_f}{R_{GB}} = 999.5 \qquad ...(1)$$

When potentiometer is at maximum position,

$$\left(1+\frac{2R_f}{50\times10^3+R_{GB}}\right)\left(\frac{1}{2}\right) = 1 \quad \text{i.e.} \quad \frac{R_f}{50\times10^3+R_{GB}} = 0.5 \qquad ...(2)$$

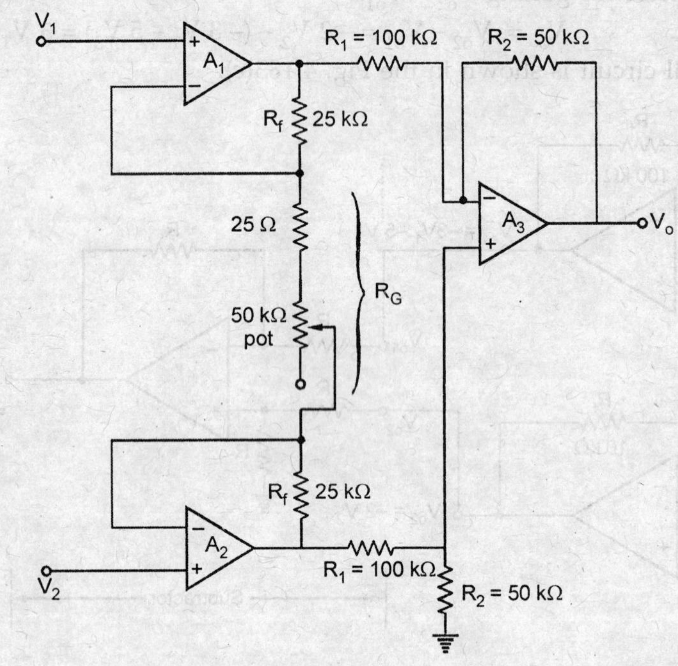

Fig. 4.184

Solving equations (1) and (2),

R_{GB} = 25 W ...Base resistance of potentiometer

∴ R_f = 25.012 kΩ

The designed circuit is shown in Fig. 2.184.

Example 4.49: Design a circuit to implement $V_o = 0.545\ V_3 + 0.273\ V_4 - 1.25\ V_1 - 2\ V_2$.

Solution: $V_o = 0.545\ V_3 + 0.273\ V_4 - 1.25\ V_1 - 2\ V_2$

Design circuit for ($0.545\ V_3 + 0.273\ V_4$) and ($1.25\ V_1 + 2\ V_2$) and then use subtractor. For ($0.545\ V_3 + 0.273\ V_4$). Let R_{f1} = 10 kΩ

$$\therefore \quad V_{o1} = -\left(\frac{R_{f1}}{R_3}V_3+\frac{R_{f1}}{R_4}V_4\right) \quad \text{i.e.} \quad \frac{R_{f1}}{R_3}=0.545, \frac{R_{f1}}{R_4}=0.273$$

∴ R_3 = 18.348 kΩ, R_4 = 36.63 kΩ

For ($1.25\ V_1 + 2\ V_2$), Let R_{f2} = 50 kΩ

$$\therefore \quad V_{o2} = -\left(\frac{R_{f2}}{R_1}V_1 + \frac{R_{f2}}{R_2}V_2\right) \quad \text{i.e.} \quad \frac{R_{f2}}{R_1} = 1.25, \quad \frac{R_{f2}}{R_2} = 2$$

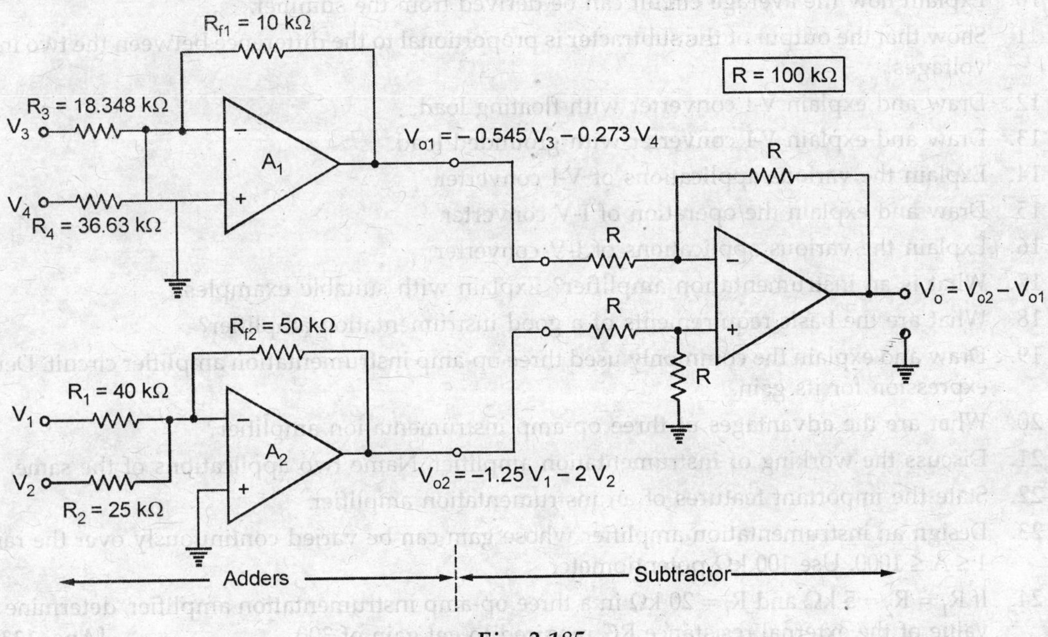

Fig. 2.185

$$\therefore \quad R_1 = 40\ \text{k}\Omega, \quad R_2 = 25\ \text{k}\Omega$$

Use the subs tractor with all resistances of same value i.e. R = 100 kΩ

Hence, $\quad V_o = V_{o2} - V_1 = -(1.25\ V_1 + 2\ V_2) - [-\,0.545\ V_3 - 0.273\ V_4]$

$\quad = 0.542\ V_3 + 0.273\ V_4 - 1.25\ V_1 - 2\ V_2$ As required

The designed circuit is shown in the Fig. 4.185.

REVIEW QUESTIONS

1. Explain the working of op-amp inverting amplifier. Derive the expression for its voltage gain.
2. Explain the working of op-amp non-inverting amplifier. Derive the expression for its voltage gain.
3. For the non-inverting operational amplifier with input resistance 100 kΩ and feedback resistance 900 kΩ. Find the effect on output voltage due to common mode voltage when the input voltage changes by 1 V. Assume CMRR as 70 dB.
4. An input of 3 V is fed to the non-inverting terminal of an operational amplifier. The amplifier has a R_i of 10 kΩ and R_f of 10 kΩ. Find the output voltage.
5. Compare inverting and non-invert ins amplifier.
6. Explain the use of voltage follower circuit.

7. Draw and explain the working of phase shifter circuit.
8. Explain thc working of inverting summer with suitable example.
9. Explain the working of lion-inverting summer with neat circuit diagram.
10. Explain how the average circuit can be derived from the summer.
11. Show that the output of the subtracter is proportional to the difference between the two input voltages.
12. Draw and explain V-I converter with floating load.
13. Draw and explain V-I converter with grounded load.
14. Explain the various applications of V-I converter.
15. Draw and explain the operation of I-V converter.
16. Explain the various applications of I-V converter.
17. What is an instrumentation amplifier? Explain with suitable examples.
18. What are the basic requirements of a good instrumentation amplifier?
19. Draw and explain the commonly used three op-amp instrumentation amplifier circuit. Derive expression for its gain.
20. What are the advantages of three op-amp instrumentation amplifier?
21. Discuss the working of instrumentation amplifier. Name two applications of the same.
22. State the important features of an instrumentation amplifier.
23. Design an instrumentation amplifier whose gain can be varied continuously over the range $1 \le A \le 1000$. Use 100 kΩ potentiometer.
24. If $R_1 = R_2 = 5$ kΩ and $R_f = 20$ kΩ in a three op-amp instrumentation amplifier, determine the value of the external resistance RC required to get gain of 300. **[Ans. 133 Ω]**
25. What is an active integrator? With neat circuit diagram explain the working of an active integrator.
26. Explain and draw the output waveforms of the ideal integrator circuit when the input is (*i*) Sine wave (*ii*) Step input (*iii*) Square wave
27. Explain the various errors in an ideal integrator circuit. How these errors are minimized?
28. Explain the practical integrator circuit. Explain its advantages.
29. Explain the working of an ideal active differentiator.
30. What are the limitations of an ideal active differentiator?
31. Explain the working of practical differentiator.
32. Explain the applications of a practical differentiator.
33. Draw the circuit diagram and explain the basic Log amplifier using (*i*) Diode (*ii*) Transistor.
34. What are the limitations of the basic Log amplifier?
35. How Log amplifier using diode is thermally compensated? Derive the output expression.
36. How Log amplifier using transistor is temperature compensated? Derive the output expression.
37. Explain the temperature compensation of Log amplifier using two op-amps.
38. Write a short note on Log amplifier.
39. Write a short note on Antilog amplifier.
40. Explain the working of Antilog amplifier.

41. Draw the circuit diagram of a temperature compensated antilog amplifier and obtain the output voltage expression.
42. Explain use of op-amp in switching circuits.
43. Draw the circuit diagram of a comparator using IC 741 op-amp. Explain its working.
44. How to move a trip point in a basic non-inverting comparator?
45. Explain zero crossing detector and draw the input-output waveforms.
46. Draw and explain the working of voltage level detector.
47. Explain how an op-amp can be used as comparator.
48. Draw and explain inverting and non-inverting comparator circuits.
49. Explain op-amp as a Schmitt trigger.
50. What is the most important applications of Schmitt trigger circuit.
51. Explain the working of Schmitt trigger.
52. Design Schmitt trigger having upper and lower thresholds of 120 mV. Input to this circuit is 1 V peak to peak triangular wave of 100 Hz. Draw the hysteresis loop.
53. Design a Schmitt trigger for UTP = + 0.5 V and LTP = – 0.5 V.
54. How to obtain different UPT and LTP levels in a Schmitt trigger?
55. What is a peak detector circuit?
56. Explain the basic principle of a peak detector circuit.
57. What do you understand by precision rectifiers ? How do they differ from conventional rectifiers?
58. Draw and explain half wave precision rectifier circuit.
59. With the help of neat diagram and waveform explain the full wave precision rectifier circuit.
60. Write a note on two output precision rectifier.
61. What is a positive clipper circuit and a negative clipper circuit?
62. Explain with neat waveforms, a positive clipper circuit.
63. Explain with neat waveforms, a negative clipper circuit.
64. What is a clamper circuit? How clamper circuits are classified ?
65. Explain with neat circuit diagram and the waveforms, the working of a negative clamper circuit.

❖❖❖

Chapter 5

Active Filters

5.1 INTRODUCTION

A filter is a circuit that is designed to pass a specified band of frequencies while attenuating all the signals outside that band. It is a frequency selective circuit.

The filters are basically classified as active filters and passive niters. The **passive filter** networks use only passive elements such as resistors, inductors and capacitors. On the other hand, **active filter** circuits use the active elements such as op-amps, transistors alongwith the resistors, inductors and capacitors. Modern active filters do not use inductors as the inductors are bulky, heavy and nonlinear. The inductors generate the stray magnetic fields. The inductors dissipate considerable amount of power.

5.1.1 Advantages of Active Filters

The active filters have the following advantges over the passive filters,

1. All the elements alongwith op-amp can be used in the integrated form. Hence there is reduction is size and weight.
2. In large quantities, the cost of the integrated circuit can be much lower than its equivalent passive network.
3. Due to availability of modern ICs, variety of cheaper op-amps are available.

4. The op-amp gain can be easily controlled in the closed loop fashion hence active filter input signals is not attenuated.
5. Due to flexibility in gain and frequency adjustments, the active filters can be easily tuned.
6. The op-amp has high input impedance and low output impedance hence the active filters using op-amp do not cause loading of the source or load.
7. The inductors are absent in the active filters hence the modern active niters are more economical.
8. Active filters can be realized under number of class of functions such as Butterworth, Thomson, Chebyshev, Cauer etc.
9. The response is improved as compared to passive filters due to ready availability of high quality components.
10. The design procedure is simpler than that for the passive niters.
11. Active niters can provide voltage gains, in contrast the passive niters often show a significant voltage loss.

Inspite of the above advantages, the active niters also have certain **limitations.** The finite bandwidth of the active devices place a limit on the highest frequency of operation. For most of the applications, with present day technology, a reasonably good filter performance can be achieved approximately upto 500 kHz. As against this, passive filters can be used upto 500 MHz.

Another important factor is the **sensitivity.** The active elements are much more sensitive to the temperature and the environmental changes than the passive elements. Hence due to changes in the parameters due to the environmental changes, the active filter performance deviates from its ideal response.

The requirement of d.c. power supply is another disadvantage of the active filters. The passive filters do not require the d.c. supply.

The most commonly used filters are,

1. Low Pass (LP) filter.
2. High Pass (HP) filter.
3. Band Pass (BP) filter.
4. Band Reject (BR) filter. This is also called Band Stop or Band Elimination filter.
5. All Pass filter.

5.2 CONCEPT OF FILTER AND FREQUENCY RESPONSE

The first basic type of filter is **low pass filter.** As the name suggests, it passes low frequency signal from input to output while it blocks high frequency signals from input.

Consider a low pass filter represented by a block as shown in the Fig. 5.1. The input voltage v_x has a low frequency signal alongwith some unwanted high frequency signals. The unwanted high frequency signals are called hash and are generated due to noise, scratching sounds etc. The function of low pass filter is to remove this hash from input and to produce low frequency signal output. This is shown in the Fig. 5.1.

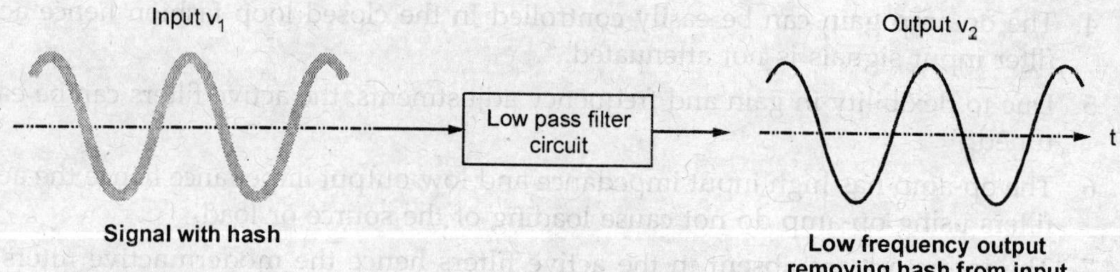

Fig. 5.1 Concept of low pass filtering

If input consists many low frequency signals then each one gets transmitted to the output, **without any change in amplitude.**

Thus **low pass filter is a circuit which passes the low frequency signals from input to output, rejecting the high frequency signals.**

5.2.1 Frequency Response

Any filter has two important characteristics,

1. **Transfer function** which is defined as the ratio of Laplace transform of output to Laplace transform of input. It is denoted as T(s). Replacing variable s by jco we get the frequency domain transfer function T(jω), This also represents **gain** of the filter.

$$\therefore \quad T(s) = \frac{V_o(s)}{V_{in}(s)}$$

$$\text{i.e.} \quad T(j\omega) = \frac{V_o(j\omega)}{V_{in}(j\omega)}$$

T(jω) need not be always gain but sometimes it may be an attenuation or loss. Depending upon whether it is an **attenuation** or **gain,** it is represented as α or **A.** Mathematically **attenuation** or **loss** is defined as,

$$\alpha = -20 \log |T| \text{ dB} \quad \text{...(1)}$$

The unit of attenuation is decibel (dB). When the circuit provides attenuation or loss, $|T| \leq 1$; $\log |T|$ is negative, hence to get α positive minus sign is introduced in equation (1).

Similarly the **gain** is defined as,

$$A = 20 \log |T| \text{ dB} \quad \text{...(2)}$$

When a circuit provides gain, $|T| > 1$; so $\log |T|$ is positive, hence A is positive. The unit of gain is also decibel (dB).

2. Consider a transfer function of the form given by,

$$T(s) = \frac{a_1(s) + a_2}{b_1(s) + b_2} = \frac{N(s)}{D(s)} \quad \text{...(3)}$$

where a_1, a_2, b_1 and b_2 are real constants; may be either positive or negative.

Then equation (3) can be modified as,

$$T(s) = \frac{a_1}{b_1} \cdot \frac{s + a_2 / a_1}{s + b_2 / b_1} = K\frac{s + z_1}{s + p_1} \qquad \text{...(4)}$$

The constant K is called **system gain factor or scale factor.**

The complex frequencies which make the value of the system function infinite are called poles. In our case, $s = -p_1$ is the pole of the system as the system function becomes infinite when it is substi hated in the denominator.

The complex frequencies which make the value of the system function zero are called zeros. In our case, $s = -z_1$ is the zero of the system as the system function becomes zero when it is substituted in the numerator. The pole-zero plot for the transfer function in equation (4) is as shown in the Fig. 5.2. Note that the scale factor K is not specified in pole-zero plot as it is a plot for frequencies only.

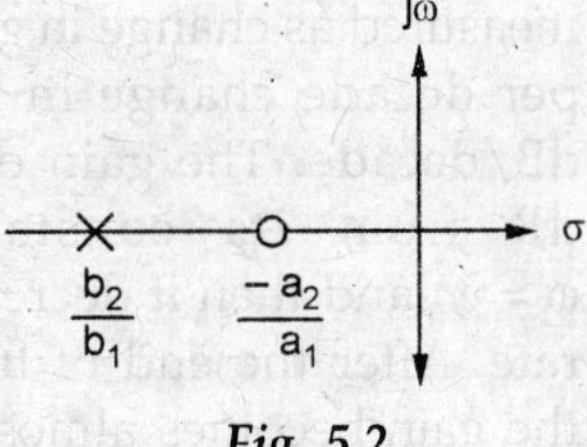

Fig. 5.2

The highest power of s in the denominator polynomial of T(s) represents the **order** of the filter.

3. The **cut-off frequency** ω_0 which indicates that frequency where filter changes its function. This means for a low pass filter, it passes all frequencies below ω_0 and rejects all frequencies above ω_0. The magnitude of transfrer function | T(jω) | remains constant for all frequencies below ω_0 for low pass filter. While | T(jωt) | decreases and becomes ideally zero for frequencies greater than ω_0.

The graph of magnitude of transfer function | T(jω) | against the frequency ω is called frequency response of a filter.

For a low pass filter, the **ideal** frequency response is shown in the Fig. 5.3.

The | T(jω) | is constant for $\omega < \omega_0$ and is zero for $\omega > \omega_0$. The frequency band $0 < \omega < \omega_0$ is called **pass band.** While the frequency band for $\omega > \omega_0$ is called **stop band.** Due to a brick like structure, the characteristics is called **brick wall,** which is **ideal** for low pass filter.

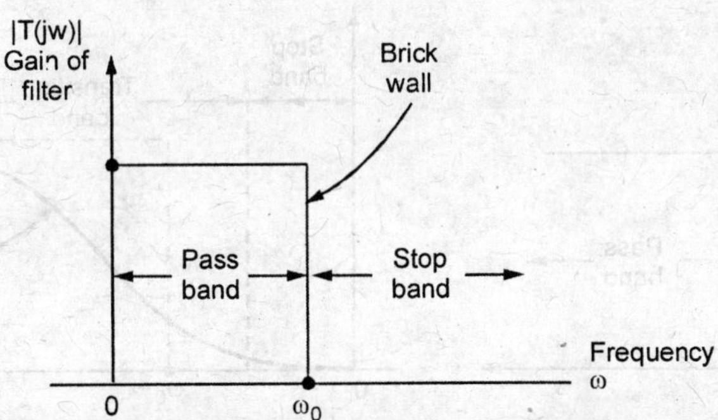

Fig. 5.3 Ideal frequency response of low pass filter

Practically such as instantaneous drop in | T(jω) | at $\omega = \omega_0$ is not possible. But the gain | T(jω) | starts decreasing from co_0 onwards and becomes almost zero after a band of frequencies. This band is called **transition band** which practically exists in all types of filters. The decrease in gain during transition band is called gain roll off. This rate depends on the order of the filler and is called **n-pole roll off.** More the value of n, more fast is the roll off and more ideal is the frequency response. This is shown in the Fig. 5.4.

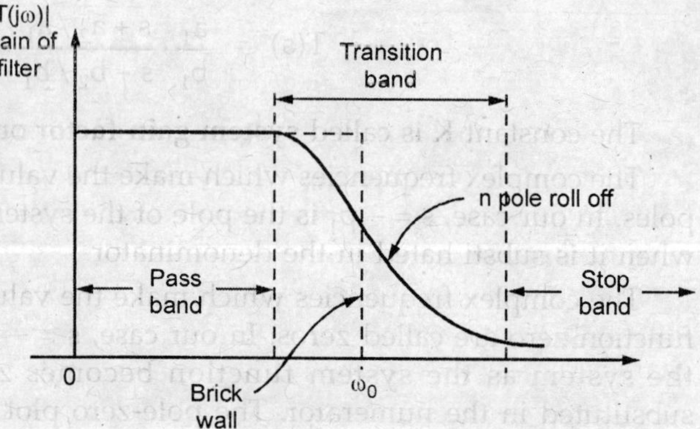

Fig. 5.4 Practical frequency response of low pass filter

Practically the n pole roll off is measured as change in gain in decibles per decade change in frequency i.e. dB/decade. The gain decreases by 3 dB from its constant value at $\omega = \omega_0$ and then it decreases by higher rate. After the end of transition band the gain becomes almost zero.

Using proper design techniques, precision components and high speed op-amps, the practical response can be obtained very close to the ideal response.

5.3 FREQUENCY RESPONSE CHARACTERISTICS OF OTHER TYPES OF FILTERS

Let us study the frequency response characteristics of other types of filters which are high pass, band pass, band elimination and all pass filters.

5.3.1 High Pass Filter

In the high pass filter, the filter rejects the frequencies which are less than cut-off frequency ω_0 and it allows to pass the frequencies which are greater than ω_0. Thus the range of frequencies $0 < \omega < \omega_0$ is stop band while $\omega > \omega_0$ is the pass band. The Fig. 5.5 shows the ideal and practical frequency response characteristics of high pass filter.

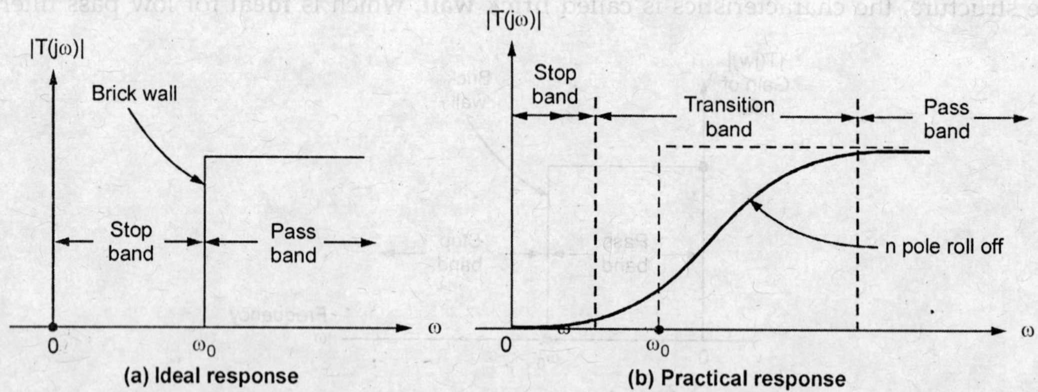

Fig. 5.5 Frequency response characteristics of high pass filter

Practically transition band is designed to be very small such that practical response almost matches with the ideal response and the two bands stop and pass are separated at cut-off frequency ω_0 itself.

5.3.2 Band Pass Filter

The band pass filter has two stop bands and two cut-off frequencies. It rejects the frequencies less than ω_1 and the frequencies higher than ω_2, while it passes the band of frequencies between ω_1 and ω_2. Thus $0 < \omega < \omega_1$ and $\omega > \omega_2$ are the stop bands while $\omega_1 < \omega < \omega_2$ is the pass band. The frequency between ω_1, and ω_2 where the gain $| T(j\omega) |$ attains maximum value is called **centre frequency** denoted as ω_0. At ω_1 and ω_2, the gain 3 dB down from its maximum value hence are called **3 dB frequencies.** The difference $\omega_2 - \omega_1$ is called **bandwidth** of the filter. The Fig. 5.6 shows the ideal and practical frequency response characteristics of band pass filter.

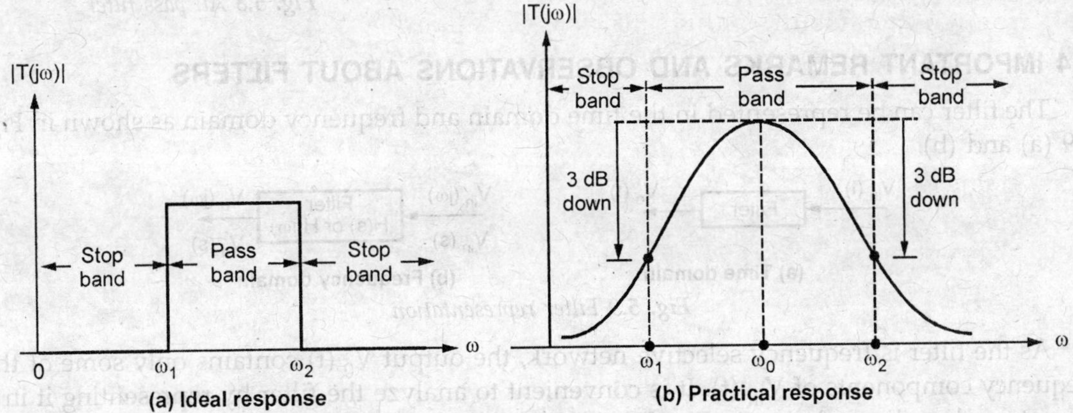

Fig. 5.6 Frequency response characteristics of band pass filter

5.3.3 Band Elimination Filter

This is also called **band rejection** filter or **band stop filter** or **notch filter.** Its characteristics are exactly opposite to that of band pass filter. There are two pass bands and one stop band. The stop band is between two 3 dB frequencies ω_1 and ω_2. For the frequency ranges $0 < \omega < \omega_1$ and $\omega_2 > \omega$ are the two pass bands while the band for $\omega_1 < \omega < \omega_2$ is the stop band. The frequency at which the gain is minimum (almost zero) is called centre frequency denoted as ω_0. At frequency ω_0, the practical response shows a notch hence this filter is also called a **notch filter.** The Fig. 5.7 shows the ideal and practical characteristics of band elimination filter.

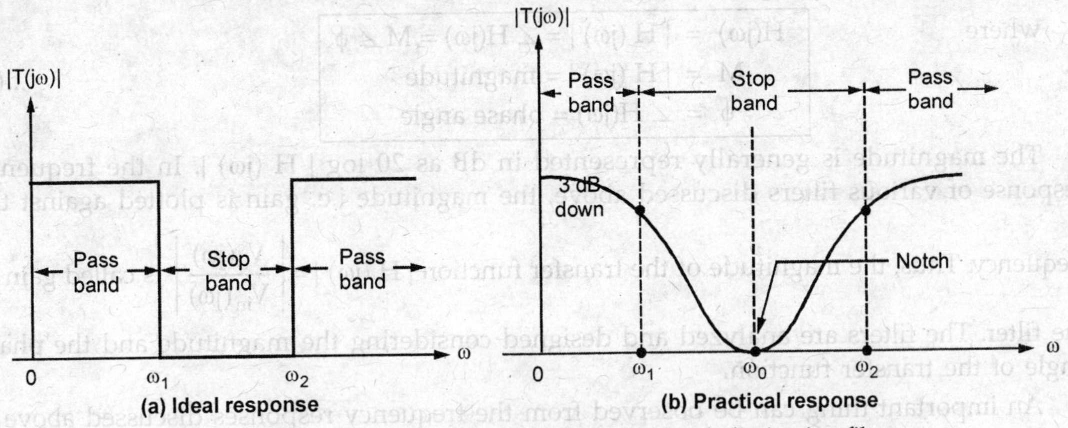

Fig. 5.7 Frequency response characteristics of band elimination filter

The all pass filter passes all the frequencies but it produces the phase shift between the input and output The output and input voltages are equal in amplitude for all the frequencies but with the phase shift between the two. The frequency upto which the input-output amplitudes remains same is decided by the **Unity Gain Bandwidth (UGB)** of the op-amp used. The characteristics showing phase shift between input and output is shown in Fig. 5.8.

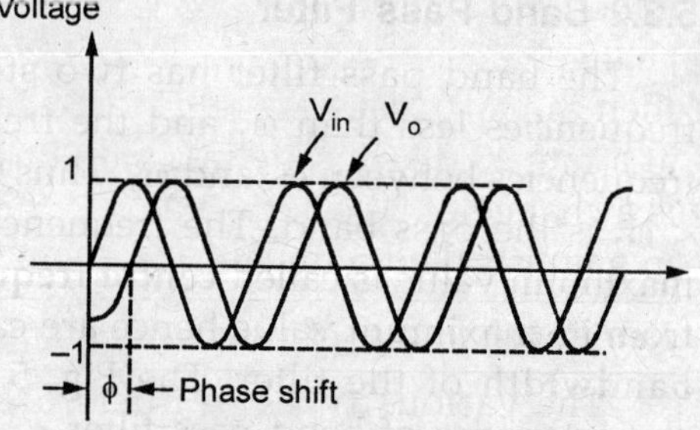

Fig. 5.8 All pass filter

5.4 IMPORTANT REMARKS AND OBSERVATIONS ABOUT FILTERS

The filter can be represented in the time domain and frequency domain as shown in Fig. 5.9 (a) and (b).

V_{in} (t) → Filter → V_o (t)

(a) Time domain

V_{in} (jω), V_{in} (s) → Filter H(s) or H(jω) → V_o (jω), V_o (s)

(b) Frequency domain

Fig. 5.9 Filter representation

As the filter is frequency selective network, the output V_0 (t) contains only some of the frequency components of V_{in} (t). It is convenient to analyze the filter by representing it in a frequency domain as shown in Fig. 5.9 (b). In the frequency domain, the filter is described by the transfer function,

$$H(s) = \frac{V_o(s)}{V_{in}(s)} \quad \text{...(1 a)}$$

or

$$H(jw) = \frac{V_o(j\omega)}{V_{in}(j\omega)} \quad \text{...(1 b)}$$

where ω = 2πf and f is the operating frequency. In the steady state, the transfer function can be represented in the polar form as,

$$H(j\omega) = |H(j\omega)| = \angle H(j\omega) = M \angle \phi$$

where

$$M = |H(j\omega)| = \text{magnitude} \quad \text{...(2)}$$

$$\phi = \angle H(j\omega) = \text{phase angle}$$

The magnitude is generally represented in dB as 20 log | H (jω) |. In the frequency response of various filters discussed above, the magnitude i.e. gain is plotted against the frequency. Thus, the magnitude of the transfer function $|H(j\omega)| = \left|\frac{V_o(j\omega)}{V_{in}(j\omega)}\right|$ is called gain of the filter. The filters are analyzed and designed considering the magnitude and the phase angle of the transfer function.

An important thing can be observed from the frequency responses discussed above is the behaviour of the gain in the stop band for the various filters. The frequency response either decreases or increases or both in the stop band. The rate at which the gain of the filter changes in the stop band is dependent on the order of the filter. If the filter is first order then gain increases at a rate 20 dB/decade in a stop band of high pass filter, the gain decreases at a rate 20 dB/decade in a stop band of low pass filter and so on. This indicates that there is a change of 20 dB in a gain per decade (10 times) change in the frequency. Such a change in gain is called **gain roll off.**

***Key Point:** In case of a second order filters, the gain roll off is at the rate of 40 dB/decade and so on.*

The various types of filters used in practice which approximately produce the ideal response are: (*i*) Butterworth filters

(*ii*) Chebyshev Filters

(*iii*) Cauer Filters.

5.4.1 Butterworth Approximation

The filter in which denominator polynomial of its transfer function is a Butterworth polynomial is called a Butterworth filter. The Butterworth polynomials of various orders are given in the Tables 5.1 and 5.2.

Table 5.1 Butterworth polynomials in factored form

n	
1	s + 1
2	$s^2 + \sqrt{2s}$ +
3	$(s^2 + s + 1)(s + 1)$
4	$(s^2 + 0.76536\,s + 1)(s^2 + 1.84776\,s + 1)$
5	$(s + 1)(s^2 + 0.6180\,s + 1)(s^2 + 1.6`80\,s + 1)$
6	$(s^2 + 0.5176\,s + 1((s^2 + \sqrt{2s} + 1)(s^2 + 1.9318\,s + 1)$
7	$(s + 1)(s^2 + 0.4450\,s + 1)(s^2 + 1.2456\,s + 1)(s^2 + 1.8022\,s + 1)$
8	$1(s^2 + 0.3986\,s + 1)(s^2 + 1.1110\,s + 1)(s^2 + 1.6630\,s + 1)(s^2 + 1.9622\,s + 1)$

The coefficient of Butterworth polynomials are as given in the Table 5.2

Table 5.2 Coefficients of Butterworth polynomials

n	a_1	a_2	a_3	a_4	a_5	a_6	a_7	a_8
1	1							
2	$\sqrt{2}$	1						
3	2	1						
4	2.613	3.414	2.613	1				
5	3.236	5.236	3.236	1				
6	3.864	9.141	7.464	2.864	1			
7	4.494	10.103	14.606	14.606	10.103	4.494	1	
8	5.126	13.138	21.848	21.691	21.848	13.138	5.126	1

$^*a_0 = 1$

Some important observations of Butterworth polynomials expressed in factored form as given in the Table 5.2 are as follows:

1. Coefficient of the highest power of s is always 1 for any order of filter.
2. Lowest order term i.e. constant term is always 1 for any order of filter.
3. For all odd ordered filters, one of the factors is always (s + 1) while the remaining factors are quadratic in nature.
4. For all even ordered filters, all factors are quadratic in nature.
5. All the poles of Butterworth polynomial are located in left half of s-plane on a circle with radius equal to one and centre at origin.

Let us discuss first and second order low and high pass Butterworth filters.

5.5 FIRST ORDER LOW PASS BUTTERWORTH FILTER

The first order low pass butterworth filter is realised by R-C circuit used alongwith an op-amp, used in the moniverting configuration. The circuit diagram is shown in Fig. 5.10.

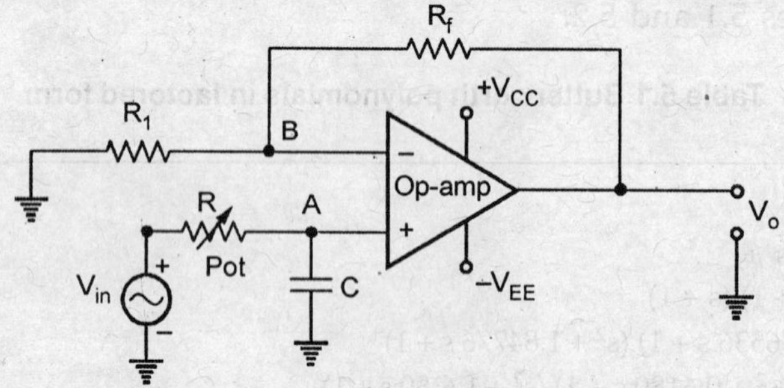

Fig. 5.10 First order low pass Butterworth filter

This also called **one pole low pass Butterworth filter.**

The resistance R_f and R_1 decide the gain of the filter in the pass band

5.5.1 Analysis of the Filter Circuit

The impedance of the capacitor C is $-jX_C$ where X_C is the capacitive reactance given by

$$X_C = \frac{1}{2\pi fC} .$$

By the potential divider rule, the voltage at the noniverting input terminal A which is the voltage across capacitor C is given by,

$$V_A = \frac{-jX_C}{R - jX_C} \cdot V_{in} \qquad ...(1)$$

$$\therefore \qquad V_A = \frac{-j\left(\frac{1}{2\pi fC}\right)}{R - j\left(\frac{1}{2\pi fC}\right)} \cdot V_{in} = \frac{-j}{2\pi fRC - j} \cdot V_{in}$$

$$= \frac{V_{in}}{1 - \frac{2\pi fRC}{j}}$$

but $$-j = \frac{1}{j} \quad \text{and} \quad -\frac{1}{j} = j$$

$$\therefore \quad V_A = \frac{V_{in}}{1 + j2\pi fRC} \qquad \text{...(2)}$$

As the op-amp is in te noniverting configuration,

$$V_o = \left(1 + \frac{R_f}{R_1}\right) V_A \qquad \text{...(3)}$$

$$\therefore \quad V_o = \left(1 + \frac{R_f}{R_1}\right) \frac{V_{in}}{(1 + j2pfRC)}$$

i.e. $$\frac{V_o}{V_{in}} = \frac{A_F}{1 + j\left(\frac{f}{f_H}\right)} \qquad \text{...(4)}$$

where $$A_F = \left(1 + \frac{R_f}{R_1}\right) = \text{Gain of filter in pass band} \qquad \text{...(5)}$$

and $$f_H = \frac{1}{2\pi RC} = \text{High cur off frequency of filter} \qquad \text{...(6)}$$

and $$f = \text{Operating frequency}$$

The $\frac{V_o}{V_{in}}$ is the transfer function of the filter and can be expressed in the polar form as,

$$\frac{V_o}{V_{in}} = \left|\frac{V_o}{V_{in}}\right| \angle \phi$$

where $$\left|\frac{V_o}{V_{in}}\right| = \frac{A_F}{\sqrt{1 + \left(\frac{f}{f_H}\right)^2}} \qquad \text{...(7)}$$

and $$\phi = -\tan^{-1}\left(\frac{f}{f_H}\right) \qquad \text{...(8)}$$

The phase angle ϕ is in degrees.

The equation (7) describes the behaviour of the low pass filter.

1. At very low frequencies, $f < f_H$

$$\left|\frac{V_o}{V_{in}}\right| \cong A_F \text{ i.e. constant}$$

2. At $f = f_H$,

$$\left|\frac{V_o}{V_{in}}\right| = \frac{A_F}{\sqrt{2}} = 0.707\, A_F \text{ i.e. 3 dB down to the level of } A_F.$$

3. At $f > f_H$

$$\left|\frac{V_o}{V_{in}}\right| = A_F$$

Thus, for the range of frequencies, $0 < f < f_H$, the gain is almost constant equal to f_H which is high cut off frequency. At $f = f_H$, gain reduces to 0.707 A_F i.e. 3 dB down from A_F And as the frequency increases than f_H, the gain decreases at a rate of 20dB/decade. The rate 20 dB/decade means decrease of 20 dB in gain per 10 times change in frequency. The same rate can be expressed as 6 dB/octave i.e. decrease of 6 dB per two times change in the frequency. The frequency f_H is called **cut off frequency, break frequency, – 3dB frequency or corner frequency.** The frequency response is shown in the Fig. 5.11.

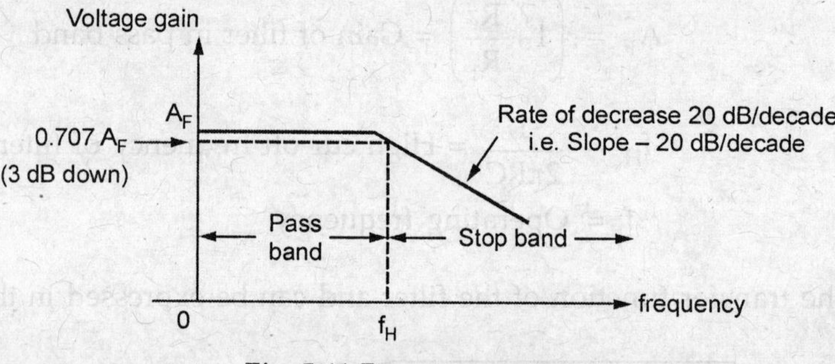

Fig. 5.11 Frequency response

Key Point: *The rate of decrease in gain is 20 dB/decade i.e. the decrease can be indicated by a negative slope in the frequency response, as – 20 dB/decade.*

5.5.2 Design Steps

The design steps for the first order low pass Butterworth filter are

1. Choose the cut off frequency, f_H.
2. Choose the capacitance C usually betwen 0.001 and 1 μF. Generally, it is selected as 1 μF or less than that. For better performance, mylar or tantalum capacitors are selected.
3. Now, for the RC circuit,

$$f_H = \frac{1}{2\pi RC}$$ refer equation (6)

Hence, as f_H and C are known, calculate the value of R. ,

4. The resistances R_f and R_1 can be selected depending on the required gain in the pass band.

$$A_F = 1 + \frac{R_f}{R_1}$$

5.5.3 Frequency Scaling

Once the filter is designed, sometimes, it is necessary to change the value of cut-off frequency f_H. The method used to change the original cut-off frequency f_H to a new cut-off frequency f_{H1} is called as frequency scaling.

To achieve such a frequency scaling, the standard value capacitor C is selected first. The required cut-off frequency can be achieved by calculating corresponding value of resistance R. But to achieve frequency scaling a potentiometer is used as shown in Fig. 5.10. Thus, the resistance R is generally a potentiometer with which required cut-off frequency f_H can be adjusted and changed later on if required.

Example 5.1: Design a low pass filter at a cut off frequency of 15.9 kHz with a passband gain 1.5.

Solution:

Step 1: The cut-off frequency is

$$f_H = 15.9 \text{ kHz}$$

Step 2: Choose $C = 0.001$ u.F

Step 3:
$$f_H = \frac{1}{2\pi RC}$$

$$\therefore \quad 15.9 \times 10^3 = \frac{1}{2\pi R \times 0.001 \times 10^{-6}}$$

$$\therefore \quad \mathbf{R = 10\ k\Omega}$$

For frequency scaling, use a pot of 20 k

Step 4:
$$A_F = 1.5$$

But
$$A_F = 1 + \frac{R_f}{R_1}$$

$$\therefore \quad \frac{R_f}{R_1} = 0.5$$

$$\therefore \quad Rf = 0.5\ R_1$$

Choose
$$\mathbf{R_1 = 10\ k\Omega}$$
$$\mathbf{R_f = 5\ k\Omega}$$

The complete circuit diagram is shown in the Fig. 5.12.

Fig. 5.12

5.6 SECOND ORDER LOW PASS BUTTERWORTH FILTER

The practical response of the filter must be very close to an ideal one. In case of low pass filter, it is always desirable that the gain rolls off very fast after the cut off frequency, in the stop band. In case of first order filter, it rolls off at a rate of 20 dB/decade. In case of second order filter, the gain rolls off at a rate of 40 dB/decade. Thus, the slope of the frequency response after $f = f_H$ is – 40 dB/decade, for a second order low pass filter.

A first order filter can be converted to second order type by using an additional RC network as shown in the Fig. 5.13.

The cut off frequency f_H for the filter is now decided by R_2, C_2, R_3 and C_3. The gain of the filter is as usual decided by op-amp i.e. the resistance R_1 and R_f.

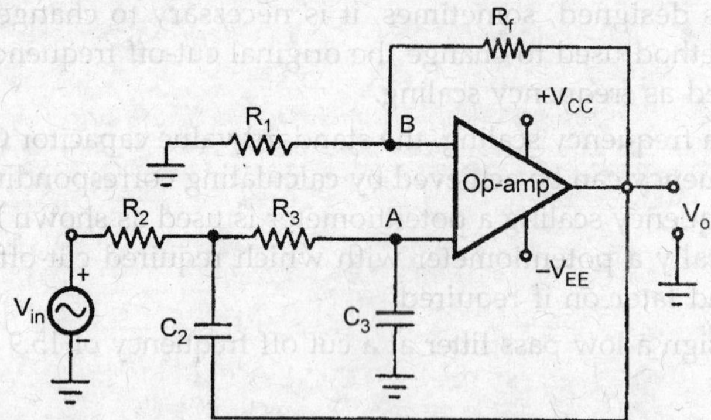

Fig. 5.13 Second order low pass Butterworth filter

5.6.1 Analysis of the Filter Circuit

For deriving the expression for the cut off frequency, let us use the Laplace transform method.

The input RC network can be represented in the Laplace domain as shown in Fig. 5.14.

R_2, V_1, R_3, V_A, $V_{in}(s)$, $I_1(s)$, $I_2(s)$, $I_3(s)$, $\frac{1}{sC_2}$, $\frac{1}{sC_3}$, $V_o(s)$, I = 0

To op-amp +ve terminal (op-amp input current is zero)

Fig. 5.14

Now $$I_1 = I_2 + I_3 \qquad ...(1)$$

i.e. $$\frac{V_{in} - V_1}{R_2} = \frac{V_1 - V_0}{\left(\frac{1}{sC_2}\right)} + \frac{V_1 - V_A}{R_3} \qquad ...(2)$$

Using potential divider rule, we can write

$$V_A = V_1 \left(\frac{\frac{1}{sC_3}}{R_3 + \frac{1}{sC_3}} \right) \quad ...(3)$$

$$V_A = \frac{V_1}{1 + sR_3 C_3} \quad ...(4)$$

$$\therefore \quad V_1 = V_A (1 + sR_3 C_3)$$

Substituting in equation (2) ans solving for V_A, we get

$$\frac{V_{in} - V_A(1 + sR_3 C_3)}{R_2} = \frac{V_A (1 + sR_3 C_3) - V_0}{\left(\frac{1}{sC_2}\right)} + \frac{V_A (1 + sR_3 C_3) - V_A}{R_3}$$

$$\frac{V_{in}}{R_2} + V_o(s C_2) = V_A \left[\frac{(1 + s R_3 C_3)}{R_2} + s C_2 (1 + s R_3 C_3) + \frac{(1 + s R_3 C_3)}{R_3} - \frac{1}{R_3} \right]$$

$$\therefore \frac{V_{in}}{R_2} + V_o (s C_2) = V_A \left[\frac{R_3 (1 + s R_3 C_3) + R_2 R_3 s C_2 (1 + s R_3 C_3) + R_2 (1 + s R_3 C_3) - R_2}{R_2 R_3} \right]$$

$$\therefore (R_3 V_{in} + V_o s R_2 R_3 C_2) = V_A [(1 + s R_3 C_3)(R_3 + R_2 R_3 s C_2 + R_2) - R_2]$$

$$V_A = \frac{R_3 V_{in} + V_o s R_2 R_3 C_2}{[(1 + s R_3 C_3)(R_3 + R_2 R_3 C_2 s + R_2) - R_2]} \quad ...(5)$$

Now, for op-amp in noninverting configuration,

$$V_o = A_F V_A \quad ...(6)$$

where $A_F = 1 + \frac{R_f}{R_1}$

and V_A = The voltage at the noninverting terminal

$$\therefore \quad V_o = A_F \left[\frac{R_3 V_{in} + V_o s R_2 R_3 C_2}{(1 + s R_3 C_3)(R_3 + R_2 R_3 C_2 s + R_2) - R_2} \right]$$

$$\therefore \frac{A_F R_3 V_{in}}{(1 + s R_3 C_3)(R_3 + R_2 R_3 C_2 s + R_2) - R_2} = V_o \left[1 - \frac{s R_2 R_3 C_2}{(1 + s R_3 C_3)(R_3 + R_3 R_2 C_2 s + R_2) - R_2} \right]$$

$$\therefore A_F R_3 V_{in} = V_o [(1 + s R_3 C_3)(R_3 + R_3 R_2 C_2 s + R_2) - R_2 - s R_2 R_3 C_2]$$

$$\therefore \frac{V_o}{V_{in}} = \frac{A_F}{s^2 + \frac{(R_3 C_3 + R_2 C_3 + R_2 C_2 - A_F R_2 C_2)s}{R_2 R_3 C_2 C_3} + \frac{1}{R_2 R_3 C_2 C_3}} \quad ...(7)$$

As the order of s in the gain expression is two, the filter is called second order filter.

The standard form of the transfer function of any second order system is

$$\boxed{\frac{V_o(s)}{V_{in}(s)} = \frac{A}{s^2 + 2\xi\,\omega_n\,s + \omega_n^2}} \quad ...(8)$$

where A = Overall gain

ξ = Damping of system

ω_n = Natural frequency of oscillations

Comparing equation (7) and equation (8), we can say that

$$\omega_n^2 = \frac{1}{R_2\,R_3\,C_2\,C_3} \quad ...(9)$$

In case of filters, this frequency is nothing but the cut-off frequency, ω_H

$$\therefore \quad \omega_H^2 = \frac{1}{R_2\,R_3\,C_2\,C_3}$$

$$\therefore \quad (2\,\pi\,f_H)^2 = \frac{1}{R_2\,R_3\,C_2\,C_3}$$

$$\therefore \quad \boxed{f_H = \frac{1}{2\pi\sqrt{R_2\,R_3\,C_2\,C_3}}} \quad ...(10)$$

This is the required cut off frequency.

Replacing s by jω, the transfer function can be written in the frequency domain and hence, finally, can be expressed in the polar form as,

$$\frac{V_o}{V_{in}} = \left|\frac{V_o}{V_{in}}\right| \angle\,\phi$$

where

$$\boxed{\left|\frac{V_o}{V_{in}}\right| = \frac{A_F}{\sqrt{1+\left(\frac{f}{f_H}\right)^4}}} \quad ...(11)$$

and A_F = Gain of filter in pass band

f = Input frequency in Hz

f_H = High cut-off frequency in Hz

The frequency response is shown in Fig. 5.15.

At the cut off frequency, f_H, the gain is 0.707 A_F i.e. 3 dB down from its 0 Hz level. After, f_H ($f > f_H$), the gain rolls off at a rate of 40 dB/decade. Hence, the slope of the response after f_H is – 40 dB/decade.

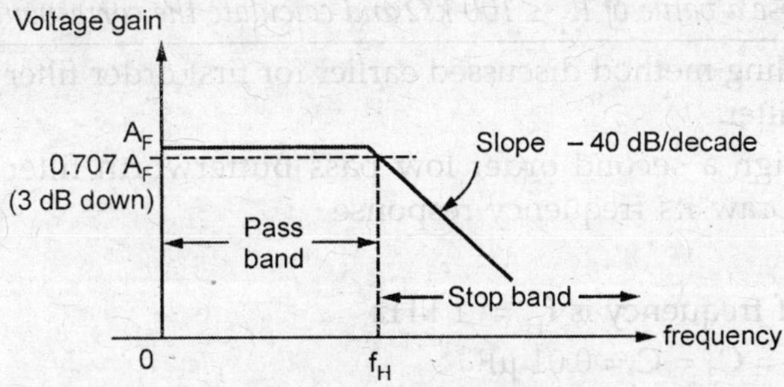

Fig. 5.15 Frequency response

5.6.2 Design Steps

The design steps for second order low pass Butterworth filter are

1. Choose the cut-off frequency f_H.
2. The design can be simplified by selecting $R_2 = R_3 = R$ and $C_2 = C_3 = C$. And choose a value of C less than or equal to 1 μF.
3. Calculate the value of R from the equation,

$$\boxed{f_H = \frac{1}{2\pi\sqrt{R_2 R_3 C_2 C_3}} = \frac{1}{2\pi RC}}$$

4. As $R_2 = R_3 = R$ and $C_2 = C_3 = C$, the pass band voltage gain $A_F = (1 + R_f/R_1)$ of the second order low pass filter has to be equal to 1.586.

Note: For $R_2 = R_3 = R$ and $C_2 = C_3 = C$, the transfer function takes the form

$$\frac{V_o(s)}{V_{in}(s)} = \frac{A_F}{s^2 + \frac{3-A_F}{RC}s + \frac{1}{R^2C^2}} \quad ...(12)$$

From this we can write that,

$$\xi = \text{Damping factor} = \frac{3-A_F}{2} \quad ...(13)$$

Now, for second order Butterworth filter, the middle term required is $\sqrt{2} = 1.414$, from the normalised Butterworth polynomial.

$$\therefore \quad 3 - A_F = \sqrt{2} = 1.414,$$

$$\therefore \quad \boxed{A_F = 1.586} \quad ...(14)$$

Thus, to ensure the Butterworth response, it is necessary that the gain A_F is 1.586.

$$\therefore \quad 1.586 = 1 + \frac{R_f}{R_1}$$

$$\therefore \quad \boxed{R_f = 0.586\,R_1} \quad ...(15)$$

***Key Point:** Hence, choose a value of $R_1 < 100$ kΩ and calculate the corresponding value of R_f.*

The frequency scaling method discussed earlier for first order filter is equally applicable to the second order filter.

Example 5.2: Design a second order low pass Butterworth filter having high cut-off frequency of 1 kHz. Draw its frequency response

Solution:

Step 1: The cut-off frequency is $f_H = 1$ kHz

Step 2: Choose $C_2 = C_3 = C = 0.01\ \mu F$

Step 3: Choose $R_2 = R_3 = R$

$$\therefore \quad f_H = \frac{1}{2\pi R C}$$

$$\therefore \quad 1 \times 10^3 = \frac{1}{2\pi R \times 0.01 \times 10^{-6}}$$

$$\therefore \quad R = 15.915\ k\Omega$$

Step 4: $R_f = 0.586\ R_1$

Choose $R_1 = 10\ k\Omega$

$$\therefore \quad R_f = 5.86\ k\Omega$$

Use a 10 kΩ potentiometer for precise adjustment of R_f.

The circuit is shown in the Fig. 5.16.

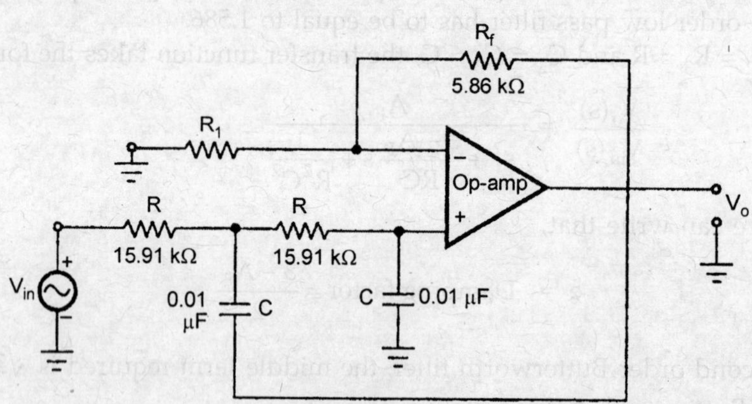

Fig. 5.16

The gain of the second order filter is given by

$$\frac{V_o}{V_{in}} = \frac{A_F}{\sqrt{1+\left(\frac{f}{f_H}\right)^4}}$$

Now, $\quad A_F = 1.586$ and $f_H = 1$ kHz

$$\therefore \quad \frac{V_o}{V_{in}} = \frac{1.586}{\sqrt{1+\left(\frac{f}{1\times10^3}\right)^4}}$$

Prepare a table to obtain the frequency response.

Table 5.3

f input frequency in Hz	$\frac{V_o}{V_{in}}$	$\frac{V_o}{V_{in}}$ in dB i.e. 20 log $\frac{V_o}{V_{in}}$
10	1.586	4 dB
100	1.585	4 dB
500	1.538	3.74 dB
1000	1.121	1 dB
5000	0.0633	−23.97 dB
7000	0.032	−29.89 dB
10000	0.015	−36.47 dB
50000	6.34×10^{-4}	−63.95 dB
100000	1.586×10^{-4}	−76 dB

The frequency response is shown in Fig. 5.17.

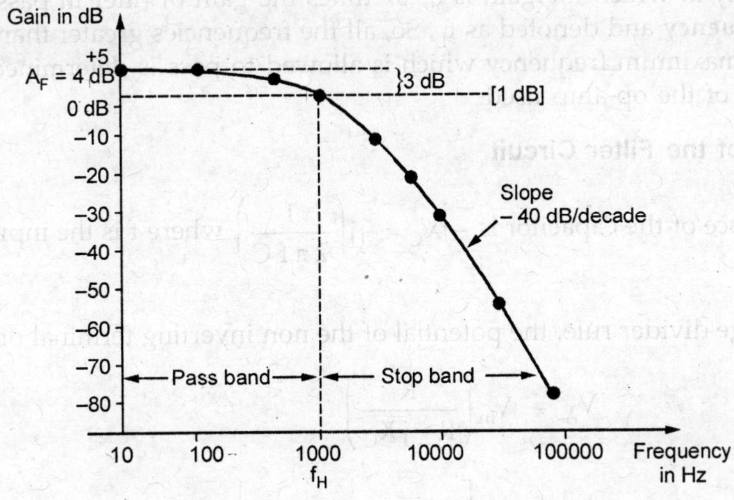

Fig. 5.17

5.7 FIRST ORDER HIGH PASS BUTTERWORTH FILTER

As mentioned earlier, a high pass filter is a circuit that attenuates all the signals below a specified cut off frequency denoted as f_L. Thus, a high pass filter performs the opposite function to that of low pass filter. Hence, the high pass filter circuit can be obtained by interchanging frequency determining resistances and capacitors in low pass filter circuit.

The first order high pass filter can be obtained by interchanging the elements R and C in a first order low pass filter circuit. The Fig. 5.18 shows the first order high pass Butterworth filter.

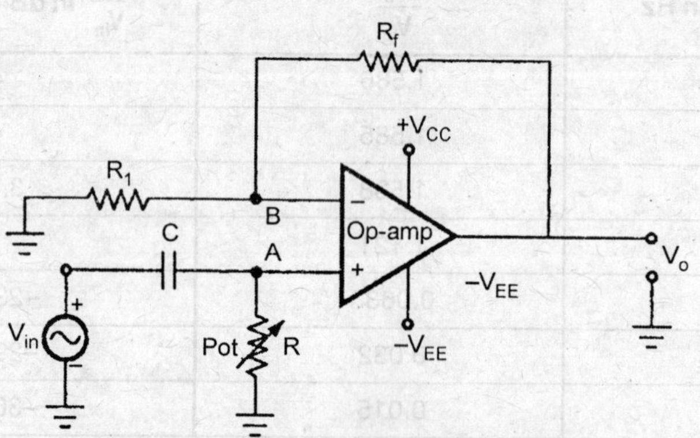

Fig. 5.18 First order high pass Butterworth filter

It can be observed that as compared to first order low pass filter (Fig. 5.10), the positions of R and C are changed in the high pass circuit shown in Fig. 5.18.

The frequency at which the gain is 0.707 times the gain of filter in pass band is called **low cut-off frequency** and denoted as f_L. So, all the frequencies greater than f_L are allowed to pass but the maximum frequency which is allowed to pass is determined by the closed loop bandwidth of the op-amp used.

5.7.1 Analysis of the Filter Circuit

The impedance of the capacitor is $-jX_C = -j\left(\frac{1}{2\pi f C}\right)$ where f is the input i.e. operating frequency.

By the voltage divider rule, the potential of the non inverting terminal of the op-amp is,

$$V_A = V_{in}\left(\frac{R}{R - jX_C}\right) \quad \text{...(1)}$$

$$\therefore \quad V_A = V_{in}\left[\frac{R}{-jX_C\left(\frac{R}{-jX_C}+1\right)}\right] \quad \text{taking } -j\,X_C \text{ outside}$$

As $-\frac{1}{j} = j$, we can write,

$$\frac{1}{-jX_C} = \frac{j}{X_C} = \frac{j}{\left(\frac{1}{2\pi fC}\right)} = j\,2\,\pi\,f\,C \quad \text{...(2)}$$

Substituting in the above expression of V_A,

$$\therefore \quad V_A = V_{in}\left[\frac{\left(-\frac{R}{jX_C}\right)}{\left(-\frac{R}{jX_C}\right)+1}\right]$$

$$\therefore \quad V_A = V_{in}\left[\frac{j2\pi fRC}{1+j2\pi fRC}\right] \quad \text{...(3)}$$

This can be represented as,

$$\therefore \quad V_A = V_{in}\left[\frac{j\left(\frac{f}{f_L}\right)}{1+j\left(\frac{f}{f_L}\right)}\right]$$

where $f_L = \frac{1}{2\pi RC}$

= Low cut-off frequency ...(4)

Now, for the op-amp in non-inverting configuration,

$$V_o = A_F V_A$$

where V_A = Voltage at the non inverting input

and $A_F = \left(1+\frac{R_f}{R_1}\right)$

= Gain of op-amp in pass band

$$\therefore \quad V_o = A_F V_{in}\left[\frac{j\left(\frac{f}{f_L}\right)}{1+j\left(\frac{f}{f_L}\right)}\right]$$

$$\therefore \quad \boxed{\frac{V_o}{V_{in}} = A_F\left[\frac{j\left(\frac{f}{f_L}\right)}{1+j\left(\frac{f}{f_L}\right)}\right]} \quad ...(5)$$

This is the required expression for the transfer function of the filter.

For the frequency response, we require the magnitude of the transfer function which is given by,

$$\left|\frac{V_o}{V_{in}}\right| = \frac{A_F\left(\frac{f}{f_L}\right)}{\sqrt{1+\left(\frac{f}{f_L}\right)^2}} \quad ...(6)$$

The equation (6) describes the behaviour of the high pass filter.

1. At low frequencies, i.e. $f < f_L$

$$\left|\frac{V_o}{V_{in}}\right| < A_F$$

2. At $f = f_1$, $\left|\frac{V_o}{V_{in}}\right| = 0.707\ A_F$ i.e. 3 dB down from the level of A_F

3. At $f > f_L$, i.e. high frequencies, 1 can be neglected as compared to $\left(\frac{f}{f_L}\right)$ from denominator

$$\therefore \quad \left|\frac{V_o}{V_{in}}\right| \cong A_F \text{ i.e. constant}$$

Thus, the circuit acts as high pass filter with a passband gain as A_F. For the frequencies, $f < f_1$, the gain increases till $f = f_L$ at a rate of +20 dB/decade. Hence, the slope of the frequency response in stop band is +20 dB/decade for first order high pass filter.

The frequency response is shown in the Fig. 5.19.

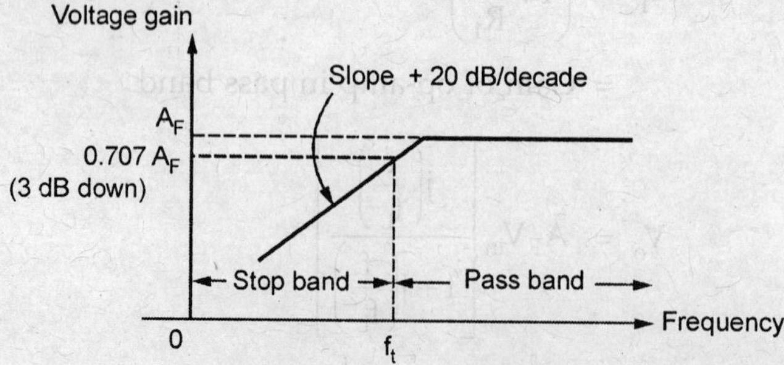

Fig. 5.19 Frequency response

***Note:** As high pass filter is basically a low pass filter circuit with positions of R and C interchanged, the design steps and the frequency scaling method discussed earlier for low pass filter is equally applicable to the first order high pass butterworth filter.*

Example 5.3: Design a high pass filter with a cut off frequency of 10 kHz with a passband gain of 1.5. Also plot the frequency response for the designed filter.

Solution:

Step 1: The lower cut-off frequency is 10 kHz.

$$\therefore \quad f_L = 10 \text{ kHz}$$

Step 2: Choose C less than 1 μR

$$\therefore \quad C = 0.02\ \mu\text{F}$$

Step 3: Calculate R

$$f_L = \frac{1}{2\pi RC}$$

$$\therefore \quad 10 \times 10^3 = \frac{1}{2\pi R \times 0.02 \times 10^{-6}}$$

$$\therefore \quad R = 795.77\ \Omega$$

For frequency scaling, use a pot of 1 kΩ

Step 4: $A_F = 1.5$

But $\quad A_F = 1 + \frac{R_f}{R_1} = 1.5$

$$\therefore \quad R_f = 0.5\ R_1$$

Choose $\quad R_1 = 10\ \text{k}\Omega$

$$\therefore \quad R_f = 5\ \text{k}\Omega$$

The complete circuit is shown in Fig. 5.20.

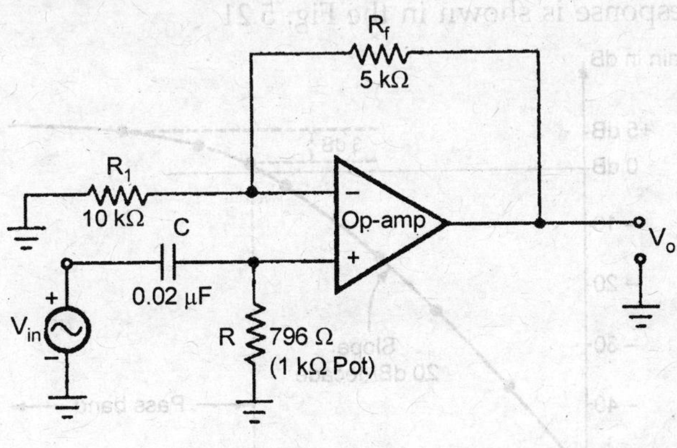

Fig. 5.20

To obtain the frequency response, consider the magnitude of the transfer function of the filter given by,

$$\left|\frac{V_o}{V_{in}}\right| = \frac{A_F\left(\frac{f}{f_L}\right)}{\sqrt{1+\left(\frac{f}{f_L}\right)^2}}$$

$$= \frac{1.5\left(\frac{f}{10\times 10^3}\right)}{\sqrt{1+\left(\frac{f}{10\times 10^3}\right)^2}}$$

Now, for various values of f, calculate $\left|\frac{V_o}{V_{in}}\right|$ in dB as shown in Table 5.4.

Table 5.4

f in Hz	$\left\|\frac{V_o}{V_{in}}\right\|$	$\left\|\frac{V_o}{V_{in}}\right\|$ in dB
10	1.5×10^{-3}	–56.48
100	0.015	–36.48
500	0.0749	–22.51
1000	0.149	–16.52
5000	0.671	–3.46
10000	1.061	0.511
30000	1.42	3.06
100000	1.49	3.47

The frequency response is shown in the Fig. 5.21

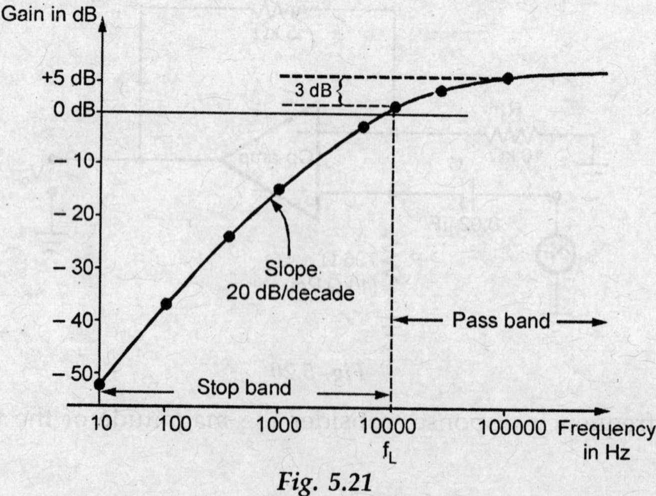

Fig. 5.21

5.8 SECOND ORDER HIGH PASS BUTTERWORTH FILTER

The second order high pass Butterworth filter produces a gain roll off at the rate of +40 dB/decadc in the stop band. This filter also can be realised by interchanging the positions of resistors and capacitors in a second order low pass Butterworth filter. The Fig. 5.22 shows the second .order high pass Butterworth filter.

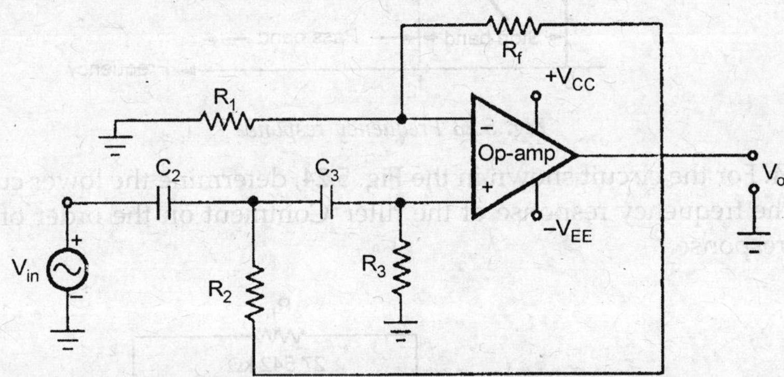

Fig. 5.22 Second order high pass Butterworth filter

The analysis, design and the scaling procedures for this filter is exactly same as that of second order low pass Butterworth filter.

The resulting expression is given here for the convenience of the reader.

The voltage gain magnitude equation for the second order high pass filter is,

$$\left|\frac{V_o}{V_{in}}\right| = \frac{A_F}{\sqrt{1+\left(\frac{f_L}{f}\right)^4}} \quad \ldots(1)$$

where, f = Input frequency in Hz

f_L = Lower cut-off frequency in Hz = $\dfrac{1}{2\pi\sqrt{R_2R_3C_2C_3}}$

For $R_2 = R_3 = R$ and $C_2 = C_3 = C$, $f_L = \dfrac{1}{2\pi RC}$

A_F = Passband gain

= 1.586 to ensure second order butterworth response

and R_f = $0.586\ R_t$

The frequency response of this filter is shown in the Fig. 5.23.

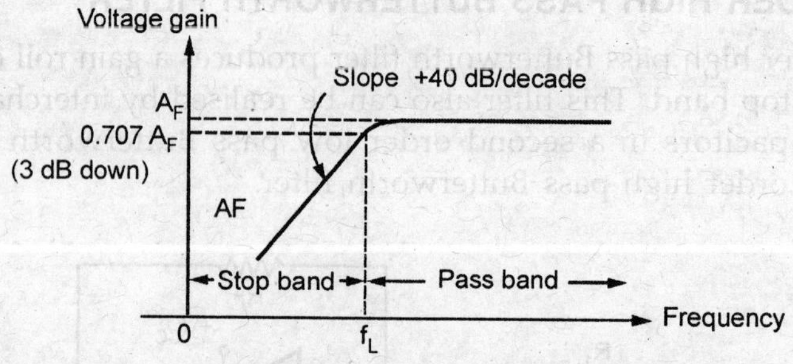

Fig. 5.23 Frequency response

Example 5.4: For the circuit shown in the Fig. 5.24, determine the lower cut off frequency and then plot the frequency response of the filter. Comment on the order of the filter from the frequency response.

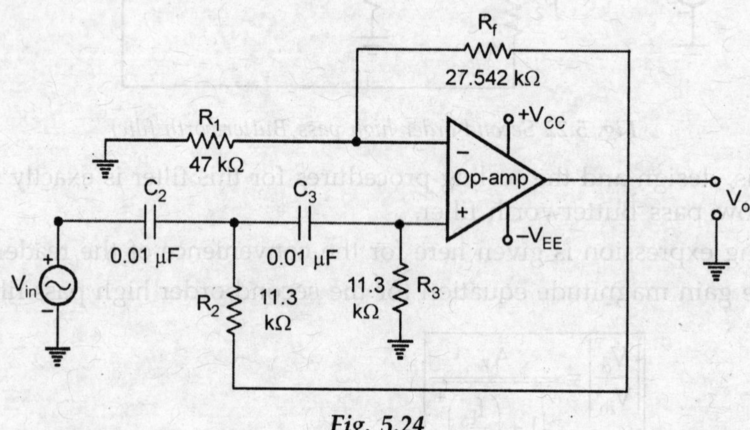

Fig. 5.24

Solution:

From the Fig. 5.24,

$$R_2 = R_3 = 11.3 \text{ k}\Omega$$
$$C_2 = C_3 = 0.01 \text{ μF}$$

$$f_L = \frac{1}{2\pi\sqrt{R_2 R_3 C_2 C_3}} = \frac{1}{2\pi\sqrt{(11.3\times10^3)^2\ (0.01\times10^{-6})^2}}$$
$$= 1.408 \text{ kHz}$$

From the two R-C sections used in the circuit, it is clear that the filter is second order high pass filter.

To plot the frequency response, let us prepare the table including the gain in dB obtained for various values of input frequency.

$$\left|\frac{V_o}{V_{in}}\right| = \frac{A_F}{\sqrt{1+\left(\frac{f_L}{f}\right)^4}}$$

where $$A_F = 1 + \frac{R_f}{R_1}$$

$$= 1 + \frac{27.542}{47} = 1.586$$

$$\therefore \quad \left|\frac{V_o}{V_{in}}\right| = \frac{1.586}{\sqrt{1 + \left(\frac{1.408 \times 10^3}{f}\right)^4}}$$

Table 5.5

f in Hz	$\left\|\frac{V_o}{V_{in}}\right\|$	$\left\|\frac{V_o}{V_{in}}\right\|$ in dB
10	8×10^{-5}	– 81.93 dB
100	8×10^{-3}	– 41.93 dB
500	0.198	– 17.04 dB
1000	0.714	– 2.93 dB
1408	1.121	0.99 dB
10000	1.585	4 dB
30000	1.585	4 dB
100000	1.586	4 dB

The frequency response is shown in the Fig. 5.25(a). (See Fig. 5.25(a) on next page).

From the frequency response, it can be observed that in the stop band i.e. $f < f_L$, as frequency is changed 10 times i.e. from 10 to 100, the gain changes from – 81.93 to – 41.93 dB. Thus, the slope of the frequency response in stop band is + 40 dB/decade. This shows that the filter is second order filter.

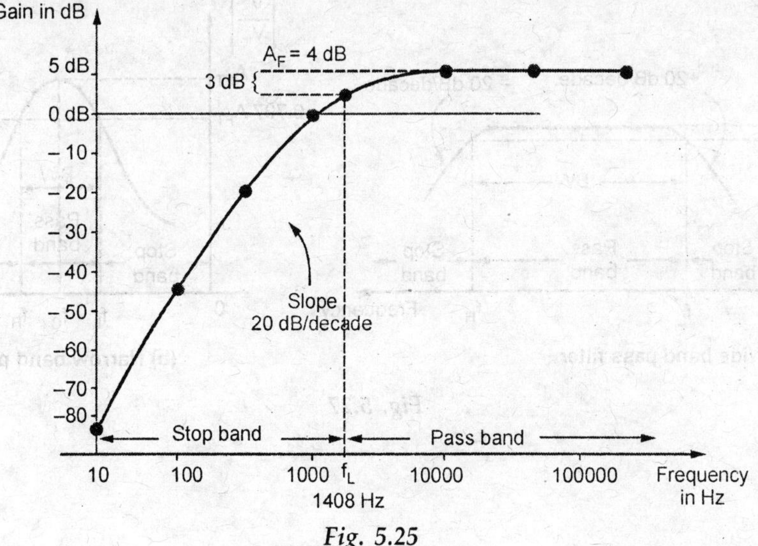

Fig. 5.25

5.9 INTRODUCTION TO BANDPASS FILTERS

A bandpass filter is basically a frequency selector. It allows one particular band of frequencies to pass. Thus, the pass band is between the two cut-off frequencies f_H and f_L where $f_H > f_L$. Any frequency outside this band gets attenuated. The frequency response of band pass filter is shown in Fig. 5.26.

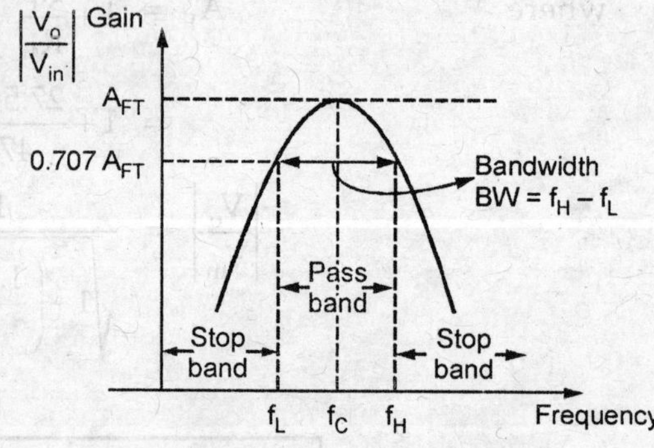

Fig. 5.26 Bandpass filter

The pass band which is between f_H and f_L is called bandwidth of the filter denoted as BW.

$$BW = f_H - f_L$$

The frequency at the centre of the pass band is called **centre frequency** denoted as f_C.

The gain is maximum at f_C and is denoted as A_{FT} called **total passband gain.**

Practically, the f_C is not exactly at the centre of the pass band hence, it is also called as resonant frequency. The gain at f_L and f_H is 0.707 A_{FT}.

There are two types of bandpass niters which are classified based on the **figure of merit or quality factor (Q).**

(*i*) **For Q < 10, the bandpass filter is called wide bandpass filter.** In this type, the bandpass is wide and we get large bandwidth. The response is shown in the Fig. 5.27 (a).

(*ii*) **For Q > 10, the bandpass filter is called narrow bandpass filter.** The bandpass is very narrow and the bandwidth is very small. Higher the value of Q, narrower is the passband and more selective is the filter. In the narrow band filter, the gain peaks at the centre frequency. The response is shown in Fig. 5.27 (b).

The gain roll off for $f < f_L$ is + 20 dB/decade while $f > f_H$ it is – 20 dB/decade.

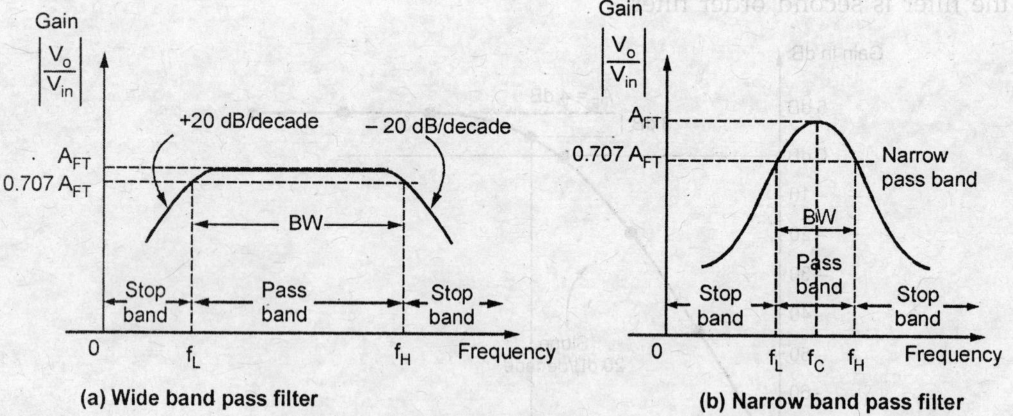

Fig. 5.27

For wide band pass filter, the centre frequency is given by,

$$f_C = \sqrt{f_L f_H} \quad ...(2)$$

The relationship between Q and 3 dB bandwidth with f_C is given by,

$$Q = \frac{f_C}{BW} = \frac{f_C}{f_H - f_L}$$

5.9.1 Wide Band Pass Filter

The wide band pass filter can be realised by simply cascading a high pass filter and low pass filter. If both high pass and low pass filters are of first order, the gain roll off in both the stop bands are ± 20 dB/dccadc and wide band pass filter is of first order. To get gain roll off ± 40 dB/decade and second order wide band pass filter, both high pass and low pass filters must be of second order and so on.

The Fig. 5.28 shows the first order wide band pass filter obtained by cascading first order high pass and low pass filter sections.

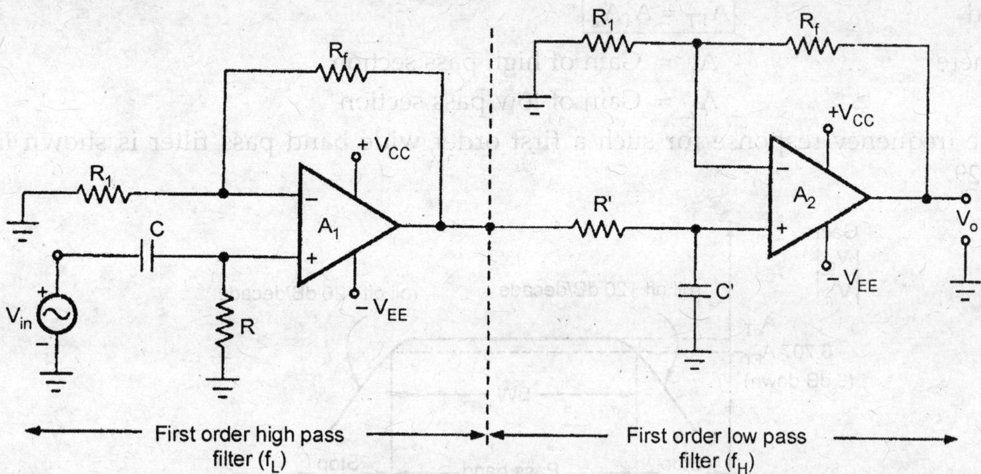

Fig. 5.28 Wide band pass filter

For wide band pass response, f_H must be greater than f_L. The voltage gain expressions for the two sections are reproduced here for the convenience.

$$\left|\frac{V_o}{V_{in}}\right| = \frac{A_F}{\sqrt{1+\left(\frac{f}{f_H}\right)^2}} \qquad \text{Low pass}$$

$$\left|\frac{V_o}{V_{in}}\right| = \frac{A_F\left(\frac{f}{f_L}\right)}{\sqrt{1+\left(\frac{f}{f_H}\right)^2}} \qquad \text{High pass}$$

The design steps discussed earlier for first order low pass and high pass filters, are to be used to design the wide band pass filter of first order.

As the two circuits are in cascade, the overall gain of wide band pass filter is the product of the two gains expressed as,

$$\left|\frac{V_o}{V_{in}}\right| = \frac{A_{FT}\left(\frac{f}{f_L}\right)}{\sqrt{\left[1+\left(\frac{f}{f_L}\right)^2\right]\left[1+\left(\frac{f}{f_H}\right)^2\right]}} \quad ...(4)$$

where
A_{FT} = Total pass band gain
f = Input frequency in Hz
f_L = Lower cut off frequency in Hz
f_H = Higher cut off frequency in Hz

and
$$A_{FT} = A_1 A_2 \quad ...(5)$$

where
A_1 = Gain of high pass section
A_2 = Gain of low pass section

The frequency response for such a first order wide band pass filter is shown in the Fig. 5.29.

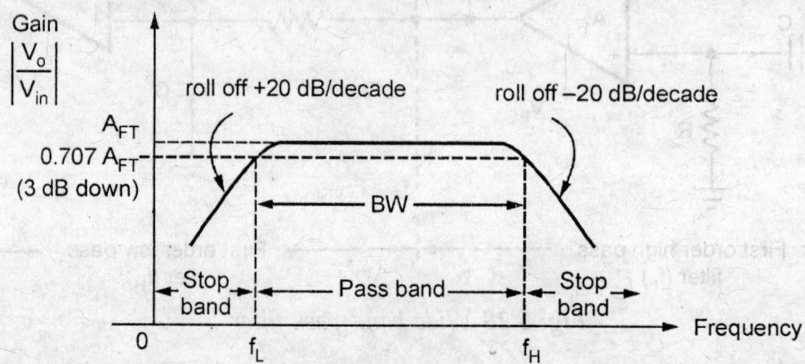

Fig. 5.29 Frequency response

5.9.2 Narrow Band Pass Filter

The narro band pass filter uses only one op-amp as against two by wide band pass filter. It has following features:

(i) It has two feedback paths.

(ii) The op-amp is in the inverting configuration.

Due to the two feedback paths, it is called **multiple feedback filter.**

The Fig. 5.30 shows the circuit diagram of narrow band pass filter.

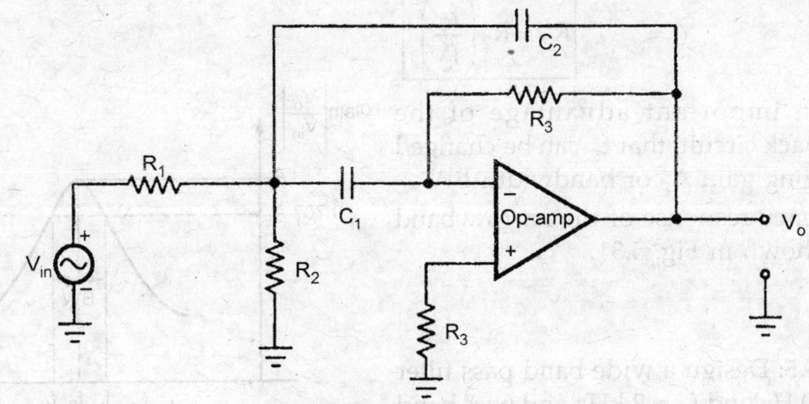

***Fig. 5.30** Narrow band pass filter*

As seen from Fig. 5.30, the input is applied to the inverting input terminal. Thus, op-amp is in inverting configuration. The resistance R_3 connected to non-inverting input terminal is offset compensating resistance.

The important parameters of the narrow band pass filter are f_L, f_H, the center frequency f_C, the gain at the center frequency A_F and the quality factor Q. The relationship of components with the various parameters are given by the following expressions.

For simplifying the calculations, choose $C_1 = C_2 = C$.

$$F_1 = \frac{Q}{2\pi f_C C A_F} \quad \text{...(6)}$$

$$R_2 = \frac{Q}{2\pi f_C C(2Q^2 - A_F)} \quad \text{...(7)}$$

$$R_3 = \frac{Q}{\pi f_C C} \quad \text{...(8)}$$

and $$A_F = \frac{R_3}{2R_1} = \text{gain at } f_C \quad \text{...(9)}$$

The gain Ap must satisfy the equation, ...(10)

$$A_F < 2\,Q^2$$

Changing the centre frequency f_c:

Let f_C = Original frequency

f'_C = New centre frequency

The new centre frequency can be achieved by changing the resistance R_2. The new value of resistance say R'_2, can be obtained as,

$$\boxed{R_2' - R_2\left(\frac{f_C}{f_C'}\right)^2} \qquad ...(11)$$

This is an important advantage of the multiple feedback circuit, that f_C can be changed without changing gain A_F or bandwidth BW.

The frequency response of the narrow band pass filter is shown m Fig. 5.31.

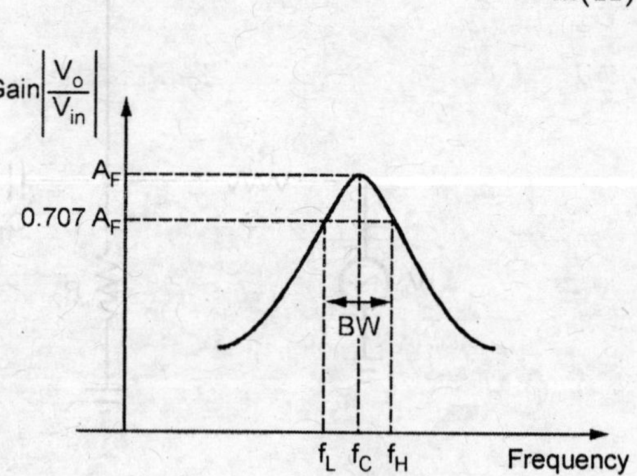

Fig. 5.31 Frequency response

Example 5.5: Design a wide band pass filter having f_L = 400 Hz and f_H = 2 kHz and pass band gain of 4. Draw the frequency response of the filter and also calculate the Q value of the filter.

Solution: Design the low pass filter,

$$f_H = 2 \text{ kHz}$$

Let $C' = 0.01\ \mu$

and $f_H = \dfrac{1}{2\pi R'C''}$

$\therefore$ $R' = \dfrac{1}{2\pi \times 2 \times 10^3 \times 0.01 \times 10^{-6}} = 7.957\ k\Omega$

Design the high pass filter,

$$f_L = 400 \text{ Hz}$$

Let $C = 0.05\ \mu F$

and $f_L = \dfrac{1}{2\pi RC}$

$\therefore$ $R = \dfrac{1}{2\pi \times 400 \times 0.05 \times 10^{-6}} = 7.95\ k\Omega$

Now $A_{FT} = A_1 A_2$

where A_1 = Gain of high pass section

A_2 = Gain of low pass section

It is given, $A_{FT} = 4$

Let $A_1 = A_2 = 2$

Now, for the non inverting op-amp,

$$A_1 = A_2 = 1 + \frac{R_f}{R_1} = 2$$

$\therefore$ $R_f = R_1$ for both the setions

$\therefore$ Let $R_f = R_1 = 100\ k\Omega$

Hence, the designed circuit is shown in Fig. 5.32.

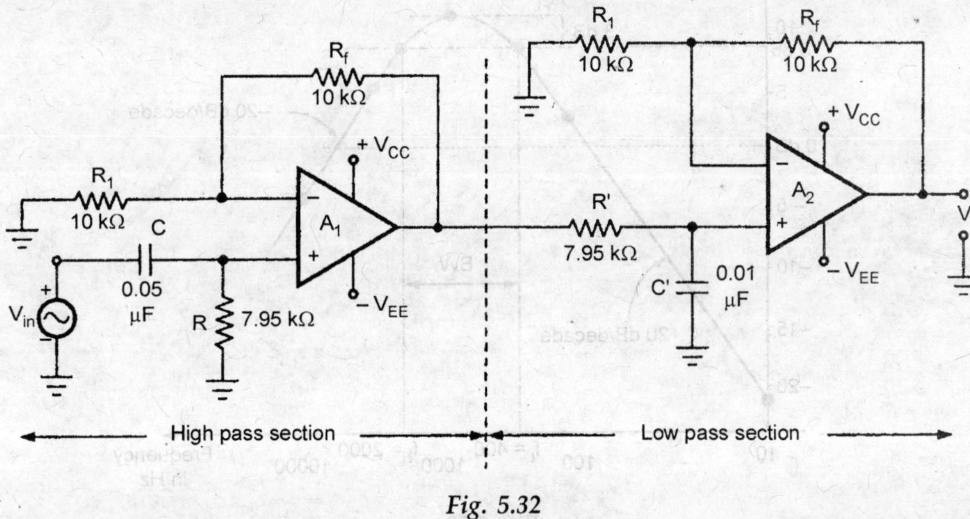

Fig. 5.32

To obtain the frequency response, use the equation (4) which is magnitude of the voltage gain of wide band pass filter.

$$\left|\frac{V_o}{V_{in}}\right| = \frac{A_{FT}\left(\frac{f}{f_L}\right)}{\sqrt{\left[1+\left(\frac{f}{f_L}\right)^2\right]\left[1+\left(\frac{f}{f_H}\right)^2\right]}} = \frac{4\left(\frac{f}{400}\right)}{\sqrt{\left[1+\left(\frac{f}{400}\right)^2\right]\left[1+\left(\frac{f}{2\times 10^3}\right)^2\right]}}$$

Now, select the frequencies and tabulate the data for the frequency response.

Table 5.6

f in Hz	$\left\|\frac{V_o}{V_{in}}\right\|$	$\left\|\frac{V_o}{V_{in}}\right\|$ **in dB**
10	0.0999	– 20.27 dB
100	0.9689	– 0.27 dB
400	2.77	8.86 dB
1000	3.32	10.42 dB
2000	2.77	8.86 dB
5000	1.48	3.41 dB
10000	0.783	– 2.11 dB

The frequency response is shown in Fig. 5.33.

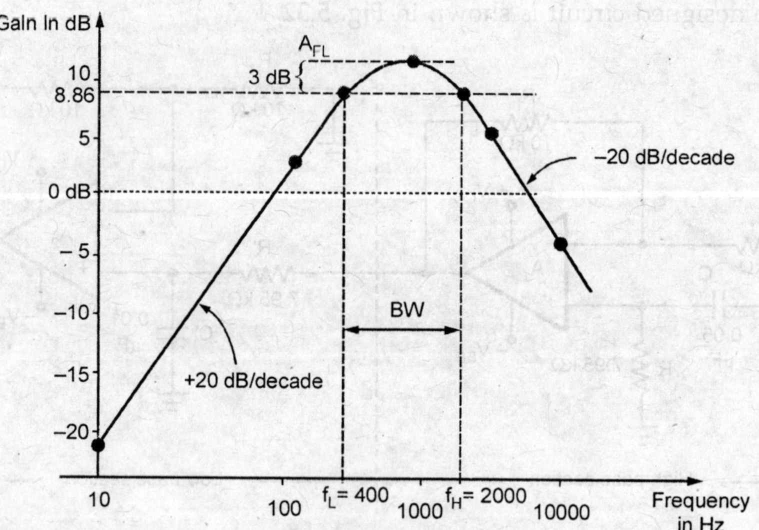

Fig. 5.33

Now,
$$f_C = \sqrt{f_l, f_{11}} = \sqrt{400 \times 200} = 894.427 \text{ Hz}$$
$$BW = f_{11} - f_1 = 2000 - 400 = 1600$$
$$\therefore \quad Q = \frac{f_C}{BW} = \frac{894.427}{1600} = \mathbf{0.559}$$

Thus, Q is less than 10, for wide band pass filter.

Example 5.6: Design the narrow band pass filter with two feed back paths with $f_C = 1.5$ kHz, $Q = 7$ and $A_F = 15$.

Calculate the new value of resistance in the circuit which will change f_C to 2 kHz.

Solution: Choose $C_1 = C_2 = C = 0.02\ \mu F$

$$R_1 = \frac{Q}{2\pi f_C C A_F} = \frac{7}{2\pi \times 1.5 \times 10^3 \times 0.02 \times 10^{-6} \times 15}$$
$$= 2.47\ k\Omega$$
$$R_2 = \frac{Q}{2\pi f_C C(2Q^2 - A_F)}$$
$$= \frac{7}{2\pi \times 1.5 \times 10^3 \times 0.02 \times 10^{-6} \times (2 \times 7^2 - 15)}$$
$$= 447.4\ \Omega$$
$$R_3 = \frac{Q}{\pi f_C C} = \frac{7}{\pi \times 1.5 \times 10^3 \times 0.02 \times 10^{-6}}$$
$$= 74.27\ k\Omega$$

The resistance R_2 can be changed to get $R'_C = 2$ kHz

$$R'_2 = R_2 \left(\frac{f_C}{f'_C}\right)^2 = 447.4 \times \left(\frac{1.5}{2}\right)^2 = 251.66\ \Omega$$

The circuit is shown in the Fig. 5.34.

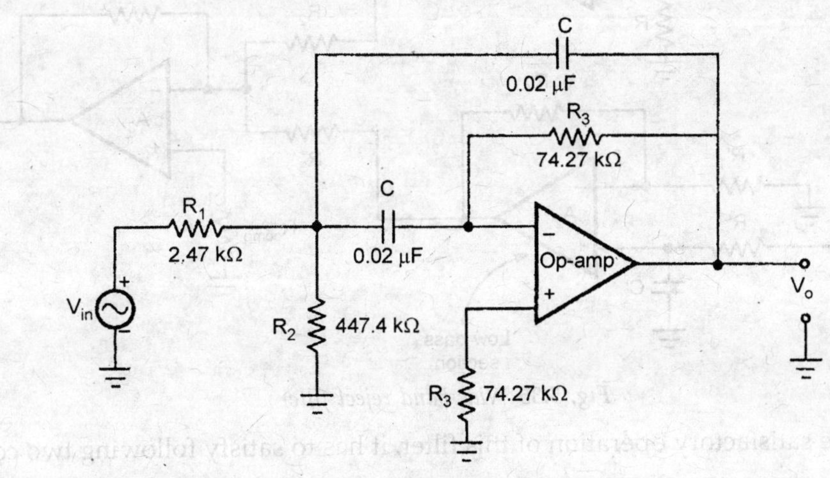

Fig. 5.34

5.10 BAND ELIMINATION FILTER

As mentioned earlier, this filter is also called **band stop filter or band reject filter.** The action of this filter is exactly opposite to the band pass filter. This filter has two pass bands and one stop band. A band of frequencies is attenuated by this filter, hence the name band elimination filter.

This filter is also classified as:

(i) Wide band reject filter and

(ii) Narow band reject filter.

The figure of merit (Q) for narrow band filter is greater than 10. In its frequency response, it shows a notch and hence, also called as notch filter. The bandwidth of the notch filter is very small as compared to wide band reject filter.

5.10.1 Wide Band Reject Filter

Similar to the wide band pass filter, this filter also consists of a high pass and low pass filter sections. Additionally, it consists of a summing amplifier. The Fig. 5.35 shows the circuit diagram of wide band reject filter.

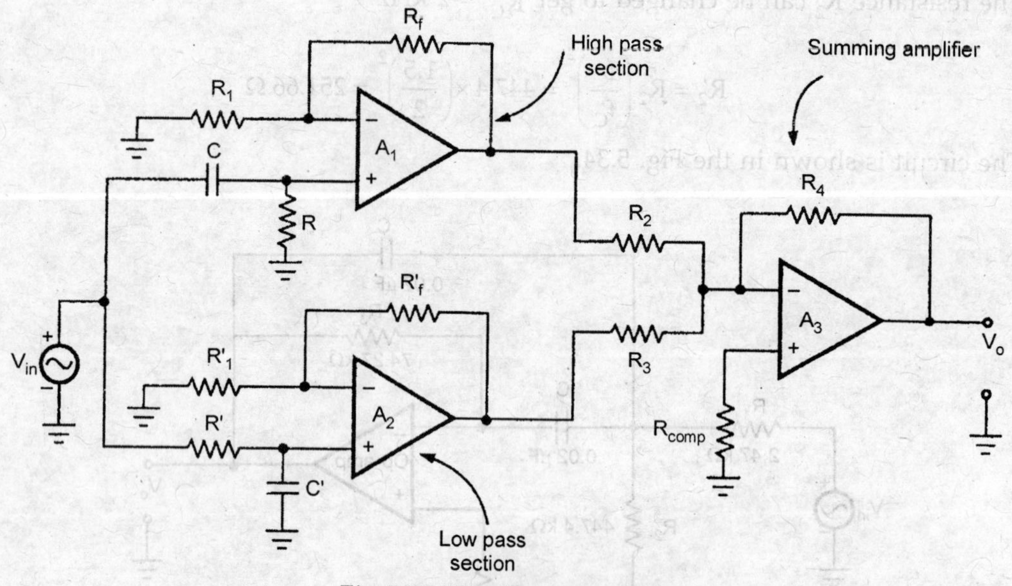

Fig. 5.35 Wide band reject filter

To have the satisfactory operation of this filter, it has to satisfy following two conditions:

(i) The low cut-off frequency, f_L of high pass filter must be greater than the high cut-off frequency, f_H of low pass filter.

(ii) The pass band gain of both high pass and low pass sections must be equal.

The design of the overall filter is based on the individual design of the various sections. The gain of the summing amplifier can be set to 1 for simplicity and thus,

$$R_2 = R_3 = R_4 = R \quad \text{...(1)}$$

Hence,

$$R_{comp} = R_2 \| R_3 \| R_4 = \frac{R}{3} \quad \text{...(2)}$$

Both high pass and low pass sections provide attenuation in the stop band between f_H and f_L. For $f < f_H$, the transmission is due to low pass section while for $f > f_L$, the transmission is due to high pass section. The frequency response for this type of filter is shown in Fig. 5.36.

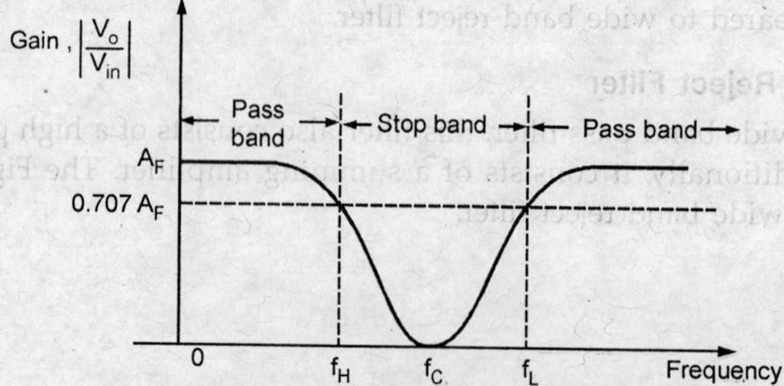

Fig. 5.36 Frequency response

The centre frequency, f_C is given by

$$f_C = \sqrt{f_H f_L}$$

Example 5.7: Design a wide band reject filter having f_H = 400 Hz and f_L = 2 kHz with a pass band gain of 2.

Solution: The given values of the frequencies are

$$f_H = 400 \text{ Hz} \quad \text{and} \quad f_L = 2 \text{ kHz}$$

For high pass section, $f_L = 2$ kHz

Choose $C = 0.01\ \mu\text{F}$

Now, $$f_L = \frac{1}{2\pi RC}$$

$$\therefore \quad 2 \times 10^3 = \frac{1}{2\pi R \times 0.01 \times 10^{-6}}$$

$$\therefore \quad R = 7.957\ \text{k}\Omega$$

For low pass section, $f_H = 400$ Hz

Choose $C' = 0.05\ \mu\text{F}$

Now, $$f_H = \frac{1}{2\pi R'C'}$$

$$\therefore \quad R' = \frac{1}{2\pi \times 400 \times 0.05 \times 10^{-6}}$$

$$= 7.957\ \text{k}\Omega$$

The gain of both the sections must be equal to 2.

$$\therefore \quad A_F = 1 + \frac{R_f}{R_1} = 2$$

$$\therefore \quad \frac{R_f}{R_1} = 1$$

$$\therefore \quad R_f = R_1 = 10\ \text{k}\Omega$$

For summing amplifier, let gain be 1.

$$\therefore \quad R_2 = R_3 = R_4 = 10\ \text{k}\Omega$$

$$\therefore \quad R_{comp} = \frac{R}{3} = \frac{10}{3} = \mathbf{3.33\ k\Omega}$$

The designed circuit is shown in Fig. 5.37.

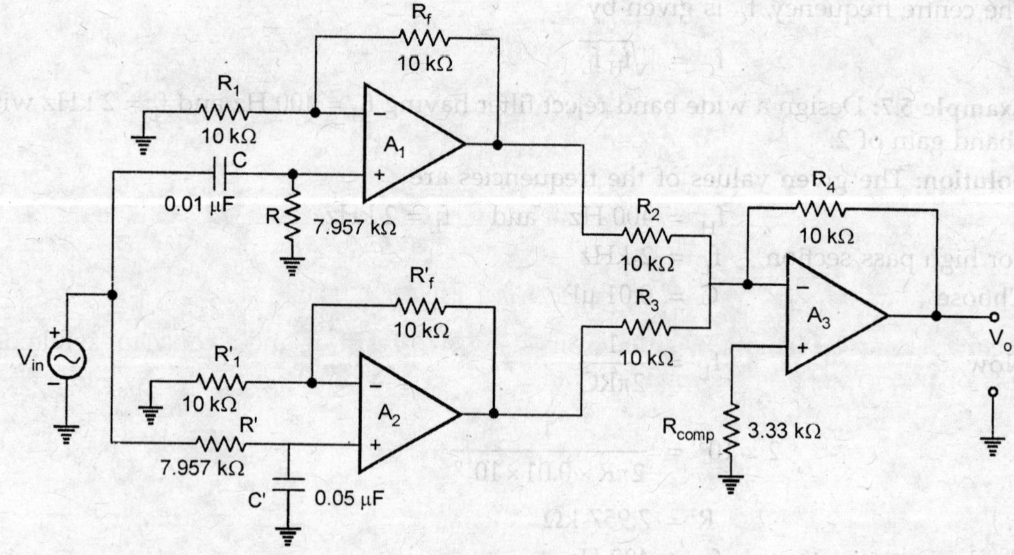

Fig. 5.37

$$f_C = \sqrt{f_{11}\, f_1} = \sqrt{400 \times 2 \times 10^3}$$

$$= \mathbf{894.427\ Hz}$$

The frequency response is shown in Fig. 5.38.

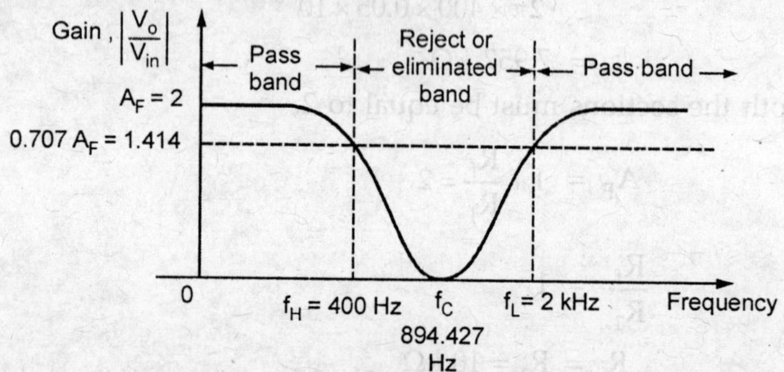

Fig. 5.38

5.10.2 Narrow Band Reject Filter (Notch Filter)

The name of the filter i.e. Notch filter is due to the characteristics shape of its frequency response curve. The stop band of this filter is very narrow. The typical application of the notch filter is the rejection of a single frequency, such as 50 Hz power line frequency hum. It is also used in the biomedical instrumentation and also blanking of control tones for telephone lines. Thus, particular unwanted frequency can be eliminated using this filter.

The **passive** circuit used to obtain the notch filter is the twin T network as shown in Fig. 5.39.

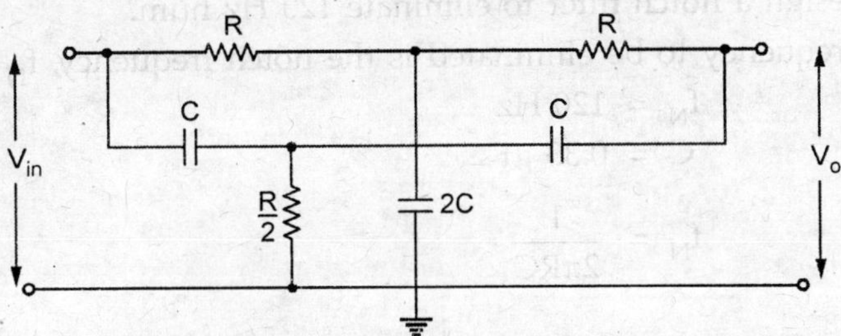

Fig. 5.39 Twin T network

It consists of two T networks. One consists of two resistors and a capacitor while other consists of two capacitors and one resistor.

The notch out frequency is the frequency at which the maximum attenuation occurs. This is given by

$$f_N = \frac{1}{2\pi RC} \quad ...(3)$$

The value of Q i.e. figure of merit for the passive network is very low, hence an **active** notch filter which uses twin T network is preferred in practice. This is shown in Fig. 5.40.

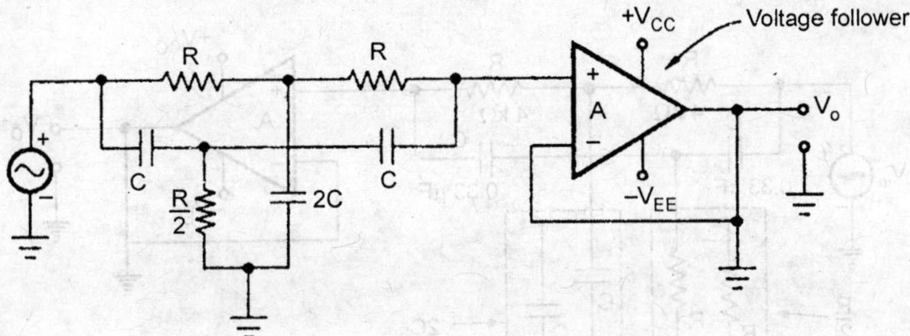

Fig. 5.40 Active notch filter

To design a notch filter, to eliminate the specific notch frequency, f_N, choose the capacitor C less than or equal to 1 μF. Then, calculate the value of R using equation (3). To obtain best performance of the circuit, the components designed should have precise values which are close to their designed values. The frequency response of the notch filter is shown in the Fig. 5.41. Fig. 5.41 (a) shows the ideal response while the Fig. 5.41 (b) shows the practical response.

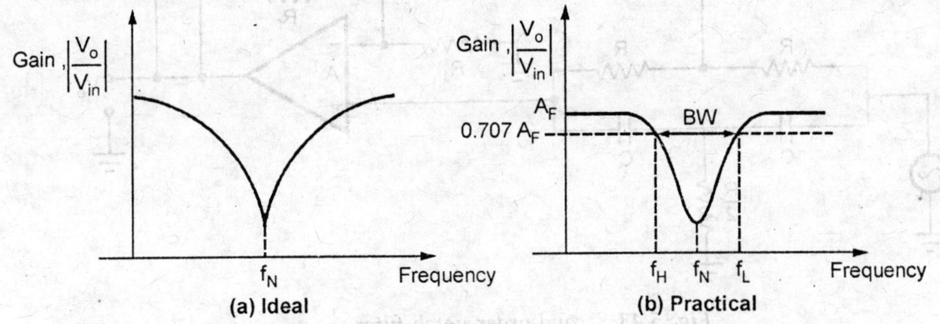

Fig. 5.41 Frequency response

Example 5.8: Design a notch filter to eliminate 120 Hz hum.

Solution: The frequency to be eliminated is the notch frequency, f_N

$$\therefore \quad f_N = 120 \text{ Hz}$$

Now, choose $\quad C = 0.33 \ \mu F$

$$\therefore \quad f_N = \frac{1}{2\pi RC}$$

$$\therefore \quad R = \frac{1}{2\pi \times 120 \times 0.33 \times 10^{-6}}$$

$$= \mathbf{4.01 \ k\Omega}$$

For R/2, use two parallel resistance of 4 kΩ while for 2 C use two 0.33 μF capacitors in parallel. The designed circuit is shown in Fig. 5.42.

For the improved response, a second order notch filter also can be used alongwith the op-amp in noninverting configuration. It also uses a twin T network. Higher the value of Q, the narrower is the width of the notch. In practice, due to mismatching of components, the infinitely deep notch cannot be obtained but using precisely matched components, reasonably deep notch can be obtained. The circuit of second order notch filter is shown in Fig. 5.43.

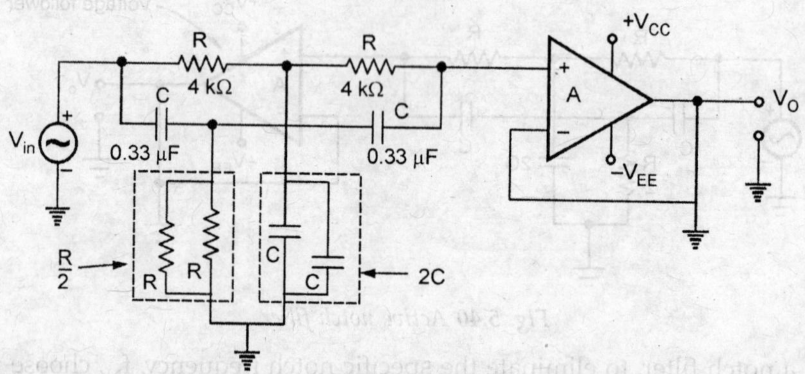

Fig. 5.42

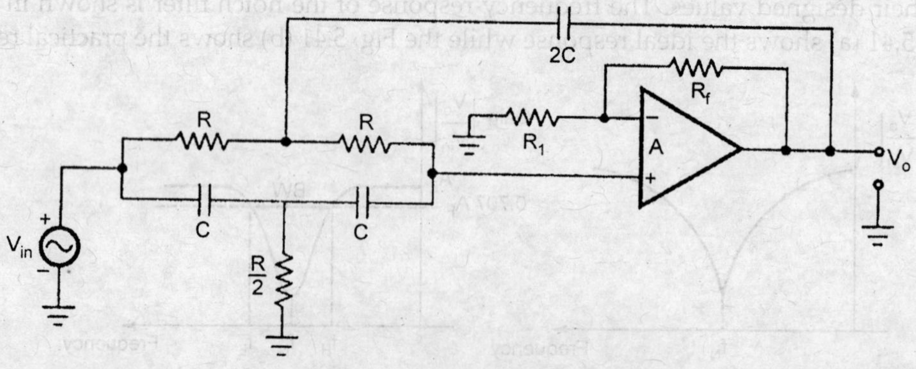

Fig. 5.43 Second order notch filter

The transfer function of this circuit is given by

$$\frac{V_0}{V_{in}} = \frac{1-\left(\frac{f}{f_N}\right)^2}{1-\left(\frac{f}{f_N}\right)^2+\left(\frac{1}{Q}\right)j\left(\frac{f}{f_N}\right)}$$

where $f_N = \frac{1}{2\pi RC}$ = Notch frequency

$$Q = \frac{1}{2(2-A_F)}$$

and $A_F = 1+\frac{R_f}{R_1}$

For the stability purpose, A_F must be less than 2.

The notch filters are used in **communication circuits** and **biomedical instruments** like ECG for removing line noise.

5.11 ALL PASS FILTER

The niters which are discussed uptil now are used to adjust the magnitude of the transfer function of the circuit. But, this also alters the phase angle characteristics of the circuit. It is many times required to control the phase response of the filter. The filter which is used to control the phase response by adding a phase shift between input and output signals is called as all pass filter. Its gain is one for all the frequencies. Thus, as the name suggests, it passes all the frequencies of the input signal. It does not produce any attenuation but provide the required phase shift for the different frequencies of the input signal.

For example, when signals are transmitted over the transmission lines, there is change in their phase. To compensate for such phase change, all pass filters are used. Hence, all pass filters are used. Hence, all pass filters arc also called as **delay equalizers** or **phase correctors.**

The Fig. 5.44 shows the simple first order all pass filter.

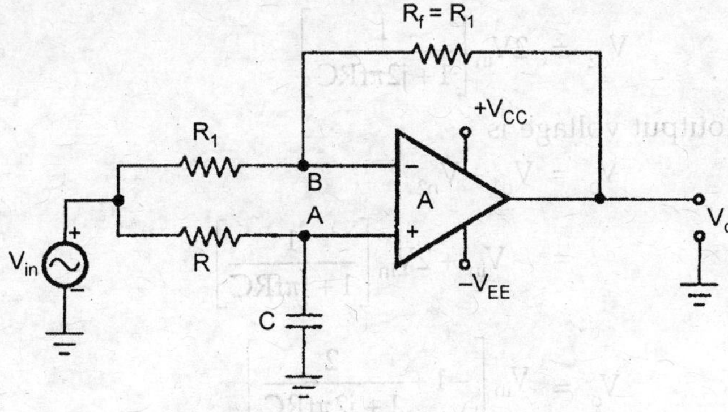

Fig. 5.44 All pass filter

5.11.1 Analysis of the Circuit

Let us use the superposition principle to obtain the expression for the output voltage V_0. Assume input to the noninverting terminal zero. The circuit acts as an inverting amplifier.

$$\therefore \quad V_{o1} = -\frac{R_f}{R_1} V_{in}$$

$$\therefore \quad V_{o1} = -V_{in} \quad \text{as} \quad R_f = R_1 \qquad ...(1)$$

Now, assume input to the inerting terminal zero. The circuit acts as a noninverting amplifier.

$$\therefore \quad V_{o2} = \left(1 + \frac{R_f}{R_1}\right) V_A$$

$$\therefore \quad V_{o2} = 2\, V_A \text{ as } R_f = R_1 \qquad ...(2)$$

and $\quad V_A = \text{Voltage at node A}$

By the potential divider rule, the voltage V_A can be obtained as

$$V_A = V_{in}\left[\frac{-jX_C}{R - jX_C}\right]$$

where $\quad -jX_C = -j\left(\frac{1}{2\pi fC}\right) = \left(\frac{1}{j2\pi fC}\right) \quad \text{as} \quad -j = \frac{1}{j}$

$$\therefore \quad V_A = V_{in}\left[\frac{\frac{1}{j2\pi fC}}{R + \frac{1}{j2\pi fC}}\right]$$

$$\therefore \quad V_A = V_{in}\left[\frac{1}{1 + j2\pi fRC}\right] \qquad ...(3)$$

Substituting in equation (2),

$$V_{o2} = 2V_{in}\left[\frac{1}{1 + j2\pi fRC}\right] \qquad ...(4)$$

Hence, the total output voltage is

$$V_o = V_{o1} + V_{o2}$$

$$= -V_{in} + 2V_{in}\left[\frac{1}{1 + j\pi fRC}\right]$$

$$\therefore \quad V_o = V_{in}\left[-1 + \frac{2}{1 + j2\pi fRC}\right] \qquad ...(5)$$

$$\therefore \quad \boxed{\frac{V_o}{V_{in}} = \frac{1 - j2\pi fRC}{1 + j2\pi fRC}} \qquad ...(6)$$

The magnitude of the transfer function as

$$\therefore \quad \boxed{\left|\frac{V_o}{V_{in}}\right| = \frac{\sqrt{1 + (2\pi fRC)^2}}{\sqrt{1 + (2\pi fRC)^2}} = 1} \qquad ...(7)$$

It is mentioned earlier that the magnitude is always 1 for all pass filter and it can pass the entire range of frequency. But the phase angle is given by

$$\therefore \quad \boxed{\phi = -2\tan^{-1}\left(\frac{2\pi fRC}{1}\right)} \qquad ...(8)$$

This is the phase angle in degrees which indicates that there is a phase shift of ϕ degrees between input and output signal. If the positions of R and C are interchanged, we get the positive phase shift. The negative phase shift indicates that the output V_o lags input Vjj, by angle §, while positive phase shift indicates that V_0 leads input V_{in} by angle ϕ.

The Fig. 5.45 shows the phase shift produced by all pass filter between input and output.

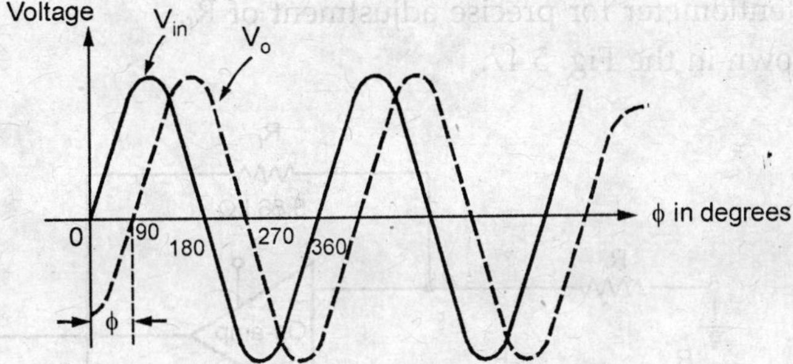

Fig. 5.45 Phase shift between V_o and V_{in}

Example 5.9: For the all pass filters, the values of R and C are 7.95 kΩ and 0.02 μF respectively. If the input frequency is 1.5 kHz, calculate the phase shift.

Solution: R = 7.95 kΩ C = 0.02 μF

f = 1.5 kHz

$$\therefore \quad \phi = -2\tan^{-1}\left(\frac{2\pi fRC}{1}\right)$$

$$= -2\tan^{-1}\left(\frac{2\pi \times 1.5 \times 10^3 \times 7.95 \times 10^3 \times 0.02 \times 10^{-6}}{1}\right)$$

$$= \mathbf{-112.56°}$$

The output voltage lags input voltage by 112.56°, but the amplitude and frequency of output will be same as that of input.

Examples with Solutions

Example 5.10: Design a second order Butterworth low pass filter having cutoff frequency of 1 kHz.

Solution:

Step 1: The cut-off frequency

is $f_H = 1$ kHz

Step 2: Choose $C_2 = C_3 = C = 0.01\ \mu F$

Step 3: Choose $R_2 = R_3 = R$

$$\therefore \quad f_H = \frac{1}{2\pi RC}$$

$$\therefore \quad 1 \times 10^3 = \frac{1}{2\pi R \times 0.01 \times 10^{-6}}$$

$$\therefore \quad R = 15.915\ k\Omega$$

Step 4: $R_f = 0.586\ R_1$

Choose $R_1 = 10\ k\Omega$

$$\therefore \quad R_f = 5.86\ k\Omega$$

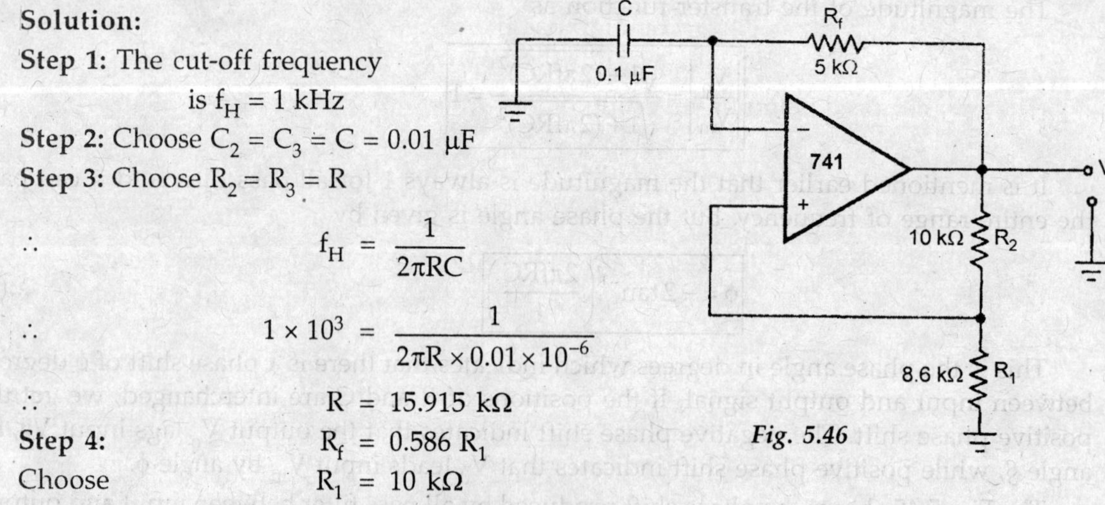

Fig. 5.46

Use a 10 kΩ potentiometer for precise adjustment of R_f.

The circuit is shown in the Fig. 5.47.

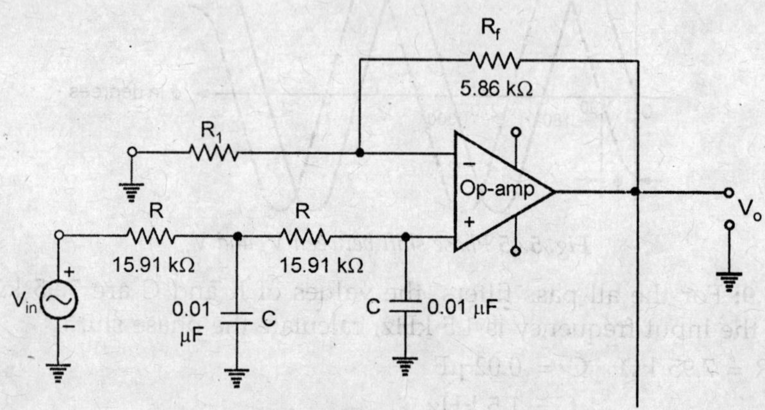

Fig. 5.47

Example 5.11: Design a fourth order Butterworth low pass filter having a upper cutoff frequency of 1 kHz.

Solution: The 4th order Butterworth low pass filter is to be obtained by cascading two second order low pass filters. For 4th order, the two second order Butterworth polynomials are,

$$(s^2 + 0.7653\ s + 1)\ (s^2 + 1.8477\ s + 1) \quad \text{... Refer Table 5.6}$$

Thus in first part of the fourth order filter, the second order filter is to be designed such that,

$$3 - A_F = 0.7653$$

$$\therefore \quad A_F = \text{op-amp pass band gain} = 3 - 0.7653 = 2.2347$$

The op-amp used is in non-inverting mode,

$$\therefore \quad A_F = 1 + \frac{R_f}{R_1}$$

$$\therefore \quad \frac{R_f}{R_1} = 2.2347 - 1 = 1.2347$$

$$\therefore \quad R_f = 1.2347\ R_1$$

While for the second part of the fourth order filter, the second order filter is to be designed such that,

$$3 - A_F = 1.8477$$

$$\therefore \quad A_F = 1.1523$$

Now $\quad A_F = 1 + \frac{R_f}{R_1}$

$$\therefore \quad \frac{R_f}{R_1} = 0.1523$$

$$\therefore \quad R_f = 0.1523\ R_1$$

Choose $\quad R_1 = 10\ k\Omega$ for both the parts.

So for first part, $\quad R_f = 1.2347 \times 10 = 12.34\ k\Omega$

And for second part, $\quad R_f = 0.1523 \times 10 = 1.523\ k\Omega$

For both the sections, $\quad f_H = 1\ kHz$

Now $\quad f_H = \frac{1}{2\pi RC}$

$$\therefore \quad 1 \times 10^3 = \frac{1}{2\pi RC}$$

Choose $\quad C = 0.1\ \mu F$

$$\therefore \quad R = \frac{1}{2\pi \times 1 \times 10^3 \times 0.1 \times 10^{-6}} = \mathbf{1.591\ k\Omega}$$

The designed fourth order low pass filter is shown in the Fig. 5.48.

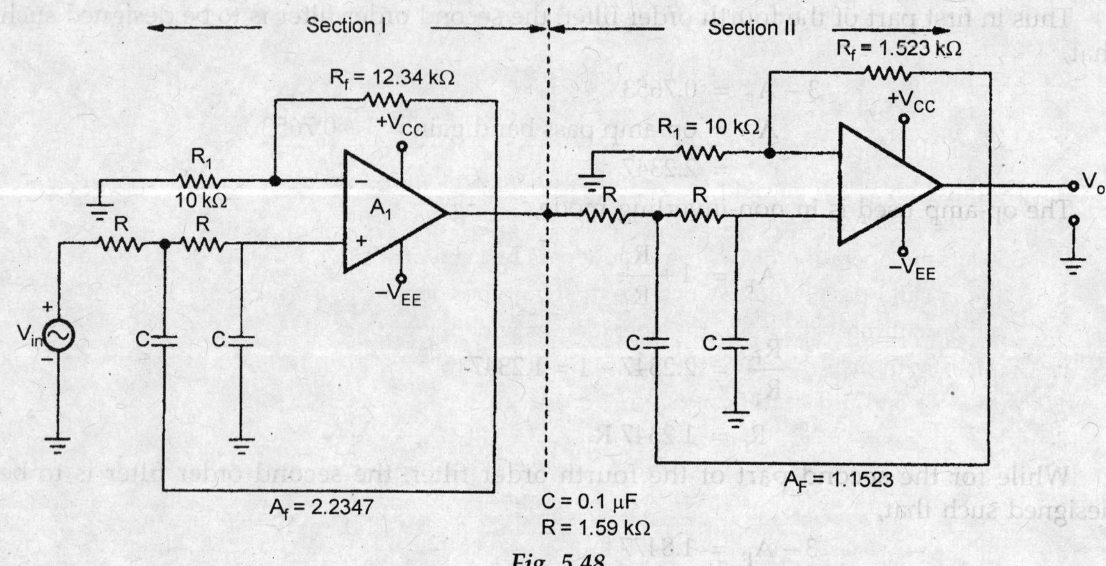

Fig. 5.48

Example 5.12: Design a second order Butterworth active high pass filter for a cut-off frequency of 5 kHz.

Solution: (i) A second order Butterworth active high pas filter for a cut-off frequency of 5 kHz : The cut-off frequency is f_0 = 5 kHz

Choose $R = 10\ \text{k}\Omega$

$\therefore$ $$f_0 = \frac{1}{2\pi RC}$$

i.e. $$C = \frac{1}{2\pi \times 10 \times 10^3 \times 5 \times 10^3} = \mathbf{3.18\ nF}$$

For second order Butterworth filter,

$$\frac{R_f}{R_1} = 1.585$$

Choose R = 10 kΩ hence R_f = 15.85 kΩ

The designed circuit is shown in the Fig. 5.49.

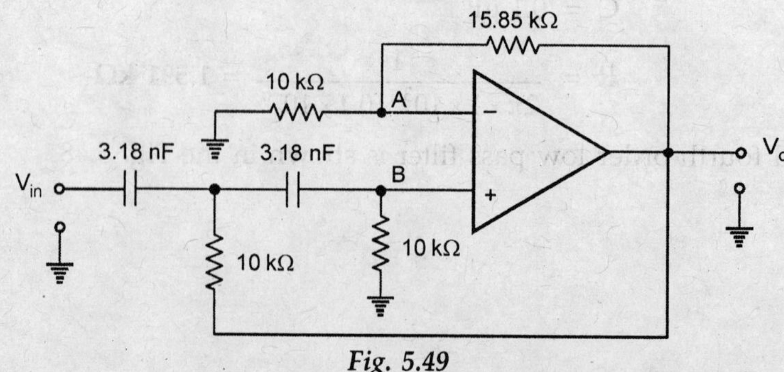

Fig. 5.49

REVIEW QUESTIONS

1. Define a filter. How filters are classified ?
2. Explain the various types of filters along with their frequency response.
3. What is the meaning of stop band and pass band of a filter ?
4. Explain the operation of second order low pass butter worth filter.
5. What are the advantages of active filter over passive filter ?
6. State the disadvantages of passive filters.
7. Explain the operation of first order low pass but tenvorth filter.
8. What is band pass filter ? Which are the two types of band pass filters ?
9. Explain with neat diagram, the operation of wide band pass filter.
10. Explain ivith neat diagram, the operation of narrow band pass filter.
11. Design second order Butterworth low pass filter having cut off frequency of 1 kHz.
12. What is band elimination filter ? Draw its frequency response.
13. Explain the operation of wide band reject filter.
14. What is a notch filter 1 Explain its operation with the help of a neat circuit diagram.
15. What is an all pass filter ? Explain its operation and application areas.
16. Derive the expression for the phase shift produced by an all pass filter.
17. Design a low pass filter at a cut off frequency of 1 kHz with a pass band gain of 2.
18. Design a high pass filter at a cut off frequency of 1 kHz with a pass band gain of 2. Plot its frequency response.
19. Design a wide band pass filter with f_L = 200 Hz and f_H = 1 kHz and a pass band gain = 4. Draw the frequency response and calculate its Q value. (Ans.: Q = 0.56)
20. Design a wide band reject filter having f_H = 200 Hz and f_L = 1 kHz.
21. For an all pass filter, the required phase shift is – 90° at a frequency of 1 kHz. Design the filter and draw the circuit diagram.
22. Explain the operation of first order high pass butterworth filter.
23. Explain the operation of second order high pass butterworth filter.

❁❁❁

Chapter

6

Phase Locked Loop

Inside the Chapter

6.1 INTRODUCTION

A phase locked loop is basically a closed loop system designed to lock the output frequency and phase to the frequency and phase of an input signal. It is commonly abbreviated as PLL. The PLL was first introduced in its discrete form in early 1930s. The high cost of realizing PLL in discrete form limited its use earlier. Now with the advanced IC technology, PLLs are available as inexpensive monolithic ICs. They are used in applications such as frequency synthesis, frequency modulation/demodulation, AM detection, tracking filters, FSK demodulator, tone detector etc.

In this chapter, we are going to discuss basic operating principle, popular PLL IC 565 and important applications of PLL. The chapter starts with the discussion of voltage controlled oscillator.

6.2 VOLTAGE CONTROLLED OSCILLATOR (VCO)

A voltage controlled oscillator is an oscillator circuit in wliich the frequency of oscillations can be controlled by an externally applied voltage. The VCO provides the linear relationship between the applied voltage and the oscillation frequency. Applied voltage is called **control voltage.** The control of frequency with the help of control voltage is also called voltage to frequency conversion.

Hence VCO is also called **voltage to frequency converter.** Practically VCO is used to produce square and triangular waveforms whose frequency is controlled by control voltage.

The Fig. 6.1 shows op-amp based voltage controlled oscillator circuit.

The circuit associated with op-amp A_1 is V-I converter i.e. it forces current through capacitor C which is linearily proportional to the input voltage V_i. For getting triangular and square waveforms, capacitor must charge and discharge and thus current direction through capacitor must kept on alternating. The direction of current through capacitor is controlled by the MOSFET switch M_1. It is driven by Schmitt trigger comparator CMP. When the output of CMP. is low, M_1 is off and resistance R remains floating while when the output of CMP is high, M_1 is on and resistance R gets connected to the ground. The square wave is obtained at terminal V_{SQ} and triangular wave is obtained at V_{TR}. When output transistor Q is on, V_{SQ} is nothing but $v_{CE(sat)}$ i.e. almost zero. While when Q is off then V_{SQ} is decided by voltage divider of R_1 and R_2.

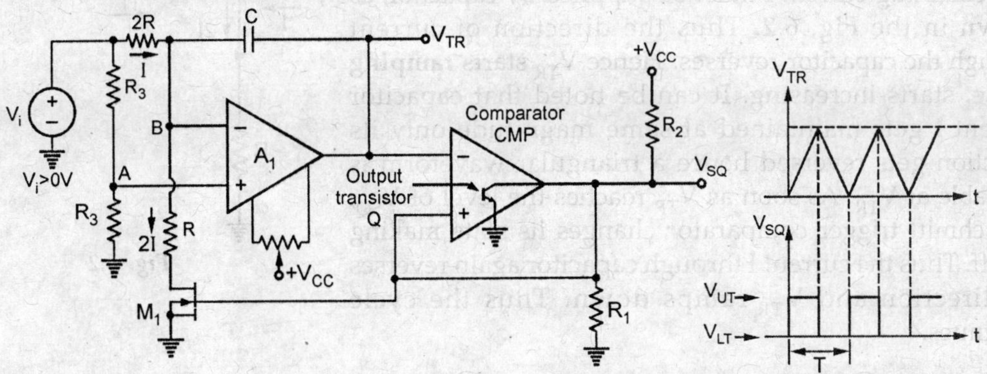

Fig. 6.1 Voltage controlled oscillator circuit

$$\therefore \quad V_{SQ} = \frac{V_{CC}R_1}{R_1 + R_2} \quad \text{... when Q is off}$$

As the output V_{SQ} is connected to the non inverting terminal of comparator CMP, the high and low values of V_{SQ} are the upper and lower threshold levels of comparator CMP.

Due to voltage divider of R_3 and R_3, the voltage of node A is,

$$V_A = \frac{V_i R_3}{R_3 + R_3} = \frac{V_i}{2} \quad \text{...(1)}$$

Due to the virtual ground, $V_B = V_A$

$$\therefore \quad V_B = \frac{V_i}{2} \quad \text{...(2)}$$

Thus the current I through 2R resistor is,

$$I = \frac{V_i - \frac{V_i}{2}}{2R} = \frac{1}{4R}(V_i) \quad \text{...(3)}$$

The equation (3) shows that op-amp A_1 acts as **voltage to current converter.**

Assume that the output V_{SQ} is low hence M_1 is off and the resistance R is floating. So entire current $I = \frac{1}{4R}(V_i)$ has to pass through the capacitor C and the voltage V_{TR} starts decreasing, **due to polarity of voltage across capacitor.** So V_{TR} **ramps down.**

When V_{TR} becomes zero then comparator CMP changes its state and output V_{SQ} becomes high. Due to this M_1 becomes on and the resistance R gets connected to the ground. The current through resistance R is thus,

$$I_R = \frac{V_B}{R} = \frac{V_i}{2R} = 2\frac{V_i}{4R} = 2I \quad ...(4)$$

Now current I out of 2I is supplied by resistance 2R but remaining current I must be supplied by capacitor, as shown in the Fig. 6.2. Thus the direction of current through the capacitor reverses. Hence V_{TR} starts **ramping up** i.e. starts increasing. It can be noted that capacitor current I gets maintained at same magnitude only its direction gets reversed hence a triangular waveform is available at V_{TR}. As soon as V_{TR} reaches the level of V_{SQ}, the schmitt trigger comparator changes its state making M_1 off. Thus the current I through capacitor again reverses its direction and V_{TR} ramps down. Thus the cycle continues.

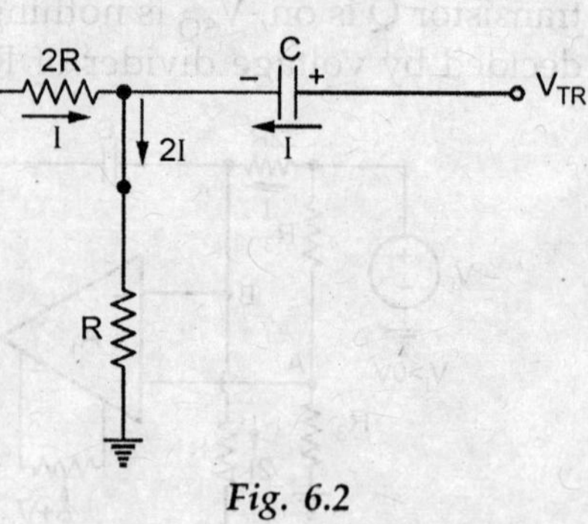

Fig. 6.2

6.2.1 Period of Oscillation

The period of oscillation is determined by equating the charge expressions,

$$C\Delta V = I\Delta t$$

During any half cycle,

$$\Delta V = V_{UT} - V_{LT}$$

$$I = \frac{V_i}{4R}$$

and

$$\Delta t = \frac{T}{2}$$

$$\therefore \quad C(V_{UT} - V_{LT}) = \frac{V_i}{4R} \times \frac{T}{2}$$

$$\therefore \quad T = \frac{4RC \times 2[V_{UT} - V_{Lt}]}{V_i}$$

$$\therefore \quad f = \frac{1}{T} = K_v V_i \quad ...(5)$$

where

$$\boxed{K_v = \frac{1}{8RC[V_{UT} - V_{LT}]}} \quad ...(6)$$

The equation (5) shows that the output frequency is linearily proportional to the input voltage V_i and thus circuit works as voltage controlled oscillator.

But the accuracy of such VCO circuit is limited because of speed of response of op-amp A_1, comparator CMP and MOSFET switch M_1. Similarly error is caused at low frequencies due to input bias current and the offset voltage of op-amp A_1 The non-zero channel resistance of M_1 when it is on i.e. r_{ds}(on) is another factor causing an error. Hence in practice integrated version of VCO is used.

Example 6.1: In theVCO circuit shown in the Fig. 6.1, the various parameters are R = 10 KΩ, C = 1.25 nF, V_{UT} = 10 V and V_{LT} = OV. If input is changed from 10 mV to 10 V calculate the range over which the output frequency can be varied.

Solution: For the VCO circuit,

$$K_v = \frac{1}{8RC[V_{UT} - V_{LT}]}$$

$$= \frac{1}{8 \times 10 \times 10^3 \times 1.25 \times 10^{-9} \times (10 - 0)}$$

$$= 1000 \text{ Hz/V}$$

And $f = K_v V_i$

For V_i = 10 mV, $f = 1000 \times 10 \times 10^{-3} = 10 \text{ Hz}$

For V_i = 10 V, $f = 1000 \times 10 = 10 \text{ kHz}$

Hence the range over which output frequency can be varied is 10 Hz to 10 kHz.

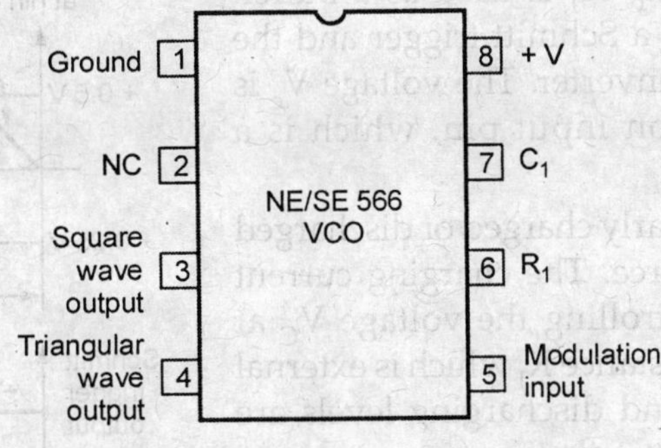

Fig. 6.3 Pin diagram of IC 566 VCO

6.3 VOLTAGE CONTROLLED OSCILLATOR (VCO) IC 566

A voltage controlled oscillator (VCO) is an oscillator circuit in which the frequency of oscillations can be controller by an externally applied voltage. The VCO provides the linear relationship between the applied voltage and the oscillation frequency. Applied voltage is called **control voltage.** The control of frequency with the help of control voltage is also called voltage to frequency conversion. Hence VCO is also called **voltage to frequency converter.** Practically VCO is used to produce square and triangular waveforms whose frequency is controlled by control voltage.

Practically VCO is available in IC form. The commonly used VCO ICs are NE/SE 566, LM 566 etc. The Fig. 6.3 shows the pin diagram of NE/SE 566 VCO manufactured by Signetics. It is 8 pin IC, which provides two output pins. Its feature is that simultaneously IC provides square and triangular wave outputs which are the functions of input voltage. The input voltage is also called modulating input voltage.

The Fig. 6.4 shows the block diagram of NE/SE 566 VCO.

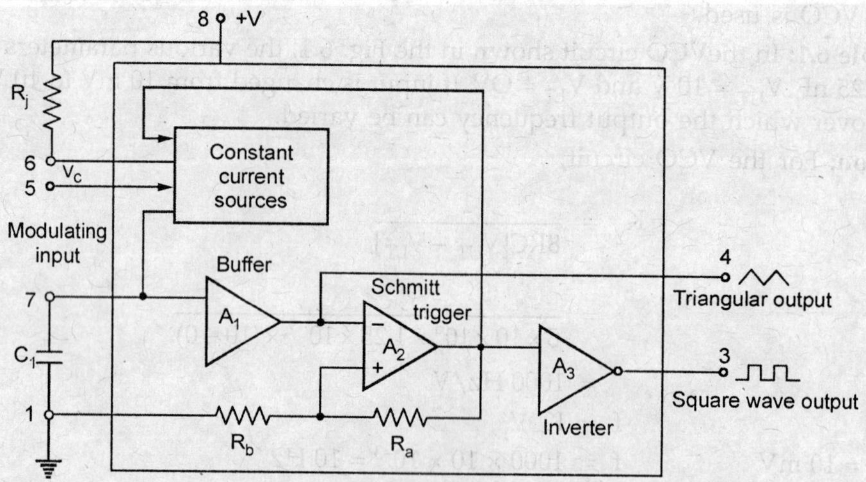

Fig. 6.4 Block diagram of NE/SE 566 VCO

Operation: The op-amp A_1 is used as a buffer. The op-amp A_2 is used as a Schmitt trigger and the op-amp A_3 is used as an inverter. The voltage V_c is applied to the modulation input pin, which is a control voltage.

The capacitor C_1 is linearly charged or discharged by a constant current source. The charging current can be controlled by controlling the voltage V_C at pin 5 or by varying the resistance R_1 which is external to the IC. The charging and discharging levels are determined by the Schmitt trigger.

The output voltage of Schmitt trigger is designed to swing between +V and 0.5 V. For $R_a = R_b$, the voltage at noninverting terminal swings between 0.5(+V) to 0.25(+V). Thus the triangular wave is generated due to alternate charging and discharging of the capacitor C_1, in linear manner. When C_1 voltage increases beyond 0.5(+V), the Schmitt trigger output goes low, and the capacitor starts discharging. When the voltage becomes less than 0.25 (+V), the output of the Schmitt trigger goes high. Due to similar

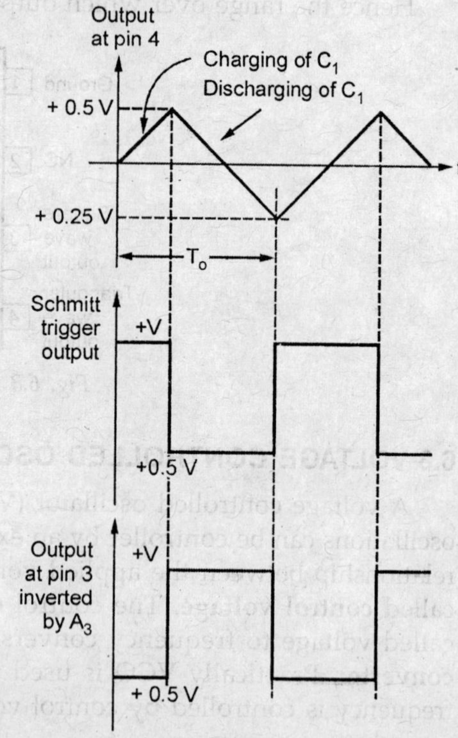

Fig. 6.5 Waveform for VCO

current sources used for charging and discharging, the time taken by C_1 to charge and discharge is same. This produces exact triangular wave. The output of the Schmitt trigger output is step response which is available at the pin 3 as a square wave output. The various waveforms are shown in the Fig. 6.5.

The frequency of the output waveform is,

$$f_o = \frac{2(+V - V_C)}{C_1 R_1 (+V)} \quad ...(1)$$

R_1 should be in the range of 2 kΩ to 20 kΩ. The equation shows that the frequency of the output oscillations is a frequency of the contol voltage V_C. The frequency is also a function of the external resistance R_1 and capacitor C_1.

6.3.1 Typical Connection Diagram of 566 VCO

Practically the modulating input voltage at the pin 5 is controlled using a potential divider formed by the resistances R_2 and R_3. Such a circuit is shown in the Fig. 6.6.

The $R_2 - R_3$ potential divider is used to set the control voltage V_C. The initial voltage V_C at the pins must be such that

$$0.75(+V) \leq V_C \leq +V$$

where +V is the supply voltage.

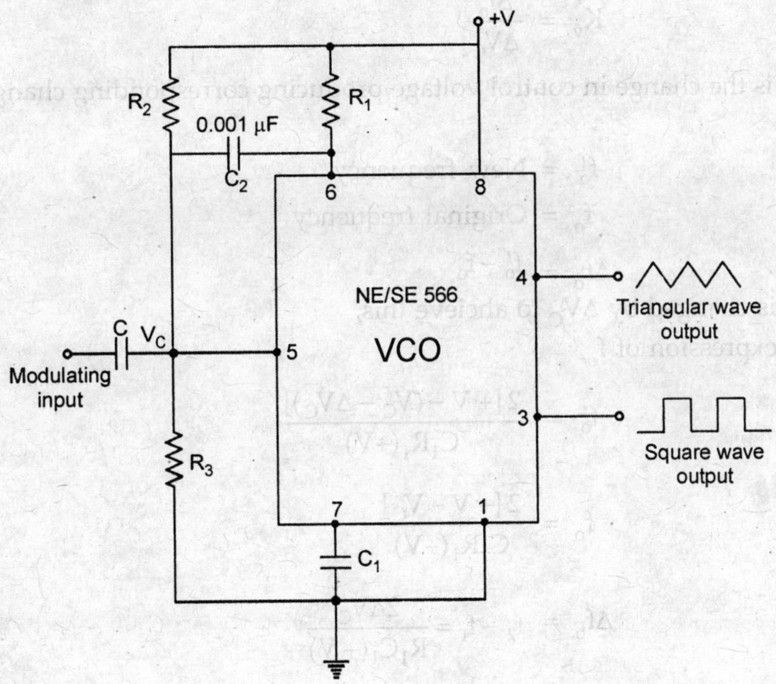

Fig. 6.6 Typical connection diagram of 566 VCO

The modulating input is connected through capacitor coupling to the pin 5 and it must be less than 3 V(p-p). For a fixed V_C and C_1 the frequency can be varied in the range 10 : 1 by varying R_1 between 2 kΩ to 20 kΩ. For fixed R_1 and C_1 the frequency can be varied in the range 10 : 1 by varying control voltage V_C. In both the cases maximum possible output frequency is 1 MHz. The capacitor C_2 connected between pins 5 and 6 is used to eliminate any oscillations present in constant current source.

6.3.2 Features of 566 VCO

1. Wide supply voltage range 10 V to 24 V.
2. Very linear modulation characteristics.
3. High temperature stability.
4. Excellent power supply rejection.
5. 10 to 1 frequency range with fixed C_1.
6. The frequency can be controlled by means of current, voltage, resistor, or capacitor.

6.3.3 Derivation of Voltage to Frequency Conversion Factor

The voltage to frequency conversion factor is an important factor of IC of 566. it is denoted as K_v, and defined as,

$$K_v = \frac{\Delta f_o}{\Delta V_C} \qquad ...(1)$$

Here ΔV_C is the change in control voltage producing corresponding change of Δf_o in the frequency.

Let f'_o = New frequency

f_o = Original frequency ...(2)

$\therefore \quad \Delta f_o = f'_o - f_o$

While V_C is changed by ΔV_C to ahcieve this,

From the expression of f_o,

$$f'_o = \frac{2[+V-(V_C-\Delta V_C)]}{C_1R_1(+V)} \qquad ...(3)$$

and

$$f_o = \frac{2[+V-V_C]}{C_1R_1(+V)} \qquad ...(4)$$

$$\therefore \quad \Delta f_o = f_2 - f_1 = \frac{2\Delta V_C}{R_1C_1(+V)} \qquad ...(5)$$

$$\therefore \quad \Delta V_C = \frac{R_1C_1\Delta f_o(+V)}{2} \qquad ...(6)$$

With no modulating input voltage,

Control voltage $V_C = (7/8)(+V)$

if f_o is original frequency then,

$$f_o = \frac{2\left[+V - \frac{7}{8}(+V)\right]}{R_1C_1(+V)} = \frac{0.25}{R_1C_1} \qquad ...(7)$$

Using value of R_1C_1 from equation (6) in equation (7),

$$f_o = \frac{0.25}{\frac{2\Delta V_C}{\Delta f_o(+V)}}$$

$$\therefore \quad K_v = \frac{\Delta f_o}{\Delta V_C} = \frac{f_o}{0.125(+V)}$$

$$\therefore \quad \boxed{K_v = \frac{8f_o}{(+V)}} \quad \text{where } f_o = \text{Original frequency}$$

This is the required voltage to frequency conversion factor.

6.3.4 Applications of VCO

The various applications of VCO are.

1. FM modulation.
2. Signal generation (Triangular or square wave)
3. Function generation.
4. Frequency shift keying i.e. FSK demodulator.
5. In frequency multipliers.
6. Converting low frequency signals such as EEC and EKG into audio frequency range signals.
7. Tone generation.

Example 6.2: For a 566 VCO shown in the Fig. 6.7, +V = 10 V, $R_2 = 1.2\ k\Omega$ and $R_1 = R_3 = 10\ k\Omega$ with $C_1 = 0.001$ uF.

(a) Calculate the frequency of output.

(b) Calculate the variation in f_0 if V_c is varied between 7.7 V and 9.5 V.

(c) Draw the square wave output if the modulating input is sine wave.

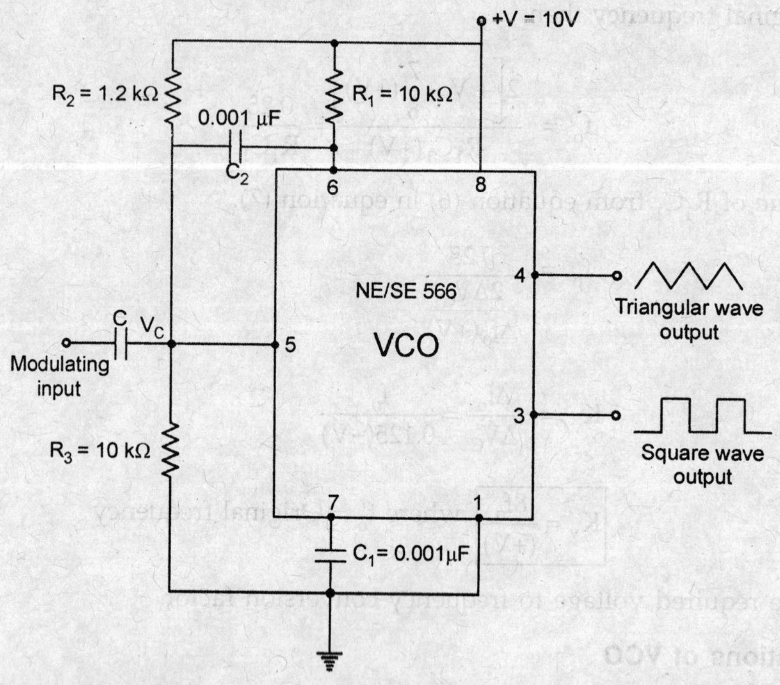

Fig. 6.7

Solution: (a) R_2 and R_3 form potential divider.

$$\therefore \quad V_C = R_3 \times \frac{+V}{R_2 + R_3}$$

$$= \frac{10 \times 10}{1.2 + 10} = 8.928 \text{ V}$$

$$\therefore \quad f_o = \frac{2(+V - V_C)}{C_1 R_1 (+V)} = \frac{2(10 - 8.928)}{0.001 \times 10^{-6} \times 10 \times 10^3 \times 10}$$

$$= 21.44 \text{ kHz}$$

(b) For $V_C = 7.7$ V,

$$f_{o1} = \frac{2(10 - 7.7)}{0.001 \times 10^{-6} \times 10 \times 10^3 \times 10}$$

$$= 46 \text{ kHz}$$

For $V_C = 9.5$ V,

$$f_{o2} = \frac{2(10 - 9.5)}{0.001 \times 10^{-6} \times 10 \times 10^3 \times 10}$$

$$= 10 \text{ kHz}$$

The change in output frequency is 46 – 10 = 36 kHz.

(c) It can be seen that as V_C increases frequency decreases. When modulating input is a sine wave then during positive half cycle V_C increases and frequency decreases. While in negative half cycle, the frequency increases. So the square wave output waveform is as shown in the Fig. 6.8.

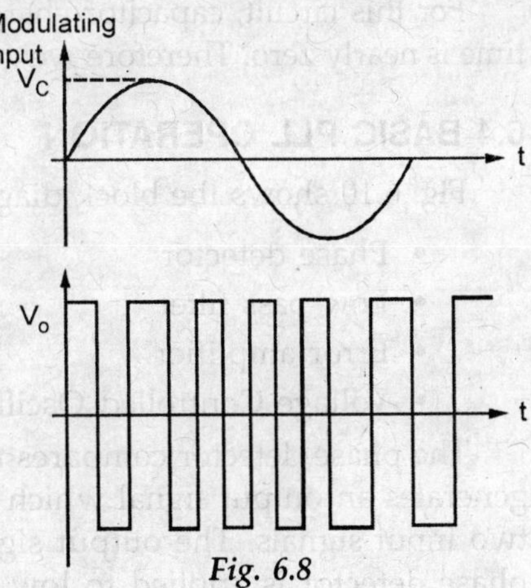

Fig. 6.8

This example explains the use of 566 to generate FM.

6.3.5 Ramp Generator using 566

The Fig. 6.9 shows the ramp generator using LM 566. Here, transistor Q_1 is used as short circuit path for capacitor. During charging the output at pin 3 is low (square wave). This makes Q_1 Off and hence capacitor charging time is decided by the R_1, C_1 +V and V_C. When capacitor voltage charges upto the upper threshold voltage of the schmitt trigger, its output goes high, making output at pin 3 high. As a result transistor Q_1 is made ON. It provides short circuit path for capacitor to discharge in no time. When capacitor voltage goes below lower threshold of schmitt trigger, the voltage at pin is made low, making transistor Q_1 OFF. The capacitor C_1 starts charging again and cycle repeats.

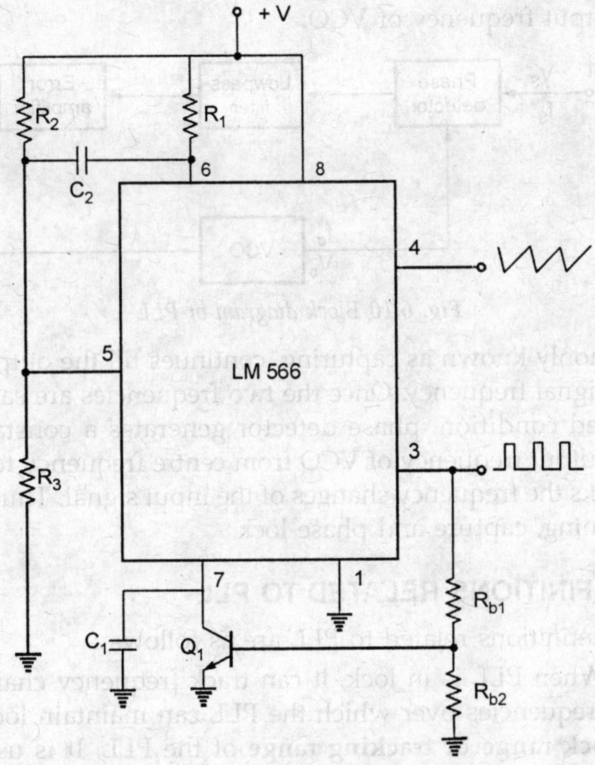

Fig. 6.9 Ramp generator using LM 566

For this circuit, capacitor charging time is decided by R_1, C_1, +V and V_C and discharge time is nearly zero. Therefore we get ramp signal at the triangular wave output of the LM 566.

6.4 BASIC PLL OPERATION

Fig. 6.10 shows the block diagram of PLL. It consists of

- Phase detector
- Low pass filter
- Error amplifier
- Voltage Controlled Oscillator (VCO)

The phase detector compares the input frequency f_s with the feedback frequency f_o and generates an output signal which is a function of the difference between the phases of the two input signals. The output signal of the phase detector is a dc voltage. The output of phase detector is applied to low pass filter to remove high frequency noise from the dc voltage. The output of low pass filter without high frequency noise is often referred to as error voltage or control voltage for VCO. When control voltage is zero, VCO is in free-running mode and its output frequency is called as center frequency f_o. The non-zero control voltage results in a shift in the VCO frequency from its free-running frequency, f_o to a frequency f, given by $f = f_o + K_v V_c$, where K_v is the voltage to frequency transfer coefficient of the VCO. The error or control voltage applied as an input to the VCO, forces the VCO to change its output frequency in the direction that reduces the difference between the input frequency and the output frequency of VCO.

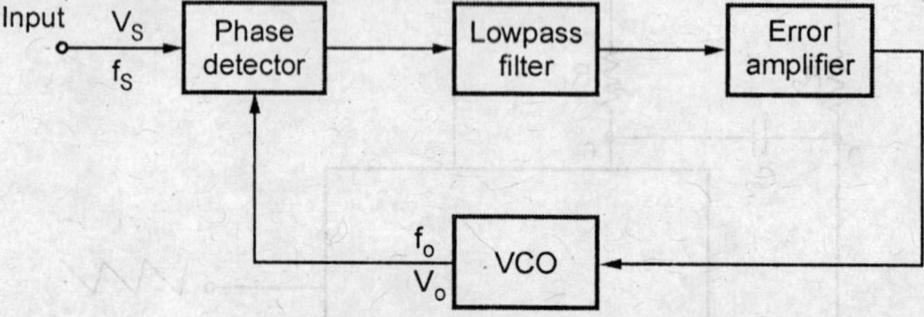

Fig. 6.10 Block diagram of PLL

This action, commonly known as capturing, continues till the output frequency of VCO is same as the input signal frequency. Once the two frequencies are same, the circuit is said to be locked. In locked condition, phase detector generates a constant dc level which is required to shift the output frequency of VCO from centre frequency to the input frequency. Once locked, PLL tracks the frequency changes of the input signal. Thus, a PLL goes through three states : free running, capture and phase lock.

6.5 IMPORTANT DEFINITIONS RELATED TO PLL

Some important definitions related to PLL are as follows :

LOCK RANGE: When PLL is in lock, it can track frequency changes in the incoming signal. The range of frequencies over which the PLL can maintain lock with the incoming signal is called the **lock range** or **tracking range** of the PLL. It is usually expressed as a percentage of f_o, the VCO frequency.

CAPTURE RANGE: The range of frequencies over which the PLL can acquire lock with an input signal is called the capture range. It is also expressed as a percentage of f_o.

PULL-IN TIME: The capture of an input signal does not take place as soon as the signal is applied, but it takes finite time. The total time taken by the PLL to establish a lock is called **pull-in time.** This depends on the initial phase and frequency difference between the two signals as well as on the overall loop again and the bandwidth of the low pass filter.

6.6 PHASE DETECTOR/COMPARATOR BLOCK OF PLL

The phase detector is an initial and most important part of PLL. There are two types of phase detectors : Analog and Digital.

6.6.1 Analog Phase Detectors

In this section we will study two types of analog phase detectors :

1. Switch Type Phase Detector.
2. Balanced Modulator Type Phase Detector.

Switch Type Phase Detector

The Fig. 6.11 shows the basic circuit and waveforms of analog phase detector. It consists of electronic switch, S. The switch is opened and closed by signal coming from VCO as shown in the Fig. 6.11(b). The switch is closed when VCO output is positive; otherwise it is open. Let us see the output of phase detector at different phase angles of input signals.

When ϕ = 0, i.e. when the input signal V_s is in phase with VCO output V_o, the output waveform V_e will be positive half sinusoids (shaded portion) shown in the Fig. 6.11(c).

When ϕ = 90°, the output waveform V_e contains half portion of negative cycle and half portion of positive cycle shown in the Fig. 6.11(d).

When ϕ = 180°, the output waveform V_e contains negative half sinusoids shown in Fig. 6.11 (e). The average output voltages (error

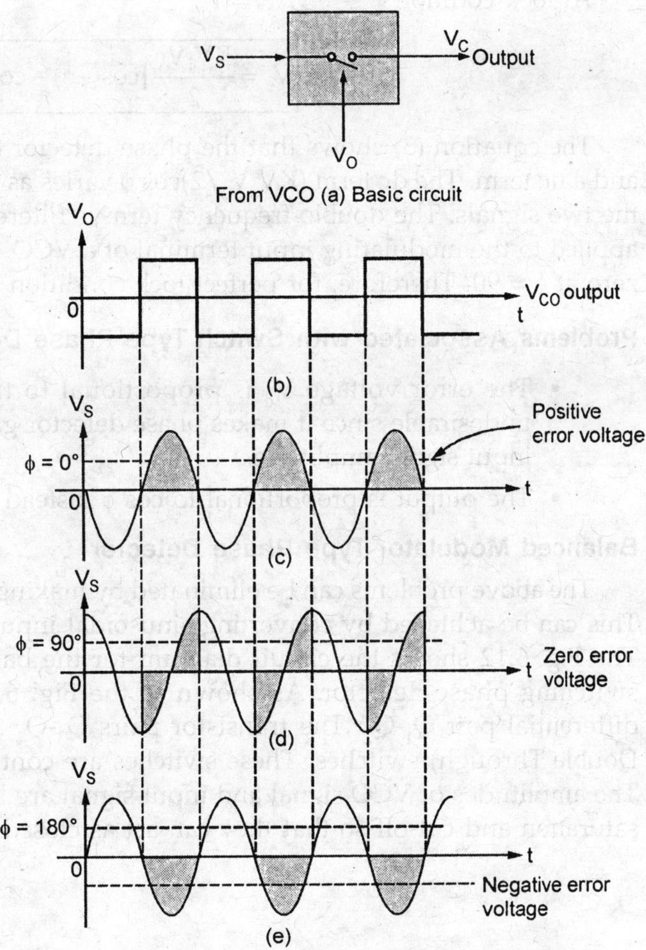

Fig. 6.11 Phase detector basic circuit and waveforms

voltage) at different phase angles of input signals are shown in respective figures by dotted line. It may be seen that the error voltage is zero when the phase shift between the two inputs (V_s and V_o) is 90°. This is a perfect lock condition. For phase shifts 0° and 180° the error voltage is positive and negative, respectively.

The detector studied above is called a half wave detector, since the phase information for only one-half of the input waveform is detected and averaged.

A phase detector actually multiplies the input signal ($V_s = V_s \sin(2\pi f_s t)$) by the VCO signal ($V_o = V_o \sin(2\pi f_o t + \phi)$). The output of multiplication is an error signal and it is given as

$$V_e = KV_sV_o \sin(2\pi f_s t) \sin(2\pi f_o t + \phi) \quad ...(1)$$

Where K : Phase detector gain or attenuation constant, the equation (1) can be simplified as

$$V_e = \frac{KV_sV_o}{2} [\cos(2\pi f_s t - 2\pi f_o t - \phi) - \cos(2\pi f_s t + 2\pi f_o t + \phi)] \quad ...(2)$$

At lock codition $f_s = f_o$,

$$\therefore \quad \boxed{V_e = \frac{KV_sV_o}{2} [\cos(-\phi) - \cos(2\pi \times 2f_o t + \phi)]} \quad ...(3)$$

The equation (3) shows that the phase detector output contains double frequency term and a dc term. The dc term $(KV_sV_o/2)\cos\phi$ varies as a function of phase ϕ, i.e. $\cos\phi$ between the two signals. The double frequency term is filtered by the low pass filter and dc term is applied to the modulating input terminal of a VCO. The dc term i.e., the dc error voltage is zero at $\phi = 90$. Therefore, for perfect lock condition $f_s = f_o$ and $0 = 90°$.

Problems Associated with Switch Type Phase Detector

- The error voltage V_e is proportional to the input signal amplitude V_s. This is undesirable since it makes phase detector gain and the loop gain dependent on the input signal amplitude.
- The output is proportional to $\cos\phi$ instead of only ϕ. This makes it non-linear.

Balanced Modulator Type Phase Detector

The above problems can be eliminated by making amplitude of the input signal constant. This can be achieved by converting sinusoidal input signal into square wave input signal. The Fig. 6.12 shows the circuit diagram for the balanced modulator used as a full-wave switching phase detector. As shown in the Fig. 6.12, the input signal is applied to the differential pair Q_1-Q_2. The transistor pairs Q_3-Q_4 and Q_5-Q_6, act as a SPDT (Single Pole Double Through) switches. These switches are controlled by the output signal from VCO. The amplitudes of VCO signal and input signal are kept such that transistor are driven into saturation and cut-off so that they can act as closed switch and open switch, respectively.

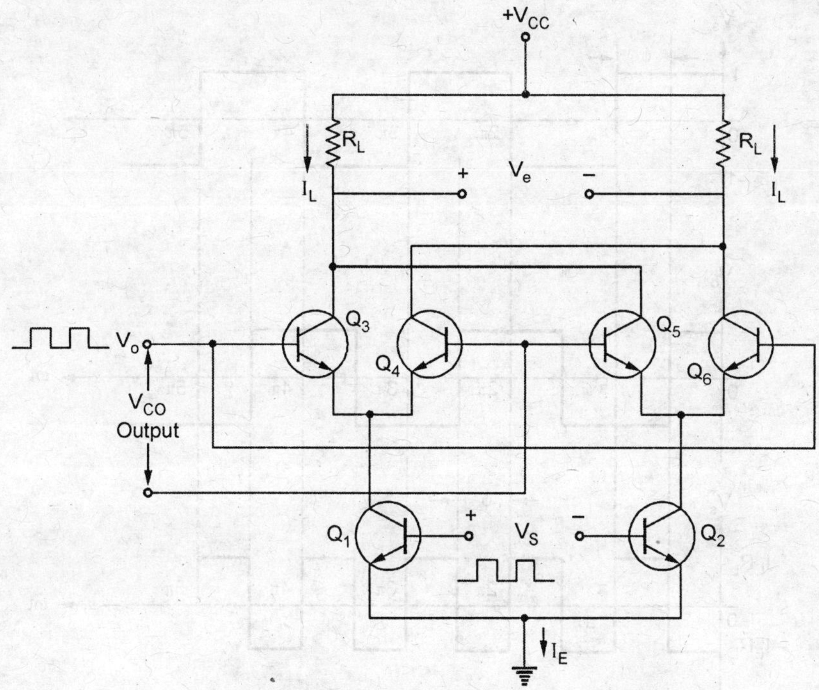

Fig. 6.12 Balanced modulator type phase detector

The Fig. 6.13 shows the waveforms for balanced modulator type phase detector with phase shift of ϕ between input signal and the VCO output signal. When V_s is positive Q_1 is ON and Q_2 is OFF. When V_o is positive Q_3 and Q_6 are ON, and Q_4 and Q_5 are OFF. Obviously, when V_s is negative Q_1 is OFF and Q_2 in ON. Similarly, when V_o is negative Q_3 and Q_6 are OFF, and Q_4 and Q_5 are ON. These conditions are listed in Table 6.1.

Table 6.1 Transistor states and output voltage at different combinations of V_s and V_o

V_s	V_o	Q_1	Q_2	Q_3-Q_6	Q_4-Q_5	V_e
Positive	Positive	ON	OFF	ON	OFF	Negative
Positive	Negative	ON	OFF	OFF	ON	Positive
Negative	Positive	OFF	ON	ON	OFF	Positive
Negative	Negative	OFF	ON	OFF	ON	Negative

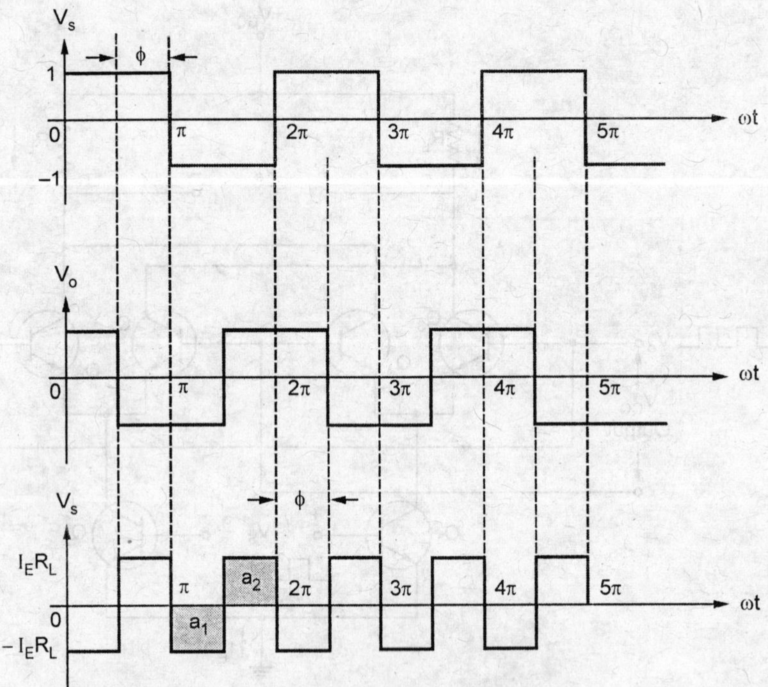

Fig. 6.13 Waveforms of balanced modulator phase detector

The output error voltage is given by

$$\boxed{V_e = I_E R_L}$$

It is positive when either Q_1 and Q_4 are ON or Q_2 and Q_6 are ON. It is negative when either Q_1 and Q_3 are ON or Q_2 and Q_5 are ON.

The average value of the output voltage of the phase detector, V_e can be obtained as,

$$V_{e(av)} = \frac{1}{\pi}(a_1 + a_2) \qquad ...(4)$$

Where a_1 : Shaded area shown in the negative half

a_2 : Shaded area shown in the positive half

$$\therefore \quad V_{e(av)} = \frac{1}{\pi}\left[(-I_E R_L)(\pi - \phi) + (I_E R_L)(\phi)\right]$$

$$= \frac{I_E R_L}{\pi}[-\pi + \phi + \phi]$$

$$= \frac{2 I_E R_L}{\pi}\left[\phi - \frac{\pi}{2}\right]$$

$$\therefore \quad V_{e(av)} = \frac{4 I_L R_L}{\pi}\left[\phi - \frac{\pi}{2}\right] \qquad \because I_E = 2 I_L$$

$$\therefore \quad V_{e(av)} = K_\phi\left(\phi - \frac{\pi}{2}\right) \qquad \because K_\phi = \frac{4I_L R_L}{\pi} \qquad ...(5)$$

Where K_ϕ is the phase angle to voltage transfer coefficient. It is the conversion ratio of the phase detector. The Fig. 6.14 shows the linear relationship between V_e and ϕ.

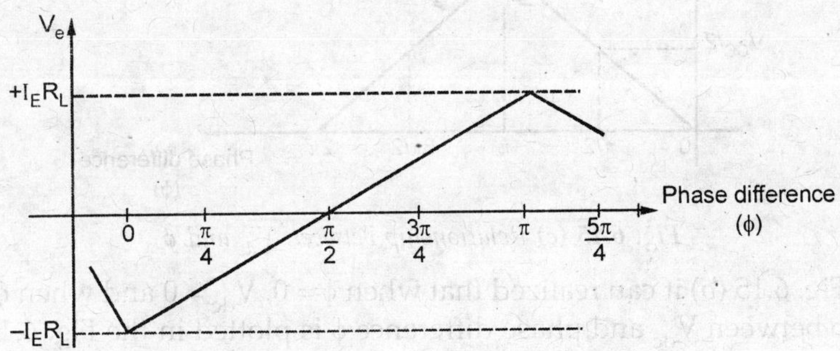

Fig. 6.14 Relationship between V_e and ϕ

6.6.2 Digital Phase Detector

Phase Detector Using XOR Gate

The XOR gate can be used as a digital phase detector when both signal f_s and f_o are square waves. This is illustrated in the Fig. 6.15. We know that the output of XOR gate is high when only one of the input signals is high. The Fig. 6.15 shows the input and output waveforms for digital phase detector. Here, f_s is leading f_o by ϕ degrees.

Fig. 6.15 (a) XOR as a digital phase detector

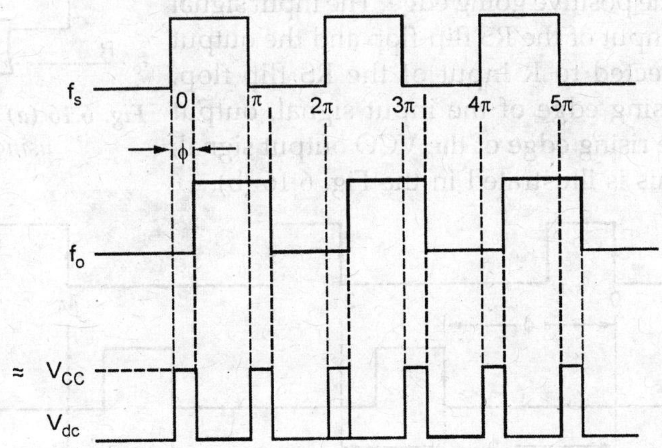

Fig. 6.15 (b) Input/output waveforms

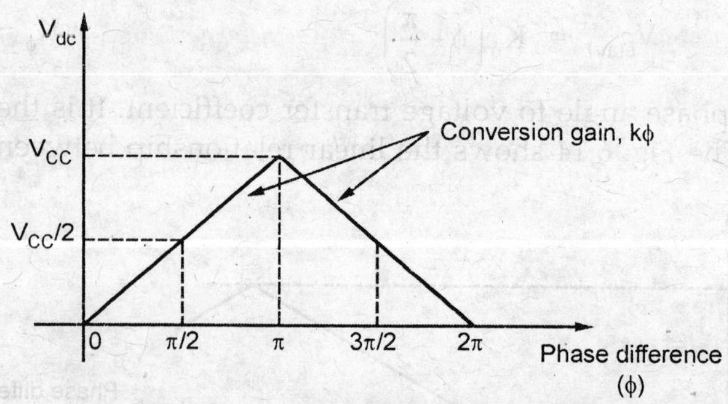

Fig. 6.15 (c) Relationship between V_{dc} and ϕ

Looking at Fig. 6.15 (b) it can realized that when $\phi = 0$, $V_{dc} = 0$ and when $\phi = \pi$, $V_{dc} \approx V_{CC}$. The relationship between V_{dc} and phase difference ϕ is plotted in the Fig. 6.15 (c). The slope of the curve shown in the Fig. 6.15 (c) gives the conversion ratio K_ϕ of the phase detector. It is given by

$$\boxed{K_\phi = \frac{V_{CC}}{\pi}}$$

for $V_{cc} = 5$ V, $K_\phi = \frac{5}{\pi} = 1.591$ V/rad

Phase Detector using RS Flip-flop

The Fig. 6.16 (a) shows the digital phase detector using edge-triggered RS flip-flop. This phase detector is used when f_s and f_o are both pulse waveforms with duty cycle less than 50 %. The flip-flop consists of NOR gates and it changes its state at the positive going edge. The input signal f_s is connected to S input of the RS flip-flop and the output of VCO, f_o is connected to R input of the RS flip-flop. Therefore, at the rising edge of the input signal, output goes high and at the rising edge of the VCO output signal, output goes low. This is illustrated in the Fig. 6.16 (b).

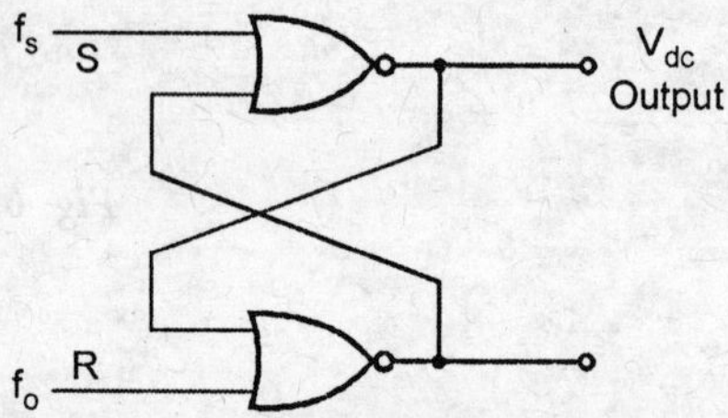

Fig. 6.16 (a) Digital phase detector using SR flip-flop

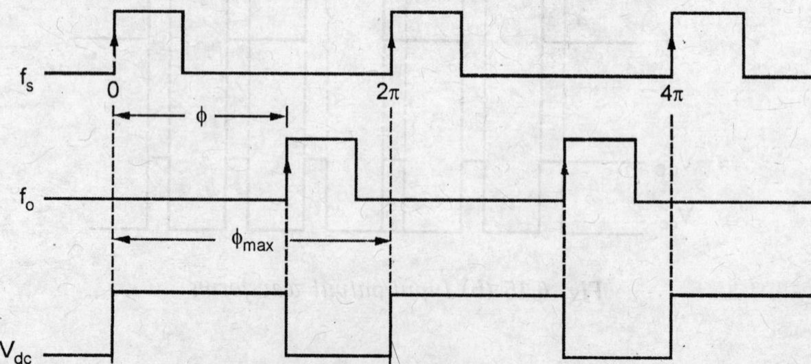

Fig. 6.16 (b) Input and output waveforms

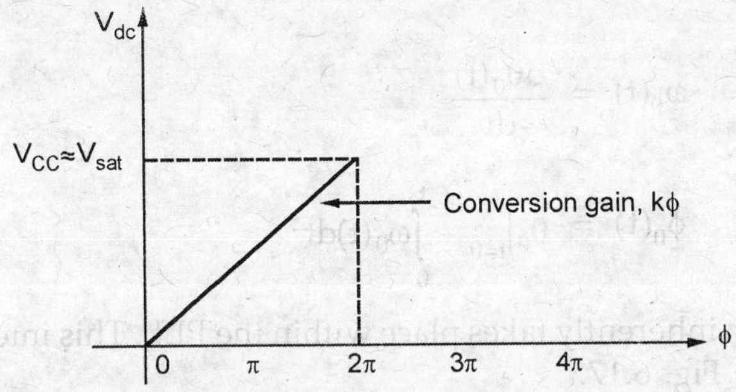

Fig. 6.16 (c) Relation between V_{dc} and ϕ

Since the duty cycle of signals is less 50 %, the maximum ϕ can approach upto 360°. As a result, the dc output voltage is linear upto 360° compared to 180° in the case of XOR detector. Thus, phase detector using RS flip-flop has better capture tracking and locking characteristics than XOR detector. This is illustrated in the Fig. 6.16 (c), which shows the relation between V_{dc} and ϕ.

Special purpose digital phase detector ICs are available. For example, IC MC4344/4044. Such ICs provide the dc output voltage linear upto 4 π radians or 720°.

6.7 CLOSE LOOP ANALYSIS OF PLL

Under locked conditions a linear relationship can exist between the output voltage of the phase detector and the phase difference between the VCO and the input signal if the later is small. The nonlinearity of phase detector makes the PLL nonlinear.

Here, we assume linear PLL in the lock condition and analyse it using standard linear feedback concept. The Fig. 6.17 shows the block diagram of PLL system,

where K_ϕ : Conversion gain of phase detector in volt/rad

F(s) : Loop filter transfer function

A : Gain of the amplifier

K_v : Voltage to frequency transfer coefficient of VCO

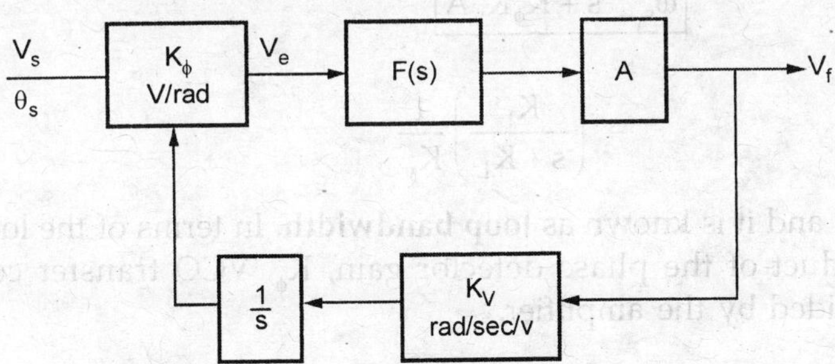

Fig. 6.17 Block diagram of PLL system

The phase of the VCO output is actually equal to the time integral of the VCO output frequency,

Since

$$\omega_0(t) = \frac{d\theta_0(t)}{dt}$$

and thus,
$$\phi_0(t) = \theta_0\big|_{t=0} + \int_0^t \omega_0(t)dt$$

Thus, an integration inherently takes place within the PLL. This integration is represented by the 1/s block in the Fig. 6.17.

In practice, when the VCO input voltage, V_f is zero, the VCO frequency is not zero. The VCO frequency when $V_f = 0$ is called VCO free **running frequency** ω_c or center frequency ω_c. The relation between the VCO output frequency ω_0 and V_f is given by

$$\boxed{\omega_0(t) = \omega_c + K_0 V_f(t)}$$

The close loop transfer function for a PLL system shown in the Fig. 6.17 is given by

$$\frac{V_f}{\theta_s} = \frac{K_\phi F(s)A}{1 + K_\phi F(s)A \cdot \dfrac{K_v}{s}}$$

$$= \frac{sK_\phi F(s)A}{s + K_\phi K_v AF(s)}$$

$$\therefore \quad \boxed{\frac{V_f}{\omega_s} = \frac{K_\phi F(s)A}{s + K_\phi K_v AF(s)}} \qquad \because \frac{V_f}{\omega_s} = \frac{1}{s}\frac{V_f}{\theta}$$

If loop-filter is not used, and F(s) = 1 then

$$\boxed{\frac{V_f}{\omega_s} = \frac{K_\phi A}{s + K_\phi K_v A}}$$

$$= \left(\frac{K_L}{s + K_L}\right)\frac{1}{K_v}$$

where $K_L = K_\phi K_v A$ and it is known as **loop bandwidth**. In terms of the loop parameters, K_L, is simply the product of the phase detector gain, K_ϕ, VCO transfer coefficient, and the electrial gain provided by the amplifier.

6.8 TRANSFER CHARACTERISTICS OF PLL

Fig. 6.18 Shows a typical frequency to voltage transfer characteristics of PLL.

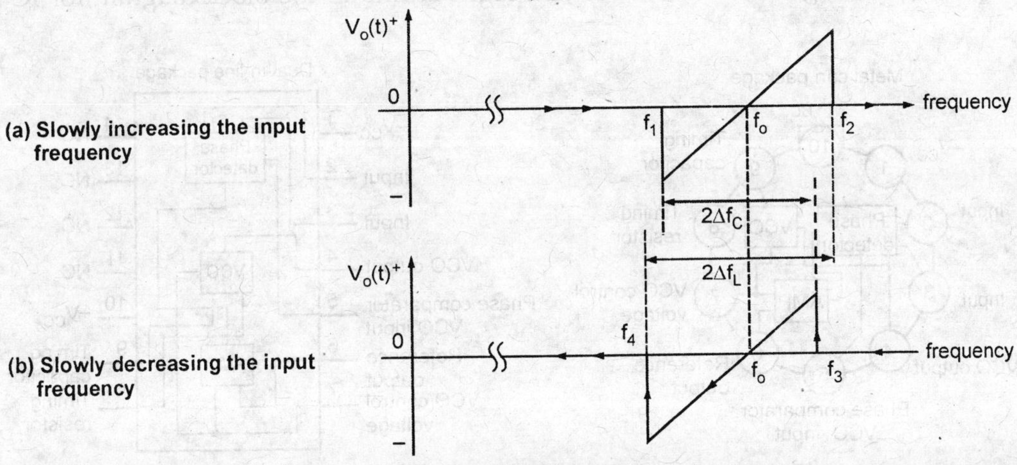

Fig. 6.18 Transfer characteristics of PLL

As shown in the Fig. 6.18, vertical axis corresponds to loop error voltage $V_o(t)$ and horizontal axis corresponds to frequency swept over a broad frequency range. Looking at Fig. 6.18 (a) we can observe that the input frequency is gradually increased. However, the PLL does not respond to the input frequency till it reaches a frequency f_1. The f_1 is a lower edge of the capture range as defined earlier. After this, PLL locks the input frequency. This causes negative jump of the error voltage to shift the output frequency of VCO. Further increase in input frequency, the error voltage $V_o(t)$ is adjusted to maintain locked condition. At $f_s = f_o$, error voltage is zero. If the input frequency is increased still further the loop tracks the input until the input frequency reaches f_2 which corresponds to the upper edge of the lock range. After this frequency the PLL loses the lock, and the error voltage drops to zero, and the VCO frequency returns to it center frequency or free running frequency. If the input frequency is reduced slowly back, the cycle repeats itself as shown in Fig. 6.18 (b). This time the PLL recaptures the signal frequency at f_3 and tracks it down upto a frequency f_4. Thus, frequency f_3 corresponds to the upper edge of the capture range and frequency f_4 is the lower edge of the lock range.

The frequency range between f_3 and f_1 is called **capture range** and **frequency range** between f_2 and f_4 is called **lock range**. The capture range is as large as lock range but in general capture range is less than the lock range as shown in the Fig. 6.18.

6.9 MONOLITHIC PHASE LOCKED LOOP IC 565

Monolithic PLLs are introduced by signetics as SE/NE 560 series and by National semiconductor as LM 560 series. The SE/NE 560 series includes SE/NE 560, 561, 562, 564, 565 and 567. These ICs are differ mainly in operating frequency range, power supply requirements, and frequency and bandwidth adjustment ranges. IC 565 PLL is most commonly used and hence discussed in this section.

6.9.1 IC 565

IC 565 is available in a 14 pin DIP package and 10 pin metal can package. Fig. 6.19 shows 14-pin package configuration for IC 565 and Fig. 6.20 shows the block diagram for IC 565.

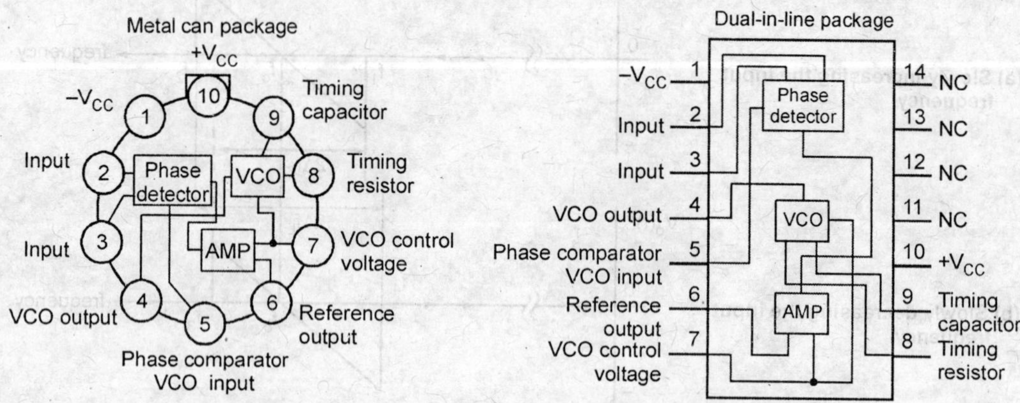

Fig. 6.19 Pin configuration of IC 565

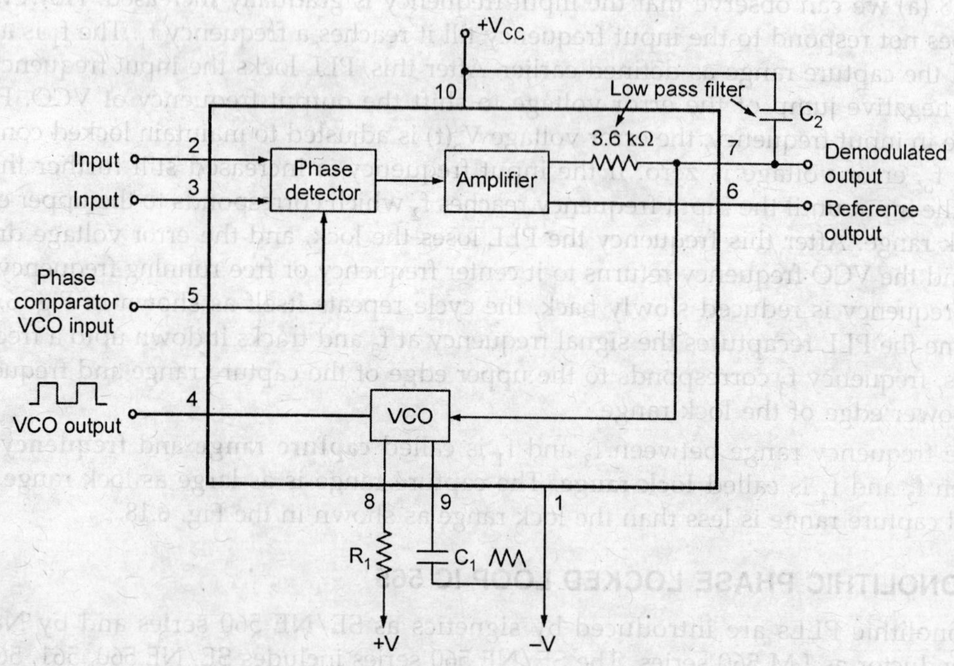

Fig. 6.20 Block diagram of IC 565 PLL

The block diagram of IC 565 PLL consists of phase detector, amplifier, low pass filter and VCO. As shown in the block diagram the phase locked feedback loop is not internally connected. Therefore, it is necessary to connect output of VCO (pin 4) to the phase comparator input (pin 5), externally. In frequency multiplication applications a digital frequency divider is inserted into the loop i.e. between pin 4 and pin 5.

The centre frequency of the PLL is determined by the free-running frequency of the VCO and it is given as

$$\boxed{f_o = \frac{1.2}{4R_1C_1}\ \text{Hz}} \qquad ...(1)$$

where R_1 and C_1 are an external resistor and a capacitor connected to pins 8 and 9, respectively. The values of R_1 and C_1 are adjusted such that the free running frequency will be at the centre of the input frequency range. The value of R_1 is restricted from 2 kΩ to 20 kΩ, but a capacitor can have any value. A capacitor C_2 connected between pin 7 and the positive supply (pin 10) forms a first order low pass filter with an internal resistance of 3.6 kΩ. The value of filter capacitor C_2 should be large enough to eliminate possible oscillations in the VCO voltage.

The lock range and capture range for IC 565 PLL are given by the following equations:

$$\boxed{f_L = \pm\frac{8f_o}{V}\ \text{Hz}} \qquad ...(2)$$

where f_o = Free running trequency of VCO in Hz

and V = (+V) – (–V) volts

$$\boxed{f_C = \pm\left[\frac{f_L}{2\pi(3.6)(10^3)(C_2)}\right]^{1/2}} \qquad ...(3)$$

where C_2 is in farads

From equation 2 we can notice that lock range increases with an increase in input voltage but decreases with increase in supply voltage. The two inputs (pin 2 and pin 3) to the phase detector allows direct coupling of an input signal, provided that there is no dc voltage difference between the pins and the dc resistances seen from pins 2 and 3 are equal. A reference voltage at pin 6 is approximately equal to the dc voltage of the demodulated output at pin 7. This reference voltage may be used as comparator input in applications like frequency shift keying will be discussed in section 7.8.

Example 6.3: For the circuit shown in the Fig. 6.21, calculate free running frequency, lock range and capture range.

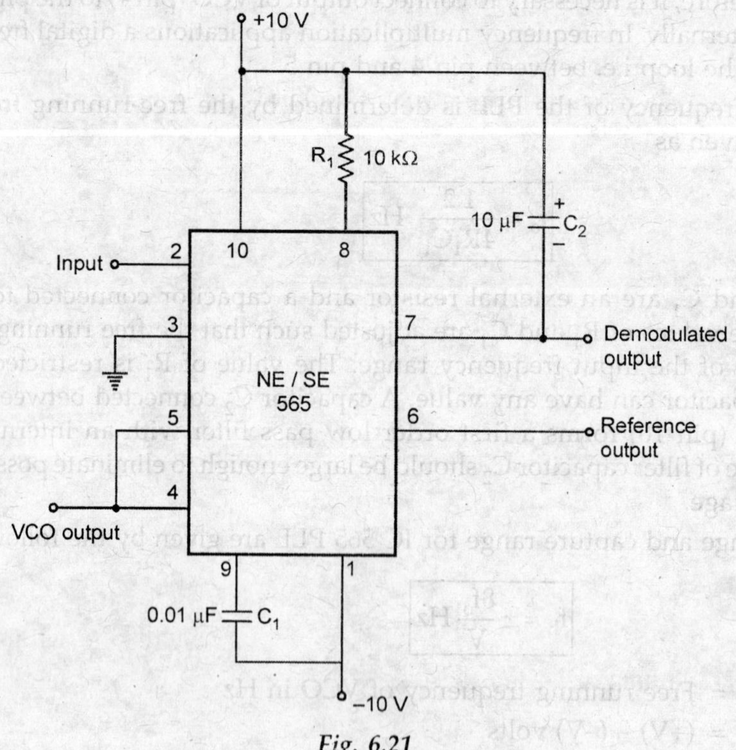

Fig. 6.21

Solution:

(i)
$$f_o = \frac{1.2}{4R_1C_1} = \frac{1.2}{4 \times 10K \times 0.01\ \mu F}$$
$$= \mathbf{3\ kHz}$$

(ii)
$$f_L = \pm\frac{8f_o}{V}$$

where $V = (+V) - (-V)$

$\therefore$ $V = (10\ V) - (-10\ V)$

$= 20\ V$

$\therefore$
$$f_L = \pm\frac{8 \times 3 \times 10^3}{20}$$
$$= \mathbf{\pm 1.2\ kHz}$$

(iii) $$f_C = \pm\left[\frac{f_L}{2\pi(3.6)(10^3)C_2}\right]^{1/2} = \pm\left[\frac{1.2\times10^3}{2\pi(3.6)(10^3)(10)(10^{-6})}\right]^{1/2}$$

$$= \pm\ \mathbf{72.83\ Hz}$$

Fig. 6.22 shows the relationship between f_o, f_L and f_C.

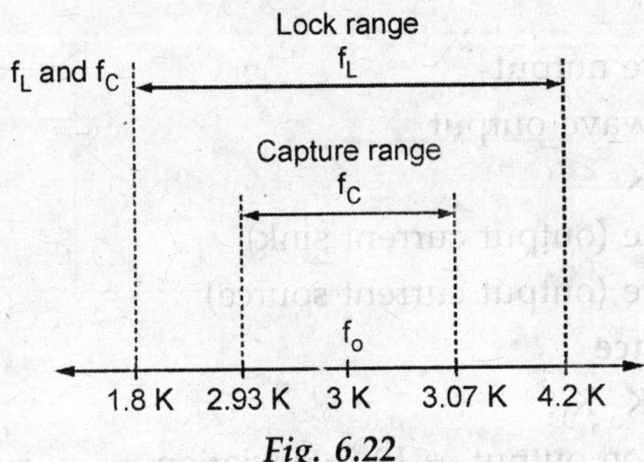

Fig. 6.22

Example 6.4: For circuit shown in Fig. 6.21, calculate the change in lock and capture range if supply voltages are changed as follows:

$$V = 12\ V \text{ and } -V = -12\ V$$

Solution: We know that, f_o = 3 kHz (calculated in the previous problem)

Now $$V = (12\ V) - (-12\ V)$$

$$= \mathbf{24\ V}$$

$$\therefore \quad f_L = \pm\frac{8\times3\times10^3}{24}$$

$$= \pm\ \mathbf{1\ kHz}$$

Therefore, the change in the capture range is 51.83 Hz.

6.9.2 Electrical Parameters of 565 PLL

Some of the important operating characteristics of the LM 565 are shown below :

$$(V_{CC} = \pm6\ V,\ T_A = 25°C)$$

Important Ratings :

Maximum operating voltage	26 V
Maximum power dissipation	300 mW
Supply current	8 mA

Phase Detector

In impedance	5 kΩ
Input level required for tracking (f_o = 50 kHz ± 10 % frequency deviation)	1 mV_{rms}
Input level for limiting	10 mV
Output resistance	3.6 kΩ

Output common mode voltage	4.5 V
Offset voltage (Between pin 6 and 7)	100 mV
Sensitivity K_D	0.68 V/rad
Voltage Control Oscillator	
Centre frequency	500 kHz
Stability	200 ppm/°C
Square wave output	5.4 V_{pp}
Triangular wave output	2.4 V_{pp}
Sensitivity K_o	4.1 f_o rad/sec/V
Square wave (output current sink)	1 mA
Square wave (output current source)	10 mA
Closed loop performance	
Loop Gain K_o K_D	2.8 f_o/sec
Demodulation output, ± 10 % deviation	300 mV

6.9.3 Derivation of LOCK Range

The Fig. 6.23 shows the block diagram to determine lock-range.

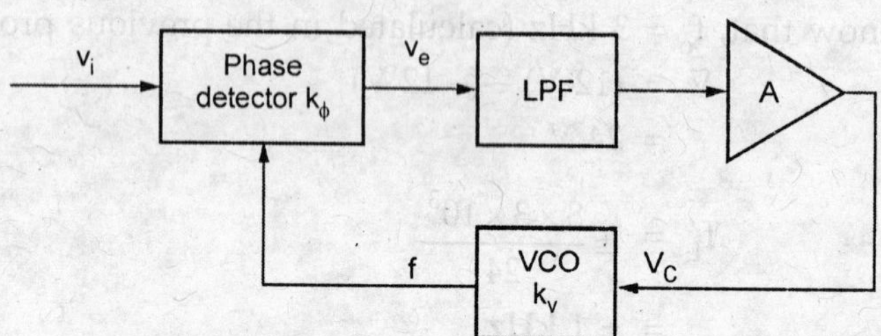

Fig. 6.23 PLL block diagram to determine lock-range

Looking at Fig. 6.23, let us assume that the output voltage of the phase detector is

$$v_e = k_\phi(\theta_e - \pi/2) \text{ where } \theta_e = \text{Phase error} \qquad ...(4)$$

The output voltage of the phase detector is filtered by the low-pass filter to remove the high frequency components. The output of the filter is amplified by a gain A and then applied as the control voltage v_C to the VCO as given by,

$$v_C = a\,v_e = k_\phi\,A(\theta_e - \pi/2) \qquad ...(5)$$

This control voltage v_C will result in a shift in the VCO frequency from its center frequency f_o to a frequency f, given by

$$f = f_o + k_v\,v_C$$

When the PLL is locked into the input signal frequency f_i, we have

$$f = f_i = f_o + k_v\,v_C \qquad ...(6)$$

Substituting value v_C from equation (5) we have,

$$f_i - f_o = k_v k_\phi A(\theta_e - \pi/2)$$

$$\therefore \quad \theta_e - \frac{\pi}{2} = \frac{f_i - f_o}{k_v k_\phi A} \quad ...(7)$$

$$\therefore \quad \theta_e = \frac{\pi}{2} + \frac{f_i - f_C}{k_v k_\phi A}$$

The maximum output voltage magnitude available from the phase detector occurs for $\phi = \pi$ and 0 radian and is

$$v_e(max) = \pm k_\phi(\pi/2)$$

The corresponding value of the maximum control voltage available to drive the VCO will be

$$v_C(max) = \pm k_\phi(\pi/2)A \quad ...(8)$$

Substituting the maximum value of v_C from equation (8) in equation (6) we have,

$$f = f_i = f_o \pm k_v k_\phi (\pi/2) A$$

$$= f_o \pm \Delta f_L$$

where $2\,\Delta f_L$ will be lock-in frequency range given by,

$$\text{lock range} = 2\Delta f_L = k_v k_\phi A \pi \therefore \Delta f_L = k_v k_\phi A \pi/2$$

The lock-in range is symmetrically located with respect to VCO free running frequency f_0. For PLL 565,

we have

$$k_v = \frac{8 f_o}{V}$$

where $\quad V = \pm V_{CC} - (V_{CC})$

For PLL 565, $\quad k_\phi = \frac{1.4}{\pi}$ and $A = 1.4$

Substituting these values we get,

$$\Delta f_L = \pm \frac{8 f_o}{V} \times \frac{1.4}{\pi} \times 1.4 \times \frac{\pi}{2}$$

$$\therefore \quad \boxed{\Delta f_L = \pm \frac{7.84 f_o}{V}} \quad ...(9)$$

6.9.4 Derivation of Capture Range

The capture range is the range of input frequencies for which the initially unlocked loop will lock on an input signal. Thus is always less than the lock range. Since capture range, $\Delta\omega_{cap}$ denotes a transient condition, it is not as readily derived as a lock-in range. However, an approximate parametric expression for the capture range will be initially derived to give an estimate of the capture range. It can be derived by employing simple lag filter. When PLL is not locked the phase angle difference between the signal and the VCO output voltage is given by

$$\theta_e = (\omega_s t + \theta_s) - (\omega_o t + \theta_o) = (\omega_o - \omega_o)\, t + \Delta\theta \quad ...(10)$$

The phase angle difference thus not be constant, but will change with time at a rate given by

$$\frac{d\theta_e}{dt} = \omega_s - \omega_o \qquad ...(11)$$

The phase detector output voltage will therefore not have a dc component, but rather will have an ac voltage with a triangular waveform of peak amplitude $k_\phi\ (\pi/2)$ and the fundamental frequency of $\omega_s - \omega_o$ i.e. $f_s - f_0 = \Delta f$.

Let us derive an approximate expression for capture range for PLL employing a simple lag filter. The transfer function for a simple lag filter is given by,

$$F(j\omega) = \frac{1}{1 + j\omega\tau_1} = \frac{1}{1 + j(\omega/\omega_1)} = \frac{1}{1 + j(f/f_1)}$$

where $\tau_1 = RC \text{ and } \omega_1 = 1/R_1C$

$$\therefore \quad f_1 = \frac{1}{2\pi R_1 C} \qquad ...(12)$$

For the condition that $(f/f_1)^2 >> 1$, the transfer function can be expressed approximately as

$$F(f) = \frac{1}{j(f/f_1)} = f_1/jf \qquad ...(13)$$

The fundamental input frequency term supplied to the low pass filter by the phase detector will be at the difference frequency $\Delta f = f_s - f_o$. If $\Delta f > 3f_1$, the transfer function of LPE will be approximately given by,

$$F(\Delta f) \cong f_1/\Delta f = f_1/(f_s - f_o) \qquad ...(14)$$

The voltage available to drive the VCO is given by,

$$v_C = v_e \times F(\Delta f) \times A \qquad ...(15)$$

or

$$v_C(max) = v_e(max) \times F(\Delta f) \times A$$

$$= \pm k_\phi\ (\pi/2) \times f_1/\Delta f \times A \qquad ...(16)$$

The corresponding value of the maximum frequency shift will be given by,

$$(f - f_o)_{max} = k_v\ v_{C(max)} = \pm\ k_v k_\phi (\pi/2)\ A(f_1/\Delta f) \qquad ...(17)$$

For the acquisition of the signal frequency f_s, we must have that $f = f_s$ so that the maximum signal frequency range can be required by the PLL will be,

$$(f_s - f_o)_{max} = \pm k_v k_\phi (\pi/2)\ A\ (f_1/\Delta f_{cap}) \qquad ...(18)$$

Since $\Delta f_{cap} = (f_s - f_0)_{max}$ we have,

$$(\Delta f_{cap})^2 = k_v\ k_\phi (\pi/2)\ A\ f_1 \qquad ...(19)$$

Since $\Delta f_L = \pm\ k_v\ k_\phi\ (\pi/2)\ A$

$$\therefore \quad (\Delta f_{cap})^2 = f_1 \Delta f_L$$

$$\therefore \quad \Delta f_{cap} = \pm\sqrt{f_1\ \Delta f_L} \qquad ...(20)$$

$$= \pm\left[\frac{\Delta f_L}{2\pi RC}\right]^{1/2} \qquad ...(21)$$

For IC PLL 565, $R = 3.6\ k\Omega$

$$\therefore \quad \boxed{\Delta f_{cap} = \pm \left[\frac{\Delta f_L}{2\pi \times 3.6 \times 10^3 \times C}\right]^{1/2}} \quad ...(22)$$

The total capture range is $2\Delta f_{cap}$.

6.9.5 Filters used in PLL

We have seen the function of filter block in the PLL. Here, we will see the types of filters used in PLL. The filter section of PLL can be made up of two types of filters :

- Loop filter and
- Sideband filter

Loop filter: Loop filter controls the fundamental loop characteristic such as loop bandwidth, locking time and transient response by establishing values for the loop natural frequency and the damping factor. The error signal from the phase detector tends to be in the form of pulse appearing at the rate of f_s. Due to this frequency, f_i and its lower harmonics present as a major spectrum component and that must be removed by filtering. Loop filter provides attenuation to these frequencies.

Sideband filter: Sideband filter provides additional filtering of the error signal to generate a stable and clean control voltage to the VCO. It provides attenuation to sidebands that are created due to disturbances on line. These disturbances frequency modulate the VCO and cause undesired frequencies or sidebands.

We can use either active or passive filters in PLL. We can also use combination of these two. Let us discuss advantages and disadvantages of active and passive filters.

Advantages of Active Filters

- They provide voltage gain and therefore, can furnish wide control voltage range to the VCO while still maintaining an optimum voltage level for the detector to interface with.
- When used as a loop filter, they are capable of providing high dc gain and low input current. This results in zero static phase error between f_s and f_o and thus very low error signal energy to be filtered exists when the loop is locked.

Disadvantages of Active Filters

- They contribute noise components that appear on the VCO signal. This is troublesome, especially in narrowband FM radios.

Advantage of Passive Filters

- They do not contribute any additional noise

Disadvantages of Passive Filters

- They result a significant variation in the value of ky for the phase detector as the VCO is tuned over its frequency range.
- They require greater static phase error between f_s and f_o to maintain lock than an active filter.

6.10 565 PLL APPLICATIONS

6.10.1 Frequency Multiplier

Fig. 6.24 shows the block diagram for a frequency multiplier using PLL 565. Here, a divide by N network is inserted between the VCO output (pin 4) and the phase comparator input (pin 5). Since the output of the divider is locked to the input frequency f_i, the VCO is actually running at a multiple of the input frequency. Therefore, in the locked state, the VCO output frequency f_o is given by,

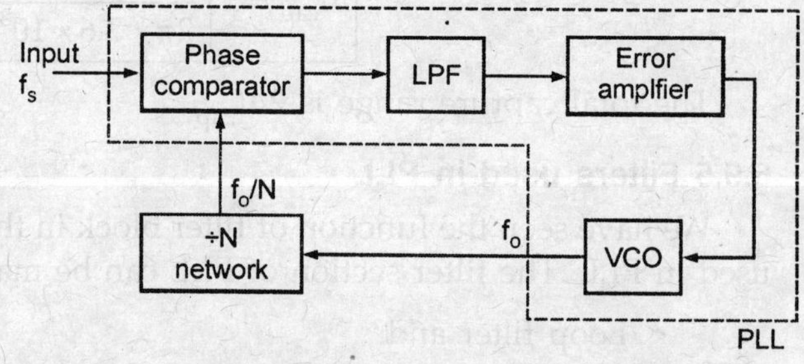

Fig. 6.24 Block diagram of frequency multiplier

$$f_o = Nf_i \quad \text{...(1)}$$

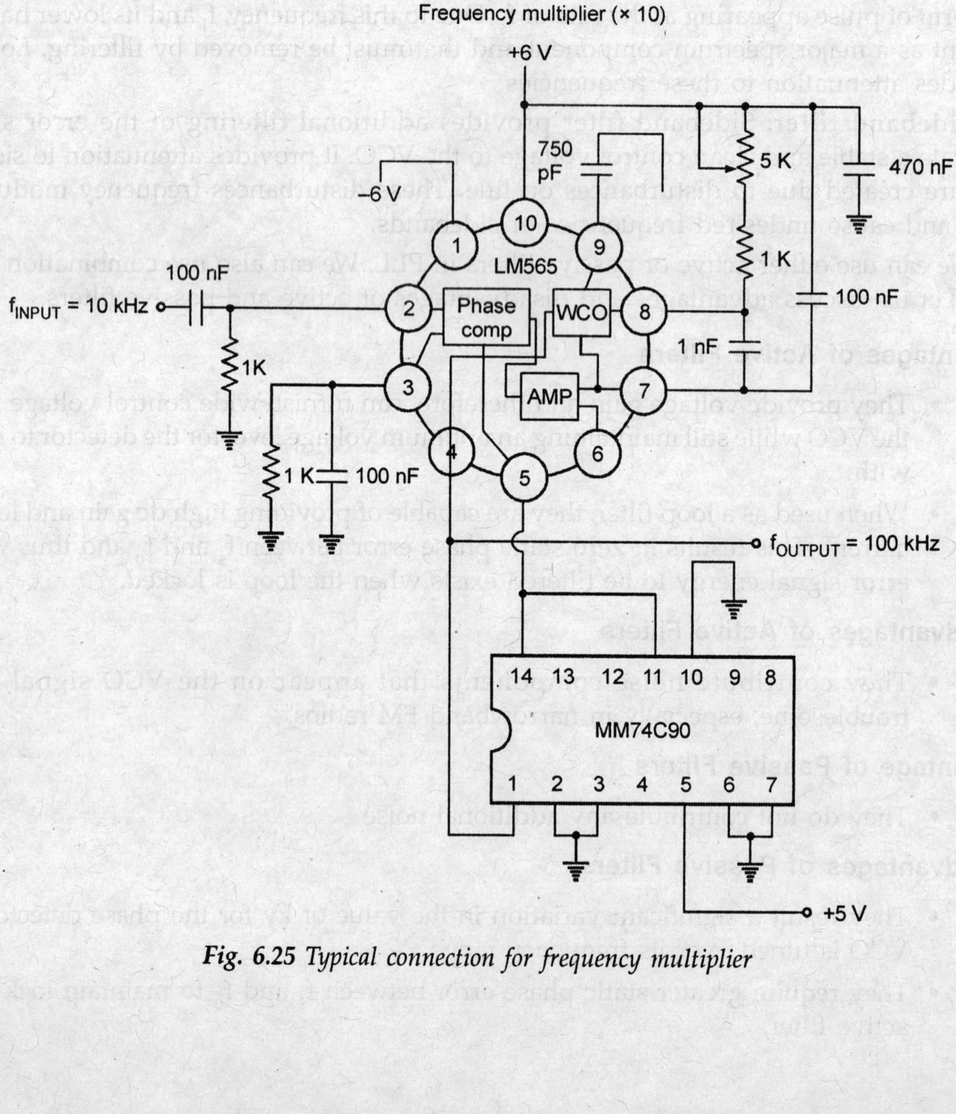

Fig. 6.25 Typical connection for frequency multiplier

By selecting proper divider by N network, we can obtain desired multiplication. For example, to obtain output frequency $f_o = 6\ f_i$, a divide by N should be equal to 6.

Fig. 6.25 shows LM 565 IC used as a frequency multiplier circuit. The IC 7490 is a 4 bit binary counter. It is configured as a divide by 10 circuit.

6.10.2 Frequency Synthesizer

The PLL can be used as the basis for frequency synthesizer that can produce a precise series of frequencies that are derived from a stable crystal controlled oscillator. Fig. 6.26 shows the block diagram of frequency synthesizer. It is similar to frequency multiplier circuit except that divided by M network is added at the input of phase lock loop. The frequency of the crystal-controlled oscillator is divided by an integer factor M by divider network to produce a frequency f_{osc}/M, where f_{osc} is the frequency of the crystal controlled oscillator. The VCO frequency f_{vco} is similarly divided by factor N by divider network to give frequency equal to f_{vco}/N. When the PLL is locked in on the divided-down oscillator frequency, we will have $f_{osc}/M = f_{vco}/N$, so that $f_{vco} = (N/M)\ f_{osc}$.

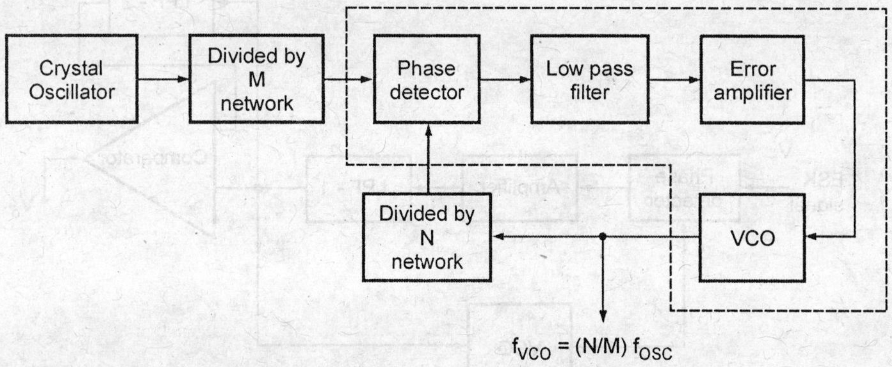

Fig. 6.26 Block diagram of frequency synthesizer

By adjusting divider counts to desired values large number of frequencies can be produced, all derived from the crystal controlled oscillator.

6.10.3 FM Demodulator

The PLL can be very easily used as an FM detector or demodulator. Fig. 6.27 shows the block diagram of FM detector.

When the PLL is locked in on the FM signal, the VCO frequency follows the instantaneous frequency of the FM signal, and the error voltage or VCO control voltage is proportional to the deviation of the input frequency from the centre frequency. Therefore, the a-c component of error voltage will represent a true replica of the

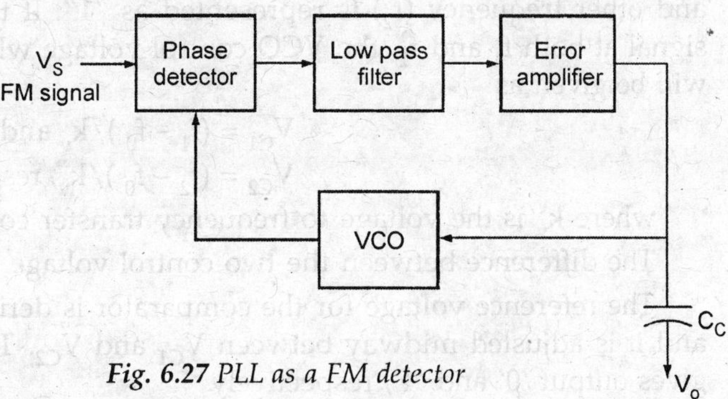

Fig. 6.27 PLL as a FM detector

modulating voltage that is applied to the FM carrier at the transmitter. The faithful reproduction of modulating voltage depends on the linearity between the instantaneous frequency deviation and the control voltage of VCO. It is also important to note that the FM frequency deviation and the modulating frequency should remain in the locking range of PLL to get the faithful replica of the modulating signal. If the product of the modulation frequency f_m and the frequency deviation exceeds the $(\Delta f_C)^2$ [Refer Fig. 6.18 (b)], the VCO will not be able to follow the instantaneous frequency variations of the FM signal.

6.10.4 Frequency Shift Keying (FSK) Demodulator

In digital data communication, binary data is transmitted by means of a carrier frequency. It uses two different carrier frequencies for logic 1 and logic 0 states of binary data signal. This type of data transmission is called frequency shift "keying (FSK). In this data transmission, on the receiving end, two carrier frequencies are converted into 1 and 0 to get the original binary data. This process is called as FSK demodulation.

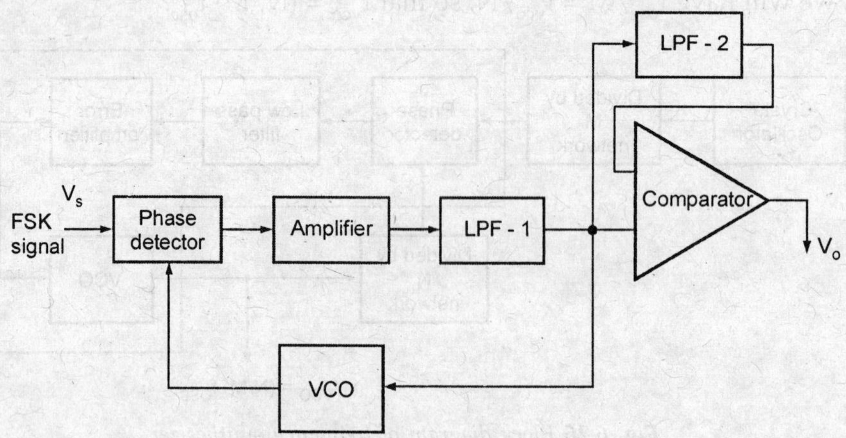

Fig. 6.28 Frequency shift keying demodulator

A PLL can be used as a FSK demodulator, as shown in the Fig. 6.28. It is similar to the PLL demodulator for analog FM signals except for the addition of a comparator to produce a reconstructed digital output signal.

Let us consider that there are two frequencies, one frequency (f_1) is represented as "0" and other frequency (f_2) is represented as "1". If the PLL remain is locked into the FSK signal at both f_1 and f_2, the VCO control voltage which is also supplied to the comparator will be given as

$$V_{C1} = (f_1 - f_0)/k_v \text{ and}$$
$$V_{C2} = (f_2 - f_0)/k_v, \text{ respectively,}$$

where k_v is the voltage to frequency transfer coefficient of the VCO.

The difference between the two control voltage levels will be $\Delta V_C = (f_2 - f_1)/k_v$.

The reference voltage for the comparator is derived from the additional low pass filter and it is adjusted midway between V_{C1} and V_{C2}. Therefore, for V_{C1} and V_{C2}, comparator gives output '0' and '1', respectively.

6.10.5 AM Detection

A PLL can be used to demodulate AM signals as shown in Fig. 6.29.

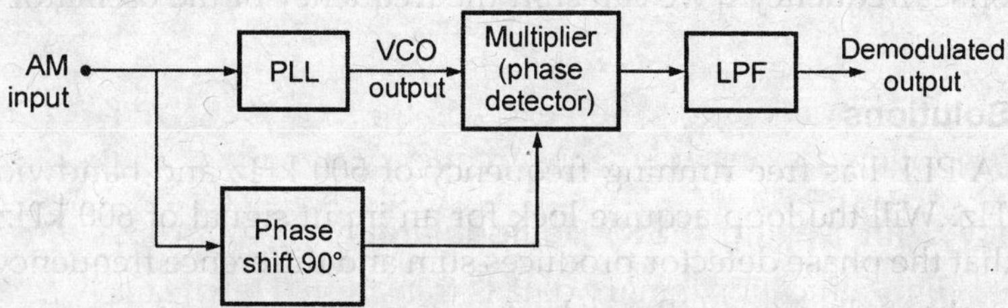

Fig. 6.29 PLL used as AM demodulator

The PLL is locked to the carrier frequency of the incoming AM signal. Once locked the output frequency of VCO is same as the carrier frequency, but it is in unmodulated form. The modulated signal with 90° phase shift and the unmodulated carrier from output of PLL are fed to the multiplier. Since VCO output is always 90° out of phase with the incoming AM signal under the locked condition, both the signals applied to the multiplier are in same phase. Therefore, the output of the multiplier contains both the sum and the difference signals. The low pass filter connected at the output of the multiplier rejects high frequency components gives demodulated output. As PLL follows the input frequencies with high accuracy, a PLL AM detector exhibits a high degree of selectivity and noise immunity which is not possible with conventional peak detector type AM modulators.

6.10.6 Frequency Translation

The frequency translation means shifting the frequency of an oscillator by a small factor. Fig. 6.30 shows the block schematic for frequency translator using PLL.

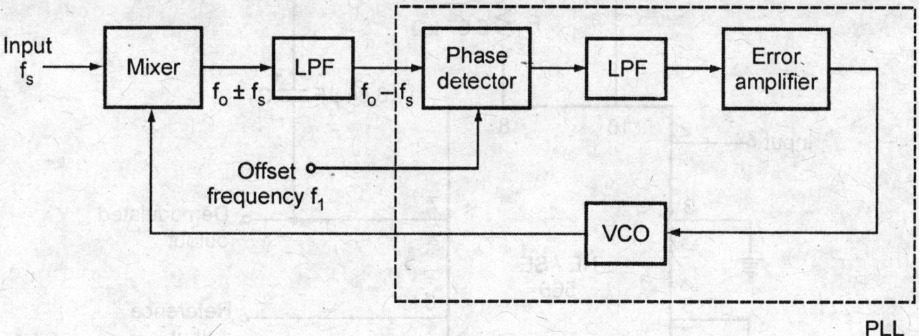

Fig. 6.30 Block schematic for frequency translator using PLL

It consists of mixer, low pass filter and the PLL. The input frequency f_s which has to be shifted is applied to the mixer. Another input to the mixer is the output voltage of VCO, f_o. Therefore, the output of mixer contains the sum and difference signal ($f_o \pm f_s$). The low pass filter connected at the output of mixer rejects the ($f_o + f_s$) signal and gives only ($f_o - f_s$) signal at the output. The ($f_o - f_s$) signal is applied to the phase detector. Another input for phase detector is the offset frequency f_1. In the locked mode, the VCO output frequency is adjusted to make two input frequencies of phase detector equal. This gives

$$f_o - f_s = f_1 \text{ and}$$

$$\therefore \quad f_o = f_s + f_1$$

By adjusting offset frequency f_1 we can shift the frequency of the oscillator to the desired value.

Examples with Solutions

Example 6.5: A PLL has free running frequency of 500 kHz and bandwidth of the low pass filter is 10 kHz. Will the loop acquire lock for an input signal of 600 kHz? Justify your answer. Assume that the phase detector produces sum and difference frequency components.

Solution: The phase detector will produce

$$f_i + f_c = 600 \text{ kHz} + 500 \text{ kHz} = 1100 \text{ kHz}$$

$$f_i - f_c = 600 \text{ kHz} - 500 \text{ kHz} = 100 \text{ kHz}$$

Both sum and different frequency components are outside the passband of low-pass filter. Hence loop will not acquire lock.

Example 6.6: Design a PLL circuit using 565 IC to get :

Free running frequency = 4.5 kHz

Lock range = 2 kHz

Capture range = 100 Hz

Assume supply voltages of ± 10 V are available. Show the circuit diagram with all component values.

Solution: We know that, free running frequency is

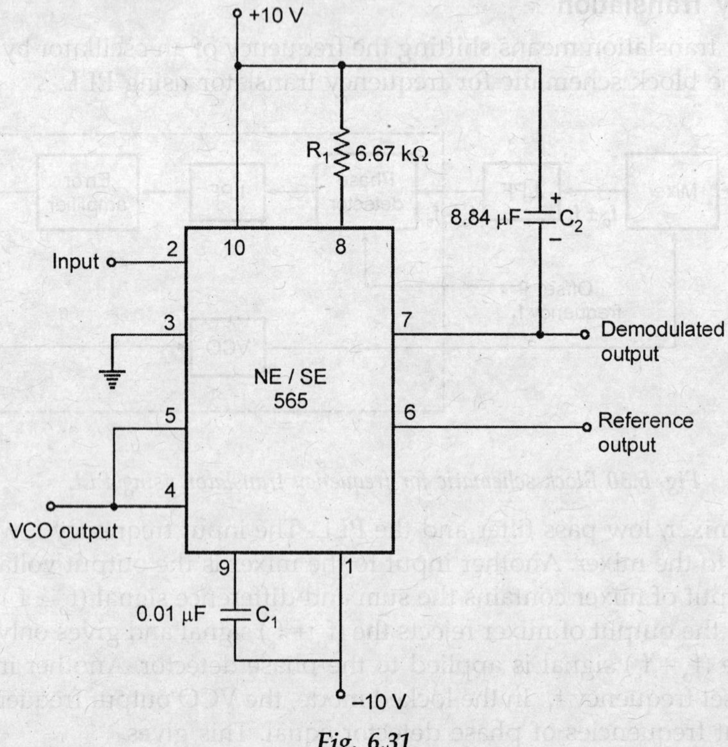

Fig. 6.31

$$f_o = \frac{1.2}{4R_1C_1}$$

Assuming $C_1 = 0.01\ \mu F$ and having $f_0 = 4.5$ kHz, we have

$$R_1 = \frac{1.2}{4 \times 4.5 kHz \times 0.01 \mu F}$$

$$= \mathbf{6.67\ K}$$

Capture range is given as,

$$f_C = \pm\left[\frac{f_L}{2\pi(3.6)(10^3)(C_2)}\right]^{\frac{1}{2}}$$

$$\therefore \quad 100\ Hz = \pm\left[\frac{2kHz}{2\pi(3.6)(10^3)(C_2)}\right]^{\frac{1}{2}}$$

$$\therefore \quad C_2 = \mathbf{8.84\ \mu F}$$

Example 6.7: Input to phased locked loop is having 2 Hz rate of variation. Design phased locked loop for 10 kHz centre frequency. Find out capture range considering the frequency deviation rate as given above i.e. 2 Hz. Find lock range also. Write values of component clearly in the circuit diagram.

Solution: $f_o = 10$ kHz

Let $V = 12$ V and $-V = -12$ V

$f_L = ?$

$f_C = ?$

$V = (+V) - (-V) = 12 + 12 = 24$ V

$$f_L = \pm\frac{8f_o}{V}$$

$$f_L = \pm\frac{8 \times 10 \times 10^3}{24}$$

$$f_L = \mathbf{\pm 3.33\ kHz}$$

$$f_C = \pm\left[\frac{f_L}{2\pi(3.6)(10^3)(C_2)}\right]^{1/2}$$

Let $C_2 = 10\ \mu F$

$$f_C = \pm\left[\frac{3.33 \times 10^3}{2\pi(3.6)(10^3)(10 \times 10^{-6})}\right]^{1/2}$$

$$f_C = \mathbf{\pm[121.33]\ Hz}$$

Example 6.8: Free running frequency is 100 kHz. Supply voltage is ± 6 V. Demodulation capacitor is 1 μF. Findout lock and capture frequencies and range of the PLL employing LM 565. Design components of this PLL for given free running frequency.

Solution: $f_L = \pm\dfrac{8f_o}{V}$ where $V = (\pm V) - (V) = 12\text{ V}$

$$\therefore \quad f_L = \pm\frac{8 \times 100 \times 10^3}{12}$$

$$= \mathbf{66.66\ kHz}$$

The lock range is 133.33 kHz

$$f_C = \pm\left[\frac{f_L}{2\pi(3.6)\,(10^3) \times 1 \times 10^{-6}}\right]^{\frac{1}{2}} = \pm\left[\frac{66.66 \times 10^3}{2 \times \pi(3.6)\,(10^3) \times 1 \times 10^{-6}}\right]^{\frac{1}{2}}$$

$$= \mathbf{\pm 1.716\ kHz}$$

The capture range is 3.432 kHz

Assuming $R_1 = 10$ K we have,

$$f_o = \frac{1.2}{4 \times 10\text{K} \times C_1}$$

$$\therefore \quad C_1 = \frac{1.2}{4 \times 10\text{K} \times 100\text{K}} = \mathbf{3 \times 10^{-10}\ F}$$

Example 6.9: Design frequency multiplier circuit using PLL IC 565 to multiply input frequency by 5 centre frequency by 50 kHz and power supply is ± 10 V DC. Find lock and capture frequency at output and state related tracking range and capture range limits for input.

Solution: Given $f_o = 50\text{ kHz},\ V = 10\text{ V} - (-10\text{ V}) = 20\text{ V}$

$$f_L = \pm\frac{8f_o}{V} = \pm\frac{8 \times 50\text{kHz}}{20}$$

$$= \mathbf{\pm 20\ kHz}$$

$$f_C = \pm\left[\frac{f_L}{2\pi(3.6)(10)^3\,C_2}\right]^{\frac{1}{2}}$$

$$= \pm\left[\frac{20\text{kHz}}{2\pi(3.6)(10)^3\,(100) \times 10^{-9}}\right]^{\frac{1}{2}}$$

$$= \mathbf{\pm 2.973\ kHz}$$

At input these limits are divided by 5

Therefore

$$f_{Li} = \frac{20\,\text{kHz}}{5} = \mathbf{4\ kHz}\ \text{and}$$

$$f_{Ci} = \frac{2.973\,\text{kHz}}{5} = \mathbf{0.5946\ kHz}$$

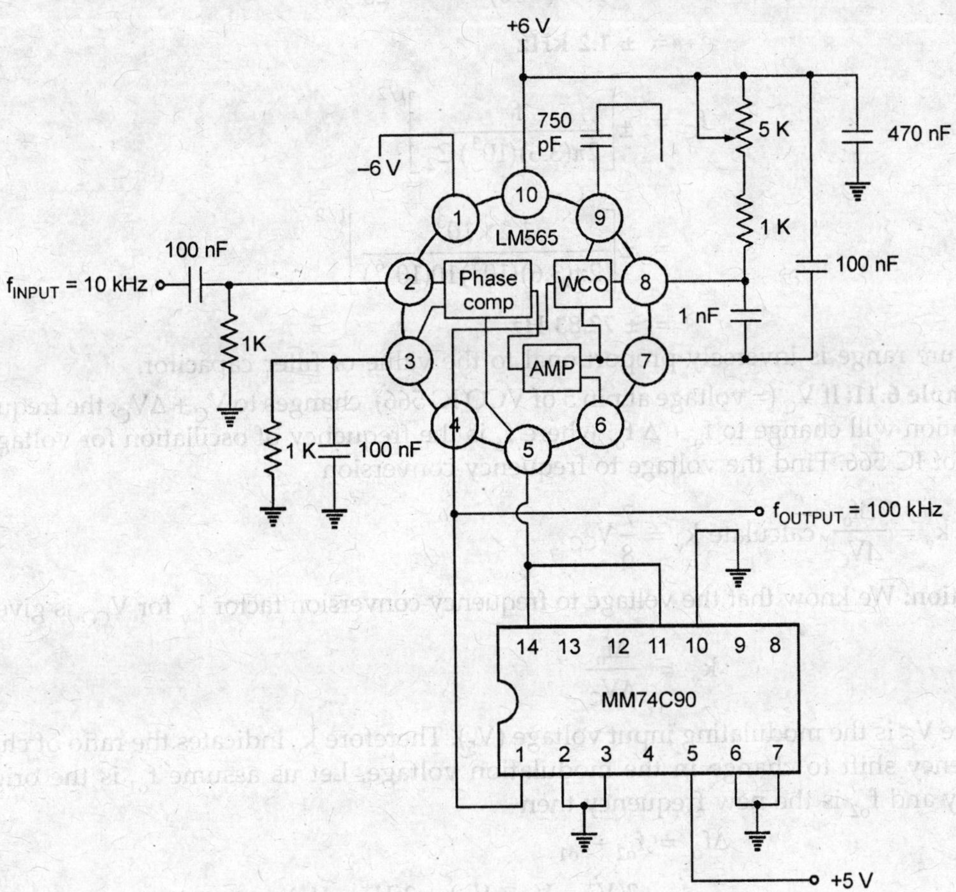

Fig. 6.32 Typical connection for frequency multiplier

Example 6.10: Calculate output frequency f_o, lock range Δf_L and capture range Δf_C of PLL if $R_T = 1$ K, $C_T = 0.1$ μF and filter capacitor C = 10 μF. What is the effect of filter capacitor on capture range?

Solution: Given $R_1 = 1$ K, $C_1 = 0.1$ μF, $C_2 = 10$ μF

We known that

(i)

$$f_o = \frac{1.2}{4R_1C_1} = \frac{1.2}{4\ 1\text{K} \times 0.1 \times 10^{-6}}$$

$$= \mathbf{3\ kHz}$$

(ii) $$f_L = \pm\frac{8f_o}{V}$$

Assuming supply voltage ±10 V we have

$$f_L = \pm\frac{8\times3\times10^3}{10-(-10)} = \pm\frac{8\times3\times10^3}{20}$$

$$= \pm\ \mathbf{1.2\ kHz}$$

(iii) $$f_C = \pm\left[\frac{f_L}{2\pi(3.6)(10^3)C_2}\right]^{1/2}$$

$$= \pm\left[\frac{1.2\times10^3}{2\pi(3.6)(10^3)10(10^{-6})}\right]^{1/2}$$

$$= \pm\ \mathbf{72.83\ Hz}$$

Capture range is inversely proportional to the value of filter capacitor.

Example 6.11: If V_C (= voltage at pin 5 of VCO IC 566), changes to $V_C + \Delta V_C$, the frequency of oscillation will change to $f_o + \Delta f_o$, where f_o is the frequency of oscillation for voltage V_C at pin 5 of IC 566. Find the voltage to frequency conversion

ratio $k_v = \frac{\Delta f_o}{\Delta V_C}$, calculate $k_v = \frac{7}{8}V_{CC}$.

Solution: We know that the voltage to frequency conversion factor k_v for V_{CO} is given by

$$k_v = \frac{\Delta f_o}{\Delta V_C}$$

where V_C is the modulating input voltage (V_5). Therefore k_v indicates the ratio of change in frequency shift to change in the modulation voltage. Let us assume f_{o1} is the original frequency and f_{o2} is the new frequency then

$$\Delta f_o = f_{o2} - f_{o1}$$

$$= \frac{2(V^+ - V_C + V_C)}{C_1R_1V^+} - \frac{2(V^+ - V_C)}{C_1R_1V^+}$$

$$= \frac{2\ V_C}{C_1R_1V^+}$$

or $$\Delta V_C = \frac{\Delta f_o C_1 R_1 V^+}{2}$$

When V_C is biased at 7/8 V^+ we have

$$f_o = \frac{2(V^+ - (7/8)V^+)}{C_1R_1V^+} = \frac{1}{4R_1C_1} = \frac{0.25}{R_1C_1}$$

Substihiting above value of f_o we have

$$\Delta V_C = \frac{\Delta f_o V^+}{8 f_o}$$

or $$k_v = \frac{\Delta f_o}{\Delta V_C} = \frac{8 f_o}{V^+} = \frac{8 f_o}{V_{CC}} \qquad \because V^+ = V_{CC}$$

Example 6.12: A PLL with a free running frequency of 1 kHz is connected to a variable frequency oscillator. The frequency of oscillator is gradually increased and when its frequency was 850 Hz the PLL got locked. The frequency of oscillator was decreased and it went out of lock for the oscillator frequency of 800 Hz. Calculate the lock range and the capture range of the PLL.

Solution: Lock range $= f_o \pm 800\text{ Hz} = 1000 \pm 800\text{ Hz} = \pm 200\text{ Hz}$

Capture range $= f_o \pm 850\text{ Hz} = 1000 \pm 850\text{ Hz}$

REVIEW QUESTIONS

1. Explain working of PLL using appropriate block diagram and explain any one application of the same.
2. With reference to a PLL, define 'capture range', 'lock range' and 'pull in time'.
3. Draw and explain the transfer characteristics of PLL.
4. Draw the block diagram of IC 565 PLL.
5. Give the typical electrical parameters for 565 PLL.
6. Using a neat circuit diagram, explain how 565 PLL can be used a FSK demodulator.
7. Derive expression for capture range of PLL IC 565.
8. Derive expression for Lock range of PLL IC 565.
9. For the circuit shown in Fig. 6.33 calculate free running frequency, lock range and capture range.

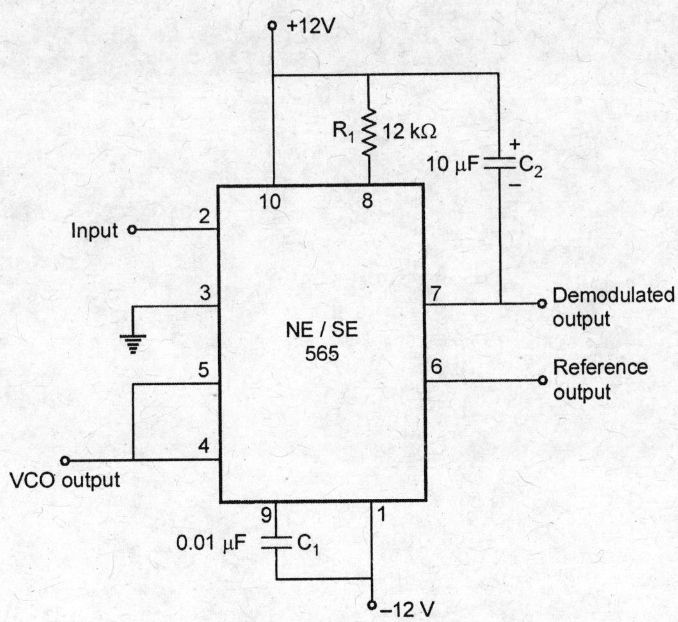

Fig. 6.33

[Ans.: f_o = 2.5 kHz; f_L = 833 Hz, f_C = 60.69 Hz]

10. Design a circuit using IC 565 to obtain a frequency multiplication of 4. The input signal is $2V_{p-p}$ with 2 kHz frequency.
11. Explain the working of PLL. Explain in detail, FSK demodulation using PLL.
12. Name two applications of PLL.
13. Explain AM demodulation using PLL.
14. Show that the lock-in range of a PLL is given by $\Delta f_L = \pm 7.8\ f_0/V$ where the symbols used have the usual meaning.
15. A PLL has free running frequency of 500 kHz and bandwidth of the low pass filter is 10 kHz. Will the loop acquire lock for an input signal of 600 kHz ? Justify your answer. Assume that the phase detector produces sum and difference frequency components.
16. Write notes on frequency synthesizers.
17. Explain the purpose and functioning of frequency synthesizer.
18. Explain PLL as a frequency synthesizer.
19. Draw and explain the frequency translator circuit using PLL.
20. What is a voltage controlled oscillator ?
21. Explain the working of a voltage controlled oscillator.
22. With reference to a VCO, define voltage to frequency conversion factor K_v.

❖❖❖

Chapter

7

D-A and A-D Converters

7.1 INTRODUCTION

Most of the information carrying signals such as voltage, current, charge, temperature, pressure and time are available in the analog form. However, for processing, transmission and storage purposes, it is often more convenient to express such signals in the digital form. When expressed in the digital form, they provide better accuracy and reduce noise.

Moreover, the development in the microprocessor technology has made it compulsory to process data in the digital form. Since digital systems such as microprocessor use a binary system of ones and zeros, we have to convert signal from analog form to digital form. The circuit that performs this conversion is called an analog to digital (A/D) converter.

On the other hand, a digital to analog (D/A) converter is used when a binary output from a digital system must be converted to some equivalent analog voltage or current. For example, if in a particular system a computer is used as a controller, the controlling signal produced by the computer is always digital. The system to be controlled requires the analog signal. Hence in between the computer and the system to be controlled the digital to analog converter is must.

Fig. 7.1 shows a typical A/D and D/ A converter application. As shown in the

Inside the Chapter

Fig. 7.1, the analog signal is converted in digital form, which is then processed or perhaps just transmitted or recorded. Once processed, received, or retrieved, the signal is D-A converted to be reused in analog form.

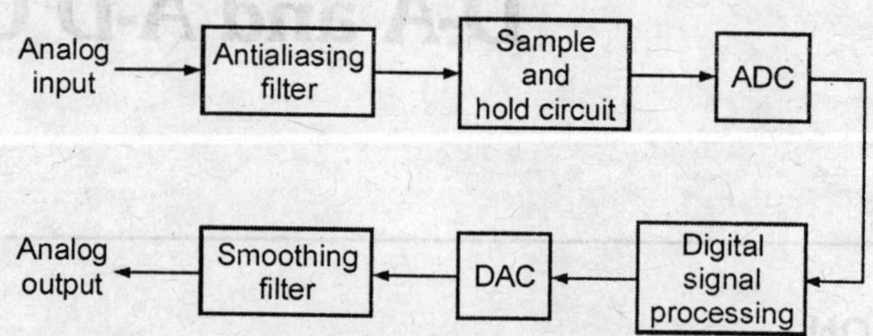

Fig. 7.1 Typical A/D and D/A converter application

In this chapter, we see the various techniques used for analog to digital conversion and digital to analog conversion alongwith the supporting circuits such as sample and hold circuits. In the later part of this chapter we see the voltage to time and voltage to frequency converters.

7.2 HIGH SPEED SAMPLE AND HOLD CIRCUITS

For accurate analog to digital conversion the analog input voltage should be held constant during the conversion cycle. If the analog input voltage changes by more than ± 1/2 LSB an error can occur in the digital output code. To illustrate the effect of a changing analog input voltage on the conversion processor, let us consider a situation of a successive approximation ADC with an analog input voltage that is initially zero, but there happen to be a large change in voltage amplitude occurring during the conversion process. Fig. 7.2 shows the changing input voltage and its effect on the successive approximation conversion process.

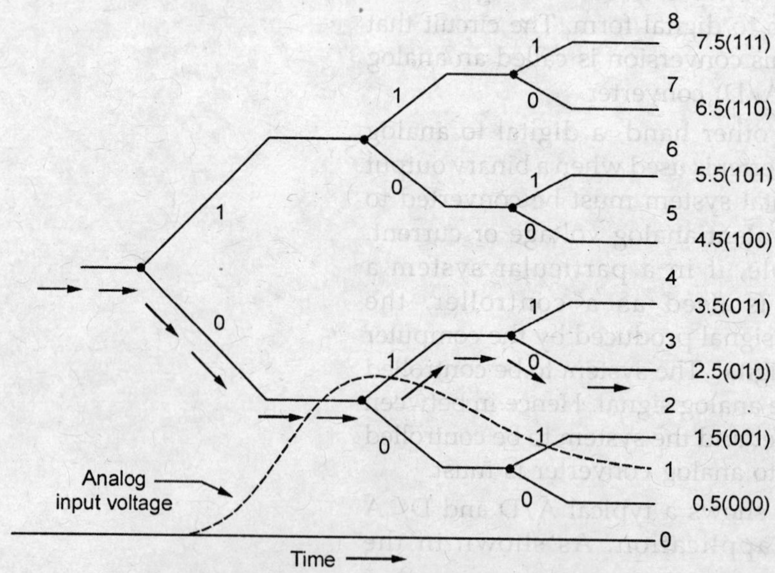

Fig. 7.2 Effect of changing input voltage.

As shown in Fig. 7.2 analog input voltage at start of conversion process is zero volts and at the end of conversion process it is near to 1.5 volts, and the conversion process result is 010_2, i.e. 2.5 V. This result does not corresponds to the analog voltage at the start of conversion or at the end of conversion. To minimise the occurrence of these errors it is necessary to hold the value of the analog input voltage constant during the conversion process. The sample and hold circuit does this task.

As its name implies, the sample and hold (S/H) circuit samples the value of the input signal in response to a sampling command and hold it at the output until arrival of the next command. It samples an analog input voltage in a very short period, generally in the range of 1 to 10 μs, and holds the sampled voltage level for an extended period, which can range from a few millisecond to several seconds. Fig. 7.3 shows input and output response of the sample and hold circuit.

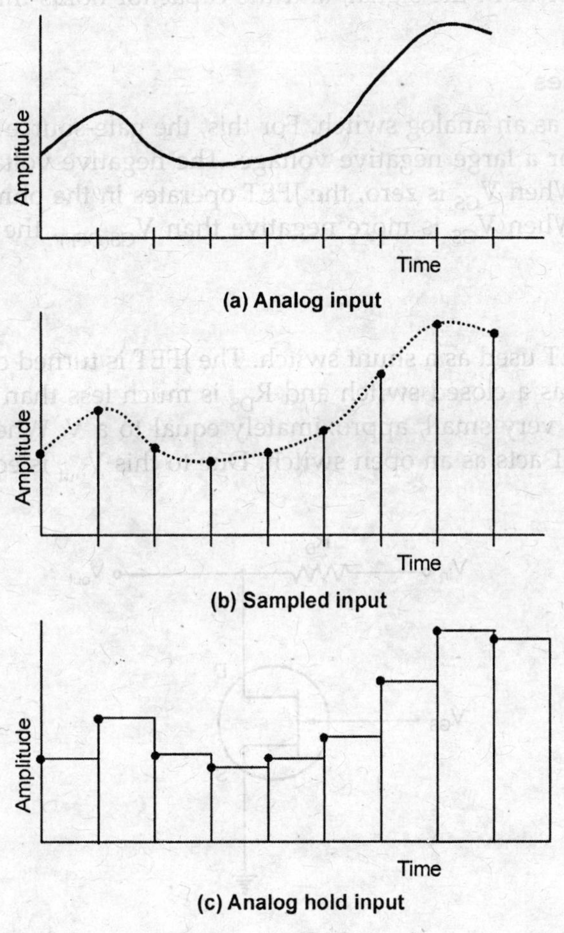

Fig. 7.3 Input and output response of sample and hold circuit.

The sample and hold circuit uses to basic components analog switch and capacitor. The Fig. 7.4 shows the basic sample and hold circuit.

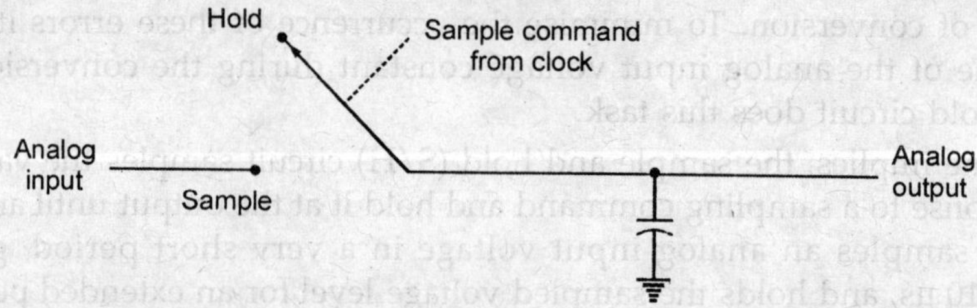

Fig. 7.4 Principle diagram for sample and hold circuit.

The circuit tracks the analog signal until the sample command causes the digital switch to isolate the capacitor from the signal, and the capacitor holds this analog voltage during A/D conversion.

7.2.1 Analog Switches

JFET can be used as an analog switch. For this, the gate-source voltage V_{GS} is restricted to two values : 0 V or a large negative voltage. The negative voltage must be equal to or more than $V_{GS(OFF)}$. When V_{GS} is zero, the JFET operates in the ohmic region and JFET acts as a closed switch. When V_{GS} is more negative than $V_{GS(OFF)}$, the JFET is cut off and the switch is open.

7.2.1.1 Shunt Switch

Fig. 7.5 shows JFET used as a shunt switch. The JFET is turned on and off by V_{GS}. When V_{GS} = 0 V, JFET acts as a closed switch and R_{DS} is much less than RD. Due to the voltage divider action V_{out} is very small, approximately equal to a V. When V_{GS} is more negative than $V_{GS(OFF)}$, the JFET acts as an open switch. Due to this V_{out} is equal to V_{in}.

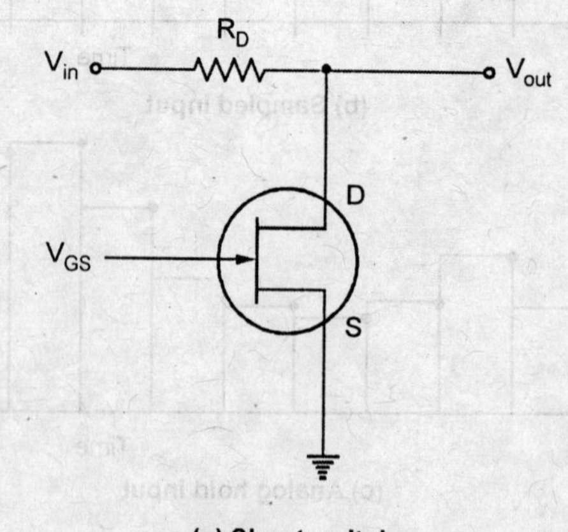

(a) Shunt switch

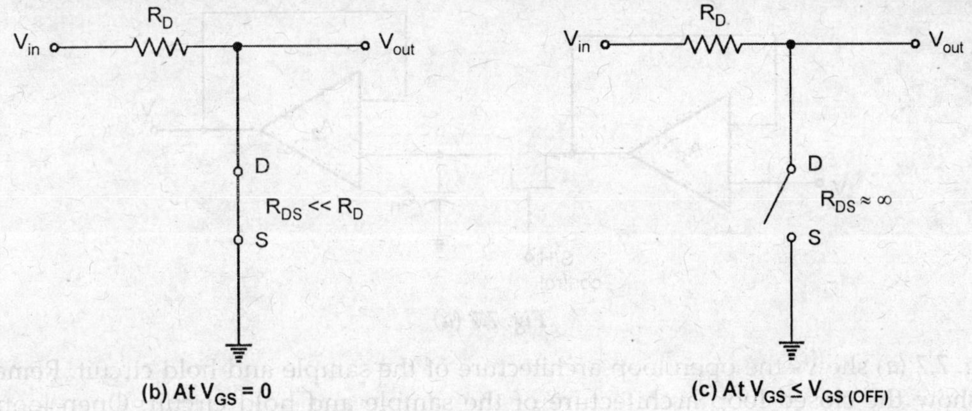

Fig. 7.5

7.2.1.2 Series Switch

Fig. 7.6 shows JFET used as a series switch. When $V_{GS} = 0$ V, the switch is closed and V_{out} equals V_{in}. When V_{GS} is equal to or more negative than $V_{GS(OFF)}$, the JFET is open and V_{out} is approximately zero.

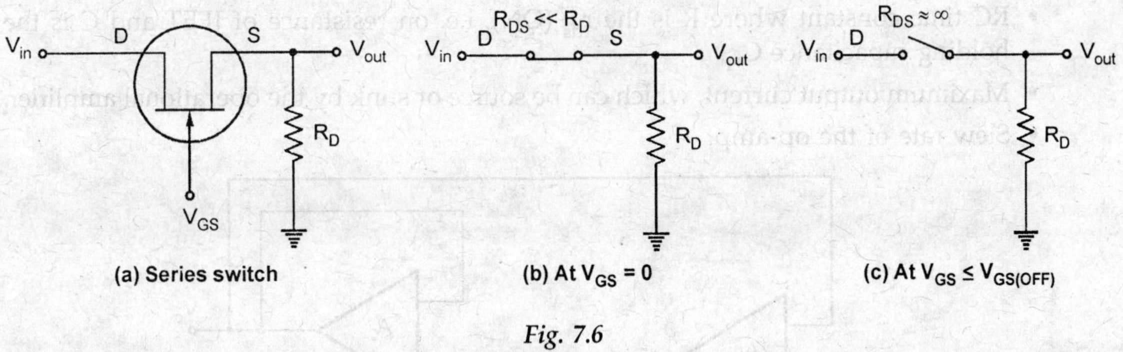

Fig. 7.6

7.2.2 Sample and Hold Circuits

Four basic sample and hold circuits are shown in Fig. 7.7. In these circuits a JFET is used as switch. During the sampling time the JFET switch is turned on, and the holding capacitor charges up to the level of the analog input voltage. At the end of this short sampling period, the JFET switch is turned off. This isolates the holding capacitor C_H from the input signal. As a result, the voltage across capacitor C_H and hence the output voltage will remain essentially constant at the value of the input voltage at the end of the sampling time. However, there will be a small drop-off or drop of the capacitor voltage during the hold period due to the various leakage currents. To avoid this, input and output buffers (voltage follower) circuits are used.

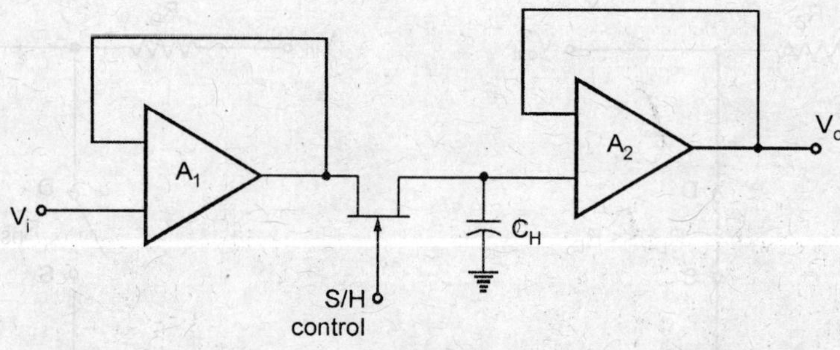

Fig. 7.7 (a)

Fig. 7.7 (*a*) shows the open loop architecture of the sample and hold circuit. Remaining Figs. show the closed loop architecture of the sample and hold circuit. Open loop type sample and hold circuits are faster than closed loop types which have delayed output feedback to the input buffer. However, closed loop architectures provide higher dc accuracy because of this feedback, cancelling the output amplifier offset errors.

The acquisition time of a S/H circuit is the time required for the holding capacitor C_H to charge up to a level close to the input voltage during sampling. The acquisition time for S/H circuit should be as low as possible. In the circuit of Fig. 7.7 (*a*) there are three principle factors that will control the acquisition time. These factors are:

- RC time constant where R is the r_{ds} (ON), i.e. on resistance of JFET and C is the holding capacitance C_H.
- Maximum output current, which can be source or sunk by the operational amplifier.
- Slew rate of the op-amp.

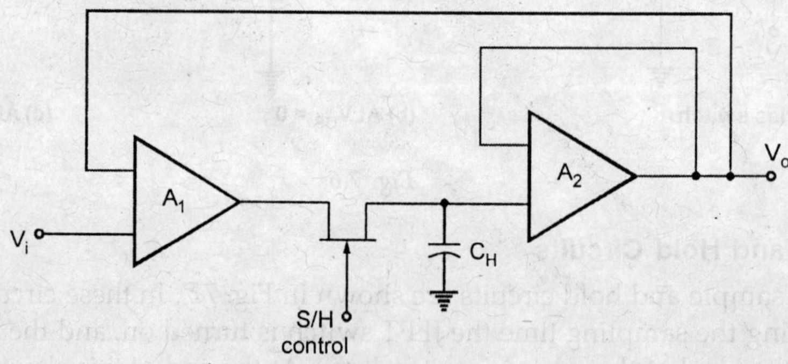

Fig. 7.7 (b)

The circuit shown in Fig. 7.7 (*b*) offers some advantage over that of Fig. 7.7 (*a*) in terms of the acquisition time since the r_{ds} (ON) of the JFET switch is inside the feedback loop of A_1 and A_2. Therefore the acquisition time for this circuit is limited by maximum output current and slew rate of the op-amp, rather than the RC time constant.

The S/H circuit of Fig. 7.7 (*c*) performs in a fashion similar to that of Fig. 7.7 (*b*) but it offers the additional feature of providing voltage gain. The voltage gain of this circuit can be given as $A = 1 + (R_F/R_1)$. Therefore, the sampled output voltage is equal to the sampled input voltage multiplied by the voltage gain factor of $1 + (R_F/R_1)$.

The S/H circuit of Fig. 7.7 (*d*) offers two advantages. The faster capacitor charging rate provides shorter acquisition time. This is because the voltage at the inverting input terminal of A_2 is equal to the capacitor voltage divided by the open loop gain of A_2. In this circuit, the summing input of A_2 remains at virtual ground. Due to this, the charge removed from the summing junction via C_{gd} is constant regardless of the input and output signal levels. This removed charge appears as a constant offset at the output. However, as it is constant, it can be nulled by any standard offset trimming technique.

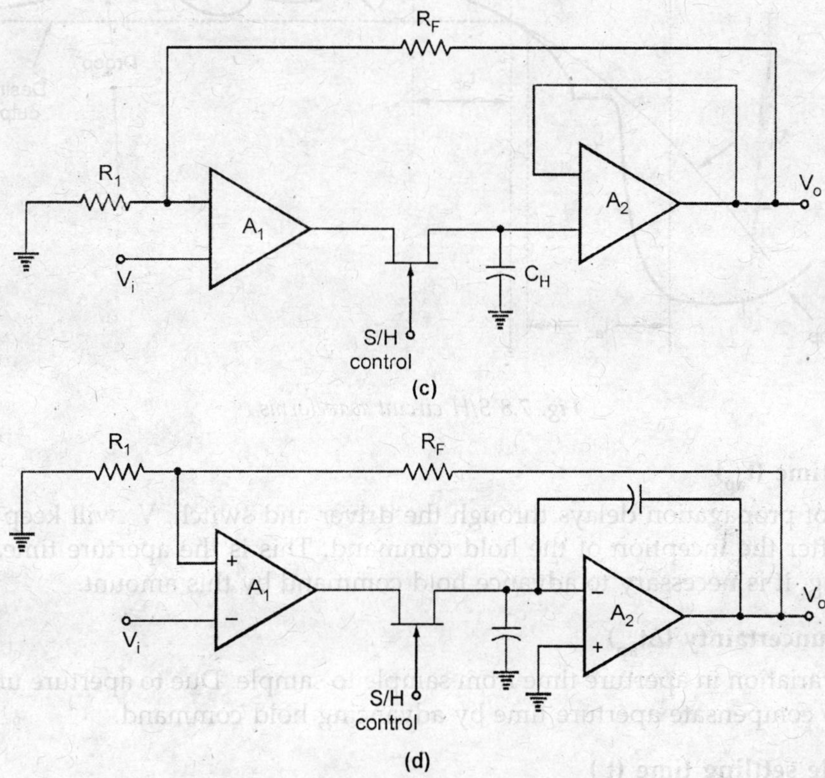

Fig. 7.7 Basic Sample and hold circuits

7.2.3 Performance Parameters of S/H Circuits

The S/H circuit performance can be characterised by specifications of the ordinary amplifier such as input offset voltage, gain error, nonlinearity etc. However, during the transitions from one mode to other (sample to hold or hold to sample), as well as in hold mode, circuit performance is characterised by specifications peculiar to S/H circuits.

These specifications are explained below with the help of S/H circuit waveform, shown in Fig. 7.8.

1. Acquisition time (t_{ac})

It is the time required for the holding capacitor C_H to charge up to a level close to the input voltage during sampling. It depends on three factors:

- RC time constant
- Maximum output current of op-amp
- Slew rate of op-amp

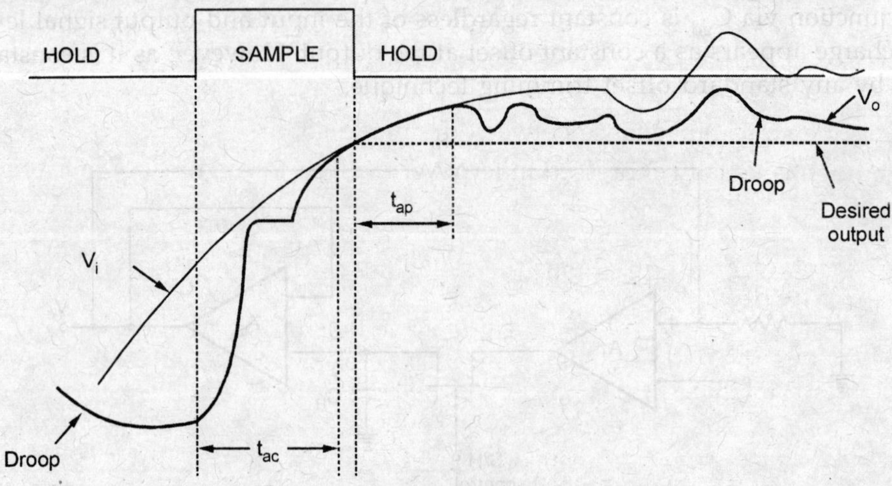

Fig. 7.8 S/H circuit waveforms

2. Aperture time (t_{ap})

Because of propagation delays through the driver and switch, V_o will keep tracking V_i some time after the inceptiorr of the hold command. This is the aperture time. To get the precise timing, it is necessary to advance hold command by this amount.

3. Aperture uncertainty (Δt_{ap})

It is the variation in aperture time from sample to sample. Due to aperture uncertainty it is difficult to compensate aperture time by advancing hold command.

4. Hold mode settling time (t_s)

After the application of hold command, it takes a certain amount of time for V_o to settle within a specified error band, such as 1%, 0.1% or 0.01%. This is the hold mode settling time.

5. Hold step

Because of the parasitic switch capacitances, at the time of switching between sample to hold mode, there is an unwanted transfer of charge between the switch driver and C_H. This changes the capacitor voltage and hence the output voltage, as shown in the Fig. 7.8. This changes in output voltage are referred to as hold step, pedestal error and sample to hold offset. The change in the output voltage can be related as given below

$$\boxed{\Delta V_o = \frac{\Delta Q}{C_H}} \quad ...(1)$$

6. Feedthrough

In the hold mode, because of stray capacitance across the switch, there is a small amount of ac coupling between V_o and V_i. This ac coupling causes output voltage to vary with variation in the input voltage. This is referred to as feed through and it is given as,

$$\boxed{\Delta V_o = \frac{C_{ds}}{C_H} \Delta V_i} \quad \text{...(2)}$$

where C_{ds} is the stray capacitance between drain and source of JFET.

Feedthrough is usually expressed in terms of the feed through rejection ratio (FRR) and it is given as,

$$\boxed{FRR = 20 \log_{10} \frac{\Delta V_i}{\Delta V_o}} \quad \text{... (3)}$$

Example 7.1: Calculate the change in the output voltage if input changes by 5 V with FRR for S/H circuit is 80 dB.

Solution : From equation (3) we can write,

$$FRR = 20 \log_{10} \frac{\Delta V_i}{\Delta V_o}$$

$$80 = 20 \log_{10} \frac{5}{\Delta V_o}$$

$$\therefore \quad \Delta V_o = \mathbf{0.5\ mV}$$

7. Voltage droop

The leakage current causes voltage of the capacitor to drop down. This is referred to as droop.

7.2.4 Advantages of Sample and Hold Circuits

1. The primary use of the sample and hold circuit to hold the sampled analog input voltage constant during conversion time of *A/D* converter.
2. In case of multichannel ADCs, synchronization can be achieved by sampling signals from all channels at the same time.
3. It also reduces the crosstalk in the multiplexer.

7.2.5 Sample and Hold ICs

For further improved performance, specially designed sample and hold ICs can be used. Such sample and hold ICs are HA 2420 by Harris Semiconductor, LF 198 and LF 398 by National Semiconductor. The LF 398 is very popularly used sample and hold IC. The functional diagram of LF 398 is shown in the Fig. 7.9.

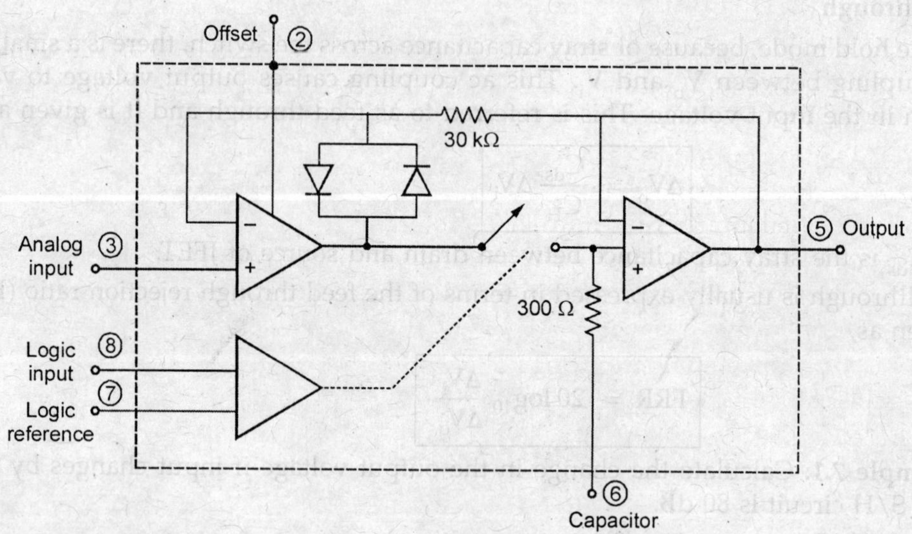

Fig. 7.9 Functional diagram of LF 398

The hold or storage capacitor is required to be connected externally. The connection diagram of LF 398 is shown in the Fig. 7.10.

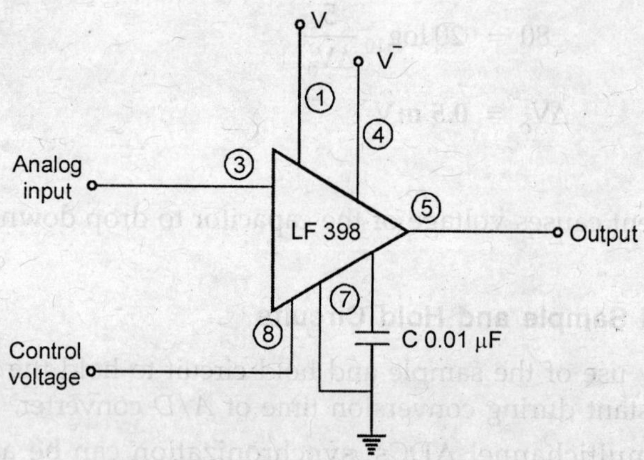

Fig. 7.10 Connection diagram of LF 398.

7.2.6 Applications of Sample and Hold Circuits

The applications of such sample and hold circuit are :

(*i*) Digital interfacing.

(*ii*) Analog to digital converter circuits.

(*iii*) Pulse modulation systems.

(*iv*) In storage of outputs of a multiplexer between updates in data distribution systems.

(*v*) In reset-stabilised op-amps.

(*vi*) In analog demultiplexers.

7.3 D/A CONVERTERS

A DAC (Digital to Analog Converter) accepts an n-bit input word $b_1, b_2, b_3, \ldots b_n$ in binary and produce an analog signal proportional to it. Fig. 7.11 shows circuit symbol and input-output characteristics of a 4-bit DAC. There are four digital inputs, indicating 4-bit DAC. Each digital input requires an electrical signal representing either a logic 1 or a logic 0. The b_n is the least significant bit, LSB, whereas b_1 is the most significant bit, MSB.

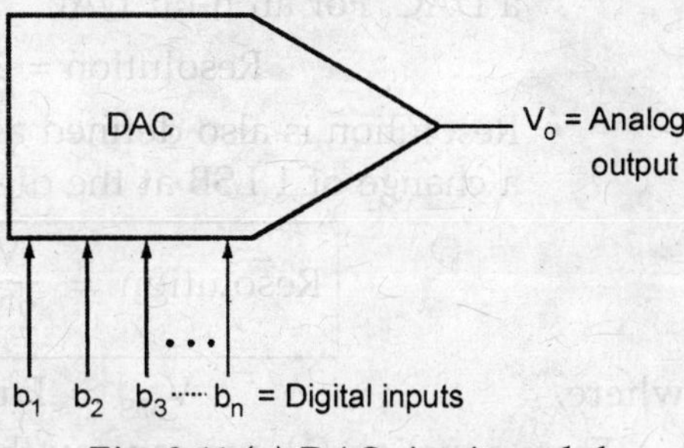

Fig. 8.11 (a) DAC circuit symbol.

Fig. 7.11 (*b*) shows analog output voltage V_o is plotted against all 16 possible digital input words.

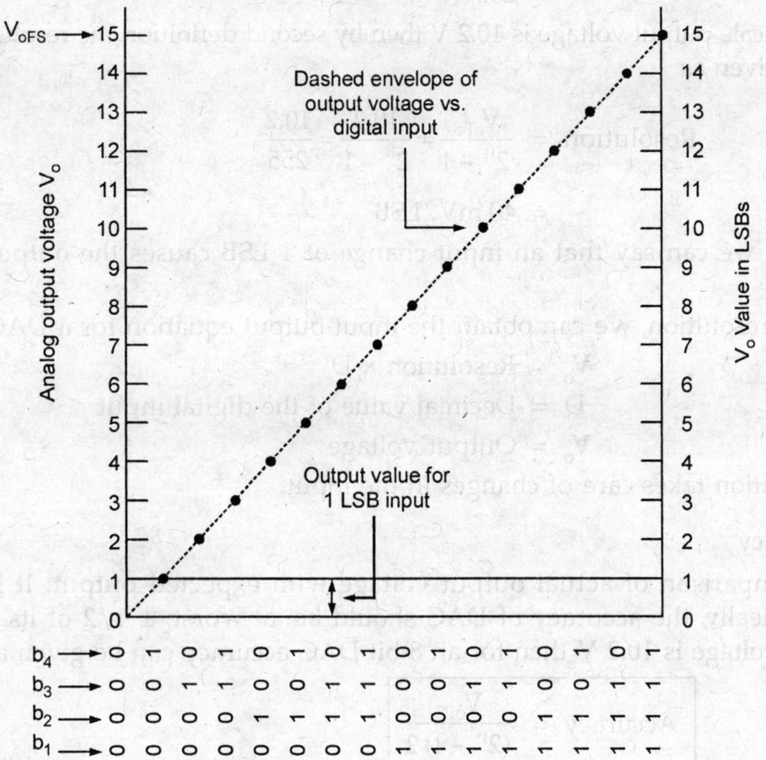

Fig. 7.11 (b)

7.3.1 Performance Parameters of DAC

The various performance parameters of DAC are,

7.3.1.1 Resolution

Resolution is defined in two ways.

- Resolution is the number of different analog output values that can be provided by a DAC. For an n-bit DAC

$$\text{Resolution} = 2^n \quad ...(1)$$

- Resolution is also defined as the ratio of a change in output voltage resulting from a change of 1 LSB at the digital inputs. For an n-bit DAC it can be given as

$$\boxed{\text{Resolution} = \frac{V_{oFS}}{2^n - 1}} \quad ...(2)$$

where, V_{oFS} = Full scale output voltage

From equation (1), we can say that, the resolution can be determined by the number of bits in the input binary word. For an 8-bit DAC resolution can be given as

$$\text{Resolution} = 2^n = 2^8 = 256$$

If the full scale output voltage is 10.2 V then by second definition the resolution for an 8-bit DAC can be given as

$$\text{Resolution} = \frac{V_{oFS}}{2^n - 1} = \frac{10.2}{2^8 - 1} = \frac{10.2}{255}$$

$$= 40 \text{ mV/LSB}$$

Therefore, we can say that an input change of 1 LSB causes the output to change by 40 mV.

From the resolution, we can obtain the input-output equation for a DAC.

Thus V_o = Resolution × D

where D = Decimal value of the digital input

and V_o = Output voltage

The resolution takes care of changes in the input.

7.3.1.2 Accuracy

It is a comparison of actual output voltage with expected output. It is expressed in percentage. Ideally, the accuracy of DAC should be, at worst, ± 1/2 of its LSB. If the full scale output voltage is 10.2 V then for an 8-bit DAC accuracy can be given as

$$\boxed{\text{Accuracy} = \frac{V_{oFS}}{(2^n - 1)\,2}} \quad ...(3)$$

$$= \frac{10.2}{255 \times 2} = 20 \text{ mV}$$

7.3.1.3 Monotonicity

A converter is said to have good monotonicity if it does not miss any step backward when stepped through its entire range by a counter.

7.3.1.4 Conversion Time

It is a time required for conversion of analog signal into its digital equivalent. It is also called setting time. It depends on the response time of the switches and the output of the amplifier.

7.3.1.5 Settling Time

This is the time required for the output of the DAC to settle to within ± 1/2 LSB of the final value for a given digital input i.e. zero to full scale.

7.3.1.6 Stability

The performance of converter changes with temperature, age and power supply variations. So all the relevant parameters such as offset, gain, linearity error and monotonicity must be specified over the full temperature and power supply ranges. These parameters represent the stability of the converter.

Example 7.2: An 8-bit DAC has an output voltage range of 0-2.55 V. Define its resolution in two ways.

Solution: For the given DAC,

$$n = \text{Number of bits} = 8$$

(*i*) $$\text{Resolution} = 2^n = 2^8 = \mathbf{256}$$

i.e. the output voltage can have 256 different values including zero.

(*ii*) $$V_{oFS} = \text{Full scale output voltage} = \mathbf{2.55\ V}$$

$$\therefore \quad \text{Resolution} = \frac{V_{oFS}}{2^n - 1} = \frac{2.55}{2^8 - 1} = \frac{\mathbf{10\ mV}}{\mathbf{1\ LSB}}$$

Thus an input change of 1 LSB causes the output to change by 10 mV.

Example 7.3: The digital input for a 4-bit DAC is 0110. Calculate its final output voltage.

Solution: For given DAC,

$$n = 4$$

$$\therefore \quad V_{oFS} = 15\ V$$

$$\therefore \quad \text{Resolution} = \frac{V_{oFS}}{2^n - 1} = \frac{15}{2^4 - 1} = \mathbf{1\ V/LSB}$$

$$\therefore \quad V_o = \text{Resolution} \times D$$

Now $$D = \text{Decimal of } (0110)_2 = 6$$

$$\therefore \quad V_o = 1\ V/LSB \times 6 = \mathbf{6\ V}$$

Example 7.4: An 8-bit DAC has resolution of 20 mV/LSB. Find V_{oFS} and V_o if the input is $(10000000)_2$.

Solution: $\text{Resolution} = \frac{V_{oFS}}{2^n - 1}$

$$\therefore \quad 20 = \frac{V_{oFS}}{2^8 - 1}$$

$$\therefore \quad V_{oFS} = \mathbf{5.1\ V}$$

$$D = \text{Equivalent of } (10000000) = \mathbf{128}$$

$$\therefore \quad V_o = \text{Resolution} \times D = 20 \times 128$$

$$= \mathbf{256\ V}$$

Example 7.5: Find out stepsize and analog output for 4-bit R-2R ladder DAC when input is 1000 and 1111. Assume $V_{ref} = +5$ V.

Solution: For given DAC, n = 4, $V_{oFS} = +5$ V

$$\text{Resolution} = \frac{V_{oFS}}{2^n - 1} = \frac{5}{2^4 - 1} = \frac{1}{3}\ \text{V/LSB}$$

$$\therefore \quad V_o = \text{Resolution} \times D$$

For $\quad D = \text{Decimal of } (1000)_2 = \mathbf{8}$

$$V_o = \frac{1}{3} \times 8 = \mathbf{2.6667\ V}$$

For $\quad D = \text{Decimal of } (1111)_2 = \mathbf{15}$

$$V_o = \frac{1}{3} \times 15 = \mathbf{5\ V}.$$

Example 7.6: A 12-bit DAC has a step size of 8 mV. Determine the full scale output voltage and percentage resolution. Also find the output voltage for the input of 010101101101?

Solution: For 12-bit DAC, step size is 8 mV.

$$V_{oFS} = 8\ \text{mV} \times 2^{12} - 1 = \mathbf{32.76\ V}$$

$$\%\ \text{Resolution} = \frac{8\ \text{mV}}{32.76\ \text{V}} \times 100 = \mathbf{0.02442\%}$$

The output voltage for the input 010101101101 is = 8 mV × $(1389)_{10}$ = 11.112 V

7.3.2 Types of D/A Converter

There are mainly two techniques used for analog to digital conversion

- Binary weighted resistor D/A converter
- R/2R ladder D/A converter

In these techniques, the shunt resistors are used to generate n binary weighted currents. These currents are added according to switch positions controlled by the digital input and then converted into voltage to give analog voltage equivalent to the digital input. Therefore, such digital to analog converters are called current driven DACs.

7.3.2.1 Binary Weighted Resistor *D/A* Converter

The binary weighted resistor DAC uses an op-amp to sum n binary weighted currents derived from a reference voltage V_R via current scaling resistors 2R, 4R, 8R, ... , 2^n R, as shown in the Fig. 7.12.

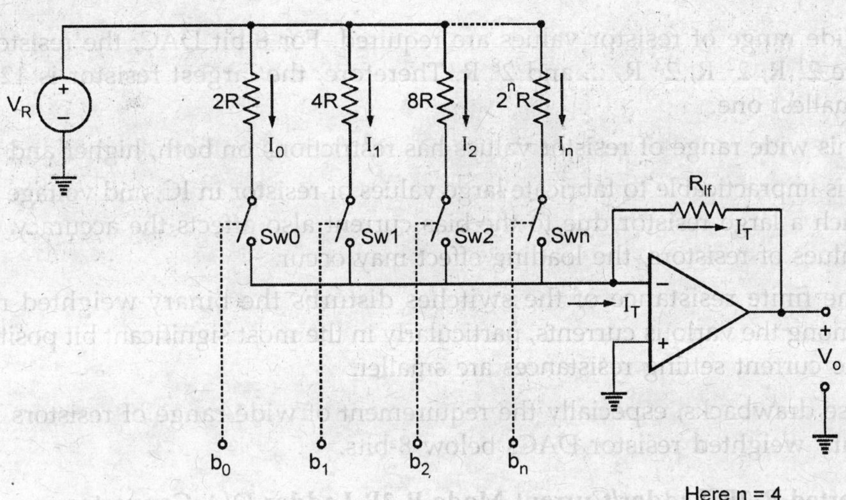

Fig. 7.12 Binary weighted resistor DAC

As shown in the Fig. 7.12, switch positions are controlled by the digital inputs. When digital input is logic 1, it connects the corresponding resistance to the reference voltage V_R; otherwise it leaves resistor open. Therefore,

For ON-switch, $I = \frac{V_R}{R}$ and

For OFF-switch, $I = 0$

Here, operational amplifier is used as a summing amplifier. Due to high input impedance of op-amp, summing current will flow through R_f. Hence the total current through R_f can be given as

$$I_T = I_0 + I_1 + I_2 + ... + I_n$$

The output voltage is the voltage across R_f and it is given as

$$V_0 = -I_T R_f = -(I_0 + I_1 + I_2 + ... + I_n) R_f$$

$$= -\left(b_0 \frac{V_R}{2R} + b_1 \frac{V_R}{4R} + b_2 \frac{V_R}{8R} + ... + b_n \frac{V_R}{2^n R}\right) R_f$$

$$= -\frac{V_R}{R} R_f (b_0 2^{-1} + b_1 2^{-2} + b_2 2^{-3} + ... + b_n 2^{-n}) \quad ...(4)$$

When $R_f = R$, V_o is given as,

$$\boxed{V_o = -V_R (b_0 2^{-1} + b_1 2^{-2} + b_2 2^{-3} + ... + b_n 2^{-n})} \quad ...(5)$$

The equation (4) indicates that the analog output voltage is proportional to the input digital word.

The simplicity of the binary weighted DAC is offset by drawbacks associated with it.

Drawbacks

1. Wide range of resistor values are required. For 8-bit DAC, the resistors required are 2^1 R, 2^2 R, 2^3 R, ... and 2^8 R. Therefore, the largest resistor is 128 times the smallest one.
2. This wide range of resistor values has restrictions on both, higher and lower ends. It is impracticable to fabricate large values of resistor in IC, and voltage drop across such a large resistor due to the bias current also affects the accuracy. For smaller values of resistors, the loading effect may occur.
3. The finite resistance of the switches disturbs the binary-weighted relationship among the various currents, particularly in the most significant bit positions, where the current setting resistances are smaller.

All these drawbacks, especially the requirement of wide range of resistors restricts the use of binary weighted resistor DACs below 8-bits.

7.3.2.2 Inverted R-2R Ladder/Current Mode R-2R Ladder D/A Converter

R/2R ladder D/A converter uses only two resistor values. This avoids resistance spread drawback of binary weighted D/A converter. Fig. 7.13 shows R/2R ladder DAC. Like binary weighted resistor DAC, it also uses shunt resistors to generate n binary weighted currents; however it uses voltage scaling and identical resistors instead of resistor scaling and common voltage reference used in binary weighted resistor DAC. Voltage scaling requires an additional set of voltage dropping series resistances between adjacent nodes, as shown in the Fig. 7.13.

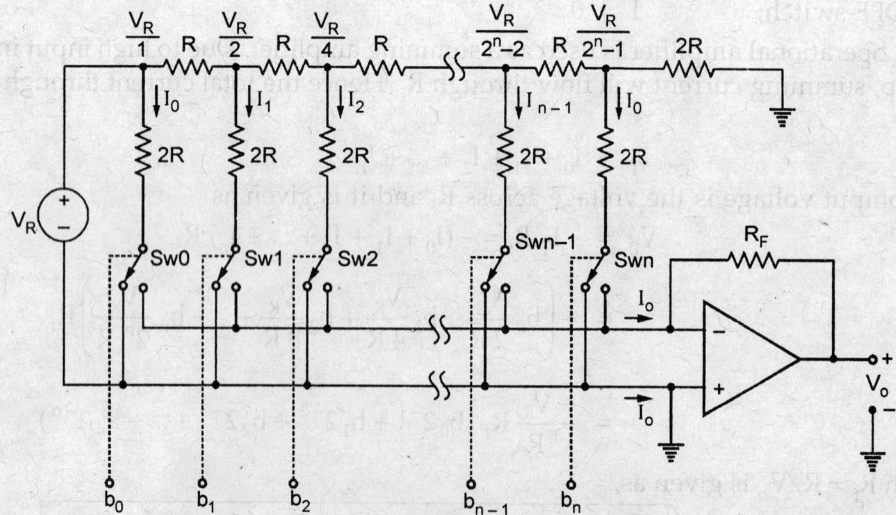

Fig. 7.13 Inverted R/2R ladder DAC.

Here, each bit of the binary word connects the corresponding switch either to ground or to the inverting input terminal of the op-amp which is at the virtual ground. Since both the positions of switches are at ground potential, the current flowing through resistances is constant and it is independent of switch position. These currents can be given as

$$I_0 = \frac{V_R}{2R} \qquad ...(6)$$

$$I_1 = \frac{V_R/2}{2R} = \frac{V_R}{4R} = \frac{I_1}{2}$$

$$I_2 = \frac{V_R/4}{2R} = \frac{V_R}{8R} = \frac{I_1}{4}$$

$$I_n = \frac{V_R/2^n - 1}{2R} = \frac{I_1}{2^n - 1} \qquad ...(7)$$

We know that, V_o is given as

$$V_o = -I_T R_f \qquad ...(8)$$

$$V_o = -R_f(b_0 \times I_0 + b_1 \times I_1 + b_2 \times I_2 + ... + b_n \times I_n)$$

$$= -R_f\left(b_0 \frac{V_R}{2R} + b_1 \frac{V_R}{4R} + b_2 \frac{V_R}{8R} + ... + b_n \frac{V_R}{2^n R}\right)$$

$$= \frac{-V_R R_f}{R}(b_0 2^{-1} + b_1 2^{-2} + b_2 2^{-3} + ... + b_n 2^{-n}) \qquad ...(9)$$

When $R_f = R$, V_0 is given as,

$$V_o = -V_R(b_0 2^{-1} + b_1 2^{-2} + b_2 2^{-3} + ... + b_n 2^{-n}) \qquad ...(10)$$

Let us consider 4-bit binary DAC with binary input 1001 and $R_f = R$, as shown in the Fig. 7.14.

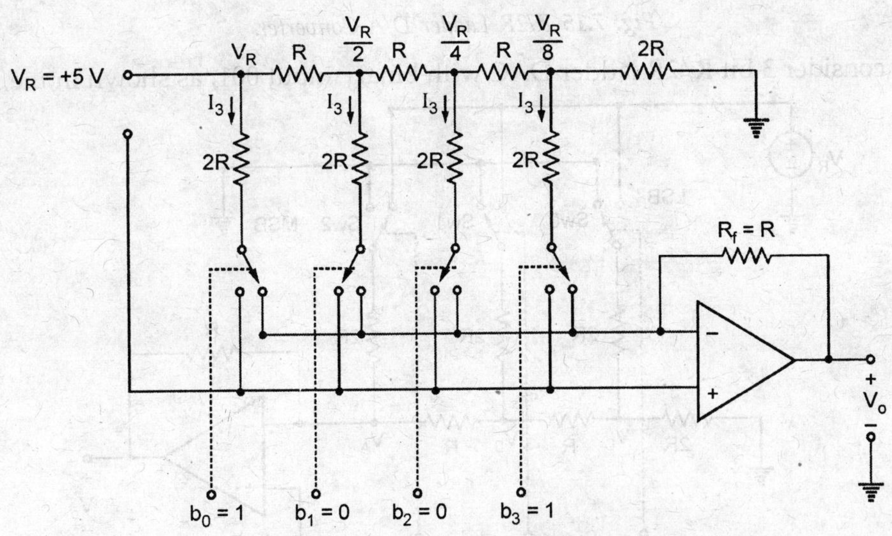

Fig. 7.14 Inverted 4-bit R/2R ladder DAC

Here, output voltage is given as,

$$V_o = -V_R(1\times 2^{-1}+0\times 2^{-2}+0\times 2^{-3}+1\times 2^{-4})$$

$$= -V_R\left(\frac{1}{2}+0+0+\frac{1}{16}\right) = -0.5626\ V_R = 2.8125\ V$$

The inverting R/2R ladder DAC works on the principle of summing currents and it is also said to operate in the **current mode**. An important advantage of the current mode is that all ladder node voltages remain constant with changing input codes, thus avoiding any shutdown effects by stray capacitances.

7.3.2.3 R-2R Ladder/Voltage Mode R-2R Ladder D/A Converter

In this type, reference voltage is applied to one of the switch positions, and other switch position is connected to ground, as shown in the Fig. 7.15.

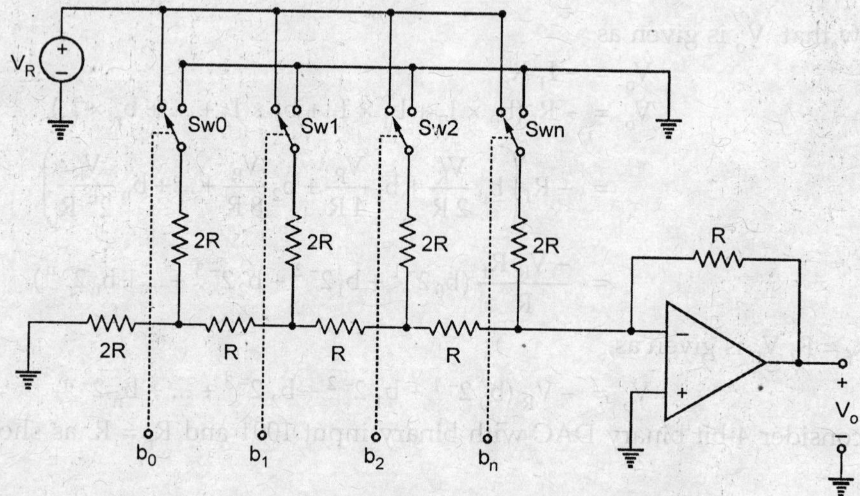

Fig. 7.15 R/2R Ladder D/A Converter.

Let us consider 3-bit R/2R ladder DAC with binary input 001, as shown in the Fig. 7.16.

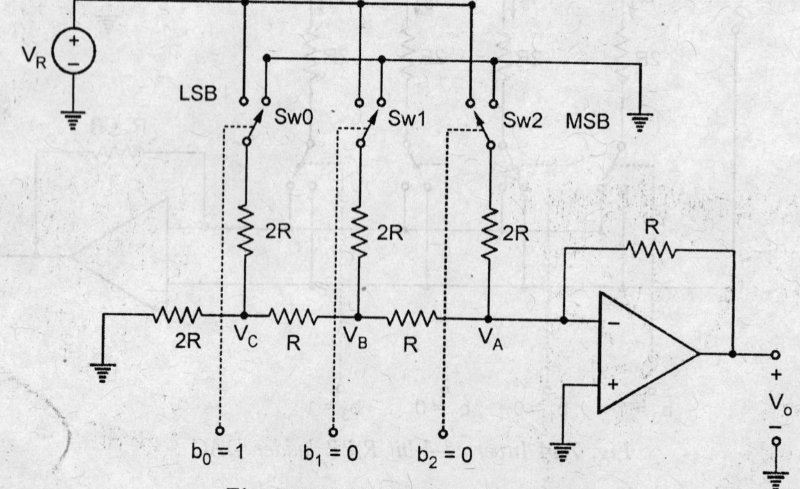

Fig. 7.16 3-bit R/2R ladder DAC.

Reducing above network to the left by Thevenin's theorem we get,

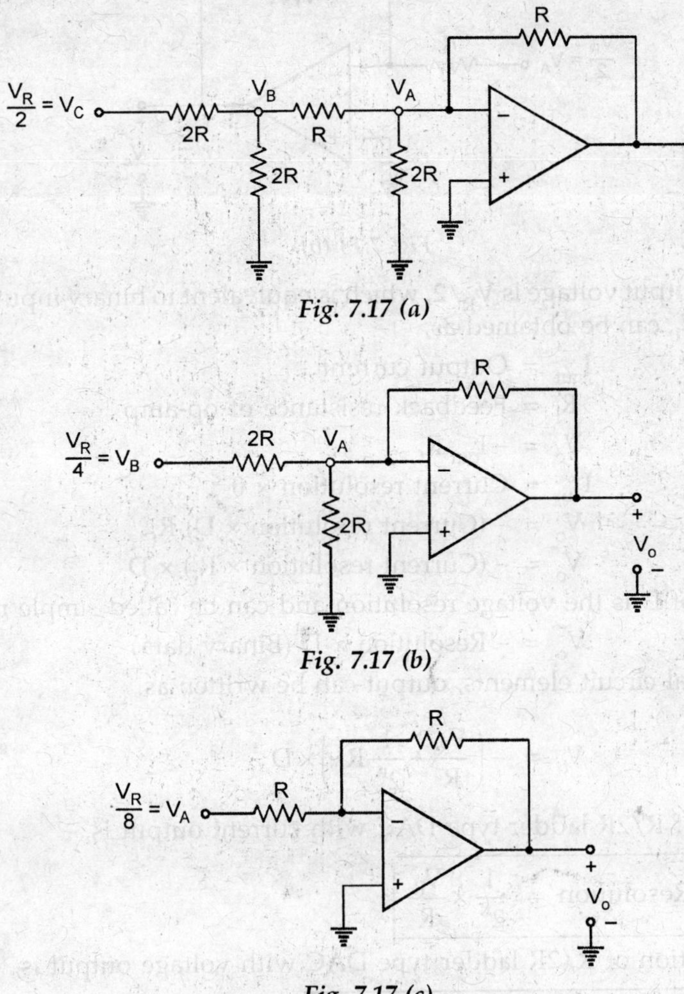

Fig. 7.17 (a)

Fig. 7.17 (b)

Fig. 7.17 (c)

Therefore, the output voltage $V_R/8$, which is equivalent to binary input 001.

For binary input 100 the network can be reduced as follows:

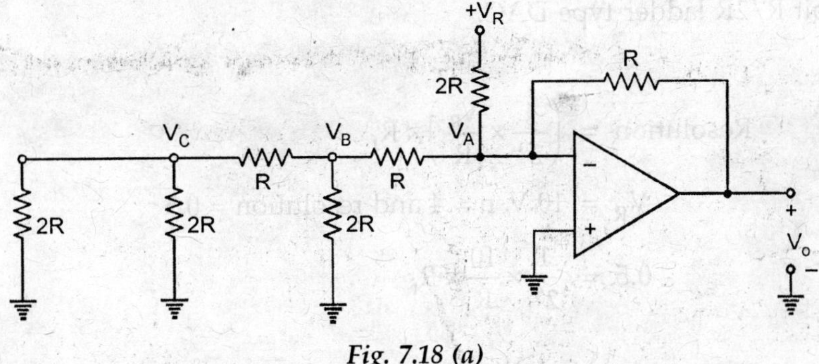

Fig. 7.18 (a)

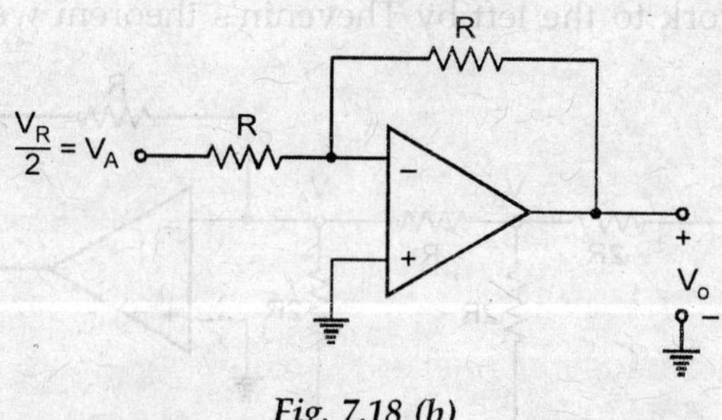

Fig. 7.18 (b)

Therefore, the output voltage is $V_R/2$, which is equivalent to binary input 100. In general, the expression for V_0 can be obtained as,

Let I_{out} = Output current

R_f = Feedback resistance of op-amp

$\therefore \quad V_o = -I_{out} R_f$

Now I_{out} = Current resolution × 0

$\therefore \quad V_o$ = – (Current resolution × D) R_f

$\therefore \quad V_o$ = – (Current resolution × R_f) × D ...(11)

The coefficient of D is the voltage resolution and can be called simple resolution.

$\therefore \quad V_o$ = – Resolution × D (Binary data) ...(12)

In terms of actual circuit elements, output can be written as,

$$V_o = -\left(\frac{V_R}{R} + \frac{1}{2^n} R_f\right) \times D \quad ...(13)$$

The resolution of R/2R ladder type DAC with current output is,

$$\boxed{\text{Resolution} = \frac{1}{2^n} \times \frac{V_R}{R}} \quad ...(14)$$

while the resolution of R/2R ladder type DAC with voltage output is,

$$\boxed{\text{Resolution} = \left(\frac{1}{2^n} \times \frac{V_R}{R}\right) \times R_f} \quad ...(15)$$

Example 7.7: Suggest the values of resistors and reference voltage if resolution required is 0.5 V for 4-bit R/2R ladder type DAC.

Solution:

$$\text{Resolution} = \left(\frac{1}{2^n} \times \frac{V_R}{R}\right) \times R_f$$

Let V_R = 10 V, n = 4 and resolution = 0.5

$$\therefore \quad 0.5 = \frac{1}{2^4} \times \frac{10}{R} \times R_f$$

$$\therefore \quad \frac{R}{R_f} = 1.25$$

Choose $\quad R_f = 10\ k\Omega$

$$\therefore \quad R = \mathbf{12.5\ k\Omega}$$

Advantages of R-2R ladder DACs

1. Easier to build accurately as only two precision metal film resistors are required.
2. Number of bits can be expanded by adding more sections of same R/2R values.
3. In inverted R/2R ladder DAC, node voltages remain constant with changing input binary words. This avoids any slowdown effects by stray capacitances.

7.3.3 Switches of DAC

CMOS Switch

The inverted ladder DACs are ideally suited to monolithic fabrication using CMOS technology. The switches used in this technology are realized with CMOS transistors. The Fig. 7.19 shows the circuit diagram of CMOS switch. The switch consists of pair of NMOS transistors, M_8 - M_9 and transistors M_1 through M_7 capable of accepting TTL and CMOS-compatible logic input. Transistors M_1 through M_7 provide antiphase drive for the gates of M_8 and M_9.

When the input is high, M_8 is OFF, M_9 is ON, and the 2R resistor is connected to the I_o bus. When the input is low, M_8 is ON, M_9 is OFF and the 2R resistor is connected to the $\overline{I}_o$ bus.

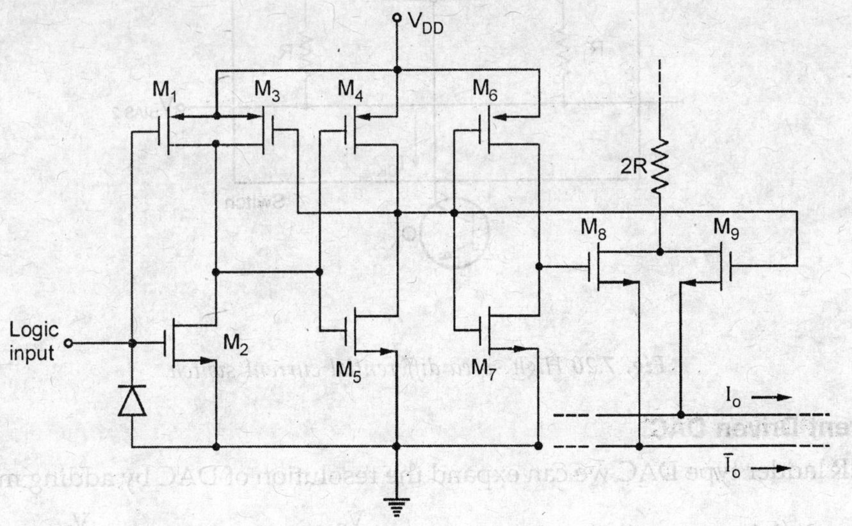

Fig. 7.19 CMOS switch.

BJT Switch

BJTs are used to provide fast current switching and true output current source or sink capability in DACs. When used in non saturating current-steering mode BJT switches provide very fast switching. Typically switching time 'is in nanoseconds. The Fig. 7.20 shows the high-speed different current switch using BJTs.

Here, when $V_K > V_{BIAS\,1}$, Q_1 is off and Q_2 in ON. As a result, Q_3 is OFF and Q_4 is ON, thus steering $Q_{K,S}$ vollector current I_{CK} to the I_o bus. When $V_K < V_{BIAS\,1}$, the conditions are reversed and I_{CK} is diverted to the $\bar{I}_o$ bus. The switching threshold is set by $V_{BIAS\,1}$, which is typically near 1.4 V to ensure compatibility with both CMOS and TTL.

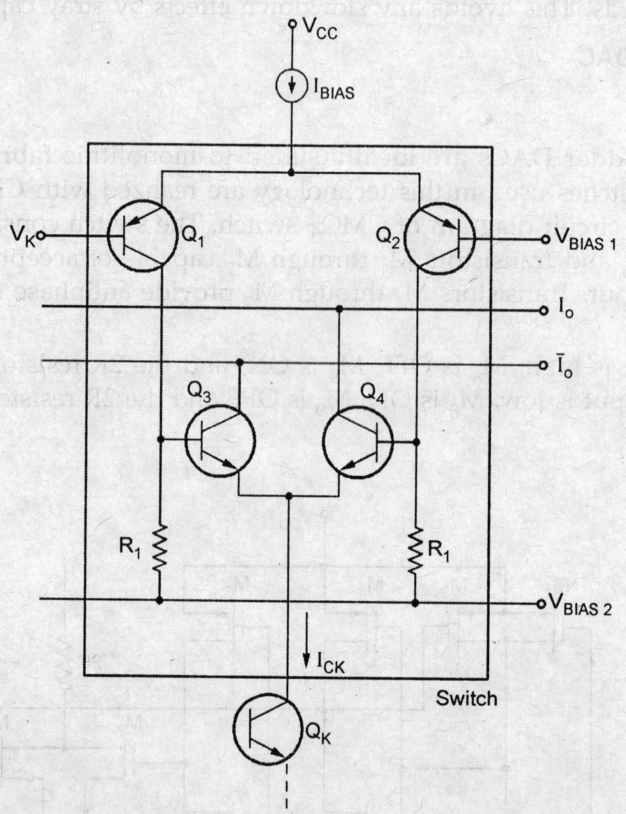

Fig. 7.20 High speed differential current switch.

7.3.4 Current Driven DAC

In R/2R ladder type DAC we can expand the resolution of DAC by adding more ladders.

We know that, the current through first ladder is $\frac{V_R}{2R}$, second ladder is $\frac{V_R}{4R}$, third ladder is

$\frac{V_R}{8R}$ and so on. When switches of such ladders are realized with BJTs, the emitter current of each BJT is current through corresponding ladder. Therefore, they are not equal. They are multiples of 2^n. When such DAC are fabricated using monolithic technology, it is necessary to maintain ratioed emitter area to pass ratioed emitter current. This leads to extravagant transistor geometries. This problem of ratioed emitter area can be eliminated altogether by using equal-value current sinks and exploiting the current-scaling capability of the inverted R-2R ladder to obtain binary-weighted contributions to the output. Such DACs are known as current driven DACs.

The Fig. 7.21 shows 4-bit current driven DAC. For the 4-bit current driven DAC shown in Fig. 7.21, the current 10 is given by

$$I_o = 2\frac{V_r}{R_r}(b_1 2^{-1} + b_2 2^{-2} + b_2 2^{-3} + b_4 2^{-4})$$

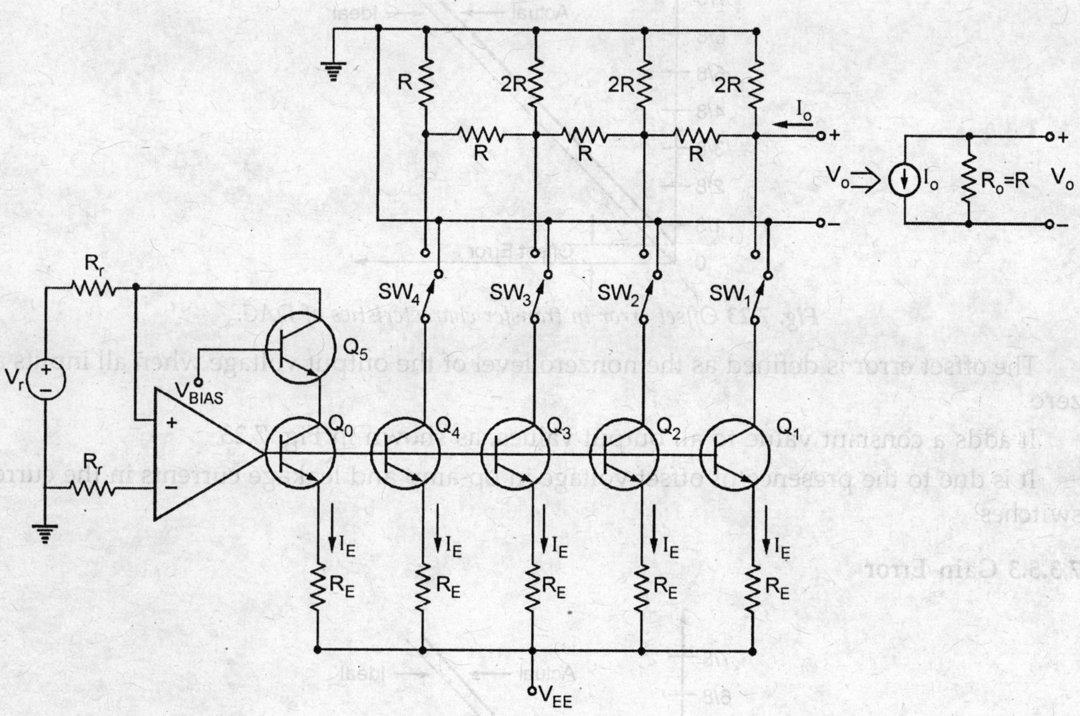

Fig. 7.21 4-bit current driven DAC.

7.3.5 Sources of Errors in DAC

There are mainly three errors in DACs : Linearity, Offset and gain errors.

7.3.5.1 Linearity Error

The error is defined as the amount by which the actual output differs from the ideal straight-line output.

Fig. 7.22 shows the linearity error in the transfer characteristics of DAC. It is mainly due to the errors in the current source resistor values.

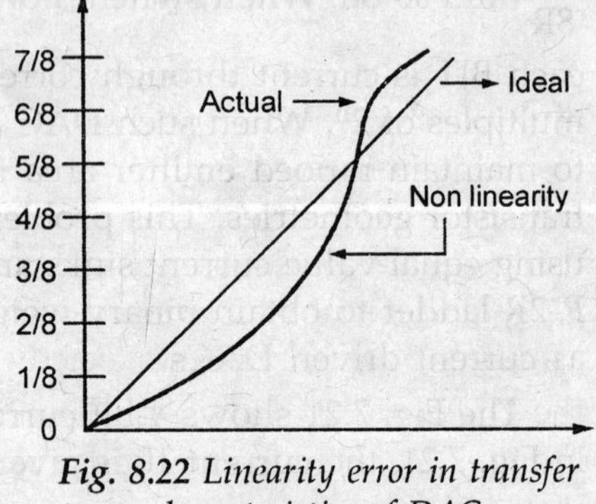

Fig. 8.22 Linearity error in transfer characteristics of DAC.

7.3.5.2 Offset Error

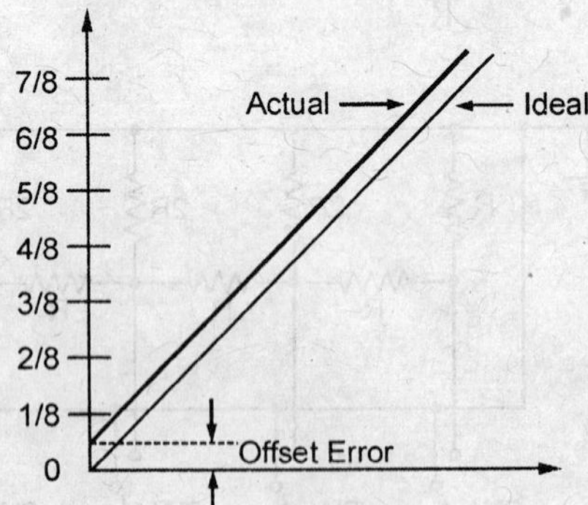

Fig. 7.23 Offset error in transfer characteristics of DAC.

The offset error is defined as the nonzero level of the output voltage when all inputs are zero.

It adds a constant value to all output values, as shown in Fig. 7.23.

It is due to the presence of offset voltage in op-amp and leakage currents in the current switches.

7.3.5.3 Gain Error

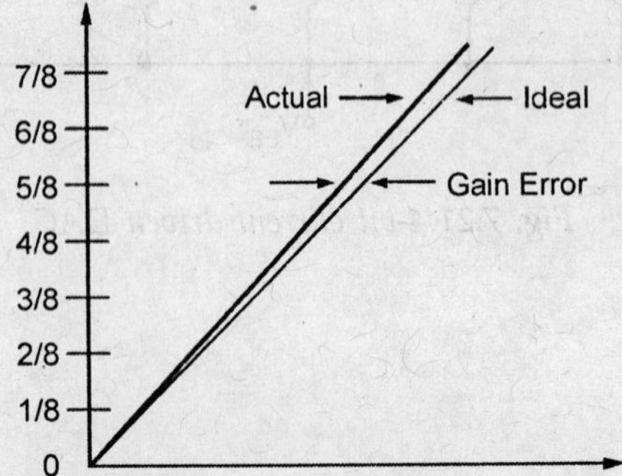

Fig. 7.24 Gain error in transfer characteristics of DAC.

The gain error is defined as the difference between the calculated gain of the current to voltage converter and the actual gain achieved. It is due to the errors in the feedback resistor on the current to voltage converter op-amp.

Fig. 7.24 Shows the gain error in transfer characteristics of DAC.

7.4 A/D CONVERTERS

The A/D conversion is a quantizing process whereby an analog signal is converted into equivalent binary word. Thus the A/D converter is exactly opposite function that of the D/A converter.

Fig. 7.25 shows symbol for A/D converter.

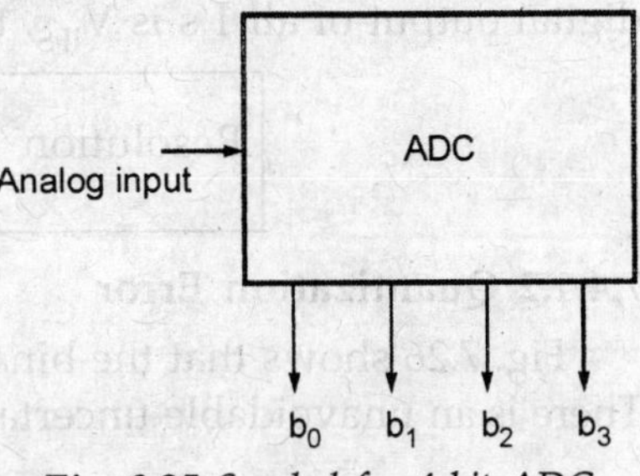

Fig. 8.25 Symbol for 4-bit ADC.

7.4.1 Performance Parameters of ADC

The Fig. 7.26 shows the digital output of an ideal 3-bit ADC plotted against the analog input voltage.

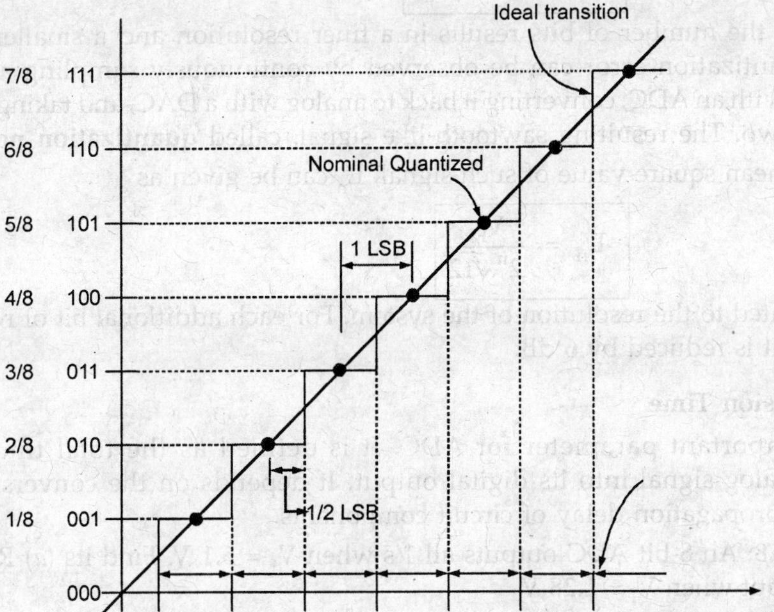

Fig. 7.26 Analog input Vs Digital output.

Let us define the performance parameters of ADC, referring to the Fig. 7.26, which shows the output-input characteristics of ADC.

7.4.1.1 Resolution

Fig. 7.26 shows eight (2^3) discrete output states from 000_2 to 111_2, each step being $\frac{1}{8}$ V apart. Therefore, we can say that expression of ADC resolution is the same as for the DAC and is repeated here:

$$\text{Resolution} = 2^n \quad \text{...(1)}$$

Resolution is also defined as the ratio of a change in value of input voltage, V_i, needed to change the digital output by 1 LSB. If the full scale input voltage required to cause a digital output of all 1's is V_{iFS}, then resolution can be given as

$$\text{Resolution} = \frac{V_{iFS}}{2^n - 1} \quad \text{...(2)}$$

7.4.1.2 Quantization Error

Fig. 7.26 shows that the binary output is 011 for all values of V_i between 1/4 and 1/2 V. There is an unavoidable uncertainty about the exact value of V_i when the output is 011. This uncertainty is specified as quantization error. Its value is $\pm\frac{1}{2}$ LSB.

It is given as,

$$Q_E = \frac{V_{iES}}{(2^n - 1)\,2} \quad \text{...(3)}$$

Increasing the number of bits results in a finer resolution and a smaller quantization error. The quantization error can be observed by continuously sampling a time-varying analog signal with an ADC, converting it back to analog with a DAC, and taking the difference between the two. The resulting sawtooth-like signal, called **quantization noise**.

The root mean square value of such signal, E_n can be given as

$$E_n = \frac{V_{FS}}{2^n\sqrt{12}}$$

This is related to the resolution of the system. For each additional bit of resolution E_n is cut in half, that is reduced by 6 dB.

7.4.1.3 Conversion Time

It is an important parameter for ADC. It is defined as the total time required to convert an analog signal into its digital output. It depends on the conversion technique used and the propagation delay of circuit components.

Example 7.8: An 8-bit ADC outputs all 1's when V_i = 5.1 V. Find its (*a*) Resolution and (*b*) Digital output when V_i = 1.28 V.

Solution: (*a*) From equation (1) we have,

$$\text{Resolution} = 2^8 = 256$$

and from equation (2) we have,

$$\text{Resolution} = \frac{5.1\text{ V}}{2^8 - 1} = \mathbf{20\ mV/LSBs}$$

Therefore, we can say that to change output by 1 LSB we have to change input by 20 mV.

(*b*) For 1.28 V analog input, digital output can be calculated as,

$$D = \frac{1.28\text{ V}}{20\text{ mV / LSB}} = \mathbf{64\text{ mV/LSBs}}$$

The binary equivalent of 64 is 01000000_2

Example 7.9: Calculate the quantizing error for 12-bit ADC with full scale input voltage 4.095 V.

Solution: From equation (3) we get

$$Q_E = \frac{4.095}{(2^{12}-1)\times 2} = \frac{4.095}{(4096-1)} = \mathbf{0.5\text{ mV}}$$

7.4.2 Basic Conversion Techniques

Analog to digital converter are classified into two general groups based on the conversion techniques. One technique involves comparing a given analog signal with the internally generated reference voltages. This group includes successive approximation, flash, delta modulated (DM), adaptive delta modulated and flash type converters. The another technique involves changing an analog signal into time or frequency and comparing these new parameters against known values. This group includes integrator converters and voltage-to-frequency converters.

In this chapter we are going to discuss following types of ADCs using various conversion techniques:

1. Single ramp or single slope
2. Dual slope
3. Successive approximation
4. Flash
5. Delta modulation
6. Adaptive delta modulation

7.4.2.1 Single Slope ADC

It consists of a ramp generator and BCD or binary counters. The Fig. 7.27 shows the single slope ADC.

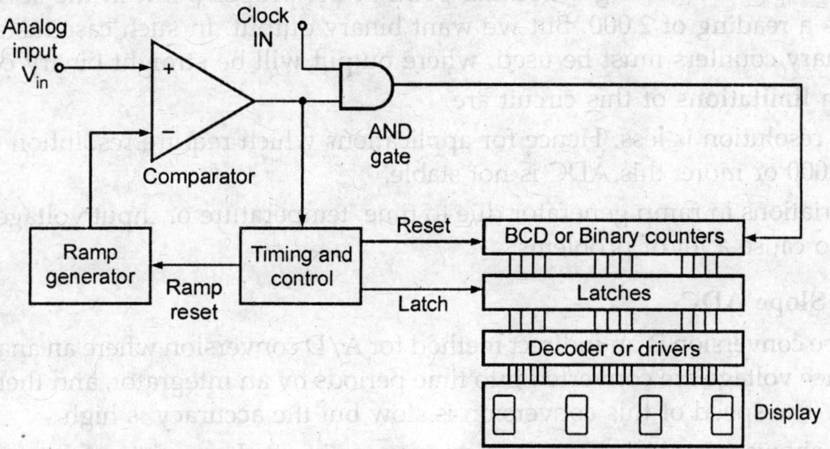

Fig. 7.27 Single slope ADC.

At the start, the reset signal is provided to the ramp generator and the counters. Thus counters are resetted to 0's. The analog input voltage V_{in} is applied to the positive terminal of the comparator. As this is more positive than the negative input, the comparator output goes high. The output of ramp generator is applied to the negative terminal of the comparator. The high output of the comparator enables the AND gate which allows clock to reach to the counters and also this high output starts the ramp.

The ramp voltage goes positive until it exceeds the input voltage. When it exceeds V_{in}, comparator output goes low. This disables AND gate which inturn stops the clock to the counters. The control circuitry provides the latch signal which is used to latch the counter data. The reset signal resets the counters to 0's and also resets the ramp generator. The latched data is then displayed using decoder and a display device.

Let us consider the practical example to understand the working. Assume that the clock frequency is 1 MHz. There are four BCD counters and the input V_{in} is 2.000 V.

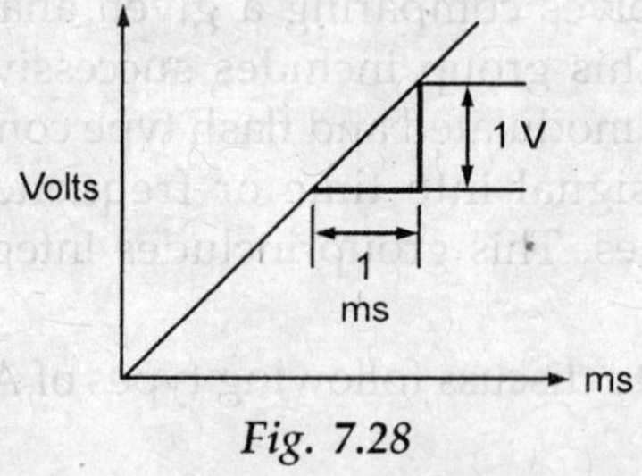

Fig. 7.28

Now let ramp has a slope of 1 V/ms as shown in the Fig. 7.28. As the input is 2.000 V, the ramp will take 2 ms to reach to 2 V and to stop the clock to the counters.

Now how many pulses will reach to the counters during 2 ms? It can be calculated from the frequency of the clock. The number of pulses reaching to the counter in 2 ms is $\frac{2 \text{ ms}}{(1/1 \text{ MHz})} = 2000$. The comparator output going high will strobe. The latches and send the count to the displays. Inserting a decimal point at the proper point in the seven segment display gives a reading of 2.000. But we want binary output. In such case instead of BCD counters, binary counters must be used, where output will be straight binary coded.

The main **limitations** of this circuit are,

(*i*) Its resolution is less. Hence for applications which require resolution of 9 part in 20,000 or more, this ADC is not stable.

(*ii*) Variations in ramp generator due to time, temperature or input voltage sensitivity also cause a lot of problems.

7.4.2.2 Dual Slope ADC

Dual slope conversion is an indirect method for A/D conversion where an analog voltage and a reference voltage are converted into time periods by an integrator, and then measured by a counter. The speed of this conversion is slow but the accuracy is high.

Fig. 7.29 shows a typical dual slope converter circuit. It consists of integrator (ramp generator), comparator, binary counter, output latch and reference voltage. The ramp generator input is switched between the analog input voltage Vi and a negative reference voltage, $- V_{REF}$. The analog switch is controlled by the MSB of the counter. When the MSB is a logic 0, the voltage being measured is connected to the ramp generator input. When MSB is logic 1, the negative reference voltage is connected to the ramp generator.

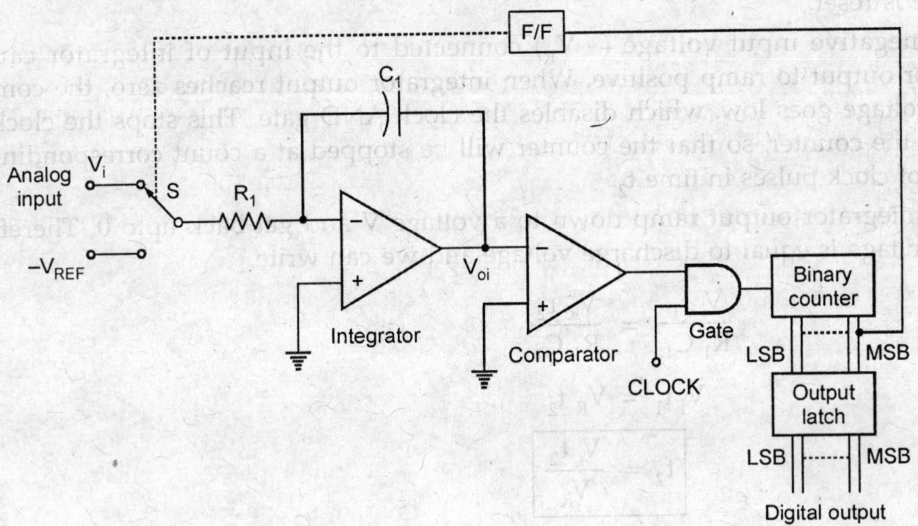

Fig. 7.29 Dual slope A/D converter.

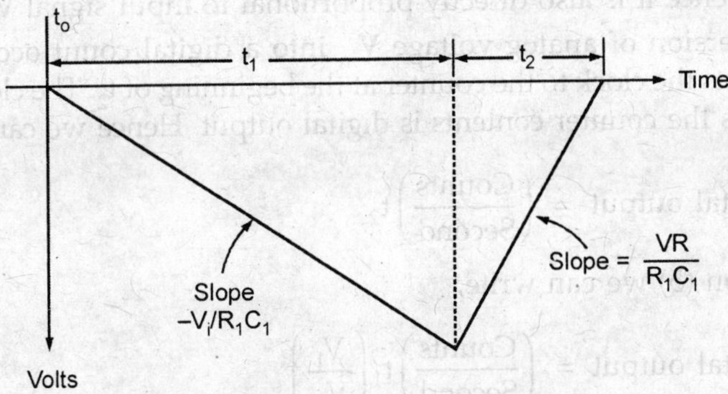

Fig. 7.30 Integrator output voltage.

At time t = 0, analog switch S is connected to the analog input voltage V_i, so that the analog input voltage integration begins. The output voltage of the integrator can be given as,

$$V_{oi} = \frac{-1}{R_1 C_1} \int_0^t V_i \, dt$$

$$= \frac{-V_i \, t}{R_1 \, C_1}$$

where $R_1 C_1$ is the integrator time constant and V_i is assumed constant over the integration time period. At the end of 2^N clock periods MSB of the counter goes high. At a result the output of the flip-flop goes high, which causes analog switch S to be switched from V_i to $-V_R$. At this very same time the binary counter which has gone through its entire count sequence is reset.

The negative input voltage ($-V_R$) connected to the input of integrator causes the integrator output to ramp positive. When integrator output reaches zero, the comparator output voltage goes low, which disables the clock AND gate. This stops the clock pulses reaching the counter, so that the counter will be stopped at a count corresponding to the number of clock pulses in time t_2.

The integrator output ramp down to a voltage V and get back upto 0. Therefore, the charge voltage is equal to discharge voltage and we can write,

$$\frac{V_i\, t_1}{R_1\, C_1} = \frac{V_R\, t_2}{R_1\, C_1}$$

$$\therefore \quad V_i\, t_1 = V_R\, t_2$$

$$\therefore \quad \boxed{t_2 = \frac{V_i\, t_2}{V_R}} \qquad \text{...(5)}$$

The above equation shows that t_2 is directly proportional only to the V_i, since V_R and t_1 are constants. The binary digital output of the counter gives corresponding digital value for time period t_2 and hence it is also directly proportional to input signal V_i.

The actual conversion of analog voltage V_{in} into a digital count occurs during t_2. The control circuit connects the clock to the counter at the beginning of t_2. The clock is disconnected at the end of t_2. Thus the counter contents is digital output. Hence we can write,

$$\text{Digital output} = \left(\frac{\text{Counts}}{\text{Second}}\right) t_2 \qquad \text{...(6)}$$

But from equation (5) we can write,

$$\text{Digital output} = \left(\frac{\text{Counts}}{\text{Second}}\right) t_1 \left(\frac{V_1}{V_R}\right) \qquad \text{...(7)}$$

The counter output can then be connected to an appropriate digital display. The advantages of dual slope AOC are,

1. It is highly accurate.
2. Its cost is low.
3. It is immune to temperature caused variations in R_1 and C_1.

The only disadvantage of this AOC is its speed which is low.

Example 7.10: For a particular dual slope ADC, t_1 is 83.33 ms and the reference voltage is 100 mV. Calculate t_2 if (*i*) V_i is 100 mV and (*ii*) 200 mV.

Solution: We know that,

$$t_2 = \left(\frac{V_i}{V_R}\right) t_1$$

(*i*) $$t_2 = \left(\frac{100}{100}\right)(83.33)$$

$$= 83.33 \text{ ms}$$

(*ii*) $$V_i = 200 \text{ mV}$$

$\therefore$ $$t_2 = \left(\frac{200}{100}\right)(83.33)$$

$$= 166.6 \text{ ms}$$

Example 7.11: Find the digital output of an ADC having t_1 as 83.33 ms and V_R as 100 mV for an input voltage of + 100 mV. The clock frequency is 12 kHz.

Solution: The digital output is given as,

$$\text{Digital output} = \left(\frac{\text{Counts}}{\text{Second}}\right) t_1 \left(\frac{V_i}{V_R}\right)$$

Now Clock frequency = 12 kHz

$$\text{i.e.} = 12000 \frac{\text{counts}}{\text{second}}$$

$\therefore$ $$\text{Digital output} = 12000 \times 83.33 \times \left(\frac{100}{100}\right) \times 10^{-3}$$

$$= 1000 \text{ counts.}$$

7.4.2.3 Successive Approximation ADC

In this technique, the basic idea is to adjust the DAC's input code such that its output is within ± 1/2 LSB of the analog input V_i to be A/D converted. The code that achieves this represents the desired ADC output. It uses very efficient code searching strategy called binary search. It completes searching process for n-bit conversion in just n clock periods.

Fig. 7.31 shows the block diagram of successive approximation A/D converter. It consists of a DAC, a comparator, and a successive approximation register (SAR). The external clock input sets the internal timing parameters. The control signal start of conversion (SOC) initiates an A/D conversion process and end of conversion signal is activated when the conversion is completed.

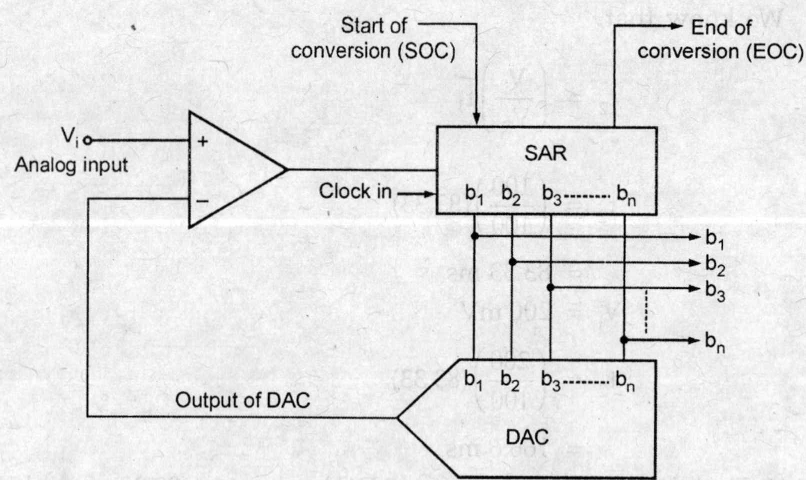

Fig. 7.31 Block diagram of successive approximation A/D converter.

Operation

The searching code process in successive approximation method is similar to weighing an unknown material with a balance scale and a set of standard weights. Let us assume that we have 1 kg, 2 kg and 4 kg weights (SAR) plus a balance scale (comparator and DAC). Now we will see the successive approximation analogy for 3-bit ADC.

Refer Fig. 7.31 and 7.32. The analog voltage V_{in} is applied at one input of comparator.

On receiving start of conversion signal (SOC) successive approximation register sets 3-bit binary code 100_2 ($b_2 = 1$) as an input of DAC. This is similar process of placing the unknown weight on one platform of the balance and 4 kg weight on the other. The DAC converts the digital word 100 and applies it equivalent analog output at the second input of the comparator. The comparator then compares two voltages just like comparing unknown weight with 4 kg weight with the help of balance scale. If the input voltage is greater than the analog output of DAC, successive approximation register keeps $b_2 = 1$ and makes $b_1 = 1$ (addition of 2 kg weight to have total 6 kg weight) otherwise it resets $b_2 = a$ and makes $b_1 = 1$ (replacing 2 kg weight). The same process is repeated for b_1 and b_0. The status of b_0, b_1 and b_2 bits gives the digital equivalent of the analog input.

Fig. 7.32 illustrates the process we have just discussed.

The dark lines in the Fig. 7.32 shows setting and resetting actions of bits for input voltage 5.2 V, on the basis of comparison. It can be seen from the Fig. 7.32 that one clock pulse is required for the successive approximation register to compare each bit. However an additional clock pulse is usually required to reset the register prior to performing a conversion.

The time for one analog to digital conversion must depend on both the clock's period T and number of bits n. It is given as,

$$T_C = T(n+1) \qquad \text{...(8)}$$

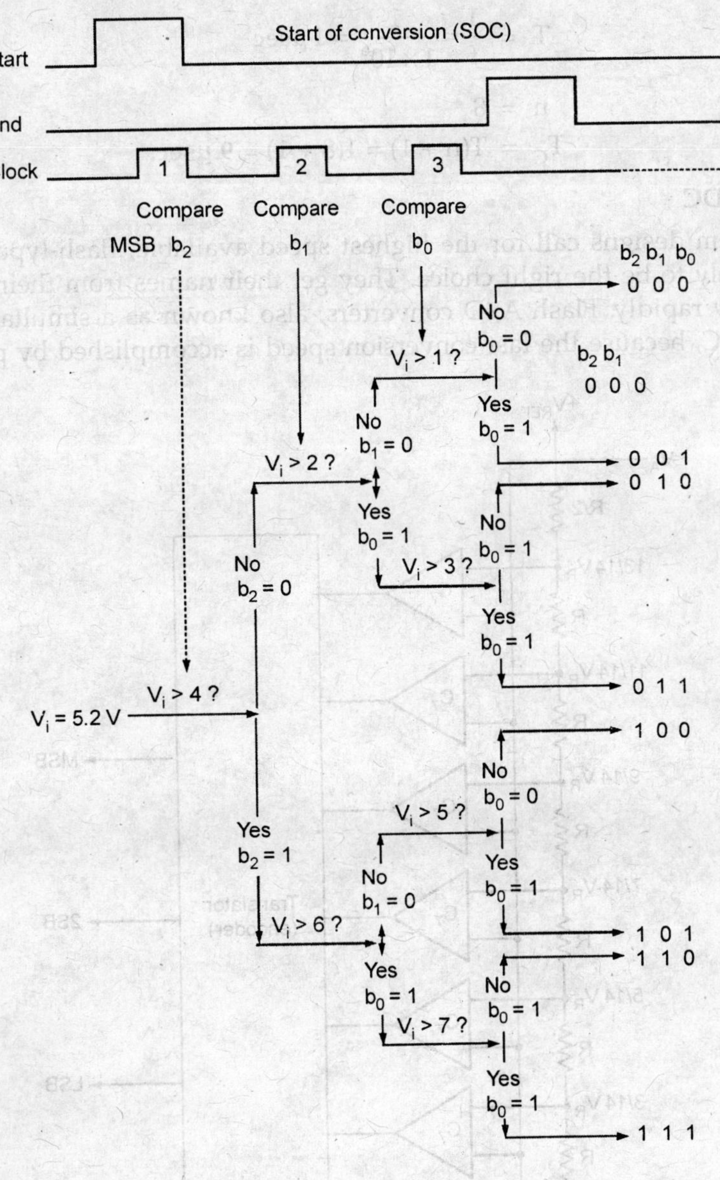

Fig. 7.32 Illustration of conversion process.

where T_C = Conversion time

T = Clock period

n = Number of bits

Example 7.12: An 8-bit successive approximation ADC is driven by a 1 MHz clock. Find its conversion time.

Solution: $f = 1 \text{ MHz}$

$$\therefore \quad T = \frac{1}{f} = \frac{1}{1 \times 10^6} = \mathbf{1\ \mu sec}$$

$$n = 8$$

$$\therefore \quad T_C = T(n+1) = 1(8+1) = \mathbf{9\ \mu sec.}$$

7.4.2.4 Flash ADC

When system designs call for the highest speed available, flash-type A/D converters (ADCs) are likely to be the right choice. They get their names from their ability to do the conversion very rapidly. Flash A/D converters, also known as a simultaneous or parallel comparator ADC, because the fast conversion speed is accomplished by providing

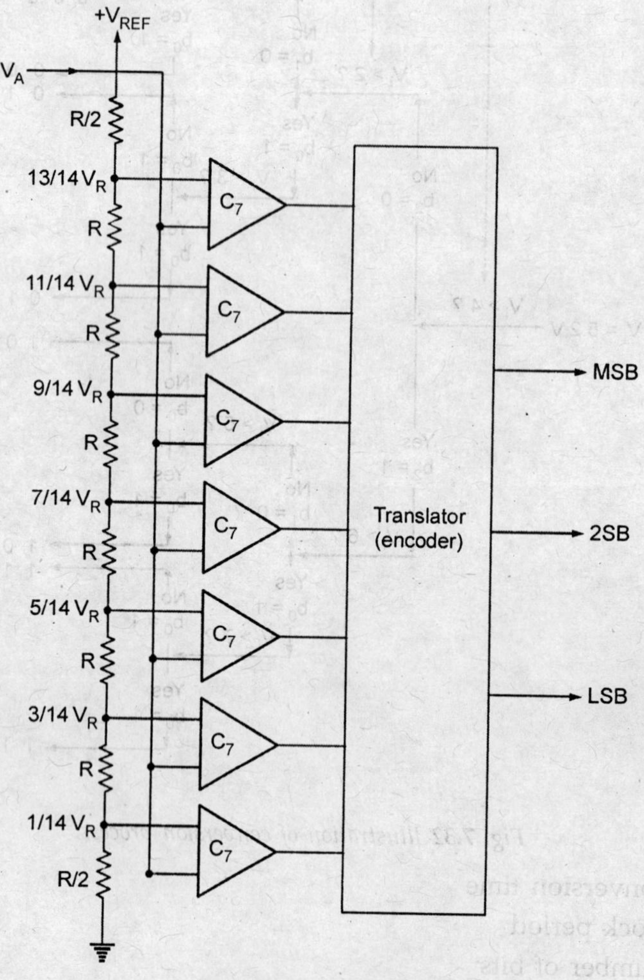

Fig. 7.33 3-bit flash converter.

$2^n - 1$ comparators and simultaneously comparing the input signal with unique reference levels spaced 1 LSB apart.

Fig. 7.33 shows 3-bit flash A/D converter. For this ADC, seven $(2^3 - 1)$ comparators are required. As shown in the Fig. 7.33, one input of each comparator is connected to the input signal and other input to the reference voltage level generated by the reference voltage divider. The reference voltage (V_{REF}) is equal to the full scale input signal voltage. The manner in which the flash A/D converter performs a quantization is relatively simple.

The comparators give output "1" or "0" state depending on whether the input signal is above or below the reference level at that instant. Those comparators referred above the input signal, remain turned-off, representing a "0" state. The comparators at or below the input signal conversely become a "1" state. The code resulting from this comparator is converted to a binary code by the encoder.

The number of comparators required for n-bit resolution is,

$$\text{Number of comparators} = 2^n - 1 \qquad \text{...(9)}$$

As seen earlier the quantization error is $\pm\frac{1}{2}$ LSB. Thus for an ADC, the maximum frequency for a sine wave V_{in} to be digitised within an accuracy of $\pm\frac{1}{2}$ LSB is,

$$\boxed{f_{max} = \frac{1}{2\pi(T_C)\,2^n}} \qquad \text{...(10)}$$

where f_{max} = Maximum input frequency

T_C = Conversion time

n = Number of bits

Example 7.13: For a particular 8-bit ADC, the conversion time is 9 μs. Find the maximum frequency of an input sine wave that can be digitized.

Solution : The maximum frequency is given by,

$$f_{max} = \frac{1}{2\pi(T_C)\,2^n} = \frac{1}{2\pi \times 9 \times 10^{-6} \times 2^8}$$

$$= \mathbf{69.07\ Hz}$$

7.4.2.5 Comparison between Flash, Dual Slope and Successive Approximation Techniques

The comparison of ADCs is given in the tabular form as below:

Parameter	Flash			Successive approximation			Dual slope		
Speed	Fastest			Fast			Slow		
Accuracy	Less			Medium			More		
Resolution	Upto 2^8			Upto 2^{16}			2^{16} or even more		
Input hold time	Very less			Depend on number of bits. It is more than flash A/D converter			It is maximum, hence sample and hold circuit is required		
Cost	Very costly			Medium			Less		
Applications	High speed fiber optic communication, Digital storage oscilloscope, Imaging and many more where high speed A/D conversion is required			The successive approximation A/D converter has the disadvantage of requiring D/A converter, but it has the advantage of high speed with excellent resolution. Hence these are most popular and used in data acquisition systems			These are used when high accuracy and resolution is required and speed is not the important criteria.		
	IC	Reso-lution	Conver-sion rate	IC	Resolu-tion	Conver-sion rate	IC	Reso-lution	Conver-sion rate
ICs	AD6020KD	6	50 MHz	ADC1103	8	1 MHz	ADC141	14	25 Hz
	AD9000SD	6	75 MHz	ADC60	10	670 kHz	7109	12	30 Hz
	TM1070	7	15 MHz	ADC1103	12	300 kHz	ADB1200	12	28 Hz
	TDC10193	9	25 MHz	ADC72	16	20 kHz	LF13300	12	28 Hz

7.4.2.6 Counter Type ADC

Principle of operation

This ADC uses DAC for A to D conversion. The output of the DAC is continuously compared with the analog input to the ADC which is to be converted into digital output. When the output of the DAC becomes greater than this analog input, the corresponding digital input to the DAC is noted which represents the analog input to the ADC.

Fig. 7.34 shows the circuit diagram of computer type ADC.

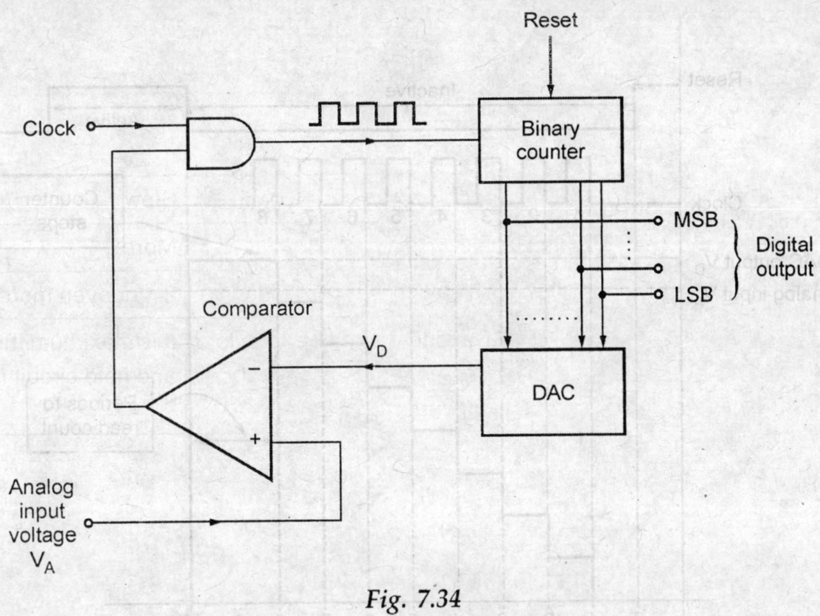

Fig. 7.34

As shown in the Fig. 7.34, the counter type ADC consists of a binary counter, DAC, comparator and AND gate. The operation of the circuit is explained below.

(*i*) Initially, the counter is reset, i.e. its output is set to zero by applying a reset pulse. The output of the counter is given as digital input to DAC. Since input to DAC is zero, its output V_D is zero.

(*ii*) When the analog input voltage V_A is applied to ADC, it becomes greater than V_D. V_A acts as input voltage for non inverting terminal and V_D acts as input voltage for inverting terminal of the comparator. Since V_A is greater than V_D, the comparator output goes high.

(*iii*) For an AND gate, one input is clock pulses and another input is the output of the comparator. Because of the high output of the comparator, the clock pulses are allowed to pass through the AND gate.

(*iv*) The counter starts counting these close pulses. According to the number of clock pulses, the output of the counter goes on increasing. This increases the output of the DAC.

(*v*) The above steps are continued till V_D is less than V_A.

(*vi*) As soon as DAC output V_D becomes greater than V_A, the comparator output goes low. This disables AND gate. So the clock pulses are not allowed to pass through the AND gate. The counting process of the binary counter is stopped.

(*vii*) The output of the binary counter which is in digital form is noted which represents the digital equivalent of the analog input voltage V_A.

(*viii*) For the next A to D conversion, the input voltage to ADC, V_A changes. The binary counter is cleared by applying a second reset pulse and all the above steps are repeated to obtain the digital equivalent of V_A.

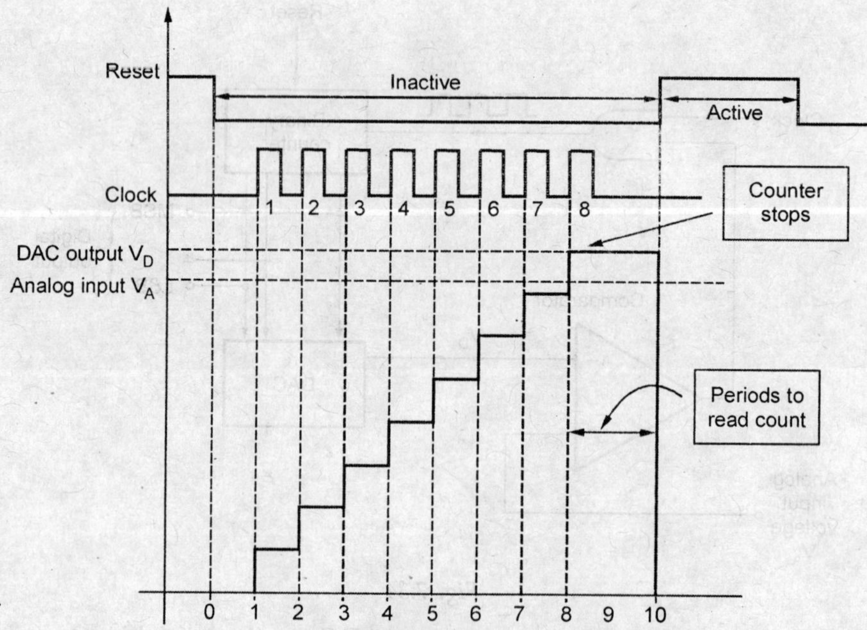

Fig. 7.35 Waveforms for counter type ADC.

Disadvantages

- It is necessary to give enough time for DAC conversion and comparator to respond. Therefore, there is a limitation on the clock frequency. As clock frequency is low, the speed of conversion is less.
- Conversion time is not constant. It increases with increase in input voltage. In other words, we can say that conversion time is high at high input voltage.

7.4.2.7 Delta Sigma Modulation

The delta modulation technique for analog to digital conversion is used in digital communication. In this technique, the analog signal is converted into pulses which can be represented by digital data.

The Fig. 7.36 shows the basic circuit for delta modulator. It consists of comparator, one bit quantizer and integrator. The comparator compares the input analog signal m(t) with the feedback signal m′(t), which is the integrated form of the output signal $P_o(t)$. The comparator output produces the difference signal. If the difference signal is positive, it is encoded as a binary 1, whereas if the difference signal is negative, it is encoded as a binary 0. A binary 1 is transmitted as a positive voltage pulse and binary 0 is transmitted as a negative voltage pulse.

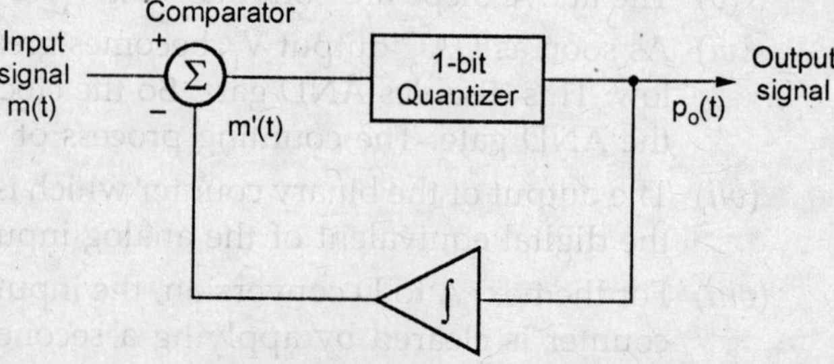

Fig. 8.36 Delta modulator.

Advantages of Delta Modulation

1. It is a simple technique.
2. It represents each analog sample by only one bit.

Disadvantages of Delta Modulation

1. Granular noise: The waveform of m′(t) is a staircase approximation to the analog signal m(t). For this staircase waveform, there will be an initial transient period labelled A. When modulator reaches steady state, the staircase waveform **"hunts"** around the analog waveform as shown at B. This hunting produces granular noise. The granular noise occurs when the step size is two large compared to small variations in the input signal. This can be observed at point B. Here, input is slowly increasing but the staircase waveform remains horizontal. The granual noise can be reduced by keeping step size small.

2. Slope overload distortion: The slope overload distortion occurs when the rate of change of the analog waveform is too high for the staircase waveform to follow. The slope overload causes a large error between the staircase approximated signal, m′(t) and the original input signal, m(t). At point C and D waveform suffers from slope overload distortion. Since the step size of delta modulator remains fixed, its maximum or minimum slopes occur along straight lines. In case of horizontal line slope overload distortion is minimum, and it is maximum along the vertical line. The slope overload distortion can be minimised by increasing the step size. But increase in step size increases granular noise. To solve this problem the delta modulation process is modified and known as **adaptive delta modulation.**

A delta modulation scheme that incorporates integration at its input is called **delta sigma modulation**. To be more precise, however, it should be called sigma delta modulation, because the integration is in fact performed before the delta modulation. The Fig. 7.36 (*a*) shows how delta sigma modulation is derived from the delta modulation.

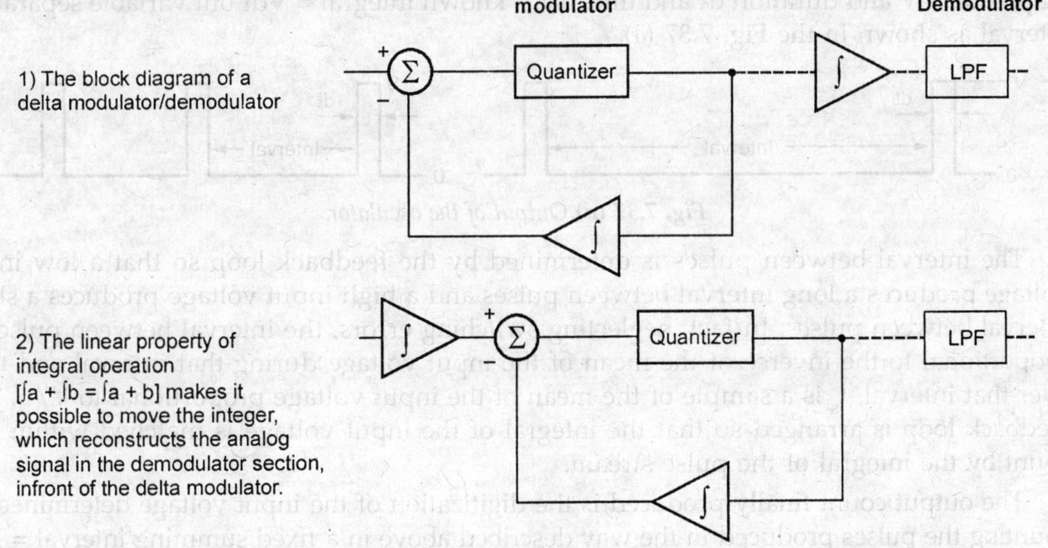

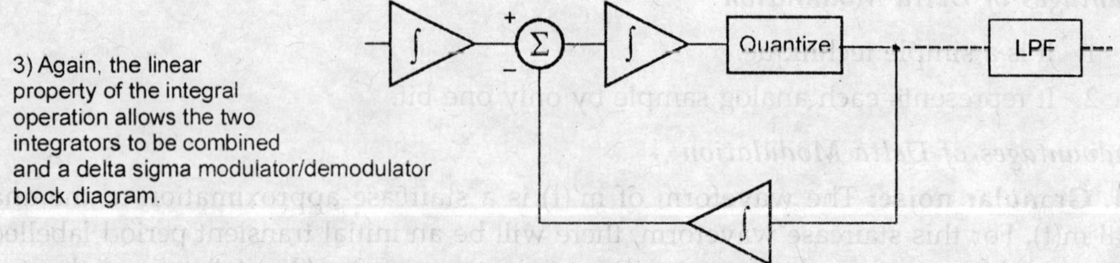

Fig. 7.36 (a) Derivation of $\Delta \sum$ *from* Δ *modulation.*

The Fig. 7.37 shows the block diagram of delta sigma modulator used to perform analog to digital conversion (ADC). The method can be thought of as a voltage-controlled oscillator, where the controlling voltage is the voltage to be measured and where linearity and proportionality are determined by a negative feedback loop.

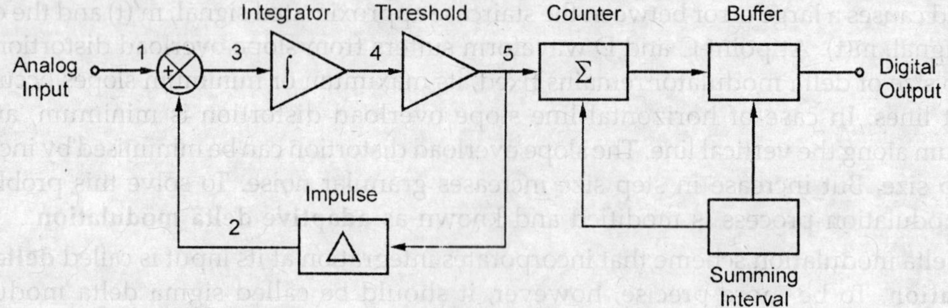

Fig. 7.37 Block diagram of delta sigma modulator.

The output of the oscillator is a pulse stream, each pulse of which is a known, constant, amplitude = V and duration dt and thus has a known integral = Vdt but variable separating interval as shown in the Fig. 7.37 (*a*).

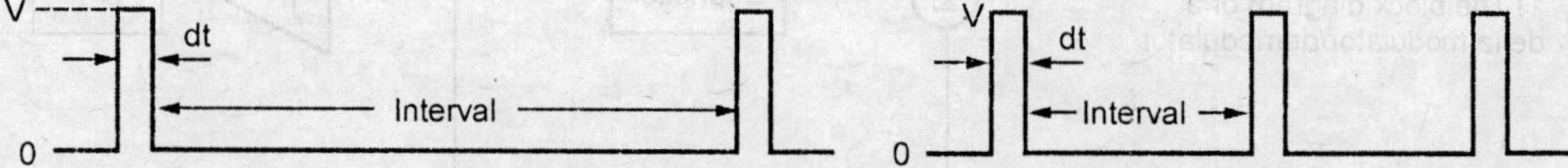

Fig. 7.37 (a) Output of the oscillator.

The interval between pulses is determined by the feedback loop so that a low input voltage produces a long interval between pulses and a high input voltage produces a short interval between pulses. In fact, neglecting switching errors, the interval between pulses is proportional to the inverse of the mean of the input voltage during that interval and thus over that interval, t_s is a sample of the mean of the input voltage proportional to V/t_s. The feedback loop is arranged so that the integral of the input voltage is matched within one count by the integral of the pulse stream.

The output count finally produced is the digitization of the input voltage determined by counting the pulses produced in the way described above in a fixed summing interval = Ndt producing a count, Σ. The integral of the pulse stream is ΣVdt which is produced over an interval Ndt and thus the average of the input voltage over the summing period is VΣ/N and is the mean of means and so subject to little variance. The accuracy achieved depends on the accuracy with which V is known and the precision or resolution is within one count in N.

Variations in scaling can be produced by either varying the fixed summing interval, Ndt or by counting down the pulses by a fixed ratio or both methods can be used.

The pulses described above may be treated as the Dirac δ (delta) function in a formal analysis and the count as Σ (sigma). It is these pulses which are transmitted for delta-sigma modulation but are counted to form sigma in the case of analog to digital conversion.

The Fig. 7.38 shows the waveforms at points designated by numbers 1 to 5 for an input of 0.2 volts on the left and 0.4 volts on the right.

Consider first the waveforms on the left

- 1 is the input and for this short interval is constant at 0.2 V.
- The stream of delta impulses is shown at 2.
- The difference between 1 and 2 is shown at 3.
- The difference shown at 3 is integrated to produce the waveform 4.
- The threshold detector generates a pulse 5 which is sustained until the waveform 4 falls below the threshold. Within the loop, 5 triggers the impulse generator and external to the loop increments the counter. The summing interval is a prefixed external to the loop increments the counter. The summing interval is a prefixed time and at its expiry the count is strobed into the buffer and the counter reset.

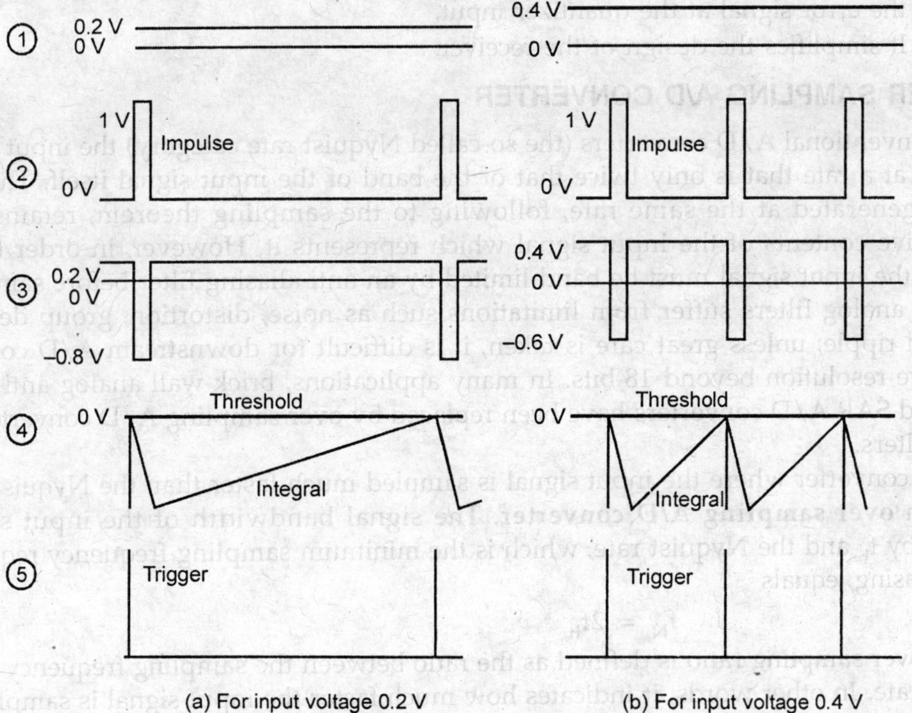

Fig. 7.38 Waveforms at points designated by numbers 1 to 5.

- It is necessary that the ratio between the impulse interval and the summing interval is equal to the maximum (full scale) count. It is then possible for the impulse duration and the summing interval to be defined by the same clock with a suitable arrangement of logic and counters. This has the advantage that neither interval has to be defined with absolute precision as only the ratio is important. Then to achieve overall accuracy it is only necessary that the amplitude of the impulse be accurately defined.

Consider first the waveforms on the right

- On the right the input is now 0.4 V and the sum during the impulse is – 0.6 V as opposed to – 0.8 V on the left. Thus the negative slope during the impulse is lower on the right than on the left.
- Also the sum is 0.4 V on the right during the interval as opposed to 0.2 V on the left. Thus the positive slope outside the impulse is higher on the right than on the left.
- The resultant effect is that the integral (4) crosses the threshold more quickly on the right than on the left. A full analysis would show that in fact the interval between threshold crossings on the right is half that on the left. Thus the frequency of impulses is doubled. Hence the count increments at twice the speed on the right to that on the left which is consistent with the input voltage being doubled.

It is important to note that in delta sigma modulation, the input signal is integrated prior to delta modulation. This gives the following benefits :

- The low frequency contents of the input signal is pre-emphasized.
- Correlation between adjacent samples of the delta modulator input is increased, which tends to improve overall system performance by reducing the variance of the error signal at the quantizer input.
- It simplifies the design of the receiver.

7.5 OVER SAMPLING A/D CONVERTER

In conventional A/D converters (the so-called Nyquist rate category) the input signal is sampled at a rate that is only twice that of the band of the input signal itself. The digital output, generated at the same rate, following to the sampling theorem, retains all the informative contents of the input signal which represents it. However, in order to avoid aliasing, the input signal must be band-limited by an anti-aliasing filter before sampling.

Such analog filters suffer from limitations such as noise, distortion, group delay, and passband ripple; unless great care is taken, it is difficult for downstream A/D converters to achieve resolution beyond 18-bits. In many applications, brick-wall analog anti-aliasing filters and SAR A/D converters have been replaced by over sampling A/D converters with digital filters.

A/D converter where the input signal is sampled much faster than the Nyquist rate, is called an **over sampling A/D converter**. The signal bandwidth of the input signal is denoted by f_b and the Nyquist rate, which is the minimum sampling frequency required to avoid aliasing, equals

$$f_N = 2f_b$$

The over sampling ratio is defined as the ratio between the sampling frequency and the Nyquist rate. In other words, it indicates how much faster the input signal is sampled than minimally required by the Nyquist theorem.

$$OSR = \frac{f_s}{f_N} = \frac{f_s}{2f_b}$$

The principle of using over sampling is that by sampling the signal many times, errors due to noise and coarse quantization are averaged out. Through the use of loop-filter and feedback, the noise is shaped to high frequencies, which can be easily removed by digital filters.

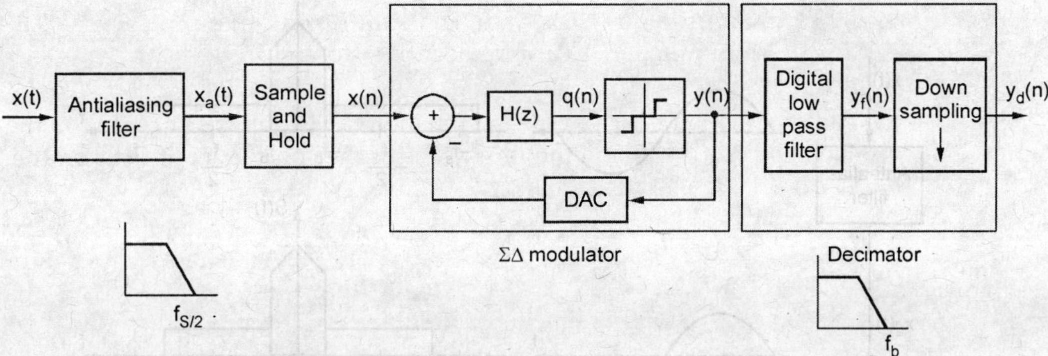

Fig. 7.39 Block diagram of over sampled ADC.

Antialiasing filter (AAF): It eliminates spectral components above half the sampling frequency from the input signal so that the modulator input signal is band-limited and the subsequent sampling operation does not alias input signals from higher frequencies into the band of interest.

ΣΔ Modulator: It performs the actual A/D conversion by means of sampling and quantizing the band-limited input signal as well as by filtering the quantization error from the internal quantizer out of the in-band. The internal feedback DAC is commonly implemented with the same low resolution as the internal quantizer and thus does not introduce an additional quantization error.

Decimation filter: After quantization, a digital low pass filter uses decimation both to reduce the sampling frequency to a nominal rate and prevent aliasing at the new, lower sampling frequency. Quantized data words are output at a lower frequency (for example, 48 or 96 kHz). The decimation low pass filter removes frequency components beyond the Nyquist frequency of the output sampling frequency to prevent aliasing when the output of the digital filter is resampled (under-sampled) at the system's sampling frequency.

The decimator typically consists of two different blocks. First, a digital low pass filter is used to remove all the frequency components above $\frac{f_N}{2}$ to avoid signal degradation due to aliasing in the down-sampling block that follows the digital filter. This digital filter also removes all the quantization noise which does not fall inside the signal band. The digital filter operates in the digital domain and its output contains N-bits words.

The next block in the decimation filter down-samples the output of the digital filter. Down-sampling by a ratio of OSR can be done by simply keeping a sample and remove the next OSR-1 samples. Since the sampling rate of signal is changed, aliasing can occur. However, the decimation process does not result in loss of information since the digital filter removes all the components that could alias in the signal band.

In an ideal over sampled converter, the anti-alias filter, the sample-block, the digital low pass filter and the decimator do not degrade the performance. Therefore, the signal-to-noise ratio is determined by the quantizer.

The Fig. 7.40 shows timing waveforms at different stages of over sampled A/D converter.

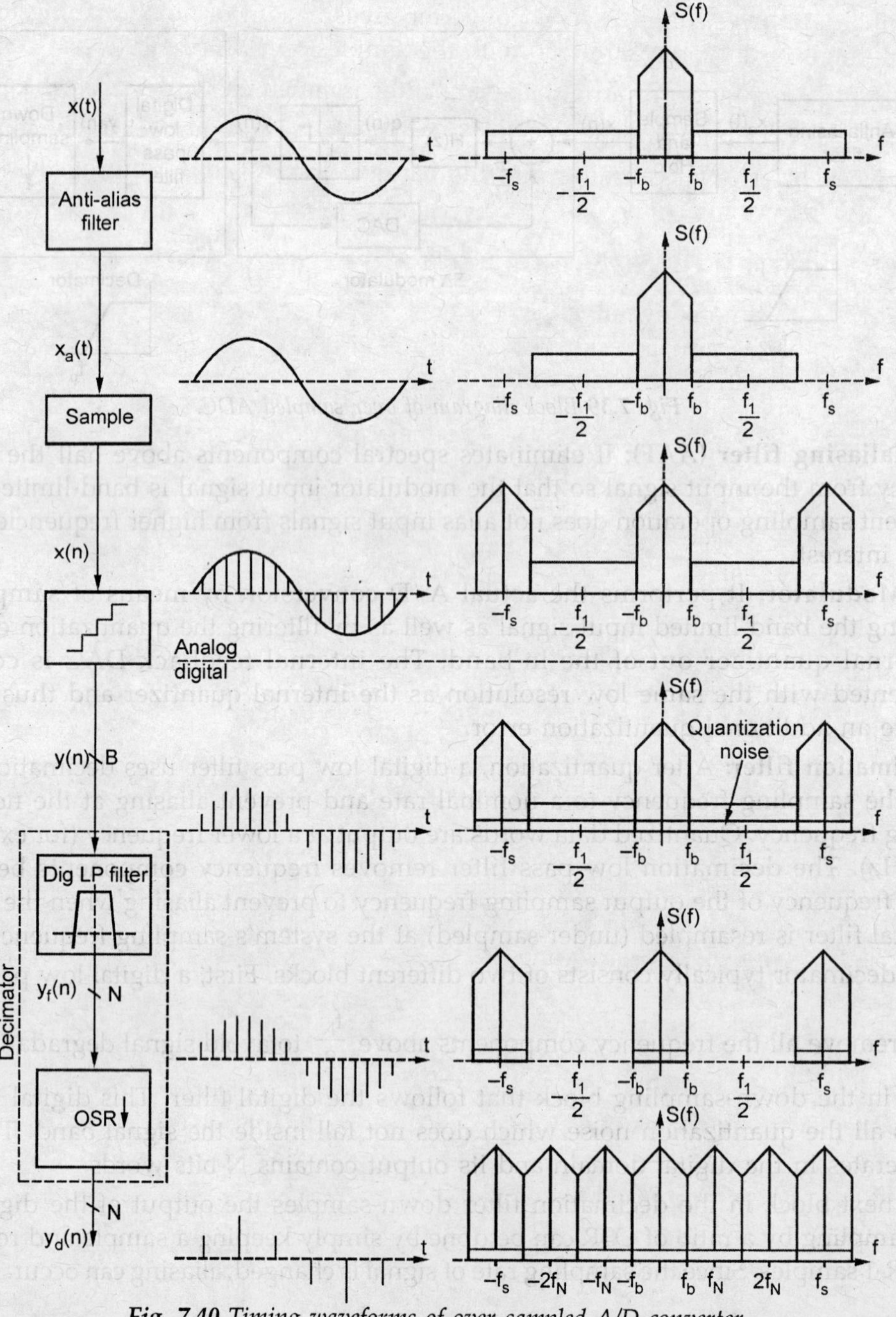

Fig. 7.40 Timing waveforms of over sampled A/D converter.

Advantages of over sampled A/D converters

- Over sampling A/D converters can achieve increased resolution compared to conventional A/D converters. Every doubling of the over sampling ratio increases the resolution by 3 dB or approximately 0.5-bit.
- They relax the requirements on analog circuits at the cost of more complex digital signal processing.
- Sigma-delta over sampled A/D converter gives benefit of noise shaping to achieve truly high resolutions (≥ 16-bits) with the simplest analog circuitry.
- The implementation of an antialiasing filter is much easier, because the transition band increases enormously for OSR ≫ 1. This is because no frequency components between $[f_B, f_s - f_B]$ can alias into the band of interest $[0, f_B]$ during the sampling procedure.

7.6 A/D CONVERTER USING VOLTAGE TO TIME CONVERSION

The voltage to time conversion can be easily obtained by using voltage to frequency converter. Infact as the time is reciprocal of the frequency, the frequency output of voltage to frequency converter can be easily converted to time using a counter, monos table multivibrator and a latch. The block diagram of voltage to time converter is shown in the Fig. 7.41.

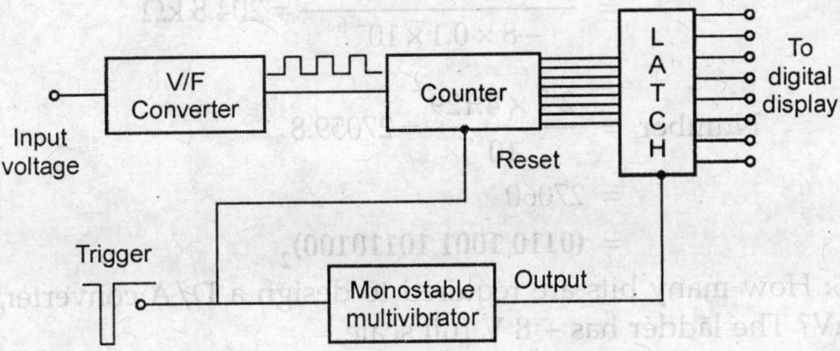

Fig. 7.41 Voltage to time converter.

A negative going pulse is used to trigger the monostable multivibrator. The same pulse is used to reset the counter. The input voltage is applied to the voltage to frequency converter. It produces the output pulses whose frequency is linearly proportional to the input.

When the trigger is applied to the monostable multivibrator its output goes high for the particular time period. At the same time the counter starts counting the pulses. After the time period of monostable multivibrator, its output goes low. This output is applied to the latch which is negative edge triggered. Hence the counter output gets latched. The number of pulses which occur during the specific time period are counted and the latched output is then displayed by the decade counting assemblies. Such a digital count when calibrated gives digital equivalent value corresponds to analog input and hence can be used as A/D converter.

EXAMPLES WITH SOLUTIONS

Example 7.14: The basic step of a 9-bit DAC is 10.3 mV. If 000000000 represents 0 V, what is the output for an input of 101101111?

Solution: The output for an input of $101101111_2 = (367_{10})$

$= 10.3 \text{ mW} \times 367$

$= \mathbf{3.7801\ V}$

Example 7.15: A dual slope ADC uses a 16-bit counter and a 4 MHz clock rate. The maximum input voltage is + 10 V. The maximum integrator output voltage should be – 8 V when the counter has cycled through 2^n counts. The capacitor used in the integrator is 0.1 μF. Find the value of the resistor R of the integrator. If the analog signal is + 4.129 V, find the corresponding binary number.

Solution: Given: $V_{oi} = -8, V_i = 10, C_1 = 0.1\ \mu F$ and

$$t = 2^N \times 1/4 \text{ MHz} = 2^{16} \times 1/4 \text{ MHz}$$

$$= 16.384 \times 10^{-3}$$

(*i*) We know that, the output voltage of the integrator is given as

$$V_{oi} = \frac{-V_i\, t}{R_1\, C_1}$$

$$\therefore \quad R_1 = \frac{-V_i\, t}{V_{oi}\, C_1}$$

$$= \frac{-10 \times 16.384 \times 10^{-3}}{-8 \times 0.1 \times 10^{-6}} = 204.8\ k\Omega$$

(*ii*) $$\text{Number} = \frac{2^{16} \times 4.129}{10} = 27059.8$$

$$= 27060$$

$$= \mathbf{(0110\ 1001\ 10110100)_2}$$

Example 7.16: How many bits are required to design a D/A converter, that can have a resolution of 5 mV? The ladder has + 8 V full scale.

Solution: $$\text{Resolution} = \frac{V_{oFS}}{2^n - 1}$$

$$\therefore \quad 2^n = \frac{V_{oFS}}{\text{Resolution}} + 1$$

$$2^n = \frac{8\text{ V}}{5\text{ mV}} + 1$$

$$\therefore \quad n = 10.644$$

$$= \mathbf{11}$$

Example 7.17: If the maximum output voltage of a 7-bit D/A converter is 25.4 V, what is the smallest change in the output as the binary count increases.

Solution: The smallest change in the output as the binary count increases (Resolution) is given as

$$= \frac{V_{oFS}}{2^n - 1} = \frac{25.4}{2^7 - 1} = \frac{25.4}{128 - 1} = \frac{25.4}{127} = \mathbf{0.2\ V}$$

Example 7.18: Determine the output voltage produced by a 4-bit DAC whose output voltage range is 0 to 10 V when the input binary number is 0110.

Solution: Given $V_{oFS} = 10\ V$

$\therefore$ Resolution $= \dfrac{10}{2^4 - 1} = 0.6667\ V$

The output voltage at 0110 = 0.6667 × 6 = **4 V**

Example 7.19: Determine the resolution of an 8-bit A/D converter for a 10 V input range.

Solution: $V_{iFS} = 10\ V$

$$R = \frac{V_{iFS}}{2^n - 1} = \frac{10}{2^8 - 1} = \frac{10}{255}$$

$= \mathbf{39.215\ mV}$

Example 7.20: Find the resolution of a 12-bit D/A converter.

Solution: Resolution $= 2^{12} = 4096.$

Example 7.21: An 8-bit D/A converter has an output voltage range of 0-2.55 V calculate its resolution.

Solution: Resolution $= \dfrac{V_{oFS}}{2^n - 1} = \dfrac{2.55}{256 - 1} = 0.01\ V$

Example 7.22: Calculate the number of comparators required for realizing a 4-bit flash A/D converter.

Solution: (2^n – 1) where n = 4 hence 24 – 1 = 15 comparators are required.

Example 7.23: A dual slope ADC has a full scale input of 2 volts. It uses an integrating time of 10 ms and integrating capacitor of 0.1 μF. The maximum magnitude of the integrator output should not exceed-S volts. Calculate the value of the integrating resistor.

Solution: $V_{oi} = \left| \dfrac{-V_i t}{R_1 C_1} \right|$

$$\therefore \quad R_1 = \frac{V_i t}{V_{oi} C_1} = \frac{2 \times 10 \times 10^{-3}}{3 \times 0.1 \times 10^{-6}}$$

$= 66.67\ k\Omega.$

The value of integrating capacitor should be greater than 66.67 K.

Example 7.24: An 12-bit D/A converter has resolution of 30 mV/LSB. Find the full scale output voltage.

Solution: Resolution $= \dfrac{V_{oFS}}{2^n - 1}$

$\therefore$ V_{oFS} = Resolution × 2^n – 1 = 30 × 10^{-3} × 2^{12} – 1

$= \mathbf{122.85\ V}$

Example 7.25: For an 8-bit successive approximation type A/D converter is driven by a 2 MHz clock. Find the conversion time.

Solution: $$T_C = T(n+1) = \frac{1}{2 \times 10^6}(8+1) = 4.5\ \mu\text{sec}$$

Example 7.26: An 8-bit DAC has a step size of 10 mV. Determine the full scale output voltage and percentage resolution. Find the output voltage for the input of 01010101.

Solution: For 8-bit DAC, step size is 10 mV

$$V_{oFS} = 10\text{ mV} \times 2^8 - 1 = 2.55\text{ V}$$

$$\%\text{ Resolution} = \frac{10\text{ mV}}{2.55\text{ V}} \times 100 = 0.392\%$$

$$(01010101)_2 = (85)_{10}$$

∴ The output voltage for the input 01010101 = 10 mV × 85 = 0.85 V.

Example 7.27: The input to a sampling gate is a sine wave. Plot its output with and without a hold capacitor.

Solution:

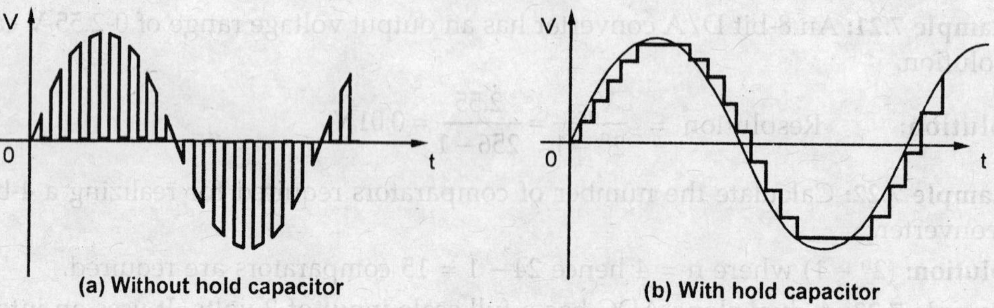

Fig. 7.42

Example 7.28: Draw the circuit of a four bit R-2R D/A converter with feedback resistance of 3R for the Op-amp. Calculate its output voltage for a digital input of 0100 if the reference voltage is 2.5 V.

Solution: Refer section 8.3.2.3.

$$V_o = -\left(\frac{2.5}{R} \times \frac{1}{2^4} \times 3R\right) \times 4 \qquad \because (0100)_2 = 4_{10}$$

$$= \mathbf{1.875\ V}$$

Example 7.29: How many comparators are required for a 4-bit flash ADC ? For an input signal in the range of 0 to + 10 V. What are the reference voltages needed ? Show how they can be generated using a 10 V reference and several 1 kohm resistors.

Solution:

$$\text{Comparator required} = 2^n - 1 = 2^4 - 1 = 15$$

See Fig. 7.43 on next page.

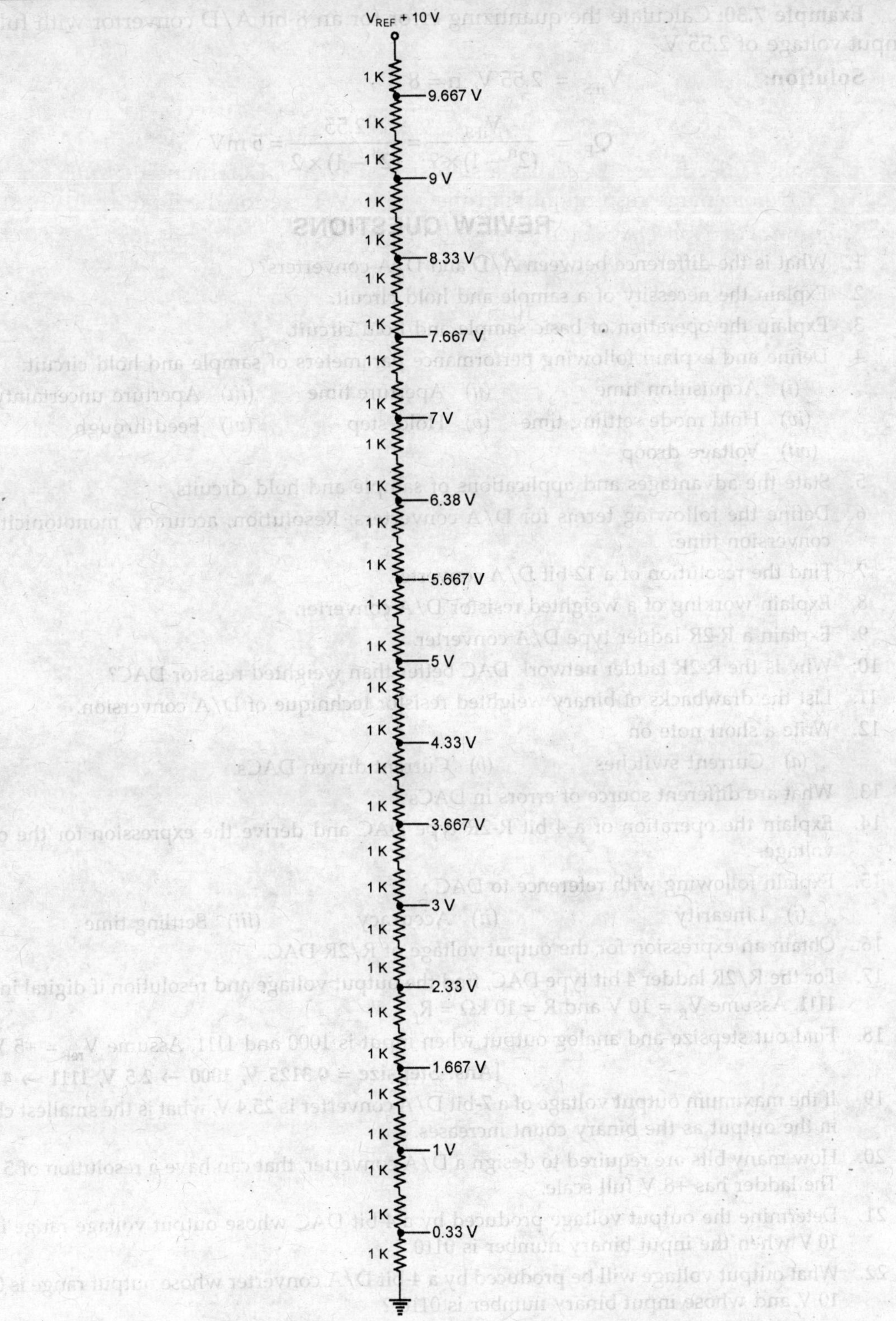

V_REF + 10 V
1 K
9.667 V
9 V
8.33 V
7.667 V
7 V
6.38 V
5.667 V
5 V
4.33 V
3.667 V
3 V
2.33 V
1.667 V
1 V
0.33 V

Fig. 7.43

Example 7.30: Calculate the quantizing error for an 8-bit A/D convertor with full scale input voltage of 2.55 V.

Solution: $V_{iFS} = 2.55\text{ V},\ n = 8$

$$Q_E = \frac{V_{iFS}}{(2^n - 1) \times 2} = \frac{2.55}{(2^8 - 1) \times 2} = 5\text{ mV}$$

REVIEW QUESTIONS

1. What is the difference between A/D and D/A converters?
2. Explain the necessity of a sample and hold circuit.
3. Explain the operation of basic sample and hold circuit.
4. Define and explain following performance parameters of sample and hold circuit:
   (*i*) Acquisition time (*ii*) Aperture time (*iii*) Aperture uncertainty
   (*iv*) Hold mode settling time (*v*) Hold step (*vi*) Feedthrough
   (*vii*) Voltage droop
5. State the advantages and applications of sample and hold circuits.
6. Define the following terms for D/A converters: Resolution, accuracy, monotonicity, and conversion time.
7. Find the resolution of a 12-bit D/A converter.
8. Explain working of a weighted resistor D/A converter.
9. Explain a R-2R ladder type D/A converter.
10. Why is the R-2R ladder network DAC better than weighted resistor DAC?
11. List the drawbacks of binary weighted resistor technique of D/A conversion.
12. Write a short note on
    (*a*) Current switches (*b*) Current driven DACs.
13. What are different source of errors in DACs ?
14. Explain the operation of a 4 bit R-2R type DAC and derive the expression for the output voltage.
15. Explain following with reference to DAC :
    (*i*) Linearity (*ii*) Accuracy (*iii*) Settling time
16. Obtain an expression for the output voltage of R/2R DAC.
17. For the R/2R ladder 4 bit type DAC, find the output voltage and resolution if digital input is 1111. Assume $V_R = 10$ V and R = 10 kΩ = R_f
18. Find out stepsize and analog output when input is 1000 and 1111. Assume $V_{ref} = +5$ V.
    [Ans. Stepsize = 0.3125 V, 1000 → 2.5 V, 1111 → 4.6875]
19. If the maximum output voltage of a 7-bit D/A converter is 25.4 V, what is the smallest change in the output as the binary count increases.
20. How many bits are required to design a D/A converter, that can have a resolution of 5 mV ? The ladder has +8 V full scale.
21. Determine the output voltage produced by a 4-bit DAC whose output voltage range is a to 10 V when the input binary number is 0110.
22. What output voltage will be produced by a 4-bit D/A converter whose output range is 0 V to 10 V and whose input binary number is 0110 ?

23. Explain the working of a weighted resistor D/A converter and state its important features.
24. List the various specifications of ADC.
25. List various A/D conversion techniques.
26. What do you mean by quantization error?
27. Explain the digital ramp A/D converter with the help of neat block diagram.
28. Using an OP AMP, draw the functional diagram of a successive approximation ADC and explain its working.
29. Explain the operation of flash A/D converter.
30. How many comparators are required to build an n bit flash type A/D converter ?
31. Which is the fastest ADC and why?
32. Explain the working of a dual slope A/D converter.
33. Explain how dual slope A/D converter provides noise rejection.
34. What is Delta modulation ?
35. Draw block diagram of delta the modulation circuit.
36. Explain ADM.
37. Explain the working of a voltage to frequency converter.
38. Explain the following two types of V/F converters:
    (*i*) Wide-sweep multitnbraior VFC
    (*ii*) Charge balancing VFC
39. Explain the working of a voltage to time converter.
40. Explain the working of a counter type A/D converter and state its important features.
41. Determine the resolution of an 8-bit A/D converter for a 10 V input range.
42. Explain the working of a high speed sample and hold circuit.
43. Write a note on over sampling A/D converter.
44. State the benefits of A/D converter.
45. Draw and explain the block diagram of over sampling A/D converter.

❀❀❀

Chapter

8

Waveform Generators

8.1 INTRODUCTION

There are many situations in which it is required to generate the signals within the system itself. Such signals have various waveforms such as square, staircase, triangular, pulse etc. The various op-amp circuits used for generating such signals are called **signal generators or waveform generators.** Waveform generators produce a time dependent signals of prescribed characteristics such as frequency, amplitude and shape. Many a times these characteristics are programmable and can be controlled externally with the help of some control mechanism.

Some waveform generator circuits using op-amp are discussed in this chapter.

Inside the Chapter

8.2 SINE WAVE GENERATORS (OSCILLATORS)

The sine wave is certainly one of the most fundamental waveforms. A variety of circuits and techniques have been developed for the generation of sine waves. The conventional sine wave oscillator circuits use phase shifting techniques that usually employ.

- Two RC tuning networks, and
- Complex amplitude limiting circuitry

Let us discuss the two commonly used oscillators for the sine wave generation.

8.3 PHASE SHIFT OSCILLATOR

RC phase shift oscillator basically consists of an amplifier and a feedback network consisting of resistors and capacitors arranged in ladder fashion. Hence such an oscillator is also called ladder type RC phase shift oscillator.

To understand the operation of this oscillator let us study RC circuit first, which is used in the feedback network of this oscillator. The Fig. 8.1 shows the basic RC circuit.

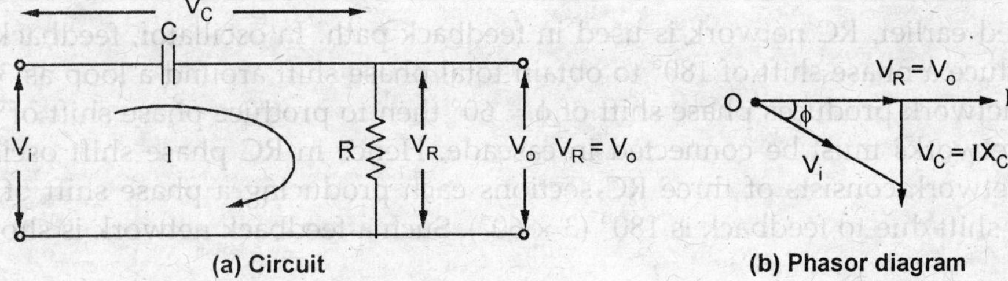

Fig. 8.1

The capacitor C and resistance R are in series. Now X_C is the capacitive reactance in ohms given by,

$$X_C = \frac{1}{2\pi f C}\ \Omega$$

The total impedance of the circuit is,

$$Z = R - jX_C = R - j\left(\frac{1}{2\pi f C}\right)\Omega = |Z| \angle -\phi^\circ\ \Omega$$

The r.m.s. value of the input voltage applied is say V_i volts. Hence the current is given by,

$$I = \frac{V_i \angle 0^\circ}{Z} = \frac{V_i \angle 0^\circ}{|Z| \angle -\phi}$$

$$\therefore \quad I = \frac{V_i}{Z} \angle +\phi\ A$$

where $$|Z| = \sqrt{R^2 + (X_C)^2}$$

and $$\phi = \tan^{-1}\left(\frac{X_C}{R}\right)$$

From expression of current it can be seen that current I leads input voltage V_i by angle ϕ. The output voltage V_o is the drop across resistance R given by,

$$V_o = V_R = IR$$

The voltage across the capacitor is,

$$V_C = I X_C$$

The drop V_R is in phase with current I while the drop V_C lags current I by 90° i.e. I leads V_C by 90°. The phasor diagram is shown in the Fig. 8.1 (*b*).

---

*Key Point: By using proper values of R and C, **the angle φ is adjusted in practice equal to 60°.***

8.3.1 RC Feedback Network

As stated earlier, RC network is used in feedback path. In oscillator, feedback network must introduce a phase shift of 180° to obtain total phase shift around a loop as 360°. Thus if one RC network produces phase shift of ϕ = 60° then to produce phase shift of 180° such three RC networks must be connected in cascade. Hence in RC phase shift oscillator, the feedback network consists of three RC sections each producing a phase shift of 60°, thus total phase shift due to feedback is 180° (3 × 60°). Such a feedback network is shown in the Fig. 8.2.

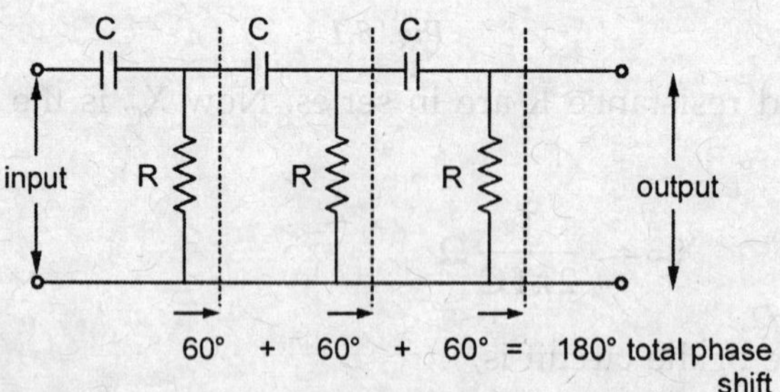

Fig. 8.2 Feedback network in RC phase shift oscillator.

The network is also called the **ladder network.** All the resistance values and all the capacitance values are same, so that for a particular frequency, each section of R and C produces a phase shift of 60°.

8.3.2 R-C Phase Shift Oscillator using Op-amp

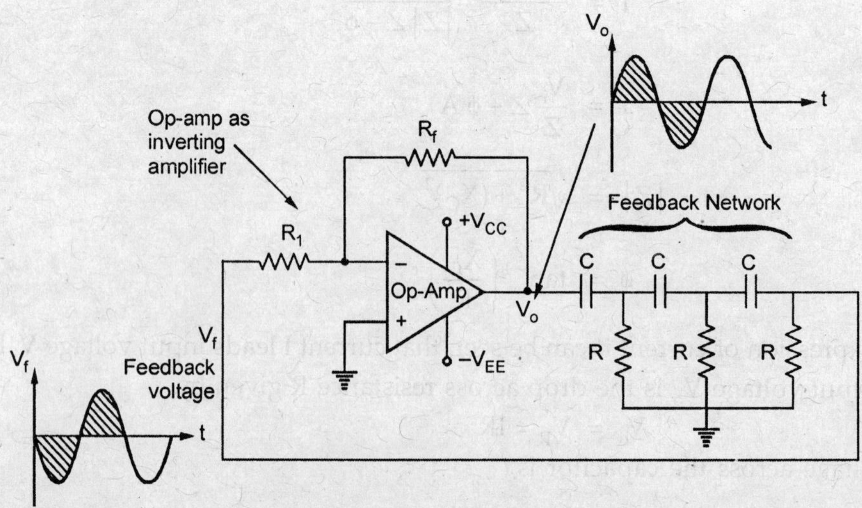

Fig. 8.3 R-C phase shift oscillator using op-amp.

R-C phase shift oscillator using op-amp uses op-amp in inverting amplifier mode. Thus it introduces the phase shift of 180° between input and output. The feedback network consists of 3 RC sections each producing 60° phase shift. Such a RC phase shift oscillator using op-amp is shown in the Fig. 8.3.

The output of amplifier is given to feedback network. The output of feedback network drives the amplifier. The total phase shift around a loop is 180° of amplifier and 180° due to 3 RC section, thus 360°. This satisfies the required condition for positive feedback and circuit works as an oscillator.

8.3.3 Derivation of Frequency of Oscillations

Let us find the transfer function of the RC feedback network.

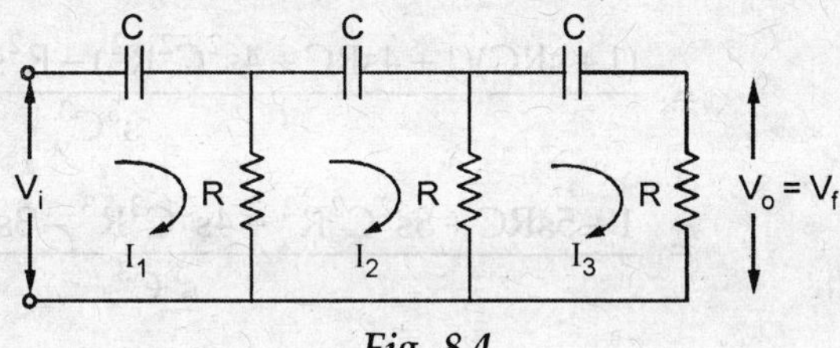

Fig. 8.4

Applying KVL to various loops we get,

$$I_1\left(R+\frac{1}{j\omega C}\right)-I_2R = V_i \qquad ...(1)$$

$$-I_1R+I_2\left(2R+\frac{1}{j\omega C}\right)-I_3R = 0 \qquad ...(2)$$

$$0-I_2R+I_3\left(2R+\frac{1}{j\omega C}\right) = 0 \qquad ...(3)$$

Replacing jω by s and writing the equations in the matrix form,

$$\begin{bmatrix} R+\frac{1}{sC} & -R & 0 \\ -R & 2R+\frac{1}{sC} & -R \\ 0 & -R & 2R+\frac{1}{sC} \end{bmatrix}\begin{bmatrix} I_1 \\ I_2 \\ I_3 \end{bmatrix} = \begin{bmatrix} V_i \\ 0 \\ 0 \end{bmatrix} \qquad ...(4)$$

Using the Cramer's rule to obtain I_3,

$$D = \begin{bmatrix} \frac{1+sRC}{sC} & -R & 0 \\ -R & \frac{1+2sRC}{sC} & -R \\ 0 & -R & \frac{1+2sRC}{sC} \end{bmatrix}$$

$$= \frac{(1+sRC)(1+2sRC)^2}{s^3C^3} - \frac{R^2(1+2sRC)}{sC} - \frac{R^2(1+sRC)}{sC}$$

$$= \frac{(1+sRC)(1+4sRC+4s^2C^2R^2) - R^2s^2C^2[1+2sRC+1+sRC]}{s^3C^3}$$

$$= \frac{1+5sRC+8s^2C^2R^2+4s^3C^3R^3-3s^3R^3C^3-2R^2s^2C^2}{s^3C^3}$$

$$= \frac{1+5sRC+6s^2C^2R^2+s^3C^3R^3}{s^3C^3} \quad \text{...(5)}$$

$$D_3 = \begin{bmatrix} \frac{1+sRC}{sC} & -R & V_i \\ -R & \frac{1+2sRC}{sC} & 0 \\ 0 & -R & 0 \end{bmatrix}$$

$$= V_i R^2 \quad \text{...(6)}$$

$$\therefore \quad I_3 = \frac{D_3}{D} = \frac{V_i R^2 s^3 C^3}{1+5sRC+6s^2C^2R^2+s^3C^3R^3}$$

Now $$V_o = V_f = I_3 R = \frac{V_i R^3 s^3 C^3}{1+5sRC+6s^2C^2R^2+s^3C^3R^3} \quad \text{...(7)}$$

$$\therefore \quad \beta = \frac{V_o}{V_i}$$

$$= \frac{R^3s^3C^3}{1+5sCR+6s^2C^2R^2+s^3C^3R^3}$$

Replacing s by jw, s^2 by $-\omega^2$, s^2 by $-j\omega^3$

$$\therefore \quad \beta = \frac{-j\omega^3 R^3C^3}{1+5j\omega CR - 6\omega^2C^2R^2 - j\omega^3C^3R^3}$$

$$\alpha = \frac{1}{\omega RC}$$

$$\therefore \quad \beta = \frac{1}{1+6j\alpha+5\alpha^2 - j\alpha^3}$$

$$\therefore \quad \beta = \frac{1}{(1-5\alpha^2)+j\alpha(6-\alpha^2)} \quad ...(9)$$

To have phase shift of 180°, the imaginary part in the denominator must be zero.

$$\therefore \quad \alpha(6-\alpha)^2 = 0$$

$$\therefore \quad \alpha^2 = 6 \quad \text{(neglecting zero value)}$$

$$\therefore \quad \alpha = \sqrt{6}$$

$$\therefore \quad \frac{1}{\omega RC} = \sqrt{6}$$

$$\therefore \quad \omega = \frac{1}{RC\sqrt{6}}$$

$$\therefore \quad \boxed{f = \frac{1}{2\pi RC\sqrt{6}}} \quad ...(10)$$

This is the frequency with which circuit oscillates.

At this frequency,

$$\beta = \frac{1}{1-5\times(\sqrt{6})^2} = -\frac{1}{29}$$

Negative sign indicates phase shift to 180°

$$\therefore \quad \boxed{|\beta| = \frac{1}{29}} \quad ...(11)$$

Now to have the oscillations, $|A\beta| \geq 1$

$$\therefore \quad |A||\beta| \geq 1$$

$$|A| \geq \frac{1}{|\beta|} \geq \frac{1}{\left(\frac{1}{29}\right)}$$

$$\boxed{|A| \geq 29} \quad ...(12)$$

Key Point: *For the oscillations to occur, the gain of the op-amp must be equal to or greater than 29, which can be adjusted using the resistances R_f and R_i.*

8.3.4 Advantages

The advantages of R-C phase shift oscillator are,

1. The circuit is simple to design.
2. Can produce output over audio frequency range.
3. Produces sinusoidal output waveform.
4. It is a fixed frequency oscillator.

8.3.5 Disadvantages

By changing the values of R and C, the frequency of the oscillator can be changed. But the values of R and C of all three sections must be changed simultaneously to satisfy the oscillating conditions. But this is practically impossible. Hence the phase shift oscillator is considered as a fixed frequency oscillator, for all practical purposes.

And the frequency stability is poor due to the changes in the values of various components, due to effect of temperature, aging etc.

8.3.6 Phase Shift Oscillator Design

Practically the resistance R_f of the inverting amplifier is designed, by making current through it, much larger than input bias current of the op-amp.

Let the current through R_f is I_1 then

$$I_1 = 100\, I_b\,(\text{max}) \qquad ...(13)$$

Without amplitude stabilization, **the output of the oscillator oscillates between the levels ± V sat.**

$$\therefore \quad V_o = +V_o(\text{sat})$$

$$\therefore \quad R_f = \frac{V_o(\text{sat})}{I_1} \qquad ...(14)$$

+ V_o (sat) can be assumed 1 V less than + V_{CC}.

Now $\quad A_{CL} \geq 29$

$$\therefore \quad \frac{R_f}{R_1} \geq 29 \qquad ...(15)$$

Design the value of R_1 from the gain requirement.

To prevent the loading of the amplifier because of RC networks, it is necessary that $R_1 \geq 10$ R.

$$\therefore \quad R = \frac{R_1}{10} \qquad ...(16)$$

Now $$C = \frac{1}{2\pi\sqrt{6}\,fR}$$ gives required value of capacitor

Note that for output frequency less than 1 kHz, op-amp 741 is used but for higher frequencies LM 318 or LF 351 op-amps are recommended due to their higher slew rates.

Example 8.1: Design the phase shift oscillator to have output frequency of 500 Hz. Use ± 12 V supply.

Solution: As f is less than 1 kHz, Use op-amp 741 with I_b (max) = 50 nA.

$$\therefore \quad I_1 = 100\, I_b(\text{max}) = 5\ \mu A$$

$$\therefore \quad R_f = \frac{V_o(\text{sat})}{I_1} \qquad V_o(\text{sat}) = 12 - 1 = 11\ V$$

$$= \frac{11}{50 \times 10^{-6}}$$

$$= \mathbf{2.2\ m\Omega} \text{ (standard value)}$$

$$A_{CL} = \frac{R_f}{R_1} \geq 29$$

$$\therefore \quad R_1 = \frac{R_f}{29}$$

$$= \frac{220 \times 10^3}{29}$$

$$= \mathbf{75.86\ k\Omega}$$

Use standard value of 75.86 kΩ

$$R = \frac{R_1}{10}$$

$$= \frac{75.8 \times 10^3}{10}$$

$$= \mathbf{7.5\ k\Omega} \text{ (standard value)}$$

$$\therefore \quad C = \frac{1}{2\pi\sqrt{6}\,fR}$$

$$= \frac{1}{2\pi\sqrt{6} \times 500 \times 7500}$$

$$= 0.017\ \mu F$$

Use standard value of 0.017 μF

The designed circuit is shown in the Fig. 8.5.

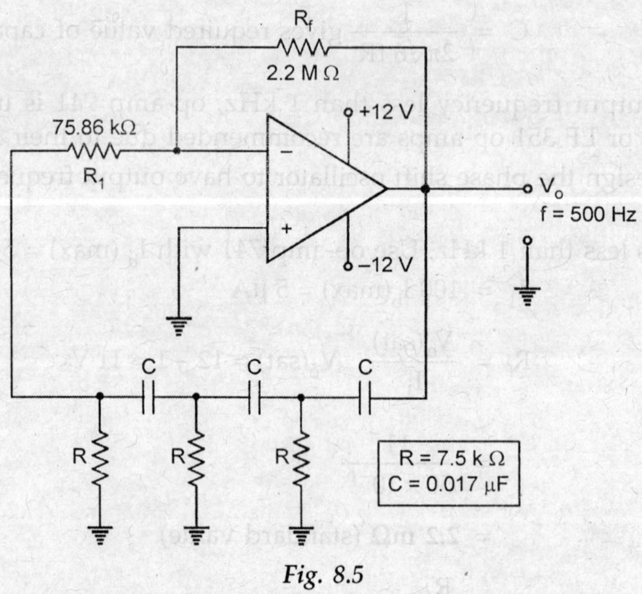

Fig. 8.5

8.4 WIEN BRIDGE OSCILLATOR

Generally in an oscillator, amplifier stage introduces 180° phase shift and feedback network introduces additional 180° phase shift, to obtain a phase shift of 360° (2π radians) around a loop. This is required condition for any oscillator. But Wien bridge oscillator uses a noninverting amplifier and hence does not provide any phase shift during amplifier stage. As total phase shift required is 0° or $2\ \pi t$ radians, in Wien bridge type no phase shift is necessary through feedback. Thus the total phase shift around a loop is 0°. Let us study the basic version of the Wien bridge oscillator and its analysis.

A basic Wien bridge used in this oscillator and an amplifier stage is shown in the Fig. 8.6.

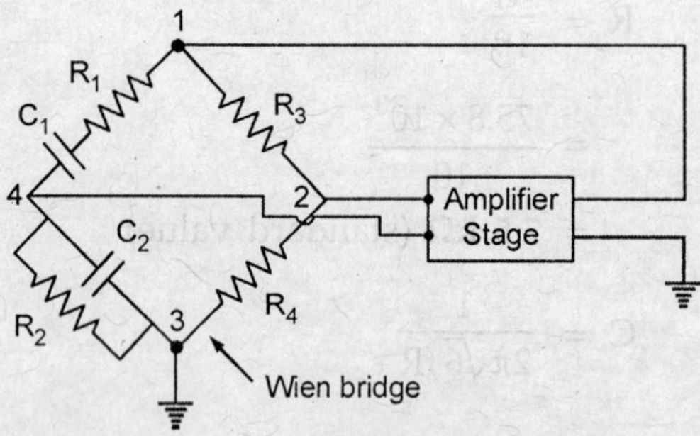

Fig. 8.6 Basic circuit of Wien bridge oscillator.

The output of the amplifier is applied between the terminals 1 and 3, which is the input to the feedback network. While the amplifier input is supplied from the diagonal terminals 2 and 4, which is the output from the feedback network. Thus amplifier supplied its own input through the Wien bridge as a feedback network.

The two arms of the bridge, namely R_1, C_1 in series and R_2, C_2 in parallel are called **frequency sensitive arms**. This is because the components of these two arms decide the frequency of the oscillator. Let us find out the gain of the feedback network. As seen earlier input V_{in} to the feedback network is between is 1 and 3 while output V_f of the feedback network is between 2 and 4. This is shown in Fig. 8.7. Such a feedback network is called **lead-lag network**. This is because at very low frequencies it acts like a lead while at very high frequencies it acts like lag network.

Fig. 8.7 Feedback network of Wien bridge oscillator.

Now from the Fig. 8.7, as shown,

$$Z_1 = R_1 + \frac{1}{j\omega C_1} = \frac{1 + j\omega R_1 C_1}{j\omega C_1}$$

$$Z_2 = R_2 || \frac{1}{j\omega C_2} = \frac{R_2 \times \frac{1}{j\omega C_2}}{R_2 + \frac{1}{j\omega C_2}}$$

$$= \frac{R_2}{1 + j\omega R_2 C_2} \quad \ldots(1)$$

Replacing jw = s,

$$Z_1 = \frac{1 + sR_1 C_1}{sC_1}$$

and

$$Z_2 = \frac{R_2}{1 + sR_2 C_2}$$

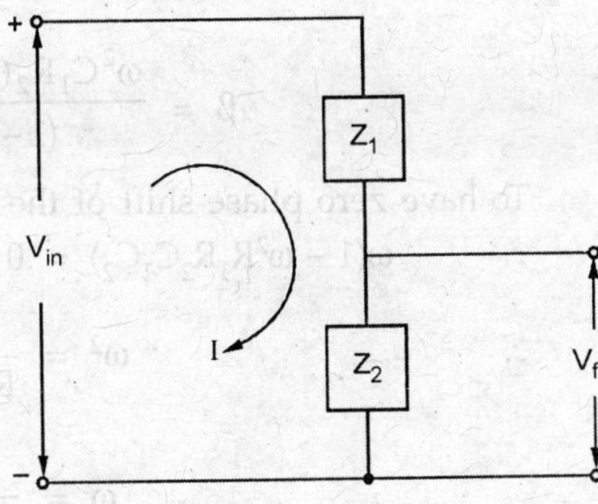

Fig. 8.8 Simplified circuit

$$I = \frac{V_{in}}{Z_1 + Z_2}$$

and

$$V_f = I\, Z_2$$

$$\therefore \quad V_f = \frac{V_{in}\, Z_2}{Z_1 + Z_2}$$

$$\therefore \quad \beta = \frac{V_f}{V_{in}} = \frac{Z_2}{Z_1 + Z_2} \quad \ldots(2)$$

Substituting the values of Z_1 and Z_2,

$$\beta = \frac{\left[\dfrac{R_2}{1+sR_2C_2}\right]}{\left[\dfrac{1+sR_1C_1}{sC_1}\right]+\left[\dfrac{R_2}{1+sR_2C_2}\right]}$$

$$\beta = \frac{sC_1R_2}{(1+sR_1C_1)(1+sR_2C_2)+sC_1R_2}$$

$$= \frac{sC_1R_2}{1+s(R_1C_1+R_2C_2)+s^2R_1R_2C_1C_2+sC_1R_2}$$

$$= \frac{sC_1R_2}{1+s(R_1C_1+R_2C_2+C_1R_2)+s^2R_1R_2C_1C_2}$$

Replacing s by $j\omega$, $s^2 = -\omega^2$

$$\therefore \quad \beta = \frac{j\omega C_1R_2}{(1-\omega^2R_1R_2C_1C_2)+j\omega(R_1C_1+R_2C_2+C_1R_2)} \quad \text{...(3)}$$

Rationalising the expression,

$$\beta = \frac{j\omega C_1R_2[(1-\omega^2R_1R_2C_1C_2)-j\omega(R_1C_1+R_2C_2+C_1R_2)]}{(1-\omega^2R_1R_2C_1C_2)^2+\omega^2(R_1C_1+R_2C_2+C_1R_2)^2}$$

$$\beta = \frac{\omega^2C_1R_2(R_1C_1+R_2C_2+C_1R_2)+j\omega C_1R_2(1-\omega^2R_1R_2C_1C_2)}{(1-\omega^2R_1R_2C_1C_2)^2+\omega^2(R_1C_1+R_2C_2+C_1R_2)^2} \quad \text{...(4)}$$

To have zero phase shift of the feedback network, its imaginary part must be zero.

$$\therefore \quad \omega(1-\omega^2R_1R_2C_1C_2) = 0$$

$$\therefore \quad \omega^2 = \frac{1}{R_1R_2C_1C_2} \text{ neglecting zero value.}$$

$$\omega = \frac{1}{\sqrt{R_1R_2C_1C_2}}$$

$$\boxed{f = \frac{1}{2\pi\sqrt{R_1R_2C_1C_2}}} \quad \text{...(5)}$$

This is the frequency of the oscillator and it shows that the components of the frequency sensitive arms are the deciding factors, from the frequency.

In practice, $R_1 = R_2 = R$ and $C_1 = C_2 = C$ are selected.

$$\therefore \qquad f = \frac{1}{2\pi\sqrt{R^2 C^2}}$$

$$\boxed{f = \frac{1}{2\pi RC}}$$

At $R_1 = R_2 = R$ and $C_1 = C_2 = C$, the gain of the feedback network becomes,

$$\beta = \frac{\omega^2 RC\,(3RC) + j\omega RC(1-\omega^2 R^2 C^2)}{(1-\omega^2 R^2 C^2) + \omega^2 (3RC)^2}$$

Substituting $\quad f = \frac{1}{2\pi RC} \quad$ i.e. $\quad \omega = \frac{1}{RC}$,

We get the magnitude of the feedback network at the resonating frequency of the oscillator as,

$$\beta = \frac{3}{0 + \frac{1}{R^2C^2} \times (3RC)^2} = \frac{3}{9}$$

$$\therefore \qquad \beta = \frac{1}{3} \qquad \text{...(7)}$$

The positive sign of β indicates that the phase shift by the feedback network is $0°$. Now to satisfy the Barkhausen criterion for the sustained oscillations, we can write,

$$|A\beta| \geq 1$$

$$\therefore \qquad |A| \geq \frac{1}{|\beta|} \geq \frac{1}{\left(\frac{1}{3}\right)}$$

$$\therefore \qquad \boxed{|A| \geq 3}$$

This is the required gain of the amplifier stage, without any phase shift.

If $R_1 \neq R_2$ and $C_1 \neq C_2$ then

$$\boxed{f = \frac{1}{2\pi\sqrt{R_1 R_2 C_1 C_2}}}$$

Substituting in the equation (7) we get,

$$\beta = \frac{C_1 R_2}{(R_1 C_1 + R_2 C_2 + C_1 R_1)}$$

$$|A\beta| \geq 1$$

$$\therefore \quad \boxed{A \geq \frac{R_1C_1 + R_2C_2 + C_1R_2}{C_1R_2}} \quad ...(8)$$

Another important **advantage** of the Wien bridge oscillator is that by varying the two capacitor values simultaneously, by mounting them on the common shaft, different frequency ranges can be provided.

8.4.1 Wien Bridge Oscillator using Op-amp

The Fig. 8.9 shows the Wien bridge oscillator using an op-amp.

The resistance R and capacitor C are the components of frequency sensitive arms of the bridge. The resistance R_f and R_1 form the part of the feedback path. The gain of noninverting op-amp can be adjusted using the resistance R_f and R_1. The gain of op-amp is,

$$A = 1 + \frac{R_f}{R_1}$$

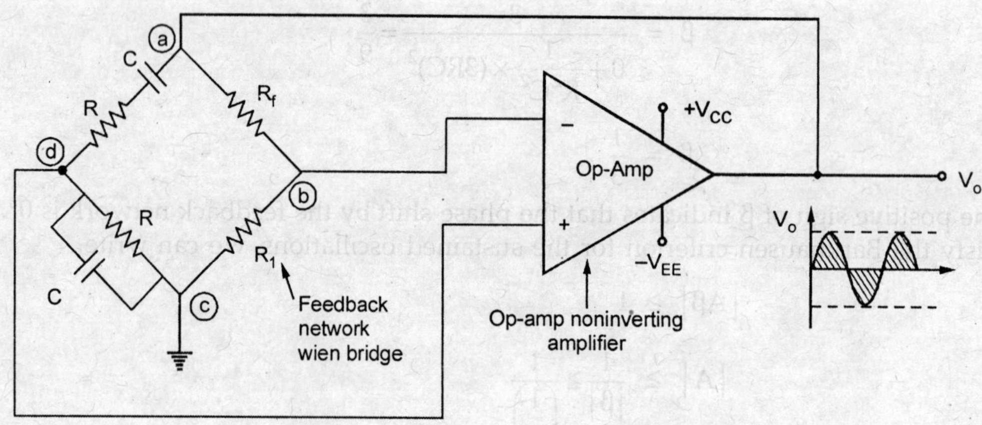

Fig. 8.9

To satisfy Barkhausen criterion that $A\beta \geq 1$ it is necessary that the gain of the noninverting op-amp amplifier must be minimum 3.

$$\therefore \quad |A| \geq 3 \quad \text{i.e.} \quad 1 + \frac{R_f}{R_1} \geq 3$$

$$\therefore \quad \frac{R_f}{R_1} \geq 2$$

Thus ratio of R_f and R_1 must be greater than or equal to 2.

The frequency of oscillations is given by,

$$\therefore \quad \boxed{f = \frac{1}{2\pi RC} \text{ Hz}}$$

The feedback is given to the noninverting terminal of op-amp which ensures zero phase shift. It is used popularly in laboratory signal generators.

If in a Wien bridge feedback network, two resistances are not equal i.e. they are R_1 and R_2 while two capacitors are not equal i.e. they are C_1 and C_2 then the frequency of oscillations is given by,

$$f = \frac{1}{2\pi\sqrt{R_1R_2C_1C_2}}$$

With $R_1 = R_2 = R$ and $C_1 = C_2 = C$ we get it as $1/2\pi$ RC as stated earlier.

The simplified circuit diagram of the Wien bridge oscillator is shown in the Fig. 8.10.

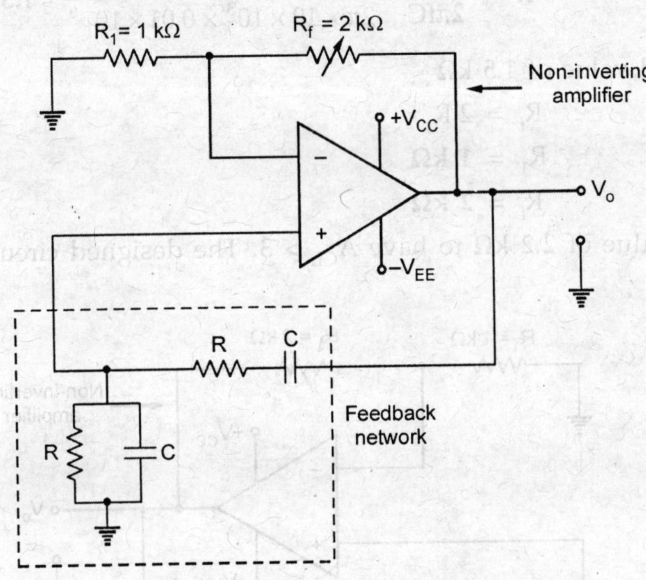

Fig. 8.10 Simplified diagram of Wien bridge oscillator.

8.4.2 Advantages

The various advantages of Wien bridge oscillator are,

1. By varying the two capacitor values simultaneously, by mounting them on the common shaft, different frequency ranges can be obtained.
2. The perfect sine wave output is possible.
3. It is useful audio frequency range i.e. 20 Hz to 100 kHz.

8.4.3 Wien Bridge Oscillator Design

Select the capacitor value much larger than the stray capacitance, about 0.01 to 0.05 μF. From the equation of frequency, obtain the value R.

$$R = \frac{1}{2\pi fC}$$

Then for noninverting amplifier,

$$R_f = 2 R_1$$

Choose R_1 and design the value of R_f. Keep R_f variable for fine adjustments.

Example 8.2: Design the Wien bridge oscillator circuit to have output frequency of 10 kHz.

Solution: Choose $C = 0.01\ \mu F$

$$\therefore \quad R = \frac{1}{2\pi fC} = \frac{1}{2\pi \times 10 \times 10^3 \times 0.01 \times 10^{-6}} = \mathbf{1.5915\ k\Omega}$$

Choose standard value of 1.5 kΩ

Now $R_f = 2 R_1$

Choose $R_1 = \mathbf{1\ k\Omega}$

$\therefore \quad R_f = \mathbf{2\ k\Omega}$

Use standard value of 2.2 kΩ to have $A_{CL} > 3$. The designed circuit is shown in the Fig. 8.11.

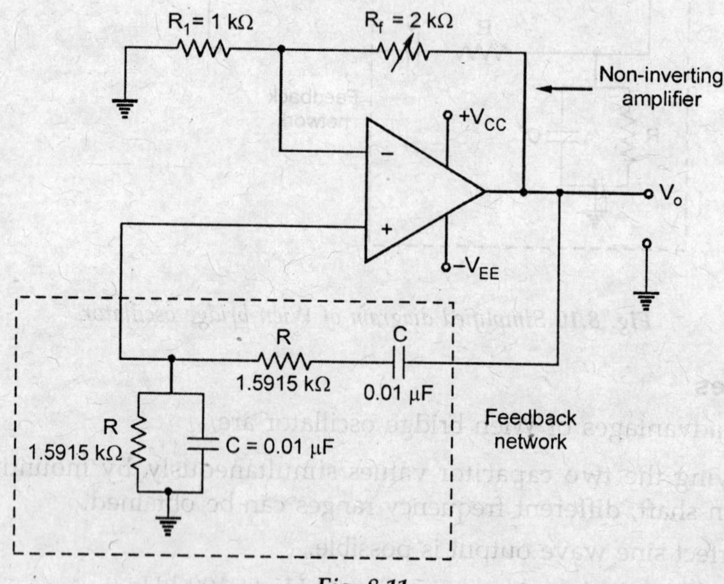

Fig. 8.11

8.5 COMPARISON OF RC PHASE SHIFT AND WIEN BRIDGE OSCILLATORS

The similarities and the differences between the two oscillators are given in the Table 8.1.

Table 8.1

Sr. No.	RC Phase Shift Oscillator	Wien Bridge Oscillator
1.	It is a phase shift oscillator used for low frequency range.	It is also a phase shift oscillator used for low frequency range.
2.	The feedback network is RC network with three RC sections.	The feedback network is lead-lag network which is called Wien bridge circuit.
3.	The feedback network introduces 180° phase shift.	The feedback network does not introduce any phase shift.
4.	Op-amp is used in an inverting mode.	Op-amp is used in non-inverting mode.
5.	Op-amp circuit introduces 180° phase shift.	Op-amp circuit does not introduce any phase shift.
6.	The frequency of oscillations is, $f = \frac{1}{2\pi RC\sqrt{6}}$	The frequency of oscillations is, $f = \frac{1}{2\pi RC}$
7.	The amplifier gain condition is, $\lvert A \rvert \geq 29$	The amplifier gain condition is, $\lvert A \rvert \geq 3$
8.	The frequency variation is difficult.	Mounting the two capacitors on common shaft and varying their values, frequency can be varied.

8.6 TYPES OF MULTIVIBRATORS

There are three types of multivibrator circuits in use, namely,

(*a*) Bistable Multivibrator.

(*b*) Monostable Multivibrator

(*c*) Astable Multivibrator

Let us see the basic concept behind the operation of these three types of. multivibrators.

8.6.1 Bistable Multivibrator

As the name suggests, the bistable multivibrator has two stable states. The multivibrator can exist indefinitely in either of the two stable states. It requires an external trigger pulse to change from one stable state to another. The circuit remains in one stable state unless an external trigger pulse is applied.

Key Point:** The bistable multivibrator is also known by variety of other names as **Eccles-Jordan circuit, trigger circuit, scale-of-2 toggle circuit, flip-flop and binary.

The bistable multivibrator is used for the performance of many digital operations such as counting and storing of the binary information. This multi vibrator circuit also finds an application in the generation and processing of pulse-type waveforms.

8.6.2 Monostable Multivibrator

The monostable multivibrator has only one **stable state**. The other state is unstable referred as quasi-stable state. When an external trigger pulse is applied to the circuit, the circuit goes into the quasi-stable state from its normal stable state. After some time interval, the circuit automatically returns to its stable state. The circuit does not require any external pulse to change from quasi-stable to stable state. The time interval for which the circuit remains in the quasi-stable state is determined by the circuit components and can be designed as per the requirement.

Key Point: The monos table multivibrator is also known by variety of other names such as ***one-shot, single-shot, a single cycle, a single swing, a single step multivibrator*** *or a* ***univibrator****. It is also called* ***gating circuit*** *or* ***delay circuit****.*

The circuit is used to generate the rectangular waveform and hence can be used to gate other circuits hence called **gating circuit**. The time between the transition from quasi-stable state to stable state can be predetermined and hence it can be used to introduce time delays with the help of fast transition. Due to this application, it is also called delay circuit.

8.6.3 Astable Multivibrator

The astable multivibrator has **both the states as quasi-stable states**. None of the states is stable state. Due to this, the multivibrator automatically makes the successive transitions from one quasi-stable state to other, without any external triggering pulse. The rate of transition from one quasi-stable state to other is determined by the circuit components.

Key Point: As this multivibrator does not require any external pulse for the transition, is called free running multivibrator.

The astable multivibrator is nothing but an oscillator. It is used as the generator of square waves. As it requires no triggering it is used as a basic source of fast waveforms.

Both monostable and astable multivibrators find extensive application in pulse circuitry.

8.7 MONOSTABLE MULTIVIBRATOR USING OP-AMP

The monostable multivibrator circuit using op-amp is shown in the Fig. 8.12.

The diode D_1 is clamping diode connected across C. The diode clamps the capacitor voltage to 0.7 V when the output is at $+V_{sat}$. A narrow negative triggering pulse V_t is applied to the noninverting input terminal, through diode D_2.

Let us see the operation of the circuit.

To understand the operation of the circuit, let us assume that the output V_o is at $+V_{sat}$ i.e. in its stable state. The diode D_1 conducts and the voltage across the capacitor C i.e. V_C gets clamped to 0.7 V. The voltage at the non-inverting input terminal is controlled by potentiometric divider of R_1 R_2 to β V 0 i.e. $+\beta V_{sat}$ in the stable state.

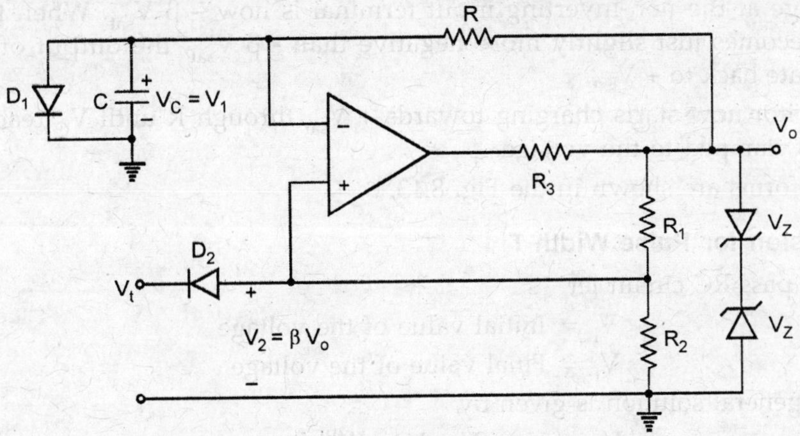

Fig. 8.12 Op-amp monostable multivibrator

Now if V_t, a negative trigger of amplitude V_t is applied to the non-inverting terminal, so that the effective voltage at this terminal is less than 0.7 V ($+\beta V_{sat} + (-V_t)$) then the output of the op-amp changes its state from $+V_{sat}$ to $-V_{sat}$.

The diode is now reverse biased and the capacitor starts charging exponentially to $-V_{sat}$ through the resistance R. The time constant of this charging is $\tau = RC$.

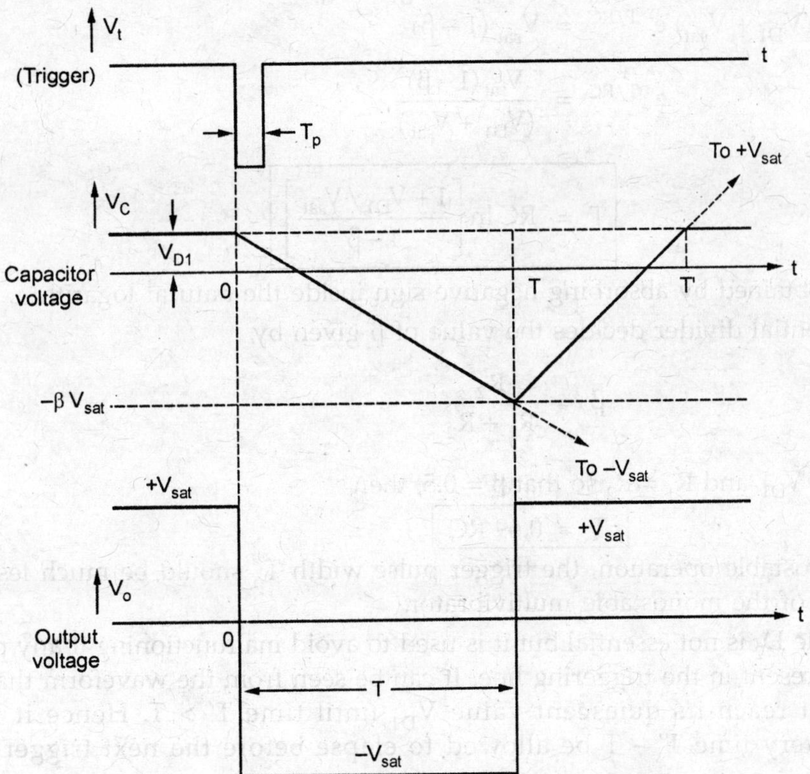

Fig. 8.13 Waveforms of monostable multivibrator

The voltage at the non-inverting input terminal is now $-\beta V_{sat}$. When the capacitor voltage V_C becomes just slightly more negative than $-\beta V_{sat}$, the output of the op-amp changes its state back to $+V_{sat}$.

The capacitor now starts charging towards $+V_{sat}$ through R until V_C reaches 0.7 V as capacitor gets clamped to the voltage.

The waveforms are shown in the Fig. 8.13.

8.7.1 Expression for Pulse Width T

For a low pass RC circuit let,

$$V_i = \text{Initial value of the voltage}$$

$$V_f = \text{Final value of the voltage}$$

Then the general solution is given by,

$$V_o = V_f + (V_i - V_f)\, e^{-t/RC} \qquad \text{...(1)}$$

Now for the monostable multivibrator discussed above, the values of V_f and V_i are,

$$V_f = -V_{sat} \text{ and } V_i = V_{D1} \quad \text{(diode forward voltage)}$$

while

$$V_o = \text{output} = \text{capacitor voltage} = V_C$$

$$\therefore \quad V_C = -V_{sat} + (V_{D1} - [-V_{sat}])\, e^{-t/RC} \qquad \text{...(2)}$$

at t = T,

$$V_C = -\beta V_{sat} \qquad \text{...(3)}$$

$$\therefore \quad -\beta V_{sat} = -V_{sat} + (V_{D1} + V_{sat})\, e^{-T/RC} \qquad \text{...(4)}$$

$$\therefore \quad (V_{D1} + V_{sat})\, e^{-T/RC} = V_{sat}(1-\beta)$$

$$\therefore \quad e^{-T/RC} = \frac{V_{sat}(1-\beta)}{(V_{D1} + V_{sat})}$$

$$\boxed{T = RC \ln\left[\frac{1 + V_{D1}/V_{sat}}{1-\beta}\right]} \qquad \text{...(5)}$$

This is obtained by absorbing negative sign inside the natural logarithm.

The potential divider decides the value of β given by,

$$\beta = \frac{R_2}{R_1 + R_2} \qquad \text{...(6)}$$

If $V_{sat} \gg V_{D1}$ and $R_1 = R_2$ so that $\beta = 0.5$, then

$$\boxed{T = 0.69\ RC} \qquad \text{...(7)}$$

For monostable operation, the trigger pulse width T_p should be much less than T, the pulse width of the monostable multivibrator.

The diode D_2 is not essential but it is used to avoid malfunctioning if any positive noise spikes are present in the triggering line. It can be seen from the waveform that the voltage V_C does not reach its quiescent value V_{D1} until time T′ > T. Hence it is necessary that a recovery time T′ – T be allowed to elapse before the next triggering signal is applied.

8.8 ASTABLE MULTIVIBRATOR USING OP-AMP

In this section we are going to study astable multivibrator operation using op-amp. Fig. 8.14 shows astable multivibrator circuit using op-amp. It looks like a comparator with Hysteresis (Schmitt trigger), except that the input voltage is replaced by a capacitor. The circuit has a time dependent elements such as resistance and capacitor to set the frequency of oscillation.

As shown in the Fig. 8.14 the comparator and positive feedback resistors R_1 and R_2 form an inverting Schmitt trigger.

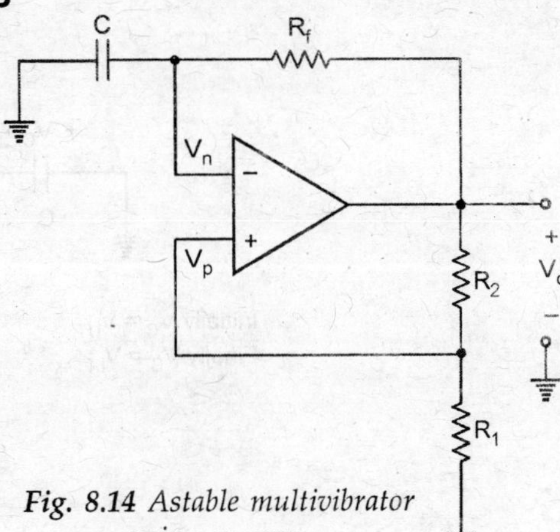

Fig. 8.14 Astable multivibrator using op-amp.

When V_o is at $+V_{sat}$, the feedback voltage is called the upper threshold voltage V_{UT} and is given as

$$V_{UT} = \frac{R_1.(+V_{sat})}{R_1 + R_2} \quad ...(1)$$

When V_o is at $-V_{sat}$, the feedback voltage is called the lower-threshold voltage V_{LT} and is given as

$$V_{LT} = \frac{R_1.(-V_{sat})}{R_1 + R_2} \quad ...(2)$$

When power is turn ON, V_o automatically swings either to $+V_{sat}$ or to $-V_{sat}$ since these are the only stable states allowed by the Schmitt trigger. Assume it swings to $+V_{sat}$. With $V_o = +V_{sat}$ we have $V_p = V_{UT}$ and capacitor starts charging towards $+V_{sat}$ through the feedback path provided by the resistor R_f to the inverting (–) input. This is illustrated in Fig. 8.15 (*a*). As long as the capacitor voltage V_C is less than V_{UT}, the output voltage remains at $+V_{sat}$.

Fig. 8.15 (a) When $V_o = +V_{sat}$, capacitor charges towards.

As soon as V_C charges to a value slightly greater than V_{UT}, the (–) input goes positive with respect to the (+) input. This switches the output voltage from $+V_{sat}$ to $-V_{sat}$ and we have $V_p = V_{LT}$, which is negative with respect to ground. As V_o switches to $-V_{sat}$, capacitor starts discharging via R_f, as shown in the Fig. 8.15 (*b*).

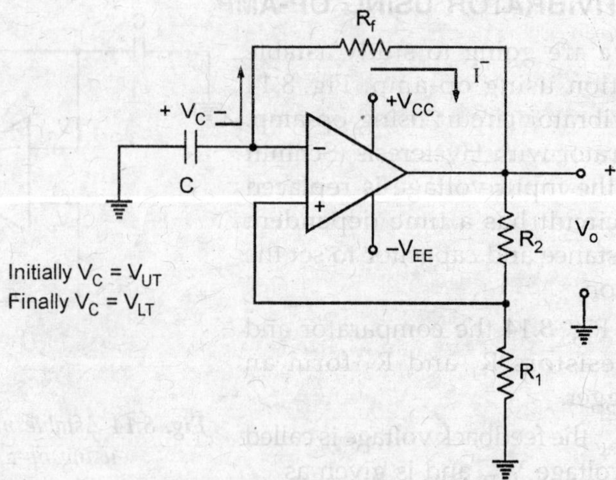

Fig. 8.15 (b) When $V_o = -V_{sat}$, capacitor charges towards V_{LT}.

The current I^- discharges capacitor to 0 V and recharges capacitor to V_{LT}. When V_C becomes slightly more negative than the feedback voltage V_{LT}, output voltage V_o switches back to $+V_{sat}$. As a result, the condition in Fig. 8.15 (*a*) is reestablished except that capacitor now has a initial charge equal to V_{LT}. The capacitor will discharge from V_{LT} to 0 V and then recharge to V_{UT}, and the process is repeating. Once the initial cycle is completed, the waveforms become periodic, as shown in the Fig. 8.15(*c*).

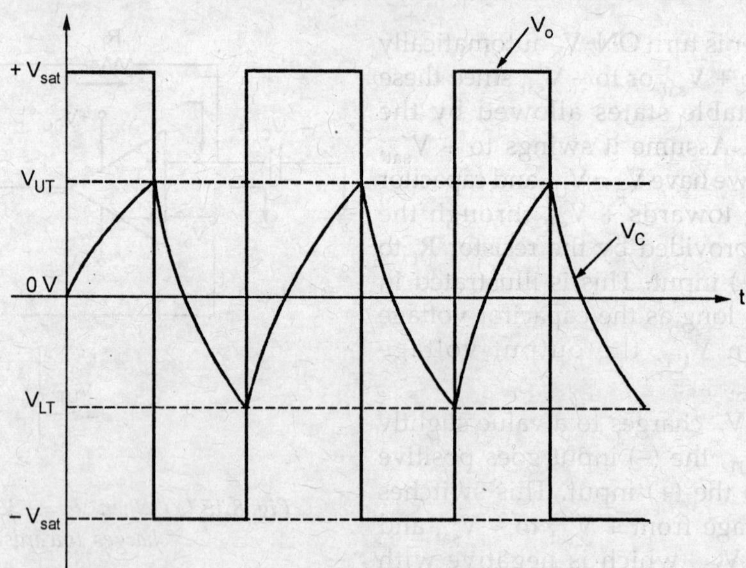

Fig. 8.15 (c) Waveforms.

8.8.1 Frequency of Oscillation

The frequency of oscillation is determined by the time it takes the capacitor to charge from V_{UT} to V_{LT} and vice versa. The voltage across the capacitor as a function of time is given as

$$V_C(t) = V_{max} + (V_{initial} - V_{max})\, e^{(-t/T)} \qquad ...(3)$$

Where $V_C(t)$ is the instantaneous voltage across the capacitor.

$V_{initial}$ is the initial voltage

V_{max} is the voltage toward which the capacitor is charging.

Let us consider the charging of capacitor from V_{LT} to V_{UT}, where V_{LT} is the initial voltage, V_{UT} is the instantaneous voltage and $+V_{sat}$ is the maximum voltage. At $t = T_1$, voltage across capacitor reaches V_{UT} and therefore equation (3) becomes

$$V_{UT} = +V_{sat} + (V_{LT} - +V_{sat})\, e^{(-T_1/R_fC)} \qquad ...(4)$$

$$\therefore -(V_{LT} - +V_{sat})\, e^{(-T_1/R_fC)} = +V_{sat} - V_{UT}$$

$$\therefore \quad e^{(-T_1/R_fC)} = \frac{(+V_{sat} - V_{UT})}{(+V_{sat} - V_{LT})}$$

$$\therefore \quad \frac{-T_1}{R_fC} = \ln\left(\frac{+V_{sat} - V_{UT}}{+V_{sat} + V_{LT}}\right)$$

$$\therefore \quad T_1 = -R_f\, C \ln\left(\frac{+V_{sat} - V_{UT}}{+V_{sat} - V_{LT}}\right)$$

$$= R_f\, C \ln\left(\frac{+V_{sat} - V_{LT}}{+V_{sat} - V_{UT}}\right) \qquad ...(5)$$

The time taken by capacitor to charge from V_{UT} to V_{LT} is same as time required for charging capacitor from V_{LT} to V_{UT}. Therefore, total time required for one oscillation is given as

$$T = 2T_1 \qquad ...(6)$$

$$\therefore \quad \boxed{T = 2R_fC \ln\left(\frac{+V_{sat} - V_{LT}}{+V_{sat} - V_{UT}}\right)} \qquad ...(7)$$

The frequency of oscillation can be determined as $f_o = 1/T$, where T represents the time required for one oscillation.

Substituting the value of T we get,

$$\boxed{f_o = \frac{1}{2R_fC \ln\left(\frac{+V_{sat} - V_{LT}}{+V_{sat} - V_{UT}}\right)}} \qquad ...(8)$$

Substituting the values of V_{UT} and V_{LT} we get,

$$T = 2R_fC \ln\left[\frac{+V_{sat} - (R_1 \times - V_{sat})/R_1 + R_2}{+V_{sat} - (R_1 \times + V_{sat})/R_1 + R_2}\right]$$

If magnitudes of $+V_{sat}$ and $-V_{sat}$ are equal,

$$T = 2R_fC\ln\left[\frac{+V_{sat}\left(1+\frac{R_1}{R_1+R_2}\right)}{+V_{sat}\left(1-\frac{R_1}{R_1+R_2}\right)}\right]$$

$$\therefore \quad \boxed{T = 2R_fC\ln\left(\frac{2R_1+R_2}{R_2}\right)}.$$

8.8.2 Non-Symmetrical Square Wave Generation

The astable multivibrator can be used to obtain non-symmetrical square wave by modifying the circuit as shown in the Fig. 8.16.

When $V_o = +V_{sat}$, the C will charge through R_3 due to forward biasing of D_1. When $V_o = -V_{sat}$, the C will discharge through R_4 due to forward biasing of D_2. Selecting different values of R_3 and R_4, charging and discharging time of C can be varied and hence nonsymmetrical square wave can be obtained.

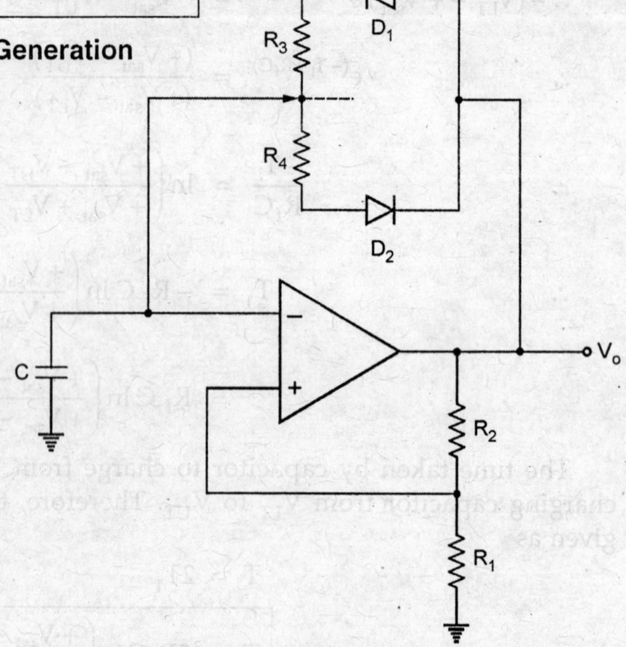

Fig. 8.16 Non-symmetrical square wave generation.

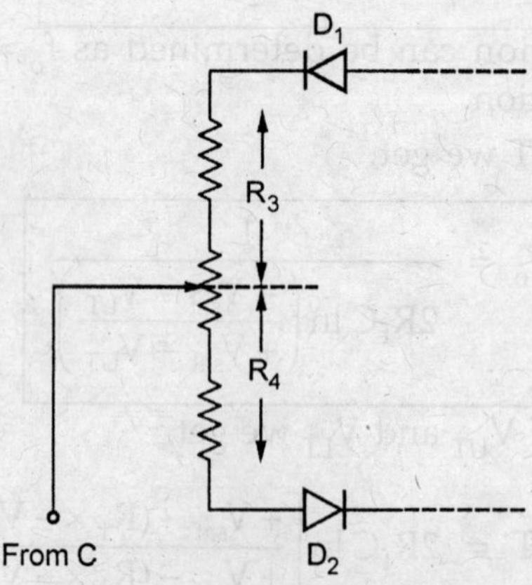

Fig. 8.17 Variable duty cycle.

Variable Duty Cycle: To vary duty cycle from say d_1% and d_2% then divide $R_3 + R_4$ resistance in the ratio of the variable duty cycle required.

For example if it is to be varied from 30% to 70% then divide $R_3 + R_4$ in the ratio 3 : 4 : 3 as shown in the Fig. 8.17.

For varying duty cycle from 10% to 90%, $R_3 + R_4$ to be divided in the ratio 1 : 8 : 1 and so on.

Example 8.3: For the circuit shown in Fig. 8.42 if R_2 = 100 kΩ, R_1 = 86 kΩ, $+V_{sat}$ = 15 V, $-V_{sat}$ = – 15 V, R_f = 100 kΩ and C = 0.1 μF, find

(*a*) VUT (*b*) V_{LT} and (*c*) frequency of oscillation

Solution: (*a*) We know that,

$$V_{UT} = \frac{R_1 \times V_{sat}}{R_1 + R_2} = \frac{86\text{ K} \times 15}{86\text{ K} + 100\text{ K}} = \mathbf{6.94\ V}$$

(*b*) We know that,

$$V_{LT} = \frac{R_1 \times -V_{sat}}{R_1 + R_2} = \frac{86\text{ K} \times (-15)}{86\text{ K} + 100\text{ K}} = \mathbf{-6.94\ V}$$

(*c*) We know that,

$$f_o = \frac{1}{2R_fC\ln\left(\dfrac{+V_{sat} - V_{LT}}{+V_{sat} - V_{UT}}\right)} = \frac{1}{2 \times 100\text{ k} \times 0.1\,\mu\text{F} \times \ln\left(\dfrac{15-(6-24)}{15-(6.94)}\right)}$$

$$= \frac{1}{0.02} = \mathbf{50\ Hz.}$$

Example 8.4: For circuit shown in Fig. 8.14 show that T = $2R_fC$ when $R_1 = 0.86\ R_2$.

Solution: We know that,

$$T = 2R_fC\ln\left(\frac{+V_{sat} - V_{LT}}{+V_{sat} - V_{UT}}\right)$$

Substituting the values of V_{UT} and V_{LT} we get,

$$T = 2R_fC\ln\left(\frac{+V_{sat} - (R_1 + R_2)/(R_1 + R_2)}{+V_{sat} - (R_1 + V_{sat})/(R_1 + R_2)}\right)$$

If the magnitude of $+V_{sat}$ and $-V_{sat}$ are equal, the term in parenthesis simplifies to

$$T = 2R_fC\ln\left(\frac{+V_{sat}\left(1 + \dfrac{R_1}{R_1 + R_2}\right)}{+V_{sat}\left(1 - \dfrac{R_1}{R_1 + R_2}\right)}\right)$$

$$= 2R_fC \ln\left(\frac{\frac{2R_1+R_2}{R_1}}{\frac{R_2}{R_1+R_2}}\right) = 2\,R_f\,C \ln\left(\frac{2R_1+R_2}{R_2}\right)$$

When $R_1 = 0.86\ R_2$, we have

$$T = 2R_fC \ln\left(\frac{2\times 0.86R_2+R_2}{R_2}\right) = 2\,R_fC \ln 2.72$$

$$= 2\ R_fC \text{ since } \ln 2.72 = 1.$$

Example 8.5: In the square wave oscillator shown, calculate the frequency of oscillations if $R_2 = 10\ k\Omega$, $R_1 = 8.6\ k\Omega$, $R_f = 100\ k\Omega$ and $C = 0.01\ \mu F$.

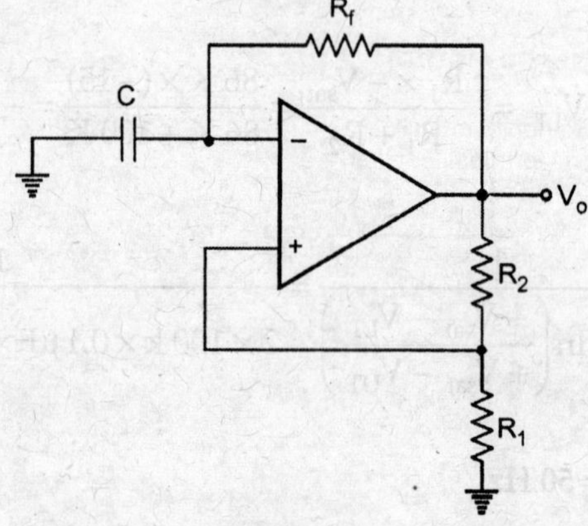

Fig. 8.18

Solution: This is astable multivibrator using op-amp. Its time period is given by,

$$T = 2R_fC \ln\left(\frac{2R_1+R_2}{R_2}\right) \quad \text{if } R_1 = 0.86\ R_2$$

$$= 2\times 100\times 10^3 \times 0.01\times 10^{-6} \ln\left[\frac{2\times 0.86\,R_2+R_2}{R_2}\right]$$

$$= 2\times 10^{-3}\ s$$

$$\therefore \qquad f = \frac{1}{T} = 499.68\ \text{Hz}$$

8.9 BISTABLE MULTIVIBRATOR USING OP-AMP

The Fig. 8.19 shows the bistable multivibrator using op-amp. The op-amp 741 can be used for this application.

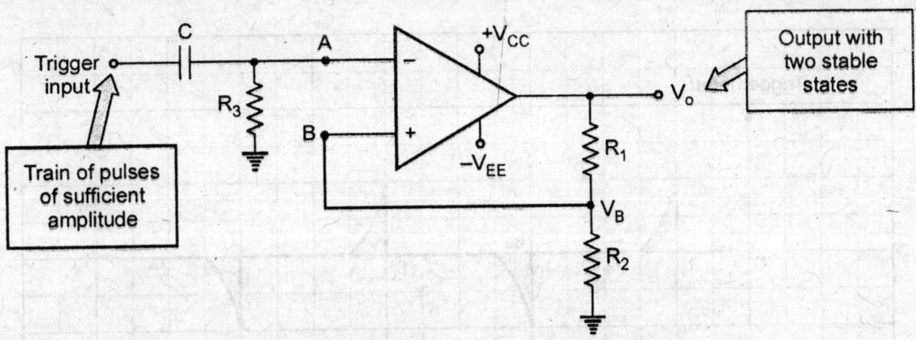

Fig. 8.19 Bistable multivibrator using op-amp.

8.9.1 Operation of the Circuit

Assume that the initial state of the output voltage is $V_o = +V_{sat}$. This state decides the voltage V_B which is voltage at the noninverting terminal of the op-amp.

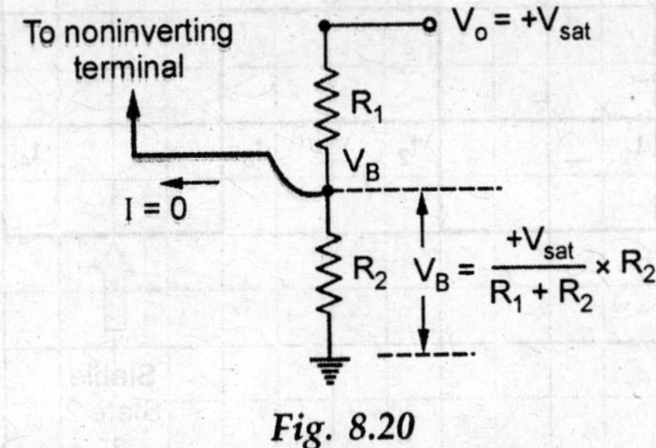

Fig. 8.20

This voltage can be obtained as shown in the Fig. 8.20.

$$\therefore \quad V_B = +V_{sat} \times \frac{R_2}{R_1 + R_2} \qquad ...(1)$$

The inverting terminal is at ground potential, through the resistance R_3.

***Key Point:** As noninverting terminal is at higher potential than inverting terminal, $V_o = +V_{sat}$ and represents the one stable state.*

In order to change this state, it is necessary to have inverting terminal potential more than the noninverting terminal.

So apply a positive going pulse to inverting terminal whose amplitude is V_{TRG} given by,

$$V_{TRG} > V_{sat} \times \frac{R_2}{R_1 + R_2} \qquad ...(2)$$

Immediately when this pulse is applied, the op-amp output changes from $+V_{sat}$ to $-V_{sat}$ as $V_A > V_B$.

Now the potential V_B is decided by this new output state which is $V_o = -V_{sat}$.

$$\therefore \quad V_B = -V_{sat} \times \frac{R_2}{R_1 + R_2} \qquad ...(3)$$

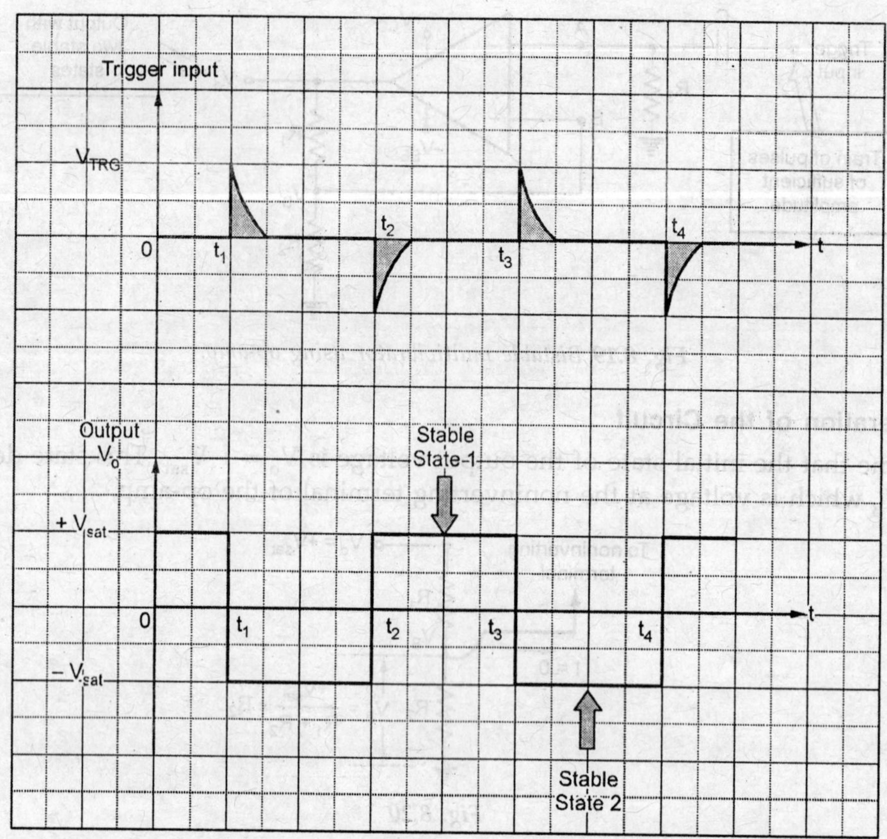

Fig. 8.21 Waveforms of bistable multivibrator using op-amp.

Key Point: *As V_B is less than inverting terminal, when trigger pulse is removed, the output remains at $-V_{sat}$ indefinitely till next pulse occurs and this state represents second stable state.*

To change the state, a negative pulse is required to be applied at the inverting terminal with the amplitude V_{TRG} given by,

$$V_{TRG} < -V_{sat} \times \frac{R_2}{R_1 + R_2} \quad \text{...(4)}$$

The output will again switch back to $+V_{sat}$, till next pulse occurs. Thus the circuit behaves as bistable multivibrator.

The trigger input and output waveforms are shown in the Fig. 8.21.

8.9.2 Applications

The various applications of bistable multivibrator using op-amp are,

1. In many digital circuits for the operations such as counting and storing the data in digital form and as a flip-flop.

2. As a memory element in shift registers and counters.
3. In processing of pulse type waveforms.
4. For generating the symmetrical square wave by controlling the time instants t_1, t_2, t_3, ... at which trigger pulses are applied.
5. As a frequency divider, as the output frequency is half of the frequency of the input triggering pulse waveform.

8.10 COMPARISON OF MULTIVIBRATORS

Sr. No.	Monostable	Astable	Bistable	
1.	Stable states	One	No	Two
2.	External trigger input	Required for changing the state.	Not required	Required for Changing the state.
3.	Other names	Mono shot, One shot, Single shot, Gating circuit or Delay circuit.	Free running circuit	Flip-flop circuit or Binary circuit.
4.	Applications	Timer, Ramp generator, Pulse width modulation, Pulse position modulation etc.	Square wave generator, voltage controlled oscillator, Flasher etc.	Flip-flop circuits, Digital circuits, Square wave generation etc.

8.11 TRIANGULAR WAVE GENERATOR

We have seen that, the output of integrator is a triangular wave if its input is a square wave. This means that a triangular wave generator can be formed by simply connecting an integrator to the square wave generator as shown in the Fig. 8.22.

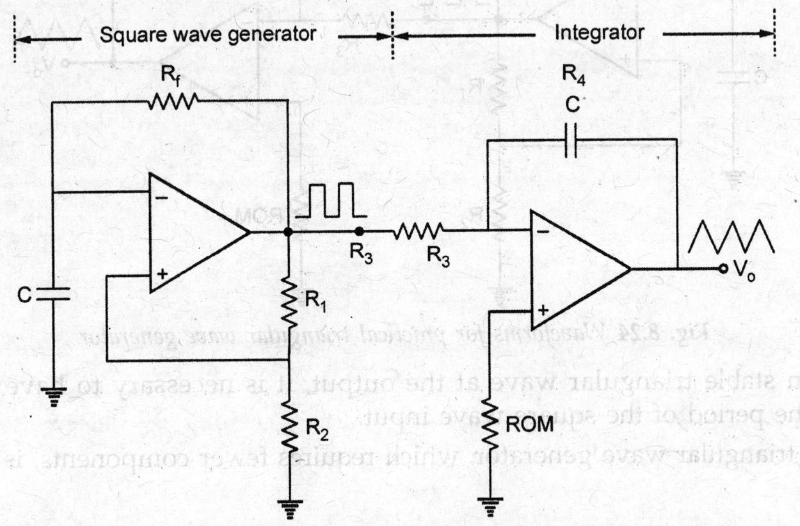

Fig. 8.22 Triangular wave generator.

Basically, triangular wave is generated by alternatively charging and discharging a capacitor with a constant current. This is achieved by connecting integrator circuit at the output of square wave generator. Assume that V′ is high at $+V_{sat}$. This forces a constant current ($+V_{sat}/R_3$) through C (left to right) to drive V_o negative, linearly. When V′ is low at $-V_{sat}$, it forces a constant current ($-V_{sat}/R_3$) through C (right to left) to drive V_o positive, linearly. The frequency of the triangular wave is same as that of square wave. This is illustrated in Fig. 8.23. Although the amplitude of the square wave is constant ($\pm V_{sat}$), the amplitude of the triangular wave decreases with an increase in its frequency, and vice versa. This is because the reactance of capacitor decreases at high frequencies and increases at low frequencies.

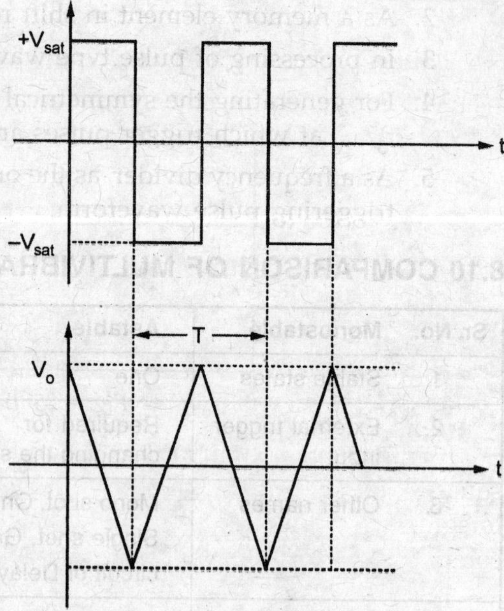

Fig. 8.23 Waveforms of triangular wave generator.

In practical circuits, resistance R_4 is connected across C to avoid the saturation problem at low frequencies as in the case of practical integrator as shown in the Fig. 8.24.

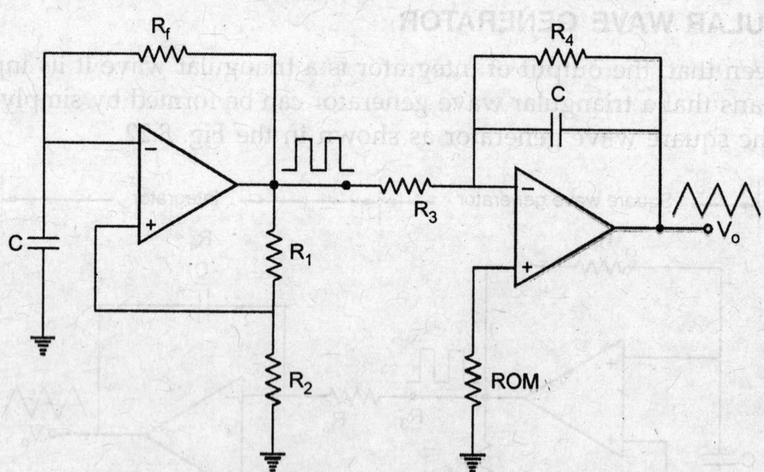

Fig. 8.24 Waveforms for practical triangular wave generator.

To obtain stable triangular wave at the output, it is necessary to have $5R_3C_2 > T/2$, where T is the period of the square wave input.

Another triangular wave generator, which requires fewer components, is shown in the Fig. 8.25.

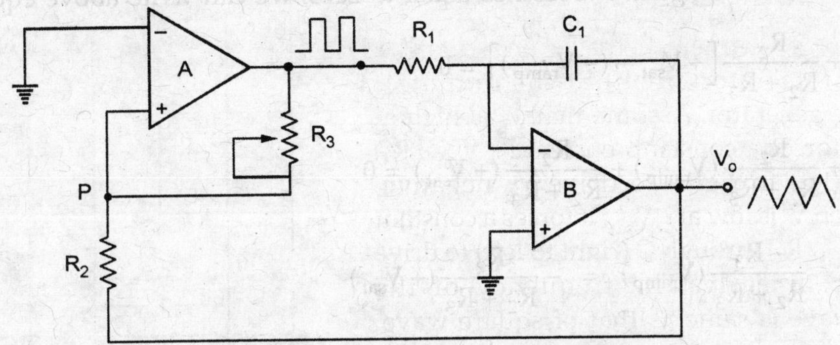

Fig. 8.25 Triangular wave generator.

It consists of a comparator (A) and an integrator (B). The output of comparator A is a square wave of amplitude ± V_{sat} and is applied to the inverting (–) input terminal of the integrator B. The output of integrator is a triangular wave and it is feedback as input to the comparator A through a voltage divider R_2R_3.

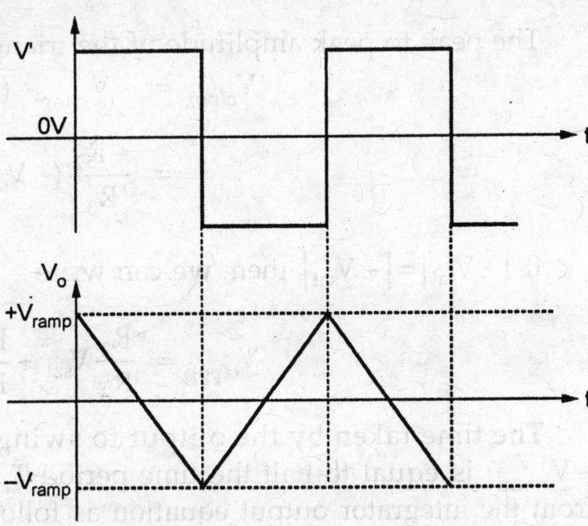

Fig. 8.26 Waveforms of triangular wave generator.

To understand circuit operation, assume that the output of comparator A is at $+V_{sat}$. This forces a constant current ($+ V_{sat}/R_1$) through C to give a negative going ramp at the output of the integrator, as shown in the Fig. 8.26. Therefore, one end of voltage divider is at a voltage $+V_{sat}$ and the other at the negative going ramp. When the negative going ramp reaches a certain value – V_{ramp}, the effective voltage at point p becomes slightly below 0 V. As a result, the output of comparator A switches from positive saturation to negative saturation ($-V_{sat}$). This forces a reverse constant current (right to left) through C to give a positive going ramp at the output of the integrator, as shown in the Fig. 8.26. When positive going ramp reaches + V_{ramp}, the effective voltage at point p becomes slightly above 0 V. As a result, the output of comparator A switches from negative saturation to positive saturation ($+ V_{sat}$). The sequence then repeats to give triangular wave at the output of integrator B.

8.11.1 Amplitude and Frequency Calculations

The frequency and amplitude of the triangular wave can be determined as follows: When comparator output is at + V_{sat}, the effective voltage at point P is given by

$$-V_{ramp} + \frac{R_2}{R_2 + R_3}\left[+V_{sat} - (-V_{ramp})\right] \qquad ...(1)$$

When effective voltage at P becomes equal to zero, we can write above equation as,

$$-V_{ramp}+\frac{R_2}{R_2+R_3}\left[+V_{sat}-(-V_{ramp})\right]=0$$

$$-V_{ramp}+\frac{R_2}{R_2+R_3}(V_{ramp})+\frac{R_2}{R_2+R_3}(+V_{sat})=0$$

$$\frac{-R_3}{R_2+R_3}(V_{ramp}) = -\frac{R_2}{R_2+R_3}(+V_{sat})$$

$$\therefore \quad -V_{ramp} = \frac{-R_2}{R_3}(+V_{sat}) \qquad ...(2)$$

Similarly, when comparator output is at $-V_{sat}$, we can write,

$$V_{ramp} = \frac{-R_2}{R_3}(-V_{sat}) \qquad ...(3)$$

The peak to peak amplitude of the triangular wave can be given as

$$V_{o(pp)} = +V_{ramp}-(-V_{ramp}) \qquad ...(4)$$

$$= \frac{-R_2}{R_3}(-V_{sat})-\left(\frac{-R_2}{R_3}\right)(+V_{sat})$$

If $|+V_{sat}|=|-V_{sat}|$ then, we can write

$$V_{o(pp)} = \frac{R_2}{R_3}V_{sat}+\frac{R_2}{R_3}V_{sat}=\frac{2R_2}{R_3}V_{sat} \qquad ...(5)$$

The time taken by the output to swing from $-V_{ramp}$ to $+V_{ramp}$ (or from $+V_{ramp}$ to $-V_{ramp}$) is equal to half the time period T/2. Refer Fig. 8.91. This time can be calculated from the integrator output equation as follows:

$$V_{o(pp)} = -\frac{1}{R_1C_1}\int_0^{\frac{T}{2}}(-V_{sat})\,dt = \left(\frac{V_{sat}}{R_1C_1}\right)\frac{T}{2} \qquad ...(6)$$

$$\therefore \quad T = \frac{2R_1C_1V_{o(pp)}}{V_{sat}} \qquad ...(7)$$

Substituting value of $V_{o(pp)}$ we get,

$$T = \frac{2R_1C_1\left(\frac{2R_2}{R_3}V_{sat}\right)}{V_{sat}} = \frac{4R_1C_1R_2}{R_3} \qquad ...(8)$$

Therefore, the frequency of oscillation can be given as,

$$\boxed{f_o = \frac{1}{T} = \frac{R_3}{4R_1 C_1 R_2}} \quad \text{...(9)}$$

Example 8.6: Design a triangular wave generator as shown in Fig. 8.25 such that $f_0 = 5$ kHz and $V_{o(pp)} = 7.5$ V. The op-amp saturation voltage is ± 15 V.

Solution: Given: $+V_{sat} = 15$ V, $-V_{sat} = -15$ V, $V_{o(pp)} = 7.5$ V, $f_o = 5$ kHz

We know that,

$$V_{o(pp)} = \frac{2R_2}{R_3} V_{sat}$$

$$\therefore \quad 7.5 = 2\frac{R_2}{R_3}15$$

$$\therefore \quad R_3 = \frac{2 \times 15R_2}{7.5} = 4R_2$$

Let $R_2 = 10$ kΩ; then $R_3 = 40$ kΩ (Use a 50 kΩ POT)

We know that,

$$f_o = \frac{R_3}{4R_1 C_1 R_2}$$

$$\therefore \quad 5 \text{ kHz} = \frac{40 \text{ k}\Omega}{(4)(R_1C_1)(10 \text{ k}\Omega)}$$

$$\therefore \quad R_1 C_1 = \frac{1}{5 \text{ kHz}}$$

$$= 0.20 \text{ ms}$$

Let $C_1 = 0.01$ µF; then

$$R_1 = \frac{0.2 \text{ ms}}{0.01 \,\mu\text{F}} = 20 \text{ k}\Omega$$

8.12 STAIRCASE AND PULSE GENERATOR

The staircase waveform is also called sawtooth waveform in which the waveform is generated by charging a capacitor at a specified rate and then discharging it very quickly through a transistor switch.

The Fig. 8.27 shows op-amp based circuit which can be used as a staircase generator and a pulse generator.

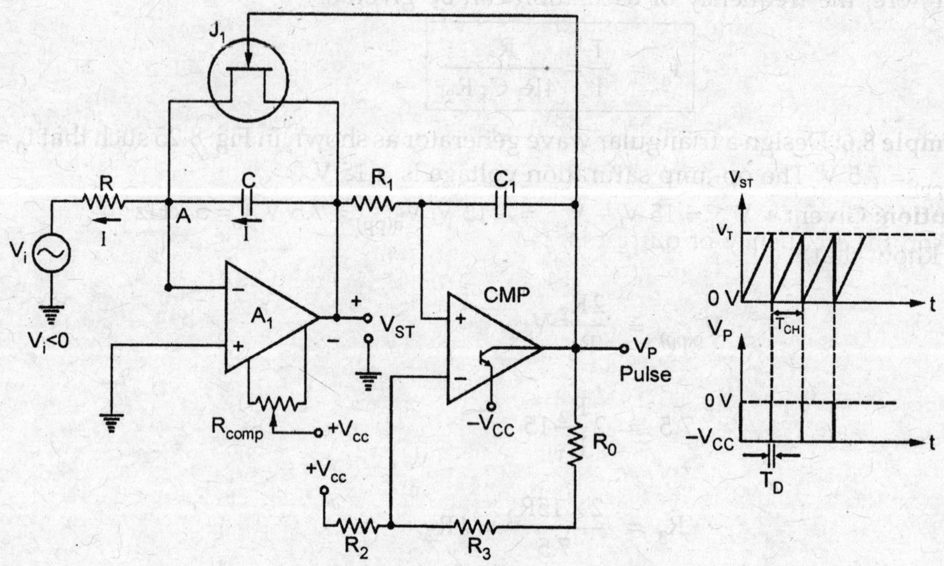

Fig. 8.27 Staircase and pulse generator.

The op-amp circuit A_1 is V-I converter which forces current I through the capacitor which is linearly proportional to the input voltage V_i.

As noninverting terminal of A_1 is grounded, the inverting terminal is also at virtual ground. Hence we can write,

$$I = \frac{|V_i|}{R} \qquad ...(1)$$

For proper operation of the circuit, this current must always flow out of node A. Hence V_i is always selected negative.

The other circuit is comparator CMP and voltage divider R_1 and R_2 decide a threshold voltage V_T given by,

$$V_T = \frac{V_{CC}R_2}{R_1 + R_2} \qquad ...(2)$$

When the capacitor charges, the CMP output is designed to be low i.e. V_p is at $-V_{CC}$. So switch J_1 is off and C charges linearly due to current I. So voltage V_{ST} increases linearly. As soon as V_{ST} becomes equal to threshold voltage V_p the output transistor of CMP becomes off and hence V_p increases from $-V_{CC}$ to 0 V giving a pulse of duration T_D. This action is instantaneous because of positive feedback provided by C_1. When V_p becomes zero, the switch J_1 becomes on and directly shorts capacitor C. This discharges capacitor C rapidly bringing V_{ST} down to 0 V instantaneously.

The design of comparator is such that due to charge accumulated in C_1, comparator does not respond immediately to the change in V_{ST} from V_T to 0 V. The duration T_D is proportional to R_1 C_1 and is designed in such a way to ensure complete discharge of C through J_1. After time T_D, V_p returns to $-V_{CC}$. This makes J_1 off and the capacitor C starts its charging again. Thus the cycle repeats. The TCH is the charging time and can be obtained as,

$$CV_T = I\,T_{CH}$$

$$\therefore \qquad T_{CH} = \frac{CV_T}{I} = \frac{CV_T}{\frac{|V_i|}{R}} = \frac{RCV_T}{|V_i|} \qquad \text{...(3)}$$

And the frequency of output is,

$$f_o = \frac{1}{T_{CH} + T_D} = \frac{1}{\frac{RCV_T}{|V_i|} + T_D} \qquad \text{...(4)}$$

T_D is generally designed to be less than 1 µsec and $T_D \ll T_{CH}$. So neglecting T_D we get,

$$\boxed{f = \frac{|V_i|}{RCV_T}} \qquad \text{...(4(a))}$$

***Key Point:** The equation shows that the output frequency is linearly proportional to the input voltage V_i. So circuit acts as VCO with the output waveforms as staircase and pulse.*

8.13 SAWTOOTH WAVE GENERATOR

Unlike triangular wave, sawtooth wave has unequal rise time and fall time. That is, it may fall negatively many times faster than it rises positively, or vice versa. The sawtooth wave generator can be implemented by slightly modifying the triangular wave generator, as shown in the Fig. 8.28.

Here, noninverting terminal (+ input) of the integrator is driven by the voltage set between $+V_{CC}$ to $-V_{EE}$ by the potentiometer. Depending on the R_4 setting, a certain dc level is added in the output of the integrator B. As output voltage of the integrator decides the effective voltage at point P, added de level in the output of the integrator will affect the duty cycle of the comparator output. When voltage at the noninverting input of the integrator is negative, the duty cycle is less than 50% resulting longer rise time than the fall time as shown in the Fig. 8.29. On the other hand, when voltage at the noninverting input of the integrator is positive, the duty cycle is greater than 50% and rise time is less than the fall time.

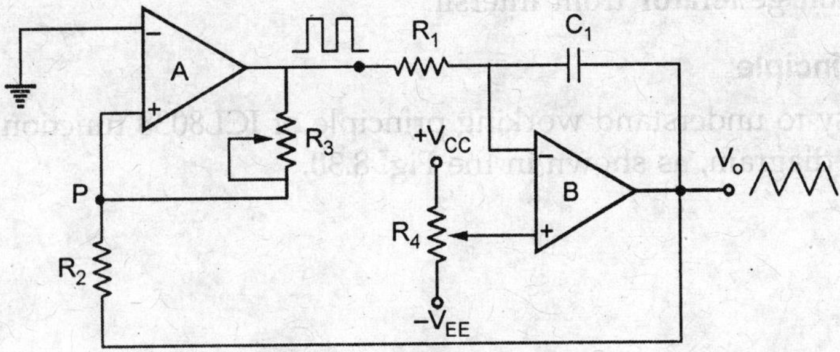

Fig. 8.28 Sawtooth wave generator.

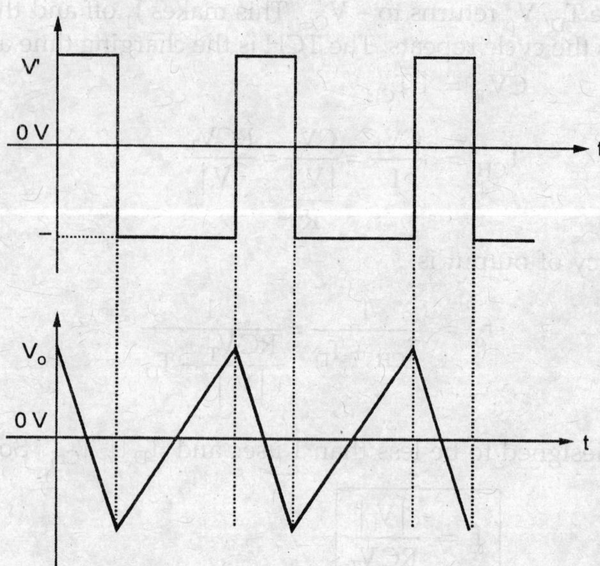

Fig. 8.29 Waveforms of sawtooth wave generator.

It is important to note that the frequency of the sawtooth waveform decreases when voltage at the noninverting input of the integrator B is adjusted towards + V_{CC} or – V_{EE}. However, the amplitude of the sawtooth wave is independent of the voltage setting at the noninverting input of the integrator B.

8.14 FUNCTION GENERATOR

Function generators are designed to provide the basic waveforms such as square wave, triangular wave and sine wave. They are also called as waveform generators. The monolithic function generators provide these basic waveforms with a minimum number of external components reducing complexity, but increasing the reliability of the circuit. They find application in communication, telemetry, electronic music, and testing and calibration in the laboratory.

In function generators, VCO (voltage controlled oscillator) generates the triangular and square waves. The triangular wave is passed through the on chip wave shaper to generate a sine wave. The sawtooth and pulse waveforms are generated by configuring the oscillator for a highly asymmetric duty cycle. In this section we will discuss the **ICL 8038 precision function generator** from Intersil.

8.14.1 Basic Principle

It is very easy to understand working principle of ICL8038 function generator using its simplified block diagram, as shown in the Fig. 8.30.

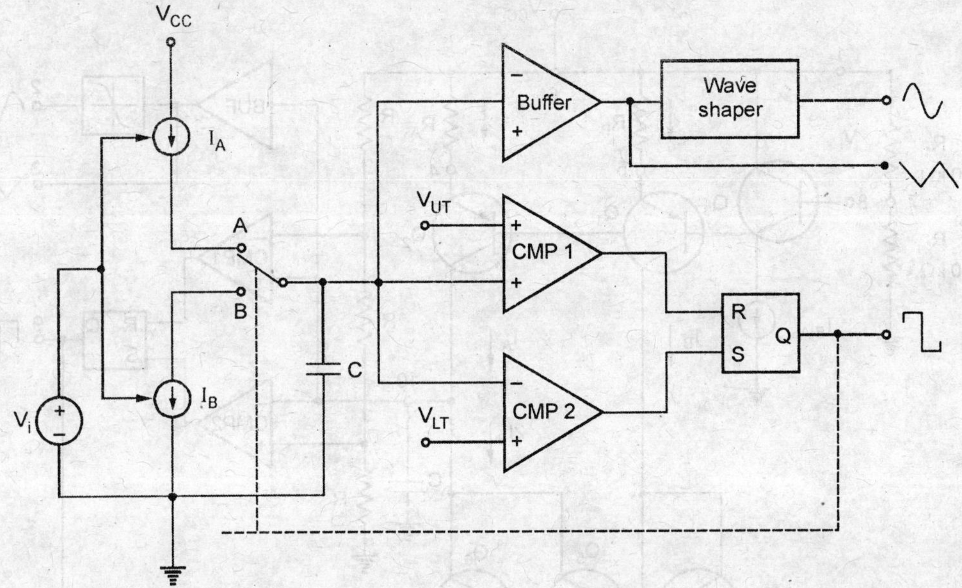

Fig. 8.30 Simplified block diagram of ICL 8038 function generator.

The operation of ICL 8038 is based on charging and discharging of a grounded capacitor C, whose charging discharging rates are controlled by programmable current generators I_A and I_B, respectively. When switch is at position A, the capacitor charges at a rate determined by current source I_A. Once the capacitor voltage reaches V_{UT}, the upper comparator (CMP1) triggers and resets the flip-flop output. This causes the switch position to change from position A to B. Now, capacitor starts discharging at the rate determined by the current sink I_B.

Once the capacitor reaches V_{LT}, the lower comparator (CMP2) triggers and sets the flip-flop output. This causes the switch position to change from position B to A. And this cycle repeats. As a result, we get square wave at the output of flip-flop and triangular wave across capacitor. The triangular wave is then passed through the on chip wave shaper to generate sine wave.

To allow automatic frequency control, currents I_A and I_B are made programmable through an external control voltage V_i. For equal magnitudes of I_A and I_B, output waveforms are symmetrical conversely, when two currents are unequal, output waveforms are asymmetrical. By making one of the currents much larger than other we can get sawtooth waveform across capacitor and rectangular wave at the output of flip-flop.

8.14.2 Circuit Diagram

The Fig. 8.31 shows the most simplified form of circuit diagram for ICL 8038.

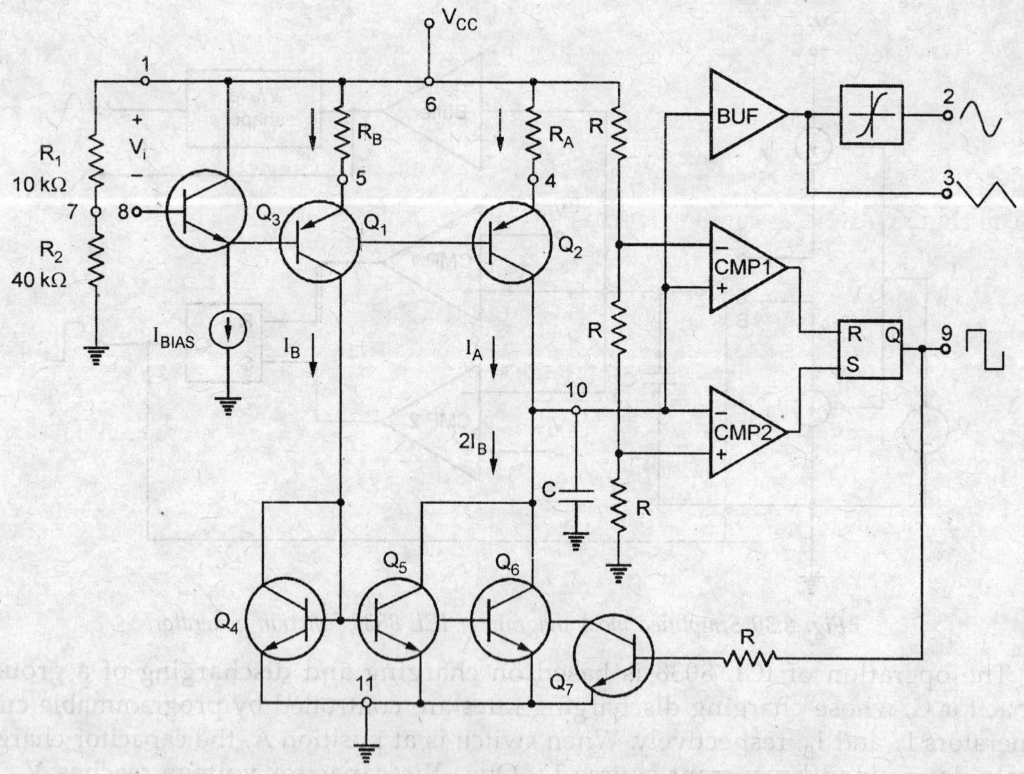

Fig. 8.31 Simplified circuit diagram for ICL 8038 function generator.

As shown in the Fig. 8.31, transistors Q_1 and Q_2 form programmable current sources whose magnitudes are set by external resistors R_A and R_B. These current sources are driven by the emitter follower (transistor Q_3). It also compensates base-emitter voltage drops for Q_1 and Q_2 to ensure $V_{RA} = V_{RB} = V_i$. Thus, $I_A = V_i/R_A$ and $I_B = V_i/R_B$. The current I_A controls the charging rate of capacitor C. The current I_B is diverted to current mirror ($Q_4 - Q_5 - Q_6$), where it undergoes polarity reversal as well as amplification by 2 due to the combined action of Q_5 and Q_6. The result is a current sink of magnitude of 2 I_B, as shown in the Fig. 8.31.

The voltage across capacitor is applied to the Schmitt trigger. The schmitt trigger shown in Fig. 8.31 is similar to that of the IC 555, with $V_{UT} = 2/3\ V_{CC}$ and $V_{LT} = 1/3V_{CC}$. Transistor Q_7 acts as a switch. When output of flip-flop is high, Q_7 saturates and pulls the bases of Q_5 and Q_6 low, thus shutting off the current sink. As a result capacitor C starts charging at a rate set by current I_A. Once the capacitor voltage reaches 2/3 $V_{CC}(V_{UT})$, CMP1 triggers and clears the flip-flop, thus turning Q_7 off. This enables current mirror to sink current equal to 2 I_B so that net current flowing out of the capacitor is $I_L = 2\ I_B - I_A$. This causes capacitor to discharge. Once capacitor voltage reaches 1/3 $V_{CC}(V_{LT})$, CMP2 triggers and sets the flip-flop and action repeats. It is important to note that net current flowing out of capacitor C should be positive i.e. $2\ I_B - I_A > 0$ discharging capacitor and hence $2\ I_B > I_A$.

8.14.3 Frequency of Output Waveform

The frequency of the output waveform can be determined as follows:

$$f_{out} = \frac{1}{T}$$

where $T = T_c + T_d$

The charging time T_c can be given as

$$T_c = \Delta V_C \times \frac{1}{\text{Charging current}} \times C = \frac{V_{CC}}{3} \times \frac{R_A}{V_i} \times C$$

$$= \frac{CR_A V_{CC}}{3V_i} \quad \text{...(1)}$$

The discharging time T_d can be given as

$$T_c = \Delta V_C \times \frac{1}{\text{Discharging current}} \times C$$

$$= \frac{V_{CC}}{3} \times \frac{1}{\left(\frac{2V_i}{R_B} - \frac{V_i}{R_A}\right)} \times C$$

$$= \frac{V_{CC}}{3V_i}\left(\frac{1}{\frac{2}{R_B} - \frac{2}{R_A}}\right) \times C \quad \text{...(2)}$$

Multiplying R_A by numerator and denominator we get,

$$= \frac{CR_A V_{CC}}{3V_i}\left(\frac{1}{\frac{2R_A}{R_B} - \frac{R_A}{R_A}}\right)$$

$$= \frac{CR_A V_{CC}}{3V_i}\left(\frac{1}{\frac{2R_A}{R_B} - 1}\right)$$

$$= \frac{CR_A V_{CC}}{3V_i}\left(\frac{R_B}{2R_A - R_B}\right) \quad \text{...(3)}$$

$$\therefore \quad T_c + T_d = \frac{CR_A V_{CC}}{3V_i} + \frac{CR_A V_{CC}}{3V_i}\left(\frac{R_B}{2R_A - R_B}\right)$$

$$= \frac{CR_A\ V_{CC}}{3V_i}\left(1 + \frac{R_B}{2R_A - R_B}\right)$$

$$= \frac{CR_A\ V_{CC}}{3V_i}\left(\frac{2R_A - R_B + R_B}{2R_A - R_B}\right)$$

$$= \frac{CR_A\ V_{CC}}{3V_i}\left(\frac{2R_A}{2R_A - R_B}\right) \quad ...(4)$$

$$f_{out} = \frac{1}{T_c + T_d} \quad ...(5)$$

$$= \frac{1}{\frac{CR_A\ V_{CC}}{3V_i}\left(\frac{2R_A}{2R_A - R_B}\right)}$$

$$= \frac{3V_i}{CR_A\ V_{CC}}\left(\frac{2R_A - R_B}{2R_A}\right)$$

$$\therefore \quad \boxed{f_{out} = \frac{3V_i}{CR_A\ V_{CC}}\left(1 + \frac{R_B}{2R_A}\right)} \quad ...(6)$$

Duty cycle is given as

$$\% d = \frac{T_c}{T_c + T_d} \times 100 \quad ...(7)$$

$$= \frac{\frac{CR_A\ V_{CC}}{3V_i}}{\frac{CR_A\ V_{CC}}{3V_i} + \frac{CR_A\ V_{CC}}{3V_i}\left(\frac{R_B}{2R_A - R_B}\right)} \times 100$$

$$= \frac{1}{1 + \frac{R_B}{2R_A - R_B}} \times 100 = \frac{1}{\frac{2R_A - R_B + R_B}{2R_A - R_B}} \times 100$$

$$= \frac{1}{\frac{2R_A}{2R_A - R_B}} \times 100$$

$$= \left(\frac{2R_A - R_B}{2R_A}\right) \times 100$$

$$\therefore \quad \boxed{\% d = \left(1 - \frac{R_B}{2R_A}\right) \times 100} \quad ...(8)$$

With $R_A = R_B$, duty cycle is 50% and we get symmetric waveforms. For symmetric waveforms i.e. $R_A = R_B$, the frequency of the waveform is given as,

$$f_{out} = \frac{3V_i}{CR_A V_{CC}}\left(1 - \frac{R_A}{2R_A}\right) \qquad \because R_A = R_B$$

$$= \frac{3V_i}{CR_A V_{CC}}\left(1 - \frac{1}{2}\right) = \frac{3V_i}{2CR_A V_{CC}}$$

$$\therefore \quad \boxed{F_{out} = \frac{1.5 V_i}{CR_A V_{CC}}} \qquad \text{...(9)}$$

8.14.4 Pin Configuration of ICL 8038

The Fig. 8.32 shows the pin configuration for ICL 8038 function generator, which is available in 14 pin DIP.

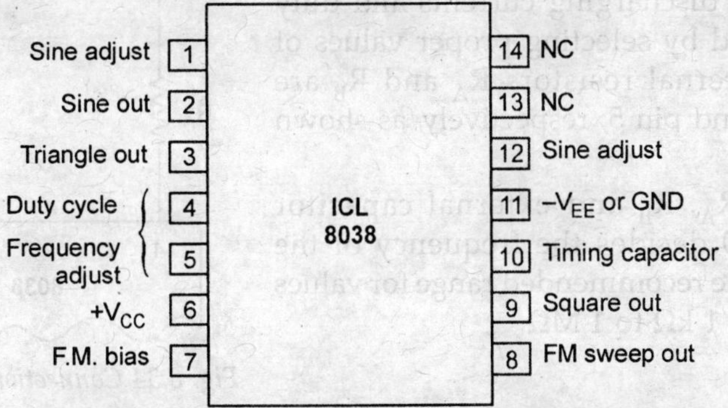

Fig. 8.32 Pin diagram of ICL 8038 function generator.

Pin 1 and Pin 12: Sine wave adjust

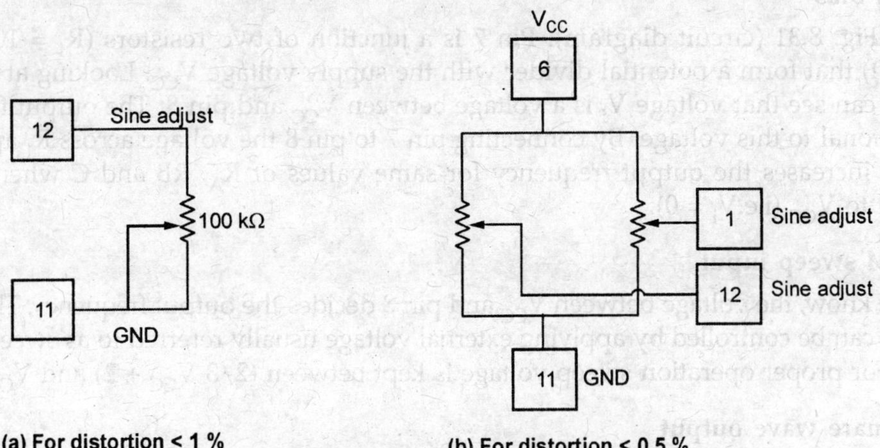

Fig. 8.33 Biasing network adjustments to reduce distortion.

The external resistor connections to these pins decides the accuracy of the sine wave. For distortion less than 1% we have to connect 100 kΩ potentiometer between pin 12 and ground or $-V_{EE}$. To achieve distortion less than 0.5% we have to connect two 100 kΩ potentiometers between V_{CC} and ground with wiper of the one potentiometer connected to pin 1 and other to pin 12, as shown in the Fig. 8.33.

Pin 2: Sine wave out

The sine wave output is available at this pin. The amplitude of sine wave is 0.22 V_{CC}, where $\pm 5\ V \leq V_{CC} \leq \pm 15\ V$.

Pin 3 : Triangular wave output

The triangular wave output is available at this pin. The amplitude of triangular wave is also function of input voltage V_{CC}. It is 0.33 V_{CC}.

Pin 4 and 5: Duty cycle/Frequency adjust

We know that frequency of output is proportional to the charging and discharging currents and duty cycle can be adjusted by selecting proper values of R_A and R_B. The external resistors R_A and R_B are connected to pin 4 and pin 5, respectively, as shown in the Fig. 8.34.

The values of R_A, R_B and external capacitor connected at pin 10 decides the frequency of the output waveform. The recommended range for values of R_A and R_B is from 1 kΩ to 1 MΩ.

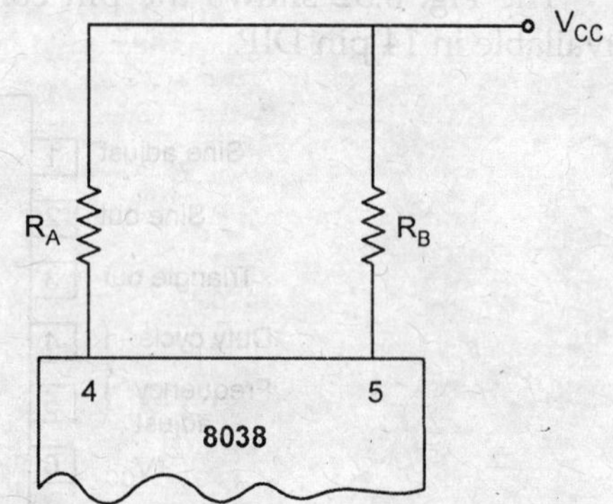

Fig. 8.34 Connections for pin 4 and pin 5.

Pin 6: $+V_{CC}$

It is a positive supply. Its voltage should be kept between 10 V to 30 V, for a single supply operation, and ± 5 V to ± 15 V for dual supply operation.

Pin 7: FM bias

Refer Fig. 8.31 (circuit diagram). Pin 7 is a junction of two resistors (R_1 = 10 kΩ and R_2 = 40 kΩ) that form a potential divider with the supply voltage V_{CC}. Looking at the same figure we can see that voltage V_i is a voltage between V_{CC} and pin 8. The output frequency is proportional to this voltage. By connecting pin 7 to pin 8 the voltage across R_1 appears as a V_i. This increases the output frequency for same values of R_A, RB and C when pin 8 is connected to V_{CC} (i.e.$V_i = 0$).

Pin 8 : FM sweep input

As we know, the voltage between V_{CC} and pin 8 decides the output frequency. The output frequency can be controlled by applying external voltage usually referred to as sweep voltage to pin 8. For proper operation sweep voltage is kept between (2/3 V_{CC} + 2) and V_{CC}.

Pin 9: Square wave output

The square wave output is available at this pin. As this output is open collector, external resistor is required to be connected between V_{CC} and pin 9 to get the square wave at pin 9.

Pin 10: Timing capacitor

The external timing capacitor C is connected at this pin.

Pin 11: – V_{EE}/Ground

If dual supply is used V_{EE} is connected to this pin. If a single is used, this pin is connected to the ground.

Pin 12 and 14: Not connected

8.14.5 Typical Connection

The Fig. 8.35 shows typical connection for ICL 8038 with fixed frequency and 50 % duty cycle.

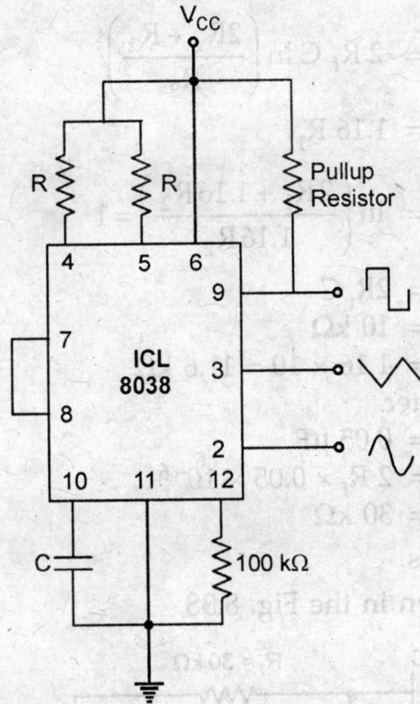

Fig. 8.35 Basic ICL 8038 connection for fixed frequency. 50 % duty cycle operation.

For the above circuit V_i is given as

$$V_i = \frac{10\text{ k}\Omega\, V_{CC}}{10\text{ k}\Omega + 40\text{ k}\Omega} = \frac{V_{CC}}{5}$$

Therefore, the output frequency is given as

$$f_{out} = \frac{1.5 \times V_{CC}/5}{C R V_{CC}} \qquad \because R_A = R_B = R \text{ and } V_i = \frac{V_{CC}}{5}$$

$$= \frac{0.3}{RC}$$

Key Point: The amplitude of square, triangular and sine waves are V_{CC}, 0.33 V_{CC} and 0.22 V_{CC}, respectively.

Examples with Solutions

Example 8.7: Design an op-amp free running multivibrator with T_{ON} = 1 msec and T_{OFF} = 2 msec.

Solution: The time period of the output waveform of free running multivibrator is given by

$$T = T_{ON} + T_{OFF} = 1 + 2$$
$$= 3 \text{ msec.}$$

The time period expression is,

$$T = 2R_f C \ln\left(\frac{2R_1 + R_2}{R_2}\right)$$

Choose $R_2 = 1.16\,R_1$

$$\therefore \quad \ln\left(\frac{2R_1 + R_2}{R_2}\right) = \ln\left(\frac{2R_1 + 1.16R_2}{1.16R_2}\right) = 1$$

$$\therefore \quad T = 2R_f C$$

Choose $R_1 = 10\ \text{k}\Omega$

$$\therefore \quad R_2 = 1.16 \times 10 = 11.6\ \text{k}\Omega$$

Use R_2 as 20 kΩ potentiometer.

Choose $C = 0.05\ \mu\text{F}$

$$\therefore \quad 3 \times 10^{-3} = 2R_f \times 0.05 \times 10^{-6}$$

$$\therefore \quad R_f = 30\ \text{k}\Omega$$

Hence the designed circuit is,

The designed circuit is shown in the Fig. 8.38.

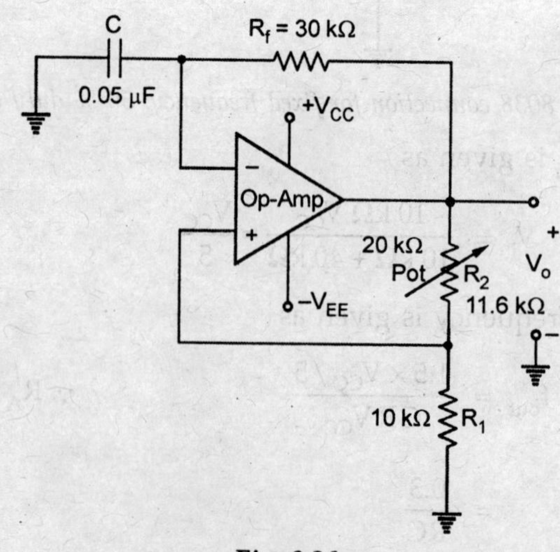

Fig. 8.36

The Fig. 8.37 shows the waveform.

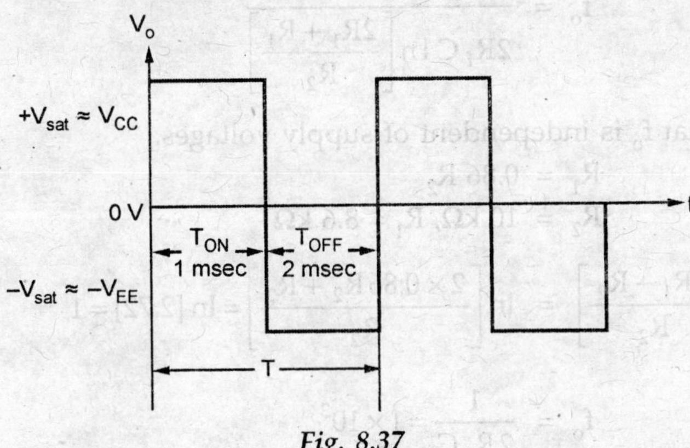

Fig. 8.37

Example 8.8: Design a square wave oscillator for f_o = 1 kHz using 741 op-amp and a DC supply voltage of ± 12 V.

Solution: The square wave oscillator means astable multivibrator.

$$f_o = 1 \text{ kHz}$$

For op-amp astable multivibrator, f_o is given by,

$$f_o = \frac{1}{2 R_f C \ln\left[\dfrac{+V_{sat} - V_{LT}}{+V_{sat} - V_{UT}}\right]} \qquad \ldots(1)$$

where

$$V_{UT} = \frac{R_1}{R_1 + R_2}[+V_{sat}]$$

$$V_{LT} = \frac{R_1}{R_1 + R_2}[-V_{sat}]$$

Substituting in equation (1),

$$f_o = \frac{1}{2 R_f C \ln\left[\dfrac{+V_{sat} - \dfrac{R_1}{R_1+R_2}[-V_{sat}]}{+V_{sat} - \dfrac{R_1}{R_1+R_2}[+-V_{sat}]}\right]}$$

$$= \frac{1}{2 R_f C \ln\left[\dfrac{+V_{sat}\left(\dfrac{2R_1+R_2}{R_1+R_2}\right)}{+V_{sat}\left(\dfrac{R_2}{R_1+R_2}\right)}\right]}$$

$$\therefore \qquad f_o = \frac{1}{2R_f C \ln\left[\dfrac{2R_1 + R_1}{R_2}\right]}$$

It can be noted that f_o is independent of supply voltages.

Let $\qquad R_1 = 0.86\ R_2$

Choose $\qquad R_2 = 10\ k\Omega,\ R_1 = 8.6\ k\Omega$

$$\therefore \qquad \ln\left[\frac{2R_1 + R_2}{R_2}\right] = \ln\left[\frac{2 \times 0.86 R_2 + R_2}{R_2}\right] = \ln[2.72] = 1$$

$$\therefore \qquad f_o = \frac{1}{2R_f C} = 1 \times 10^3$$

Choose $\qquad C = 0.1\ mF$

$$\therefore \qquad R_f = \frac{1}{2 \times 0.1 \times 10^{-6} \times 1 \times 10^3} = 5\ k\Omega$$

The designed circuit is shown in the Fig. 8.38.

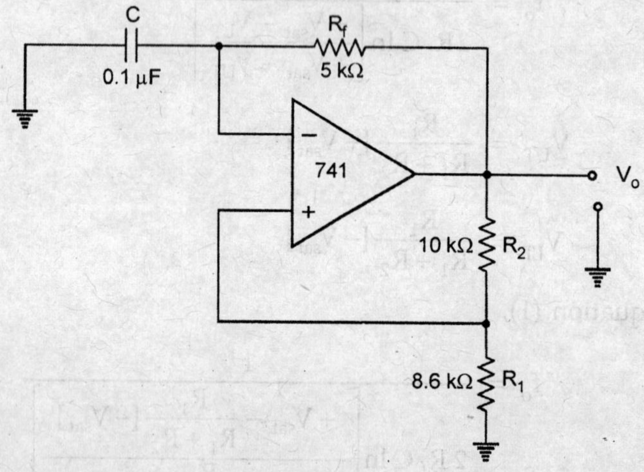

Fig. 8.38

Example 8.9: Design a square wave generator using op-amp to have output frequency of 2.5 kHz with variable duty cycle from 30 % to 70 %.

Solution: The circuit for non-symmetrical square wave is shown in the Fig. 8.39.

Let us design the circuit for f = 2.5 kHz.

Assume $\qquad R_1 = R_2 = 10\ k\Omega$

And $V_o = \pm V_{sat} = \pm 15$ V. When $V_o = +V_{sat} = +15$ V then C will charge through R_3 due to forward biased diode D_1.

And when $V_o = -V_{sat} = -15$ V. C will discharge through R_4 due to forward biased diode D_2. Selecting different combinations of R_3 and R_4, charging and discharging time can be varied and non-symmetrical square wave can be obtained.

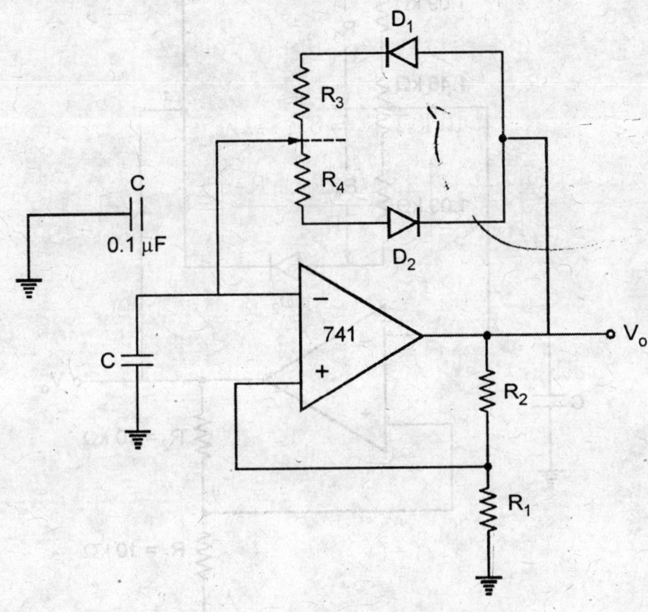

Fig. 8.39

Now
$$V_{UT} = \frac{R_1(+V_{sat})}{R_1 + R_2}$$
$$= 0.5 \times 15 = 7.5 \text{ V}$$
$$V_{LT} = \frac{R_1(-V_{sat})}{R_1 + R_2}$$
$$= 0.5 \times -15 = \mathbf{-7.5\ V}$$

Choose $C = 0.1\ \mu F$

The charging equation is,
$$V_{UT} = +V_{sat} + (V_{LT} - V_{sat})\, e^{-T_{ON}/R_3C}$$
$$\therefore \quad 7.5 = 15 + (-7.5 - 15)\, e^{-T_{ON}/R_3C}$$
$$\therefore \quad -\frac{T_{ON}}{R_3C} = \ln (0.333)$$

Now assume 50% duty cycle and design hence $R_3 = R_4$ and,
$$T_{ON} = \frac{1}{2}T = \frac{1}{2} \times \frac{1}{f}$$

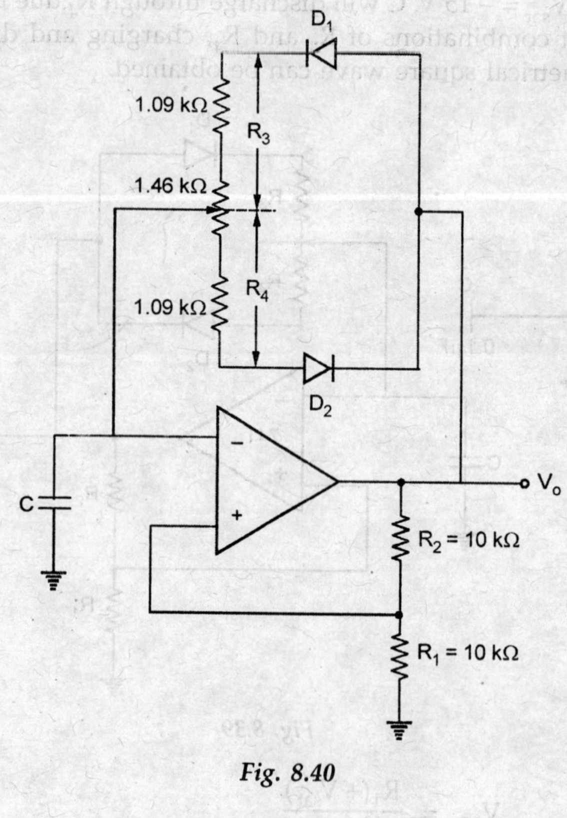

Fig. 8.40

$$= \frac{1}{2} \times \frac{1}{25 \times 10^3} = 2 \times 10^{-4} \text{ sec}$$

$$\therefore \quad R_3 = \frac{-2 \times 10^{-4}}{0.1 \times 10^{-6} \times \ln(0.333)}$$

$$= 1.8203 \text{ k}\Omega$$

$$\therefore \quad R_4 = \mathbf{1.8203 \text{ k}\Omega}$$

Now to vary the duty cycle from 30% to 70% divide $R_3 + R_4$ in the ratio 3 : 4 : 3 as shown in the Fig. 8.40. The modified circuit will give f = 2.5 kHz with variable duty cycle from 30% to 70%.

REVIEW QUESTIONS

1. Write the notes on:
   (1) Triangular wave generator
   (2) Monostable multivibrator using op-amp
   (3) Astable multivibrator using op-amp
   (4) RC phase shift oscillator using op-amp
   (5) Wien bridge oscillator using op-amp

2. Discuss the working of a RC phase shift oscillator.
3. Using an Op-amp, design a phase shift oscillator to oscillate at 100 Hz.
4. Design an Op-amp phase shift oscillator for f_0 = 500 Hz.
5. Explain the working of sine wave oscillator.
6. Draw and explain the operation of triangular/rectangular wave generator.
7. With a neat circuit diagram explain the working of RC phase shift oscillator.
8. What is the expression for the frequency of a RC phase shift oscillator? State the op-amp gain required for the oscillations.
9. In a RC phase shift oscillator, the phase shift network uses the resistances each of 4.7 kΩ and the capacitors each of 0.47 μF. Find the frequency of oscillations. **(Ans. : 29.413 Hz)**
10. Explain the basic circuit of Wien bridge oscillator.
11. Draw and explain the operation of Wien bridge oscillator using op-amp.
12. State the frequency of oscillation and minimum gain of op-amp required in Wien bridge oscillator.
13. Use ± 15 V power supply and 741. Design Wien bridge oscillator for 1 kHz frequency.
14. Compare RC phase shift oscillator and Wien bridge oscillator.
15. What is a multivibrator circuit ? What is its use ? State its types.
16. Draw and explain the operation of monostable multivibrator using op-amp.
17. Derive the expression for pulse width of monostable multivibrator.
18. Draw and explain the operation of astable multivibrator using op-amp.
19. Derive the expression for frequency of oscillations in astable multivibrator.
20. Write a note on ICL 8038 function generator.

❖❖❖

Chapter

9

555 Timer

9.1 INTRODUCTION TO TIMER IC 555

In most of the industries, operations are scheduled according to specific time requirements. In process industry, raw material is processed in different stages. In each stage, raw material is processed for a particular time period. For example, process may be the heating process and the heat may be required to be applied for, say, 5 minutes. There are number of applications where event must be delayed for specific delay periods. For example, one can snap by setting proper time period in automatic cameras. To achieve such timing requirements, the timer IC 555 is popularly used to build the timing circuits. This chapter explains the basic operation of timer IC 555 and its use as monostable and astable multivirator circuits.

9.2 BASIC CIRCUITS

When d.c. voltage is applied directly to a capacitor, it charges almost instantly to the value of the source voltage. The time required to charge a capacitor can be controlled by connecting a resistance in series with the capacitor as shown in Fig. 9.1.

The charging time increases with increase in resistance value. A high-value resistance will take more time to charge capacitor whereas a low-value resistance with the same capacitor will take shorter time to charge capacitor.

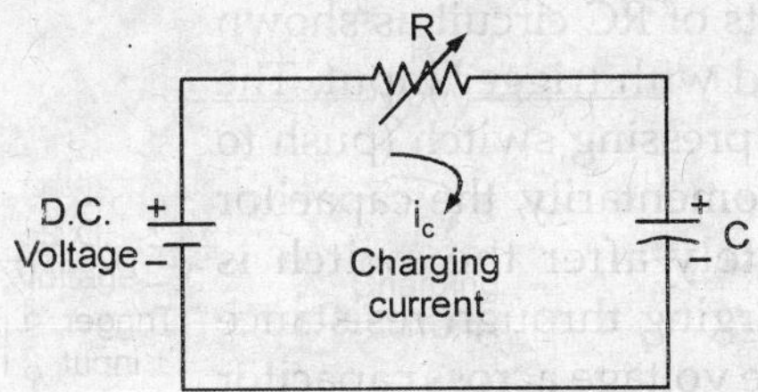

Fig. 9.1 Charging of Capacitor Through Series Resistance

The rate at which the capacitor charges is determined by the time constant. The **time constant** of a series RC circuit is known as RC-time constant and it is a product of the resistance and the capacitance. The symbol for the time constant is τ.

Mathematically, $\boxed{\tau = RC}$

where τ : Time constant in second

R : Resistance in MΩ, normally.

C : Capacitance in μF, normally.

Thus, the charging time (time constant) can be adjusted by selecting proper values of R and C.

So time constant of the R-C series circuit is defined **as time required by the capacitor voltage to rise from zero to 0.632 of its final steady state value during charging.**

Time constant is the time required for the charging current of the capacitor to drop to 0.368 of its initial maximum value.

$$\boxed{\tau = RC}$$

So **time constant** can be defined as,

(*i*) The time required for capacitor voltage to fall to 0.368 of its initial maximum value, while discharging from its initial maximum value.

(*ii*) The time required during which the capacitor discharge current falls to 0.368 of its initial maximum value.

9.3 BLOCK SCHEMATIC OF TIMER

The Fig. 9.2 shows the block schematic of timer. It consists of three input signals, one output signal and three units: Charging unit, comparator and output unit.

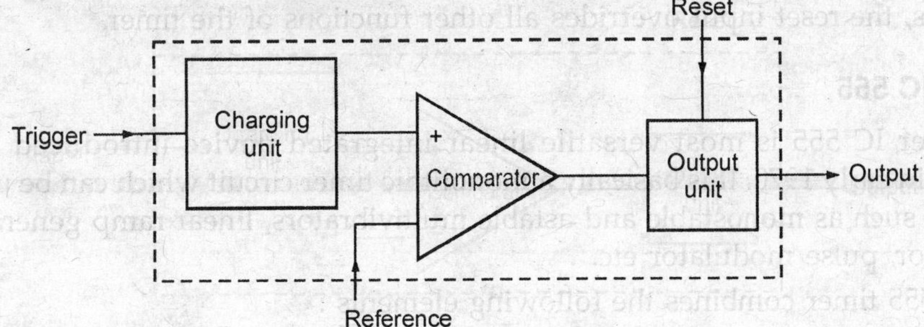

Fig. 9.2 Block Schematic of timer

The charging unit consists of RC circuit as shown in the Fig. 9.3. It is associated with trigger input. The trigger input is activated by pressing switch (push to ON). Pressing this switch momentarily, the capacitor can be discharged completely after the switch is released capacitor starts charging through resistance R, with time constant RC. The voltage across capacitor is applied to the +ve input of the comparator. The comparator compares this voltage with reference voltage applied at the reference input.

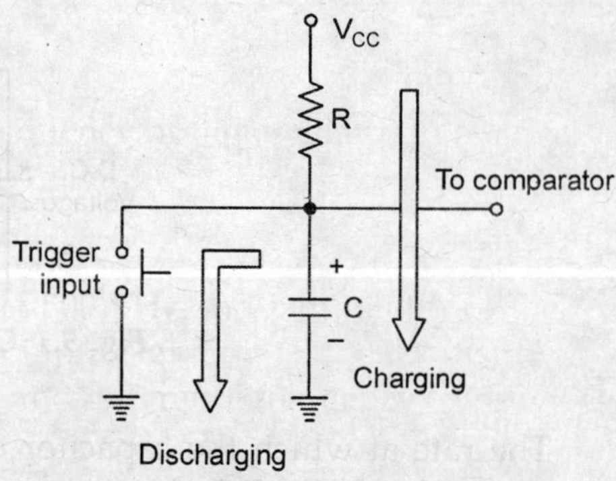

Fig. 9.3 Charging unit

When the voltage across capacitor is greater than reference voltage is greater than reference voltage, output is low; otherwise output is high. Immediately after trigger, output jumps to its high level and remains high until capacitor charges above level of reference voltage, as shown in the Fig. 9.4.

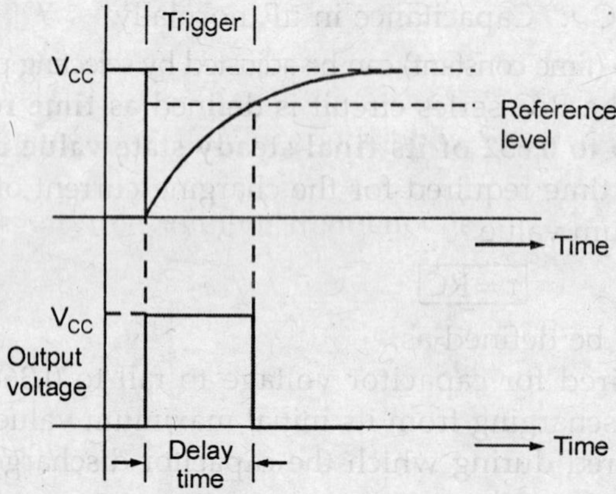

Fig. 9.4 Charging of capacitor and output waveform

The time period for which output remains high is controlled by adjusting time constant RC (R and C values).

The reset input to the timer is used to interrupt the timing interval and to reset the output. Thus, the reset input overrides all other functions of the timer.

9.4 TIMER IC 555

The timer IC 555 is most versatile linear integrated device introduced by Signetics corporation in early 1970. It is basically a monolithic timer circuit which can be used in many applications such as monostable and astable multivibrators, linear ramp generator, missing pulse detector, pulse modulator etc.

The IC 555 timer combines the following elements :

1. A relaxation oscillator
2. R-S flip-flop
3. Two comparators
4. Discharge transistor

9.4.1 R-S Flip-Flop

The Fig. 9.5 shows the part of the basic R-S flip-flop circuit. It uses a pair of cross-coupled transistors. Each collector drives the base of opposite transistor through a resistance R_B. In such a circuit, one transistor operates in saturation while the other in cut-off region. Thus if the transistor Q_2 is in saturation then its collector voltage is zero. This its collector voltage is zero. This voltage is the base drive of transistor Q_1. So there is no base drive for Q_1 and it goes into cut-off. Thus its collector voltage is approximately $+V_{CC}$. This drives the base of Q_2 which ensures that the Q_2 operates in saturation region.

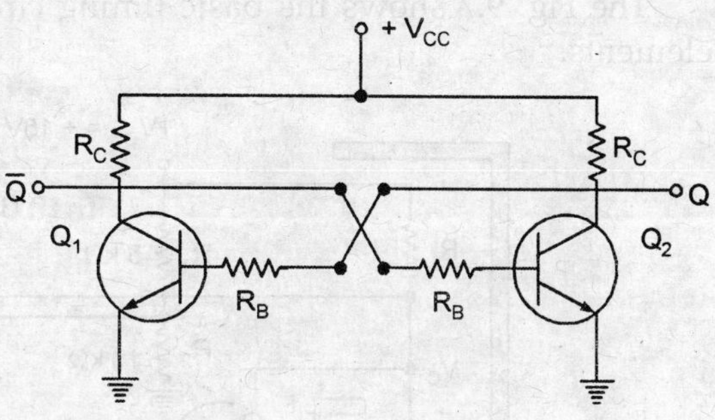

Fig. 9.5 R-S flip-flop

Now if transistor Q_2 goes into cut-off, its collector voltage approaches to $+V_{CC}$ which drives the transistor Q_1 into saturation. This makes collector voltage of Q_1 almost zero which keeps the transistor Q2 in cut-off. So all the time, **circuit ensures that one transistor operates in saturation and other in cut-off.**

The two outputs Q and $\bar{Q}$ are taken from two collectors. Depending upon the transistor operation, the output value gets decided.

Now this basic circuit along with some additional components gives us a popular R-S flip-flop circuit. It has two inputs set (S) and reset (R). With the help of these inputs, the output Q can be made high or low. Depending on Q, complementary will be the value of output $\bar{Q}$ i.e. low or high.

The Fig. 9.6 shows the symbol of R-S flip-flop circuit. When input set (S) is high, the output Q is high and $\bar{Q}$ is low. The high reset (R) input resets the output Q to low. The output Q remains in its state of high or low unless and untill it is triggered externally into the opposite state.

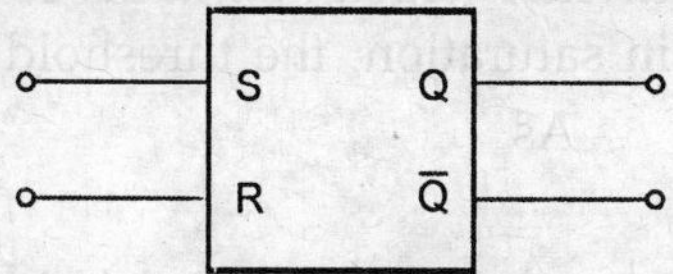

Fig. 9.6 Symbol of R-S flip-flop

The Table 9.1 shows the state of output Q relates to inputs S and R.

Table 9.1

	S	R	Output Q
No change →	0	0	NC
Flip-flop reset →	0	1	0 (low)
Flip-flop set →	1	0	1 (high)
Avoid this →	1	1	-

S = Set, R = Reset, NC = No Change, * = Race

To understand the operation of IC 555 timer, let us study the basic timing concept using R-S flip-flop.

9.4.2 Basic Timing Circuit

The Fig. 9.7 shows the basic timing circuit, which uses R-S flip-flop along some other elements.

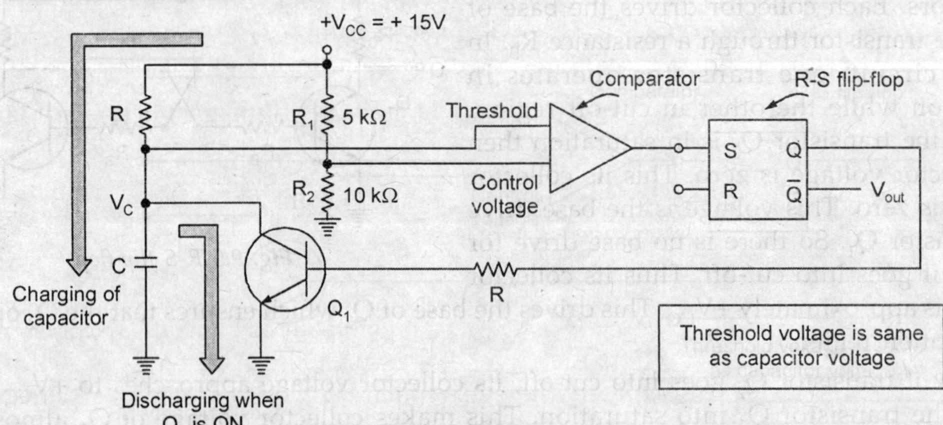

Fig. 9.7 Basic timing circuit

To understand the operation, consider that the output Q is high. This drives the base of Q1 and as it is high it drives Q_1 into saturation. It makes the capacitor voltage zero and as other end of capacitor is grounded, the capacitor is shorted. In this condition it cannot be charged.

The circuit uses a comparator. The noninverting input of comparator is called threshold voltage. While its inverting input is called **control voltage**. The R_1 and R_2 forms a potential divider which maintains control voltage constant at +10 V. As Q is high and transistor Q_1 is in saturation, the threshold voltage is zero.

As $R_1 = 5\ k\Omega$ and $R_2 = 10\ k\Omega$

$$\text{Control Voltage} = \frac{R_2}{R_1 + R_2} \times V_{CC} = \frac{10}{5+10} \times 15 = 10V \text{ constant}$$

Now if high voltage is applied to the reset (R) input of flip-flop then it resets R-S flip-flop and output Q goes low. This drives the transistor Q_1 in cut-off. Now the capacitor is free to charge and starts charging through resistance R. The threshold voltage thus starts increasing. When it becomes just greater than +10 V which is the control voltage, the comparator output goes high. This signal is driving the set (S) input of R-S flip-flop.

This changes the state of output Q back to high. This drives transistor Q_1 into saturation which quickly discharges the capacitor C.

The Fig. 9.8 shows the waveforms of threshold voltage and output voltage V_{out}. The charging of capacitor is exponential hence the threshold voltage is also exponential in nature. When Q goes low, the $\bar{Q}$ becomes high and positive going pulse appears at V_{out}. Similarly when capacitor voltages increases more than the control voltage, Q becomes high and $\bar{Q}$ becomes low. This brings V_{out} to zero instantly. Thus a rectangular output gets produced.

It can be observed that output remains high for the time which is required by the capacitor to charge upto control voltage, through R. Thus by varying R or C, the output pulse width can be varied. This is the working principle of timer IC 555.

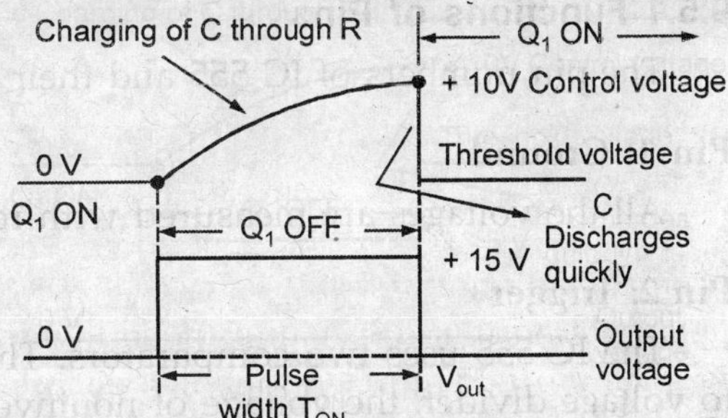

Fig. 9.8 Waveforms of basic timing circuit

9.5 BLOCK DIAGRAM OF IC 555

The Fig. 9.9 (a) and (b) show the pin diagram and the block diagram of the IC NE 555 timer. This is 8 pin IC timer.

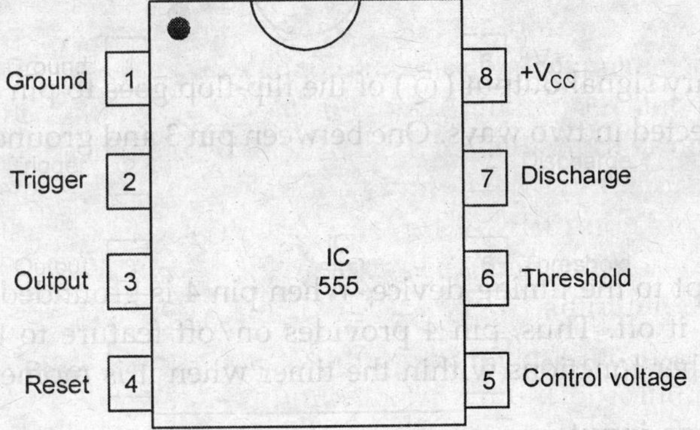

Fig. 9.9 (a) Pin diagram

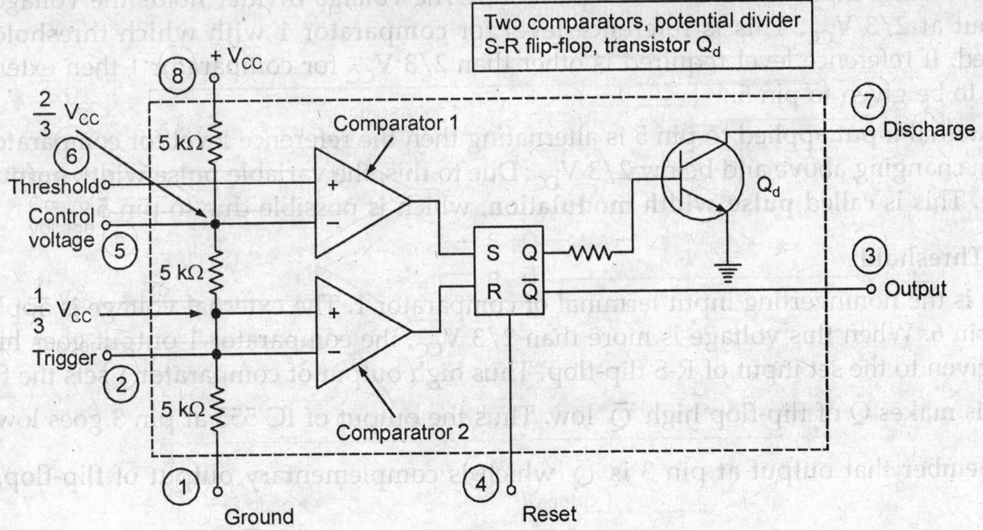

Fig. 9.9 (b) Block diagram of IC 555 timer

9.5.1 Functions of Pins

The pin numbers of IC 555 and their functions are discussed below:

Pin 1: Ground

All the voltages are measured with respect to this terminal.

Pin 2: Trigger

The IC 555 uses two comparators. The divider consists of three equal resistances. Due to voltage divider, the voltage of noninverting terminal of comparator 2 is fixed at $V_{CC}/3$. The inverting input of comparator 2 which is compared with $V_{CC}/3$, is nothing but trigger input brought out as pin number 2. When the trigger input is slightly less than $V_{CC}/3$ the comparator 2 output goes high. This output is given to reset input of R-S flip-flop. So high output of comparator 2 resets the flip-flop.

Pin 3 : Output

The complementary signal output ($\bar{Q}$) of the flip-flop goes to pin 3 which is the output. The load can be connected in two ways. One between pin 3 and ground while other between pin 3 and pin 8.

Pin 4 : Reset

This is an interrupt to the timing device. When pin 4 is grounded, it stops the working of device and makes it off. Thus, pin 4 provides on/off feature to the IC 555. This reset input overrides all other functions within the timer when it is momentarily grounded.

Pin 5 : Control Voltage input

In most of the applications, external control voltage input is not used. This pin is nothing but the inverting input terminal of comparator 1. The voltage divider holds the voltage of this input at $2/3\ V_{CC}$. This is reference level for comparator 1 with which threshold is compared. If reference level required is other than $2/3\ V_{CC}$ for comparator 1 then external input is to be given to pin 5.

If external input applied to pin 5 is alternating then the reference level for comparator 1 keeps on changing above and below $2/3\ V_{CC}$. Due to this, the variable pulse width output is possible. This is called **pulse width modulation**, which is possible due to pin 5.

Pin 6 : Threshold

This is the noninverting input terminal of comparator 1. The external voltage is applied to this pin 6. When this voltage is more than $2/3\ V_{CC}$, the comparator 1 output goes high. This is given to the set input of R-S flip-flop. Thus high output of comparator 1 sets the flip-flop. This makes Q of flip-flop high $\bar{Q}$ low. Thus the output of IC 555 at pin 3 goes low.

Remember that output at pin 3 is $\bar{Q}$ which is complementary output of flip-flop. In short,

For threshold $> \frac{2}{3} V_{CC}$, flip-flop → Q → high, output at pin 3 → low

For trigger $< \frac{1}{3} V_{CC}$, flip-flop → reset, Q → low, output at pin 3 → high

Pin 7 : Discharge

This pin is connected to the collector of the discharge transistor Q_d. When the output is high then Q is low and transistor Q_d is off. It acts as an open circuit to the external capacitor C to be connected across it, so capacitor C can charge as described earlier. When output is low, Q is high which drives the base of Q_d high, driving transistor Q_d in saturation. It acts as short circuit, shorting the external capacitor C to be connected across it.

Pin 8 : Supply $+V_{CC}$

The IC 555 timer can work with any supply voltage between 4.5 V and 16 V.

9.6 MONOSTABLE MULTIVIBRATOR USING IC 555

The IC 555 timer can be operated as a monostable multivibrator by connecting an external resistor and a capacitor as shown in the Fig. 9.10.

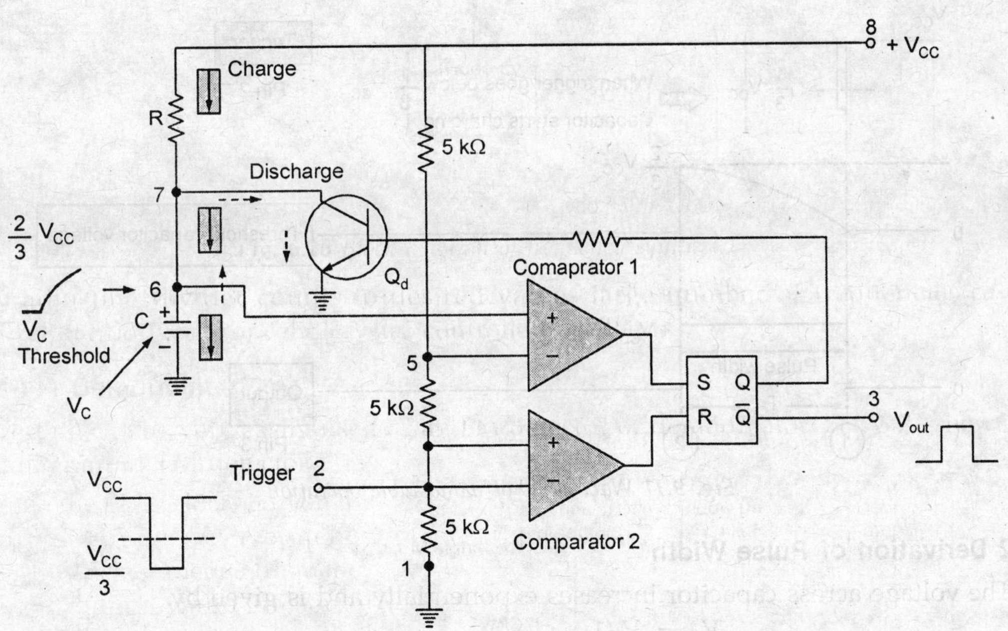

Fig. 9.10 Monostable operation of 555

The circuit has only one **stable state**. When trigger is applied, it produces a pulse at the output and returns back to its stable state. The duration of the pulse depends on the values of R and C. As it has only one stable state, it is called one shot multivibrator.

9.6.1 Operation

The flip-flop is initially set i.e. Q is high. This drives the transistor Q_d in saturation. The capacitor discharge completely and voltage across it is nearly zero. The output at pin 3 is low.

When a trigger input, a low going pulse is applied, then circuit state remains unchanged till trigger voltage is greater than 1/3 V_{CC}. When it becomes less than 1/3 V_{CC}, then comparator 2 output goes high. This resets the flip-flop so Q goes low and $\bar{Q}$ goes high. Low Q makes the transistor Q_d off. Hence capacitor starts charging through resistance R, as shown by dark arrows in the Fig. 9.10.

The voltage across capacitor increases exponentially. This voltage is nothing but the threshold voltage at pin 6. When this voltage becomes more than 2/3 V_{CC}, then comparator 1 output goes high. This set the flip-flop i.e. Q becomes high and $\bar{Q}$ low. This high Q drives the transistor Q_d in saturation. Thus capacitor C quickly discharges through Qd as shown by dotted arrows in the Fig. 9.11.

So it can be noted that V_{out} at pin 3 is low at start, when trigger is less than 1/3 V_{CC} it becomes high and when threshold is greater than 2/3 V_{CC} again becomes low, till next trigger pulse occurs. So a rectangular wave is produced at the output. The pulse width of this rectangular pulse is controlled by the charging time of capacitor. This depends on the time constant RC. Thus RC controls the pulse width. The waveforms are shown in the Fig. 9.11.

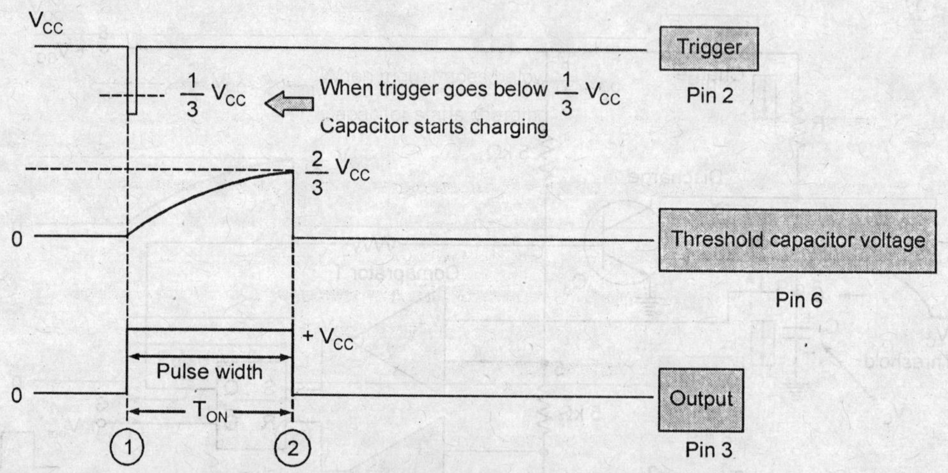

Fig. 9.11 Waveforms of monostable operation

9.6.2 Derivation of Pulse Width

The voltage across capacitor increases exponentially and is given by,

$$V_C = V(1 - e^{-t/CR})$$

If

$$V_C = 2/3\ V_{CC}$$

then

$$\frac{2}{3} V_{CC} = V_{CC}(1 - e^{-t/CR})$$

$$\frac{2}{3} - 1 = -e^{-t/CR}$$

$$\frac{1}{3} = e^{-t/CR}$$

$$\therefore \quad \frac{1}{CR} = -1.0986$$

$$\therefore \quad t = +1.0986 \text{ CR}$$

$$\therefore \quad t = 1.1 \text{ CR}$$

where C in farads, R in ohms, t in seconds.

Thus, we can say that voltage across capacitor will reach $2/3\ V_{CC}$ in approximately 1.1 times, time constant i.e. 1.1 RC.

Thus the pulse width denoted as W is given by,

$$\boxed{W = 1.1\ RC}$$

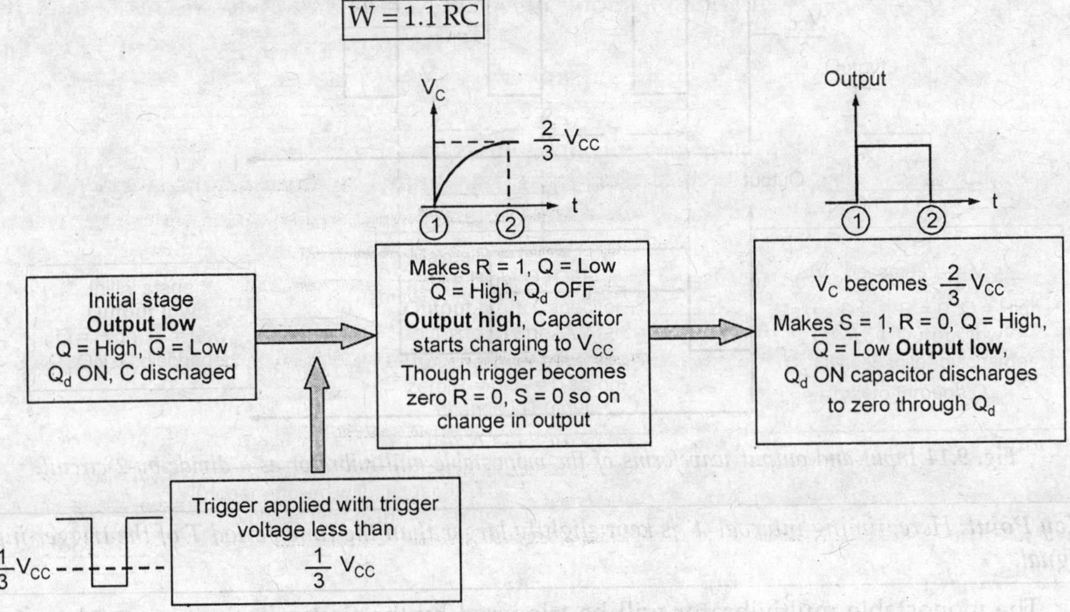

Fig. 9.12 Summary of monostable operation

9.6.3 Schematic Diagram

Generally a schematic diagram of the IC 555 circuits is shown which does not include comparators, flip-flop etc. It only shows the external components to be connected to the 8 pins of IC 555. Thus, the schematic diagram of IC 555 as a monostable multivibrator is shown in the Fig. 9.13.

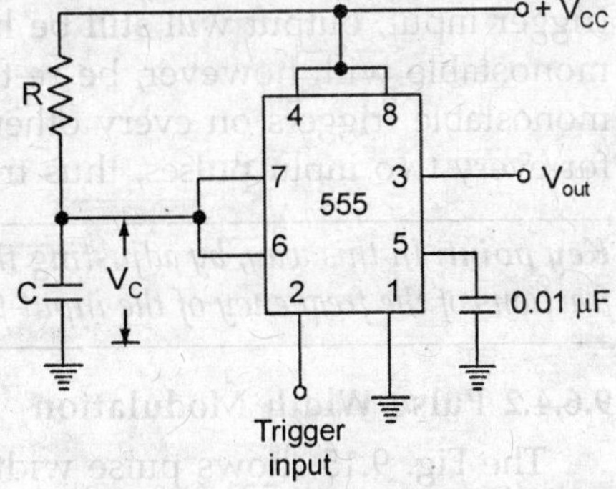

Fig. 9.13 555 timer as monostable multivibrator

The external components R and C are shown. To avoid accidental reset, pin 4 is connected to pin 8 which is supply $+V_{CC}$. To have the noise filtering of control voltage, the pin 5 is grounded through a small capacitor of 0.01 μF.

9.6.4 Application of Monostable Multivibrator

9.6.4.1 Frequency Divider

We know that, in monostable multivibrator, application of trigger pulse gives a positive going pulse on the output. The same monostable circuit can be used as a frequency divider if the timing interval is adjusted to be longer than the period of the input signal, as shown in the Fig. 9.14.

The Fig. 9.14 shows monostable multivibrator as a **divider-by-2 circuit**.

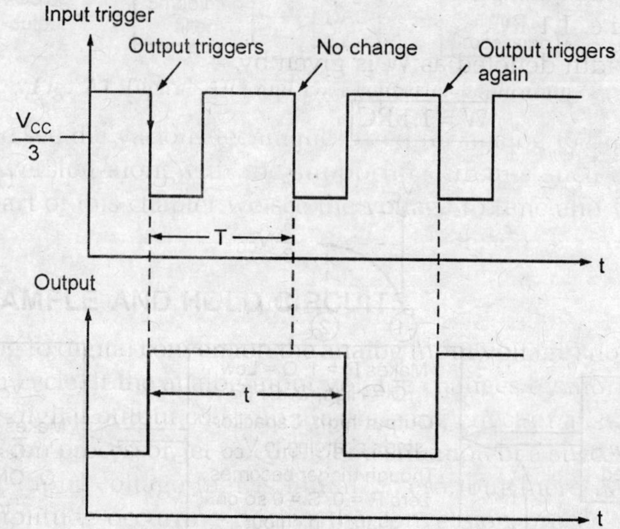

Fig. 9.14 Input and output waveforms of the monostable multivibrator as a divide-by-2 circuit.

***Key Point:** Here, timing interval 't' is kept slightly larger than the time period T of the trigger input signal.*

The monostable multivibrator will be triggered by the first negative going edge of the trigger input, which will make output to go in its high state. The output will remain high for the period equal to 'timing interval'. As timing interval is greater than time period of the trigger input, output will still be high when the second negative going pulse occurs. The monostable will, however, be re-triggered on the third negative-going pulse. Therefore, monostable triggers on every other pulse of the trigger input, so there is only one output for every two input pulses, thus trigger signal is, divided by 2.

***Key point:** In this way, by adjusting the timing interval, the monostable circuit can be made integral fractions of the frequency of the input trigger signal.*

9.6.4.2 Pulse Width Modulation

The Fig. 9.15 shows pulse width modulator. It is basically a monostable multivibrator with a modulating input signal applied at the control voltage input (pin 5). Internally, the control voltage is adjusted to the 2/3 V_{CC}. Externally applied modulating signal changes the control voltage and hence the threshold voltage level of the upper comparator (comparator 1). As a result, time period required to charge the capacitor upto threshold voltage level changes, giving pulse width modulated signal at the output as shown in the Fig. 9.16.

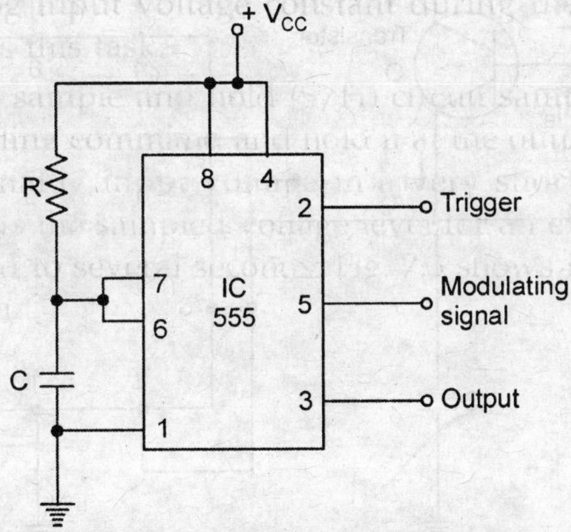

Fig. 9.15 Pulse width modulator

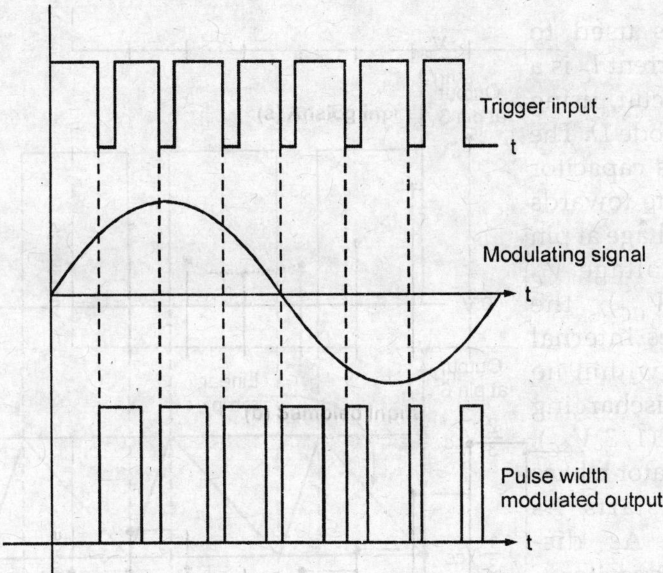

Fig. 9.16 Pulse width modulator waveforms

***Key Point:** It may be noted from the output waveform that the pulse duration varies according to the modulating signal level, but the frequency of the output pulses is same as that of the trigger input signal.*

9.6.4.3 Linear Ramp Generator

When a capacitor is charged with a constant current source then linear ramp is obtained. This concept is used in linear ramp generator. The circuit is shown in the Fig. 9.17.

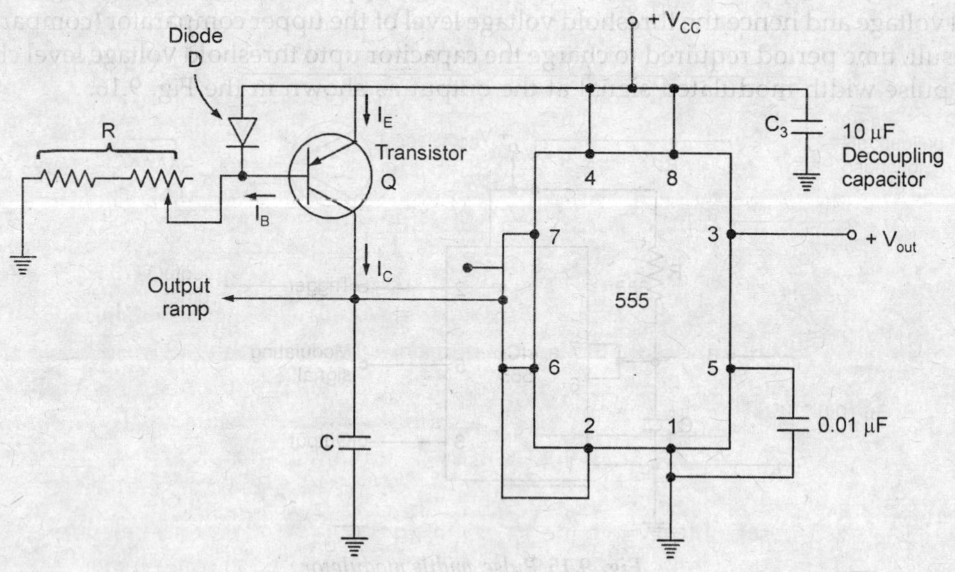

Fig. 9.17 Linear ramp generator

The circuit is used to obtain constant current I_C is a current mirror circuit, using transistor Q and diode D. The current I_C, charges capacitor C at a constant rate towards $+V_{CC}$. But when voltage at pin 6 *i.e.* capacitor voltage V_C becomes $(2/3\ V_{CC})$, the comparator makes internal transistor Q_1 ON within no time. But while discharging when V_C becomes $(1/3\ V_{CC})$, the second comparator makes Q_1 OFF and C starts its charging again. As discharging time of capacitor C is very small, the time period of ramp is assumed practically same as that of charging time of capacitor.

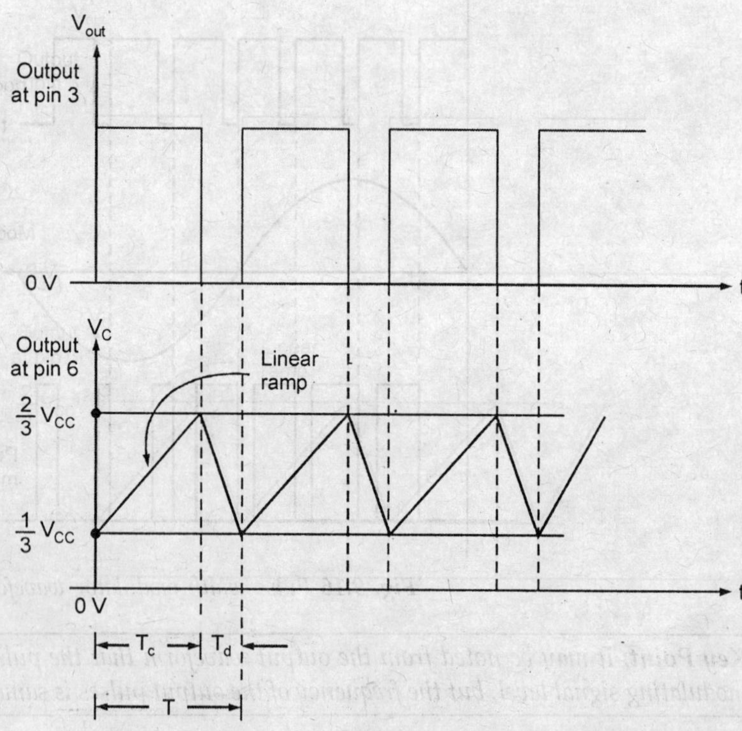

Fig. 9.18 Waveforms of ramp generator

The waveforms are shown in the Fig. 9.18.

The time period of the ramp is approximately given by,

$$T = \frac{V_{CC} C}{3 I_C} \text{ sec}$$

where $\quad I_C = \text{Charging current} = \frac{V_{CC} - V_D}{R} = \frac{V_{CC} - V_{BF}}{R}$

Hence the frequency is given by,

$$\boxed{f = \frac{1}{T} = \frac{3 I_C}{V_{CC} C} \text{ Hz}}$$

9.6.4.4 Missing Pulse Detector

The Fig. 9.19 (a) shows the circuit diagram for missing pulse detector. When signal input is at ground level (0 V), the emitter diode of transistor T_1 forward biases and clamps capacitor voltage V_C to 0.7 V. This forces output voltage to stay in its high state. When signal input goes high, the transistor cuts off and capacitor C begins to charge. If input signal again goes low before the 555 **completes** its timing cycle, the voltage across C is reset to 0.7 V. If, however, input signal does not go low before the 555 completes its timing cycle, the 555 enters its normal state and output voltage goes low. This is illustrated in Fig. 9.19 (b).

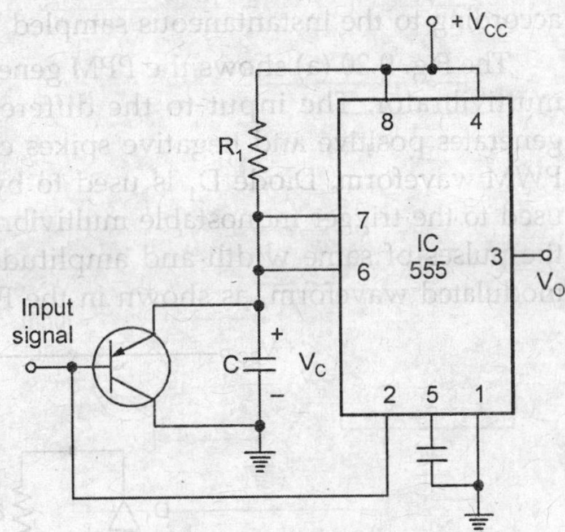

Fig. 10.19 (a) Missing pulse detector

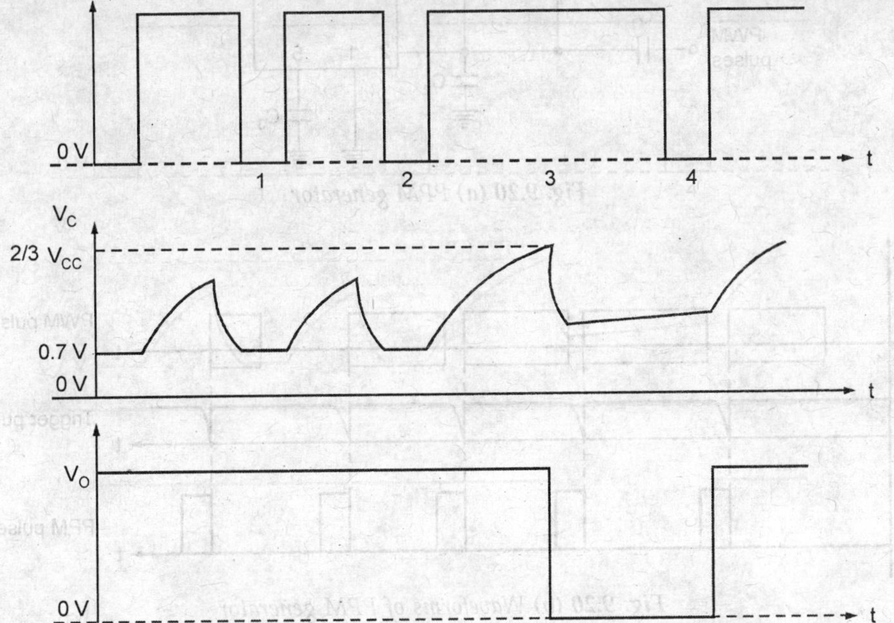

Fig. 9.19 (b) Waveforms for missing pulse detector

For this circuit timing interval is adjusted such that it is slightly longer than the period of input signal. The continuous low going pulses of the period less than the timing interval do not allow capacitor to charge upto 2/3 V_{CC}. As a result, output voltage remains high. In case of missing pulse (pulse 4), capacitor charges upto 2/3 V_{CC} and forces output voltage in to its low state, as shown in the Fig. 9.19 (b). This type of circuit can be used to detect a missing heart beat.

9.6.4.5 Pulse Position Modulation (PPM)

In pulse position modulation, the amplitude and width of the pulses are kept constant, while the position of each pulse with reference to the position of a reference pulse, is changed according to the instantaneous sampled value of the modulating signal.

The Fig. 9.20 (a) shows the PPM generator. It consists of differentiator and a monostable multivibrator. The input to the differentiator is a PWM waveform. The differentiator generates positive and negative spikes corresponding to leading and trailing edges of the PWM waveform. Diode D_1 is used to bypass the positive spikes. The negative spikes are used to the trigger monostable multivibrator. The monostable multivibrator then generates the pulses of same width and amplitude with reference to trigger to give pulse position modulated waveform, as shown in the Fig. 9.20 (b).

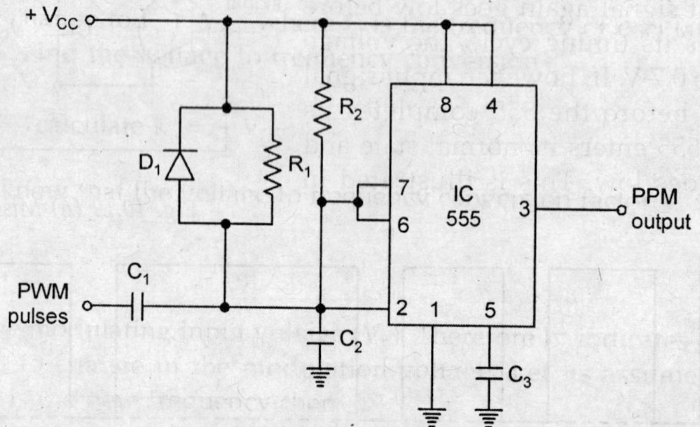

Fig. 9.20 (a) PPM generator

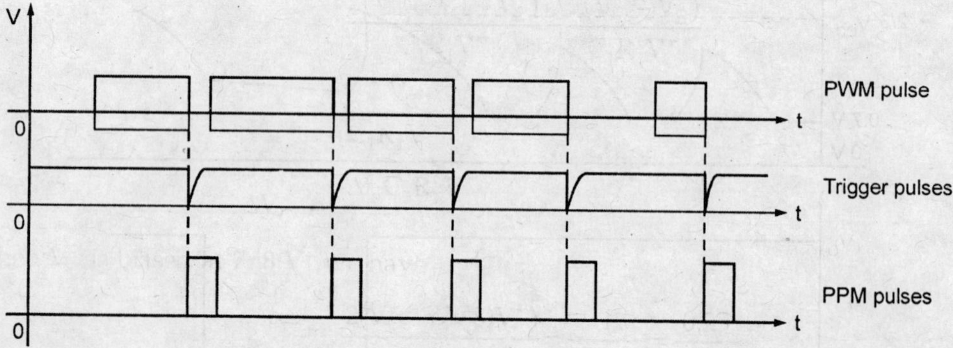

Fig. 9.20 (b) Waveforms of PPM generator

Example 9.1: Design a timer, which should turn ON heater immediately after pressing a push button and should hold heater in 'ON-state' for 5 seconds.

Solution: Fig. 9.21 shows monostable circuit used to drive the relay.

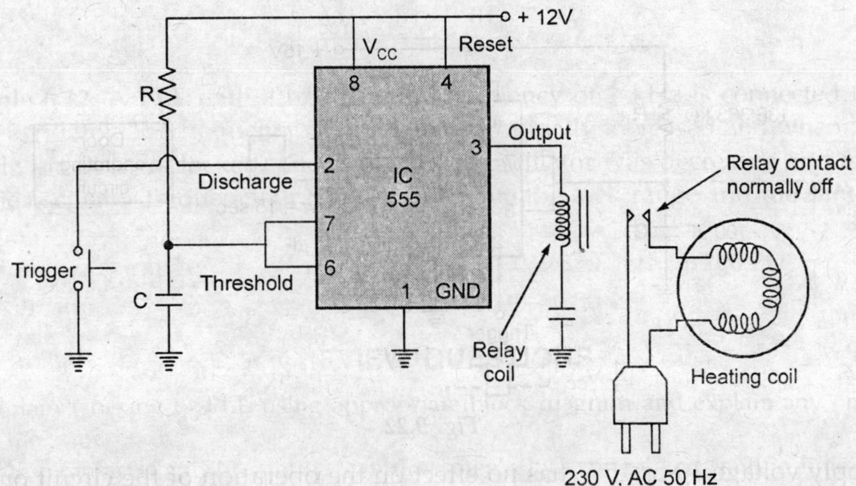

Fig. 9.21 Monostable multivibrator used to switch 'ON' relay for specific time

This relay should be energized for 5 seconds to hold heater 'ON' for 5 seconds. Thus, T_{ON} for monostable is 5 seconds.

We know that the pulse width is given by,

$$W = 1.1\ RC$$

$$\therefore \quad 5 = 1.1\ RC$$

Now, there are two unknowns. In this case, we have to select value for capacitor and with the selected value we have to find the value of resistance from the formula.

$\therefore$ If capacitor value is 10 μF

then

$$5 = 1.1 \times R \times 10\ \mu F$$

$$\therefore \quad R = \frac{5}{1.1 \times 10\ \mu F}$$

$$= 454545.45\ \Omega = \mathbf{454.54\ k\Omega}$$

The calculated value is not standard value, but we can adjust this value by connecting variable resistance i.e. potentiometer.

Example 9.2 : Draw the circuit diagram of timer using IC 555. Calculate the component values if the controlled door should remain open for 15 secs after a trigger signal is received. The d.c. voltage available is either 10 or 15 volts.

Solution : The requirements is that the door must be open 15 sec after receiving a trigger signal and then gets shut down automatically. This requires IC 555 in a monostable mode with a pulse width of 15 sec.

$$\therefore \quad W = 15\ sec$$

Now

$$W = 1.1\ RC$$

$$\therefore \quad 15 = 1.1\ RC$$

Choose $C = 100\ \mu F$

$\therefore$ $R = 136.363\ k\Omega$

The designed circuit is shown in the Fig. 9.22.

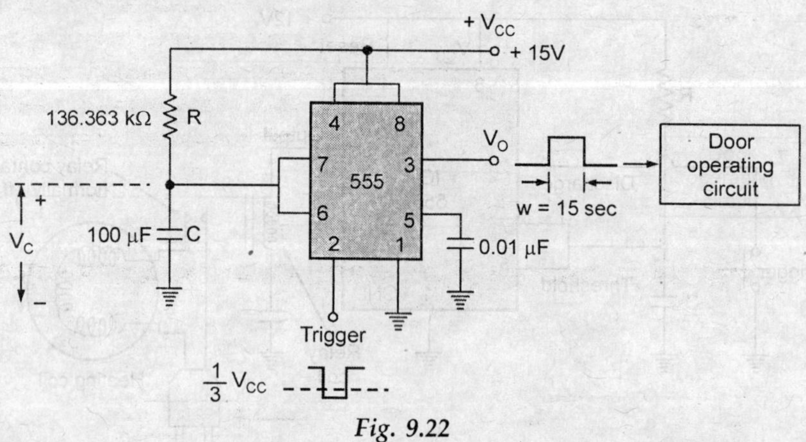

Fig. 9.22

The supply voltage 10 or 15 V has no effect on the operation of the circuit or the values of R and C selected.

9.7 ASTABLE MULTIVIBRATOR USING IC 555

The Fig. 9.23 shows the IC 555 connected as an astable multivibrator. The threshold input is connected to the trigger input. Two external resistance R_A, R_B and a capacitor C is used in the circuit.

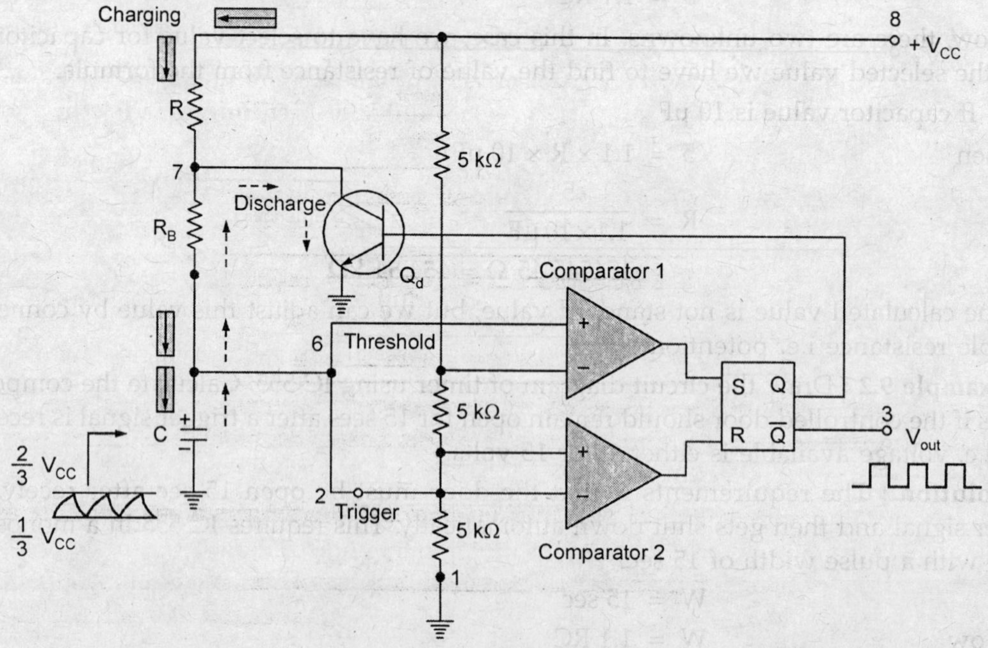

Fig. 9.23 Astable operation of 555

This circuit has no stable state. The circuits changes its state alternately. Hence the operation is also called free running nonsinusoidal oscillator.

9.7.1 Operation

When the flip-flop is set, Q is high which drives the transistor Q_d in saturation and the capacitor gets discharged. Now the capacitor voltage is nothing but the trigger voltage. So while discharging, when it becomes less than $1/3\ V_{CC}$, comparator 2 output goes high. This resets the flip-flop hence Q goes low and $\bar{Q}$ goes high.

The low Q makes the transistor off. Thus capacitor starts charging through the resistances R_A, R_B and V_{CC}. The charging path is shown by thick arrows in the Fig. 9.23. As total resistance in the charging path is $(R_A + R_B)$, the charging time constant is $(R_A + R_B)$ C.

Now the capacitor voltage is also a threshold voltage. While charging, capacitor voltage increases i.e. the threshold voltage increases. When it exceeds $2/3\ V_{CC}$, then the comparator 1 output goes high which sets the flip-flop. The flip-flop output Q becomes high and output at pin 3 i.e. $\bar{Q}$ becomes low. High Q drives transistor Q_d in saturation and capacitor starts discharging through resistance R_B and transistor Q_d. This path is shown by dotted arrows in the Fig. 9.23. Thus the discharging time constant is $R_B C$. When capacitor voltage becomes less than $1/3\ V_{CC}$, comparator 2 output goes high, resetting the flip-flop. This cycle repeats.

Thus when capacitor is charging, output is high while when it is discharging the output is low. The output is a rectangular wave. The capacitor voltage is exponentially rising and falling. The waveforms are shown in the Fig. 9.24.

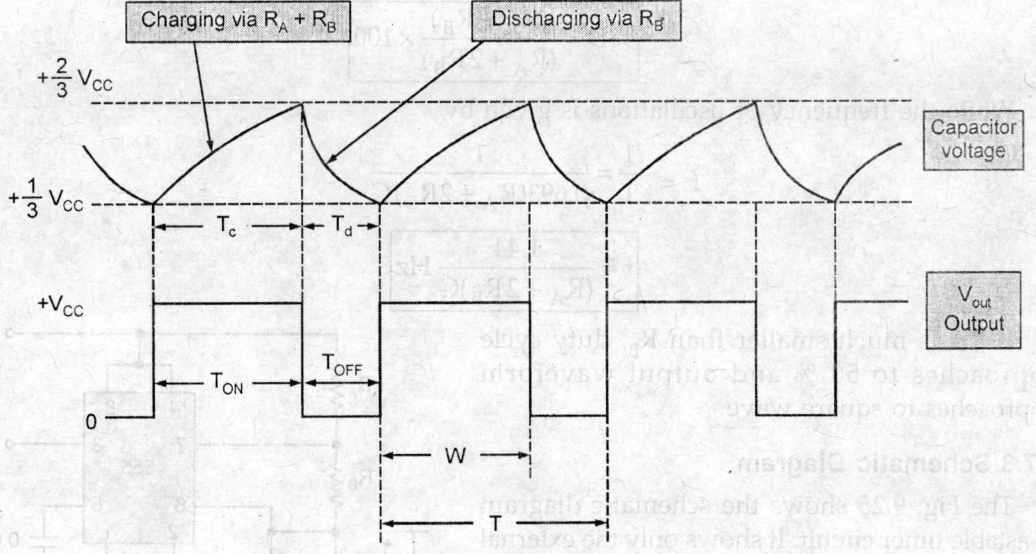

Fig. 9.24 Waveforms of astable operation

9.7.2 Duty Cycle

Generally the charging time constant is greater than the discharging time constant. Hence at the output, the waveform is not symmetric. The high output remains for longer period than low output. The ratio of high output period and low output period is given by a mathematical parameter called duty cycle. It is defined as the ratio of ON time i.e. high output to the total time of one cycle. As shown in the Fig. 9.24.

$$W = \text{Time for output is high} = \text{TON}$$

$$T = \text{Time of one cycle}$$

$$\therefore \quad D = \text{Duty cycle}$$

$$\therefore \quad \boxed{\% D = \frac{W}{T} \times 100 \ \%}$$

The charging time for the capacitor is given by,

$$T_c = \text{Charging time} = 0.693 \ (R_A + R_B) \ C$$

While the discharge time is given by,

$$T_d = \text{Discharging time} = 0.693 \ R_B C$$

Hence the time for one cycle is,

$$T = T_c + T_d = 0.693 \ (R_A + R_B) \ C + 0.0693 \ R_B \ C$$

$$\therefore \quad \boxed{T = 0.693 \ (R_A + 2R_B) \ C}$$

while $\quad W = Tc = 0.693 \ (R_A + R_B) \ C$

$$\therefore \quad \% D = \frac{W}{T} \times 100 = \frac{0.693 (R_A + R_B) C}{0.693 \ (R_A + 2 \ R_B) C} \times 100$$

$$\therefore \quad \boxed{\% D = \frac{(R_A + R_B)}{(R_A + 2R_B)} \times 100}$$

While the frequency of oscillations is given by,

$$f = \frac{1}{T} = \frac{1}{0.693 (R_A + 2R_B) C}$$

$$\therefore \quad \boxed{f = \frac{1.44}{(R_A + 2R_B) C} \ \text{Hz}}$$

If RA is much smaller than R_B, duty cycle approaches to 50 % and output waveform approaches to square wave.

9.7.3 Schematic Diagram

The Fig. 9.25 shows the schematic diagram of astable timer circuit. It shows only the external components R_A, R_B and C. The pin 4 is tied to pin 8 and pin 5 is grounded through a small capacitor.

The important application of astable multivibrator is voltage controlled oscillator (VCO).

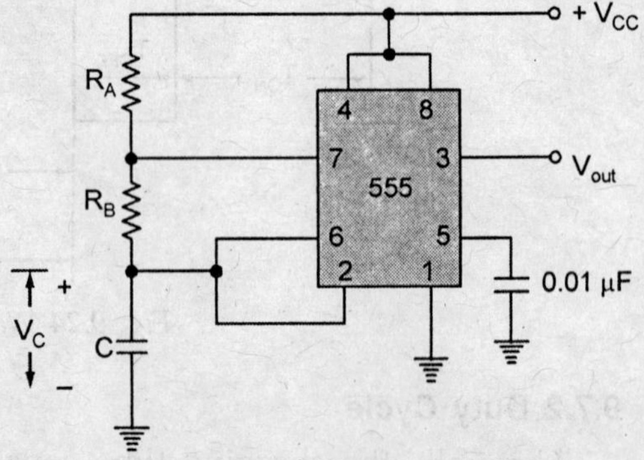

Fig. 9.25 555 timer as astable multivibrator

Example 9.3: A 555 timer is configured to run in astable with RA = 4 kΩ, R_B = 4 kΩ and C = 0.01 μF. Determine the frequency of the output and duty cycle.

Solution : The frequency of output is given by,

$$f = \frac{1.44}{(R_A + 2R_B)C} = \frac{1.44}{(4+2\times4)\times10^3\times0.01\times10^{-6}}$$

$$= \mathbf{12\ kHz}$$

The duty cycle is given by,

$$D = \frac{R_A + R_B}{R_A + 2R_B}$$

$$= \frac{4+4}{4+(2\times4)} = 0.6667$$

Thus the duty cycle is **66.67 %**

9.7.4 Application of Astable Multivibrator

9.7.4.1 Square Wave Generator

It can be observed from the expression of duty cycle that in astable exact 50 % duty cycle is not possible to achieve. To get exactly 50 % duty cycle i.e. square wave output it is necessary to modify the astable timer circuit.

The modified astable circuit used to obtain the square wave output is shown in the Fig. 9.26.

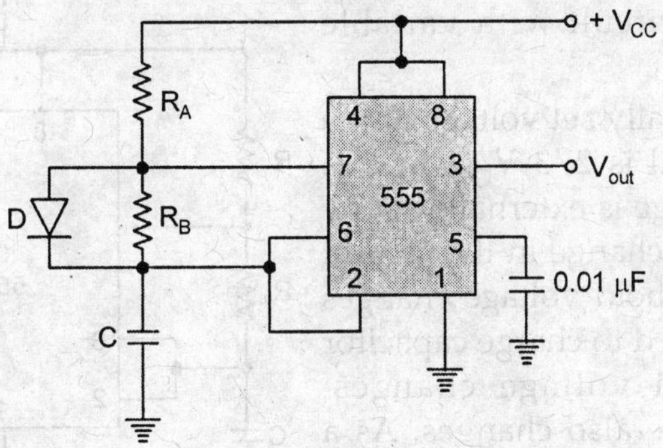

Fig. 9.26 Square wave generator

In the modified circuit, the capacitor C discharges through R_B. To obtain square wave (50 % duty cycle) resistance R_B is adjusted such that it is equal to the summation of resistance RA and the forward resistance of diode D. Usually, potentiometer is used for exact adjustment of resistors.

The waveforms of square wave generator are shown in the Fig. 9.27.

The other application are,

1. Voltage controlled oscillator
2. FSK generator
3. Flasher circuit.

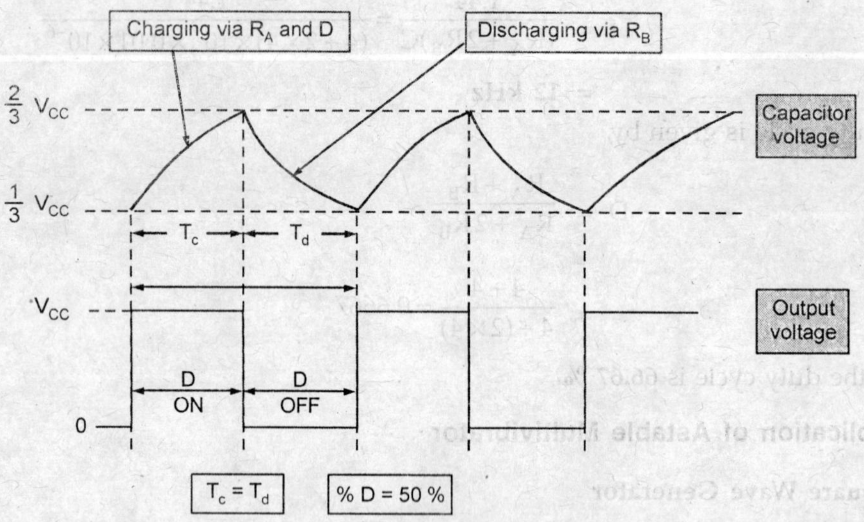

Fig. 9.27 Waveforms of square wave generator

9.7.4.2 Voltage Controlled Oscillator (VCO)

The Fig. 9.28 shows the circuit diagram for voltage controlled oscillator. It is basically an astable multivibrator circuit with variable control voltage.

We know that internally set voltage at the control voltage terminal is 2/3 V_{CC}. In this circuit, the control voltage is externally set by the potentiometer. With change in the control voltage, the upper threshold voltage changes and thus the time required to charge capacitor upto upper threshold voltage changes. Similarly, discharge time also changes. As a result, the frequency of the output voltage changes.

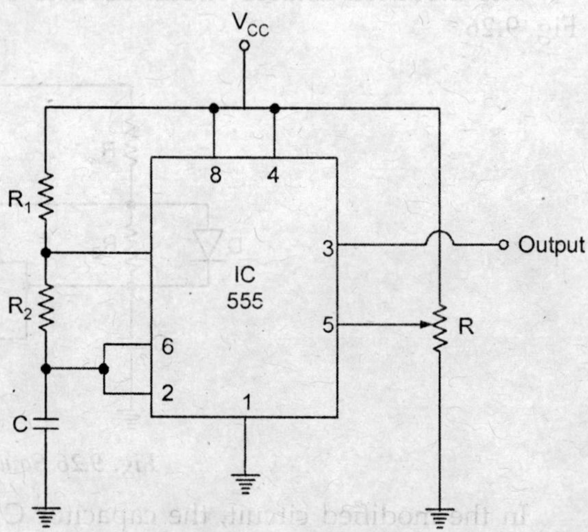

Fig. 9.28 Voltage controlled oscillator

If control voltage is increased, the capacitor will take more time to charge and discharge and therefore frequency will decrease. On the otherhand, if control voltage is decreased, the capacitor will take less time to charge and discharge, increasing the frequency of the output signal. Thus by varying control voltage we can change the frequency.

9.7.4.3 FSK Generator

Binary code consists of 1's and 0's. It can be transmitted by shifting a carrier frequency. One fix frequency represents one and other represents zero. This type of transmission if called frequency shift keying (FSK) technique. Astable multivibrator using 555 can be used to generate FSK signal.

The circuit for FSK generation is as shown in the Fig. 9.29.

When digital input is HIGH (logic 1), transistor T_1 is OFF and 555 timer works in a normal astable mode. The frequency of the output waveform can be given as

$$\boxed{f_o = \frac{1.45}{(R_1 + 2R_2)C}}$$

When input is LOW (logic 0), transistor T_1 is ON and connects the resistance R_p in parallel with R_1. With this connecting effective R_{left} becomes $R_1 \parallel R_p$, and output frequency is now given by

$$\boxed{f_o = \frac{1.45}{[(R_1 \parallel +2R_2)]C}}$$

With this connection effective resistance Reff becomes $R_1 \; | \; | \; R_p$.

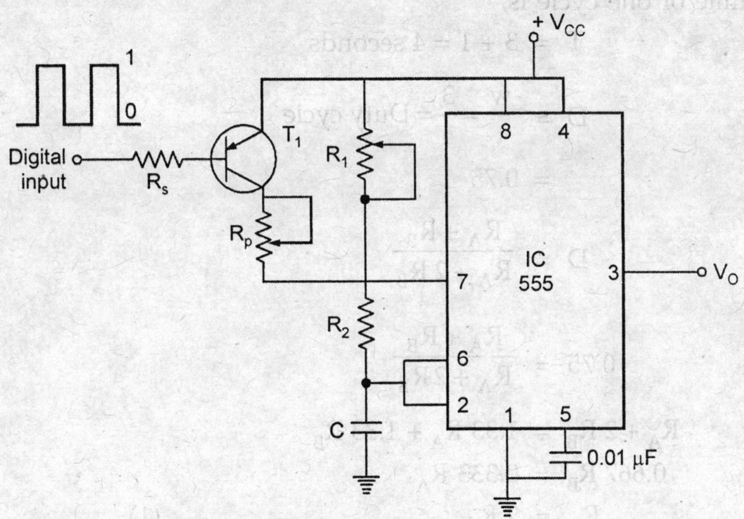

Fig. 9.29 FSK generator

Example 9.4 : Design an astable multivibrator which will flash the electric bulb such that its ON time will be 3 seconds and off time will be 1 seconds.

Solution : Fig. 9.30 shows astable circuit used to drive relay.

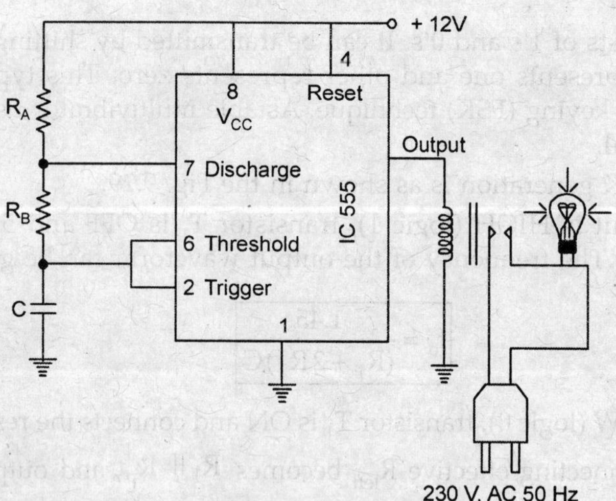

Fig. 9.30 Astable multivibrator used to flash electric bulb with specific frequency

This relay should be energized for 3 seconds and then de-energize by 1 seconds. Hence charging time of capacitor is 3 seconds which is a pulse width W.

$$\therefore \quad W = 3 \text{ seconds}$$

While total time of one cycle is,

$$\therefore \quad T = 3 + 1 = 4 \text{ seconds}$$

$$\therefore \quad D = \frac{w}{t} = \frac{3}{4} = \text{Duty cycle}$$

$$= 0.75$$

But
$$D = \frac{R_A + R_B}{R_A + 2\,R_B}$$

$$\therefore \quad 0.75 = \frac{R_A + R_B}{R_A + 2R_B}$$

$$\therefore \quad R_A + 2\,R_B = 1.33\,R_A + 1.33\,R_B$$

$$\therefore \quad 0.667\,R_B = 0.333\,R_A$$

$$\therefore \quad R_A = 2\,R_B \qquad \text{... (1)}$$

Now the charging time is given by,

$$Tc = 0.693\,(R_A + R_B)\,C$$

$$\therefore \quad 3 = 0.693\,(R_A + R_B)\,C$$

Choose
$$C = 10\,\mu F$$

$$\therefore \quad R_A + R_B = \frac{3}{0.693 \times 10 \times 10^{-6}}$$

$\therefore \quad R_A + R_B = 4.329 \times 10^5$ ... (2)

Solving equation (1) and (2) we get,

$$R_A = \mathbf{288.6\ k\Omega}$$

$$R_B = \mathbf{144.3\ k\Omega}$$

The values are not standard but can be adjusted using the potentiometers.

Example 9.5 : Design a 555 based square wave generator to produce a symmetrical square wave of 1 kHz. If VCC = 12 V, draw the voltage across timing capacitor and the output.

Solution : The circuit used for the square wave generator is shown in the Fig. 9.31.

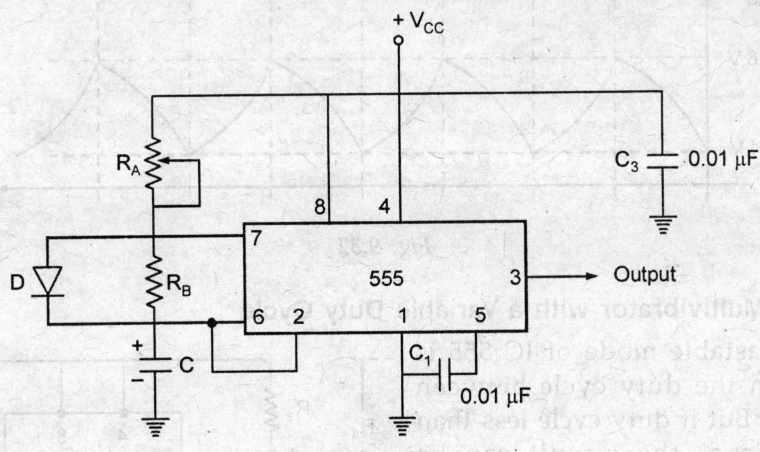

Fig. 9.31

For square wave, charging and discharging resistors must be equal.

The resistance R_1 is a combination of a fixed resistor and a potentiometer. So that potentiometer can be adjusted for the exact square wave.

$$f = 1\text{ kHz}$$

$$T = \frac{1}{f} = 1 \times 10^{-3}\text{ s}$$

$\therefore \quad T_d = \text{Discharge} = \frac{T}{2} = 0.5\text{ msec}$

Now $\quad T_d = 0.69\text{ RB C and } T_c = T_d \text{ for D} = 0.5 \text{ i.e. } 50\ \%$

Choose $\quad C = 0.1\ \mu F$

$\therefore \quad 0.5 \times 10^{-3} = 0.69\ R_B \times 0.1 \times 10^{-6}$

$\therefore \quad R_B = 7.25\ k\Omega$

For square wave, $\quad R_A = R_B$

$\therefore \quad R_A = 100\ \Omega + 10\ k\Omega \text{ pot}$

With pot precise square wave can be obtained.

The output voltage and capacitor voltage waveforms are shown in the Fig. 9.32.

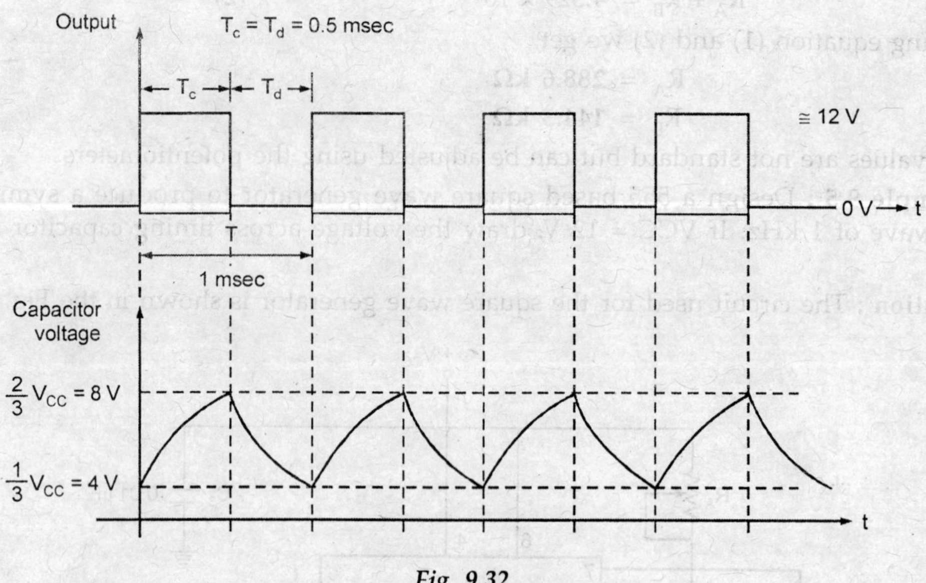

Fig. 9.32

9.7.5 Astable Multivibrator with a Variable Duty Cycle

Generally astable mode of IC 555 is used to obtain the duty cycle between 50 % to 100 %. But if duty cycle less than 50 % is required, the circuit can be modified as shown in the Fig. 9.33

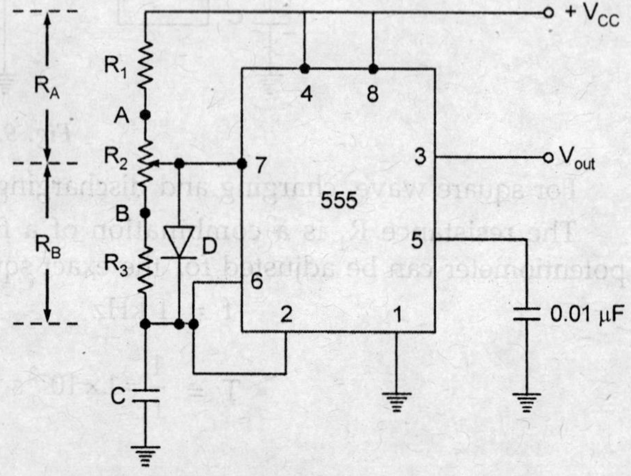

Fig. 9.33

The circuit is similar to square wave generator, but instead of connecting diode across fixed R_B it is connected across a combination of variable R_2 and R_3. The resistance R_2 is variable and it can be added to R_1 or R_3 to change the charging and discharging resistance. Thus a variable duty cycle can be achieved.

The charging time for the capacitor is,

$$T_c = (R_1 + \text{Part of } R_2)\ 0.693\ C$$

The discharging time for the capacitor is,

$$T_d = (R_3 + \text{Remaining part of } R_2)\ 0.693\ C$$

$$\therefore \quad T = T_c + T_d = 0.693\ C\ [R_1 + R_2 + R_3]$$

$$\therefore \quad \boxed{f = \frac{1}{T} = \frac{1.44}{(R_1 + R_2 + R_3)C}}$$

Range of duty possible:

Let position of R_2 potentiometer is A i.e. part of R_2 in series with R_1 is zero and entirely R_2 gets added to R_3

$\therefore$ $$T_c = R_1\, 0.693\, C$$

and $$T_d = 0.693\,(R_2 + R_3)\, C$$

$\therefore$ $$T = 0.693\, C\,(R_1 + R_2 + R_3)$$

So there is no change in frequency but,

$$\%\, \text{Dmin} = \frac{T_c}{T} \times 100 = \frac{0.693\, R_1\, C}{0.693\, C(R_1 + R_2 + R_3)} \times 100$$

$\therefore$ $$\boxed{\%D_{min} = \frac{R_1}{R_1 + R_2 + R_3} \times 100}$$

Thus minimum duty cycle, less than 50 % is possible.

For R_2 potentiometer position at B, maximum duty cycle is possible.

$\therefore$ $$T_c = 0.693\,(R_1 + R_2)\, C$$

and $$T_d = 0.693\, R_3\, C$$

and $$T = 0.693\, C(R_1 + R_2 + R_3)$$

$\therefore$ $$\boxed{\%D_{max} = \frac{R_1 + R_2}{R_1 + R_2 + R_3} \times 100}$$

Example 9.6 : Design an astable multivibrator for an output frequency of 1 kHz but a variable duty cycle of 30% to 70%. Assume VCC = 12 V.

Solution : For variable cycle, use the modified circuit as shown in the Fig. 9.34.

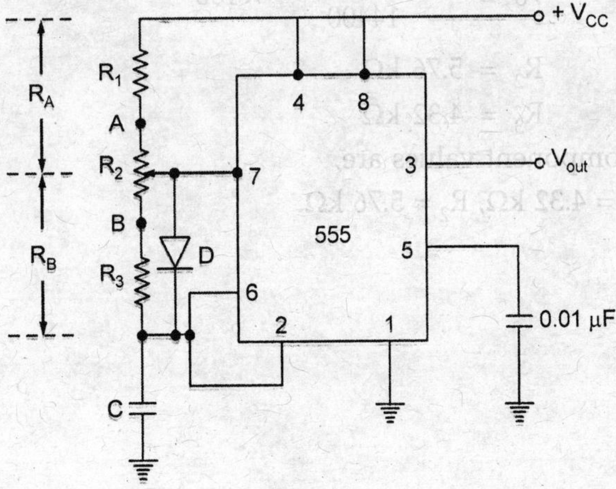

Fig. 9.34

Now $$f = \frac{1.44}{C(T_1 + R_2 + R_3)}$$

Choose $$C = 0.1\ \mu F$$

$$\therefore \quad 1 \times 10^3 = \frac{1.44}{0.1 \times 10^{-6}(R_1 + R_2 + R_3)}$$

$$\therefore \quad R_1 + R_2 + R_3 = 14400 \qquad \ldots (1)$$

$$\% D_{min} = \frac{R_1}{14400} \times 100$$

$$\therefore \quad 30 = \frac{R_1}{R_1 + R_2 + R_3} \times 100$$

Using equation (1) $$30 = \frac{R_1}{14400} \times 100 \qquad \ldots (2)$$

$$\therefore \quad R_1 = 4.32\ k\Omega$$

Now $$\%D_{max} = \frac{R_1 + R_2}{R_1 + R_2 + R_3} \times 100 \qquad \ldots (3)$$

$$\therefore \quad 70 = \frac{R_1 + R_2}{R_1 + R_2 + R_3} \times 100$$

Using equation (1) and (3),

$$70 = \frac{4.32 \times 10^3 + R_2}{14400} \times 100$$

$$\therefore \quad R_2 = 5.76\ k\Omega$$

Using equation (1) $R_3 = 4.32\ k\Omega$

Thus the various component values are,

$C = 0.1\ \mu F$, $R_1 = R_3 = 4.32\ k\Omega$, $R_2 = 5.76\ k\Omega$.

9.8 COMPARISON OF MULTIVIBRATOR CIRCUITS

Sr. No.	*Monostable multivibrator*	*Astable multivibrator*
1.	It has only one stable state.	There is no stable state.
2.	Trigger is required for the operation, to change the state	Trigger is not required to change the state, hence called free running.
3.	Two components R and C are necessary with IC 555 to obtain the circuit.	Three components RA, RB and C are necessary with IC 555 to obtain the circuit.
4.	The pulse width is given by, W = 1.1 RC seconds	The frequency is given by $f = \frac{1.44}{(R_A + 2R_B)} \text{Hz}$
5.	The frequency of operation is controlled by frequency of trigger pulses applied.	The frequency of operation is controlled by R_A. R_B and C.
6.	The applications are, timer, frequency divider, pulse width modulation etc.	The applications are square wave generator, flasher, voltage controlled oscillator, FSK generator etc.

Example 9.7 : Draw the circuits diagram of an astable multivibrator to generate the output signal with frequency of 1 kHz and the duty cycle of 75%.

Solution :

$$f = 1 \text{ kHz}$$

$$D = 75\ \% = 0.75$$

Now $$f = \frac{1.44}{(R_A + 2R_B)C} \text{Hz}$$

$$\therefore \quad 1 \times 10^3 = \frac{1.44}{(R_A + 2R_B)C}$$

$$\therefore \quad (R_A + 2\,R_B)\,C = 1.44 \times 10^{-3}$$

$$\therefore \text{ while} \quad \%\,D = \frac{(R_A + R_B)}{(R_A + 2R_B)} \times 100 \qquad \dots (1)$$

$$\therefore \quad 0.75 = \frac{R_A + R_B}{R_A + 2R_B}$$

$$\therefore \quad R_A + 2\,R_B = \frac{(R_A + R_B)}{0.75}$$

$$\therefore \quad RA + 2\,R_B = 1.33\,(R_A + R_B)$$

$$\therefore \quad 0.66\,R_B = 0.33\,R_A$$

$$\therefore \quad R_B = 0.5\,R_A$$

Choose $$C = 0.1\ \mu F \qquad \dots (2)$$

Substituting in equation (1),

$$(R_A + 2R_B) \times 0.1 \times 10^{-6} = 1.44 \times 10^{-3}$$

$$\therefore \quad R_A + 2\,R_B = 14400 \qquad \text{... (3)}$$

Substituting equation (2) in equation (3),

$$R_A + 2(\,0.5\,R_A) = 14400$$

$$\therefore \quad R_A = 7.2\ k\Omega$$

$$\therefore \quad R_B = 3.6\ k\Omega$$

and $\quad C = 0.1\ \mu F$

Hence the circuit diagram is as shown in the Fig. 9.35.

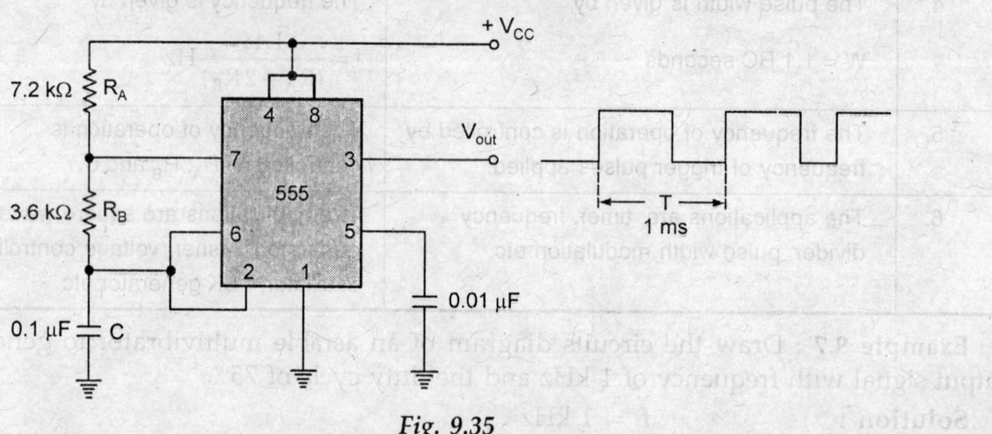

Fig. 9.35

Example 9.8 : An astable multivibrator is to be designed for getting rectangular waveform with t_{ON} = 0.3 ms. Draw the circuit diagram with various component values. Also calculate frequency of oscillations and duty cycle. Assume total time period (T) to be 1 ms.

Solution : T_{ON} = 0.6 ms, T = 1 ms

$$\therefore \quad D = \frac{t_{ON}}{T} = \frac{0.6}{1} = 60\%$$

Now $$D = \frac{R_A + R_B}{R_A + 2R_B} = 0.6$$

$$\therefore \quad R_A + R_B = 0.6\,R_A + 1.2\,R_B$$

$$\therefore \quad 0.4\,R_A = 0.2\,R_B$$

$$\therefore \quad R_B = 2\,R_A$$

$$f = \frac{1.44}{(R_A + 2R_B)C} = \frac{1}{T} = 1000$$

Choose $\quad C = 0.1\ \mu F$

$$\therefore \quad R_A + 2R_B = 14400$$

Using equation (1), $\quad 5R_A = 14400$

$$\therefore \quad R_A = 2.88\ k\Omega,\ R_B = 5.77\ k\Omega$$

The circuit is shown in the Fig. 9.36.

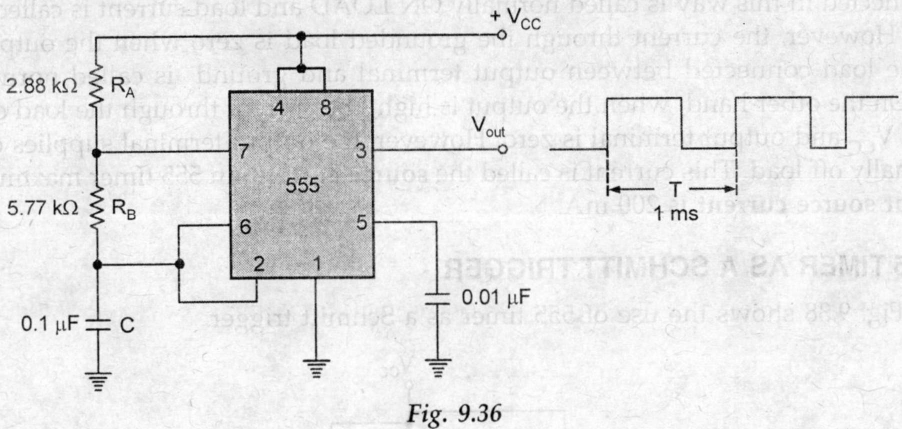

Fig. 9.36

9.9 FEATURES OF IC 555 TIMER

The various features of the IC 555 timer are :

1. The 555 is a monolithic timer device which can be used to produce accurate and highly stable time delays or oscillation. It can be used to produce time delays ranging from few microseconds to several hours.
2. It has two basic operating modes : monostable and astable.
3. It is available in three packages : 8-pin metalcan, 8-pin mini DIP or a 14-pin DIP. A 14-pin package is IC 556 which consists of two 555 timers.
4. The NE 555 (Signetics) can operate with a supply voltage in the range of 4.5 V to 18 V and is capable of sourcing and sinking output currents of 200 mA. Its CMOS version (TLC 555) can operate over a supply range of 2 V to 18 V and has output current sinking and sourcing capabilities of 100 mA and 10 mA, respectively.
5. It has a very high temperature stability, as it is designed to operate in the temperature range of –55°C to 125°C.
6. Its output is capatible with TTL, CMOS and Op-amp circuits.

9.10 CONNECTING LOAD TO IC 555

The output of 555 is used to drive load (controlling devices) such as transistors and relays. Therefore are two ways of connecting load to output terminal : either between output terminal (pin 3) and ground (pin 1) or between output terminal (pin 3) and supply voltage (pin 8) as shown in Fig. 9.37.

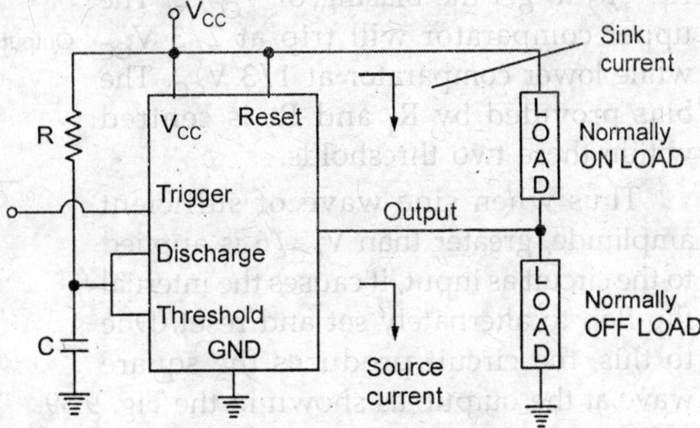

Fig. 9.37 Connection of normally OFF load and normally ON load

Normally output is low. Thus, load current flows through the load connected between VCC and output terminal. The load connected in this way is called normally ON LOAD and load current is called the **sink current**. However, the current through the grounded load is zero when the output is low. Thus, the load connected between output terminal and ground is called normally OFF LOAD. On the other hand, when the output is high, the current through the load connected between V_{CC} and output terminal is zero. However, the output terminal supplies current to the normally off load. This current is called the **source current.** In 555 timer maximum value of **sink or source current** is 200 mA.

9.11 555 TIMER AS A SCHMITT TRIGGER

The Fig. 9.38 shows the use of 555 timer as a Schmitt trigger.

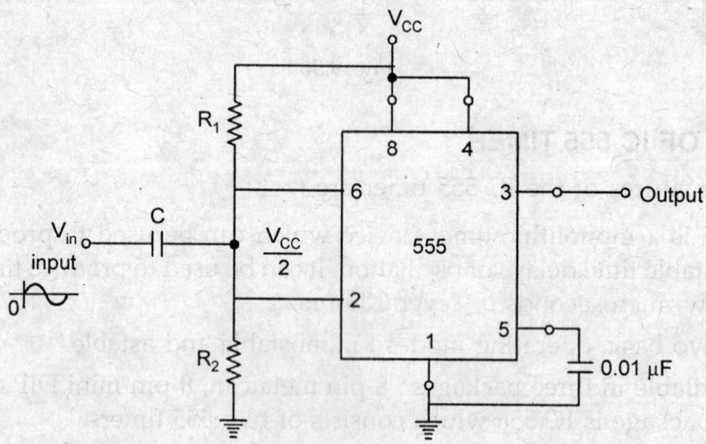

Fig. 9.38 555 as a Schmitt trigger

The input is given to the pins 2 and 6 which are tied together. Pins 4 and 8 are connected to supply voltage $+V_{CC}$. The common point of two pins 2 and 6 is externally biased at $V_{CC}/2$ through the resistance network R_1 and R_2. Generally $R_1 = R_2$ to get the biasing of $V_{CC}/2$. The upper comparator will trip at $2/3$ V_{CC} while lower comparator at $1/3$ V_{CC}. The bias provided by R_1 and R_2 is centred within these two thresholds.

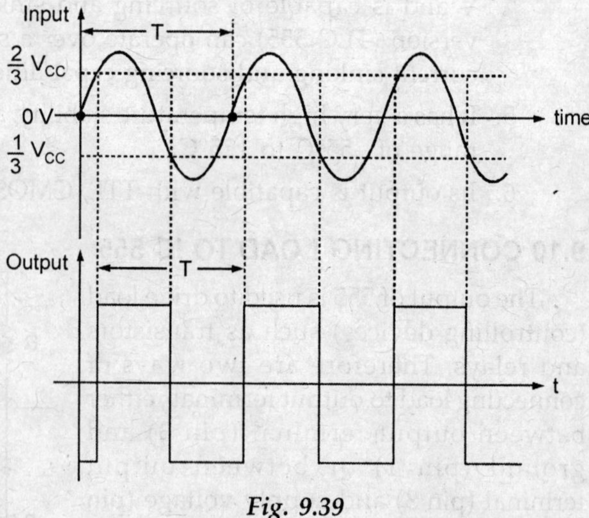

Fig. 9.39

Thus when sine wave of sufficient amplitude, greater than $V_{CC}/6$ is applied to the circuit as input, it causes the internal flip-flop to alternately set and reset. Due to this, the circuit produces the square wave at the output, as shown in the Fig. 9.39.

The frequency of square wave remains same as that of input. The Schmitt trigger can operate with the input frequencies upto 50 kHz.

Example 9.9 : Draw the circuit diagram of IC 555 used as the astable mode to generate square wave of 1 kHz frequency, giving output equal to 5 V for $\frac{1}{2}$ msec and output equal to 0 V for next $\frac{1}{2}$ msec. Connect one red lamp and one green lamp so that for $\frac{1}{2}$ msec red lamp is ON and green is OFF and for next $\frac{1}{2}$ msec green lamp is ON and red is OFF. Lamp have ratings of 5 V and 50 mA.

Solution : $T_{ON} = T_{OFF} = 0.5$ msec

$\therefore$ $T = T_{ON} + T_{OFF} = 1$ msec

i.e. $f = \frac{1}{T} = 1\,\text{kHz}$

Now $Td = TOFF = 0.69\ R_B C$

Choose $C = 0.1\ \mu F$

$\therefore$ $0.5 \times 10^{-3} = 0.69\ R_B \times 0.1 \times 10^{-6}$

$\therefore$ $R_B = 7.246\ k\Omega$

Now duty cycle is 50 % so RA = RB = 7.246 kΩ

Practically a modified circuit is required for 50 % duty cycle where diode is connected across R_B and charging takes place through R_A and diode. And R_B must be equal to sum of R_A and diode forward resistance. So to have perfect square wave, R_A is kept variable i.e. pot of say 10 kΩ, in this case. It is then adjusted to obtain precise square wave. The resistance required in series with LED to be connected is,

$$R = \frac{V_o - V_{LED}}{I_{LEd}}$$

Assuming $V_{LED} = 0.7$ V

Current limiting $R = \frac{5-0.7}{50\times 10^{-3}} = 86\,\Omega \approx 100\,\Omega$

The wattage of R is,

$$P = (50 \times 10^{-3})^2 \times 100$$

$$= 0.25 \text{ W}$$

Both resistors R can be of $\frac{1}{4}$ W.

The required circuit is shown in the Fig. 9.40.

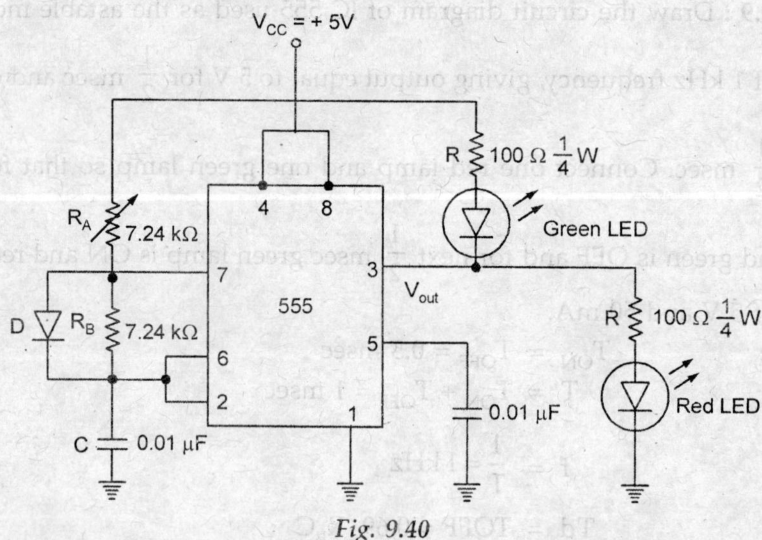

Fig. 9.40

Example 9.10 : Design a ramp generator using 555 timer having output frequency of 5 kHz, with V_{CC} = + 5 V.

Solution : f = 5 kHz, V_{CC} = + 5 V

Assume $\quad V_D = V_{BE} = 0.7\text{ V}$

$$\therefore \quad I_C = \frac{V_{CC} - V_D}{R} = \frac{V_{CC} - 0.7}{R} = \frac{4.3}{R} \quad \text{... (1)}$$

and $$f = \frac{3I_C}{V_{CC}C}$$

$$\therefore \quad 5 \times 10^3 = \frac{3 \times 4.3}{RC \times 5}$$

$$\therefore \quad R_C = 5.16 \times 10^{-4} \quad \text{... (2)}$$

Choose $\quad C = 0.05\ \mu F$

$$\therefore \quad R = \frac{5.16 \times 10^{-4}}{0.05 \times 10^{-6}} = 10.32\text{ k}\Omega$$

Use 10 kΩ resistance and 1 kΩ pot for precise adjustment.

$$I_C = \frac{4.3}{10.32 \times 10^3} = 0.416\text{ mA}$$

Example 9.11 : Design a ramp generator using 555 timer having output frequency of 5 kHz, with V_{CC} = +5 V.

Solution : f = 5 kHz, V_{CC} = +5V

Assume $\quad V_D = V_{BE} = 0.7\text{ V}$

$$\therefore \quad I_C = \frac{V_{CC} - V_D}{R} = \frac{V_{CC} - 0.7}{R} = \frac{4.3}{R} \quad ...(1)$$

and
$$f = \frac{3I_C}{V_{CC}C}$$

$$\therefore \quad 5 \times 10^3 = \frac{3 \times 4.3}{RC \times 5}$$

$$\therefore \quad RC = 5.13 \times 10^{-4}$$

Choose
$$C = 0.05\ \mu F \quad ...\ (2)$$

$$\therefore \quad R = \frac{5.13 \times 10^{-4}}{0.05 \times 10^{-6}} = 10.32\ k\Omega$$

Use 10 kΩ resistance and 1 kΩ pot for precise adjustment.

$$I_C = \frac{4.3}{10.32 \times 10^3} = 0.416\ mA$$

Example 9.12 : Design an astable multivibrator having an output frequency of 10 kHz with a duty cycle of 25 %, using IC 555.

Solution : f = 10 kHz and % D = 25 %

$$\therefore \quad T = \frac{1}{f} = \frac{1}{10 \times 10^3} = 0.1\ ms$$

$$\therefore \quad D = \frac{T_{ON}}{T} \text{ i.e. } T_{ON} = 0.25\ T = 0.025\ ms$$

and
$$T_{OFF} = T - T_{ON} = 0.25\ T = 0.025\ ms$$

Duty cycle is less than 50 % hence modified circuit must be used.

The modified circuit is shown in the Fig. 9.41.

Choose
$$C = \mathbf{0.01\ \mu F}$$

The charging of C takes place through R_A and diode D while the discharging of C takes place through R_B only.

Assuming ideal diode,

$$T_{ON} = 0.693\ R_{AC} \text{ and } T_{OFF} = 0.693\ R_BC$$

$$\therefore \quad R_A = \frac{0.025 \times 10^{-3}}{0.01 \times 10^{-6} \times 0.693} = 3.607\ k\Omega$$

and
$$R_B = \frac{0.075 \times 10^{-3}}{0.01 \times 10^{-6} \times 0.693} = 10.82\ k\Omega$$

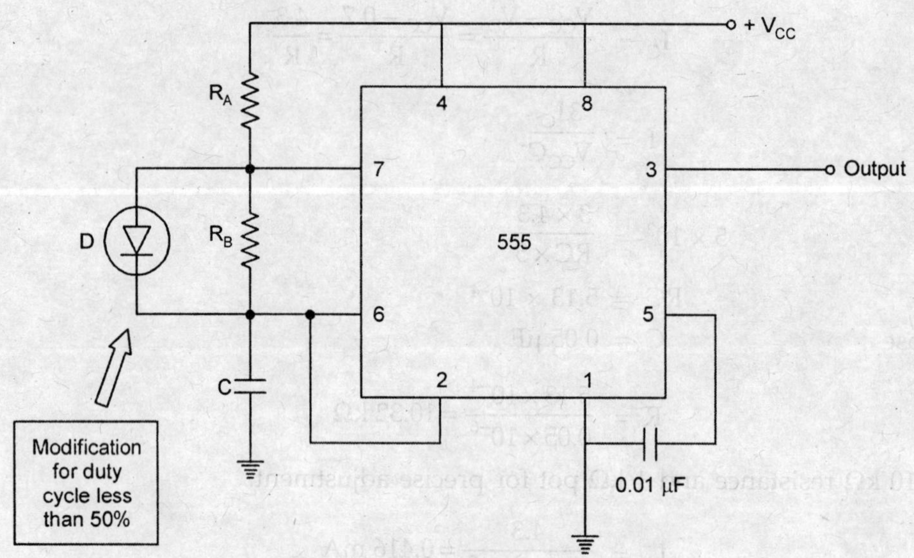

Fig. 9.41

Thus for C = 0.01 μF, R_A = 3.607 kΩ and R_B = 10.82 kΩ the modified circuit gives duty cycle of 25%.

Example 9.13 : A 555 timer configured in astable mode with R_A = 2 kohm, R_B = 4 kohm and C = 0.1 μF. Determine the frequency of the output and duty cycle.

Solution : This is astable mode of operation.

$$\therefore \quad f = \frac{1.44}{(R_A + 2R_B)C} = \frac{1.44}{[2 \times 10^3 + 2 \times 4 \times 10^3] \times 0.1 \times 10^{-6}}$$

$$= \mathbf{1.44\ kHz}$$

$$\%\, D = \frac{R_A + R_B}{R_A + 2R_B} = \frac{2 \times 10^3 + 4 \times 10^3}{2 \times 10^3 + 2 \times 4 \times 10^3} \times 100 = 60\%$$

REVIEW QUESTIONS

1. Explain the functional block diagram of IC 555.
2. State the functions of each pin of IC 555.
3. Name the two packages in which IC 555 timer is available.
4. Explain the working of a timer circuit.
5. Explain the use of IC 555 as a monostable multivibrator.
6. Derive the expression for the pulse width of a monostable multivibrator using IC 555.
7. State any two applications of monostable multivibrator.
8. Determine the output pulse width of the monostable multivibrator using IC 555, if R = 12 kΩ and C = 0.1 μF. (Ans. : 1.32 ms)
9. A IC 555 timer used as a monostable has R = 20 kΩ and C = 0.01 μF. What is the duration of output pulse ? (Ans. : 0.22 ms)

10. Estimate the values of timing elements to generate a time delay of 1 ms using IC 555 timer as a monostable multivibrator.
11. Explain with neat sketch, the working of IC 555 as an astable multivibrator.
12. Define duty cycle. Derive the expression for the duty cycle of an astable multivibrator using 555.
13. Derive the expression for the frequency of the output of an astable multivibrator.
14. How astable mode of 555 can be modified to get a square wave generator ?
15. List the features of IC 555 timer.
16. A 555 timer is configured to run in astable mode with R_1 = 20 kΩ and R_2 = 8 kΩ and C = 0.1 μF. Determine the output frequency and duty cycle. (Ans.: 400 Hz, 77.77 %)
17. Design an astable multivibrator using 555 timer to produce a 1 kHz square waveform for duty cycle D = 0.50.
18. State the applications of astable multivibrator using IC 555.
19. Design a 555 based square wave generator to produce a symmetrical square wave of 2 kHz. If V_{CC} = 12 volts, draw the voltage across timing capacitor and the output?
20. Explain the internal structure of IC μA 78S40 switching voltage regulator.
21. List the important features of IC μA 78S40.
22. Draw and explain the functional block diagram of the LM317 three terminal adjustable regulator.
23. Define ripple rejection with respect to voltage regulators.
24. What is switched capacitor ? Explain its need ?
25. How switched capacitor can simulate a resistor ?
26. Draw and explain switched capacitor integrator.
27. Write a note on MF5 universal filter, stating its features.
28. Draw the functional block diagram of IC switched capacitor filter and explain the function of each block.
29. Write a note switched capacitor filter.
30. Why are switched capacitor filters ideal for MOS technology ?
31. Draw the pin diagram of IC MF10 and explain the function of each pin.
32. Write a note on modes of operation of MF10.
33. Explain the internal circuit diagram of LM380 audio amplifier.
34. List down the features of LM380 audio amplifier.
    (i) Audio amplifier
    (ii) Phono amplifier
    (iii) Siren/Alarm system
    (iv) Intercom system
    (v) Bridge amplifier
    (vi) Audio amplifier with variable gain
36. Explain the use of IC 810 as :
    (i) Portable radio receiver
    (ii) Television receiver
37. List down the features of IC 810.

38. Explain the working of typical frequency to voltage converter and draw the waveforms.
39. List the features of TC 9400.
40. Explain the function of important pins of TC 9400.
41. Explain with the help of block diagram working of TC 9400 as a voltage to frequency converter.
42. Explain with the help of block diagram working of TC 9400 as a frequency to voltage converter.
43. What do you mean by video amplifier ?
44. What are the design considerations to be followed for video amplifiers ?
45. Explain the cascade video amplifier with a neat circuit diagram.
46. Explain the working of a video amplifier circuit.
47. Draw the frequency response of a video amplifier.
48. What are the features of CA 3040 video amplifier ?
49. Explain the internal circuit diagram of CA 3040 video amplifier.
50. Write a short note on fiber optic ICs.
51. Write a note on fibre optic IC.
52. Write a short note on optoisolators.
53. Give various performance parameters for optoisolators.
54. Write a note on IC optocoupler and packages.
55. Write a note on optocoupler.
56. Discuss the advantages of optocouplers. How is it superior to an electrical coupling ?
57. Write a note on isolation amplifier.
58. Explain the working of the following circuits :
    (i) Isolation amplifier,
    (ii) Voltage regulator.
59. Explain any one isolation amplifier IC with the help of block diagram.
50. Name two applications of isolation amplifier.
61. Write a short note on ISO100, an isolation amplifier.

❖❖❖

BIBLIOGRAPHY

- *Burr-Brown General Catalog.* Tucson, Ariz.: Burr-Brown, 1979.
- *Datel-Intersil Data Acquisition Component Handbook.* Mansfield, Mass.: Datel-Intersil, Inc., 1980.
- *Fairchild Linear Op-Amp Data Book.* Mountain View, Calif.: Fairchild Camera and Instrument Corporation, 1979.
- *Intersil Data Book.* Cupertino, Calif.: Intersil Inc., 1979.
- *Motorola Linear Integrated Circuit.* Motorola CMOS Data. Phoenix, Ariz.; Motorola Semiconductor Products, 1979.
- *National Linear Data Book.* Santa Clara, Calif.: National Semiconductor Corporation, 1978.
- *RCA Linear Integrated Circuits.* Somerville, N. J.: RCA Solid State, 1978.
- *Signetics Analog Data Manual.* Sunny vale, Calif.: Signetics Corporation, 1979.
- *Teledyne Data Conversion Design Manual.* Mountain View, Calif.: Teledyne, Semiconductor, 1980.
- *The TTL Data Book for Design Engineers.* Texas Instruments, 1976..

  BERLIN, H. M., *The Design of Active Filters with Experiments.* Derby, Conn.: E and L Instruments, Inc., 1977.
- BYLERLY, J. E., and M. VANDER Kooi, LM380 Power Audio Amplifier. *National Semiconductor Linear Applications*, Vol. 1, 1976.
- KAHN, M., *The Versatile Op-Amp.* New York: Holt, Rinehart and Winston, Inc., 1970.
- MALVINO, A. P., *Electronic Principles.* New York: McGraw-Hill Book Company, 1979.
- *Phase-Locked Loops.* Benton Harbor, Mich.: Heathkit Continuing Education, 1980.
- PRENSKY, S. D., *Manual of Linear Integrated Circuits: Operational Amplifiers and Analog ICs.* Reston, Va.: Reston Publishing Company, Inc., 1974.
- RUTKOWSKI, G. B., *Handbook of Integrated Circuit Operational Amplifier.* Englewood Cliffs, N. J.: Prentice-Hall, Inc., 1975.
- STANLEY, WILLIAM D., *Operational Amplifiers with Linear Integrated Circuits.* Columbus, Ohio: Merrill Publishing Company, 1989.
- BAR-LEV, A, Semiconductors, and electronic Devices, Practice-Hall, Inc., 1975.

- GAYAKWAD, R.A., Op-amps and Linear Integrated Circuit Technology, Prentice-Hall India Ltd., New Delhi, 1987.
- BOTKAR, K. R., Integrated Circuits, Khanna Publishers, 2011.
- SZE, S. M., Physics of Semiconductor Devices, Wiley, 1969.
- Millman, J., Microelectronics, McGraw Hill, 1979.
- Gandhi, S.K., the Theory and Practice of Microelectronics, Wiley, 1968.
- U.A. Bakshi., A.P. Godse, Linear Integrated Circuits, Technical Publications, Pune, 2008.

❑❑❑